THE COLLECTED PAPERS OF

Albert Einstein

VOLUME 2

THE SWISS YEARS:

WRITINGS, 1900–1909

THE COLLECTED PAPERS OF

Albert Einstein

VOLUME 2

THE SWISS YEARS: WRITINGS, 1900–1909

John Stachel, EDITOR

DAVID C. CASSIDY, JÜRGEN RENN, AND
ROBERT SCHULMANN, ASSOCIATE EDITORS

DON HOWARD, ASSISTANT EDITOR

A. J. KOX, CONTRIBUTING EDITOR

ANN LEHAR, EDITORIAL ASSISTANT

Princeton University Press
1989

Copyright © 1989 by Princeton University Press
Published by Princeton University Press, 41 William Street, Princeton, New Jersey 08540
In the United Kingdom: Princeton University Press, Oxford

LIBRARY OF CONGRESS CATALOGING-IN-PUBLICATION DATA
(Revised for vol. 2)

Einstein, Albert, 1879-1955.
The collected papers of Albert Einstein.

Includes bibliographies and indexes.
Contents: v. 1. The early years, 1879-1902—
v. 2. The Swiss years, writings, 1900-1909.
1. Physics. 2. Physicists—Biography.
I. Stachel, John J., 1928-   . II. Title.
QC16.F5A2   1987   530   86-43132
ISBN 0-691-08407-6 (v. 1)

This book has been composed in Linotron Times Roman

Printed in the United States of America by Princeton University Press
Princeton, New Jersey

CONTRIBUTORS

The Hebrew University of Jerusalem and Princeton University Press wish to express their appreciation to the Contributors for their generous financial support of the editorial work that has made this edition of *The Collected Papers of Albert Einstein* possible.

ENDOWMENT

Harold W. McGraw, Jr.

INSTITUTIONAL CONTRIBUTORS

The National Science Foundation (U.S.A.)
The Alfred P. Sloan Foundation (U.S.A.)
The National Endowment for the Humanities (U.S.A.)
The Swiss National Science Foundation
The Dr. Tomalla Foundation, Vaduz, Liechtenstein
The Canton of Aargau, Switzerland

INDIVIDUAL CONTRIBUTORS

The Estate of Louise R. Berman
The Estate of Albert Einstein
Robert A. Hefner III
The Estate of William H. Schwartz

CONTENTS

LIST OF TEXTS

The works included in this volume established Einstein as a central figure in twentieth-century theoretical physics. He submitted the first paper as a recent, still-unemployed graduate of the Swiss Federal Polytechnical School (ETH) in Zurich. The last major paper is the published version of a lecture that Einstein as newly appointed nontenured professor (Außerordentlicher Professor) in theoretical physics at the University of Zurich gave to the Physics Section of the Gesellschaft Deutscher Naturforscher und Ärzte at their 1909 meeting.[1] The invitation to attend and speak at the Salzburg meeting as an honored guest ("Ehrengast") of the German Physical Society constituted official recognition of Einstein's standing in the German-speaking physics community. The recommendation for his Zurich appointment states: "Einstein currently ranks among the most important theoretical physicists" ("Einstein gehört gegenwärtig zu den bedeutendsten theoretischen Physikern").[2]

By 1909 Einstein was actively corresponding with such well-known physicists as H. A. Lorentz, Max Planck, Arnold Sommerfeld, Johannes Stark, and Wilhelm Wien. Einstein's correspondence from this period, to be published in Volume 5, provides important additional documentation of the development of his ideas. Insofar as these letters are relevant to an understanding of Einstein's writings and their context, they are cited in the editorial notes and footnotes in this volume.[3]

The work that led to Einstein's widespread recognition as a major figure in theoretical physics is embodied in three papers written in 1905, while Einstein was a technical expert in the Swiss Patent Office. Two of the papers dealt with important problems of contemporary physics, while the third formulated a problem that could not be solved without overturning the foundations of that physics. Einstein's first paper on relativity resolves paradoxes that had arisen in his attempt to integrate the results of numerous experiments on the electrodynamics and optics of moving bodies into Lorentz's electron theory.[4] It did so by introducing a new

---

[1] *Einstein 1909c* (Doc. 60).

[2] Otto Stoll, Dekan, Philosophical Faculty of the University of Zurich, to Heinrich Ernst, Erziehungsdirektor, Canton of Zurich, 4 March 1909, SzZSa, U 110 b 2 (44).

[3] The Introduction and the editorial notes in this volume refer to Einstein's activities only when they are directly relevant to his writings.

Einstein's career during this period will be discussed in Vol. 5, the *Correspondence* for *The Swiss Years (1902–1914)*.

[4] See *Einstein 1905r* (Doc. 23). For a discussion of Einstein's work in this field, see the editorial note, "Einstein on the Theory of Relativity," pp. 253–274.

kinematics that could readily be incorporated into what is now called the classical tradition in physics,[5] however difficult some contemporary physicists may have found it to accept the theory of relativity. Einstein's first paper on Brownian motion led to experiments that helped persuade the remaining skeptics of the validity of the kinetic-molecular theory of heat, and initiated the theoretical study of stochastic processes in physics.[6] His first paper on the quantum hypothesis shows that electromagnetic radiation interacts with matter as if the radiation has a granular structure, a structure that could not be explained on the basis of either Maxwell's theory or classical mechanics.[7]

By 1909 the theory of relativity was well on its way to general acceptance, if not full comprehension, while Einstein's views on the nature of radiation were becoming the subject of intense controversy. In his Salzburg lecture Einstein made explicit a number of interconnections between his work on the quantum hypothesis, on relativity, and on Brownian motion and statistical physics, interconnections that had been generally only implicit in his earlier articles. The lecture espouses a thesis that did not find ready acceptance.[8] In spite of the general belief that the wave theory of light had completely triumphed over the emission theory, Einstein maintained that "the next phase in the development of theoretical physics will bring us a theory of light that may be conceived of as a sort of fusion of the wave and of the emission theory of light" ("die nächste Phase der Entwickelung der theoretischen Physik uns eine Theorie des Lichtes bringen wird, welche sich als eine Art Verschmelzung von Undulations- und Emissionstheorie des Lichtes auffassen läßt").[9] While acknowledging fully the successes of Maxwell's electromagnetic theory of light in explaining numerous optical effects, Einstein reviewed the growing body of evidence for his light quantum hypothesis. He stressed the existence of a symmetry between the elementary processes of emission and absorption of light, inexplicable in terms of any wave theory, but characteristic of an emission theory. He also showed that the theory of relativity makes it possible to avoid the dilemma of an exclusive choice between wave and emission theories. Light no longer need be regarded "as a sequence of states of a hypothetical medium" ("als Folge von Zuständen eines hypothetischen Mediums"); rather, it can be treated "as something existing independently, like

---

[5] This usage of "classical" was already coming into currency. Einstein referred to "classical thermodynamics" in 1908 (the passage is quoted below). For references to "classical electrodynamics" by Planck and to "classical mechanics" by Herglotz dating from about the same time, see *Jungnickel and McCormmach 1986b*, p. 313.

[6] See *Einstein 1905k* (Doc. 16). For a discussion of his work in this field, see the editorial

note, "Einstein on Brownian Motion," pp. 206–222.

[7] See *Einstein 1905i* (Doc. 14). For a discussion of Einstein's work in this field, see the editorial note, "Einstein's Early Work on the Quantum Hypothesis," pp. 134–148.

[8] See, e.g., Planck's comments in the discussion that followed Einstein's talk, *Einstein et al. 1909c* (Doc. 61), pp. 825–826.

[9] *Einstein 1909c* (Doc. 60), pp. 482–483.

matter'' (''als etwas wie die Materie selbständig Bestehendes''), something which transfers mass between emitters and absorbers of light.[10] To put the capstone on his argument that radiation exhibits particulate as well as undulatory aspects, Einstein analyzed fluctuations in the energy and momentum of the radiation field, drawing upon ideas and methods he developed earlier in the course of his work on statistical physics and Brownian motion.[11]

Einstein's concern with radiation dates back to his student years.[12] As early as 1899 he was pondering its nature; two years later he was reading some of Planck's papers on this subject. Einstein's investigation of the structure of radiation forms part of his search for what, a few years later, he called ''elementary foundations'' (''elementare Grundlagen'') for physics,[13] a search that began while he was still a student. He does not appear ever to have doubted the atomic structure of matter, and he accepted the new electron theory of electricity very quickly.[14] In his first preserved comments on electrodynamics, he suggested that it is ''the theory of the motions in empty space of moving electric and magnetic charges'' (''die Lehre von den Bewegungen bewegter Elektrizitäten und Magnetismen . . . im leeren Raum'').[15] He welcomed Drude's electron theory of metals, and explored new models to explain the thermal, electrical, and optical properties of matter and their interconnections.[16]

Attempts to account for a number of phenomena on the basis of detailed atomistic assumptions also characterize Einstein's first two papers, on the nature of intermolecular forces, published in 1901 and 1902.[17] Taking for granted the existence of atoms, molecules, and ions, Einstein attempted to determine the strength of the attractive forces between them with the help of empirical data concerning such phenomena as capillarity and electrolytic conduction. The papers and

[10] *Einstein 1909c* (Doc. 60), p. 490.

[11] Einstein's analysis of fluctuations is discussed below. See also the editorial notes, ''Einstein on the Foundations of Statistical Physics,'' pp. 41–55, and ''Einstein on Brownian Motion,'' pp. 206–222.

[12] See Vol. 1, the editorial note, ''Einstein on Thermal, Electrical, and Radiation Phenomena,'' pp. 235–237.

[13] In 1908 Einstein emphasized that he considered a physical theory as satisfactory only ''if it builds its structures out of *elementary* foundations'' (''wenn sie aus *elementaren* Grundlagen ihre Gebilder zusammensetzt''), noting ''that we are still far away from having satisfactory elementary foundations for electrical and mechanical processes'' (''dass wir noch weit davon entfernt sind, befriedigende elementare Grundlagen

für die elektrischen und mechanischen Vorgänge zu besitzen'') (Einstein to Arnold Sommerfeld, 14 January 1908).

[14] See Vol. 1, the editorial notes, ''Einstein on the Electrodynamics of Moving Bodies,'' pp. 223–225; ''Einstein on Thermal, Electrical, and Radiation Phenomena,'' pp. 235–237; and ''Einstein on Molecular Forces,'' pp. 264–266.

[15] Einstein to Mileva Marić, 10 August 1899 (Vol. 1, Doc. 52).

[16] See Vol. 1, the editorial note, ''Einstein on Thermal, Electrical, and Radiation Phenomena,'' pp. 235–237.

[17] See Vol. 1, the editorial note, ''Einstein on Molecular Forces,'' pp. 264–266; and, in this volume, the editorial note, ''Einstein on the Nature of Molecular Forces,'' pp. 3–8.

contemporary correspondence demonstrate a familiarity with physical chemistry, in particular the theory of solutions, as well as a concern for finding experimental evidence to support his hypotheses. At the end of 1901, Einstein attempted unsuccessfully to obtain a doctorate from the University of Zurich with a thesis on intermolecular forces.[18]

Einstein's interest in a wide variety of problems involving the atomic constitution of matter and electricity led him to study the relationship between thermodynamics and the microstructure of various systems.[19] In his work on intermolecular forces in liquids, for example, he made extensive use of thermodynamical methods, and even postulated an extension of the range of validity of the second law of thermodynamics. Einstein's next three papers, published between 1902 and 1904, are on the foundations of statistical physics; their aim is to establish the principal thermodynamic properties that any macroscopic system must have, with as few assumptions as possible about the constituent elements of the system, or the nature of their interactions.

Presumably as a consequence of his papers for the *Annalen der Physik* on topics in thermal physics, Einstein started to contribute to the *Beiblätter zu den Annalen der Physik*, reviewing papers in the category "Theory of Heat."[20] His services as a reviewer did not last long, however; all but two of his contributions appeared in 1905.

In his last paper on statistical physics, Einstein showed that fluctuations around the equilibrium values of thermodynamical quantities must occur as a consequence of the kinetic theory of heat. He derived an expression for the mean square fluctuations in the energy of a system represented by a canonical ensemble. The expression, which depends only on the system's thermodynamical state, enabled him to interpret Boltzmann's constant as fixing the scale for the thermal stability of a system. Einstein's formulae for fluctuations became an important tool in his later research. In addition to the thermodynamic approach, based on equilibrium statistical mechanics, he elaborated a second, distinct approach to fluctuation phenomena: the stochastic approach, based on the analysis of fluctuation-dissipation mechanisms.

Einstein first developed the latter approach in 1905, when he realized that the motions of microscopic particles suspended in a liquid must manifest visible fluctuations owing to the molecular structure of the fluid. The methods Einstein used

---

[18] For a discussion of Einstein's first attempt to obtain a doctorate, see the editorial note, "Einstein's Dissertation on the Determination of Molecular Dimensions," section III, pp. 173–176.

[19] See the editorial note, "Einstein on the

Foundations of Statistical Physics," pp. 45–46.

[20] See the editorial note, "Einstein's Reviews for the *Beiblätter zu den Annalen der Physik*," pp. 109–111.

in his work on Brownian motion combine statistical physical concepts he had recently investigated with concepts from the theory of diffusion familiar to him from his work on molecular forces.[21] As a consequence, he not only gave the first satisfactory explanation of Brownian motion as a stochastic process, but also developed a new method for the determination of molecular dimensions, based on related techniques, and completed a doctoral dissertation on this topic.[22] Einstein emphasized that the existence and properties of stochastic fluctuations represented a critical test of the kinetic theory of heat, which was still challenged by a number of scientists. His theoretical studies helped to stimulate much experimental work by Perrin and others that effectively served to put such doubts to rest.

Initially, Einstein believed black-body radiation to be the only system for which energy fluctuations are empirically significant. In 1904 he applied the thermodynamic approach developed for mechanical systems to the energy fluctuations of "a radiation-filled space" ("ein Strahlungsraum"), with encouraging results.[23] The following year, Einstein returned to the problem of black-body radiation, using thermodynamic methods combined with Boltzmann's principle to show that, in the high v/T limit, the entropy of radiation varies with its volume in the same way as does the entropy of a system of particles. Such considerations led him to formulate the heuristic viewpoint that radiant energy, like matter and electricity, is quantized.[24]

In 1907 he extended the quantum hypothesis to the vibrational energy of solids. He was thus able to explain a long-standing anomaly in the theory of specific heats of solids, which had led to serious doubts about the kinetic theory of heat. Beginning about 1911, with the first experimental confirmations of Einstein's theory of specific heats, this work played an important role in persuading many physicists of the basic significance of the quantum hypothesis.[25] In 1909 he applied both of his methods for treating fluctuations in an investigation of black-body radiation based on Planck's law. He used a thermodynamic approach, based on the inversion of Boltzmann's principle, to investigate energy fluctuations; and a stochastic approach, examining the Brownian motion of a small mirror in equilibrium with the radiation, to investigate pressure fluctuations. As noted above, the results of

---

[21] See the editorial notes, "Einstein on Brownian Motion," pp. 206–222, and "Einstein on the Nature of Molecular Forces," pp. 3–8.

[22] See the editorial note, "Einstein's Dissertation on the Determination of Molecular Dimensions," pp. 170–182.

[23] See *Einstein 1904* (Doc. 5), pp. 360–362. This success of methods developed for systems with a finite number of degrees of freedom when

applied to radiation may have encouraged Einstein to treat radiation itself as such a system (see *Einstein 1905i* [Doc. 14], pp. 132–133).

[24] See the editorial note, "Einstein's Early Work on the Quantum Hypothesis," pp. 134–148.

[25] For a discussion of the important role of the quantum theory of specific heats, see *Klein 1965*, and *Kuhn 1978*, pp. 210–220.

both calculations indicated the need for a fusion of wave and emission theories of radiation.

In spite of his many successes, Einstein felt that he had come no closer to finding the "elementary foundations" of an adequate theory of matter, radiation, and electricity. He was quite convinced of the inability of either classical mechanics or Maxwell's electrodynamics to explain the phenomena that had led him to propose the quantum hypothesis,[26] but "with no firm foundation to be seen anywhere upon which one could have built" ("ohne dass sich irgendwo fester Grund zeigte, auf dem man hätte bauen können").[27] Many years later he characterized his outlook during this period:

> Gradually I despaired of the possibility of discovering the true laws by means of constructive efforts based on known facts. The longer and more desperately I tried, the more I came to the conviction that only the discovery of a universal formal principle could lead us to assured results. The example I saw before me was thermodynamics.

> Nach und nach verzweifelte ich an der Möglichkeit die wahren Gesetze durch auf bekannte Tatsachen sich stützende konstruktive Bemühungen herauszufinden. Je länger und verzweifelter ich mich bemühte, desto mehr kam ich zu der Überzeugung, dass nur die Auffindung eines allgemeinen formalen Prinzipes uns zu gesicherten Ergebnissen führen könnte. Als Vorbild sah ich die Thermodynamik vor mir.[28]

Einstein had developed great confidence in the laws of thermodynamics. His work on the foundations of statistical physics convinced him that, within their limits, the principles of thermodynamics are valid for all systems with many degrees of freedom, regardless of their detailed microstructure, and are applicable even to

[26] Einstein later recollected the origins of his work on the quantum hypothesis: "Stimulated by Wien's and Planck's investigations, I realized that mechanics and electrodynamics stand in insoluble contradiction to the facts of experience" ("Angeregt durch Wiens und Plancks Forschungen erkannte ich, daß Mechanik und Elektrodynamik in einem unlösbaren Widerspruch zu den Erfahrungstatsachen stehen") (from a recording made 6 February 1924, text printed in *Herneck 1966a*, p. 134).

[27] *Einstein 1979*, p. 42; translation, p. 43.

[28] *Einstein 1979*, p. 48; translation, p. 49. A decade later, Einstein characterized both the theory of relativity and thermodynamics as "theories of principle," as opposed to "constructive theories." Of the latter he wrote: "These attempt to construct a picture of more complex phenomena from a relatively simple underlying formalism. . . . When we say that we have succeeded in understanding a group of natural processes, we always mean by this that a constructive theory has been found, which embraces the processes in question" ("Diese suchen aus einem relativ einfachen zu grunde gelegten Formalismus ein Bild der komplexeren Erscheinungen zu konstruieren. . . . Wenn man sagt, es sei gelungen, eine Gruppe von Naturvorgängen zu begreifen, so meint man damit immer, dass eine konstruktive Theorie gefunden sei, die die betreffenden Vorgänge umfasst") (*Einstein 1919*, German text cited from the manuscript).

thermal radiation.[29] Were there other such principles upon which one could rely in the search for "elementary foundations"?

In retrospect, whether or not Einstein saw it in such terms at the time, one can view the development of the theory of relativity as Einstein's first great success in finding a "universal formal principle." For some years, he had been trying to develop an electrodynamics of moving bodies that did not invoke an ether frame of reference.[30] After vain attempts to modify the Maxwell-Lorentz equations or replace them with an emission theory of light, he found that the correct way to approach the problem was to start from principles, whose role is analogous to that of the laws of thermodynamics. In 1907 he summarized the nature of the resulting theory:

> One is not dealing here in any way . . . with a "system," in which the individual laws would implicitly be contained and from which they could be found merely by deduction, but only with a principle that (in a way similar to the second law of thermodynamics) permits certain laws to be deduced from others.

> Es handelt sich hier . . . keineswegs um ein "System," in welchem implizite die einzelnen Gesetze enthalten wären, und nur durch Deduktion daraus gefunden werden könnten, sondern nur um ein Prinzip, das (ähnlich wie der zweite Hauptsatz der Wärmetheorie) gewisse Gesetze auf andere zurückzuführen gestattet.[31]

The theory demonstrates that the relativity principle, in conjunction with the principle of the constancy of the velocity of light, requires a new kinematical foundation for all of physics. By applying this kinematics to the Maxwell-Lorentz equations for bodies at rest, Einstein was able to formulate an electrodynamics of moving bodies that is in accord with experiment.

Einstein's 1905 paper discusses Maxwell's equations in empty space. The problem of formulating relativistically invariant equations for electromagnetic fields in moving media, first discussed by Minkowski in 1907,[32] was addressed by Einstein in 1908, in collaboration with Jakob Laub.[33]

---

[29] Describing many years later the deep impression that thermodynamics had made upon him, Einstein stated: "It is the only physical theory of universal content concerning which I am convinced that, within the framework of the applicability of its basic concepts, it will never be overthrown" ("Es ist die einzige physikalische Theorie allgemeinen Inhaltes, von der ich überzeugt bin, dass sie im Rahmen der Anwendbarkeit ihrer Grundbegriffe niemals umgestossen keit ihrer Grundbegriffe niemals umgestossen werden wird") (*Einstein 1979*, p. 30, translation, p. 31).

[30] See Vol. 1, the editorial note, "Einstein on the Electrodynamics of Moving Bodies," pp. 223–225.

[31] *Einstein 1907g* (Doc. 44), p. 207.

[32] See *Minkowski 1908*.

[33] See the editorial note, "Einstein and Laub on the Electrodynamics of Moving Media," pp. 503–507.

Einstein's work on relativity is remarkable not only for its successful application of the method of formal principles, but also for its explicit concern with foundational issues.[34] The introduction to his first paper on the subject concludes:

> The theory to be developed—like every other electrodynamics—is based upon the kinematics of rigid bodies, since the assertions of any such theory concern relations between rigid bodies (coordinate systems), clocks, and electromagnetic processes. Insufficient consideration of this circumstance is at the root of the difficulties with which the electrodynamics of moving bodies currently has to contend.

> Die zu entwickelnde Theorie stützt sich—wie jede andere Elektrodynamik—auf die Kinematik des starren Körpers, da die Aussagen einer jeden Theorie Beziehungen zwischen starren Körpern (Koordinatensystemen), Uhren und elektromagnetischen Prozessen betreffen. Die nicht genügende Berücksichtigung dieses Umstandes ist die Wurzel der Schwierigkeiten, mit denen die Elektrodynamik bewegter Körper gegenwärtig zu kämpfen hat.[35]

Einstein's sensitivity to such methodological questions in physics can be attributed in some part to extensive readings and discussions in the area now called the philosophy of science.[36] Discussing "the axiom of the absolute character of time, or rather of simultaneity" ("das Axiom des absoluten Charakters der Zeit, bzw. der Gleichzeitigkeit") Einstein later stated:

> To recognize clearly this axiom and its arbitrary character is actually to solve the problem. The critical reasoning required for the discovery of this central point was decisively promoted, in my case, especially by the reading of David Hume's and Ernst Mach's philosophical writings.

> Dies Axiom und seine Willkür klar erkennen bedeutet eigentlich schon die Lösung des Problems. Das kritische Denken, dessen es zur Auffindung dieses zentralen Punktes bedurfte, wurde bei mir entscheidend

[34] Einstein later recalled: "From youth onwards my entire scientific effort was directed to deepening the foundations of physics" ("Von der Jugend an war mein ganzes wissenschaftliches Streben auf die Vertiefung der Grundlagen der Physik gerichtet") (from a recording made 6 February 1924, text in *Herneck 1966a*, p. 134).

[35] *Einstein 1905r* (Doc. 23), p. 892

[36] Emphasizing his early concern with foundational questions in physics (see note 34), Einstein immediately added: "Philosophical viewpoints and requirements in the narrower sense only affected me secondarily" ("Philosophische Gesichtspunkte und Bedürfnisse in engerem Sinne wirkten nur sekundär auf mich") (from a recording made 6 February 1924, text in *Herneck 1966a*, p. 134).

gefördert insbesondere durch die Lektüre von David Humes und Ernst Machs philosophischen Schriften.[37]

Einstein read two of Ernst Mach's influential historical-critical studies, the *Mechanik* and the *Wärmelehre*, while a student at the ETH.[38] Other classics of theoretical physics that he studied during his student years, such as the works of Boltzmann, Helmholtz, Kirchhoff, and Hertz,[39] also explicitly treat many foundational and methodological questions. The wide-ranging ''regular philosophical reading and discussion evenings'' (''regelmässige philosophische Lese- und Diskussionsabende'')[40] of Einstein and his friends, Maurice Solovine and Conrad Habicht, who banded together in 1902 to form the ''Olympia Academy'' (''Akademie Olympia''),[41] also played a significant role in directing his attention to such issues. In his reminiscences of Einstein, Solovine gives an extensive list of the readings that formed the basis of their discussions.[42]

Solovine and Einstein started to read Pearson's *Grammar of Science*[43] before Habicht joined them. After that, the three of them read and discussed Mach's

[37] *Einstein 1979*, p. 50. Elsewhere he included Poincaré together with Hume and Mach in a similar list, noting that he read Hume ''in a quite good German edition'' (''in einer recht guten deutschen Ausgabe'') (Einstein to Michele Besso, 6 March 1952), possibly *Hume 1895*. In 1915, Einstein wrote to Schlick: ''You have also correctly seen that this trend of thought [positivism] was of great influence on my efforts, and specifically E. Mach and still much more Hume, whose treatise on understanding I studied with fervor and admiration shortly before the discovery of the theory of relativity. It is very well possible that without these philosophical studies I would not have arrived at the solution'' (''Auch darin haben Sie richtig gesehen, dass diese Denkrichtung von grossem Einfluss auf meine Bestrebungen gewesen ist, und zwar E. Mach und noch viel mehr Hume, dessen Traktat über den Verstand ich kurz vor Auffindung der Relativitätstheorie mit Eifer und Bewunderung studierte. Es ist sehr gut möglich, dass ich ohne diese philosophischen Studien nicht auf die Lösung gekommen wäre'') (Einstein to Moritz Schlick, 14 December 1915). In spite of Einstein's garbled reference to the work that he read, it appears to have been *A Treatise of Human Nature* (see the reading list of the Olympia Academy, cited below, and Einstein to Michele Besso, 6 January 1948).

[38] See Einstein to Mileva Marić, 10 September 1899 (Vol. 1, Doc. 54), especially note 8.

Commenting later on the prevailing mechanistic outlook in physics at this time, Einstein stated: ''It was Ernst Mach who, in his *History of Mechanics*, shook this dogmatic faith; this book exercised a profound influence upon me precisely in this regard while I was a student'' (''Ernst Mach war es, der in seiner *Geschichte der Mechanik* an diesem dogmatischen Glauben rüttelte; dies Buch hat gerade in dieser Beziehung einen tiefen Einfluss auf mich als Student ausgeübt'') (*Einstein 1979*, p. 18). Einstein probably read the third edition, *Mach 1897*, while a student.

[39] See Einstein to Mileva Marić, 10 August 1899, 10 September 1899, 1 August 1900, 13 September 1900, and 15 April 1901 (Vol. 1, Docs. 52, 54, 69, 75, and 101).

[40] Einstein to Michele Besso, 6 March 1952.

[41] The earliest record of the name and membership occurs in a postcard of 30 November 1903 sent by Einstein to Conrad Habicht.

[42] *Solovine 1956*, pp. vii–viii. The works are listed in the order in which Solovine mentions them. He indicates that they read many other works on similar themes, as well as a number of literary works. In a letter of 14 April 1952 to Carl Seelig (SzZE Bibliothek, Hs 304:1006), Solovine includes *Poincaré 1905a* in a similar, but otherwise less complete list of the readings of the Olympia Academy.

[43] *Pearson 1900*.

Analyse der Empfindungen[44] and *Mechanik*,[45] Mill's *Logic*,[46] Hume's *A Treatise of Human Nature*,[47] Spinoza's *Ethics*,[48] several unspecified memoirs and lectures by Helmholtz,[49] several chapters of Ampère's *L'essai sur la philosophie des sciences*,[50] Riemann's "Ueber die Hypothesen, welche der Geometrie zu Grunde liegen,"[51] several chapters of Avenarius's *Kritik der reinen Erfahrung*,[52] Clifford's "On the Nature of Things-in-Themselves,"[53] Dedekind's *Was sind und was sollen die Zahlen?*,[54] and Poincaré's *La science et l'hypothèse*.[55]

By directing Einstein's attention to the problem of the formation of scientific concepts (such as the concepts of space and time), to the role of conventions in scientific thought (such as that required for defining the simultaneity of distant events), and to the place of formal principles in the structure of scientific theories (such as the principle of relativity),[56] to name only a few relevant topics, his foundational readings helped to prepare him for the task of revising the kinematical foundations of physics.

Einstein considered the theory of relativity only a first step toward a satisfactory resolution of the difficulties facing physics, toward "a complete world view that is in accord with the principle of relativity" ("ein vollständiges, dem Relativitätsprinzip entsprechendes Weltbild").[57] He stated that

> The theory of relativity is just as little ultimately satisfactory as, for example, classical thermodynamics was before Boltzmann had inter-

[44] Three editions came out within a few years of the founding of the Olympia Academy: *Mach 1900a*, *1902*, and *1903*.

[45] For Einstein's earlier reading of Mach, see note 38. The Olympia Academy members may have used the fourth or fifth editions, *Mach 1901*, *1904*.

[46] The 8th edition, *Mill 1872*, was translated into German twice (see *Mill 1877*, *1884–1887*).

[47] The first part of *Hume 1739* was translated as *Hume 1895*.

[48] *Spinoza 1677*. Einstein may have used one of several German translations, possibly *Spinoza 1887* or *1893*.

[49] Helmholtz's scientific papers are reprinted in three volumes: *Helmholtz 1882*, *1883*, *1895*; the epistemological papers are included in *Helmholtz 1883*. His more popular writings are printed in *Helmholtz 1884*. For evidence of Einstein's earlier reading of Helmholtz, see Einstein to Mileva Marić, early August 1899 and 10 August 1899 (Vol. 1, Docs. 50 and 52).

[50] *Ampère 1834*.

[51] *Riemann 1854*.

[52] *Avenarius 1888*, *1890*.

[53] Clifford's essay was issued in German as a book, *Clifford 1903*, which Einstein presumably used.

[54] *Dedekind 1893*.

[55] *Poincaré 1902*. Einstein may have used the German edition, *Poincaré 1904a*.

[56] It is well known that elements of Einstein's distinction between principle and constructive theories occur in Poincaré's writings (see, e.g., *Poincaré 1904b*). Two other sources that may have influenced Einstein's emphasis on the role of principles in physics are Violle, whose textbook Einstein studied (see Vol. 1, "Albert Einstein—Beitrag für sein Lebensbild," p. lxiv), and Alfred Kleiner, with whom Einstein had been in contact since 1901 when he began to discuss physics with him (see Einstein to Mileva Marić, 19 December 1901, Vol. 1, Doc. 130). For Violle's comments on the role of principles, see *Violle 1892*, p. 90; for Kleiner's comments, see, e.g., *Kleiner 1901*, pp. 21–23.

[57] *Einstein 1907h* (Doc. 45), p. 372.

preted entropy as probability. If the Michelson-Morley experiment had not put us in the most awkward position, no one would have accepted the theory of relativity as a (partial) salvation.

Die Relativitätstheorie ist ebensowenig endgültig befriedigend, wie es z. B. die klassische Thermodynamik war, bevor Boltzmann die Entropie als Wahrscheinlichkeit gedeutet hatte. Wenn uns nicht das Michelson-Morley'sche Experiment in die grösste Verlegenheit gebracht hätte, hätte niemand die Relativitätstheorie als eine (halbe) Erlösung empfunden.[58]

While searching for adequate elementary foundations of physics, Einstein continued to use formal principles as tools to derive new connections between established laws, and as criteria to help delimit the possibilities in the search for a more fundamental theory. In 1911 he stated this strategy:

We are all agreed that the quantum theory, in its present form, can be a useful tool, but does not truly constitute a theory in the ordinary sense of the word. . . . The question arises of knowing what are the general principles of physics on which we may rely for the solution of the problems that concern us.

Nous sommes tous d'accord que la théorie des quanta, sous sa forme actuelle, peut être d'un emploi utile, mais ne constitue pas véritablement une théorie au sens ordinaire du mot. . . . La question se pose de savoir quels sont les principes généraux de la Physique sur lequels nous pouvons compter pour la solution des questions qui nous occupent.[59]

He singled out the energy principle and Boltzmann's principle as such principles; elsewhere, he similarly utilized the relativity principle and the second law of thermodynamics.[60]

At the end of the decade Einstein believed that he had still not achieved any real understanding of quantum phenomena. His dissatisfaction was caused by the lack of an interpretation of Planck's constant "in an intuitive manner" ("in anschaulicher Weise"),[61] and by the role of the quantum of electric charge, which remained "an alien" ("ein Fremdling") in Maxwell's theory.[62] He was convinced that the quantum of electricity and of radiation should not simply be postulated,

---

[58] Einstein to Arnold Sommerfeld, 14 January 1908.

[59] *Solvay 1911*, p. 436.

[60] See, e.g., *Einstein 1909c* (Doc. 60) for the relativity principle, and *Einstein 1912a* for the laws of thermodynamics.

[61] Einstein to Arnold Sommerfeld, 14 January 1908.

[62] *Einstein 1909b* (Doc. 56), p. 192.

but should be constructed, emerging from a satisfactory theory of matter and radiation. He anticipated that "the same modification of the [Maxwell-Lorentz] theory which contains the elementary quantum [of electricity] will also contain the quantum structure of radiation" ("die gleiche Modifikation der Theorie, welche das Elementarquantum als Konsequenz enthält, auch die Quantenstruktur der Strahlung als Konsequenz enthalten wird").[63]

In 1909 Einstein made his first known attempt to find a field theory that would explain the structure of both matter and radiation. After investigating relativistically invariant, nonlinear generalizations of Maxwell's equations, he wrote:

> I have not yet succeeded . . . in finding a system of equations that I could see was suited to the construction of the elementary quantum of electricity and the light quantum. The manifold of possibilities does not seem to be so large, however, that one should shrink from the task.

> Es ist mir noch nicht gelungen, ein . . . Gleichungssystem zu finden, von dem ich hätte einsehen können, daß es zur Konstruktion des elektrischen Elementarquantums und der Lichtquanten geeignet sei. Die Mannigfaltigkeit der Möglichkeiten scheint aber nicht so groß zu sein, daß man vor der Aufgabe zurückschrecken müßte.[64]

The attempt to find such a system of equations constitutes a forerunner of his later search for a unified field theory.

Of course, the search for a unified theoretical foundation for all of physics did not originate with Einstein. During his formative years as a physicist, three such foundational programs were current, each with universal, and hence exclusive, ambitions: the traditional program based upon classical mechanics, the energetics program based upon thermodynamics, and the electromagnetic program based upon Maxwell's theory.[65] Einstein was apparently attracted to the idea of a unified foundation for all of physics quite early in his career.[66] While he does not appear to have been attracted by the energetics program,[67] he was profoundly

---

[63] *Einstein 1909b* (Doc. 56), pp. 192–193.

[64] *Einstein 1909b* (Doc. 56), p. 193.

[65] For a survey of these three programs, see *Jungnickel and McCormmach 1986b*, chap. 24, pp. 212–245.

[66] In 1901, in commenting on his first paper, Einstein wrote: "It is a glorious feeling to recognize the unity of a complex of phenomena, which appear to direct sense perception as quite distinct things" ("Es ist ein herrliches Gefühl, die Einheitlichkeit eines Komplexes von Erscheinungen zu erkennen, die der direkten sinn-

lichen Wahrnehmung als ganz getrennte Dinge erscheinen") (Einstein to Marcel Grossmann, 14 April 1901, Vol. 1, Doc. 100). Einstein may have picked up this theme from Humboldt's *Kosmos*, which he is reported to have read as a youth (see note 8 to ibid.).

[67] In 1913, he endorsed Planck's judgment (see *Planck 1896*) "that energetics is worthless as a heuristic method, indeed, that it even operates with untenable concepts" ("daß die Energetik als heuristische Methode wertlos ist, ja, daß sie sogar mit unhaltbaren Begriffen oper-

impressed by the successes of classical mechanics and of Maxwell's theory.[68] However much he may have regretted the resultant dualism, he accepted the need to employ a mixture of both mechanical and electromagnetic concepts in his work.[69]

Einstein was the first physicist to argue that the fundamental concepts of both mechanics and electrodynamics require drastic modification.[70] Yet, even while seeking ways to modify them, he was not disposed to jettison the concepts of either theory prematurely.[71] He recognized the existence of a broad domain of phenomena, to the understanding of which both theories continue to provide reliable guidance.[72]

His work on molecular forces, the foundations of statistical physics, molecular dimensions, and Brownian motion all involve efforts to extend and perfect the classical mechanical (Galilei-Newtonian) approach, including its statistical extensions. Similarly, his work on the theory of relativity and the electrodynamics of moving media represent efforts to extend and perfect Maxwell's electrodynamics, as incorporated in Lorentz's electron theory, while modifying classical mechanics to make it cohere with this theory.

His work on the quantum hypothesis, on the other hand, consisted of demonstrations of the limited validity of both classical mechanics and Maxwell's electromagnetic theory, and efforts to comprehend phenomena that cannot be explained

---

iert'') (*Einstein 1913*, p. 1077). For a possible influence of energetics on Einstein, see the editorial note, "Einstein on the Foundations of Statistical Physics," p. 46.

[68] See *Einstein 1979*, pp. 18, 30, 32.

[69] In 1914, Einstein presumably was recalling his own acceptance of such a dualistic outlook when he recalled that "about fifteen years ago, no one yet doubted that a correct account of the electrical, optical, and thermal properties of bodies was possible on the basis of Galilei-Newtonian mechanics, applied to molecular motions, and of Maxwell's theory of the electromagnetic field" ("vor etwa fünfzehn Jahren zweifelte man noch nicht daran, daß auf der Grundlage der auf die Molekülbewegungen angewendeten Galilei-Newtonschen Mechanik und der Maxwellschen Theorie des elektromagnetischen Feldes eine richtige Darstellung der elektrischen, optischen und thermischen Eigenschaften der Körper möglich sei") (*Einstein 1914b*, p. 740).

For a contemporary critical comment on this dualism, see *Einstein 1905i* (Doc. 14), pp. 132–133; for a later one, see *Einstein 1979*, p. 34.

[70] See *Einstein 1905i* (Doc. 14), *1906d* (Doc. 34).

[71] In 1909 he noted "that it will only be a matter of a modification of our present theories, and not a complete abandonment of them" ("Daß es sich nur um eine Modifikation unserer heutigen Theorien, nicht um ein vollständiges Verlassen derselben handeln wird") (*Einstein 1909b* [Doc. 56], p. 192).

[72] See *Einstein 1907h* (Doc. 45), pp. 372–373, for Einstein's comments on the range of applicability of Maxwell's equations. In the *Autobiographical Notes*, he commented: "The success of the theory of Brownian motion showed again clearly that classical mechanics always yielded reliable results when it was applied to motions in which the higher time derivatives of the velocity are negligibly small" ("Der Erfolg der Theorie der Brown'schen Bewegung zeigte wieder deutlich, dass die klassische Mechanik stets dann zuverlässige Resultate lieferte, wenn sie auf Bewegungen angewandt wurde, bei welchen die höheren zeitlichen Ableitungen der Geschwindigkeit vernachlässigbar klein sind") (*Einstein 1979*, p. 46).

on the basis of these theories by means of both principle and constructive theories. At first, he seemed to suggest that some version of a generalized mechanical system, involving only a finite number of degrees of freedom, might provide a new basis for all of physics, including radiation.[73] But by the end of the decade, as we have seen, he hoped to derive the granular structure of radiation and electricity from some form of nonlinear field theory.

Einstein's efforts to incorporate gravitation into the theory of relativity led him in 1907 to formulate a new formal principle, later named the principle of equivalence.[74] He stressed that, when gravitational effects are taken into account, it is impossible to maintain the privileged role that inertial frames of reference still have in the original relativity theory. He concluded that, if gravitation is to be included, it is necessary to extend the relativity principle. The search for a group of transformations, wider than the Lorentz group, under which the laws of physics remain invariant when gravitation is included, lasted from 1907 until the end of 1915, leading finally to what Einstein considered his greatest achievement, the general theory of relativity.

[73] See *Einstein 1905i* (Doc. 14), pp. 132–133, and *Einstein 1907h* (Doc. 45), p. 372.

[74] See the editorial note, "Einstein on the Theory of Relativity," pp. 273–274. The similarity that he saw between relativity and thermodyamics, discussed above, may have prepared him to expect that, like the second law of thermodynamics, the original principle of relativity would have its limits.

SUPPLEMENT TO THE
EDITORIAL METHOD IN
VOLUME ONE

PRESENTATION OF TEXTS

Published papers are dated by the date of completion, if known; otherwise, by the earliest known of the dates of submission, reception, or publication.

Previously published texts are presented in facsimile, preserving all relevant features of the original publication, including pagination and running heads. Parts of texts not relevant to this edition but appearing on the same page as text by Einstein are, where possible, "screened" to reproduce them in lighter type. On the page preceding the text the reader will find the document number, English title, the document's short title from the list of Literature Cited, dates of completion, submission, reception and publication, if known, as well as the place of completion and the source for each text. Republished versions of an Einstein text with revisions or annotations that may be ascribed to Einstein are also noted on that page. Some of the information on the title page is repeated in the unnumbered descriptive note at the end of the document.

An English running head and a running page number appear above the rule on every page of text. Numbers indicating editorial footnotes are presented in square brackets in the margins of texts reproduced in facsimile, directly adjacent to the line containing the passage or equation upon which the note comments. Annotation begins after the last page of facsimile text, or on the bottom of the page of very short texts, if space allows.

If a text of a later supplement to a paper is, in the original, printed immediately following the paper, then this presentation is retained. Both dates are noted in the Table of Contents, on the relevant title page, and in a textual note.

In the case of Einstein's reviews, the original title of the article or book under review is given.

Translations are provided in the editorial apparatus for all non-English citations.

In this volume and in future volumes, subjects of biographies in an earlier *Correspondence* volume are identified in the Index by printing the subject's last name in small capitals, and following the name by the volume number in parentheses (e.g., "BESSO, Michele Angelo [Vol. 1]").

ANNOTATION

Notes correcting textual errors in facsimile texts are provided only where these errors might prove confusing.

Page references to documents reproduced in facsimile always refer to the pagination in the original publication.

A note appearing in an original text is cited as "fn." An editorial footnote is cited as "note."

In cases where readers are more familiar with a later version of a document, the occurrence of page breaks in the more familiar version is noted on the document and is referred to on the title page.

A NOTE ON SECONDARY LITERATURE

In preparing the editorial notes and other annotations for this volume, the editors have made extensive use of a large body of secondary literature commenting on the primary sources contained or cited in this volume. Since these secondary sources are cited only for specific purposes, the extent of our general debt to the secondary literature on Einstein and on the history of modern physics is not always evident. We acknowledge it here, with an expression of gratitude to the authors. Works which have been particularly useful include: *Bernhardt 1971*; *Brush 1976*; *Goldberg 1984*; *Hermann 1969*; *Hirosige 1976*; *Holton 1973* (Part 2), *1986* (Part 1); *Jammer 1966*; *Jungnickel and McCormmach 1986a, 1986b*; *Kangro 1976*; *Kerker 1976*; *Klein 1963b, 1965, 1967, 1970, 1972, 1974b, 1977, 1979, 1980, 1982a*; *Kuhn 1978*; *McCormmach 1967, 1970a, 1970b, 1976*; *Mehra and Rechenberg 1982*; *Miller 1981b, 1986*; *Nye 1972*; *Pais 1982*; *Pauli 1921, 1958*; *Pyenson 1985*; *Torretti 1983*; *Wheaton 1978b, 1983*.

Some of the Einstein correspondence cited in this volume has been previously published. In particular, the Einstein-Stark correspondence has been published in *Hermann 1966*; the Einstein-Mach correspondence in *Herneck 1966b*; the Einstein-Besso correspondence in *Einstein/Besso 1972*; and those parts of the Einstein-Sommerfeld correspondence cited in this volume in *Eckert and Pricha 1984*. Since the letters cited will ultimately be included in this edition, citations of these sources are not included in our annotations.

Unless otherwise noted, biographical information in the annotations is drawn from *Debus 1968*; *Gillispie 1970-1980*; or *J. C. Poggendorffs biographisch-literarisches Handwörterbuch*.

ACKNOWLEDGMENTS

We extend our thanks to the following persons, who made written comments on earlier versions of the editorial material in this volume: Professors Peter Bergmann, Yehuda Elkana, Gerald Holton, Res Jost, Horst Melcher, and Arthur I. Miller; and to Professor Nathan Reingold for extensive oral comments. We thank the members of the Editorial Committee, who discussed plans for the volume and an earlier draft at several lengthy committee meetings. The editor thanks the members of the Editorial Advisory Board for sharing their comments on the draft at a Board meeting he attended. Thanks are also extended to the Hebrew University of Jerusalem.

We also wish to thank Rhoda Bilansky and the other members of the staff of the Interlibrary Loan Office at Mugar Library of Boston University; Carolyn Fawcett of the Widener Library of Harvard University; Keith Glavash of the Microlab of MIT Library, who made the production of facsimiles possible, and Richard Newton, who assembled the material from a variety of sources; Klaus Hentschel, who assisted in preparing the index; Rudolf Hertz, who provided us with copies from the *Nachlaß* of his father, Paul Hertz; Professor Gerald Holton, who kindly permitted the editor to examine Einstein's offprint collection of his papers, given to Professor Holton by Einstein's secretary, Miss Helen Dukas; Bruno Jech, who made available a copy of an unpublished paper on and annotations of Einstein's dissertation; Peter McLaughlin, who assisted in the translation of material in the editorial apparatus; Annette Pringle, who shared major responsibility for preparing the Literature Cited, including extensive library research; and Clara Berry, Emmanuelle Cesari, Neil Schnepf, and Shawn Smith, who assisted in producing this volume. Anne Kox gratefully acknowledges financial support from the Netherlands Organization for Scientific Research (NWO).

TRANSLATION

The National Science Foundation has generously sponsored a translation project of *The Collected Papers of Albert Einstein*. This work is carried out independent of the documentary-edition project, by an independent translator, Dr. Anna Beck, and consultant, Professor Peter Havas. The volumes, produced in paperback in a typescript format, should be read only in conjunction with the documentary edition, as they contain none of the editorial commentary of that edition.

LOCATION SYMBOLS

Cz-Ar	Státní Ústřední Archív, Prague, Czechoslovakia
IsReW	Wix Library, Weizmann Institute, Rehovoth, Israel
NeLR	Museum Boerhaave (Rijksmuseum voor de Geschiedenis van de Natuurwetenschappen en van de Geneeskunde), Leiden, The Netherlands
SzBe	Stadt- und Universitätsbibliothek, Bern, Switzerland
SzBeSa	Staatsarchiv des Kantons Bern, Bern, Switzerland
SzZE	Eidgenössische Technische Hochschule, Zurich, Switzerland
SzZSa	Staatsarchiv des Kantons Zürich, Zurich, Switzerland
SzZU	Archiv der Universität Zürich, Zurich, Switzerland

TEXTS

EINSTEIN ON THE NATURE OF
MOLECULAR FORCES

Einstein's first two papers have a common theme: an investigation of the nature of molecular forces by means of the effect of such forces on various observable phenomena in liquids. *Einstein 1901* (Doc. 1) concentrates on capillary phenomena in neutral liquids; *Einstein 1902a* (Doc. 2) deals with dilute salt solutions.

Popular-scientific books and school texts no doubt familiarized Einstein at an early age with the general ideas of the atomic-molecular theory and the concept of cohesive forces between molecules.[1] Violle's textbook of physics, which Einstein studied in 1895, argues strongly for the molecular outlook, and contains a lengthy chapter on capillarity, based on Laplace's theory of short-range forces between the molecules of a fluid (see below).[2] Mach's *Mechanik*, which Einstein read about 1897, also introduces such forces to explain the shape of fluid bodies.[3] A record of his nascent interest in studying the nature of these forces dates from Einstein's ETH years. His notes on Weber's physics lectures contain the marginal comment "Investigate! Holidays." ("Untersuchen! Ferien.") next to his summary of Weber's comment on the unknown function of the distance between two identical molecules that characterizes the cohesive force between them.[4]

In his first published paper, *Einstein 1901* (Doc. 1), Einstein chose the phenomenon of capillarity to study the nature of intermolecular forces. Before him, many physicists had sought to explain capillarity on the basis of cohesive forces acting between the constituent particles of matter.[5] Laplace had proposed an ambitious program for unifying physics by postulating the existence of various central forces between molecules in order to explain a number of physical phenomena.[6] The most detailed and successful example of this approach is his theory of capillarity, based on the hypothesis of an attractive intermolecular

[1] These concepts are mentioned in both *Bernstein 1853–1857* and *Büchner 1855* (for evidence that Einstein read these books as a boy, see Vol. 1, "Albert Einstein—Beitrag für sein Lebensbild," p. lxii). The physics text that Einstein used during his last term at the Luitpold Gymnasium (*Krist 1891*; see Luitpold-Gymnasium, Curriculum, Vol. 1, Appendix B, p. 353), discusses the molecular structure of matter (ibid., p. 5) and forces between molecules, including cohesive forces (ibid., pp. 13–14).

[2] Einstein used Violle's textbook to prepare for the ETH entrance examination (see Vol. 1, "Albert Einstein—Beitrag für sein Lebensbild," p. lxiv). *Violle 1892* argues that "the assumption of molecules is . . . not just a hypothesis, but the simple expression of a fact" ("Die Annahme von Molecülen ist . . . nicht nur eine Hypothese, sondern der einfache Ausdruck einer Thatsache") (p. 339). The section on capillarity, *Violle 1893*, chap. III, pp. 575–663, includes extensive historical, experimental, and theoretical discussions.

[3] See *Mach 1897*. Mach's discussion of molecular forces is in sec. 10 of chap. III. For evidence that Einstein read Mach's *Mechanik*, see Einstein to Mileva Marić, 10 September 1899 (Vol. 1, Doc. 54), especially note 8.

[4] Vol. 1, Doc. 37, p. 130. Einstein took the course notes during the winter semester of 1897–1898, but it is possible that some of his marginal comments were added later. See Vol. 1, the editorial note, "Einstein on Molecular Forces," pp. 264–266, for a discussion of Einstein's early interest in this subject.

[5] For reviews of eighteenth and nineteenth-century theories of capillarity, see *Minkowski 1907a*; *Weber, R. H. 1916*; *Bikerman 1975, 1978*; *Rowlinson and Widom 1982*, chap. 1; *Rüger 1985*.

[6] For Laplace's program, see *Fox 1974*.

force with a range so short that the force vanishes at microscopic distances. On the basis of this assumption, Laplace was able to explain the existence of surface tension and several other capillary phenomena.[7] His theory of capillarity was elaborated by Poisson and Gauss, among others. Such static molecular models could not be used for the investigation of the thermal aspects of capillary phenomena. In the latter part of the nineteenth century a strictly thermodynamical approach to capillarity was developed, which relates various capillary phenomena without any assumptions about the underlying cause of the assumed surface tension. The concurrent successes of the kinetic theory of gases led to attempts to develop a kinetic theory of liquids, and thus to a revival of interest in the molecular theory of capillarity. Van der Waals, in particular, used the study of capillarity as a way of investigating molecular cohesive forces.[8] His work was the first to combine the methods of the Laplacian theory with those of kinetic theory. Maxwell, Rayleigh, William Thomson, and Boltzmann are other leading physicists who worked on the theory of capillarity during the last quarter of the nineteenth century. However, no substantial progress toward a kinetic theory of liquids was made.

A fellow ETH student later reported that Einstein was very impressed by a lecture on capillarity given by Minkowski during Einstein's last term at the ETH (April–July 1900).[9] In September 1900 he finished reading Boltzmann's *Gastheorie*, which contains a discussion of capillarity, based on Van der Waals's approach that may have further stimulated Einstein's interest.[10] The first evidence that he was actually at work on a theory of molecular forces dates from the following month. In a letter to Mileva Marić he refers to results on capillarity that he had recently found.[11] He believes the results to be completely new, and suggests that they look for empirical data to test them. "If a law of nature results" ("wenn sich dabei ein Naturgesetz ergibt"), he wrote, a paper would be submitted to the *Annalen der Physik*. Two months later, on 16 December 1900, the manuscript of *Einstein 1901* (Doc. 1) was received by the *Annalen*.

Einstein studied the dependence of capillary phenomena, and hence of molecular forces, on the chemical composition of neutral liquids. Like many contemporary treatments of capillarity, Einstein's paper combines the use of both thermodynamic and molecular-theoretical methods. The paper starts with an ingenious thermodynamical argument, based on the experimentally observed, approximately linear decrease in surface tension with increasing temperature, to prove that the surface energy of a fluid consists almost exclusively of potential energy. Einstein then calculated this surface potential energy on the basis of an assumption about the nature of molecular forces. Guided by a

---

[7] See *Laplace 1806*.

[8] See *Van der Waals 1873*. For discussions of Van der Waals's work, see *Rowlinson 1973; Klein 1974a*.

[9] See *Kollros 1956*, p. 21. Minkowski was already at work on his review article on capillarity, *Minkowski 1907a*.

[10] See Einstein to Mileva Marić, 13 September and 19 September 1900 (Vol. 1, Docs. 75 and 76). Boltzmann discusses Van der Waals's theory of fluids and its relationship with Laplace's theory of capillarity in sec. I of *Boltzmann 1898a*. Einstein sent a manuscript copy of his first paper to Boltzmann (see Mileva Marić to Helene Savić, 20 December 1900, Vol. 1, Doc. 85).

[11] Einstein to Mileva Marić, 3 October 1900 (Vol. 1, Doc. 79).

presumed analogy with gravitational forces, he postulated a similar form for the potential between two molecules:

$$P = P_\infty - c_1 c_2 \varphi(r),$$

where $\varphi(r)$ is a universal function of the distance r between the molecules. At the time, such an expression for the intermolecular potential was not uncommon; the masses of the interacting molecules were usually chosen as the constants c.[12] In Einstein's paper, however, the constants c_1 and c_2 are assumed to depend on the chemical nature of the molecules. The atoms of a chemical element are characterized by a certain value of c; the c of a molecule is assumed to be the sum of the c's of its constituent atoms. Einstein did not discuss the range of the intermolecular force in the paper, but his assumptions about the form of the potential imply that the range is the same for all molecules.[13] Einstein derived relations between the constants c for the molecules of a neutral liquid and several measurable properties of the liquid. Comparison with experimental values of these properties for various liquids allowed him to establish (relative) values of c for a number of elements.[14] From these, he calculated the values for a number of compounds, which he then compared with the experimentally established values for the latter. On the whole, he found the agreement between observed and calculated values to be satisfactory. Einstein continued to work on the topic of liquid surfaces during 1901, apparently without any further success.[15] He also derived a consequence from his theory that appeared to be in contradiction with the Van der Waals theory of liquids.[16]

After the paper was published on 1 March 1901, offprints were sent to several prominent physicists, notably Wilhelm Ostwald (who is cited in the paper), as part of Einstein's unsuccessful attempt during that spring to find a position as *Assistent* at some university.[17] In a letter to Ostwald accompanying the offprint, Einstein expressed his indebtedness to Ostwald's work.[18] If Ostwald read the paper, he was presumably not sympathetic to Einstein's molecular approach. Ostwald had been an ardent energeticist for some time, and was quite hostile to atomistic explanations.[19] In 1891 he already wrote some-

[12] See, e.g., *Boltzmann 1898a*, p. 56.

[13] A decade later he noted that this assumption is not tenable (see *Einstein 1911a*, and the discussion of it below).

[14] One of the constants he calculated turned out to be negative, implying the possibility of a repulsive force between molecules. Einstein did not comment on this question.

[15] See Einstein to Mileva Marić, 22 July 1901 (Vol. 1, Doc. 119).

[16] See Einstein to Mileva Marić, 12 December 1901 (Vol. 1, Doc. 127). The problem arose from Einstein's tacit assumption that the range of molecular forces is the same for all molecules.

[17] See Einstein to Otto Wiener, 9 March 1901 (Vol. 1, Doc. 90); Einstein to Wilhelm Ostwald, 19 March 1901 (Vol. 1, Doc. 92); and Einstein to Heike Kamerlingh Onnes, 12 April 1901 (Vol. 1, Doc. 98). In his letter to Wiener, Einstein only mentions his paper, presumably because he had not yet received offprints.

[18] Einstein to Wilhelm Ostwald, 19 March 1901 (Vol. 1, Doc. 92). Einstein states that reading *Ostwald 1891* had stimulated *Einstein 1901* (Doc. 1); but he seems to have used Ostwald's book more as a source of data on capillary phenomena than as a source of theoretical ideas on molecular forces.

[19] For a discussion of the development of Ostwald's views on energetics, see *Deltete 1983*, chap. V and appendix I.

what disparagingly about Laplace's work.[20] Two years later, he stated that atomistic explanations of capillarity depend on so many arbitrary assumptions that the topic is "not so much clarified as obscured" ("nicht sowohl aufgeklärt, als verdunkelt") by such explanations.[21] In spite of a second note by Einstein, and a letter from his father appealing for a word of encouragement, Ostwald apparently did not respond.[22]

A review of Einstein's paper comments on the "not altogether clear and unobjectionable derivations" ("nicht durchweg klaren und einwandfreien Herleitungen"), and notes that Einstein overlooked a thermodynamical relation between two equations he derived.[23] *Einstein 1901* (Doc. 1) is cited with more favorable comments several times in the literature on capillarity,[24] and one of its thermodynamical results was later called Einstein's equation.[25]

Einstein evidently was highly gratified by the ability of his theory of molecular forces to explain several apparently unrelated phenomena. A letter to Marcel Grossmann enthuses: "It is a glorious feeling to recognize the unity of a complex of phenomena, which appear to direct sense perception as quite distinct things." ("Es ist ein herrliches Gefühl, die Einheitlichkeit eines Komplexes von Erscheinungen zu erkennen, die der direkten sinnlichen Wahrnehmung als ganz getrennte Dinge erscheinen.")[26] Even more ambitious hopes for the theory are suggested by Einstein's allusions to gravitation in connection with it.[27] Although he acknowledged in *Einstein 1901* (Doc. 1) that his results leave the question of whether molecular forces are related to gravitation completely open, he continued to hope. In the letter to Grossmann just quoted, Einstein asserted that a generalization of his theory to gases would enable him to evaluate c for almost all the elements, which would take him a large step closer to settling this question. The next day he wrote to Marić expressing similar hopes.[28] However, there is no further mention of the question in his letters,[29] nor in his second paper on molecular forces, *Einstein 1902a* (Doc. 2).

Toward the end of 1901, Einstein hoped to obtain a doctorate with a dissertation based

---

[20] After praising Laplace's phenomenological theory of capillarity, Ostwald stated that Laplace's molecular forces "have not remained without contradiction and have not led to a knowledge of the nature of these forces" ("sind nicht ohne Widerspruch geblieben und haben zu einer Erkenntnis der Natur dieser Kräfte nicht geführt") (*Ostwald 1891*, p. 515).

[21] *Ostwald 1893*, p. 28.

[22] See Einstein to Wilhelm Ostwald, 3 April 1901 (Vol. 1, Doc. 95), and Hermann Einstein to Wilhelm Ostwald, 13 April 1901 (Vol. 1, Doc. 99). For Ostwald's failure to reply, see Einstein to Mileva Marić, 10 April 1901 (Vol. 1, Doc. 97).

[23] See *Wiedeburg 1901*; see also *Einstein 1901* (Doc. 1), note 24.

[24] See *Pockels 1908*, pp. 1123; *Freundlich 1909*, pp. 41, 43–45; *Weber, R. H. 1916*, p.

110; and *Schottky 1929*, p. 116. *Kleeman 1909* tries to develop Einstein's theory by assuming, contrary to an argument in Einstein's paper, that the surface energy is temperature dependent.

[25] See *Defay and Prigogine 1951*, pp. 38–40; and *Einstein 1901* (Doc. 1), note 3.

[26] Einstein to Marcel Grossmann, 14 April 1901 (Vol. 1, Doc. 100).

[27] Speculations on a common origin of gravitation and molecular forces were not uncommon at the time. See, e.g., *Ostwald 1891*, which attributes the idea of a relation between the two forces to Van 't Hoff (see ibid., p. 1142).

[28] Einstein to Mileva Marić, 15 April 1901 (Vol. 1, Doc. 101).

[29] However, Michele Besso refers to the question in a letter to Einstein of 7–11 February 1903.

on all of his work on molecular forces.[30] He failed to obtain the doctorate at this time, but continued his study of molecular forces. His second paper includes an application of his theory to metallic ions in dilute salt solutions.

Einstein had been interested in physical chemistry for some time. Reporting to Marić in October 1900 on his study of this subject, he expressed enthusiasm about the results obtained during the last 30 years, calling the ion theory "by far the most splendid" ("das aller prachtvollste").[31] A reference to Nernst's theory of electrolytes in *Einstein 1902a* (Doc. 2) indicates that he had previously studied it.[32] Einstein first mentioned his plan to extend his theory of molecular forces to salt solutions in a letter to Marić.[33] He apparently had hoped to treat solutions of arbitrary strength, but now realized that he would have to confine himself to infinitely dilute solutions in order to avoid the need to consider interactions between the solute molecules. In *Einstein 1902a* (Doc. 2), published in July 1902, Einstein considered systems consisting of metallic electrodes immersed in dilute solutions of various salts of the same metal as the electrodes. The ions and solvent molecules are subject to external conservative forces. A purely thermodynamical argument is used to prove that the potential difference between the electrodes and the salt solution does not depend on the nature of the negative salt ions, or on the external forces. Next, Einstein's theory of molecular forces is used to determine how this potential difference depends on the nature of the solvent, assuming that the force between solute ions and solvent molecules in dilute salt solutions does not depend on the charge of the ions. Einstein proposed an experimental method to both test the validity of this assumption and evaluate the constants c for the ions and solvent molecules. He concluded by expressing the hope that someone would undertake the "cumbersome experiment" ("mühevolle Untersuchung") described, since he is not in a position to do so; but no one seems to have carried out this program.

Although he published nothing further on his theory of molecular forces for the remainder of the decade, Einstein remained actively interested in the subject until at least 1903. In January of that year he wrote: "In the near future I will occupy myself with molecular forces in gases" ("In der nächsten Zeit will ich mich mit den Molekularkräften in Gasen abgeben").[34] In March he wrote: "I am having great difficulty collecting the material for my work on molecular forces" ("Ich habe grosse Mühe, bis ich für meine Arbeit über Molekularkräfte das Material zusammenkriege").[35] A few years later, Einstein wrote

[30] He actually submitted a dissertation on molecular forces in gases to Professor Alfred Kleiner of the University of Zurich, but later withdrew it (see Mileva Marić to Helene Savić, ca. 23 November–mid-December 1901, Vol. 1, Doc. 125; Einstein to Mileva Marić, 28 November 1901, 12 December 1901, 17 December 1901, Vol. 1, Docs. 126–128; Einstein to Swiss Patent Office, 18 December 1901, Vol. 1, Doc. 129; and Receipt for Return of Doctoral Fees, 1 February 1902, Vol. 1, Doc. 132). For a discussion of Einstein's first attempt to obtain a doc-torate, see the editorial note, "Einstein's Dissertation on the Determination of Molecular Dimensions," pp. 174–175.

[31] Einstein to Mileva Marić, 3 October 1900 (Vol. 1, Doc. 79).

[32] See, e.g., *Nernst 1898*, pp. 659–678.

[33] Einstein to Mileva Marić, 15 April 1901 (Vol. 1, Doc. 101).

[34] Einstein to Michele Besso, 22 January 1903.

[35] Einstein to Michele Besso, 17 March 1903.

deprecatingly about his two papers on molecular forces. He sent Stark a collection of his offprints, omitting "my two worthless beginners' works" ("meine zwei wertlosen Erstlingsarbeiten").[36]

Einstein published one final paper on capillarity and molecular forces, *Einstein 1911a*. In this paper, he demonstrated, without mentioning his earlier papers, that the assumption of a universal range for the molecular cohesive force is incompatible with certain empirically established laws obeyed by the surface tension. He stressed the need to introduce a range for the force between two molecules that is dependent on the nature of the molecules.

Einstein 1902a (Doc. 2) opens with a discussion of the conditions for the validity of the second law of thermodynamics, a discussion that proved to be of some significance for Einstein's work. He points out that the theories of dissociation and of dilute solutions are based on the application of the second law of thermodynamics to processes involving idealized semipermeable membranes that can separate any two (or more) substances. Although even the approximate realizability of such processes is often doubtful, the predictions of the theories are confirmed by experiment.[37] It seems, therefore, that one can draw valid conclusions from the consideration of highly idealized processes. In particular, it seems that one may apply the second law to mixtures acted on by external conservative forces, which produce the same effects as semipermeable membranes. Generalizing this conclusion, Einstein formulated the hypothesis that the second law may be applied to mixtures, the components of which are subjected to arbitrary conservative forces. The arguments in the body of the paper make extensive use of this hypothesis. Einstein's next paper, which initiated his study of the statistical foundations of thermodynamics, may have been stimulated by the need he felt to justify this hypothesis. In any case, the final section of that paper is devoted to providing such a justification.[38]

[36] Einstein to Stark, 7 December 1907. He did not mention his theory of molecular forces in his *Autobiographical Notes* (see *Einstein 1979*).

[37] See *Planck 1891* for a discussion of the role of ideal processes in thermodynamics. *Nernst 1898*, pp. 102–103, discusses the thermodynamical use of idealized semipermeable membranes. See also *Einstein 1902a* (Doc. 2), note 1.

[38] See *Einstein 1902b* (Doc. 3), p. 433, and the editorial note, "Einstein on the Foundations of Statistical Physics," p. 46.

1. "Conclusions Drawn from the Phenomena of Capillarity"

[Einstein 1901]

DATED Zurich, 13 December 1900
RECEIVED 16 December 1900
PUBLISHED 1 March 1901

IN: *Annalen der Physik* 4 (1901): 513–523.

5. *Folgerungen aus den Capillaritätserscheinungen;* *von Albert Einstein.*

Bezeichnen wir mit γ diejenige Menge mechanischer Arbeit, welche wir der Flüssigkeit zuführen müssen, um die freie Oberfläche um die Einheit zu vergrössern, so ist γ nicht etwa die gesamte Energiezunahme des Systems, wie folgender Kreisprocess lehrt. Sei eine bestimmte Flüssigkeitsmenge vorliegend von der (absoluten) Temperatur T_1 und der Oberfläche O_1. Wir vermehren nun isothermisch die Oberfläche O_1 auf O_2, erhöhen die Temperatur auf T_2 (bei constanter Oberfläche), vermindern dann die Oberfläche auf O_1 und kühlen dann die Flüssigkeit wieder auf T_1 ab. Nimmt man nun an, dass dem Körper ausser der ihm vermöge seiner specifischen Wärme zukommenden keine andere Wärmemenge zugeführt wird, so ist bei dem Kreisprocess die Summe der dem Körper zugeführten Wärme gleich der Summe der ihm entnommenen. Es muss also nach dem Princip von der Erhaltung der Energie auch die Summe der zugeführten mechanischen Arbeiten gleich Null sein.

Es gilt also die Gleichung:

[1]
$$(O_2 - O_1)\gamma_1 - (O_2 - O_1)\gamma_1 = 0 \quad \text{oder} \quad \gamma_1 = \gamma_2.$$

[2] Dies widerspricht aber der Erfahrung.

Es bleibt also nichts anderes übrig als anzunehmen, dass mit der Aenderung der Oberfläche auch ein Austausch der Wärme verbunden sei, und dass der Oberfläche eine eigene specifische Wärme zukomme. Bezeichnen wir also mit U die Energie, mit S die Entropie der Oberflächeneinheit der Flüssigkeit, mit s die specifische Wärme der Oberfläche, mit w_0 die zur Bildung der Oberflächeneinheit erforderliche Wärme in mechanischem Maass, so sind die Grössen:

$$dU = s.O.dT + \{\gamma + w_0\}dO$$

und

$$dS = \frac{s.O.dT}{T} + \frac{w_0}{T}dO$$

vollständige Differentiale. Es gelten also die Gleichungen:

$$\frac{\partial\,(s\,.\,O)}{\partial\,O} = \frac{\partial\,(\gamma + w_0)}{\partial\,T},$$

$$\frac{\partial}{\partial\,O}\left(\frac{s\,O}{T}\right) = \frac{\partial}{\partial\,T}\left(\frac{w_0}{T}\right).$$

Aus diesen Gleichungen folgt:

$$\gamma + w_0 = \gamma - T\frac{\partial\,\gamma}{\partial\,T}.$$

Dies aber ist die gesamte Energie, welche zur Bildung der Einheit der Oberfläche nötig ist.

Bilden wir noch:

$$\frac{d}{d\,T}\,(\gamma + w_0) = -\,T\,\frac{d^2\,\gamma}{d\,T^2}.$$

[3]

Die Experimentaluntersuchungen haben nun ergeben, dass sich stets sehr nahe γ als lineare Function der Temperatur darstellen lässt, d. h.:

[4]

Die zur Bildung der Oberflächeneinheit einer Flüssigkeit nötige Energie ist unabhängig von der Temperatur.

Ebenso folgt:

$$s = \frac{d\,\gamma}{d\,T} + \frac{d\,w_0}{d\,T} = \frac{d\,\gamma}{d\,T} - \frac{d\,\gamma}{d\,T} - T\frac{d^2\,\gamma}{d\,T^2} = 0,$$

also: Der Oberfläche als solcher ist kein Wärmeinhalt zuzuschreiben, sondern die Energie der Oberfläche ist potentieller Natur. Man sieht schon jetzt, dass

[5]

$$\gamma - T\frac{d\,\gamma}{d\,T}$$

eine zu stöchiometrischen Untersuchungen sich geeignetere Grösse ist, als das bisher benutzte γ bei Siedetemperatur. Die Thatsache, dass die zur Bildung der Oberflächeneinheit erforderliche Energie kaum mit der Temperatur variirt, lehrt uns aber auch, dass die Configuration der Molecüle in der Oberflächenschicht mit der Temperatur nicht variiren wird (abgesehen von Aenderungen von der Grössenordnung der thermischen Ausdehnung).

Um nun für die Grösse

$$\gamma - T\frac{d\,\gamma}{d\,T}$$

eine stöchiometrische Beziehung aufzufinden, ging ich von den einfachsten Annahmen über die Natur der molecularen Anziehungskräfte aus, und prüfte deren Consequenzen auf ihre Uebereinstimmung mit dem Experiment hin. Ich liess mich [6] dabei von der Analogie der Gravitationskräfte leiten.

Sei also das relative Potential zweier Molecüle von der Form:

$$P = P_\infty - c_1 \cdot c_2 \cdot \varphi(r),$$

wobei c eine für das betreffende Molecül charakteristische Constante ist, $\varphi(r)$ aber eine vom Wesen der Molecüle un- [7] abhängige Function ihrer Entfernung. Wir nehmen ferner an, dass

$$\frac{1}{2} \sum_{\alpha=1}^{n} \sum_{\beta=1}^{n} c_\alpha c_\beta \varphi(r_{\alpha,\beta})$$

der entsprechende Ausdruck für n Molecüle sei. Sind speciell alle Molecüle gleich beschaffen, so geht dieser Ausdruck in

$$\frac{1}{2} c^2 \sum_{\alpha=1}^{n} \sum_{\beta=1}^{n} \varphi(r_{\alpha,\beta})$$

über. Wir machen ferner noch die Annahme, dass das Potential der Molecularkräfte ebenso gross sei, wie wenn die Materie homogen im Raume verteilt wäre; es ist dies allerdings eine Annahme, von der wir nur angenähert die Richtigkeit erwarten dürfen. Mit ihrer Hülfe verwandelt sich der obige Ausdruck in:

[8]
$$P = P_\infty - \frac{1}{2} c^2 N^2 \iint d\tau \cdot d\tau' \varphi(r_{d\tau, d\tau'}),$$

wobei N die Anzahl der Molecüle in der Volumeneinheit ist. Ist das Molecül unserer Flüssigkeit aus mehreren Atomen zusammengesetzt, so soll analog wie bei den Gravitationskräften $c = \sum c_\alpha$ gesetzt werden können, wobei die c_α den Atomen der Elemente charakteristische Zahlen bedeuten. Setzt man noch $1/N = v$, wobei v das Molecularvolum bedeutet, so erhält man die endgültige Formel:

$$P = P_\infty - \frac{1}{2} \frac{\left(\sum c_\alpha\right)^2}{v^2} \iint d\tau \cdot d\tau' \varphi(r_{d\tau, d\tau'}).$$

Setzen wir nun noch voraus, dass die Dichte der Flüssigkeit bis zu deren Oberfläche constant ist, was ja durch die Thatsache wahrscheinlich gemacht wird, dass die Energie der Oberfläche von der Temperatur unabhängig ist, so sind wir nun im stande die potentielle Energie der Volumeneinheit im

516   *A. Einstein.*

Inneren der Flüssigkeit und die der Oberflächeneinheit zu berechnen.

Setzen wir nämlich

$$\frac{1}{2}\int_{x=-\infty}^{+\infty}\int_{y=-\infty}^{+\infty}\int_{z=-\infty}^{+\infty} dx\,dy\,dz\,.\,\varphi\,(\sqrt{x^2+y^2+z^2}) = K,$$

so ist die potentielle Energie der Volumeneinheit

$$P = P_\infty - K\frac{(\sum c_a)^2}{v^2}.$$

Denken wir uns eine Flüssigkeit vom Volumen V und von der Oberfläche S, so erhalten wir durch Integration

$$P = P_\infty - K\frac{(\sum c_a)^2}{v^2}\cdot V - K'\frac{(\sum c_a)^2}{v^2}\cdot O, \qquad [9]$$

wobei die Constante K' bedeutet:

$$\int_{x'=0}^{x'=1}\int_{y'=0}^{y'=1}\int_{z'=\infty}^{z'=0}\int_{x=-\infty}^{z=\infty}\int_{y=-\infty}^{y=\infty}\int_{z=0}^{z=\infty} dx.dy.dz.dx'.dy'.dz \qquad [10]$$
$$\varphi\,\sqrt{(x-x')^2+(y-y')^2+(z-z')^2}.$$

Da über φ nichts bekannt ist, bekommen wir natürlich keine Beziehung zwischen K und K'.

Dabei ist zunächst im Auge zu behalten, dass wir nicht wissen können, ob das Flüssigkeitsmolecül nicht die n-fache Masse des Gasmolecüles besitzt, doch folgt aus unserer Herleitung, dass dadurch unser Ausdruck der potentiellen Energie der Flüssigkeit nicht geändert wird. Für die potentielle Energie der Oberfläche bekommen wir, auf Grund der eben gemachten Annahme, den Ausdruck:

$$P = K'\frac{(\sum c_a)^2}{v^2} = \gamma - T\frac{d\gamma}{dT}, \qquad [11]$$

oder

$$\sum c_a = v\cdot\sqrt{\gamma - T\frac{d\gamma}{dT}\cdot\frac{1}{\sqrt{K'}}}.$$

Da die rechts stehende Grösse für Siedetemperatur für viele Stoffe aus den Beobachtungen von R. Schiff berechenbar ist, so bekommen wir reichlichen Stoff zur Bestimmung der Grössen c_a: Ich entnahm das gesamte Material dem Buch

[12]   über Allgemeine Chemie von **W. Ostwald**. Ich gebe hier zunächst das Material an, mittels dessen ich das c_α für C, H, O nach der Methode der kleinsten Quadrate berechnete. In der mit $\sum c_{\alpha_{\text{ber.}}}$ überschriebenen Columne sind die $\sum c_\alpha$ angegeben, wie sie mit Hülfe der so gewonnenen c_α aus den chemischen Formeln sich ergeben. Isomere Verbindungen wurden zu einem Wert vereinigt, weil die ihnen zugehörigen Werte der linken Seite nur unbedeutend voneinander abwichen. Die Einheit wurde willkürlich gewählt, weil, da K' unbekannt ist, eine absolute Bestimmung der c_α nicht möglich ist.

Ich fand:

$$c_H = -1{,}6, \quad c_C = 55{,}0, \quad c_O = 46{,}8 .$$

Formel	$\sum c_\alpha$	$\sum c_{\alpha_{\text{ber.}}}$	Name der Verbindung
$C_{10}H_{16}$	510	524	Citronenterpen
CO_2H_2	140	145	Ameisensäure
$C_2H_4O_2$	193	197	Essigsäure
$C_3H_6O_2$	250	249	Propionsäure
$C_4H_8O_2$	309	301	Buttersäure und Isobuttersäure
$C_5H_{10}O_2$	365	352	Valeriansäure
$C_4H_6O_3$	350	350	Acetanhydrid
$C_6H_{10}O_4$	505	501	Aethyloxalat
$C_8H_8O_2$	494	520	Methylbenzoat
$C_9H_{10}O_2$	553	562	Aethylbenzoat
$C_6H_{10}O_3$	471	454	Acetessigäther
C_7H_8O	422	419	Anisol
$C_8H_{10}O$	479	470	Phenetol und Methylcresolat
$C_8H_{10}O_2$	519	517	Dimethylresorcin
$C_5H_4O_2$	345	362	Furfurol
$C_5H_{10}O$	348	305	Valeraldehyd
$C_{10}H_{14}O$	587	574	Carvol

Man sieht, dass die Abweichungen in fast allen Fällen die Versuchsfehler wohl kaum übersteigen und keinerlei Gesetzmässigkeit zeigen.

Hierauf berechnete ich gesondert die Werte für Cl, Br und J, welchen Bestimmungen natürlich eine geringere Sicherheit zukommt, und fand:

$$c_{Cl} = 60, \quad c_{Br} = 152, \quad c_J = 198 .$$

Ich lasse nun in gleicher Weise wie oben das Material folgen:

518   *A. Einstein.*

Formel	$\sum c_a$	$\sum c_{a_{ber.}}$	Name der Verbindung
C_6H_5Cl	385	379	Chlorbenzol
C_7H_7Cl	438	434	Chlortoluol
C_7H_7Cl	450	434	Benzychlorid
C_3H_5OCl	270	270	Epichlorhydrin
C_2OHCl_3	358	335	Chloral
C_7H_5OCl	462	484	Benzoylchlorid
$C_7H_6Cl_2$	492	495	Benzylidenchlorid
Br_2	217	304	Brom
C_2H_5Br	251	254	Aethylbromid
C_3H_7Br	311	306	Propylbromid
C_3H_7Br	311	306	Isopropylbromid
C_3H_5Br	302	309	Allylbromid
C_4H_5Br	353	354	Isobutylbromid
$C_5H_{11}Br$	425	410	Isoamylbromid
C_6H_5Br	411	474	Brombenzol
C_7H_7Br	421	526	o-Bromtoluol
$C_2H_4Br_2$	345	409	Aethylenbromid
$C_3H_6Br_2$	395	461	Propylenbromid
C_2H_5J	288	300	Aethyljodid
C_3H_7J	343	352	Propyljodid
C_3H_7J	357	352	Isopropyljodid
C_3H_5J	338	355	Allyljodid
C_4H_9J	428	403	Isobutyljodid
$C_5H_{11}J$	464	455	Isoamyljodid

Es scheint mir, dass grössere Abweichungen von unserer Theorie bei solchen Stoffen eintreten, welche verhältnismässig grosse Molecularmaasse und kleines Molecularvolum haben.

Wir haben aus unseren Annahmen gefunden, dass die potentielle Energie der Volumeneinheit den Ausdruck besitzt:

$$P = P_\infty - K \frac{(\sum c_a)^2}{v^2},$$

dabei bedeutet K eine bestimmte Grösse, welche wir aber nicht berechnen können, da es überhaupt erst durch die Wahl der c_a vollkommen definirt wird. Wir können daher $K = 1$ setzen und gewinnen so eine Definition für die absoluten Werte der c_a. Berücksichtigen wir dies von nun an, so erhalten wir für die Grösse des Potentiales, welche dem Aequivalent (Molecül) zukommt, den Ausdruck:

$$P = P_\infty - K \frac{(\sum c_a)^2}{v},$$

wobei natürlich P_∞ eine andere Constante bedeutet. Nun könnten wir aber das zweite Glied der rechten Seite dieser Gleichung der Differenz $D_m J - A v_d$ gleich setzen — wobei D_m die moleculare Verdampfungswärme (Dampfwärme × Molecularmasse), J das mechanische Aequivalent der Calorie, A den Atmosphärendruck in absolutem Maass und v_d das Molecular-

[13] volum des Dampfes ist —, wenn die potentielle Energie des Dampfes Null wäre und wenn für Siedetemperatur der Inhalt an kinetischer Energie beim Uebergang vom flüssigen in den Gaszustand ungeändert bliebe. Die erste dieser Annahmen scheint mir unbedenklich. Da wir aber zu der letzteren Annahme keinen Grund haben, aber auch keine Möglichkeit die fragliche Grösse abzuschätzen, so bleibt uns nichts anderes übrig, als die obige Grösse selbst zur Rechnung zu benutzen.

[14]   In die erste Spalte der folgenden Tabelle habe ich die Grössen $\sqrt{D_m' . v}$ im Wärmemaass eingetragen, wobei D_m' die um die äussere Verdampfungsarbeit (in Wärmemaass) verminderte Verdampfungswärme bedeutet. In die zweite setzte ich die Grössen $\sum c_a$, wie sie aus den Capillaritätsversuchen ermittelt sind; in der dritten finden sich die Quotienten beider Werte. Isomere Verbindungen sind wieder zu einer Zeile vereinigt.

Name der Verbindung	Formel	$\sqrt{D_m' . v}$	$\sum c_{a_{\text{ber.}}}$	Quotient
Isobutylpropionat	$C_7H_{14}O_2$	1157	456	2,54
Isoamilacetat	„			
Propylacetat	„			
Isobutylisobutyrat	$C_8H_{16}O_2$	1257	510	2,47
Propylvalerat	„			
Isobutylbutyrat	„			
Isoamylpropionat	„			
Isoamylisobutyrat	$C_9H_{18}O_2$	1367	559	2,45
Isobutylvalerat	„			
Isoamylvalerat	$C_{10}H_{10}O_2$	1464	611	2,51
Benzol	C_6H_6	795	310	2,57
Toluol	C_7H_8	902	372	2,48
Aethylbenzol	C_8H_{10}	1005	424	2,37
m-Xylol	„			
Propylbenzol	C_9H_{12}	1122	475	2,36
Mesitylen	„			
Cymol	$C_{10}H_{14}$	1213	527	2,30
Aethylformiat	$C_3H_6O_2$	719	249	2,89
Methylacetat	„			

Name der Verbindung	Formel	$\sqrt{D_m' \cdot v}$	$\sum c_{\alpha \text{ber.}}$	Quotient
Aethylacetat	$C_4H_8O_2$	837	301	2,78
Methylpropionat	„			
Propylformiat	„			
Methylisobutyrat	$C_5H_{10}O_2$	882	353	2,50
Isobutylformiat	.,			
Aethylpropionat	.,			
Propylacetat	„			
Methylbutyrat	„			
Aethylisobutyrat	$C_6H_{12}O_2$	971	405	2,40
Methylvalerat	,,			
Isobutylacetat	,,			
Aethylbutyrat	,,			
Propylpropionat	,.			
Isoamylformiat	„			

Trotzdem der in der fünften Columne eingetragene Quotient keineswegs eine Constante ist, sondern vielmehr deutlich von der Constitution der Stoffe abhängt, so können wir das vorliegende Material doch dazu benutzen, diejenige Zahl, wenigstens der Grössenordnung nach, zu ermitteln, mit der unsere c_α multiplicirt werden müssen, damit wir sie in der von uns gewählten absoluten Einheit erhalten. Der gesuchte Multiplicator ergiebt sich im Mittel:

$$2,51 \cdot \sqrt{4,17 \cdot 10^7} = 1,62 \cdot 10^4. \qquad [15]$$

Da die vorhergehende Betrachtung zeigt, dass sich bei der Verdampfung die kinetischen Verhältnisse der Molecüle verändern (wenigstens wenn unser Ausdruck für die potentielle Energie richtig ist), unternahm ich es die absolute Grösse c_α noch auf eine andere Weise aufzusuchen. Dabei ging ich von der folgenden Idee aus:

Comprimirt man eine Flüssigkeit isothermisch und ändert [16] sich dabei ihr Wärmeinhalt nicht, was wir nun voraussetzen wollen, so ist die bei der Compression entweichende Wärme gleich der Summe der Compressionsarbeit und der von den Molecularkräften geleisteten Arbeit. Wir können also letztere Arbeit berechnen, wenn wir die bei der Compression entweichende Wärmemenge eruiren können. Dazu aber verhilft uns das Carnot'sche Princip.

Sei nämlich der Zustand der Flüssigkeit durch den Druck p in absoluten Einheiten und die absolute Temperatur T bestimmt; ist nun bei einer unendlich kleinen Zustandsänderung dQ die dem Körper zugeführte Wärme in absolutem Maass, dA die ihm zugeführte mechanische Arbeit, und setzen wir

$$dQ = X\,dp + S\,.\,dT,$$

$$dA = -p\,.\,dv = -p\left\{ \frac{\partial v}{\partial p}\,dp + \frac{\partial v}{\partial T}\,dT \right\}$$

[17]
$$= p\,.\,v\,.\,\varkappa\,dp - p\,.\,v\,.\,\alpha\,dT,$$

so liefert uns die Bedingung, dass dQ/T und $dQ + dA$ vollständige Differentiale sein müssen, die Gleichungen

$$\frac{\partial}{\partial T}\left(\frac{X}{T} \right) = \frac{\partial}{\partial p}\left(\frac{S}{T} \right)$$

und

[18]
$$\frac{\partial}{\partial T}(X + p\,\varkappa) = \frac{\partial}{\partial p}(S - p\,\alpha)$$

hierbei bedeuten, wie man sieht, X die bei isothermischer Compression durch den Druck $p = 1$ dem Körper zugeführte Wärme in mechanischem Maass, S die specifische Wärme bei constantem Druck, $\varkappa$ den Compressibilitätscoefficienten, α den thermischen Ausdehnungscoefficienten. Aus diesen Gleichungen findet man:

[19]
$$X\,dp = -T\left\{ \alpha + p\,\frac{\partial \alpha}{\partial p} + p\,\frac{\partial \varkappa}{\partial T} \right\}dp\,.$$

Nun ist daran zu erinnern, dass der Atmosphärendruck, unter dem sich unsere Körper gewöhnlich finden, für Compressionserscheinungen von Flüssigkeiten unbedenklich als unendlich klein zu betrachten ist, ebenso sind die Compressionen in unseren Experimenten sehr nahe proportional den angewandten Compressionskräften. Die Erscheinungen gehen also so vor sich, wie wenn die Compressionskräfte unendlich klein wären. Berücksichtigt man dies, so geht unsere Gleichung über in:

[20]
$$X\,.\,dp = -T\,.\,\alpha\,.\,dp\,.$$

Wenden wir nun die Voraussetzung an, dass bei isothermischer Compression die kinetische Energie des Systems nicht geändert wird, so erhalten wir die Gleichung

$$X\,.\,dp + \text{Compressionsarbeit} + \text{Arbeit der Molecularkräfte} = 0.$$

Ist P das Potential der Molecularkräfte, so ist die letzte Arbeit:

$$\frac{\partial P}{\partial v} \cdot \frac{\partial v}{\partial p} \cdot dp\,.$$

[21]

Setzt man unseren Ausdruck für die Grösse des Potentiales der Molecularkräfte hierin ein und berücksichtigt, dass die Compressionsarbeit von der Ordnung dp^2 ist, so erhält man bei Vernachlässigung dieser unendlich kleinen Grösse zweiter Ordnung

$$\frac{T_a}{\varkappa} = \frac{(\sum c_a)^2}{v^2}\,,$$

[22]

wobei $\varkappa$ den Compressibilitätscoefficienten in absolutem Maasse bezeichnet. Wir erhalten so abermals ein Mittel, den gesuchten Proportionalitätscoefficienten für die Grössen c_a zu bestimmen. Die Grössen α und $\varkappa$ für die Temperatur des Eises entnahm ich den Tabellen von Landolt und Börnstein. Man erhält so [23] für den gesuchten Factor die Werte:

Xylol	$1{,}71 \cdot 10^4$	Aethylalkohol	$1{,}70 \cdot 10^4$
Cymol	$1{,}71 \cdot 10^4$	Methylalkohol	$1{,}74 \cdot 10^4$
Terpentinöl	$1{,}73 \cdot 10^4$	Propylalkohol	$1{,}82 \cdot 10^4$
Aethyläther	$1{,}70 \cdot 10^4$	Amylalkohol	$2{,}00 \cdot 10^4$

Zunächst ist zu bemerken, dass die beiden durch verschiedene Methoden erlangten Coefficienten recht befriedigend übereinstimmen, trotzdem sie aus ganz verschiedenen Phenomenen hergeleitet sind. Die letzte Tabelle zeigt sehr befriedigende Uebereinstimmung der Werte, nur die kohlenstoffreicheren Alkohole weichen ab. Es ist dies auch zu erwarten, denn aus den Abweichungen, welche die Alkohole von dem thermischen Ausdehnungsgesetz von Mendelejew und von dem [25] stöchiometrischen Capillaritätsgesetz von R. Schiff zeigen, hat [26] man schon früher geschlossen, dass bei diesen Verbindungen mit Temperaturänderungen Aenderungen der Grösse der Flüssigkeitsmolecüle verbunden sind. Es ist also auch zu erwarten, dass bei isothermischer Compression solche moleculare Veränderungen auftreten, sodass für solche Stoffe bei gleicher Temperatur der Wärmeinhalt Function des Volums sein wird.

Zusammenfassend können wir also sagen, dass sich unsere fundamentale Annahme bewährt hat: Jedem Atom entspricht ein moleculares Anziehungsfeld, welches unabhängig von der Temperatur und unabhängig von der Art ist, wie das Atom mit anderen Atomen chemisch verbunden ist.

Schliesslich ist noch darauf hinzuweisen, dass mit steigendem Atomgewicht im allgemeinen auch die Constanten c_α steigen, doch nichts stets und nicht in proportionaler Art. Die Frage, ob und wie unsere Kräfte mit den Gravitationskräften verwandt sind, muss also noch vollkommen offen gelassen werden. Es [27] ist auch hinzuzufügen, dass die Einführung der Function $\varphi(r)$, welche unabhängig von der Natur der Molecüle sein sollte, nur [28] als Näherungsannahme aufzufassen ist, ebenso die Ersetzung [29] der Summen durch Integrale; in der That scheint sich unsere Theorie für Stoffe von kleinem Atomvolum nicht zu bewähren, wie das Beispiel des Wassers darthut. Ueber diese Fragen sind erst von eingehenden Specialforschungen Aufschlüsse zu hoffen.

Zürich, den 13. December 1900.

(Eingegangen 16. December 1900.)

Published in *Annalen der Physik* 4 (1901): 513–523. Dated Zurich, 13 December 1900, received 16 December 1900, published 1 March 1901.

[1] The second γ_1 in the first equation should be γ_2.

[2] Experiment shows that the surface tension γ decreases as the temperature increases (see note 4). Einstein's argument here is similar to one first given by William Thomson (Lord Kelvin) in *Thomson, W. 1858*.

[3] This equation has been called Einstein's equation (*Defay and Prigogine 1951*, pp. 38–40). The use of total derivatives here is justified since γ is a function of temperature only, if the curvature of the surface is neglected.

[4] *Ostwald 1891*, which Einstein used as a source for many of the experimental data cited in this paper (see the editorial note, "Einstein on the Nature of Molecular Forces," p. 5), cites experimental evidence for such a linear dependence of γ on temperature (see pp. 523–525, 526–530).

[5] "Wärmeinhalt" is used here in the sense of

internal kinetic energy (see H. F. Weber's Lectures on Physics, ca. December 1897–ca. June 1898, Vol. 1, Doc. 37, note 34).

[6] For a discussion of Einstein's comments on a possible relation between gravitational and molecular forces, see the editorial note, "Einstein on the Nature of Molecular Forces," p. 6. See also p. 523 of the present paper.

[7] Later in this paper, Einstein notes that his assumption about $\varphi(r)$ is to be regarded as only an approximation (see p. 523). *Einstein 1911a* reexamines this assumption without mentioning his earlier work. Einstein shows that assuming the range of the intermolecular forces to be the same for all molecules leads to consequences that are incompatible with empirically established relations between γ, v, and T. These relations imply that the range must be approximately equal to the distance between the molecules of the fluid.

[8] Since Laplace, the replacement of sums by integrals in the theory of capillarity traditionally had been justified by assuming that the range of

the molecular force was much greater than the molecular radius. However, by the end of the nineteenth century there were reasons to question this assumption. For a historical review of this mean-field approximation in capillary theory, see *Rowlinson and Widom 1982*, pp. 17–21. See also note 7.

[9] *O* denotes the surface area. The sign of the last term in this equation should be positive.

[10] This expression for K' is incorrect, as is clear even from dimensional considerations. *Einstein 1911a*, without mentioning the present paper, gives the correct expression (see the formula for K_2 on p. 167).

[11] This expression is actually the surface potential energy per unit area.

[12] See *Ostwald 1891*, pp. 528–530, for Schiff's data on surface tension; pp. 376–385 list values for the molar volumes of a large number of chemical compounds at their boiling point.

[13] *Boltzmann 1898a*, pp. 59–60, calculates an expression for the heat of vaporization of a liquid, based on the intermolecular force law, to which Einstein's is equivalent. In calculating the external work done against atmospheric pressure, Einstein neglected the volume of the liquid in comparison with that of the vapor.

[14] *Ostwald 1891*, pp. 354–356, gives values, due to Schiff, for the heat of vaporization of a number of compounds; pp. 376–385 lists molar volumes at the boiling point. Einstein uses the molar, not the molecular values for both D'_m and v; the external work is put equal to $RT = 1.991\,T$ cal/mol (see, e.g., *Nernst 1898*, p. 61).

[15] Here 2,51 is the average of the quotients given in the table, and $4{,}17 \cdot 10^7$ is the value of the mechanical equivalent of heat used by Einstein.

[16] For the meaning of "Wärmeinhalt," see note 5.

[17] Here, κ and α denote the standard coefficients of isothermal compressibility and isobaric thermal expansion.

[18] The terms $p\kappa$ and $p\alpha$ should be $pv\kappa$ and $pv\alpha$, respectively.

[19] The right-hand side of this equation should be multiplied by v.

[20] The right-hand side of this equation should be multiplied by v.

[21] *P* here is the potential energy per mole, given on the last line of p. 518. This expression should be multiplied by v.

[22] The numerator of the left-hand side should be $T\alpha$.

[23] See *Landolt and Börnstein 1894*, pp. 107–109. The values for v are again taken from *Ostwald 1891*, pp. 376–385 (Einstein uses the molar, not the molecular volume).

[24] The two results are thermodynamically equivalent, however. If the volume of the fluid is neglected, the Clausius-Clapeyron equation for vaporization is

$$\left(\frac{\partial p}{\partial T}\right)_v = \frac{D_m}{Tv},$$

with v the volume of the vapor and D_m the heat of vaporization. Neglecting the atmospheric pressure, as did Einstein, we may set $D_m = D'_m$. The identity

$$\left(\frac{\partial p}{\partial T}\right)_v \left(\frac{\partial v}{\partial p}\right)_T \left(\frac{\partial T}{\partial v}\right)_p = -1$$

then gives

$$D'_m = \frac{Tv\alpha}{k}.$$

The existence of a thermodynamic relation between Einstein's two results is suggested in a review of Einstein's paper (*Wiedeburg 1901*).

[25] See *Ostwald 1891*, pp. 279, 398–399, for a discussion of Mendeleev's formula.

[26] Possibly a reference to Ostwald's use of Schiff's data to test Eötvös's law (*Ostwald 1891*, pp. 541–543).

[27] See note 6.

[28] See note 7.

[29] See note 8.

2. "On the Thermodynamic Theory of the Difference in Potentials between Metals and Fully Dissociated Solutions of Their Salts and on an Electrical Method for Investigating Molecular Forces"

[*Einstein 1902a*]

DATED Bern, April 1902
RECEIVED 30 April 1902
PUBLISHED 10 July 1902

IN: *Annalen der Physik* 8 (1902): 798–814.

798

5. Ueber die thermodynamische Theorie der Potentialdifferenz zwischen Metallen und vollständig dissociirten Lösungen ihrer Salze und über eine elektrische Methode zur Erforschung der Molecularkräfte; von A. Einstein.

§ 1. Eine hypothetische Erweiterung des zweiten Hauptsatzes der mechanischen Wärmetheorie.

Der zweite Hauptsatz der mechanischen Wärmetheorie kann auf solche physikalische Systeme Anwendung finden, die im stande sind, mit beliebiger Annäherung umkehrbare Kreisprocesse zu durchlaufen. Gemäss der Herleitung dieses Satzes aus der Unmöglichkeit der Verwandlung latenter Wärme in mechanische Energie, ist hierbei notwendige Voraussetzung, dass jene Processe realisirbar seien. Bei einer wichtigen Anwendung der mechanischen Wärmetheorie ist es aber zweifelhaft, ob dieses Postulat erfüllt ist, nämlich bei der Vermischung zweier oder mehrerer Gase mit Hülfe von semipermeabeln Wänden. Auf der Voraussetzung der Realisirbarkeit dieses Vorganges basirt die thermodynamische Theorie der Dissociation der Gase und die Theorie der verdünnten Lösungen. [1]

Die einzuführende Voraussetzung ist bekanntlich folgende: Zu je zwei Gasen A und B sind zwei Scheidewände herstellbar, sodass die eine durchlässig für A, nicht aber für B, die andere durchlässig für B, nicht aber für A ist. Besteht die Mischung aus mehreren Componenten, so gestaltet sich diese Voraussetzung noch complicirter und unwahrscheinlicher. Da nun die Erfahrung die Resultate der Theorie vollständig bestätigt hat, trotzdem wir mit Processen operirt haben, deren Realisirbarkeit wohl bezweifelt werden kann, so erhebt sich die Frage, ob nicht vielleicht der zweite Hauptsatz auf ideale Processe gewisser Art angewendet werden kann, ohne dass man mit der Erfahrung in Widerspruch gerät.

In diesem Sinne können wir auf Grund der gewonnenen Erfahrung jedenfalls den Satz aussprechen: **Man** bleibt im Einklang mit der Erfahrung, wenn man den zweiten Haupt-

Thermodynamische Theorie der Potentialdifferenz etc. 799

satz auf physikalische Gemische ausdehnt, deren einzelne Componenten durch in gewissen Flächen wirkende conservative Kräfte auf gewisse Teilräume beschränkt werden. Diesen Satz verallgemeinern wir hypothetisch zu folgendem:

Man bleibt im Einklange mit der Erfahrung, wenn man den zweiten Hauptsatz auf physikalische Gemische anwendet, auf deren einzelne Componente beliebige conservative Kräfte [2] wirken.

Auf diese Hypothese werden wir uns im Folgenden stets stützen, auch wo es nicht absolut notwendig erscheint.

§ 2. Ueber die Abhängigkeit der elektrischen Potentialdifferenz einer vollkommen dissociirten Salzlösung und einer aus dem Lösungsmetall bestehenden Elektrode, von der Concentration der Lösung und vom hydrostatischen Druck.

In einem cylindrischen Gefässe, dessen Axe zusammen-falle mit der z-Axe eines cartesischen Coordinatensystems befinde sich ein vollkommen dissociirtes Salz in Lösung. $v\,do$ sei die Anzahl der Grammmolecüle des Salzes, welche sich im Volumenelemente do gelöst finden, $v_m\,do$ die Anzahl der [3] Metallionen, $v_s\,do$ die Anzahl der Säureionen daselbst, wobei v_m und v_s ganzzahlige Vielfache von v sind, sodass die Gleichungen bestehen:

$$v_m = n_m \cdot v,$$
$$v_s = n_s \cdot v.$$

Ferner sei $n \cdot v \cdot E \cdot do$ die Grösse der gesamten positiven elektrischen Ionenladung in do, also auch, bis auf unendlich Kleines, die Grösse der negativen. n ist dabei die Summe der Wertigkeiten der Metallionen des Molecüls, E die Elektricitätsmenge, welche zur elektrolytischen Ausscheidung eines Grammmolecüles eines einwertigen Ions erforderlich ist. [4] eines einwertigen Ions erforderlich ist.

Diese Gleichungen gelten jedenfalls, da die Anzahl der überzähligen Ionen einer Gattung zu vernachlässigen sein wird.

Wir wollen ferner annehmen, dass auf die Metall- bez. Säureionen eine äussere conservative Kraft wirke, deren Potential [5] pro Ion die Grösse P_m bez. P_s besitze. Wir vernachlässigen ferner die Veränderlichkeit der Dichte des Lösungsmittels mit dem Druck und der Dichte des gelösten Salzes, und nehmen

A. Einstein.

an, dass auf die Teile des Lösungsmittels ebenfalls eine conservative Kraft wirke, deren Potential pro Grammäquivalent des Lösungsmittels die Grösse P_0 besitze, wobei $\nu_0\, d\,o$ Grammmolecüle des Lösungsmittels in $d\,o$ vorhanden seien.

Alle die Kräftefunctionen seien lediglich von der z-Coordinate abhängig, und das System befinde sich im elektrischen, thermischen und mechanischen Gleichgewicht. Es werden dann die Grössen: Concentration ν, das elektrische Potential π, osmotische Drucke der beiden Ionengattungen p_m und p_s, hydrostatischer Druck p_0 nur Functionen von z sein.

Es müssen nun an jeder Stelle des Elektrolyten die beiden Elektronengattungen für sich im Gleichgewicht sein, was durch die Gleichungen ausgedrückt wird:

$$-\frac{d\,p_m}{d\,z}\cdot\frac{1}{\nu}-n_m\frac{d\,P_m}{d\,z}-n\,E\,\frac{d\,\pi}{d\,z}=0,$$

$$-\frac{d\,p_s}{d\,z}\cdot\frac{1}{\nu}-n_s\,\frac{d\,P_s}{d\,z}+n\,E\,\frac{d\,\pi}{d\,z}=0,$$

dabei ist:

$$p_m = \nu\,.\,n_m\,.\,R\,T\,.$$

$$p_s = \nu\,.\,n_s\,.\,R\,T,$$

wo R eine für alle Ionenarten gemeinsame Constante ist. Die Gleichungen nehmen also die Form an:

$$(1)\quad\left\{\begin{array}{l} n_m\,R\,T\dfrac{d\lg\nu}{d\,z}+n_m\dfrac{d\,P_m}{d\,z}+n\,E\dfrac{d\,\pi}{d\,z}=0,\\[2mm] n_s\ R\,T\dfrac{d\lg\nu}{d\,z}+n_s\dfrac{d\,P_s}{d\,z}-n\,E\dfrac{d\,\pi}{d\,z}=0,\end{array}\right.$$

Sind P_m und P_s für alle z, sowie ν und π für ein bestimmtes z bekannt, so liefern die Gleichungen (1) ν und π als Functionen von z. Auch ergäbe die Bedingung, dass sich die Lösung als Ganzes im Gleichgewicht befindet, eine Gleichung zur Bestimmung des hydrostatischen Druckes p_0, die nicht angeschrieben zu werden braucht. Wir bemerken nur, dass $d\,p_0$ von $d\,\nu$ [6] und $d\,\pi$ deshalb unabhängig ist, weil es uns freisteht, beliebige conservative Kräfte anzunehmen, welche auf die Molecüle des Lösungsmittels wirken.

Wir denken uns nun in $z = z_1$ und $z = z_2$ Elektroden in die Lösung eingeführt, welche aus dem Lösungsmetalle bestehen, und nur einen verschwindend kleinen Teil des Querschnittes des cylindrischen Gefässes ausfüllen sollen. Lösung

Thermodynamische Theorie der Potentialdifferenz etc. 801

und Elektroden zusammen bilden ein physikalisches System, welches wir folgenden umkehrbaren isothermischen Kreisprocess ausführen lassen:

1. Teilprocess: Wir lassen die Elektricitätsmenge $n\,E$ *unendlich langsam* durch die Lösung passiren, indem wir die in $z = z_1$ bez. $z = z_2$ befindliche Elektrode als Anode bez. Kathode verwenden.

2. Teilprocess: Wir bewegen die hierbei elektrolytisch von z_1 nach z_2 bewegte Metallmenge mechanisch in der Lösung unendlich langsam wieder von z_2 nach z_1.

Man ersieht zunächst, dass der Process strenge umkehrbar ist, da alle Vorgänge unendlich langsam vor sich gehend gedacht werden, derselbe also aus (idealen) Gleichgewichtszuständen zusammengesetzt ist. Der zweite Hauptsatz verlangt für einen solchen Process, dass die Summe der dem System während des Kreisprocesses zugeführten Wärmemengen verschwinde. Der erste Hauptsatz verlangt in Verbindung mit dem zweiten, dass die Summe der übrigen Energien, welche dem System während des Kreisprocesses zugeführt werden, verschwinde.

Während des ersten Teilprocesses wird die elektrische Arbeitsmenge zugeführt:

$$- n\,E\,(\Pi_2 - \Pi_1),$$

wobei Π_2 und Π_1 die elektrischen Potentiale der Elektroden bedeuten.

Während des zweiten Teilprocesses wird:

$$\int_{z_2}^{z_1} K\,d z$$

zugeführt, wobei K die in der positiven z-Richtung wirkende Kraft bedeutet, welche notwendig ist, um die zu bewegenden n_m Metallionen, welche sich jetzt im metallischen Zustande befinden, an der beliebigen Stelle z in Ruhe zu erhalten. Für K gilt, wie leicht ersichtlich die Gleichung:

$$K - n_m \frac{d\,P_m}{d\,z} - n_m v_m \frac{d\,p_o}{d\,z} = 0.$$

Dabei bedeutet v_m das Volumen eines Metallions im metallischen Zustande. Jene Arbeit erhält also den Wert:

$$\int_{z_2}^{z_1} K \cdot dz = - \int_{z_1}^{z_2} \left(n_m \frac{dP_m}{dz} + n_m v_m \frac{dp_o}{dz} \right) dz$$

$$= - n_m \left[(P_{m_2} - P_{m_1}) + v_m (p_{o_2} - p_{o_1}) \right],$$

wobei der zweite Index die Coordinate der Elektrode bezeichnet.

Wir erhalten also die Gleichung:

(2) $\qquad n \cdot E \cdot (\Pi_2 - \Pi_1) = - n_m (P_{m_2} - P_{m_1}) - n_m v_m (p_{o_2} - p_{o_1})$.

Bezeichnet man mit π_1 und π_2 die elektrischen Potentiale, welche in den Elektrodenquerschnitten im Innern der Lösung herrschen, so erhält man durch Integration aus der ersten Gleichung (1):

$$- n \cdot E (\pi_2 - \pi_1) = n_m \left[P_{m_2} - P_{m_1} \right] + n_m R T \log \left(\frac{v_2}{v_1} \right),$$

wobei sich v_1 und v_2 wieder auf die Elektrodenquerschnitte beziehen. Durch Addition dieser Gleichungen erhält man:

(3) $\quad \begin{cases} (\Pi_2 - \pi_2) - (\Pi_1 - \pi_1) = (\varDelta \Pi)_2 - (\varDelta \Pi)_1 \\ \qquad\qquad = \dfrac{n_m R T}{n E} \log \left(\dfrac{v_2}{v_1} \right) - \dfrac{n_m v_m}{n E} (p_{o_2} - p_{o_1}). \end{cases}$

Da die v und p_o vollständig unabhängig voneinander sind, so enthält diese Gleichung die Abhängigkeit der Potentialdifferenz $\varDelta \Pi$ zwischen Metall und Lösung von Concentration und hydrostatischem Druck. Es ist zu bemerken, dass die angenommenen Kräfte im Resultat nicht mehr vorkommen. Kämen sie vor, so wäre die § 1 aufgestellte Hypothese ad absurdum geführt. Die gefundene Gleichung lässt sich in zwei zerlegen, nämlich:

(4) $\quad \begin{cases} (\varDelta \Pi)_2 - (\varDelta \Pi)_1 = \dfrac{n_m}{n} \dfrac{R T}{E} \log \left(\dfrac{v_2}{v_1} \right) \text{ bei const. Druck,} \\ (\varDelta \Pi)_2 - (\varDelta \Pi)_1 = - \dfrac{n_m v_m}{n} \dfrac{1}{E} (p_{o_2} - p_{o_1}) \text{ bei const. Concentration.} \end{cases}$

Man hätte die Endformel (3) auch erhalten, ohne die in § 1 vorgeschlagene Hypothese, wenn man die äusseren Kräfte mit der Erdschwere identificirt hätte. Dann wären aber v und p_o nicht unabhängig voneinander und eine Zerlegung in die Gleichungen (4) wäre nicht erlaubt.

Es soll noch kurz erwähnt werden, dass die Nernst'sche Theorie der elektrischen Kräfte im Innern dissociirter Elektro-

Thermodynamische Theorie der Potentialdifferenz etc.   803

lyte in Verbindung mit der ersten der Gleichungen (4) die elektromotorische Kraft des Concentrationselementes zu berechnen gestattet. Man gelangt so zu einem bereits mehrfach geprüften Resultat, welches bis jetzt aus speciellen Annahmen [7] hergeleitet wurde.

§ 3. Ueber die Abhängigkeit der Grösse $\varDelta\,II$ von der Natur der Säure.

Wir betrachten folgenden idealen Gleichgewichtszustand: Sei wieder ein cylindrisches Gefäss vorhanden. In den Teilen I und II mögen sich vollständig dissociirte Salzlösungen befinden mit identischem Metallion (gleiches Metall und gleiche elektrische Ladung), aber verschiedenem Säureion. Zwischen den beiden befinde sich der Verbindungsraum V, in welchem

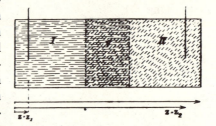

beide Salze gelöst vorkommen. In V mögen auf die Säureionen Kräfte wirken, deren Potentiale $P_s^{(1)}$ und $P_s^{(2)}$ nur von z abhängen, welche Kräfte bewirken sollen, dass nur unendlich wenig Säureionen erster Art in II, zweiter Art in I gelangen. Ausserdem seien $P_s^{(1)}$ und $P_s^{(2)}$ so gewählt, dass die Concentration der Metallionen in den beiden Teilen I und II die gleiche sei. Ebenso sei $p_{o_1} = p_{o_2}$.

Es seien $v_m^{(1)}$ Metallionen in der Volumeneinheit, welche der ersten, $v_m^{(2)}$, welche der zweiten Satzart entsprechen, dann ist:

$$(1) \qquad v_{m_1}^{(1)} = v_{m_2}^{(2)}, \quad v_{s_1}^{(2)} = 0, \quad v_{s_2}^{(1)} = 0,$$

wobei die unteren Indices die Zugehörigkeit zu Raum I bez. Raum II bezeichnet.

In V erhält man aber als Gleichgewichtsbedingung der Metallionen:

$$- R T \frac{d \log \left(v_m^{(1)} + v_m^{(2)} \right)}{d z} - \varepsilon E \frac{d \pi}{d z} = 0,$$

wobei ε die Wertigkeit des Metallions bedeutet.

804   *A. Einstein.*

Durch Integration über V und Berücksichtigung der Gleichungen (1) ergiebt sich:

(2) $\pi_2 = \pi_1.$

Wir bilden ferner, nachdem wir in I und II Elektroden aus Lösungsmetall eingesetzt denken, folgenden idealen Kreisprocess:

1. Teilprocess: Wir schicken durch das System unendlich langsam die Elektricitätsmenge εE, indem wir die im Raum I befindliche Elektrode als Anode, die andere als Kathode betrachten.

2. Teilprocess: Wir führen das so durch Elektrolyse von $z = z_1$ nach $z = z_2$ transportirte Metall, welches die Masse eines Grammäquivalentes besitzt, mechanisch wieder nach der in $z = z_1$ befindlichen Elektrode zurück.

Durch Anwendung der beiden Hauptsätze der mechanischen Wärmetheorie folgert man wieder, dass die Summe der dem System während des Kreisprocesses zugeführten mechanischen und elektrischen Energie verschwindet. Da, wie leicht ersichtlich, der zweite Teilprocess keine Energie erfordert, so erhält man die Gleichung

(3) $\Pi_2 = \Pi_1,$

wobei Π_2 und Π_1 wieder die Elektrodenpotentiale bedeuten. Durch Subtraction der Gleichungen (3) und (2) erhält man:

$$(\Pi_2 - \pi_2) - (\Pi_1 - \pi_1) = (\varDelta \Pi)_2 - (\varDelta \Pi)_1 = 0$$

und also folgenden Satz:

Die Potentialdifferenz zwischen einem Metall und einer vollständig dissociirten Lösung eines Salzes dieses Metalles in einem bestimmten Lösungsmittel ist unabhängig von der Natur des elektronegativen Bestandteiles, sie hängt lediglich von der Concentration der Metallionen ab. Voraussetzung ist dabei jedoch, dass bei den Salzen das Metallion mit derselben Elektricitätsmenge geladen ist.

§ 4.

Bevor wir dazu übergehen, die Abhängigkeit von $(\varDelta \Pi)$ von der Natur des Lösungsmittels zu studiren, wollen wir kurz die Theorie der conservativen Molecularkräfte in Flüssigkeiten entwickeln. Ich entnehme dabei die Bezeichnungsweise einer

Thermodynamische Theorie der Potentialdifferenz etc. 805

früheren Abhandlung über diesen Gegenstand[1]), welche zugleich die einzuführenden Hypothesen einstweilen rechtfertigen soll.

Jedem Molecüle einer Flüssigkeit oder einer in einer Flüssigkeit gelösten Substanz komme eine gewisse Constante c zu, sodass der Ausdruck für das relative Potential der Molecularkräfte zweier Molecüle, welche durch die Indices $\ldots_1$ und $\ldots_2$ charakterisirt seien, lautet:

(a) $$P = P_\infty - c_1 \, c_2 \, \varphi \, (r),$$

wobei $\varphi(r)$ eine für alle Molecülarten gemeinsame Function der Entfernung sei. Jene Kräfte sollen sich einfach superponiren, sodass der Ausdruck des relativen Potentiales von n Molecülen die Form habe:

(b) $$\text{Const.} - \tfrac{1}{2} \sum_{\alpha=1}^{\alpha=n} \sum_{\beta=1}^{\beta=n} c_\alpha \, c_\beta \, \varphi \, (r_{\alpha\beta}).$$

Wären speciell alle Molecüle gleich beschaffen, so erhielten wir den Ausdruck:

(c) $$\text{Const.} - \tfrac{1}{2} c^2 \sum_{\alpha=1}^{\alpha=n} \sum_{\beta=1}^{\beta=n} \varphi \, (r_{\alpha\beta})$$

Ferner sei das Wirkungsgesetz und das Verteilungsgesetz der Molecüle so beschaffen, dass die Summen in Integrale verwandelt werden dürfen, dann geht dieser Ausdruck über in:

$$\text{Const.} - \tfrac{1}{2} c^2 N^2 \int \int d\tau \cdot d\tau' \, \varphi \, (r_{d\tau, d\tau'}).$$

N bedeutet dabei die Zahl der Molecüle in der Volumeneinheit. Bszeichnet N_0 die Anzahl der Molecüle in einem Grammäquivalent, so ist $N_0/N = v$ das Molecularvolumen der Flüssigkeit, und nehmen wir an, dass ein Grammäquivalent zur Untersuchung vorliegt, so geht, wenn wir den Einfluss der Flüssigkeitsoberfläche vernachlässigen, unser Ausdruck über in:

[9]

$$\text{Const.} - \tfrac{1}{2} \frac{c^2}{v} N_0^2 \int_{-\infty}^{\infty} d\tau' \cdot \varphi \, (r_{0, d\tau'}).$$

----

[8]   1) A. Einstein, Ann. d. Phys. **4.** p. 513. 1901.

806   *A. Einstein.*

Wir wollen nun die Einheit der c so wählen, dass dieser Ausdruck übergeht in

(d)   Const. $- \dfrac{c^2}{v}$,   also $\frac{1}{2} N_0^2 \displaystyle\int_{-\infty}^{\infty} d\tau' \cdot \varphi \left(r_{0, d\tau'}\right) = 1$.

Durch diese Festsetzung gewinnt man für die Grössen c ein absolutes Maass. In jener Abhandlung ist gezeigt, dass man mit der Erfahrung in Uebereinstimmung bleibt, wenn man setzt $c = \sum c_\alpha$, wo sich die Grössen c_α auf die Atome beziehen, aus denen das Molecül zusammengesetzt ist.

Wir wollen nun das relative Anziehungspotential des Grammmolecüls eines Ions in Bezug auf sein Lösungsmittel berechnen, wobei wir ausdrücklich die Annahme machen, dass die Anziehungsfelder der Molecüle des Lösungsmittels nicht auf die elektrischen Ladungen der Ionen wirken. Später zu entwickelnde Methoden werden ein Mittel an die Hand geben, welches über die Zulässigkeit dieser Voraussetzung zu entscheiden gestattet.

Sei c_j die moleculare Constante des Ions, c_l die des Lösungsmittels, so hat das Potential eines Molecüles des Ions gegen das Lösungsmittel die Form:

$$\text{Const.} - \sum_l c_j c_l \cdot \varphi\,(r) = \text{const.} - c_j \cdot c_l N_l \int d\tau \cdot \varphi\,(r_{0, d\tau}),$$

wobei N_l die Zahl der Molecüle des Lösungsmittels pro Volumeneinheit bedeutet. Da $N_0/N_l = v_l$ ist, so geht dieser Ausdruck über in:

$$\text{Const.} - c_j \cdot c_l \cdot \frac{N_0}{v_l} \int d\tau \cdot \varphi\,(r_{0, d\tau}).$$

Das aber das Grammäquivalent N_0 Molecüle des Ions enthält, so erhalten wir für das relative Potential des Grammäquivalentes des Ions:

$$\text{Const.} - \frac{c_j \cdot c_l}{v_l} N_0^2 \int d\tau \cdot \varphi\,(r_{0, d\tau}) = \text{const.} - 2 \frac{c_j \cdot c_l}{v_l}.$$

Führt man die Concentration des Lösungsmittels $1/v_l = \nu_l$ ein, so erhält man die Form:

(e) $P_{jl} = \text{const.} - 2\,c_j \cdot c_l \,\nu_l .$

Thermodynamische Theorie der Potentialdifferenz etc.   807

Ist das Lösungsmittel eine Mischung mehrerer Flüssigkeiten, welche wir durch Indices unterscheiden wollen, erhalten wir

(e′) $$P_{jl} = \text{const.} - 2\,c_j \sum c_l v_l,$$

wobei die v_l die Anzahl der Grammmolecüle der einzelnen Componenten des Lösungsmittels pro Volumeneinheit bedeuten. Die Formel (e′) gilt angenähert auch in dem Falle, dass die Grössen v_l mit dem Orte variiren.

§ 5. Ueber die Abhängigkeit der zwischen einem Metall und einer vollständig dissociirten Lösung eines Salzes dieses Metalles herrschenden elektrischen Potentialdifferenz von der Natur des Lösungsmittels.

Ein cylindrisches Gefäss zerfalle wieder, wie im § 3 angegeben wurde, in die Räume *I, II* und den Verbindungsraum *V*. In *I* befinde sich ein erstes, in *II* ein zweites Lösungsmittel, in *V* mögen beide gemischt vorkommen und es mögen in diesem Raume auf die Lösungsmittel Kräfte wirken, welche eine Diffusion verhindern. In dem Gefässe befinde sich ein gelöstes Salz im Zustande vollständiger Dissociation. Auf die Säureionen desselben sollen in *V* Kräfte wirken, deren Potential P_s heisse und so gewählt sei, dass das Salz in *I* und *II* gleiche Concentration besitze. Wir stellen nun die Bedingung für das Gleichgewicht der Metallionen auf. Die z-Axe führen wir wieder ∥ der Cylinderaxe von *I* nach *II*.

Als Ausdruck der auf das Grammäquivalent wirkenden Kraft elektrischen Ursprunges ergiebt sich:

$$-\frac{r}{n_m}\,E\,\frac{d\pi}{dz}.$$

Die auf das Aequivalent vom osmotischen Druck ausgeübte Kraft ist:

$$-R\,T\,\frac{d\log v}{dz}.$$

Die auf das Aequivalent ausgeübte Wirkung der Molecularkräfte ist:

$$-\frac{d}{dz}\left\{-2\,c_m\,c_l^{(1)}\,v_l^{(1)} - 2\,c_m\,c_l^{(2)}\,v_l^{(2)}\right\},$$

wobei sich die oberen Indices auf die Lösungsmittel beziehen. Die gesuchte Gleichgewichtsbedingung ist also:

$$- \frac{n}{n_m} E \frac{d\pi}{dz} - RT \frac{d \log \nu}{dz} + \frac{d}{dz} \{2 c_m c_i^{(1)} \nu_i^{(1)} + 2 c_m c_i^{(2)} \nu_i^{(2)}\} = 0.$$

Integrirt man durch V hindurch und berücksichtigt, dass ν in I und II identisch ist, und dass $\nu_i^{(2)}$ in I und $\nu_i^{(1)}$ in II nach unserer Voraussetzung verschwindet, so erhält man:

$$\pi_2 - \pi_1 = \frac{n_m}{n} \frac{2 c_m}{E} \{c_i^{(2)} \nu_i^{(2)} - c_i^{(1)} \nu_i^{(1)}\},$$

wobei sich die oberen Indices auf Raum I bez. II beziehen.

Wir denken uns nun in I und II Elektroden angebracht, welche aus dem gelösten Metall bestehen, und bilden einen Kreisprocess, indem wir die Electricitätsmenge $n/n_m E$ durch das System schicken, und dann die transportirte Metallmenge mechanisch wieder zurückbewegen, was keine Arbeit erfordert, wenn wir annehmen, dass in I und II der hydrostatische Druck der nämliche sei. Durch Anwendung der beiden Hauptsätze der Wärmetheorie erhält man:

$$\Pi_2 - \Pi_1 = 0.$$

Durch Subtraction beider Resultate ergiebt sich:

$$(\Pi_2 - \pi_2) - (\Pi_1 - \pi_1) = (\Delta \Pi)^{(2)} - (\Delta \Pi)^{(1)}$$
$$= - \frac{n_m}{n} \frac{2 c_m}{E} \{c_i^{(2)} \nu_i^{(2)} - c_i^{(1)} \nu_i^{(1)}\}.$$

Ist jedes der beiden Lösungsmittel eine Mischung mehrerer nichtleitender Flüssigkeiten, so erhält man etwas allgemeiner:

$$(\Delta \Pi)^{(2)} - (\Delta \Pi)^{(1)} = - \frac{n_m}{n} \frac{2 c_m}{E} \{\sum c_i^{(2)} \nu_i^{(2)} - \sum c_i^{(1)} \nu_i^{(1)}\},$$

in welcher Formel ν_i die Zahl der Grammmolecüle einer Componente des Lösungsmittels in einem Volumelemente des gemischten Lösungsmittels bezeichnet.

Die Potentialdifferenz $\Delta \Pi$ ist also von der Natur des Lösungsmittels abhängig. Auf diese Abhängigkeit lässt sich eine Methode zur Erforschung der Molecularkräfte gründen.

§ 6. Methode zur Bestimmung der Constanten c für Metallionen und Lösungsmittel.

In einem cylindrischen Gefässe seien zwei vollständig dissociirte Salzlösungen in Diffusion begriffen; diese Salze

Thermodynamische Theorie der Potentialdifferenz etc.   **809**

seien durch untere Indices bezeichnet. Das Lösungsmittel sei im ganzen Gefäss dasselbe und werde durch den oberen Index bezeichnet. Das Gefäss zerfalle wieder in die Räume *I*, *II* und den Verbindungsraum *V*. Im Raume *I* sei nur das erste, im Raume *II* nur das zweite Salz vorhanden; im Raume *V* finde Diffusion beider Salze statt. In die Räume *I* und *II* seien Elektroden eingeführt, welche aus dem betreffenden Lösungsmetalle bestehen und die elektrischen Potentiale Π_1 bez. Π_2' besitzen; an die zweite Elektrode sei ein Stück des ersten Elektrodenmetalles angelötet, dessen Potential Π_2 sei. Wir bezeichnen ausserdem die elektrischen Potentiale im Innern der unvermischten, in *I* und *II* befindlichen Lösungen, mit π_1 und π_2, dann ist:

$$(\Pi_2 - \Pi_1)^{(1)} = (\Pi_2 - \Pi_2') + (\Pi_2' - \pi_2)^{(1)} + (\pi_2 - \pi_1)^{(1)} - (\Pi_1 - \pi_1)^{(1)}.$$

Stellt man ganz dieselbe Anordnung her, mit dem einzigen Unterschiede, dass man ein anderes Lösungsmittel benutzt, das durch den oberen Index [2] bezeichnet werde, so hat man:

$$(\Pi_2 - \Pi_1)^{(2)} = (\Pi_2 - \Pi_2') + (\Pi_2' - \pi_2)^{(2)} + (\pi_2 - \pi_1)^{(2)} - (\Pi_1 - \pi_1)^{(2)}.$$

Durch Subtraction dieser beiden Ausdrücke erhält man mit Berücksichtigung des in § 5 gefundenen Resultates:

$$(\Pi_2 - \Pi_1)^{(2)} - (\Pi_2 - \Pi_1)^{(1)} = \{(\pi_2 - \pi_1)^{(2)} - (\pi_2 - \pi_1)^{(1)}\}$$
$$- \frac{2}{E}\left\{\left(\frac{c_m n_m}{n}\right)_2 - \left(\frac{c_m n_m}{n}\right)_1\right\} \cdot \{c_i^{(2)} v_i^{(2)} - c_i^{(1)} v_i^{(1)}\}.$$

Die erforderliche Erweiterung für den Fall, dass die Lösungsmittel Mischungen sind, erhält man leicht wie in § 5.

Die Werte der linken Seite dieser Gleichung ergeben sich unmittelbar durch das Experiment. Mit der Bestimmung des ersten Gliedes der rechten Seite werden wir uns im folgenden Paragraph beschäftigen; es sei einstweilen gesagt, dass man dies Glied aus den angewandten Concentrationen und den molecularen Leitfähigkeiten der betreffenden Ionen für das betreffende Lösungsmittel berechnen kann, wenn man die Anordnung in geeigneter Weise wählt. Die Gleichung erlaubt daher die Berechnung des zweiten Gliedes der rechten Seite.

Dies benutzen wir zur Bestimmung der Constanten *c* für Metallionen und zur Prüfung unserer Hypothesen. Wir benutzen zu einer Reihe von Experimenten der geschilderten

810 *A. Einstein.*

Art immer dieselben beiden Lösungsmittel. Für die ganze Untersuchungsreihe ist dann die Grösse

$$\frac{2}{E}\{c_l^{(2)}\, v_l^{2} - c_l^{(1)}\, v_l^{(1)}\} = k = \text{const.}$$

Setzt man $n_1/n_{m_1} = E_1$ etc. gleich der Wertigkeit des ersten etc. Metallions, so ist also das berechnete letzte Glied der rechten Seite ein relatives **Maass** für die Grösse [10]

$$\left(\frac{c_{m_2}}{\varepsilon_2} - \frac{c_{m_1}}{\varepsilon_1}\right).$$

Untersucht man so Combinationen aller Elektrodenmetalle zu Paaren, so erhält man in relativem Maass die Grössen

$$\left\{\frac{c_{m_j}}{\varepsilon_j} - \frac{c_{m_k}}{\varepsilon_k}\right\}.$$

Man erhält in demselben **Maasse** die Grössen c_m/ε selbst, wenn man bei einem Metall eine analoge Untersuchung in der Weise ausführt, dass man Salze und Elektroden in *I* und *II* von demselben Metall wählt, sodass jedoch ε, d. h. die Wertigkeit (elektrische Ladung) des Metallions auf beiden Seiten verschieden ist. Es sind dann in jenem **Maasse** die Werte für die Grössen c_m der einzelnen Metalle selbst ermittelbar. Eine Reihe von solchen Untersuchungen führt also auf die Verhältnisse der c_m, d. h. der Constanten für die Molecularattraction der Metallionen. Diese Reihe der c_m muss unabhängig sein von der Natur der benutzten Salze, und die Verhältnisse der so erhaltenen c_m unabhängig von der Natur der beiden Lösungsmittel, welche wir für die Untersuchung zu Grunde legten. Ferner muss verlangt werden, dass c_m unabhängig von der elektrischen Ladung (Wertigkeit), in welcher ein Ion auftritt, sich herausstelle. Ist dies der Fall, so ist die oben gemachte Voraussetzung richtig, dass die Molecularkräfte nicht auf die elektrischen Ladungen der Ionen wirken.

Will man den Wert der Grössen c_m wenigstens angenähert absolut bestimmen, so kann man dies, indem man die Grösse k angenähert für die beiden Lösungsmittel aus den Resultaten der oben angeführten Abhandlung entnimmt, indem man die Formel $c = \sum c_\alpha$ anwendet. Freilich ist hier zu bemerken, dass sich gerade für die als Lösungsmittel am meisten naheliegenden Flüssigkeiten, Wasser und Alkohol, die Gültigkeit

Thermodynamische Theorie der Potentialdifferenz etc.   811

des Attractionsgesetzes aus den Erscheinungen der Capillarität,
[11] Verdampfung und Compressibilität nicht hat darthun lassen.

Es lässt sich auf Grund unseres Ergebnisses aber ebenso-
gut eine Erforschung der Constanten c_l von Lösungsmitteln
gründen, indem man der Untersuchung zwei Metallionen zu
Grunde legt und das Lösungsmittel variiren lässt, sodass nun
die Grösse

$$\frac{2}{E}\left\{\left(\frac{c_m\,n_m}{n}\right)_2 - \left(\frac{c_m\,n_m}{n}\right)_1\right\}$$

als constant zu betrachten ist. Indem man auch Mischungen
als Lösungsmittel zulässt, kann so die Untersuchung auf alle
elektrisch nicht leitenden Flüssigkeiten ausgedehnt werden.
Es lassen sich aus solchen Versuchen relative Werte für die
Grössen c_a herausrechnen, welche den die Flüssigkeitsmolecüle
bildenden Atomen zukommen. Auch hier bietet sich eine Fülle
von Prüfungen für die Theorie, indem die c_a beliebig über-
bestimmt werden können. Ebenso muss das Resultat unabhängig sein von der Wahl der Metallionen.

§ 7. Berechnung von $(\pi_2 - \pi_1)$.

Wir haben nun noch den Diffusionsvorgang im Raume V
genauer zu studiren. Die variabeln Grössen seien nur von z
abhängig, wobei die z-Axe des von uns gewählten cartesischen
Coordinatensystems mit der Richtung der Axe unseres Ge-
fässes zusammenfalle. v_{m_1}, v_{s_1}, v_{m_2}, v_{s_2} seien die von z abhängigen Concentrationen (Grammäquivalente pro Volumen-
einheit) der vier Ionengattungen, $\varepsilon_{m_1}E$, $-\varepsilon_{s_1}E$, $\varepsilon_{m_2}E$, $-\varepsilon_{s_2}E$
die elektrischen Ladungen, welche dieselben tragen; π sei das
elektrische Potential. Da nirgends beträchtliche elektrische
Ladungen auftreten, so ist für alle z nahezu:

(α) $\qquad v_{m_1}\varepsilon_{m_1} - v_{s_1}\varepsilon_{s_1} + v_{m_2}\varepsilon_{m_2} - v_{s_2}\varepsilon_{s_2} = 0.$

Ausserdem erhalten wir für jede Ionenart eine Gleichung,
welche ausdrückt, dass die Vermehrung der Zahl der in einem
Volumenelement befindlichen Ionen bestimmter Gattung pro
Zeiteinheit gleich ist der Differenz der in dieser Zeit ins
Volumenelement eintretenden und der in derselben Zeit aus
ihm austretenden Molecüle:

$$(\beta) \begin{cases} v_{m_1} \cdot \dfrac{\partial}{\partial z} \left\{ R\,T \dfrac{\partial \nu_{m_1}}{\partial z} + \varepsilon_{m_1}\, \nu_{m_1}\, E\, \dfrac{\partial \pi}{\partial z} \right\} = \dfrac{\partial \nu_{m_1}}{\partial t}, \\[2ex] v_{s_1} \cdot \dfrac{\partial}{\partial z} \left\{ R\,T \dfrac{\partial \nu_{s_1}}{\partial z} - \varepsilon_{s_1}\, \nu_{s_1}\, E\, \dfrac{\partial \pi}{\partial z} \right\} = \dfrac{\partial \nu_{s_1}}{\partial t}, \\[2ex] \cdot\quad\cdot\quad\cdot\quad\cdot\quad\cdot\quad\cdot\quad\cdot\quad\cdot\quad\cdot\quad\cdot\quad\cdot \\ \cdot\quad\cdot\quad\cdot\quad\cdot\quad\cdot\quad\cdot\quad\cdot\quad\cdot\quad\cdot\quad\cdot\quad\cdot \end{cases}$$

[12]

wobei v mit dem betreffenden Index die constante Geschwindigkeit bedeutet, welche die mechanische Krafteinheit
dem Grammäquivalent des betreffenden Ions in der Lösung
erteilt.

Diese vier Gleichungen bestimmen im Verein mit den
Grenzbedingungen den stattfindenden Vorgang vollständig, da
sie für jeden Zeitmoment die fünf Grössen

$$\frac{\partial \pi}{\partial z}, \quad \frac{\partial \nu_{m_1}}{\partial t} \cdots \frac{\partial \nu_{s_2}}{\partial t}$$

in eindeutiger Weise zu berechnen gestatten. Die allgemeine
Behandlung des Problemes wäre aber mit sehr grossen
Schwierigkeiten verknüpft, zumal Gleichungen (β) nicht linear
in den Unbekannten sind. Uns kommt es aber nur auf die
Bestimmung von $\pi_2 - \pi_1$ an. Wir multipliciren daher die
Gleichungen (β) der Reihe nach mit $\varepsilon_{m_1}, -\varepsilon_{s_1}, \varepsilon_{m_2}, -\varepsilon_{s_2}$ und
erhalten mit Rücksicht auf (α)

$$\frac{\partial \varphi}{\partial z} = 0.$$

wobei

$$\varphi = R\,T \left\{ v_{m_1} \varepsilon_{m_1} \frac{\partial \nu_{m_1}}{\partial z} - v_{s_1} \varepsilon_{s_1} \frac{\partial \nu_{s_1}}{\partial z} + \cdot - \cdot \right\} \\ + \left\{ v_{m_1} \varepsilon_{m_1}^2 \nu_{m_1} + v_{s_1} \varepsilon_1^2 \nu_{s_1} + \cdot + \cdot \right\} \frac{\partial \pi}{\partial z}.$$

[13]

Durch Integration dieser Gleichung nach z ergiebt sich
unter Berücksichtigung des Umstandes, dass überall, wo keine
Diffusion stattfindet,

$$\frac{\partial \nu_{m_1}}{\partial z}, \quad \frac{\partial \nu_{s_1}}{\partial z} \cdots \frac{\partial \pi}{\partial z}$$

verschwinden:

$$\varphi = 0.$$

Thermodynamische Theorie der Potentialdifferenz etc. 813

Da die Zeit als constant zu betrachten ist, lässt sich schreiben:

$$[14] \qquad d\pi = -\frac{RT\{v_{m_1}\varepsilon_{m_1} dv_{m_1} - v_{s_1}\cdot\varepsilon_{s_1}\cdot dv_{s_1} + v_{m_2}\varepsilon_{m_2} dv_{m_2} - v_{s_2}\varepsilon_{s_2} dv_{s_2}\}}{v_{m_1}\varepsilon_{m_1}^2 v_{m_1} + v_{s_1}\varepsilon_{s_1}^2 v_{s_1} + v_{m_2}\varepsilon_{m_2}^2 v_{m_2} + v_{s_2}\cdot\varepsilon_{s_2}^2 v_{s_2}}.$$

Der Ausdruck rechts ist im allgemeinen kein vollständiges Differential, was bedeutet, dass $\Delta\Pi$ nicht nur durch die an den diffusionslosen Bereichen herrschenden Concentrationen, sondern auch durch den Charakter des Diffusionsvorganges bestimmt wird. Es gelingt indessen durch einen Kunstgriff in der Anordnung, die Integration zu ermöglichen.

Wir denken uns den Raum V in drei Teile, Raum (1), Raum (2) und Raum (3) eingeteilt und dieselben vor Beginn des Experimentes durch zwei Scheidewände voneinander getrennt. (1) communicire mit I, (3) mit II, in (2) seien beide Salze gleichzeitig gelöst, mit genau denselben Concentrationen wie in I bez. II. Vor Beginn des Experimentes befindet sich also in I und (1) nur das erste, in II und (3) nur das zweite Salz in Lösung, in (2) eine Mischung beider. Die Concentration ist dabei allenthalben constant. Bei Beginn des Experimentes werden die Scheidewände weggenommen und gleich darauf die Potentialdifferenz zwischen den Elektroden gemessen. Für diese Zeit ist aber die Integration über die diffundirenden Schichten möglich, da in der ersten diffundirenden Schicht v_{m_1} und v_{s_1}, in der zweiten v_{m_2} und v_{s_2} constant sind. Die Integration liefert:

$$\pi_2 - \pi_1 = RT\left\{\frac{v_{m_1} - v_{s_1}}{v_{m_1}\varepsilon_{m_1} + v_{s_1}\varepsilon_{s_1}} \lg\left[1 + \frac{v_{m_1}\varepsilon_{m_1}^2 v_{m_1} + v_{s_1}\varepsilon_{s_1}^2 v_{s_1}}{v_{m_2}\varepsilon_{m_2}^2 v_{m_2} + v_{s_2}\varepsilon_{s_2}^2 v_{s_2}}\right]\right.$$

$$\left. - \frac{v_{m_2} - v_{s_2}}{v_{m_2}\varepsilon_{m_2} + v_{s_2}\varepsilon_{s_2}} \lg\left[1 + \frac{v_{m_2}\varepsilon_{m_2}^2 v_{m_2} + v_{s_2}\varepsilon_{s_2}^2 v_{s_2}}{v_{m_1}\varepsilon_{m_1}^2 v_{m_1} + v_{s_1}\varepsilon_{s_1}^2 v_{s_1}}\right]\right\}.$$

Eine Vereinfachung der Methode lässt sich erzielen, wenn es möglich ist, in I und II gleiches Säureion von gleicher Concentration zu wählen. Verbindet man nämlich in diesem Falle Raum I mit Raum II direct, so ist für den Anfang des Diffusionsvorganges zu setzen:

$$\frac{\partial(v_{s_1} + v_{s_2})}{\partial z} = 0; \quad v_{s_1} + v_{s_2} = v_s = \text{const.}$$

814   *A. Einstein. Thermodynamische Theorie etc.*

Ebenso ist nach Voraussetzung:

$$\varepsilon_{s_1} = \varepsilon_{s_2} = \varepsilon_s \quad \text{und} \quad v_{s_1} = v_{s_2} = v_s.$$

Gleichung (1) geht dann über in

[15]

(1′) $$v_{m_1} \varepsilon_{m_1} + v_{m_2} \varepsilon_{m_2} - v_s \varepsilon_s = 0.$$

Von den Gleichungen (2) bleibt die erste und dritte unverändert bestehen, aus der zweiten und vierten ergiebt sich durch Addition:

$$v_s \frac{\partial}{\partial x} \left\{ R T \frac{\partial v_s}{\partial x} - \varepsilon_s v_s E \frac{\partial \pi}{\partial x} \right\} = \frac{\partial v_s}{\partial t}.$$

Eliminirt man aus den so veränderten Gleichungen (2) vermittelst der Gleichung (1′) die Ableitungen nach der Zeit, so erhält man wie vorhin einen Ausdruck für $d\pi$, welcher ein vollständiges Differential ist. Durch Integration desselben erhält man:

$$\pi_2 - \pi_1 = -\frac{R T}{E} \frac{v_{m_2} - v_{m_1}}{v_{m_2} \varepsilon_2 - v_{m_1} \varepsilon_1} \, \lg \frac{\varepsilon_{m_2}^2 \, v_{m_2} \, v_{m_2} + \varepsilon_s^2 \, v_s \, v_s}{\varepsilon_{m_1}^2 \, v_{m_1} \, v_{m_1} + \varepsilon_s^2 \, v_s \, v_s},$$

wobei sich jetzt die Zahlenindices auf die Integrationsgrenzen beziehen. Infolge der Beziehung

$$\varepsilon_{m_1} v_{m_1} = \varepsilon_s v_s = \varepsilon_{m_2} v_{m_2}$$

erhalten wir noch einfacher

$$\pi_2 - \pi_1 = -\frac{R T}{E} \frac{v_{m_2} - v_{m_1}}{v_{m_2} \varepsilon_2 - v_{m_1} \varepsilon_1} \, \lg \frac{\varepsilon_{m_2} \, v_{m_2} + \varepsilon_s \, v_s}{\varepsilon_{m_1} \, v_{m_1} + \varepsilon_s \, v_\varepsilon}.$$

Zum Schlusse empfinde ich noch das Bedürfnis, mich zu entschuldigen, dass ich hier nur einen dürftigen Plan für eine mühevolle Untersuchung entwerfe, ohne selbst zur experimentellen Lösung etwas beizutragen; ich bin jedoch dazu nicht in der Lage. Doch hat diese Arbeit ihr Ziel erreicht, wenn sie einen Forscher veranlasst, das Problem der Molecularkräfte von dieser Seite her in Angriff zu nehmen.

[16]

Bern, April 1902.

(Eingegangen 30. April 1902.)

Published in *Annalen der Physik* 8 (1902): 798–814. Dated Bern, April 1902, received 30 April 1902, published 10 July 1902.

[1] See *Planck 1891* for a general discussion of the role of ideal processes in thermodynamics. *Nernst 1898*, pp. 102–103, discusses the use of semipermeable membranes in thermodynamics. Nernst stresses that processes involving semipermeable membranes must be physically realizable for the second law to be applicable to them. Applying an argument first developed by Boltzmann (*Boltzmann 1878b*), he concludes that the second law can be used in the case of mixing or separation of any two chemically different gases by means of semipermeable membranes. There is no mention of external forces mimicking the membranes.

[2] See *Einstein 1902b* (Doc. 3) for Einstein's proof of this hypothesis, based on the "mechanical theory of heat."

[3] The symbols usually denote the number of moles per unit volume.

[4] "*E*" is known as Faraday's constant.

[5] P_m and P_s are actually the potentials per mole.

[6] The relation:

$$RT \frac{d \log \nu_0}{dz} + \frac{dp_0}{dz} = 0,$$

where $p_0 = \nu_0 RT$.

[7] In 1889 Nernst developed a general thermodynamic theory for electromotive phenomena in electrolytes (*Nernst 1889*); *Nernst 1898*, pp. 659–678, gives a "mechanical" theory for these phenomena, based on a molecular model involving moving ions. The expression for the electromotive force Einstein refers to is

$$E = \frac{RT}{n} \log \frac{c_2}{c_1},$$

where c_1 and c_2 are the concentrations of the salt ions at the two electrodes, and n is the valency of the metal ions. *Nernst 1898*, pp. 667–668, cites experimental evidence in support of this equation.

[8] *Einstein 1901* (Doc. 1).

[9] ν is actually the molar volume.

[10] "E_1", etc., should be "ϵ_1", etc.

[11] See *Einstein 1901* (Doc. 1), pp. 522–523.

[12] Note that v_m and v_s are velocities per unit force.

[13] In the second term of the right-hand side of this equation a factor "*E*" is missing.

[14] The factor "*RT*" should be "*RT/E*" in both equations on this page.

[15] The reference is to equation (α). References to (2) in the following are to (β).

[16] There are no indications that anyone performed an experiment based on Einstein's work.

EINSTEIN ON THE FOUNDATIONS OF
STATISTICAL PHYSICS

I

Einstein wrote only three papers devoted exclusively to the foundations of statistical physics: *Einstein 1902b* (Doc. 3), *Einstein 1903* (Doc. 4), and *Einstein 1904* (Doc. 5). Nevertheless, these papers develop a number of ideas that play a prominent role in Einstein's subsequent work; and they represent the beginning of a life-long interest in the foundations of statistical physics, an interest amply demonstrated in other publications;[1] in his correspondence with, among others, Michele Besso, Paul Hertz, Paul Ehrenfest, H. A. Lorentz, and D.K.C. MacDonald; in the design of instruments for the measurement of electrical fluctuations related to Brownian motion;[2] in his university lectures;[3] and in papers and diaries of friends and acquaintances.[4]

Einstein's papers on statistical physics start from within the Maxwell-Boltzmann tradition in kinetic theory. His aim was to fill what he considered a gap in the mechanical foundations of thermodynamics by deriving the laws of thermal equilibrium and the second law of thermodynamics from the most general possible mechanical assumptions and the probability calculus. Although there was very little contemporary response to his three papers,[5] and Einstein later downplayed their significance,[6] the harvest of results used in his later work is considerable. These include: the derivation of an energy fluctuation formula that presaged Einstein's work on Brownian motion and other fluctuation phenomena;[7] the derivation of an expression for the entropy that Einstein used repeatedly in his

[1] See, e.g., *Einstein 1910c, 1911h, 1915a, 1916a, 1916b, 1924, 1925a, 1925b*; *Einstein and Hopf 1910a, 1910b*.

[2] See *Einstein 1908a* (Doc. 48), and the editorial notes, "Einstein on Brownian Motion," pp. 221–222, and Vol. 5, "Einstein's 'Maschinchen' for the Measurement of Small Quantities of Electricity."

[3] Einstein lectured on statistical physics during the summer semester of 1908, University of Bern (see the Kreisschreiben of 6 July 1908, SzBeSa); summer semester of 1910, University of Zurich (see *Zürich Verzeichnis 1910a*, p. 22); summer semester of 1912, German University of Prague (see *Prag Ordnung 1912a*, p. 34); summer semester of 1913, ETH (see *ETH Programm 1913a*, p. 14); winter semester of 1915/1916, University of Berlin (see *Berlin Verzeichnis 1915b*, p. 48); and winter semester of 1917/1918, University of Berlin (see *Berlin Verzeichnis 1917b*, p. 44).

[4] For example, Paul Ehrenfest recorded in his diary that when he met Einstein for the first time in Prague in February 1912, they immediately started to discuss the ergodic hypothesis (*Klein 1970*, p. 176). And Karl F. Herzfeld reported conversations with Einstein concerning his objections to a derivation of the Boltzmann principle in *Herzfeld 1913*, p. 1553.

[5] *Einstein 1902b* (Doc. 3) and *Einstein 1903* (Doc. 4) are cited in *Boltzmann and Nabl 1907*, p. 549. Martin Klein (*Klein 1970*, p. 46) reported that the young Ehrenfest read Einstein's papers during a visit to Leiden in the spring of 1903. Ehrenfest later cited *Einstein 1902b* (Doc. 3) and *Einstein 1903* (Doc. 4) in *Ehrenfest and Ehrenfest 1911*, pp. 6–7, 80. *Einstein 1903* (Doc. 4) is discussed at some length in *Ornstein 1910* and *Lorentz 1916*. Paul Hertz subjected all three papers to a careful analysis in *Hertz, P. 1910a*, and they are cited frequently in his other papers on the foundations of statistical mechanics, *Hertz, P. 1910b, 1912, 1913a, 1913b*, as well as in his *Repertorium* article, *Hertz, P. 1916*.

[6] See *Einstein 1911c*, p. 176.

[7] See the editorial note, "Einstein on Brownian Motion," pp. 206–222.

papers on Brownian motion and on the quantum hypothesis;[8] the proof of the equipartition theorem for canonical ensembles, a result crucial to Einstein's dispute with Planck over the interpretation of the quantum hypothesis;[9] and the definition of probabilities as time averages, along with their equation to ensemble averages, which lies behind Einstein's conflict with Planck and others over the manner in which probabilities are introduced in physics.[10]

II

Einstein probably first read about the theory of heat in the popular-scientific literature and textbooks that he read as an adolescent.[11] But his formal introduction to the subject came in his fourth-year physics class at the Aargau Kantonsschule, and above all in H. F. Weber's physics lectures at the ETH.[12] In this respect, as in others, however, Einstein's ETH physics education failed to acquaint him with recent developments in the field.[13] Weber's lectures presented a survey of the theory of heat in the form in which it had been developed by Clausius.[14] The only more recent developments covered were experimental studies of topics such as diffusion, specific heats, thermal conductivity, and black-body radiation—all of them research interests of Weber.[15] The course included nothing about developments in kinetic theory in the last third of the nineteenth century owing to Maxwell, Boltzmann, Helmholtz, and others,[16] nor about the controversies surrounding this theory.[17]

[8] See *Einstein 1905k* (Doc. 16), p. 551, and the editorial note, "Einstein on Brownian Motion," pp. 206–222; see also *Einstein 1906d* (Doc. 34), p. 201, and the editorial note, "Einstein's Early Work on the Quantum Hypothesis," pp. 134–148.

[9] See especially *Einstein 1909b* (Doc. 56), p. 186.

[10] For Einstein's comments on Planck, see the papers cited in note 63.

[11] Einstein's early readings included *Büchner 1855, Bernstein 1853–1857, Krist 1891*, and *Violle 1892, 1893* (see Vol. 1, the editorial note, "Einstein's First Scientific Essay," pp. 5–6).

[12] For evidence that Einstein studied the theory of heat at the Aargau Kantonsschule, see his Aargau Kantonsschule Record (Vol. 1, Doc. 10) and his Aargau Kantonsschule Curriculum (Vol. 1, Appendix D, p. 361). For his notes on Weber's lectures on heat, see H. F. Weber's Lectures on Physics, Vol. 1, Doc. 37, pp. 63–147.

[13] For shortcomings in Einstein's physics education at the ETH, in particular the absence of a course on Maxwell's electrodynamics, see Vol. 1, the editorial note, "Einstein as a Student of Physics, and His Notes on H. F. Weber's Course," pp. 60–62.

[14] See *Clausius 1879–1891*. Einstein may have studied Clausius sometime before 1905. Asked about relevant readings prior to his work on Brownian motion (see Carl Seelig to Einstein, 11 September 1952), Einstein wrote: "Naturally I was familiar with Clausius's general investigations of kinetic theory" ("Clausius' allgemeine Untersuchungen über Kinetik kannte ich natürlich") (Einstein to Carl Seelig, 15 September 1952).

[15] See Vol. 1, the editorial note, "Einstein as a Student of Physics, and His Notes on H. F. Weber's Course," pp. 60–62, and the editorial note in this volume, "Einstein's Early Work on the Quantum Hypothesis," p. 135.

[16] See *Maxwell 1871, 1877, 1878; Boltzmann 1896, 1898a; Helmholtz 1903; Kirchhoff 1894*; and *Meyer, O. E. 1877, 1895, 1899*. For a survey of the history of kinetic theory in the nineteenth century, see *Brush 1976*.

[17] The earliest evidence of Einstein's acquaintance with any of these controversies is a discussion of Poincaré cycles in Einstein's notes for his lectures on the kinetic theory of heat at the University of Zurich, summer semester of 1910. See also *Discussion/Einstein 1911*, pp. 436–439.

Just as he studied Maxwell's electrodynamics on his own, Einstein pursued his interest in thermodynamics and kinetic theory through independent reading. Among the first books that he studied was Mach's *Wärmelehre*, which he probably read in 1897 or shortly thereafter.[18] But as Einstein's papers on statistical physics themselves make clear, Boltzmann's *Gastheorie* was the most immediate influence on his work in this field.[19] He began an intensive study of Boltzmann, perhaps in the summer of 1899,[20] and late the following summer he wrote:

> I am firmly convinced of the correctness of the principles of the theory, that is, I am convinced that in the case of gases it is really a matter of the movement of discrete mass points of definite, finite magnitude that move in accordance with certain conditions. . . . It is a step forward in the dynamical explanation of physical phenomena.

> Ich bin fest von der Richtigkeit der Prinzipien der Theorie überzeugt, das heißt ich bin überzeugt, daß es sich wirklich um Bewegung diskreter Massenpunkte von bestimmter endlicher Größe bei den Gasen handelt, die sich gemäß gewissen Bedingungen bewegen. . . . Es ist ein Schritt weiter in der dynamischen Erklärung der physikalischen Erscheinungen.[21]

Einstein continued to study the *Gastheorie* at least through the summer of 1901.[22] In the spring of that year, he mentioned O. E. Meyer's widely read text, *Die kinetische Theorie der Gase*, as a possible source of empirical data for an investigation of intermolecular forces that he was then carrying out with Marić's assistance,[23] and also asked Marić to send him a copy of Kirchhoff's *Vorlesungen über die Theorie der Wärme*.[24] He seems to

[18] See *Mach 1896*. Besso recalled recommending Mach's writings to Einstein in 1897 or 1898 (Michele Besso to Einstein, 12 October–8 December 1947); Einstein recalled that Besso had recommended both the *Mechanik* (*Mach 1897*) and the *Wärmelehre* "during my first years of [university] studies" ("während meiner ersten Studienjahre") (Einstein to Besso, 6 January 1948), and "about the year 1897" ("etwa im Jahre 1897") (Einstein to Carl Seelig, 8 April 1952). A copy of the second edition of the *Wärmelehre*, *Mach 1900b*, is in Einstein's personal library and contains a few short marginalia in Einstein's hand. The earliest extant mention of Mach in Einstein's correspondence is in Einstein to Mileva Marić, 10 September 1899 (Vol. 1, Doc. 54).

[19] See *Boltzmann 1896, 1898a*. A copy, with the label of a Milan book store, is in Einstein's personal library and contains a few, mostly insignificant marginalia and inclusions in Einstein's hand.

[20] See Einstein to Mileva Marić, 10 September 1899 (Vol. 1, Doc. 54).

[21] Einstein to Mileva Marić, 13 September 1900 (Vol. 1, Doc. 75). Shortly thereafter, Einstein sent Boltzmann a manuscript copy of *Einstein 1901* (Doc. 1) (see Mileva Marić to Helene Savić, 20 December 1900, Vol. 1, Doc. 85); for a discussion of the possible influence of *Boltzmann 1898a* on this paper, see the editorial note, "Einstein on the Nature of Molecular Forces," p. 4. Einstein's sister Maja claimed that Einstein continued to correspond with Boltzmann until the latter's death; but she stated that their correspondence concerned the theory of relativity (*Winteler-Einstein 1924*, p. 18).

[22] See Einstein to Marcel Grossmann, 6 September 1901 (Vol. 1, Doc. 122).

[23] *Meyer, O. E. 1877, 1895, 1899* (see Einstein to Mileva Marić, 30 April 1901, Vol. 1, Doc. 102). For a discussion of Einstein's work on molecular forces, see the editorial note, "Einstein on the Nature of Molecular Forces," pp. 3–8.

[24] See *Kirchhoff 1894*. For the request, see Einstein to Mileva Marić, 15 April 1901 (Vol. 1, Doc. 101).

have read a book by Planck, possibly the *Vorlesungen über Thermodynamik*, sometime before the fall of 1901.[25]

Many years later, Einstein recalled that, when he began work on the foundations of statistical physics, he was not familiar with the work of Gibbs and Boltzmann, which he characterized as "totally exhaustive of the subject" ("den Gegenstand tatsächlich erschöpfenden").[26] It is, indeed, improbable that Einstein knew Gibbs's *Elementary Principles in Statistical Mechanics*[27] when he wrote his three papers on statistical physics.[28] And in 1911 he went so far as to write: "Had I known Gibbs's book at that time, I would not have published those papers at all, but would have confined myself instead to the treatment of just a few points" ("Wenn mir das Gibbssche Buch damals bekannt gewesen wäre, hätte ich jene Arbeiten überhaupt nicht publiziert, sondern mich auf die Behandlung einiger weniger Punkte beschränkt").[29] But the extent of his knowledge of Boltzmann's work is harder to determine. As noted above, he certainly had studied Boltzmann's *Gastheorie*, which is cited in two of the papers. Thus, the question is whether he knew any of Boltzmann's earlier papers on the subject, the contents of which are not always covered in the *Gastheorie*.[30] A letter of Einstein's from September 1901, in which he refers to his recent study of Boltzmann's "works,"[31] suggests that he may have read some of these, as do certain features of his papers.[32] His first published mention of Gibbs, as well as of a pre-*Gastheorie* work by Boltzmann, is in 1909.[33] He was clearly well acquainted with Gibbs's research by the time he lectured at the University of Zurich in the summer semester of 1910.[34]

[25] See Mileva Marić to Einstein, early November 1901 (Vol. 1, Doc. 123). Only two books by Planck had been published by this date, *Planck 1887* and *Planck 1897*. In *Einstein 1913*, Einstein praised *Planck 1897* very highly and indicated that a copy was in every physicist's library.

[26] *Einstein 1979*, p. 44.

[27] See *Gibbs 1902*.

[28] It is unlikely that Einstein read the English edition, *Gibbs 1902*, and the German translation did not appear until 1905. A copy of *Gibbs 1905* is in Einstein's personal library. It is possible that Einstein learned something about Gibbs's approach from Planck's review of *Gibbs 1902* in the *Beiblätter zu den Annalen der Physik* (*Planck 1903a*), which was not published, however, until after *Einstein 1903* (Doc. 4). He may also have read Planck's paper (*Planck 1904*) in the Boltzmann *Festschrift* (*Meyer, S. 1904*) comparing Boltzmann's and Gibbs's conceptions of entropy. Einstein reviewed several papers from this *Festschrift* for the *Beiblätter* (see the editorial note, "Einstein's Reviews for the *Beiblätter zu den Annalen der Physik*"), which also published a review of Planck's paper (see *Valentiner 1905*).

[29] *Einstein 1911c*, p. 176. This paper was a reply to *Hertz, P. 1910a*, which criticizes certain features of Einstein's papers on statistical physics (see *Einstein 1902b* [Doc. 3], note 20, and *Einstein 1903* [Doc. 4], note 17). Hertz was one of the few physicists at the time who took Einstein's approach to the foundations of statistical physics seriously as an alternative to Gibbs's work. Einstein and Hertz were corresponding by the summer of 1910 (see Einstein to Paul Hertz, 14 August 1910), and Rudolf Hertz, Paul's son, reports that they had met while Einstein was still working in Bern (private communication).

[30] For a survey of the development of Boltzmann's views, see *Klein 1973*.

[31] Einstein to Marcel Grossmann, 6 September 1901 (Vol. 1, Doc. 122).

[32] These features are indicated in the notes to Docs. 3, 4, and 5.

[33] See *Einstein 1909b* (Doc. 56), p. 186, fn. 3, for a mention of Gibbs's investigations; and p. 187 for a reference to *Boltzmann 1877*.

[34] Einstein's lecture notes on kinetic theory incorporate both the vocabulary and notation of *Gibbs 1902*.

Aside from his reading of material directly concerned with kinetic theory, Einstein was studying at least three other topics in 1901 and 1902 that may have suggested the need to extend the foundations of thermodynamics and kinetic theory.[35] First, since at least the spring of 1901, Einstein had been reading Planck's papers on irreversible radiation processes, in which Planck sought to extend the concept of entropy to radiation.[36]

Second, Einstein was studying the work of Drude and others on the electron theory of metals, in which the apparatus of kinetic theory is employed to explain such phenomena as electrical and thermal conductivity.[37] Especially important for Einstein's conception of the foundations of statistical physics was his reading in late May 1901 of a paper by Reinganum that convinced him completely of the validity of the electron theory.[38] Reinganum showed that the principal results of Drude's electron theory can be derived without any assumptions about the nature of the elementary charge carriers or the mechanism of their interactions with the atoms of a metal. The only crucial assumption is that the equipartition theorem applies to a gas of moving charge carriers. Reinganum concluded:

> It appears therefore . . . that in metals, too, electricity moves in discrete quantities with the magnitude of the charges on electrolytic ions, and that the principles of the gas theory are applicable to the masses moving with the charges.

> Es erscheint daher . . . dass auch in den Metallen die Elektricität in discreten Mengen von der Grösse elektrolytischer Ionenladungen sich bewegt, und dass für die mit den Ladungen bewegten Massen die Principien der Gastheorie anzuwenden sind.[39]

Reinganum's results helped to confirm Einstein's commitment to an atomistic view of electricity, as well as of matter; but they may also have suggested that the equipartition theorem, as a vital principle of statistical physics, needed a more general mechanical foundation than it had hitherto been given. In fact, one of the major results of *Einstein 1902b* (Doc. 3) is precisely a demonstration that the equipartition theorem is valid not only for a

---

[35] Anticipations of Einstein's later distinction between "theories of principle" and "constructive theories" (see *Einstein 1919*) may also have inclined him to seek more general foundations for thermodynamics and the kinetic theory. For a discussion of the distinction, see the Introduction, pp. xxi–xxii.

[36] Einstein apparently studied one or both of *Planck 1900a, 1900b*; see Einstein to Mileva Marić, 4 April 1901 and 10 April 1901 (Vol. 1, Docs. 96 and 97). For a discussion of Einstein's interest in Planck's work, see Vol. 1, the editorial note, "Einstein on Thermal, Electrical, and Radiation Phenomena," pp. 235–237. See also Einstein to Marić, second half of May 1901, 4

June 1901, and 7 July 1901 (Vol. 1, Docs. 110, 112, and 114), as well as Einstein to Jost Winteler, 8 July 1901 (Vol. 1, Doc. 115).

[37] His reading included at least *Drude 1900a, 1900b* (see Einstein to Mileva Marić, 4 April 1901, and 28 May 1901, Vol. 1, Docs. 96 and 111). For a discussion of Einstein's interest in the electron theory of metals, see Vol. 1, the editorial note, "Einstein on Thermal, Electrical, and Radiation Phenomena," pp. 235–237.

[38] See *Reinganum 1900*. For Einstein's comments on it, see Einstein to Mileva Marić, 28 May 1901 (Vol. 1, Doc. 111).

[39] *Reinganum 1900*, p. 403. Reinganum's comments concern primarily *Drude 1900a*.

microcanonical ensemble, as Boltzmann proved in the *Gastheorie*,[40] but also for a canonical ensemble.[41]

Third and most important, Einstein was working on a theory of molecular forces.[42] The second of his two papers on this subject, *Einstein 1902a* (Doc. 2), rests upon what he calls a hypothetical extension of the second law of thermodynamics to mixtures, the individual components of which are subject to arbitrary external conservative forces.[43] At issue is the employment in thermodynamics and kinetic theory of idealized devices, such as semipermeable membranes, which had been used quite successfully in studies of diffusion, dissociation, and dilute solutions, in spite of doubts about even the approximate feasibility of such devices.[44] Einstein proposed to replace semipermeable membranes by the equally idealized, but theoretically more tractable, device of arbitrary external conservative forces. A proof of the validity of the second law when such forces are employed is one of the chief results of his first paper on statistical physics.[45]

One additional source may have impelled Einstein to seek a more general foundation for statistical physics, independent of special assumptions about the nature of the elementary constituents of physical systems or the interactions among them: his reading of Ostwald and Mach, both of whom were well known to be skeptical regarding hasty assumptions concerning the existence of molecules or atoms.[46] Given Einstein's firm commitment to the atomic theory (at least in the form of the assumption that mechanical systems possess only a finite number of degrees of freedom) and to Boltzmann's approach to statistical physics, the skepticism of Ostwald and Mach could well have been an inducement to free the foundations of statistical physics from dependence upon detailed and debatable hypotheses about the constitution of the systems that statistical physics aims to describe.[47]

[40] See *Boltzmann 1898a*, p. 101.

[41] See *Einstein 1902b* (Doc. 3), § 6. It is possible that, in the doctoral dissertation that he submitted in 1901, Einstein's interest in Boltzmann's work on kinetic theory and in Drude's work on electron theory may have coalesced with his work on molecular forces; for a discussion of the 1901 dissertation, see the editorial note, "Einstein's Dissertation on the Determination of Molecular Dimensions," pp. 174–175.

[42] For discussions of this work, see Vol. 1, the editorial note, "Einstein on Molecular Forces," pp. 264–266, and, in this volume, the editorial note, "Einstein on the Nature of Molecular Forces," pp. 3–8.

[43] See *Einstein 1902a* (Doc. 2), p. 799.

[44] For a review of this topic, see *Planck 1891*.

[45] See *Einstein 1902b* (Doc. 3), p. 433.

[46] See especially *Ostwald 1893* and *Mach 1896*. For evidence that Einstein read *Ostwald 1893*, see the apparent references to it in Ein-

stein to Mileva Marić, 10 April 1901 (Vol. 1, Doc. 97), and in Einstein to Wilhelm Ostwald, 19 March 1901 (Vol. 1, Doc. 92), in which Einstein claimed that his work on capillarity had been influenced by Ostwald. For evidence that Einstein read Mach, see note 18 (for an early expression of Einstein's high opinion of Mach, see *Einstein 1916c*).

[47] Many years later, Einstein said about his papers on statistical physics: "My principal aim in this was to find facts that would guarantee as much as possible the existence of atoms of definite finite size" ("Mein Hauptziel dabei war es, Tatsachen zu finden, welche die Existenz von Atomen von bestimmter endlicher Grösse möglichst sicher stellten") (*Einstein 1979*, p. 44, translation, p. 45). Einstein went on to say that it was his work on Brownian motion, together with Planck's determination of Avogadro's constant from the Rayleigh-Jeans radiation law, that convinced the "numerous skeptics" ("zahlreichen Skeptiker"), including Ostwald and Mach, of the reality of atoms (*Einstein 1979*,

The only evidence of personal influences on Einstein's early work on statistical physics concerns his correspondence with Michele Besso from 1903, and his conversations with a patent-office colleague, Joseph Sauter. Einstein's correspondence with Besso is discussed in section III below, and in the annotations to one of the papers.[48] A lengthy discussion with Sauter is mentioned in a letter from Besso to Einstein of 13 June 1952:

> Lately he found again your papers on thermodynamics of 26.VI.1902 [Doc. 3] and 26.I.1903 [Doc. 4]. He remembers having discussed one of them at length at that time, and having to a certain extent saved what was essential in it, in spite of a mistake that he discovered, this in the face of a pessimistic attitude on your part; but he cannot recall which of the two it concerned.

> Er fand letztlich wieder Deine Arbeiten über Thermodynamik vom 26.VI.1902 [Doc. 3] und 26.I.1903 [Doc. 4]. Er erinnert sich über die eine, seinerzeit, lange diskutiert zu haben, und das Wesentliche davon, trotz eines damals von ihm entdeckten Fehlers, entgegen pessimistischer Einstellung von Dir, gewissermassen gerettet zu haben; kann sich aber nicht besinnen welche von den beiden es betraf.[49]

III

As noted above, Einstein's contributions to the foundations of statistical physics grew out of the Maxwell-Boltzmann tradition in kinetic theory, which was then also called the "molecular theory of heat" ("molekulare Wärmetheorie"), more specifically, the aspect of the tradition that emphasizes the general statistical foundations of thermodynamics. Einstein's aim was to derive the basic concepts and laws of thermodynamics from mechanics and the probability calculus. In doing so, Einstein improved upon Boltzmann's arguments in several important respects, so that his three papers on the subject form a bridge, parallel to that in *Gibbs 1902*, between Boltzmann's work and the modern approach to statistical mechanics.[50] Among the features that distinguish Einstein's approach from Boltzmann's are: (1) his attempt at generality—he tried to proceed with a minimum of assumptions about the nature of the systems under consideration; (2) his conception of what needs to be derived from these assumptions—not only the second law of thermodynamics, but also

---

pp. 44 and 46). See also *Einstein 1915a*, pp. 260–261, and the editorial note, "Einstein on Brownian Motion," p. 218.

[48] See *Einstein 1903* (Doc. 4), notes 4, 6, and 7.

[49] Einstein replied on 13 July 1952 that he recalled many discussions with Sauter about the papers, but could not remember the details. Sauter himself described this episode in a radio broadcast in 1955, in which he recalled discussing Einstein's early papers during walks home

with Einstein after work at the patent office (for the text of this talk, see *Flückiger 1974*, pp. 155–156). An error that may be the one discovered by Sauter is discussed in *Einstein 1903* (Doc. 4), note 7.

[50] Another important transitional work is *Ehrenfest and Ehrenfest 1911*. Two of the earliest texts to incorporate the modern statistical point of view are *Wassmuth 1915* and *Hertz, P. 1916*.

an observable measure of temperature and the associated laws of thermal equilibrium; (3) his method of carrying out the derivation—he independently formulated the distinction between what are now, following *Gibbs 1902*, called the microcanonical and the canonical ensembles; (4) his appreciation of certain crucial conceptual problems attendant upon the introduction of probabilities in physics, which led him to assume what would later be called an ergodic hypothesis, in the form of the equality of time and ensemble averages; and (5) his extraction of important new physical consequences from his foundational studies, including most notably the energy fluctuation formula for the canonical ensemble.

Einstein 1902b (Doc. 3) was submitted to the *Annalen* in June 1902 and published on 18 September. It opens with the statement that Einstein intends to fill the "gap" ("Lücke") in the mechanical foundations of "the general theory of heat" ("die allgemeine Wärmetheorie") by deriving the laws of thermal equilibrium and the second law of thermodynamics exclusively from the equations of mechanics and the probability calculus. Einstein noted that Maxwell's and Boltzmann's theories "have already come near to this goal" ("diesem Ziele bereits nahe gekommen sind"). In a letter written nine months earlier, he gave a similar description of a work that may have been a draft of part of this paper:

> Recently I have been thoroughly occupied with Boltzmann's works on the kinetic theory of gases & have, in the last few days, myself written a small paper, which provides the keystone to a chain of derivations begun by him. . . . [I will] probably publish it in the Annalen.

> Ich habe mich in letzter Zeit gründlich mit Boltzmanns Arbeiten über kinetische Gastheorie befaßt & in den letzten Tagen selbst eine kleine Arbeit geschrieben, welche den Schlußstein einer von ihm begonnenen Beweiskette liefert. . . . [Ich werde] sie wahrscheinlich in den Annalen publizieren.[51]

It is not clear, however, just what shortcomings Einstein saw in Boltzmann's derivations. The chief novelties of *Einstein 1902b* (Doc. 3) are Einstein's derivation of the equilibrium distribution for a virtual canonical ensemble of systems from the microcanonical distribution, and his subsequent employment of the canonical distribution to derive the equipartition theorem, the laws of thermal equilibrium, and the second law for systems in equilibrium. This suggests that Einstein saw one shortcoming as Boltzmann's failure to understand the significance of the canonical ensemble and its connection with the microcanonical ensemble.

Support for this hypothesis comes from a comment by Einstein in 1909 on Planck's

---

[51] Einstein to Marcel Grossmann, 6 September 1901 (Vol. 1, Doc. 122). Parts of this earlier paper may survive in § 8–10 of *Einstein 1902b* (Doc. 3). These sections provide the proof of a thermodynamical assumption crucial to the argument of *Einstein 1902a* (for a discussion of this assumption, see the editorial note, "Einstein on the Nature of Molecular Forces," p. 8). And, quite independently of the rest of the paper, § 8 begins with a reiteration of the characterization of a "mechanical system" ("mechanisches System") given in § 1.

attempt to deny the applicability of the equipartition theorem to radiation.[52] Planck, citing Boltzmann, argued that a necessary condition for the validity of the equipartition theorem is that the state distribution be an ergodic one, in the sense that the probability of a system's occupying a given cell of phase space is proportional to the size of the cell, however small the cell may be; but that an ergodic distribution is not guaranteed if the phase space is partitioned into cells of finite size, as in Planck's version of the quantum hypothesis.[53] Einstein wrote:

> The attempt . . . to call into question the general validity of equation II [the equipartition theorem] rests—it seems to me—solely on a gap in Boltzmann's inquiries, which has been filled in the meantime by Gibbs's investigations.

> Der . . . Versuch, die Allgemeingültigkeit der Gleichung II in Frage zu stellen, beruht—wie mir scheint—nur auf einer Lücke in Boltzmanns Betrachtungen, welche unterdessen durch die Gibbsschen Untersuchungen ausgefüllt wurde.[54]

Gibbs, like Einstein, showed that the equipartition theorem holds not only for an ergodic or microcanonical ensemble (which is what Planck apparently took Boltzmann to have asserted), but also for a canonical ensemble.[55]

Einstein may have considered another shortcoming of Boltzmann's approach to be that, as a consequence of his failure to grasp the significance of the canonical ensemble, he could not construct a link between the fundamental concepts of the kinetic theory and the measurement of crucial macroscopic thermodynamical quantities like temperature. Such measurements require physical interactions of a kind that can only be treated theoretically with the introduction of the canonical ensemble. In fact, one of the main topics of *Einstein*

---

[52] See *Einstein 1909b* (Doc. 56), pp. 186–187. This comment was made in the context of Einstein's criticism of Planck's derivation of the radiation law. For a discussion of this criticism, see the editorial note, "Einstein's Early Work on the Quantum Hypothesis," p. 138.

[53] See *Planck 1906c*, p. 178. Planck cites *Boltzmann 1898a*, p. 101, where one reads: "Of course this equality of the mean value of the kinetic energy corresponding to each momentoid has only been proved for the assumed (ergodic) state distribution. . . . But in general there can and will be other stationary state distributions, for which these theorems are not valid" ("Selbstverständlich ist diese Gleichheit des Mittelwerthes der jedem Momentoide entsprechenden lebendigen Kraft nur für die vorausgesetzte (ergodische) Zustandsvertheilung bewiesen. . . . Es kann und wird im Allgemeinen aber auch andere stationäre Zustandsvertheilungen geben, für welche diese Sätze nicht gelten"). In

Boltzmann 1898a, § 32, and earlier in *Boltzmann 1885*, an "Ergode" is defined as a virtual ensemble of systems with identical energies, where the energy is the only time-independent conserved quantity. "Momentoide" is Boltzmann's term for "degrees of freedom"; see *Boltzmann 1898a*, § 33. For a discussion of Planck's argument, see the editorial note, "Einstein's Early Work on the Quantum Hypothesis," p. 138.

[54] *Einstein 1909b* (Doc. 56), p. 186; see also *Einstein and Hopf 1910b*.

[55] See *Einstein 1902b* (Doc. 3), pp. 427–428, and *Gibbs 1902*, p. 49, which corresponds to *Gibbs 1905*, p. 48. A few pages after the passage cited by Planck, Boltzmann himself proved the equipartition theorem for a *real* ensemble of canonically distributed systems whose interactions are assumed to be negligible (see *Boltzmann 1898a*, § 42), but nowhere did he prove it for a canonically distributed virtual ensemble.

1902b (Doc. 3) is the establishment of a connection between a constant in Einstein's formula for the canonical distribution (defined below) and an "observable measure of temperature" ("beobachtbare[s] Temperaturmaass") (p. 425).

Following the programmatic remark about the "gap" ("Lücke"), *Einstein 1902b* (Doc. 3) defines a very general kind of "mechanical system" ("mechanisches System"), possessing a large but finite number of degrees of freedom, the energy of which is the sum of a potential term and a kinetic term that is quadratic in the velocities. A microcanonical ensemble—a virtual ensemble of N such systems with energies between E and $E + \delta E$— is introduced. Using Liouville's theorem, and assuming that the energy is the only explicitly time-independent conserved quantity for a mechanical system, Einstein derived an equilibrium distribution formula for this ensemble. A more special ensemble satisfying these conditions is then introduced, consisting of coupled systems S and Σ (thermometer and heat bath), with the energy of Σ taken to be infinitely great compared to that of S. The equilibrium distribution of the ensemble of subsystems S (a canonical ensemble) is derived:

$$ dN = A'' e^{-2hE} dp_1 \ldots dq_n. $$

In the course of this derivation, Einstein first introduced the structure function

$$ \omega(E) \, \delta E = \int_{E}^{E + \delta E} dp_1 \ldots dp_n, $$

which later played an important role in his papers on the quantum hypothesis.[56]

After establishing the above-mentioned connection between the constant h and an "observable measure of temperature," used to prove the laws of thermal equilibrium, Einstein derived the equipartition theorem for the canonical ensemble and the second law of thermodynamics for reversible processes, as well as an expression for the entropy of a mechanical system at equilibrium,

$$ \epsilon = E/T + 2 \kappa \log\{ \int e^{-2hE} dp_1 \ldots dq_n \} + \text{const.}, $$

that he employed several times in later papers.[57] Einstein pointed out that the expression for the entropy is noteworthy,

because it depends solely on E and T, while no longer allowing the special form

---

[56] See, e.g., *Einstein 1907a* (Doc. 38), and the editorial note, "Einstein's Early Work on the Quantum Hypothesis," p. 141. The structure function is cited here in the form first given in *Einstein 1904* (Doc. 5), p. 355, which is the version that Einstein cited in his later papers.

[57] See, e.g., *Einstein 1905k* (Doc. 16), p. 551, and *Einstein 1906d* (Doc. 34), p. 201.

of E as the sum of potential energy and kinetic energy to come to the fore. This fact suggests that our results are more general than the mechanical representation utilized.

weil er lediglich von E und T abhängt, die specielle Form von E als Summe potentieller Energie und lebendiger Kraft aber nicht mehr hervortreten lässt. Diese Thatsache lässt vermuten, dass unsere Resultate allgemeiner sind als die benutzte mechanische Darstellung.[58]

Einstein's next paper on the foundations of statistical physics, *Einstein 1903* (Doc. 4), was received by the *Annalen* on 26 January 1903 and published on 16 April. Einstein summarized its contents in a letter to Besso of 22 January 1903:

The concepts of temperature and entropy follow from the assumption of the energy principle and the atomistic theory; as the second law also does in its most general form, namely the impossibility of a *perpetuum mobile* of the second kind, by using the hypothesis that state distributions of isolated systems never evolve into more improbable ones.

Unter Voraussetzung des Energieprinzips und der atomistischen Theorie folgen die Begriffe Temperatur und Entropie, sowie mit Benützung der Hypothese, daß Zustandsverteilungen isol. Systeme niemals in unwahrscheinlichere übergehen, auch der zweite Hauptsatz in seiner allgemeinsten Form, nämlich die Unm[öglichkeit] eines perpetuum mobile zweiter Art.

In a letter to Jakob Ehrat of the last week of March, Einstein emphasized the generality of the foundations for the theory of heat that he employed in this paper: "It assumes only the foundations of atomic physics, no further physical hypothesis" ("Sie setzt nur die Grundlagen der Atomphysik voraus, keine weitere physikalische Hypothese").[59]

After noting that his previous paper had employed assumptions from the kinetic theory of heat, the paper goes on to raise the question, already suggested in that paper:

whether the kinetic theory is really necessary . . . or whether assumptions of a more general kind are, perhaps, already sufficient. That the latter is the case . . . is to be shown in this article.

ob die kinetische Theorie auch wirklich notwendig ist . . . oder ob vielleicht bereits Voraussetzungen allgemeinerer Art dazu genügen können. Daß dieses letztere der Fall ist . . . soll in dieser Abhandlung gezeigt werden.[60]

---

[58] *Einstein 1902b* (Doc. 3), p. 433. On possible echoes of *Hertz, H. 1894* in this remark, see *Einstein 1902b* (Doc. 3), note 42.

[59] Einstein to Jakob Ehrat, last week of March 1903. Einstein did not specify what he meant by the expressions "atomistic theory" ("atomistische Theorie") and the "foundations of atomic physics" ("Grundlagen der Atomphysik"). The only assumption in *Einstein 1903* (Doc. 4) fitting this description is that the systems under consideration possess a finite number of degrees of freedom, which excludes, most importantly, continuous fields.

[60] *Einstein 1903* (Doc. 4), p. 170.

The paper begins with the definition of "physical system" ("physikalisches System"), a definition more general than that of "mechanical system" given in the previous paper. Although the energy is still assumed to be the only explicitly time-independent conserved quantity, no assumption is made about its form, in particular about a distinction between potential and kinetic energy. Aside from the assumption of a large but finite number of degrees of freedom, all that is explicitly required is that the change in a system's state at any time be fully determined by its state at that time. A microcanonical ensemble of N systems is again introduced, but in this paper the derivation of the equilibrium distribution for such an ensemble is preceded by a discussion of time averages. Einstein argued that (a) the macroscopic thermodynamical stability of a physical system entails the existence of a well-defined limit, as $t \to \infty$, for the proportion of time that the system spends in a given region of phase space; (b) the time average thus defined for an individual physical system is numerically equivalent to an ensemble average, that is, to the average fraction of an ensemble of N identical systems occupying the region at any time.

The subsequent derivation of the microcanonical equilibrium distribution rests upon an implicit hydrodynamical analogy with the time evolution of the ensemble of systems in phase space. Einstein started from the equation of continuity for the ensemble density function in phase space, but—in contrast to the now standard derivation—he did not explicitly invoke either Liouville's theorem or the incompressibility condition, a reflection, perhaps, of his desire to achieve maximum generality. He soon came to believe, however, that the assumptions actually necessary for his derivation are equivalent to the assumption that Hamilton's equations are satisfied, and hence that the only advances in generality are the elimination of the force concept and of the distinction between potential and kinetic energy.[61]

Particularly noteworthy in this derivation is Einstein's equating of time and ensemble averages. It represents an advance in his understanding of the problem of ergodicity that became a hallmark of his approach to the foundations of thermodynamics, distinguishing it from Gibbs's approach.[62] All of his later critical remarks about Planck's way of introducing probabilities into statistical physics concern the latter's failure to provide some kind of physical justification for his assumptions about equiprobable cases, such as that provided by the concept of time averages.[63]

After rederiving a number of results in *Einstein 1902b* (Doc. 3), including the canonical

---

[61] See Einstein to Michele Besso, 17 March 1903. For a discussion of this letter, see *Einstein 1903* (Doc. 4), note 7.

[62] For discussions of Einstein's equating of time and ensemble averages, see *Ornstein 1910*, pp. 805–808, and *Lorentz 1916*; see also *Einstein 1903* (Doc. 4), note 5.

[63] See, e.g., *Einstein 1905i* (Doc. 14), p. 140; *Einstein 1909b* (Doc. 56), p. 187, where Einstein criticizes both Planck and Boltzmann for introducing probabilities into their calculations without defining these probabilities; *Dis-cussion/Planck 1911*, p. 115, where Einstein criticizes Planck for trying to apply the Boltzmann equation, $S = k \cdot \log W$, without giving "a physical definition" ("une définition physique") of the probabilities, W; *Einstein 1910c*, pp. 1276–1279, which makes essentially the same point without mentioning Planck; and *Discussion/Einstein 1911*, pp. 437–438, a general discussion of the definition of probabilities in terms of time averages, relating this approach to the problem of the statistical character of the second law of thermodynamics.

distribution and the equipartition theorem, Einstein gave a more general derivation of his expression for the entropy, which no longer requires the distinction between kinetic and potential energy and is supposed to be valid for nonequilibrium distributions.

The paper concludes with a derivation of the second law that also claims validity for the nonequilibrium case. However, the derivation depends upon the assumption that "ever more probable state distributions will follow more improbable ones" ("immer wahrscheinlichere Zustandsverteilungen auf unwahrscheinlichere folgen werden");[64] here, the term "state distributions" ("Zustandsverteilungen") refers to the distribution of microstates of a virtual ensemble of systems over a hypersurface of constant energy. Einstein was later criticized for his failure to prove this assumption,[65] but his next paper modifies it in a way that makes clear that he did not intend to deny or to ignore thermodynamical fluctuations.[66] The main difficulty with which Einstein was grappling concerns the kind of microstate or state distribution that should be related to the thermodynamic state of a system in attempting a derivation of such thermodynamical principles as the second law.[67]

On 6 April 1904 Einstein wrote to his friend Marcel Grossmann that he was sending him a paper he had just submitted to the *Annalen*, in which he treated "the atomistic theory of heat without the kinetic hypothesis" ("die atomistische Wärmelehre ohne die kinetische Hypothese"). Einstein's last paper devoted exclusively to the foundations of statistical physics, *Einstein 1904* (Doc. 5), was submitted for publication in late March and appeared in early June 1904. This paper is the culmination of Einstein's efforts to generalize and extend the foundations of statistical physics. It is distinguished from his two previous papers by a more polished and direct style, and by the concrete and novel physical consequences drawn from foundational investigations, which foreshadow Einstein's later employment of the principles of statistical physics in such fields as black-body radiation, specific heats, Brownian motion, and critical opalescence. Most of the paper's formal apparatus is taken from the previous paper, but there are a number of new results. Among these are (1) an improved derivation of the second law, replacing the problematic assumption about the monotonic increase in the probability of a distribution of microstates with the assumption of a monotonic increase in the probability that the energies, $E_1, E_2, \ldots$, E_l, of the individual members of a set of systems will each lie in a given range, E_1 to $E_1 + \delta E_1$, of energies; such a distribution of energies is now designated as the "state" of the set of systems; (2) a determination of the constant κ ($\kappa = k/2$, where k is Boltzmann's constant), which establishes a relation between the values of Avogadro's number and of Boltzmann's constant. Most important, however, are results that anticipate some of

[64] *Einstein 1903* (Doc. 4), p. 184.

[65] See *Hertz, P. 1910a*, p. 552. Einstein replied to Hertz in *Einstein 1911c*. See also *Einstein 1902b* (Doc. 3), note 20, and *Einstein 1903* (Doc. 4), note 17.

[66] See *Einstein 1904* (Doc. 5), pp. 355–357. For a more complete discussion of the questions posed by this assumption, see *Einstein 1903*

(Doc. 4), note 17. For Einstein's later views on the statistical character of the second law, see, e.g., *Discussion/Einstein 1911*, pp. 436–443, and *Einstein 1915a*, pp. 262–263.

[67] See *Hertz, P. 1916*, pp. 547–558, for a discussion of the relations between microstates, phenomenological states, and entropy informed by Einstein's views.

Einstein's later interests: (3) the interpretation of the constant κ (again, in effect, Boltzmann's constant) as a measure of the thermodynamical stability of a system; (4) the derivation of an energy fluctuation formula that heralds Einstein's subsequent investigations of Brownian motion and other fluctuation phenomena; and (5) the application of this formula to fluctuations in the energy of black-body radiation within a small cavity, resulting in the derivation of an analog to Wien's displacement law, with a numerical factor that agrees to an order of magnitude with the empirical results.[68]

The last of Einstein's papers devoted solely to the foundations of statistical physics by no means represents the last of Einstein's contributions to the discussion of the issues raised in these papers. In the years after 1904, Einstein continued to comment upon these issues, but always in the context of specific problems, or applications in other areas. For example, Einstein's use of the Boltzmann principle, $S = k \cdot \log W$, in his papers on the quantum hypothesis, led him to think about the principle's proper interpretation, and to disagree with Planck about the right way to count "complexions" ("Komplexionen").[69] Many of Einstein's arguments for defining probabilities in physics as time averages are found in his papers on the quantum hypothesis.[70] His continuing interest in fluctuations, and in the attendant question of the limits of validity of macroscopic thermodynamics, is evident in his papers on Brownian motion and critical opalescence.[71] In the latter paper, Einstein also laid the foundations of what later became known as statistical thermodynamics.[72] Finally, his papers on quantum statistics in 1916–1917 and 1924–1925 are still investigating the changes in the foundations of statistical physics required by the quantum hypothesis.[73]

Further evidence of Einstein's interest in the foundations of statistical physics is provided by the lectures on statistical physics that he delivered at various universities between 1908 and 1918.[74] It is noteworthy that the point of view presented in these lectures is virtually identical to that developed in his first three papers on the subject (Docs. 3, 4, and 5). The presentation is more sophisticated, incorporates the Gibbsian terminology in discussions of the microcanonical and canonical ensembles, and displays a clearer understanding of the nature of the required ergodicity assumptions.[75] But, in spite of the fact

[68] For a discussion of black-body radiation and the Wien displacement law, see the editorial note, "Einstein's Early Work on the Quantum Hypothesis," p. 135.

[69] See especially *Einstein 1907a* (Doc. 38), § 1, and *Einstein 1909b* (Doc. 56), p. 187. For a discussion of this dispute with Planck, and of Einstein's views on probability in this context, see the editorial note, "Einstein's Early Work on the Quantum Hypothesis," pp. 137–139.

[70] See the papers cited above in note 63.

[71] See, e.g., *Einstein 1905k* (Doc. 16), *Einstein 1906b* (Doc. 32), *Einstein 1907b* (Doc. 39), *Einstein 1910c*, and the editorial note, "Einstein on Brownian Motion," pp. 206–222.

[72] See *Einstein 1910c*, which was partially

anticipated in *Einstein 1907b* (Doc. 39). For a discussion of statistical thermodynamics, see *Tisza and Quay 1963*.

[73] See *Einstein 1916a, 1916b, 1924, 1925a,* and *1925b*.

[74] See note 3. In Berlin, Einstein was not obliged to lecture on any topic (see *Kayser 1930*, p. 124), yet chose to lecture several times on topics in statistical physics.

[75] See, e.g., Walter Dällenbach's notes for Einstein's lectures on "Molekulartheorie der Wärme" at the ETH, summer semester of 1913 (SzZE Bibliothek, Hs 304:1224), p. 15, where the equivalence of time and ensemble averages is discussed.

that Gibbs's approach had gained a wide following among physicists by the end of the first decade of this century,[76] Einstein treated the essential issues in much the same way as in his earlier papers.[77]

[76] See *Ehrenfest and Ehrenfest 1911* for a contemporary survey.

[77] For further discussion of Einstein's work on the foundations of statistical physics, see *Byrne 1980, 1981*; *Klein 1967, 1974b, 1982a*; *Kuhn 1978*, pp. 171–182; *Mehra and Rechenberg 1982*, pp. 64–70; and *Pais 1982*, chap. 4.

3. "Kinetic Theory of Thermal Equilibrium and of the Second Law of Thermodynamics"

[*Einstein 1902b*]

DATED Bern, June 1902
RECEIVED 26 June 1902
PUBLISHED 18 September 1902

IN: *Annalen der Physik* 9 (1902): 417–433.

417

6. *Kinetische Theorie des Wärmegleichgewichtes und des zweiten Hauptsatzes der Thermodynamik; von A. Einstein.*

So gross die Errungenschaften der kinetischen Theorie der Wärme auf dem Gebiete der Gastheorie gewesen sind, so ist doch bis jetzt die Mechanik nicht im stande gewesen, eine hinreichende Grundlage für die allgemeine Wärmetheorie zu liefern, weil es bis jetzt nicht gelungen ist, die Sätze über das Wärmegleichgewicht und den zweiten Hauptsatz unter alleiniger Benutzung der mechanischen Gleichungen und der Wahrscheinlichkeitsrechnung herzuleiten, obwohl Maxwell's und Boltzmann's Theorien diesem Ziele bereits nahe gekommen sind. Zweck der nachfolgenden Betrachtung ist es, [1] diese Lücke auszufüllen. Dabei wird sich gleichzeitig eine Erweiterung des zweiten Hauptsatzes ergeben, welche für die [2] Anwendung der Thermodynamik von Wichtigkeit ist. Ferner wird sich der mathematische Ausdruck für die Entropie vom mechanischen Standpunkt aus ergeben.

§ 1. Mechanisches Bild für ein physikalisches System.

Wir denken uns ein beliebiges physikalisches System darstellbar durch ein mechanisches System, dessen Zustand durch sehr viele Coordinaten $p_1, \ldots p_n$ und die dazu gehörigen Ge- [3] schwindigkeiten

$$\frac{d p_1}{d t}, \ldots \frac{d p_n}{d t}$$

eindeutig bestimmt sei. Die Energie E derselben bestehe aus zwei Summanden, der potentiellen Energie V und der lebendigen Kraft L. Erstere sei eine Function der Coordinaten allein, letztere eine quadratische Function der

$$\frac{d p_\nu}{d t} = p'_\nu,$$

deren Coefficienten beliebige Function der p sind. Auf die Massen des Systems sollen zweierlei äussere Kräfte wirken.

418   *A. Einstein.*

Die einen seien von einem Potentiale V_a ableitbar und sollen
die äusseren Bedingungen (Schwerkraft, Wirkung von festen
Wänden ohne thermische Wirkung etc.) darstellen; ihr Potential
kann die Zeit explicite enthalten, doch soll seine Ableitung
nach derselben sehr klein sein. Die anderen Kräfte seien
nicht von einem Potential ableitbar und seien schnell veränderlich. Sie sind als diejenigen Kräfte aufzufassen, welche
die Wärmezufuhr bewirken. Wirken solche Kräfte nicht, ist
aber V_a explicite von der Zeit abhängig, so haben wir einen
adiabatischen Process vor uns.

Wir werden auch statt der Geschwindigkeiten, lineare
Functionen derselben, die Momente $q_1, \ldots q_n$ als Zustandsvariable des System einführen, welche durch n Gleichungen
von der Form

$$q_\nu = \frac{\partial L}{\partial p'_\nu}$$

definirt sind, wobei L als Function der $p_1, \ldots p_n$ und
$p_1', \ldots p_n'$ zu denken ist.

§ 2. Ueber die Verteilung der möglichen Zustände unter N identischen adiabatischen stationären Systemen, bei nahezu gleichem Energieinhalt.

Seien unendlich viele (N) Systeme gleicher Art vorhanden,
deren Energieinhalt zwischen den bestimmten sehr wenig ver-
[4]   schiedenen Werten $\overline{E}$ und $E + \delta E$ continuirlich verteilt sind.
Aeussere Kräfte, welche nicht von einem Potential ableitbar
sind, sollen nicht vorhanden sein und V_a möge die Zeit nicht
explicite enthalten, sodass das System ein conservatives System
ist. Wir untersuchen die Zustandsverteilung, von welcher wir
voraussetzen, dass sie stationär sei.

Wir machen die Voraussetzung, dass ausser der Energie
[5] $E = L + V_a + V_i$ oder einer Function dieser Grösse, für das
einzelne System keine Function der Zustandsvariabeln p und q
[6]   allein vorhanden sei, welche mit der Zeit sich nicht ändert;
auch fernerhin seien nur Systeme betrachtet, welche diese
Bedingung erfüllen. Unsere Voraussetzung ist gleichbedeutend
mit der Annahme, dass die Zustandsverteilung unserer Systeme
durch den Wert von E bestimmt sei, und sich aus jeden beliebigen Anfangswerten der Zustandsvariabeln, welche nur

unserer Bedingung für den Wert der Energie Genüge leisten, von selbst herstelle. Existirte nämlich für das System noch eine Bedingung von der Art $\varphi(p_1 \ldots q_n) = $ const., welche [7] nicht auf die Form $\varphi(E) = $ const. gebracht werden kann, so wäre offenbar durch geeignete Wahl der Anfangsbedingungen zu erzielen, dass für jedes der N Systeme φ einen beliebigen vorgeschriebenen Wert hätte. Da sich diese Werte aber mit der Zeit nicht ändern, so folgt z. B., dass der Grösse $\sum \varphi$, erstreckt über alle Systeme, bei gegebenem Werte von E, durch geeignete Wahl der Anfangsbedingungen, jeder beliebige Wert erteilt werden könnte. $\sum \varphi$ ist nun andererseits aus der Zustandsverteilung eindeutig berechenbar, sodass anderen Werten von $\sum \varphi$ andere Zustandsverteilungen entsprechen. Man ersieht also, dass die Existenz eines zweiten solchen Integrals φ notwendig zur Folge hat, dass durch E allein die Zustandsverteilung noch nicht bestimmt wäre, sondern dass dieselbe notwendig vom Anfangszustande der Systeme abhängen müsste.

Bezeichnet man mit g ein unendlich kleines Gebiet aller Zustandsvariabeln $p_1, \ldots p_n,\ q_1, \ldots q_n$, welches so gewählt sein soll, dass $E(p_1, \ldots q_n)$ zwischen $\bar{E}$ und $\bar{E} + \delta E$ liegt, wenn die Zustandsvariabeln dem Gebiete g angehören, so ist die Verteilung der Zustände durch eine Gleichung von folgender Form zu charakterisiren

$$dN = \psi(p_1, \ldots q_n) \int_g dp_1 \ldots dq_n,$$

dN bedeutet die Anzahl der Systeme, deren Zustandsvariable zu einer bestimmten Zeit dem Gebiete g zugehören. Die Gleichung sagt die Bedingung aus, dass die Verteilung stationär ist.

Wir wählen nun ein solches unendlich kleines Gebiet G. Die Anzahl der Systeme, deren Zustandsvariable zu irgend einer bestimmten Zeit $t = 0$ dem Gebiete G angehören, ist dann

$$dN = \psi(P_1, \ldots Q_n) \int_G dP_1 \ldots dQ_n,$$

[8]

wobei die grossen Buchstaben die Zugehörigkeit der abhängigen Variabeln zur Zeit $t = 0$ andeuten sollen.

Wir lassen nun die beliebige Zeit t verstreichen. Besass ein System in $t = 0$ die bestimmten Zustandsvariabeln $P_1 . \ldots Q_n$, so besitzt es zur Zeit $t = t$ die bestimmten Zustandsvariabeln $p_1, \ldots q_n$. Die Systeme, deren Zustandsvariabeln in $t = 0$ dem Gebiete G angehörten, und zwar nur diese, gehören zur Zeit $t = t$ einem bestimmten Gebiete g an, sodass also die Gleichung gilt:

[9]
$$d N = \psi(p_1, \ldots q_n) \int_g \cdot$$

Für jedes derartige System gilt aber der Satz von Liouville, welcher die Form hat:

$$\int d P_1, \ldots d Q_n = \int d p_1, \ldots d q_n.$$

Aus den drei letzten Gleichungen folgt

$$\psi(P_1, \ldots Q_n) = \psi(p_1, \ldots q_n).\,^{1)}$$

ψ ist also eine Invariante des Systems, welche nach dem obigen die Form haben muss $\psi(p_1, \ldots q_n) = \psi^*(E)$. Für alle betrachteten Systeme ist aber $\psi^*(E)$ nur unendlich wenig verschieden von $\psi^*(\overline{E}) = \text{const.}$, und unsere Zustandsgleichung lautet einfach

$$d N = A \int_g d p_1, \ldots d q_n.$$

wobei A eine von den p und q unabhängige Grösse bedeutet.

§ 3. **Ueber die (stationäre) Wahrscheinlichkeit der Zustände eines Systems S, das mit einem System Σ von relativ unendlich grosser Energie mechanisch verbunden ist.**

Wir betrachten wieder unendlich viele (N) mechanische Systeme, deren Energie zwischen zwei unendlich wenig verschiedenen Grenzen $\overline{E}$ und $\overline{E} + \delta \overline{E}$ liege. Jedes solche mechanische System sei wieder eine mechanische Verbindung eines Systems S mit den Zustandsvariabeln $p_1, \ldots q_n$ und eines [11] Systems Σ mit den Zustandsvariabeln $\pi_1 \ldots \chi_n$. Der Ausdruck für die Gesamtenergie beider Systeme soll so beschaffen sein, dass jene Terme der Energie, welche durch Einwirkung der Massen eines Teilsystems auf die des anderen Teilsystems

[10]   1) Vgl. L. Boltzmann, Gastheorie, II. Teil. § 32 u. § 37.

Kinetische Theorie des Wärmegleichgewichtes etc.   **421**

hinzukommen, gegen die Energie E des Teilsystems S zu vernachlässigen seien. Ferner sei die Energie H des Teilsystems Σ [12]
unendlich gross gegen E. Bis auf unendlich Kleines höherer
Ordnung lässt sich dann setzen:

$$\mathsf{E} = H + E.$$

Wir wählen nun ein in allen Zustandsvariabeln $p_1 \ldots q_n$,
$\pi_1 \ldots \chi_n$ unendlich kleines Gebiet g, welches so beschaffen sei,
dass E zwischen den constanten Werten $\bar{\mathsf{E}}$ und $\bar{\mathsf{E}} + \delta\bar{\mathsf{E}}$ liege.
Die Anzahl dN der Systeme, deren Zustandsvariabeln dem
Gebiet g angehören, ist dann nach dem Resultate des vorigen
Paragraphen:

$$dN = A \int\limits_g dp_1 \ldots d\chi_n.$$

Wir bemerken nun, dass es in unserem Belieben steht, statt A
irgend eine stetige Function der Energie zu setzen, welche
für $\mathsf{E} = \bar{\mathsf{E}}$ den Wert A annimmt. Dadurch ändert sich nämlich
unser Resultat nur unendlich wenig. Als diese Function wählen
wir $A' . e^{-2h\mathsf{E}}$, wobei h eine vorläufig beliebige Constante
bedeutet, über welche wir bald verfügen werden. Wir
schreiben also:

$$dN = A' \int\limits_g e^{-2h\mathsf{E}} dp_1 \ldots d\chi_n.$$ [13]

Wir fragen nun: Wie viele Systeme befinden sich in Zuständen,
sodass p_1 zwischen p_1 und $p_1 + dp_1$, p_2 bez. p_2 und $p_2 + dp_2 \ldots q_n$
zwischen q_n und $q_n + dq_n$, $\pi_1 \ldots \chi_n$ aber beliebige, mit den
Bedingungen unserer Systeme verträgliche Werte besitzen?
Nennt man diese Anzahl dN', so erhält man:

$$dN' = A' e^{-2hE} dp_1 \ldots dq_n \int e^{-2hH} d\pi_1 \ldots d\chi_n.$$

Die Integration erstreckt sich dabei auf jene Werte der Zu-
standsvariabeln, für welche H zwischen $\bar{\mathsf{E}} - E$ und $\bar{\mathsf{E}} - E + \delta\bar{\mathsf{E}}$
liegt. Wir behaupten nun, der Wert von h sei auf eine und
nur eine Weise so zu wählen, dass das in unserer Gleichung
auftretende Integral von E unabhängig wird.

Das Integral $\int e^{-2hH} d\pi_1 \ldots d\chi_n$, wobei die Grenzen der
Integration durch die Grenzen $\bar{\mathsf{E}}$ und $\bar{\mathsf{E}} + \overline{\delta\mathsf{E}}$ bestimmt sein   [14]
mögen, ist nämlich bei bestimmtem $\delta\bar{\mathsf{E}}$ offenbar lediglich

422   *A. Einstein.*

Function von E allein; nennen wir dieselbe $\chi(\mathsf{E})$. Dass in dem Ausdruck für dN' auftretende Integral lässt sich dann in der Form schreiben:

$$\chi(\bar{\mathsf{E}} - E).$$

Da nun E gegen $\bar{\mathsf{E}}$ unendlich klein ist, so lässt sich dies bis auf unendlich Kleines höherer Ordnung in der Form schreiben:

$$\chi(\bar{\mathsf{E}} - E) = \chi(\bar{\mathsf{E}}) - E\chi'(\bar{\mathsf{E}}).$$

Die notwendige und hinreichende Bedingung dafür, dass jenes Integral von E unabhängig ist, lautet also

$$\chi'(\bar{\mathsf{E}}) = 0.$$

Nun lässt sich aber setzen

$$\chi(\mathsf{E}) = e^{-2h\mathsf{E}} \cdot \omega(\mathsf{E}),$$

wobei $\omega(\mathsf{E}) = \int d\pi_1 \ldots d\chi_n$, erstreckt über alle Werte der Variabeln, deren Energiefunction zwischen E und $\mathsf{E} + \delta\mathsf{E}$ liegt.

Die gefundene Bedingung für h nimmt also die Form an:

$$e^{-2h\bar{\mathsf{E}}} \cdot \omega(\bar{\mathsf{E}}) \cdot \left\{ -2h + \frac{\omega'(\bar{\mathsf{E}})}{\omega(\bar{\mathsf{E}})} \right\} = 0,$$

oder

$$h = \tfrac{1}{2} \frac{\omega'(\bar{\mathsf{E}})}{\omega(\bar{\mathsf{E}})}.$$

Es giebt also stets einen und nur einen Wert für h, welcher die gefundenen Bedingungen erfüllt. Da ferner, wie im nächsten Paragraphen gezeigt werden soll, $\omega(\mathsf{E})$ und $\omega'(\mathsf{E})$ stets positiv sind, ist auch h stets eine positive Grösse.

Wählen wir h in dieser Weise, so reducirt sich das Integral auf eine von E unabhängige Grösse, sodass wir für die Zahl der Systeme, deren Variabeln $p_1 \ldots q_n$ in den bezeichneten Grenzen liegen, den Ausdruck erhalten

[15]
$$dN' = A'' e^{-2h E} \cdot dp_1 \ldots dq_n.$$

Dies ist also auch bei anderer Bedeutung von A'' der Ausdruck für die Wahrscheinlichkeit, dass die Zustandsvariabeln eines mit einem System von relativ unendlich grosser Energie mechanisch verbundenen Systems zwischen unendlich nahen [16] Grenzen liegen, wenn der Zustand stationär geworden ist.

§ 4. Beweis dafür, dass die Grösse h positiv ist.

Sei $\varphi(x)$ eine homogene, quadratische Function der Variabeln $x_1 \ldots x_n$. Wir betrachten die Grösse $z = \int dx_1 \ldots dx_n$, wobei die Integrationsgrenzen dadurch bestimmt sein mögen, dass $\varphi(x)$ zwischen einem gewissen Wert y und $y + \varDelta$ liege, wobei $\varDelta$ eine Constante sei. Wir behaupten, dass z, welches allein von y Function ist, stets mit wachsendem y zunimmt, wenn $n > 2$.

Führen wir die neuen Variabeln ein $x_1 = \alpha\, x_1' \ldots x_n = \alpha\, x_n'$, wobei $\alpha = $ const., dann ist:

$$z = \alpha^n \int dx_1' \ldots dx_n'.$$

Ferner erhalten wir $\varphi(x) = \alpha^2 \varphi(x')$.

Die Integrationsgrenzen des gewonnenen Integrals lauten also für $\varphi(x')$

$$\frac{y}{\alpha^2} \quad \text{und} \quad \frac{y}{\alpha^2} + \frac{\varDelta}{\alpha^2}.$$

Ist ferner $\varDelta$ unendlich klein, was wir annehmen, so erhalten wir

$$z = \alpha^{n-2} \int dx_1' \ldots dx_n'.$$

Hierbei ist y' zwischen den Grenzen   [17]

$$\frac{y}{\alpha^2} \quad \text{und} \quad \frac{y}{\alpha^2} + \varDelta.$$

Obige Gleichung lässt sich auch schreiben

$$z(y) = \alpha^{n-2}\, z\left(\frac{y}{\alpha^2}\right).$$

Wählt man α positiv und $n > 2$, so ist also stets   [18]

$$\frac{z(y)}{z\left(\dfrac{y}{\alpha^2}\right)} > 1,$$

was zu beweisen war.

Dieses Resultat benutzen wir, um zu beweisen, dass h positiv ist.

Wir fanden

$$h = \tfrac{1}{2}\, \frac{\omega'(\mathsf{E})}{\omega(\mathsf{E})},$$

wobei

$$\omega(\mathsf{E}) = \int dp_1 \ldots dq_n,$$

424   *A. Einstein.*

und E zwischen E und $E + \delta \overline{E}$. ω (E) ist der Definition nach notwendig positiv, wir haben nur zu zeigen, dass auch ω' (E) stets positiv ist.

Wir wählen E_1 und E_2, sodass $E_2 > E_1$, und beweisen, dass $\omega (E_2) > \omega (E_1)$ und zerlegen $\omega (E_1)$ in unendlich viele Summanden von der Form

$$d \left(\omega (E_1) \right) = d p_1 \ldots d p_n \int d q_1 \ldots d q_n.$$

Bei dem angedeuteten Integral besitzen die p bestimmte und zwar solche Werte, dass $V \leqq E_1$. Die Integrationsgrenzen des Integrals sind so charakterisirt, dass L zwischen $E_1 - V$ und $E_1 + \delta \overline{E} - V$ liegt.

Jedem unendlich kleinen derartigen Summanden entspricht aus $\omega (E_2)$ ein Term von der Grösse

$$d \left[\omega (E_2) \right] = d p_1 \ldots d p_n \int d q_1 \ldots d q_n,$$

wobei die p und $d p$ die nämlichen Werte haben wie in $d \left[\omega (E_1) \right]$, L aber zwischen den Grenzen $E_2 - V$ und $E_2 - V + \delta \overline{E}$ liegt.

Es ist also nach dem eben bewiesenen Satze

$$d \left[\omega (E_2) \right] > d \left[\omega (E_1) \right].$$

Folglich

$$\sum d \left[\omega (E_2) \right] > \sum d \left[\omega (E_1) \right],$$

wobei $\sum$ über alle entsprechende Gebiete der p zu erstrecken ist.

Es ist aber

$$\sum d \left[\omega (E_1) \right] = \omega (E_1),$$

wenn das Summenzeichen über alle p erstreckt wird, sodass

$$V \leqq E_1.$$

Ferner ist

[19]
$$\sum d \left[\omega (E_2) \right] < \omega (E_2),$$

weil das Gebiet der p, welches durch die Gleichung

$$V \leqq E_2$$

bestimmt wird, das durch die Gleichung

$$V \leqq E_1$$

definirte Gebiet vollständig in sich einschliesst.

Kinetische Theorie des Wärmegleichgewichtes etc.   **425**

§ 5. Ueber das Temperaturgleichgewicht.

Wir wählen nun ein System S von ganz bestimmter Beschaffenheit und nennen es Thermometer. Es stehe mit dem System Σ von relativ unendlich grosser Energie in mechanischer Wechselwirkung. Ist der Zustand des Ganzen stationär, so ist der Zustand des Thermometers durch die Gleichung definirt

$$dW = A e^{-2hE} dp_1 \ldots dq_n,$$

wobei dW die Wahrscheinlichkeit dafür bedeutet, dass die Werte der Zustandsvariabeln des Thermometers innerhalb der angedeuteten Grenzen liegen. Dabei besteht zwischen den Constanten A und h die Gleichung

$$1 = A \cdot \int e^{-2hE} dp_1 \ldots dq_n,$$

wobei die Integration über alle möglichen Werte der Zustandsvariabeln erstreckt ist. Die Grösse h bestimmt also den Zustand des Thermometers vollkommen. Wir nennen h die Temperaturfunction, indem wir bemerken, dass nach dem Gesagten jede an dem System S beobachtbare Grösse H Function von h allein sein muss, solange V_a unverändert bleibt, was wir angenommen haben. Die Grösse h aber hängt lediglich vom Zustande des Systems Σ ab (§ 3), ist also unabhängig davon, wie Σ mit S thermisch verbunden ist. Es folgt daraus unmittelbar der Satz: Ist ein System Σ mit zwei unendlich kleinen Thermometern S und S' verbunden, so kommt diesen beiden Thermometern dieselbe Grösse h zu. Sind S und S' identische Systeme, so kommt ihnen auch noch derselbe Wert der beobachtbaren Grösse H zu.

Wir führen nun nur identische Thermometer S ein und nennen H das beobachtbare Temperaturmaass. Wir erhalten also den Satz: Das an S beobachtbare Temperaturmaass H ist unabhängig von der Art, wie Σ mit S mechanisch verbunden ist; die Grösse H bestimmt h, dieses die Energie E des Systems Σ und diese dessen Zustand nach unserer Voraussetzung.

Aus dem Bewiesenen folgt sofort, dass zwei Systeme Σ_1 und Σ_2 im Falle mechanischer Verbindung kein im statio-

nären Zustand befindliches System bilden können, wenn nicht
zwei mit ihnen verbundene Thermometer S gleiches Temperaturmaass oder, was dasselbe bedeutet, sie selbst gleiche
Temperaturfunction besitzen. Da der Zustand der Systeme
Σ_1 und Σ_2 durch die Grössen h_1 und h_2 oder H_1 und H_2
vollständig definirt wird, so folgt, dass das Temperaturgleichgewicht lediglich durch die Bedingungen $h_1 = h_2$ oder $H_1 = H_2$
bestimmt sein kann.

　　Es bleibt jetzt noch übrig, zu zeigen, dass zwei Systeme
von gleicher Temperaturfunction h (oder gleichem Temperaturmaass H) mechanisch verbunden werden können zu einem
einzigen System von gleicher Temperaturfunction.

　　Seien zwei mechanische Systeme Σ_1 und Σ_2 mechanisch
zu einem System verschmolzen, so jedoch, dass die Terme
der Energie unendlich klein sind, welche Zustandsvariabeln
beider Systeme enthalten. Sowohl Σ_1 als Σ_2 seien verknüpft
mit einem unendlich kleinen Thermometer S. Die Angaben
H_1 und H_2 desselben sind bis auf unendlich Kleines jedenfalls dieselben, weil sie sich nur auf verschiedene Stellen, eines
einzigen, im stationären Zustande befindlichen Systems beziehen. Ebenso natürlich die Grössen h_1 und h_2. Wir denken
uns nun unendlich langsam die beiden Systemen gemeinsame
Terme der Energie gegen Null hin abnehmen. Hierbei ändern
sich sowohl die Grössen H und h, als auch die Zustandsverteilungen beider Systeme unendlich wenig, da diese allein
durch die Energie bestimmt sind. Ist dann die vollständige
mechanische Trennung von Σ_1 und Σ_2 ausgeführt, so bleiben
gleichwohl die Beziehungen

$$H_1 = H_2, \quad h_1 = h_2$$

bestehen und die Zustandsverteilung ist unendlich wenig ver-
[20] ändert. H_1 und h_1 beziehen sich aber nur mehr auf Σ_1,
H_2 und h_2 nur mehr auf Σ_2. Unser Process ist streng umkehrbar, da er sich aus einer Aufeinanderfolge von stationären
Zuständen zusammensetzt. Wir erhalten also den Satz:

　　Zwei Systeme von der gleichen Temperaturfunction h
lassen sich zu einem einzigen System von der Temperaturfunction h verknüpfen, sodass sich deren Zustandsverteilung
unendlich wenig ändert.

Kinetische Theorie des Wärmegleichgewichtes etc.   **427**

Gleichheit der Grössen h ist also die notwendige und hinreichende Bedingung für die stationäre Verknüpfung (Wärmegleichgewicht) zweier Systeme. Daraus folgt sofort: Sind die Systeme Σ_1 und Σ_2, und die Systeme Σ_1 und Σ_3 stationär mechanisch verknüpfbar (im Wärmegleichgewichte), so sind es auch Σ_2 und Σ_3.

Ich will hier bemerken, dass wir bis jetzt von der Voraussetzung, dass unsere Systeme mechanische seien, nur insofern Gebrauch gemacht haben, als wir den Liouville'schen Satz und das Energieprincip verwendet haben. Wahrscheinlich lassen sich die Fundamente der Wärmetheorie für noch weit allgemeiner definirte Systeme entwickeln. Solches wollen wir hier jedoch nicht versuchen, sondern uns auf die mechanischen Gleichungen stützen. Die wichtige Frage, inwiefern sich der Gedankengang von dem benutzten Bilde loslösen und verallgemeinern lässt, werden wir hier nicht behandeln.   [21]

§ 6. Ueber die mechanische Bedeutung der Grösse h.[1]

Die lebendige Kraft L eines Systems ist eine homogene quadratische Function der Grössen q. Durch eine lineare Substitution lassen sich stets Variable r einführen, sodass die lebendige Kraft in der Form erscheint

$$L = \tfrac{1}{2}\left(\alpha_1 r_1^2 + \alpha_2 r_2^2 + \ldots + \alpha_n r_n^2\right)$$

und dass

$$\int d q_1 \ldots d q_n = \int d r_1 \ldots d r_n,$$

wenn man die Integrale über entsprechende unendlich kleine Gebiete ausdehnt. Die Grössen r nennt Boltzmann Momentoiden. Die mittlere lebendige Kraft, welche einer Momentoide entspricht, wenn das System mit einem anderen, von viel grösserer Energie, ein System bildet, nimmt die Form an:

$$\frac{\int A'' e^{-2h\left[V + a_1 r_1^2 + a_2 r_2^2 + \ldots + a_n r_n^2\right]} \cdot \dfrac{\alpha_\nu r_\nu^2}{2} \cdot d p_1 \ldots d p_n \cdot d r_1 \ldots d r_n}{\int A'' e^{-2h\left[V + a_1 r_1^2 + a_2 r_2^2 + \ldots + a_n r_n^2\right]} \cdot d p_1 \ldots d p_n \, d r_1 \ldots d r_n} = \frac{1}{4h}\cdot \quad \text{[23]}$$

---

1) Vgl. L. Boltzmann, Gastheorie, II. Teil, §§ 33, 34, 42.   [22]

28*

428   *A. Einstein.*

Die mittlere lebendige Kraft aller Momentoiden eines Systems ist also dieselbe und gleich:

[24]
$$\frac{1}{4\,h} = \frac{L}{n},$$

wobei L die lebendige Kraft des Systems bedeutet.

§ 7. Ideale Gase. Absolute Temperatur.

Die entwickelte Theorie enthält als speciellen Fall die **Maxwell**'sche Zustandsverteilung der idealen Gase. Verstehen wir nämlich in § 3 unter dem System S ein Gasmolecül, unter Σ die Gesamtheit aller anderen, so folgt für die Wahrscheinlichkeit, dass die Werte der Variabeln $p_1 \ldots q_n$ von S in einem in Bezug auf alle Variabeln unendlich kleinen Gebiet g liegen, der Ausdruck

$$dW = A\,e^{-2\,h\,E} \int\limits_{g} dp_1 \ldots dq_n.$$

Auch erkennt man sogleich aus unserem, für die Grösse h in § 3 gefundenen Ausdruck, dass die Grösse h bis auf unendlich Kleines die nämliche wäre für ein Gasmolecül anderer Art, welches in dem Systeme vorkommt, in dem die Systeme Σ, welche h bestimmen, für beide Molecüle bis auf unendlich Kleines identisch sind. Damit ist die verallgemeinerte **Maxwell**'sche Zustandsverteilung für ideale Gase erwiesen. —
Ferner folgt sofort, dass die mittlere lebendige Kraft der Schwerpunktsbewegung eines Gasmolecüles, welches in einem System S vorkommt, den Wert $3/4\,h$ besitzt, weil dieselbe drei Momentoiden entspricht. Nun lehrt die kinetische Gastheorie, dass diese Grösse proportional dem vom Gase bei constanten Volumen ausgeübten Druck ist. Setzt man diesen definitionsgemäss der absoluten Temperatur T proportional, so hat man eine Beziehung von der Form

$$\frac{1}{4\,h} = \varkappa \,.\, T = \tfrac{1}{2}\,\frac{\omega\,(\bar{E})}{\omega'\,(\bar{E})},$$

[25] wobei $\varkappa$ eine universelle Constante, ω die in § 3 eingeführte Function bedeutet.

Kinetische Theorie des Wärmegleichgewichtes etc.   **429**

§ 8. Der zweite Hauptsatz der Wärmetheorie als Folgerung der mechanischen Theorie.

Wir betrachten ein gegebenes physikalisches System S als mechanisches System mit den Coordinaten $p_1 \ldots p_n$. Als Zustandsvariable in demselben führen wir ferner die Grössen

$$\frac{d\,p_1}{d\,t} = p_1{}' \ldots \frac{d\,p_n}{d\,t} = p_n{}'$$

ein. $P_1 \ldots P_n$ seien die äusseren Kräfte, welche die Coordinaten des Systems zu vergrössern streben. V_i sei die potentielle Energie des Systems, L dessen lebendige Kraft, welche eine homogene quadratische Function der p_ν' ist. Die Bewegungsgleichungen von Lagrange nehmen für ein solches System die Form an

$$\frac{\partial (V_i - L)}{\partial p_\nu} + \frac{d}{d\,t}\left[\frac{\partial L}{\partial p_\nu'}\right] - P_\nu = 0, \quad (\nu = 1, \ldots \nu = n).$$

Die äusseren Kräfte setzen sich aus zweierlei Kräften zusammen. Die einen, $P_\nu^{(1)}$, sind diejenigen Kräfte, welche die Bedingungen des Systems darstellen, und von einem Potential ableitbar sind, welches nur Function der $p_1 \ldots p_n$ ist (adiabatische Wände, Schwerkraft etc.):

$$P_\nu^{(1)} = \frac{\partial V_a}{\partial p_\nu}. \qquad\qquad [26]$$

Da wir Processe zu betrachten haben, welche mit unendlicher Annäherung aus stationären Zuständen bestehen, haben wir anzunehmen, dass V_a die Zeit zwar explicite enthalte, dass aber die partiellen Ableitungen der Grössen $\partial V_a / \partial p_\nu$ nach der Zeit unendlich klein seien.

Die anderen Kräfte, $P_\nu^{(2)} = \Pi_\nu$, seien nicht von einem Potential ableitbar, welches nur von den p_ν abhängt. Die Kräfte Π_ν stellen die Kräfte dar, welche die Wärmezufuhr vermitteln.

Setzt man $V_a + V_i = V$, so gehen die Gleichungen (1) [27] über in

$$\Pi_\nu = \frac{\partial (V - L)}{\partial p_\nu} + \frac{d}{d\,t}\left\{\frac{\partial L}{\partial p_\nu'}\right\}.$$

Die Arbeit, welche durch die Kräfte Π_ν in der Zeit $d\,t$ dem System zugeführt wird, ist dann die Darstellung der vom

430   *A. Einstein.*

System S während dt aufgenommenen Wärmemenge dQ, welche wir im mechanischen Maass messen wollen.

$$dQ = \sum \Pi_\nu \, dp_\nu = \sum \frac{\partial V}{\partial p_\nu} dp_\nu - \sum \frac{\partial L}{\partial p_\nu} dp_\nu$$
$$+ \sum \frac{dp_\nu}{dt} \frac{d}{dt}\left\{\frac{\partial L}{\partial p_\nu}\right\} dt.$$

[28]

Da aber

[29]
$$\sum p_\nu' \frac{d}{dt}\left\{\frac{\partial L}{\partial p_\nu'}\right\} dt = d \sum p_\nu' \frac{\partial L}{\partial p_\nu'} - \sum \frac{\partial L}{\partial p_\nu'} dp_\nu.$$

ferner

[30]
$$\sum \frac{\partial L}{\partial p_\nu'} p_\nu' = 2L, \quad \sum \frac{\partial L}{\partial p_\nu} dp_\nu + \sum \frac{\partial L}{\partial p_\nu'} dp_\nu' = dL,$$

so ist

[31]
$$dQ = \sum \frac{\partial V}{\partial p_\nu'} dp_\nu + dL.$$

Da ferner

[32]
$$T = \frac{1}{4 \varkappa h} = \frac{L}{n \varkappa},$$

so ist

[33] **(1)**
$$\frac{dQ}{T} = n \varkappa \frac{dL}{L} + 4 \varkappa h \sum \frac{\partial V}{\partial p_\nu} dp_\nu.$$

Wir beschäftigen uns nun mit dem Ausdruck

$$\sum \frac{\partial V}{\partial p_\nu} dp_\nu.$$

Derselbe stellt die Zunahme des Systems an potentieller Energie dar, welche stattfinden würde während der Zeit dt, wenn V nicht explicite von der Zeit abhängig wäre. Das Zeitelement dt sei so gross gewählt, dass an die Stelle jener Summe deren Mittelwert für unendlich viele gleichtemperirte Systeme S gesetzt werden kann, aber doch so klein, dass die expliciten [34] Aenderungen von h und V nach der Zeit unendlich klein seien.

Unendlich viele Systeme S im stationären Zustande, welche alle identische h und V_a besitzen, mögen übergehen in neue stationäre Zustände, welche durch die allen gemeinsamen Werte $h + \delta h$, $V + \delta V$ charakterisirt sein mögen. „δ" bezeichne allgemein die Aenderung einer Grösse beim Uebergang des Systems in den neuen Zustand; das Zeichen „d" bezeichne nicht mehr die Aenderung mit der Zeit, sondern Differentiale bestimmter Integrale. —

Die Anzahl der Systeme, deren Zustandsvariable vor der Aenderung innerhalb des unendlich kleinen Gebietes g sich befinden, ist durch die Formel gegeben

$$dN = A \, e^{-2h(V+L)} \int dp_1 \ldots dp_n, \qquad \text{[35]}$$

dabei steht es in unserer Willkür, für jedes gegebene h und V_a die willkürliche Constante von V so zu wählen, dass die Constante A der Einheit gleich wird. Wir wollen dies thun, um die Rechnung einfacher zu gestalten, und die so genauer definirte Function V^* nennen.

Man sieht nun leicht, dass die von uns gesuchte Grösse den Wert erhält:

$$(2) \qquad \sum \frac{\partial V^*}{\partial p_n} dp_n = \frac{1}{N} \int \delta \{e^{-2h(V+L)}\} \cdot V^* \, dp_1 \ldots dq_n, \qquad \text{[36]}$$

wobei die Integration über alle Werte der Variabeln zu erstrecken ist. Dieser Ausdruck stellt nämlich die Vermehrung der mittleren potentiellen Energie des Systems dar, welche einträte, wenn zwar die Zustandsverteilung sich gemäss δV^* und δh änderte, V aber sich nicht explicite veränderte.

Ferner erhalten wir:

$$(3) \quad \begin{cases} 4\varkappa h \sum \dfrac{\partial V}{\partial p_\nu} dp_\nu = 4\varkappa \dfrac{1}{N} \int \delta\{e^{-2h(V^*+L)}\} \cdot h \cdot V \cdot dp_1 \ldots dq_n \\[2mm] \qquad = 4\varkappa \, \delta[h\,\overline{V}] - \dfrac{4\varkappa}{N} \int e^{-2h(V^*+L)} \delta[h\,V] \\[2mm] \qquad\qquad\qquad\qquad\qquad dp_1 \ldots dq_n. \end{cases}$$

Die Integrationen sind hier und im Folgenden über alle möglichen Werte der Variabeln zu erstrecken. Ferner hat man zu bedenken, dass die Anzahl der betrachteten Systeme sich nicht ändert. Dies liefert die Gleichung:

$$\int \delta(e^{-2h(V^*+L)}) \, dp_1 \ldots dq_n = 0,$$

oder

$$\int e^{-2h(V^*+L)} \delta(h\,V) \, dp_1 \ldots dq_n + \delta h \int e^{-2h(V^*+L)} \delta(L) \qquad \text{[37]}$$
$$dp_1 \ldots dq_n = 0,$$

oder

$$(4) \quad \frac{4\varkappa}{N} \int e^{-2h(V^*+L)} \delta(h\,V) \, dp_1 \ldots dq_n + 4\varkappa\,\overline{L}\,\delta h = 0.$$

432   *A. Einstein.*

$\overline{V}$ und $\overline{L}$ bezeichnen die Mittelwerte der potentiellen Energie und der lebendigen Kraft der N-Systeme. Durch Addition von (3) und (4) erhält man:

$$4 \varkappa h \sum \frac{\partial V^*}{\partial p_\nu} \, dp_\nu = 4 \varkappa \, \delta \, [h \, \overline{V}] + 4 \varkappa \, \overline{L} \cdot \delta \, h,$$

oder, weil

[38] $$h = \frac{n}{4 \overline{L}}, \qquad \delta \, h = - \frac{n}{4 \, L^2} \cdot \delta \, L,$$

$$4 \varkappa h \sum \frac{\partial V}{\partial p_\nu} \, dp_\nu = 4 \varkappa \, \delta \, [h \, \overline{V}] - n \varkappa \frac{\delta \, L}{L}.$$

Setzt man diese Formel in (1) ein, so erhält man

[39] $$\frac{d \, Q}{T} = \delta \, [4 \varkappa h \, \overline{V}^*] = \delta \left[\frac{\overline{V}^*}{T} \right].$$

$d \, Q / T$ ist also ein vollständiges Differential. Da

$$\frac{\overline{L}}{T} = n \varkappa, \quad \text{also} \quad \delta \left(\frac{L}{T} \right) = 0$$

ist, so lässt sich auch setzen

$$\frac{d \, Q}{T} = \delta \left(\frac{E^*}{T} \right).$$

E^* / T ist also bis auf eine willkürliche additive Constante der Ausdruck für die Entropie des Systems, wobei $E^* = V^* + L$ gesetzt ist. Der zweite Hauptsatz erscheint also als notwendige Folge des mechanischen Weltbildes.

§ 9. Berechnung der Entropie.

Der für die Entropie ε gefundene Ausdruck $\varepsilon = E^* / T$ ist nur scheinbar so einfach, da E^* aus den Bedingungen des mechanischen Systems erst berechnet werden muss. Es ist nämlich

$$E^* = E + E_0,$$

wobei E unmittelbar gegeben, E_0 aber durch die Bedingung

[40] $$\int e^{- 2 \, h \, (E - E_0)} \, dp_1 \ldots dq_n = N$$

als Function von E und h zu bestimmen ist. Man erhält so:

[41] $$\varepsilon = \frac{E^*}{T} = \frac{E}{T} + 2 \varkappa \log \left\{ \int e^{- 2 \, h \, E} \, dp_1 \ldots dq_n \right\} + \text{const.}$$

Kinetische Theorie des Wärmegleichgewichtes etc. **433**

In dem so gefundenen Ausdruck ist die der Grösse E zuzufügende willkürliche Constante ohne Einfluss auf das Resultat, und das als „const" bezeichnete dritte Glied ist von V und T unabhängig.

Der Ausdruck für die Entropie ε ist darum merkwürdig, weil er lediglich von E und T abhängt, die specielle Form von E als Summe potentieller Energie und lebendiger Kraft aber nicht mehr hervortreten lässt. Diese Thatsache lässt vermuten, dass unsere Resultate allgemeiner sind als die benutzte mechanische Darstellung, zumal der in § 3 für h gefundene Ausdruck dieselbe Eigenschaft aufweist. [42]

§ 10. Erweiterung des zweiten Hauptsatzes.

Ueber die Natur der Kräfte, welche dem Potential V_a entsprechen, brauchte nichts vorausgesetzt zu werden, auch nicht, dass solche Kräfte in der Natur vorkommen. Die mechanische Theorie der Wärme verlangt also, dass wir zu richtigen Resultaten gelangen, wenn wir das Carnot'sche Princip auf ideale Processe anwenden, welche aus den beobachteten durch Einführung beliebiger V_a erzeugt werden können. Natürlich haben die aus der theoretischen Betrachtung jener Processe gewonnenen Resultate nur dann reale Bedeutung, wenn in ihnen die idealen Hülfskräfte V_a nicht mehr vorkommen. [43]

Bern, Juni 1902.

(Eingegangen 26. Juni 1902.)

Published in *Annalen der Physik* 9 (1902): 417–433. Dated Bern, June 1902, received 26 June 1902, published 18 September 1902.

[1] For a discussion of what Einstein might have regarded as the remaining shortcomings in the theories of Maxwell and Boltzmann, see the editorial note, "Einstein on the Foundations of Statistical Physics," pp. 48–50.

[2] See § 10 for this extension of the second law.

[3] Einstein's notation—using the p_v to designate generalized coordinates and the q_v their associated momenta—is same as that used in *Boltzmann 1898a*, § 25.

[4] Such a group of systems is now, following *Gibbs 1902*, p. 115, called a microcanonical ensemble. While Einstein later adopted Gibbs's terminology, at this time he was unaware of the work of Gibbs (see the editorial note, "Einstein on the Foundations of Statistical Physics," p. 44). However, the idea of a virtual ensemble of systems, all with identical, constant energies, had been used in kinetic theory at least since *Maxwell 1879* (which is cited in *Boltzmann*

1898a, § 29, p. 82). It is implicit in *Boltzmann 1871a*, and it appears explicitly in *Boltzmann 1885*, pp. 78–79, where Boltzmann first defined the term "Ergode" (see note 6), and in *Boltzmann 1898a*, § 32, p. 89. The striking locution, "unendlich viele (N) Systeme" ("infinitely many (N) systems") may have been taken from Boltzmann; exactly the same words are used, e.g., in *Boltzmann 1887*, p. 207.

[5] V_i denotes the internal potential energy of the system.

[6] This condition on the energy is implicit in *Boltzmann 1871a*, p. 707, and its significance is discussed in *Maxwell 1879*, pp. 548–560. It is employed explicitly in *Boltzmann 1885*, pp. 78–79, to define the concept of an "Ergode," a virtual ensemble of systems with identical energies, for which the energy is the only explicitly time-independent conserved quantity. The definition of an "Ergode" in *Boltzmann 1898a*, § 32, p. 89, does not include this condition.

[7] The same notation for constants of motion is employed on pp. 78–79 of *Boltzmann 1885*, where the concept of an "Ergode" (see note 6) was first defined. In *Boltzmann 1898a* the same notation is used, more generally, to represent the integrals of motion of a system (see § 30), but it is not used in the discussion of an "Ergode" (§ 32).

[8] In *Boltzmann 1898a*, § 25, uppercase P_ν and Q_ν are used to represent coordinates and momenta at $t = 0$, with lowercase p_ν and q_ν representing coordinates and momenta at $t = t$; G and g are used there to designate the associated regions of phase space.

[9] This equation should be

$$dN = \psi(p_1, \ldots q_n) \int_g dp_1 \ldots dq_n.$$

[10] *Boltzmann 1898a*.

[11] The idea of a virtual ensemble of composite systems is anticipated in *Boltzmann 1871a*, pp. 707–711, and in *Boltzmann 1898a*, § 35, where Boltzmann considered an ensemble of systems, each of which is divided into two parts separated by a heat-conducting wall.

[12] The symbol H represents the uppercase Greek eta.

[13] Cf. *Boltzmann 1898a*, § 37, p. 108, eq. (115). The constant h was first introduced by Boltzmann in *Boltzmann 1868*, p. 523, and used regularly thereafter; see, e.g., *Boltzmann 1896*, § 7, p. 48.

[14] $\overline{\delta E}$ should be $\delta \overline{E}$.

[15] Such a distribution is now called a canonical distribution, following *Gibbs 1902*, pp. 33–34. The concept of such an ensemble is already implicit in *Boltzmann 1872*, pp. 368–370. For a discussion of Einstein's later adoption of Gibbs's terminology, see the editorial note, "Einstein on the Foundations of Statistical Physics," pp. 54–55.

[16] Boltzmann referred to Kirchhoff as the source of such a probabilistic interpretation of this distribution (see *Boltzmann 1898a*, § 38, p. 112, and § 37). Boltzmann was probably referring to *Kirchhoff 1894*, Lecture 13, pp. 134–141.

[17] The limits of integration are defined by $\phi(x')$ taking a value between y' and $y' + \Delta$, where $y' = y/\alpha^2$.

[18] In fact, α must be greater than 1.

[19] The inequality should be

$$\Sigma \, d[\omega(E_1)] < \omega(E_2).$$

[20] Einstein's proof of what Paul Hertz dubbed the "Trennungssatz" (the assertion that after a composite system Σ is separated into two parts, Σ_1 and Σ_2, $h = h_1 = h_2$) is strongly criticized in *Hertz, P. 1910a*, pp. 247–255. Hertz argued that, in contrast to the proof of the "Vereinigungssatz" (the assertion that, after two separate systems, Σ_1 and Σ_2, are joined to form a composite system Σ, $h_1 = h_2 = h$), the proof of the "Trennungssatz" requires the assumption "that all phases of the composite system are explored" ("daß alle Phasen des zusammengesetzten Systems durchwandert werden") (*Hertz, P. 1910a*, p. 254). In his reply to Hertz, Einstein wrote: "The . . . objections against an observation on thermal equilibrium contained in my first essay on the subject rest upon a misunderstanding that was brought about by an all too terse and insufficiently careful formulation of that observation" ("Die . . . Bemerkungen gegen eine in meiner ersten einschlägigen Abhandlung enthaltene Betrachtung über das Temperaturgleichgewicht beruht auf einem Mißverständnis, das durch eine allzu knappe und nicht genügend sorgfältige Formulierung jener Betrachtung hervorgerufen wurde") (*Einstein 1911c*, p. 175; see also Einstein to Paul Hertz, 14 August 1910). Einstein did not explain the nature of the misunderstanding. For another discussion of this topic by Einstein, see *Einstein 1903* (Doc. 4), § 4. See also *Hertz, P. 1910b*, as well as *Ornstein 1910*, *1911*, for further discus-

sion of the problems posed by these theorems.

[21] In his next paper, Einstein attempted a generalization of the concept of a "mechanisches System" (see *Einstein 1903* [Doc. 4], § 1).

[22] *Boltzmann 1898a*. In § 33 Boltzmann defined "Momentoiden," and in § 33 and § 34 he proved the equipartition theorem for a microcanonical ensemble, or, in Boltzmann's terminology, an "Ergode" (see note 6). In § 42 he proved the equipartition theorem by a method quite similar to Einstein's, but only for a *real* ensemble of canonically distributed systems whose interactions are assumed to be negligible; nowhere did he prove it for a canonically distributed virtual ensemble.

[23] The exponents of e in both numerator and denominator should be $-h[2V + a_1r_1^2 + a_2r_2^2 + \ldots + a_nr_n^2]$. Cf. *Boltzmann 1898a*, § 42, p. 124.

[24] Here and in the next line, L should be $\overline{L}$, the mean kinetic energy of the system as a whole.

[25] $\kappa = k/2$, where k is now called Boltzmann's constant.

[26] A minus sign should be added to the right-hand side of the equation.

[27] By "Gleichungen (1)" Einstein evidently meant Lagrange's equations as given above on this page.

[28] The denominator of the expression in curly brackets should be $\partial p'_\nu$.

[29] On the right-hand side, dp_ν should be dp'_ν.

[30] On the left-hand side of the first equation, p_ν should be p'_ν. On the left-hand side of the second equation, the denominator of the first term should be ∂p_ν.

[31] On the right-hand side, the denominator of the first term should be ∂p_ν.

[32] L should be $\overline{L}$; see note 24.

[33] The dL in this equation is reinterpreted as $\delta\overline{L}$ below; see note 39.

[34] This assumption regarding dt anticipates Einstein's explicit identification of time and ensemble averages in *Einstein 1903* (Doc. 4), § 2.

[35] The integral should be $\int dp_1 \ldots dq_n$.

[36] The subscripts on the left-hand side should be ν rather than n. From here to the end of § 8, every V or $\overline{V}$ should be V^* or $\overline{V}^*$, respectively.

[37] In the second integral, $\delta(L)$ should be just L.

[38] From here to the end of § 8, every occurrence of L should be $\overline{L}$.

[39] Einstein tacitly assumed that dL in eq. (1) is to be replaced by $\delta\overline{L}$, in accord with his earlier remark that differentials corresponding to dt are to be replaced by corresponding ensemble averages derived from a variation in the state distribution, here δh, δV^*.

[40] The exponent on the left-hand side of the equation should be $-2h(E + E_0)$.

[41] The E/T should be $\overline{E}/T$. In *Einstein 1905k* (Doc. 16), p. 551, and in all subsequent citations of this equation, Einstein was careful to write $\overline{E}$ (or some equivalent) for the average energy, whereas in all previous citations he neglected to do so. Einstein's expression for the entropy is essentially identical to that derived in *Boltzmann 1871b*, pp. 725–728.

[42] Just such a generalization is attempted in Einstein's next paper on the foundations of statistical physics, where no assumption is made about the distinction between potential and kinetic energy in defining a physical system (see *Einstein 1903* [Doc. 4], § 1). The avoidance of a distinction between kinetic and potential energy is characteristic of Heinrich Hertz's program in mechanics (see *Hertz, H. 1894*, pp. 25–27). Einstein is reported to have studied at least the opening portion of *Hertz, H. 1894* as a student (see the transcript of the radio talk by Joseph Sauter in *Flückiger 1974*, p. 154).

[43] In effect, Einstein here claimed to have given a justification for his hypothetical extension of the second law, stated in *Einstein 1902a* (Doc. 2), p. 799.

4. "A Theory of the Foundations of Thermodynamics"

[Einstein 1903]

DATED Bern, January 1903
RECEIVED 26 January 1903
PUBLISHED 16 April 1903

IN: *Annalen der Physik* 11 (1903): 170–187.

170

9. *Eine Theorie der Grundlagen der Thermodynamik; von A. Einstein.*

In einer neulich erschienenen Arbeit habe ich gezeigt, [1]
daß die Sätze vom Temperaturgleichgewicht und der Entropie-
begriff mit Hülfe der kinetischen Theorie der Wärme hergeleitet werden können. Es drängt sich nun naturgemäß die
Frage auf, ob die kinetische Theorie auch wirklich notwendig
ist, um jene Fundamente der Wärmetheorie herleiten zu können,
oder ob vielleicht bereits Voraussetzungen allgemeinerer Art
dazu genügen können. Daß dieses letztere der Fall ist, und
durch welche Art von Überlegungen man zum Ziele gelangen
kann, soll in dieser Abhandlung gezeigt werden.   [2]

§ 1. Über eine allgemeine mathematische Darstellung der Vor- gänge in isolierten physikalischen Systemen.

Der Zustand irgend eines von uns betrachteten physikalischen Systems sei eindeutig bestimmt durch sehr viele (n)
skalare Größen $p_1, p_2 \ldots p_n$, welche wir Zustandsvariabeln [3]
nennen. Die Änderung des Systems in einem Zeitelement dt
ist dann durch die Änderungen $dp_1, dp_2 \ldots dp_n$ bestimmt,
welche die Zustandsvariabeln in jenem Zeitelement erleiden.

Das System sei isoliert, d. h. das betrachtete System stehe
mit anderen Systemen nicht in Wechselwirkung. Es ist dann
klar, daß der Zustand des Systems in einem bestimmten Zeitmoment in eindeutiger Weise die Veränderung des Systems
im nächsten Zeitelement dt, d. h. die Größen $dp_1, dp_2 \ldots dp_n$
bestimmt. Diese Aussage ist gleichbedeutend mit einem System
von Gleichungen von der Form:

$$(1) \qquad \frac{dp_i}{dt} = \varphi_i(p_1 \ldots p_n) \quad (i = 1 \ldots i = n),$$

wobei die φ eindeutige Funktionen ihrer Argumente sind.

Für ein solches System von linearen Differentialgleichungen
existiert im allgemeinen keine Integralgleichung von der Form

$$\psi(p_1 \ldots p_n) = \text{konst.},$$

welche die Zeit nicht explizite enthält. Für das Gleichungssystem aber, welches die Veränderungen eines nach außen abgeschlossenen, physikalischen Systems darstellt, müssen wir annehmen, daß mindestens eine solche Gleichung besteht, nämlich die Energiegleichung:

$$E(p_1 \ldots p_n) = \text{konst.}$$

Wir nehmen zugleich an, daß keine weitere, von dieser unabhängige Integralgleichung solcher Art vorhanden sei.

[4]

§ 2. Über die stationäre Zustandsverteilung unendlich vieler isolierter physikalischer Systeme, welche nahezu gleiche Energie besitzen.

Die Erfahrung zeigt, daß ein isoliertes physikalisches System nach einer gewissen Zeit einen Zustand annimmt, in welchem sich keine wahrnehmbare Größe des Systems mehr mit der Zeit ändert; wir nennen diesen Zustand den **stationären**. Es wird also offenbar nötig sein, daß die Funktionen q_i eine gewisse Bedingung erfüllen, damit die Gleichungen (1) ein solches physikalisches System darstellen können.

Nehmen wir nun an, daß eine wahrnehmbare Größe stets durch einen zeitlichen Mittelwert einer gewissen Funktion der Zustandsvariabeln $p_1 \ldots p_n$ bestimmt sei, und daß diese Zustandsvariabeln $p_1 \ldots p_n$ immer wieder dieselben Wertsysteme mit stets gleichbleibender Häufigkeit annehmen, so folgt aus dieser Bedingung, welche wir zur Voraussetzung erheben wollen, mit Notwendigkeit die Konstanz der Mittelwerte aller Funktionen der Größen $p_1 \ldots p_n$; nach dem obigen also auch die Konstanz jeder wahrnehmbaren Größe.

Diese Voraussetzung wollen wir genau präzisieren. Wir betrachten ein physikalisches System, welches durch die Gleichungen (1) dargestellt und dessen Energie E sei, von einem beliebigen Zeitpunkte an die Zeit T hindurch. Denken wir uns ein beliebiges Gebiet Γ der Zustandsvariabeln $p_1 \ldots p_n$ gewählt, so werden in einem bestimmten Zeitpunkt der Zeit T die Werte der Variabeln $p_1 \ldots p_n$ in diesem Gebiete Γ gelegen sein, oder sie liegen außerhalb desselben; sie werden also während eines Bruchteiles der Zeit T, welchen wir τ nennen wollen, in dem gewählten Gebiete Γ liegen. Unsere Bedingung lautet dann folgendermaßen: Wenn $p_1 \ldots p_n$ Zu-

standsvariable eines physikalischen Systems sind, also eines
Systems, welches einen stationären Zustand annimmt, so besitzt die Größe τ/T für $T=\infty$ für jedes Gebiet Γ einen bestimmten Grenzwert. Dieser Grenzwert ist für jedes unendlich kleine Gebiet unendlich klein.

Auf diese Voraussetzung kann man folgende Betrachtung
gründen. Seien sehr viele (N) unabhängige physikalische
Systeme vorhanden, welche sämtlich durch das nämliche Gleichungssystem (1) dargestellt seien. Wir greifen einen beliebigen
Zeitpunkt t heraus und fragen nach der Verteilung der möglichen Zustände unter diesen N Systemen, unter der Voraussetzung, daß die Energie E aller Systeme zwischen E^* und
dem unendlich benachbarten Werte $E^* + \delta E^*$ liege. Aus
der oben eingeführten Voraussetzung folgt sofort, daß die
Wahrscheinlichkeit dafür, daß die Zustandsvariabeln eines zufällig herausgegriffenen der N Systeme in der Zeit t innerhalb
des Gebietes Γ liegen, den Wert

$$\lim_{T=\infty} \frac{\tau}{T} = \text{konst.}$$

habe. Die Zahl der Systeme, deren Zustandsvariable in der
Zeit t innerhalb des Gebietes Γ liegen, ist also:

$$N \cdot \lim_{T=\infty} \frac{\tau}{T},$$

also eine von der Zeit unabhängige Größe. Bezeichnet g ein
in allen Variabeln unendlich kleines Gebiet der Koordinaten
$p_1 \ldots p_n$, so ist also die Anzahl der Systeme, deren Zustandsvariable zu einer beliebigen Zeit das beliebig gewählte unendlich kleine Gebiet g erfüllen:

$$(2) \qquad d N = \varepsilon(p_1 \ldots p_n)\int_g d p_1 \ldots d p_n. \qquad [5]$$

Die Funktion ε gewinnt man, indem man die Bedingung
in Zeichen faßt, daß die durch die Gleichung (2) ausgedrückte
Zustandsverteilung eine stationäre ist. Es sei im speziellen
das Gebiet g so gewählt, daß p_1 zwischen den bestimmten
Werten p_1 und $p_1 + d p_1$, p_2 zwischen p_2 und $p_2 + d p_2 \ldots p_n$
zwischen p_n und $p_n + d p_n$ gelegen ist, dann ist für die Zeit t

$$d N_t = \varepsilon(p_1 \ldots p_n) \cdot d p_1 \cdot d p_2 \ldots d p_n.$$

wobei der Index von $d\,N$ die Zeit bezeichnet. Mit Berücksichtigung der Gleichung (1) erhält man ferner für die Zeit $t + d\,t$ und dasselbe Gebiet der Zustandsvariabeln

$$d\,N_{t+dt} = d\,N_t - \sum_{\nu=1}^{\nu=n} \frac{\partial(\varepsilon\,\varphi_\nu)}{\partial p_\nu} \cdot d\,p_1 \ldots d\,p_n \cdot d\,t.$$

Da aber $d\,N_t = d\,N_{t+dt}$ ist, da die Verteilung eine stationäre ist, so ist

[6]
$$\sum \frac{\partial(\varepsilon\,\varphi_\nu)}{\partial p_\nu} = 0.$$

Daraus ergibt sich

$$-\sum \frac{\partial\,\varphi_\nu}{\partial p_\nu} = \sum \frac{\partial(\log\varepsilon)}{\partial p_\nu}\cdot\varphi_\nu = \sum \frac{\partial(\log\varepsilon)}{\partial p_\nu}\cdot\frac{d\,p_\nu}{d\,t} = \frac{d\,(\log\varepsilon)}{d\,t},$$

wobei $d\,(\log\varepsilon)/d\,t$ die Veränderung der Funktion $\log\varepsilon$ für ein einzelnes System nach der Zeit unter Berücksichtigung der zeitlichen Veränderung der Größen p_ν bezeichnet.

Man erhält ferner:

$$\varepsilon = e^{\displaystyle -\int d\,t \sum_{\nu=1}^{\nu=n} \frac{\partial\,\varphi_\nu}{\partial p_\nu} + \psi\,(E)} = e^{-n + \psi\,(E)}.$$

Die unbekannte Funktion ψ ist die von der Zeit unabhängige Integrationskonstante, welche von den Variabeln $p_1 \ldots p_n$ zwar abhängen, sie jedoch, nach der im § 1 gemachten Voraussetzung, nur in der Kombination, wie sie in der Energie E auftreten, enthalten kann.

Da aber $\psi\,(E) = \psi\,(E^*) = $ konst. für alle N betrachteten Systeme ist, reduziert sich für unseren Fall der Ausdruck für ε auf:

$$\varepsilon = \text{konst.}\; e^{\displaystyle -\int d\,t \sum_{\nu=1}^{\nu=n} \frac{\partial\,\varphi_\nu}{\partial p_\nu}} = \text{konst.}\; e^{-n}.$$

Nach dem obigen ist nun:

$$d\,N = \text{konst.}\; e^{-n} \int_g d\,p_1 \ldots d\,p_n.$$

174   *A. Einstein.*

Der Einfachheit halber führen wir nun neue Zustandsvariabeln für die betrachteten Systeme ein; sie mögen mit π_r bezeichnet werden.   Es ist dann:

$$dN = \frac{e^{-m}}{\frac{D(\pi_1 \ldots \pi_n)}{D(p_1 \ldots p_n)}_g} \int d\pi_1 \ldots d\pi_n,$$

wobei das Symbol D die Funktionaldeterminante bedeutet. — Wir wollen nun die neuen Koordinaten so wählen, daß

$$e^{-m} = \frac{D(\pi_1 \ldots \pi_n)}{D(p_1 \ldots p_n)}$$

werde.   Diese Gleichung läßt sich auf unendlich viele Arten befriedigen, z. B. wenn man setzt:

$$\pi_2 = p_2$$
$$\pi_3 = p_3 \qquad \pi_1 = \int e^{-m} \cdot dp_1.$$
$$\cdots$$
$$\pi_n = p_n$$

Wir erhalten also unter Benutzung der neuen Variabeln

$$dN = \text{konst.} \int d\pi_1 \ldots d\pi_n.$$

Im folgenden wollen wir uns stets solche Variabeln eingeführt denken.   [7]

§ 3. Über die Zustandsverteilung eines Systems, welches ein System von relativ unendlich großer Energie berührt.

Wir nehmen nun an, daß jedes der N isolierten Systeme, aus zwei Teilsystemen Σ und σ, welche in Wechselwirkung stehen, zusammengesetzt sei. Der Zustand des Teilsystems Σ möge durch die Werte der Variabeln $\Pi_1 \ldots \Pi_\lambda$, der Zustand des Systems σ durch die Werte der Variabeln $\pi_1 \ldots \pi_l$ bestimmt sein. Ferner setze sich die Energie E, welche für jedes System zwischen den Werten E^* und $E^* + \delta E^*$ liegen mag, also bis auf unendlich kleines gleich E^* sein soll, bis auf unendlich kleines, aus zwei Termen zusammen, von denen der erste H nur durch die Werte der Zustandsvariabeln von Σ, der zweite η nur durch die der Zustandsvariabeln von σ bestimmt sei, sodaß bis auf relativ unendlich kleines gilt:

$$E = H + \eta.$$ [8]

Zwei in Wechselwirkung stehende Systeme, welche diese Bedingung erfüllen, nennen wir zwei sich berührende Systeme. Wir setzen noch voraus, daß η gegen H unendlich klein sei.

Für die Anzahl dN_1 der N-Systeme, deren Zustandsvariabeln $\Pi_1 \ldots \Pi_\lambda$ und $\pi_1 \ldots \pi_\iota$ in den Grenzen zwischen Π_1 und $\Pi_1 + d\Pi_1$, Π_2 und $\Pi_2 + d\Pi_2 \ldots \Pi_\lambda$ und $\Pi_\lambda + d\Pi_\lambda$ und π_1 und $\pi_1 + d\pi_1$, π_2 und $\pi_2 + d\pi_2 \ldots \pi_\iota$ und $\pi_\iota + d\pi_\iota$ liegen, ergibt sich der Ausdruck:

$$dN_1 = C \cdot d\Pi_1 \ldots d\Pi_\lambda \cdot d\pi_1 \ldots d\pi_\iota,$$

wobei C eine Funktion von $E = H + \eta$ sein kann.

Da aber nach der obigen Annahme die Energie eines jeden betrachteten Systems bis auf unendlich kleines den Wert E^* besitzt, so können wir, ohne an dem Resultat etwas zu ändern, C durch konst. $e^{-2hE^*} = $ konst. $e^{-2h(H+\eta)}$ ersetzen, wobei h eine noch näher zu definierende Konstante bedeutet. Der Ausdruck für dN_1 geht also über in:

$$dN_1 = \text{konst.} \, e^{-2h(H+\eta)} \cdot d\Pi_1 \ldots d\Pi_\lambda \cdot d\pi_1 \ldots d\pi_\iota.$$

Die Anzahl der Systeme, deren Zustandsvariabeln π zwischen den angedeuteten Grenzen liegen, während die Werte der Variabeln Π keiner beschränkenden Bedingung unterworfen sind, wird sich also in der Form

$$dN_2 = \text{konst.} \, e^{-2h\eta} \cdot d\pi_1 \ldots d\pi_\iota \int e^{-2hH} d\Pi_1 \ldots d\Pi_\lambda$$

darstellen lassen, wobei das Integral über alle Werte der Π auszudehnen ist, denen Werte der Energie H zukommen, welche zwischen $E^* - \eta$ und $E^* + \delta E^* - \eta$ gelegen sind. Wäre die Integration ausgeführt, so hätten wir die Zustandsverteilung der Systeme σ gefunden. Dies ist nun tatsächlich möglich. Wir setzen:

$$\int e^{-2hH} \cdot d\Pi_1 \ldots d\Pi_\lambda = \chi(E),$$

wobei die Integration auf der linken Seite über alle Werte der Variabeln zu erstrecken ist, für welche H zwischen den bestimmten Werten E und $E + \delta E^*$ liegt. Das Integral, welches im Ausdruck dN_2 auftritt, nimmt dann die Form an

$$\chi(E^* - \eta),$$

oder, da η gegen E^* unendlich klein ist:

$$\chi(E^*) - \chi'(E^*) \cdot \eta.$$

176   *A. Einstein.*

Läßt sich also h so wählen, daß $\chi'(E^*) = 0$, so reduziert sich das Integral auf eine vom Zustand von σ unabhängige Größe.

Es läßt sich bis auf unendlich kleines setzen:

$$\chi(E) = e^{-2hE} \int d\Pi_1 \ldots d\Pi_\lambda = e^{-2hE} \cdot \omega(E),$$

wo die Grenzen der Integration gleich sind wie oben, und ω [9]
eine neue Funktion von E bedeutet.

Die Bedingung für h nimmt nun die Form an:

$$\chi'(E^*) = e^{-2hE^*} \cdot \{\omega'(E^*) - 2h\,\omega(E^*)\} = 0.$$

folglich:

$$h = \tfrac{1}{2}\frac{\omega'(E^*)}{\omega(E^*)}.$$

Es sei h in dieser Weise gewählt, dann wird der Ausdruck für dN_2 die Form annehmen:

(3) $dN_2 = \text{konst.}\, e^{-2h\eta}\, d\pi_1 \ldots d\pi_l.$

Bei geeigneter Wahl der Konstanten stellt dieser Ausdruck   [10]
die Wahrscheinlichkeit dafür dar, daß die Zustandsvariabeln eines Systems, welches ein anderes von relativ unendlich großer Energie berührt, innerhalb der angedeuteten Grenzen liegen. Die Größe h hängt dabei lediglich vom Zustande jenes Systems Σ von relativ unendlich großer Energie ab.

§ 4. Über absolute Temperatur und Wärmegleichgewicht.

Der Zustand des Systems σ hängt also lediglich von der   [11]
Größe h ab, und diese lediglich vom Zustande des Systems Σ. Wir nennen die Größe $1/4\,h\varkappa = T$ die absolute Temperatur des Systems Σ, wobei $\varkappa$ eine universelle Konstante bedeutet.   [12]
Nennen wir das System σ „Thermometer", so können wir sofort die Sätze aussprechen:

1. Der Zustand des Thermometers hängt nur ab von der absoluten Temperatur des Systems Σ, nicht aber von der Art der Berührung der Systeme Σ und σ.

2. Erteilen zwei Systeme Σ_1 und Σ_2 einem Thermometer σ gleichen Zustand im Falle der Berührung, so besitzen sie gleiche absolute Temperatur, und erteilen folglich

einem anderen Thermometer σ' im Falle der Berührung ebenfalls gleichen Zustand.

Seien ferner zwei Systeme Σ_1 und Σ_2 in Berührung miteinander und Σ_1 außerdem in Berührung mit einem Thermometer σ. Es hängt dann die Zustandsverteilung von σ lediglich von der Energie des Systems $(\Sigma_1 + \Sigma_2)$, bez. von der Größe $h_{1,2}$ ab. Denkt man sich die Wechselwirkung von Σ_1 und Σ_2 unendlich langsam abnehmend, so ändert sich dadurch der Ausdruck für die Energie $H_{1,2}$ des Systems $(\Sigma_1 + \Sigma_2)$ nicht, wie leicht aus unserer Definition von der Berührung und dem im letzten Paragraphen aufgestellten Ausdruck für die Größe h zu ersehen ist. Hat endlich die Wechselwirkung ganz aufgehört, so hängt die Zustandsverteilung von σ, welche sich während der Trennung von Σ_1 und Σ_2 nicht ändert, nunmehr von Σ_1 ab, also von der Größe h_1; wobei der Index die Zugehörigkeit zum System Σ_1 allein andeuten soll. Es ist also:

$$h_1 = h_{12}.$$

Durch eine analoge Schlußweise hätte man erhalten können:

$$h_2 = h_{12},$$

also

$$h_1 = h_2,$$

oder in Worten: Trennt man zwei sich berührende Systeme Σ_1 und Σ_2, welche ein isoliertes System $(\Sigma_1 + \Sigma_2)$ von der absoluten Temperatur T bilden, so besitzen nach der Trennung die nunmehrigen isolierten Systeme Σ_1 und Σ_2 gleiche Temperatur. [13] Wir denken uns ein gegebenes System mit einem idealen Gase in Berührung. Dieses Gas sei unter dem Bilde der kinetischen Gastheorie vollkommen darstellbar. Als System σ betrachten wir ein einziges einatomiges Gasmolekül von der Masse μ, dessen Zustand durch seine rechtwinkligen Koordinaten x, y, z und die Geschwindigkeiten ξ, η, ζ vollkommen bestimmt sei. Wir erhalten dann nach § 3 für die Wahrscheinlichkeit, daß die Zustandsvariabeln dieses Moleküles zwischen den Grenzen x und $x + dx \ldots \zeta$ und $\zeta + d\zeta$ liegen, den bekannten **Maxwell**schen Ausdruck:

$$dW = \text{konst. } e^{-h\mu(\xi^2 + \eta^2 + \zeta^2)} \cdot dx \ldots d\zeta.$$

178   *A. Einstein.*

Daraus erhält man durch Integration für den Mittelwert der
lebendigen Kraft dieses Moleküles

$$\overline{\frac{\mu}{2}\,(\xi^2 + \eta^2 + \zeta^2)} = \frac{1}{4\,h}\,\cdot$$ [14]

Die kinetische Gastheorie lehrt aber, daß diese Größe bei
konstantem Volumen des Gases proportional dem vom Gase
ausgeübten Drucke ist. Dieser ist definitionsgemäß der in
der Physik als absolute Temperatur bezeichneten Größe proportional. Die von uns als absolute Temperatur bezeichnete
Größe ist also nichts anderes als die mit dem Gasthermo-
meter gemessene Temperatur eines Systems.

§ 5. Über unendlich langsame Prozesse.

Wir haben bisher nur Systeme ins Auge gefaßt, welche
sich im stationären Zustande befanden. Wir wollen nun auch
Veränderungen von stationären Zuständen untersuchen, jedoch
nur solche, welche sich so langsam vollziehen, daß die in einem
beliebigen Momente herrschende Zustandsverteilung von der
stationären nur unendlich wenig abweicht; oder genauer gesprochen, daß in jedem Momente die Wahrscheinlichkeit, daß
die Zustandsvariabeln in einem gewissen Gebiete G liegen, bis
auf unendlich kleines durch die oben gefundene Formel dargestellt sei. Eine solche Veränderung nennen wir einen unendlich langsamen Prozeß.

Wenn die Funktionen φ_ν (Gleichung (1)) und die Energie E
eines Systems bestimmt sind, so ist nach dem vorigen auch
seine stationäre Zustandsverteilung bestimmt. Ein unendlich
langsamer Prozeß wird also dadurch bestimmt sein, daß sich
entweder E ändert oder die Funktionen φ_ν die Zeit explizite
enthalten, oder beides zugleich, jedoch so, daß die entsprechenden Differentialquotienten nach der Zeit sehr klein sind.

Wir haben angenommen, daß die Zustandsvariabeln eines
isolierten Systems sich nach Gleichungen (1) verändern. Umgekehrt wird aber nicht stets, wenn ein System von Gleichungen (1) existiert, nach denen sich die Zustandsvariabeln
eines Systems ändern, dieses System ein isoliertes sein müssen.
Es kann nämlich der Fall eintreten, daß ein betrachtetes
System derart unter dem Einfluß anderer Systeme sich be-

Theorie der Grundlagen der Thermodynamik.   **179**

findet, daß dieser Einfluß lediglich von Funktionen von veränderlichen Koordinaten beeinflussender Systeme abhängt, die
sich bei konstanter Zustandsverteilung der beeinflussenden
Systeme nicht ändern. In diesem Falle wird die Veränderung
der Koordinaten p_ν des betrachteten Systems auch durch ein
System von der Form der Gleichungen (1) darstellbar sein.
Die Funktionen φ_ν werden aber dann nicht nur von der
physikalischen Natur des betreffenden Systems, sondern auch
von gewissen Konstanten abhängen, welche durch die beeinflussenden Systeme und deren Zustandsverteilungen definiert
sind. Wir nennen diese Art von Beeinflussung des betrachteten
Systems eine adiabatische. Es ist leicht einzusehen, daß für
die Gleichungen (1) auch in diesem Falle eine Energiegleichung
existiert, solange die Zustandsverteilungen der adiabatisch
beeinflussenden Systeme sich nicht ändern. Ändern sich die
[15]   Zustände adiabatisch beeinflussender Systeme, so ändern sich
die Funktionen φ_ν des betrachteten Systems explizite mit der
Zeit, wobei in jedem Moment die Gleichungen (1) ihre Gültigkeit behalten. Wir nennen eine solche Änderung der Zustandsverteilung des betrachteten Systems eine adiabatische.

 Wir betrachten nun eine zweite Art von Zustandsver-
änderungen eines Systems Σ. Es liege ein System Σ zu
Grunde, welches adiabatisch beeinflußt sein kann. Wir nehmen
an, daß das System Σ in der Zeit $t = 0$ mit einem System P
von verschiedener Temperatur in solche Wechselwirkung trete,
wie wir sie oben als „Berührung" bezeichnet haben, und entfernen das System P nach der zum Ausgleich der Temperaturen von Σ und P nötigen Zeit. Es hat sich dann die
Energie von Σ geändert. Während des Prozesses sind die
Gleichungen (1) von Σ ungültig, vor und nach dem Prozesse
aber gültig, wobei die Funktionen φ_ν vor und nach dem
Prozesse dieselben sind. Einen solchen Prozeß nennen wir
[16]   einen „isopyknischen" und die Σ zugeführte Energie „zugeführte Wärme".

 Bis auf relativ unendlich kleines läßt sich nun offenbar
jeder unendlich langsame Prozeß eines Systems Σ aus einer
Aufeinanderfolge von unendlich kleinen adiabatischen und isopyknischen Prozessen konstruieren, sodaß wir, um einen Gesamt-
überblick zu erhalten, nur die letzteren zu studieren haben.

180   *A. Einstein.*

§ 6. Über den Entropiebegriff.

Es liege ein physikalisches System vor, dessen momentaner Zustand durch die Werte der Zustandsvariabeln $p_1 \ldots p_n$ vollkommen bestimmt sei. Dieses System mache einen kleinen, unendlich langsamen Prozeß durch, indem die das System adiabatisch beeinflussenden Systeme eine unendlich kleine Zustandsveränderung erfahren, und außerdem dem betrachteten System durch berührende Systeme Energie zugeführt wird. Wir tragen den adiabatisch beeinflussenden Systemen dadurch Rechnung, daß wir festsetzen, die Energie E des betrachteten Systems sei außer von $p_1 \ldots p_n$ noch von gewissen Parametern $\lambda_1, \lambda_2 \ldots$ abhängig, deren Werte durch die Zustandsverteilungen der das System adiabatisch beeinflussenden Systeme bestimmt seien. Bei rein adiabatischen Prozessen gilt in jedem Moment ein Gleichungssystem (1), dessen Funktionen φ_ν außer von den Koordinaten p_ν auch von den langsam veränderlichen Größen λ abhängen; es gilt dann auch bei adiabatischen Prozessen in jedem Moment die Energiegleichung, welche die Form besitzt:

$$\sum \frac{\partial E}{\partial p_\nu} \, \varphi_\nu = 0 \, .$$

Wir untersuchen nun die Energiezunahme des Systems während eines beliebigen unendlich kleinen, unendlich langsamen Prozesses.

Für jedes Zeitelement dt des Prozesses gilt:

(4) $$d E = \sum \frac{\partial E}{\partial \lambda} \, d\lambda + \sum \frac{\partial E}{\partial p_\nu} \, dp_\nu \, .$$

Für einen unendlich kleinen isopyknischen Prozeß verschwinden in jedem Zeitelement sämtliche $d\lambda$, mithin auch das erste Glied der rechten Seite dieser Gleichung. Da aber dE nach dem vorigen Paragraphen für einen isopyknischen Prozeß als zugeführte Wärme zu betrachten ist, so ist für einen solchen Prozeß die zugeführte Wärme dQ durch den Ausdruck:

$$d Q = \sum \frac{\partial E}{\partial p_\nu} \, dp_\nu$$

dargestellt.

Für einen adiabatischen Prozeß aber, während dessen stets die Gleichungen (1) gelten, ist nach der Energiegleichung

$$\sum \frac{\partial E}{\partial p_\nu} \, dp_\nu = \sum \frac{\partial E}{\partial p_\nu} \, \varphi_\nu \, dt = 0 \, .$$

Andererseits ist nach dem vorigen Paragraphen für einen adiabatischen Prozeß $dQ = 0$, sodaß auch für einen adiabatischen Prozeß

$$dQ = \sum \frac{\partial E}{\partial p_r}\, dp_r$$

gesetzt werden kann. Diese Gleichung muß also für einen beliebigen Prozeß in jedem Zeitelement als gültig betrachtet werden. Die Gleichung (4) geht also über in

$$(4')\qquad dE = \sum \frac{\partial E}{\partial \lambda}\, d\lambda + dQ.$$

Dieser Ausdruck stellt auch bei veränderten Werten von $d\lambda$ und von dQ die während des ganzen unendlich kleinen Prozesses stattfindende Veränderung der Energie des Systems dar.

Am Anfang und am Ende des Prozesses ist die Zustandsverteilung des betrachteten Systems eine stationäre und wird, wenn das System vor und nach dem Prozesse mit einem Systeme von relativ unendlich großer Energie in Berührung steht, welche Annahme nur von formaler Bedeutung ist, durch die Gleichung definiert von der Form:

$$dW = \text{konst.}\, e^{-2hE}.dp_1 \ldots dp_n$$
$$= e^{c-2hE}.dp_1 \ldots dp_n,$$

wobei dW die Wahrscheinlichkeit dafür bedeutet, daß die Werte der Zustandsvariabeln des Systems in einem beliebig herausgegriffenen Zeitmoment zwischen den angedeuteten Grenzen liegen. Die Konstante c ist durch die Gleichung definirt:

$$(5)\qquad \int e^{c-2hE}.dp_1 \ldots dp_n = 1,$$

wobei die Integration über alle Werte der Variabeln zu erstrecken ist.

Gelte Gleichung (5) speziell vor dem betrachteten Prozesse, so gilt nach demselben:

$$(5')\qquad \int e^{(c+dc)-2(h+dh)\left(E+\sum \frac{\partial E}{\partial \lambda} d\lambda\right)}.dp_1 \ldots dp_n = 1$$

und aus den beiden letzten Gleichungen ergibt sich:

$$\int \left(dc - 2E\,dh - 2h \sum \frac{\partial E}{\partial \lambda} . d\lambda\right).e^{c-2hE}.dp_1 \ldots dp_n = 0,$$

182   *A. Einstein.*

oder, da bei der Integration der Klammerausdruck als eine Konstante gelten kann, da die Energie E des Systems vor und nach dem Prozesse sich nie merklich von einem bestimmten Mittelwerte unterscheidet, und unter Berücksichtigung von Gleichung (5):

$$(5'') \qquad dc - 2\,E\,dh - 2\,h \sum \frac{\partial E}{\partial \lambda}\,d\lambda = 0.$$

Nach Gleichung (4') ist aber:

$$-2\,h\,dE + 2\,h \sum \frac{\partial E}{\partial \lambda}\,d\lambda + 2\,h\,dQ = 0$$

und durch Addition dieser beiden Gleichungen erhält man:

$$2\,h\,.\,dQ = d(2\,h\,E - c)$$

oder, da $1/4\,h = \varkappa\,.\,T$

$$\frac{dQ}{T} = d\left(\frac{E}{T} - 2\,\varkappa\,c\right) = dS.$$

Diese Gleichung sagt aus, das dQ/T ein vollständiges Differential einer Größe ist, welche wir die Entropie S des Systems nennen wollen. Unter Berücksichtigung von Gleichung (5) erhält man:

$$S = 2\,\varkappa\,(2\,h\,E - c) = \frac{E}{T} + 2\,\varkappa \log \int e^{-2\,h\,E}\,dp_1 \ldots dp_{\mathbf{n}}.$$

wobei die Integration über alle Werte der Variabeln zu erstrecken ist.

§ 7. Über die Wahrscheinlichkeit von Zustandsverteilungen.

Um den zweiten Hauptsatz in seiner allgemeinsten Form herzuleiten, müssen wir die Wahrscheinlichkeit von Zustandsverteilungen untersuchen.

Wir betrachten eine sehr große Zahl (N) isolierte Systeme, welche alle durch das nämliche Gleichungssystem (1) darstellbar seien, und deren Energie bis auf unendlich kleines übereinstimme. Die Zustandsverteilung dieser N Systeme läßt sich dann jedenfalls darstellen durch eine Gleichung von der Form:

$$(2') \qquad dN = \varepsilon\,(p_1 \ldots p_{\mathbf{n}},\,t)\,dp_1 \ldots dp_{\mathbf{n}}.$$

wobei ε im allgemeinen von den Zustandsvariabeln $p_1 \ldots p_{\mathbf{n}}$ und außerdem von der Zeit explizite abhängt. Die Funktion ε charakterisiert hierbei die Zustandsverteilung vollständig.

Aus § 2 geht hervor, daß, wenn die Zustandsverteilung konstant ist, was bei sehr großen Werten von t nach unseren

Voraussetzungen stets der Fall ist, $\varepsilon =$ konst. sein muß, sodaß also für eine stationäre Zustandsverteilung

$$d N = \text{konst.}\, d p_1 \ldots d p_n$$

ist.

Daraus folgt sofort, daß die Wahrscheinlichkeit $d W$ dafür, daß die Werte der Zustandsvariabeln eines zufällig heraus-gegriffenen der N Systeme, in dem unendlich kleinen, innerhalb der angenommenen Energiegrenzen gelegenen Gebiete g der Zustandsvariabeln gelegen sind, der Ausdruck:

$$d W = \text{konst.} \int_g d p_1 \ldots d p_n .$$

Dieser Satz läßt sich auch so aussprechen: Teilt man das ganze in Betracht kommende, durch die angenommenen Energie-grenzen bestimmte Gebiet der Zustandsvariabeln in l Teil-gebiete $g_1, g_2 \ldots g_l$, derart, daß

$$\int_{g_1} = \int_{g_2} = \ldots = \int_{g_l} ,$$

und bezeichnet man mit W_1, W_2 etc. die Wahrscheinlichkeiten dafür, daß die Werte der Zustandsvariabeln des beliebig heraus-gegriffenen Systems in einem gewissen Zeitpunkt innerhalb $g_1 . g_2 \ldots$ liegen, so ist

$$W_1 = W_2 = \ldots = W_l = \frac{1}{l} .$$

Das momentane Zugehören des betrachteten Systems zu einem bestimmten dieser Gebiete $g_1 \ldots g_l$ ist also genau ebenso wahr-scheinlich, als das Zugehören zu irgend einem anderen dieser Gebiete.

Die Wahrscheinlichkeit dafür, daß von N betrachteten Systeme zu einer zufällig herausgegriffenen Zeit ε_1 zum Ge-biete g_1, ε_2 zum Gebiete $g_2 \ldots \varepsilon_l$ zum Gebiete g_l gehören, ist also

$$W = \left(\frac{1}{l}\right)^N \frac{N!}{\varepsilon_1! \, \varepsilon_2! \ldots \varepsilon_n!} ,$$

oder auch, da $\varepsilon_1, \varepsilon_2 \ldots \varepsilon_n$ als sehr große Zahlen zu denken sind:

$$\log W = \text{konst.} - \sum_{\varepsilon = 1}^{\varepsilon = l} \varepsilon \log \varepsilon .$$

184 *A. Einstein.*

Ist l groß genug, so kann man hierfür ohne merklichen Fehler setzen:

$$\log W = \text{konst.} - \int \varepsilon \log \varepsilon \, dp_1 \ldots dp_n.$$

In dieser Gleichung bedeutet W die **Wahrscheinlichkeit dafür,** daß die bestimmte, durch die Zahlen $\varepsilon_1, \varepsilon_2 \ldots \varepsilon_l$, bez. durch eine bestimmte Funktion ε von $p_1 \ldots p_n$ gemäß Gleichung (2') ausgedrückte Zustandsverteilung zu einer bestimmten Zeit herrscht.

Wäre in dieser Gleichung $\varepsilon = \text{konst.}$, d. h. von den p_ν unabhängig zwischen den betrachteten Energiegrenzen, so wäre die betrachtete Zustandsverteilung stationär, und, wie leicht zu beweisen, der Ausdruck für die Wahrscheinlichkeit W der Zustandsverteilung ein Maximum. Ist ε von den Werten der p_ν abhängig, so läßt sich zeigen, daß der Ausdruck für $\log W$ für die betrachtete Zustandsverteilung kein Extremum besitzt, d. h. es gibt dann von der betrachteten Zustandsverteilung unendlich wenig verschiedene, für welche W größer ist.

Verfolgen wir die betrachteten N Systeme eine beliebige Zeit hindurch, so wird sich die Zustandsverteilung, also auch W beständig mit der Zeit ändern, und wir werden anzunehmen haben, daß immer wahrscheinlichere Zustandsverteilungen auf unwahrscheinliche folgen werden, d. h. daß W stets zunimmt, bis die Zustandsverteilung konstant und W ein Maximum geworden ist. [17]

In den folgenden Paragraphen wird gezeigt, daß aus diesem Satze der zweite Hauptsatz der Thermodynamik gefolgert werden kann.

Zunächst ist:

$$-\int \varepsilon' \log \varepsilon' \, dp_1 \ldots dp_n \geqq -\int \varepsilon \log \varepsilon \, dp_1 \ldots dp_n,$$

wobei durch die Funktion ε die Zustandsverteilung der N Systeme zu einer gewissen Zeit t, durch die Funktion ε' die Zustandsverteilung zu einer gewissen späteren Zeit t' bestimmt, und die Integration beiderseits über alle Werte der Variabeln zu erstrecken ist. Wenn ferner die Größen $\log \varepsilon$ und $\log \varepsilon'$ der einzelnen unter den N Systemen sich nicht merklich von einander unterscheiden, so geht, da

$$\int \varepsilon \, d p_1 \ldots d p_n = \int \varepsilon' \, d p_1 \ldots d p_n = N,$$

die letzte Gleichung über in:

(6) $- \log \varepsilon' \gneqq - \log \varepsilon$.

§ 8. Anwendung der gefundenen Resultate auf einen bestimmten Fall.

Wir betrachten eine endliche Zahl von physikalischen Systemen σ_1, σ_2 ..., welche zusammen ein isoliertes System bilden, welches wir Gesamtsystem nennen wollen. Die Systeme σ_1, σ_2 ... sollen thermisch nicht merklich in Wechselwirkung stehen, wohl aber können sie sich adiabatisch beeinflussen. Die Zustandsverteilung eines jeden der Systeme σ_1, σ_2 ..., die wir Teilsysteme nennen wollen, sei bis auf unendlich kleines eine stationäre. Die absoluten Temperaturen der Teilsysteme können beliebig und voneinander verschieden sein.

Die Zustandsverteilung des Systems σ_1 wird sich nicht merklich von derjenigen Zustandsverteilung unterscheiden, welche gelten würde, wenn σ_1 mit einem physikalischen System von derselben Temperatur in Berührung stände. Wir können daher dessen Zustandsverteilung durch die Gleichung darstellen:

$$d w_1 = e^{c_{(1)} - 2 h_{(1)} \, E_{(1)}} \int_g d p_1^{(1)} \ldots d p_{(n)}^{(1)},$$

wobei die Indizes (1) die Zugehörigkeit zum Teilsystem σ_1 andeuten sollen.

Analoge Gleichungen gelten für die übrigen Teilsysteme. Da die augenblicklichen Werte der Zustandsvariabeln der einzelnen Teilsysteme von denen der anderen unabhängig sind, so erhalten wir für die Zustandsverteilung des Gesamtsystems eine Gleichung von der Form:

[18]   (7) $$d w = d w_1 . d w_2 \ldots = e^{\sum c_\nu - 2 h_\nu \, E_\nu} \int_g d p_1 \ldots d p_n,$$

wobei die Summation über alle Systeme, die Integration über das beliebige in allen Variabeln des Gesamtsystems unendlich kleine Gebiet g zu erstrecken ist.

Wir nehmen nun an, daß die Teilsysteme σ_1, σ_2 ... nach einer gewissen Zeit in beliebige Wechselwirkung zueinander treten, bei welchem Prozesse aber das Gesamtsystem stets ein isoliertes bleiben möge. Nach Verlauf einer gewissen Zeit möge ein Zustand des Gesamtsystems eingetreten sein, bei welchem die Teilsysteme σ_1. σ_2 ... einander thermisch nicht beeinflussen und bis auf unendlich kleines sich im stationären Zustand befinden.

Es gilt dann für die Zustandsverteilung des Gesamtsystems eine Gleichung, welche der vor dem Prozesse gültigen vollkommen analog ist:

$$(7')\quad dw' = dw_1' \cdot dw_2' \ldots = e^{\sum \left(c_\nu' - 2h_\nu' E_\nu'\right)} \int_g dp_1 \ldots dp_n.$$

Wir betrachten nun N solcher Gesamtsysteme. Für jedes derselben gelte bis auf unendlich kleines zur Zeit t die Gleichung (7), zur Zeit t' die Gleichung (7'). Es wird dann die Zustandsverteilung der betrachteten N Gesamtsysteme zu den Zeiten t und t' gegeben sein durch die Gleichungen:

$$dN_t = N \cdot e^{\sum \left(c_\nu - 2h_\nu E_\nu\right)} \cdot dp_1 \ldots dp_n.$$

$$dN_{t'} = N \cdot e^{\sum \left(c_\nu' - 2h_\nu' E_\nu'\right)} \cdot dp_1 \ldots dp_n.$$

Auf diese beiden Zustandsverteilungen wenden wir nun die Resultate des vorigen Paragraphen an. Es sind hier sowohl die

$$\varepsilon = N \cdot e^{\sum \left(c_\nu - 2h_\nu E_\nu\right)}$$

als auch die

$$\varepsilon' = N \cdot e^{\sum \left(c_\nu' - 2h_\nu' E_\nu'\right)}$$

für die einzelnen der N Systeme nicht merklich verschieden, sodaß wir Gleichung (6) anwenden können, welche liefert

$$\sum (2\,h'\,E' - c') \geqq \sum (2\,h\,E - c),$$

oder indem man beachtet, daß die Größen $2\,h_1\,E_1 - c_1$. $2\,h_2\,E_2 - c_2, \ldots$ nach § 6 bis auf eine universelle Konstante mit den Entropien S_1, S_2 ... der Teilsysteme übereinstimmen:

$$(8)\qquad S_1' + S_2' + \ldots \geqq S_1 + S_2 + \ldots,$$

d. h. die Summe der Entropien der Teilsysteme eines isolierten Systems ist nach einem beliebigen Prozesse gleich oder größer als die Summe der Entropien der Teilsysteme vor dem Prozesse.

Theorie der Grundlagen der Thermodynamik.   **187**

§ 9. Herleitung des zweiten Hauptsatzes.

Es liege nun ein isoliertes Gesamtsystem vor, dessen Teil-systeme W, M und Σ_1, Σ_2 ... heißen mögen. Das System W, welches wir Wärmereservoir nennen wollen, besitze gegen das System M (Maschine) eine unendlich große Energie. Ebenso sei die Energie der miteinander in adiabatischer Wechsel-wirkung stehenden Systeme Σ_1, Σ_2 ... gegen diejenige der Maschine M unendlich groß. Wir nehmen an, daß die sämt-lichen Teilsysteme M, W, Σ_1, Σ_2 ... sich im stationären Zu-stand befinden.

Es durchlaufe nun die Maschine M einen beliebigen Kreis-prozeß, wobei sie die Zustandsverteilungen der Systeme Σ_1, Σ_2 ... durch adiabatische Beeinflussung unendlich langsam ändere, d. h. Arbeit leiste, und von dem Systeme W die Wärme-menge Q aufnehme. Am Ende des Prozesses wird dann die gegenseitige adiabatische Beeinflussung der Systeme Σ_1, Σ_2 ... eine andere sein als vor dem Prozesse. Wir sagen, die Maschine M hat die Wärmemenge Q in Arbeit verwandelt.

Wir berechnen nun die Zunahme der Entropie der ein-zelnen Teilsysteme, welche bei dem betrachteten Prozeß ein-tritt. Die Zunahme der Entropie des Wärmereservoirs W be-trägt nach den Resultaten des § 6 $-Q/T$, wenn T die absolute Temperatur bedeutet. Die Entropie von M ist vor und nach dem Prozeß dieselbe, da das System M einen Kreisprozeß durchlaufen hat. Die Systeme Σ_1, Σ_2 ... ändern ihre Entropie während des Prozesses überhaupt nicht, da diese Systeme nur unendlich langsame adiabatische Beeinflussung erfahren. Die Entropievermehrung $S' - S$ des Gesamtsystems erhält also den Wert

$$S' - S = - \frac{Q}{T} \, .$$

Da nach dem Resultate des vorigen Paragraphen diese Größe $S' - S$ stets $\geqq 0$ ist, so folgt

$$Q \leqq 0 \, .$$

Diese Gleichung spricht die Unmöglichkeit der Existenz eines Perpetuum mobile zweiter Art aus.

Bern, Januar 1903.

(Eingegangen 26. Januar 1903.)

Published in *Annalen der Physik* 11 (1903): 170–187. Dated Bern, January 1903, received 26 January 1903, published 16 April 1903.

[1] *Einstein 1902b* (Doc. 3).

[2] It is not entirely clear what Einstein meant by the "kinetische Theorie der Wärme" ("kinetic theory of heat"); consequently, it is not clear exactly what kind of generalization Einstein was aiming for here. In the cited paper Einstein described his aim as being "to derive the laws of thermal equilibrium and the second law by using exclusively the equations of mechanics and the probability calculus" ("die Sätze über das Wärmegleichgewicht und den zweiten Hauptsatz unter alleiniger Benutzung der mechanischen Gleichungen und der Wahrscheinlichkeitsrechnung herzuleiten") (p. 417). The principal difference between that paper and the present one is that here no distinction is made between coordinates and momenta in the definition of what Einstein called a "physical system" ("physikalisches System") (p. 170), rather than a "mechanical system" ("mechanisches System") (*Einstein 1902b* [Doc. 3], p. 417). In particular, therefore, there is no distinction here between kinetic and potential energy; and avoidance of this distinction was precisely the kind of generalization entertained near the end of the previous paper (see *Einstein 1902b* [Doc. 3], p. 433).

[3] The state variables, $p_1, p_2, \ldots, p_n$, here replace both the coordinates and momenta used in *Einstein 1902b*. *Boltzmann 1871a*, p. 679, employs a similar set of variables to characterize the state of a system. The temporal evolution of these variables is governed by a set of first-order differential equations similar to eqs. (1) in the text.

[4] This is the same assumption made in *Einstein 1902b* (Doc. 3), p. 418 (see especially note 6). In a letter to Michele Besso of 17 March 1903, Einstein wrote: "The condition that E be the *only* integral of the equations of the given form is no limitation, since I free myself of it when considering 'adiabatically' influenced systems" ("Die Bedingung, daß E das *einzige* Integral der Gleichungen sei von der bewußten Form ist keine Einschränkung, weil ich mich bei Betrachtung der sich 'adiabatisch' beeinflussenden Systeme davon befreie"). The concept of adiabatic influence is defined below, on p. 179. L. S. Ornstein questioned whether Einstein's assumption is sufficient to guarantee that the system would explore with equal frequency *all* of the phase points lying on the hypersurface of phase space defined by the energy's being the only conserved quantity; see *Ornstein 1910*, pp. 805–808.

[5] The definition of probabilities by time averages, and the equating of time and ensemble averages, is implicit in *Boltzmann 1868*, pp. 517–518, *Boltzmann 1871a*, pp. 691, 708, and *Boltzmann 1898a*, § 35, p. 103. Einstein's explicit equating of time and ensemble averages is foreshadowed in *Einstein 1902b* (Doc. 3), p. 430 (see note 34). It was later criticized by Ornstein, who objected that "[i]t is impossible for a system to pass rigidly in a finite time (whatever length it may have) through all the possible phases; and in calculating a time-average such a finite time must be assumed" (*Ornstein 1910*, p. 805). Ornstein's criticism of the equating of time and ensemble averages is related to his criticism of the proofs given in *Hertz, P. 1910a* of the "Vereinigungssatz" and the "Trennungssatz," theorems discussed by Einstein here (§ 4) and earlier in *Einstein 1902b* (Doc. 3), § 5 (see especially note 20). There is no record of a reply by Einstein to Ornstein; see, however, Hertz's reply, *Hertz, P. 1910b*. See also *Lorentz 1916*.

[6] Presumably in response to a question from Besso, Einstein explained his reasoning in this part of the paper in his letter to Besso of 17 March 1903: "First of all about the $\Sigma \frac{\partial \epsilon \varphi_\nu}{\partial p_\nu}$. If one interprets $p_1 \ldots p_n$ as coordinates in an n-dimensional space, then the system corresponds to a point. ϵ is the density of points, $\epsilon \varphi_\nu$ are the components of the material flow, and the above expression the solenoidal condition" ("Zuerst zu der $\Sigma \frac{\partial \epsilon \varphi_\nu}{\partial p_\nu}$. Interpretiere $p_1 \ldots p_n$ als Koordinaten in einem n-dimensionalen Raume, dann entspricht dem Systeme ein Punkt. ϵ ist die Dichte der Punkte, $\epsilon \varphi_\nu$ sind die Komponenten der materiellen Strömung, und der obige Ausdruck die Solenoidalitätsbedingung"). The first equation on p. 173 is a form of the equation of continuity for the points.

[7] The validity of the preceding argument requires that $\Sigma \frac{\partial \varphi_\nu}{\partial p_\nu} = 0$ (the incompressibility condition for the points of the state space), since $d\epsilon/dt$ must vanish for a stationary distribution. Einstein evidently had second thoughts about the argument even before the paper was published. In the previously cited letter to Besso of 17

March, he wrote: "If you look at my paper more closely, you will find that the assumption of the energy principle and of the fundamental atomistic idea alone does not suffice for an explanation of the second law; instead, coordinates p must exist for the representation of things, such that for every conceivable total system $\Sigma \, \partial\varphi_\nu/\partial p_\nu = 0$" ("Wenn Du meine Arbeit genauer ansiehst, dann wirst Du finden, daß die Voraussetzung des Energieprinzips & des atomistischen Grundgedankens allein nicht hinreicht zur Erklärung des zweiten Hauptsatzes, sondern es muß zur Darstellung der Dinge Koordinaten p geben, so daß für jedes denkbare Gesammtsystem $\Sigma \, \partial\varphi_\nu/\partial p_\nu = 0$"). After stating his belief that the conjunction of the incompressibility and stationarity conditions is equivalent to the assumption that Hamilton's equations are satisfied, he added: "If that is true, then the entire generalization attained in my last paper consists in the elimination of the concept of force as well as in the fact that E can possess an arbitrary form (yet not completely)?" ("Wenn das wahr ist, dann besteht die ganze durch meine letzte Arbeit erzielte Verallgemeinerung in der Eliminierung des Kraftbegriffs sowie darin, daß E beliebige Form besitzen kann (doch nicht ganz)?"). Compare this comment to Einstein's remark in the letter to Besso of 22 January 1903 to the effect that the "energy principle" ("Energieprinzip") and the "atomistic theory" ("atomistische Theorie") are alone sufficient for the derivation (see the editorial note, "Einstein on the Foundations of Statistical Physics," p. 51). In later papers, Einstein regularly imposed the incompressibility condition, citing the present paper, without noting that the condition is not stated explicitly here (see, e.g., *Einstein 1904* [Doc. 5], p. 358, and *Einstein 1907a* [Doc. 38], p. 180). It is possible that Einstein's initial neglect of the incompressibility condition is the error that his colleague, Sauter, claimed to have discovered (see the editorial note, "Einstein on the Foundations of Statistical Physics," p. 47).

[8] The interaction term is assumed to be vanishingly small. See *Einstein 1902b* (Doc. 3), pp. 420–421.

[9] See *Einstein 1904* (Doc. 5), p. 355, for another definition of $\omega(E)$.

[10] On the probabilistic interpretation, see *Einstein 1902b* (Doc. 3), p. 422 and note 16.

[11] By "der Zustand des Systems σ" ("the state of the system σ") Einstein did not mean a specification of the instantaneous values of the $p_1 \ldots p_n$, but rather the canonical distribution of microstates (Zustandsverteilung) in a virtual ensemble of subsystems σ. See p. 177, line 5. The aim of the present section (§ 4) is to establish a connection between this state distribution and an observable measure of temperature for the systems Σ.

[12] For Einstein's introduction of the constant κ, see *Einstein 1902b* (Doc. 3), p. 428 and note 25.

[13] Compare this proof of this theorem on the separation of a system into two ("Trennungssatz") with that in *Einstein 1902b* (Doc. 3), § 5, and see note 20 to that paper for Hertz's criticism of Einstein's proofs.

[14] The term on the right-hand side should be $3/4h$.

[15] By "states" ("Zustände"), Einstein means again canonical distributions of microstates ("Zustandsverteilungen") (see note 11).

[16] The term "isopyknisch" ("isopycnic") is recommended in *Boltzmann 1898b*, p. 68, to designate a change of state at constant volume. The term is used with the same meaning in *Boltzmann 1898a*, § 20.

[17] This assumption was later criticized by Paul Hertz, who wrote: "If one assumes, like Einstein, that more probable distributions follow more improbable ones, one thereby introduces a special assumption that is not at all self-evident and is very much in need of proof" ("Wenn man wie Einstein annimmt, daß wahrscheinlichere Verteilungen auf unwahrscheinlichere folgen, führt man damit eine besondere Annahme ein, die keinerlei Evidenz besitzt und durchaus des Beweises bedarf") (*Hertz, P. 1910a*, p. 552). Hertz remarked that the assumption is proved in *Gibbs 1902*, p. 150, and defended this proof against the criticisms in *Zermelo 1906*, p. 238, and *Ehrenfest and Ehrenfest 1906*. After a personal discussion with Hertz, which, on Einstein's account, resulted in complete agreement between them, Einstein wrote: "I regard this criticism as completely correct. My derivation had not satisfied me even at the time, which is why I gave a second derivation shortly thereafter, one that is also cited by Hertz." ("Ich halte diese Kritik für vollkommen zutreffend. Meine Ableitung hatte mich schon damals nicht befriedigt, weshalb ich kurz darauf eine zweite Ableitung gab, die auch von Hrn. Hertz zitiert ist.") (*Einstein 1911c*, p. 175). For the "second derivation," see *Einstein 1904*

(Doc. 5), pp. 355–357. Einstein's way of introducing his assumption about the probabilities of state distributions—"wir werden anzunehmen haben"—suggests that he may not have regarded the assumption as a putatively true assertion about the actual behavior of the systems constituting the ensemble, but rather as a hypothesis necessary for the derivation of the second law in its strict classical thermodynamic form from statistical premises, which is exactly how Einstein described it in his letter to Besso of 22 January 1903 (see the editorial note, "Einstein on the Foundations of Statistical Physics," p. 51). For later comments on the statistical character of the second law, see *Discussion/Einstein 1911*, pp. 436–443, and *Einstein 1915a*, pp. 261–262.

[18] Here v, which was used above to index the members of a virtual ensemble of N replicas of a single system, is used to index the subsystems $\sigma_1, \sigma_2, \ldots$ of the composite system.

5. "On the General Molecular Theory of Heat"

[Einstein 1904]

DATED Bern, 27 March 1904
RECEIVED 29 March 1904
PUBLISHED 2 June 1904

IN: *Annalen der Physik* 14 (1904): 354–362.

354

6. *Zur allgemeinen molekularen Theorie der Wärme; von A. Einstein.*

Im folgenden gebe ich einige Ergänzungen zu einer letztes Jahr von mir publizierten Abhandlung.[1]

Wenn ich von „allgemeiner molekularer Wärmetheorie" spreche, so meine ich damit eine Theorie, welche im wesentlichen auf den in § 1 der zitierten Abhandlung genannten Voraussetzungen beruht. Ich setze jene Abhandlung als bekannt voraus, um unnütze Wiederholungen zu vermeiden, und bediene mich der dort gebrauchten Bezeichnungen.

Zuerst wird ein Ausdruck für die Entropie eines Systems abgeleitet, welcher dem von Boltzmann für ideale Gase gefundenen und von Planck in seiner Theorie der Strahlung vorausgesetzten vollständig analog ist. Dann wird eine einfache Herleitung des zweiten Hauptsatzes gegeben. Hierauf wird die Bedeutung einer universellen Konstanten untersucht, welche in der allgemeinen molekularen Theorie der Wärme eine wichtige Rolle spielt. Schließlich folgt eine Anwendung der Theorie auf die Strahlung schwarzer Körper, wobei sich zwischen der erwähnten, durch die Größen der Elementarquanta der Materie und der Elektrizität bestimmten universellen Konstanten und der Größenordnung der Strahlungswellenlängen, ohne Zuhilfenahme spezieller Hypothesen, eine höchst interessante Beziehung ergibt.

[2]
[3]
[4]
[5]

§ 1. Über den Ausdruck der Entropie.

Für ein System, welches Energie nur in Form von Wärme aufnehmen kann, oder mit anderen Worten, für ein System, welches von anderen Systemen nicht adiabatisch beeinflußt wird, gilt zwischen der absoluten Temperatur T und der Energie E, nach § 3 und § 4, l. c., die Gleichung:

$$(1) \qquad h = \tfrac{1}{2}\,\frac{\omega'(E)}{\omega(E)} = \frac{1}{4 \times T}\,, \qquad [6]$$

---

1) A. Einstein, Ann. d. Phys. 11. p. 170. 1903.   [1]

wobei z eine absolute Konstante bedeutet und ω (etwas ab-
[7] weichend von der zitierten Abhandlung) durch die Gleichung
definiert sei:

$$\omega(E).\delta E = \int\limits_{E}^{E+\delta E} dp_1 \ldots dp_n.$$

Das Integral rechts ist hierbei über alle Werte der den momentanen Zustand des Systems vollkommen und eindeutig definieren-
den Zustandsvariabeln zu erstrecken, denen Werte der Energie
entsprechen, die zwischen E und $E+\delta E$ liegen.

Aus Gleichung (1) folgt:

$$S = \int \frac{dE}{T} = 2\,z \log[\omega(E)].$$

Der Ausdruck stellt also (unter Weglassung der willkürlichen
Integrationskonstanten) die Entropie des Systems dar. Dieser
Ausdruck für die Entropie eines Systems gilt übrigens keineswegs nur für Systeme, welche nur rein thermische Zustandsänderungen erfahren, sondern auch für solche, welche beliebige
[8] adiabatische und isopyknische Zustandsänderungen durchlaufen.

Der Beweis kann aus der letzten Gleichung von § 6, l. c.,
geführt werden; ich unterlasse dies, da ich hier keine Anwendung des Satzes in seiner allgemeinen Bedeutung zu machen
beabsichtige.

§ 2. Herleitung des zweiten Hauptsatzes.

Befindet sich ein System in einer Umgebung von bestimmter konstanter Temperatur T_0 und steht es mit dieser
Umgebung in thermischer Wechselwirkung („Berührung"), so
nimmt es ebenfalls erfahrungsgemäß die Temperatur T_0 an
und behält die Temperatur T_0 für alle Zeiten bei.

Nach der molekularen Theorie der Wärme gilt jedoch
dieser Satz nicht streng, sondern nur mit gewisser — wenn
auch für alle der direkten Untersuchung zugänglichen Systeme
mit sehr großer — Annäherung. Hat sich vielmehr das betrachtete System unendlich lange in der genannten Umgebung
befunden, so ist die Wahrscheinlichkeit W dafür, daß in einem

A. Einstein.

beliebig herausgegriffenen Zeitpunkt der Wert der Energie des Systems sichz wischen den Grenzen E und $E+1$ befindet (§ 3, 1 c.):

$$W = C e^{-\frac{E}{2 \varkappa T_0}} \omega(E),$$
                                                                    [9]

wobei C eine Konstante bedeutet. Dieser Wert ist für jedes E ein von Null verschiedener, hat jedoch für ein bestimmtes E ein Maximum und nimmt — wenigstens für alle der direkten Untersuchung zugänglichen Systeme — für jedes merklich größere oder kleinere E einen sehr kleinen Wert an. Wir nennen das System „Wärmereservoir" und sagen kurz: obiger Ausdruck stellt die Wahrscheinlichkeit dafür dar, daß die Energie des betrachteten Wärmereservoirs in der genannten Umgebung den Wert E hat. Nach dem Ergebnis des vorigen Paragraphen kann man auch schreiben:

$$W = C e^{\frac{1}{2\varkappa}\left(s - \frac{E}{T_0}\right)},$$

wobei S die Entropie des Wärmereservoirs bedeutet.

Es mögen nun eine Anzahl Wärmereservoirs vorliegen, welche sich sämtlich in der Umgebung von der Temperatur T_0 befinden. Die Wahrscheinlichkeit dafür, daß die Energie des ersten Reservoirs den Wert E_1, des zweiten den Wert $E_2 \ldots$ des letzten den Wert E_l besitzt, ist dann in leicht verständlicher Bezeichnung: [10]

(a) $\mathfrak{W} = W_1 . W_2 \ldots W_l = C_1 . C_2 \ldots C_l\, e^{\frac{1}{2\varkappa}\left\{\sum\limits_1^l s - \frac{\sum\limits_1^l E}{T_0}\right\}}.$ [11]

Diese Reservoirs mögen nun in Wechselwirkung treten mit einer Maschine, wobei letztere einen Kreisprozeß durchläuft. Bei diesem Vorgange finde weder zwischen Wärmereservoirs und Umgebung noch zwischen Maschine und Umgebung ein Wärmeaustausch statt. Nach dem betrachteten Vorgange seien die Energien und Entropien der Systeme:

$$E_1', \; E_2' \ldots E_l',$$

bez.

$$S_1', \; S_2' \ldots S_l'.$$

Allgemeine molekulare Theorie der Wärme. **357**

[12] Dem Gesamtzustande der Wärmereservoirs, welcher durch diese Werte definiert ist, kommt die Wahrscheinlichkeit zu:

(b) $$\mathfrak{W}' = C_1 \cdot C_2 \ldots C_l\, e^{\frac{1}{2\varkappa}\left(\sum\limits_1^l S' - \frac{\sum\limits_1^l E'}{T_0} \right)}.$$

Bei dem Vorgange hat sich weder der Zustand der Umgebung noch der Zustand der Maschine geändert, da letztere einen Kreisprozeß durchlief.

Nehmen wir nun an, daß nie unwahrscheinlichere Zustände auf wahrscheinlichere folgen, so ist:

[13] $$\mathfrak{W}' \geqq \mathfrak{W}.$$

Es ist aber auch nach dem Energieprinzip:

$$\sum_1^l E = \sum_1^l E'.$$

Berücksichtigt man dies, so folgt aus Gleichungen (a) und (b):

$$\sum S' \geqq \sum S.$$

§ 3. Über die Bedeutung der Konstanten $\varkappa$ in der kinetischen Atomtheorie.

[14]

Es werde ein physikalisches System betrachtet, dessen momentaner Zustand durch die Werte der Zustandsvariabeln

$$p_1, p_2 \ldots p_n$$

vollständig bestimmt sei.

Wenn das betrachtete System mit einem System von relativ unendlich großer Energie und der absoluten Temperatur T_0 in „Berührung" steht, so ist dessen Zustandsverteilung durch die·Gleichung bestimmt:

$$dW = C\, e^{-\frac{E}{2\varkappa T_0}}\, dp_1 \ldots dp_n.$$

In dieser Gleichung ist $\varkappa$ eine universelle Konstante, deren Bedeutung nun untersucht werden soll.

Unter Zugrundelegung der kinetischen Atomtheorie gelangt man auf folgendem, aus Boltzmanns Arbeiten über Gas-[15] theorie geläufigen Wege zu einer Deutung dieser Konstanten.

Es seien die p_ν die rechtwinkligen Koordinaten $x_1\, y_1\, z_1,$ $x_2\, y_2 \ldots, x_n\, y_n\, z_n$ und $\xi_1\, \eta_1\, \zeta_1, \xi_2\, \eta_2 \ldots, \xi_n\, \eta_n\, \zeta_n$ die Geschwindigkeiten

358 *A. Einstein.*

der einzelnen (punktförmig gedachten) Atome des Systems. Diese Zustandsvariabeln können gewählt werden, weil sie der Bedingung $\sum \partial \varphi_\nu / \partial p_\nu = 0$ Genüge leisten (l. c., § 2). Man hat dann: [16]

$$E = \Phi(x_1 \ldots z_n) + \sum_1^n \frac{m_\nu}{2}(\xi_\nu^2 + \eta_\nu^2 + \zeta_\nu^2),$$

wobei der erste Summand die potentielle Energie, der zweite die lebendige Kraft des Systems bezeichnet. Sei nun ein unendlich kleines Gebiet $d x_1 \ldots d z_n$ gegeben. Wir finden den Mittelwert der Größe

$$\frac{m_\nu}{2}(\xi_\nu^2 + \eta_\nu^2 + \zeta_\nu^2),$$

welcher diesem Gebiete entspricht:

$$\overline{L_\nu} = \overline{\frac{m}{2}(\xi_\nu^2 + \eta_\nu^2 + \zeta_\nu^2)}$$

$$= \frac{e^{-\frac{(\Phi x_1 \ldots z_n)}{4 \varkappa T_0}} dx_1 \ldots dz_n \int \frac{m_\nu}{2}(\xi_\nu^2 + \eta_\nu^2 + \zeta_\nu^2) e^{\frac{\sum_1^n \frac{m_\nu}{2}(\xi_\nu^2 + \eta_\nu^2 + \zeta_\nu^2)}{2 \varkappa T_0}} d\xi_1 \ldots d\zeta_n}{e^{-\frac{\Phi(x_1 \ldots z_n)}{4 \varkappa T_0}} dx_1 \ldots dz_n \int e^{\frac{\sum \frac{m_\nu}{2}(\xi_\nu^2 + \eta_\nu^2 + \zeta_\nu^2)}{2 \varkappa T_0}} d\xi_1 \ldots d\zeta_n}$$ [17]

$$= 3 \frac{\int\limits_{-\infty}^{+\infty} m_\nu \xi_\nu^2 e^{\frac{m_\nu \xi_\nu^2}{4 \varkappa T_0}} d\xi_\nu}{\int\limits_{-\infty}^{+\infty} e^{\frac{m_\nu \xi_\nu^2}{4 \varkappa T_0}} d\xi_\nu} = 3 \varkappa T_0.$$

Diese Größe ist also unabhängig von der Wahl des Gebietes und von der Wahl des Atoms, ist also überhaupt der Mittelwert des Atoms bei der absoluten Temperatur T_0. Die Größe $3\varkappa$ ist gleich dem Quotienten aus der mittleren lebendigen Kraft eines Atoms in die absolute Temperatur.[1]

Die Konstante $\varkappa$ ist ferner aufs engste verknüpft mit der Anzahl N der wirklichen Moleküle, welche in einem Molekül

---

1) Vgl. L. Boltzmann, Vorl. über Gastheorie 2. § 42. 1898. [18]

im Sinne des Chemikers (Äquivalentgewicht bezogen auf 1 g Wasserstoff als Einheit) enthalten sind.

Liege nämlich eine solche Quantität eines idealen Gases vor, so ist bekanntlich, wenn Gramm und Zentimeter als Einheiten benutzt werden

[19]
$$p\,v = R\,T, \quad \text{wobei} \quad R = 8{,}31 \cdot 10^7.$$

Nach der kinetischen Gastheorie ist aber:

$$p\,v = \tfrac{2}{3} N \overline{L},$$

wobei $\overline{L}$ den Mittelwert der lebendigen Kraft der Schwerpunktsbewegung eines Moleküles bedeutet. Berücksichtigt man noch, daß

$$\overline{L} = \overline{L}_r,$$

so erhält man:

$$N \cdot 2\varkappa = R.$$

Die Konstante $2\varkappa$ ist also gleich dem Quotienten der Konstanten R in Anzahl der in einem Äquivalent enthaltenen Moleküle.

[20] Setzt man mit O. E. **Meyer** $N = 6{,}4 \cdot 10^{23}$, so erhält
[21] man $\varkappa = 6{,}5 \cdot 10^{-17}$.

§ 4. Allgemeine Bedeutung der Konstanten $\varkappa$.

Ein gegebenes System berühre ein System von relativ unendlich großer Energie und der Temperatur T. Die Wahrscheinlichkeit dW dafür, daß der Wert seiner Energie in einem beliebig herausgegriffenen Zeitpunkte zwischen E und $E + dE$ liegt, ist:

[22]
$$dW = C e^{-\frac{E}{2\varkappa T}} \omega E \, dE.$$

Für den Mittelwert $\bar{E}$ von E erhält man:

$$\bar{E} = \int\limits_0^\infty C E e^{-\frac{E}{2\varkappa T}} \omega E \, dE.$$

Da ferner

$$1 = \int\limits_0^\infty C e^{-\frac{E}{2\varkappa T}} \omega E \, dE,$$

360   *A. Einstein.*

so ist

$$\int_0^\infty (\bar{E} - E)\, e^{-\frac{E}{2 \varkappa T}}\, \omega(E)\, dE = 0 .$$

Differenziert man diese Gleichung nach T, so erhält man:

$$\int_0^\infty \left(2\,\varkappa\, T^2 \frac{d\bar{E}}{dT} + \bar{E}E - E^2 \right) e^{-\frac{E}{2 \varkappa T}}\, \omega\, E\, dE = 0 . \qquad [23]$$

Diese Gleichung besagt, daß der Mittelwert der Klammer verschwindet, also:

$$2\,\varkappa\, T^2 \frac{d\bar{E}}{dT} = \bar{E^2} - \bar{E}\bar{E} . \qquad [24]$$

Im allgemeinen unterscheidet sich der Momentanwert E der Energie von $\bar{E}$ um eine gewisse Größe, welche wir „Energieschwankung" nennen; wir setzen:

$$E = \bar{E} + \varepsilon .$$

Man erhält dann

$$\bar{E^2} - \bar{E}\bar{E} = \overline{\varepsilon^2} = 2\,\varkappa\, T^2 \frac{d\bar{E}}{dT} . \qquad [25]$$

Die Größe $\overline{\varepsilon^2}$ ist ein Maß für die thermische Stabilität des Systems; je größer $\overline{\varepsilon^2}$, desto kleiner diese Stabilität.

Die absolute Konstante $\varkappa$ bestimmt also die thermische Stabilität der Systeme. Die zuletzt gefundene Beziehung ist darum interessant, weil in derselben keine Größe mehr vorkommt, welche an die der Theorie zugrunde liegenden Annahmen erinnert.

Durch wiederholtes Differenzieren kann man ohne Schwierigkeit die Größen $\overline{\varepsilon^3}$, $\overline{\varepsilon^4}$ etc. berechnen.

§ 5. Anwendung auf die Strahlung.

Die zuletzt gefundene Gleichung würde eine exakte Bestimmung der universellen Konstanten $\varkappa$ zulassen, wenn es möglich wäre, den Mittelwert des Quadrates der Energieschwankung eines Systems zu bestimmen; dies ist jedoch bei dem gegenwärtigen Stande unseres Wissens nicht der Fall.

Wir können überhaupt nur bei einer einzigen Art physikalischer Systeme aus der Erfahrung vermuten, daß ihnen eine Energieschwankung zukomme; es ist dies der mit Temperaturstrahlung erfüllte leere Raum.

Ist nämlich ein mit Temperaturstrahlung erfüllter Raum von Lineardimensionen, welche sehr groß gegen die Wellenlänge ist, der das Energiemaximum der Strahlung bei der betreffenden Temperatur zukommt, so wird offenbar der Betrag der Energieschwankung im Mittel im Vergleich zur mittleren Strahlungsenergie dieses Raumes sehr klein sein. Wenn dagegen der Strahlungsraum von der Größenordnung jener Wellenlänge ist, so wird die Energieschwankung von derselben Größenordnung sein, wie die Energie der Strahlung des Strahlungsraumes.

Es ist allerdings einzuwenden, daß wir nicht behaupten können, daß ein Strahlungs*raum* als ein *System* von der von uns vorausgesetzten Art zu betrachten sei, auch dann nicht, wenn die Anwendbarkeit der allgemeinen molekularen Theorie zugestanden wird. Vielleicht müßte man zum Beispiel die Grenzen des Raumes als mit den elektromagnetischen Zuständen desselben veränderlich annehmen. Diese Umstände kommen indessen hier, wo es sich nur um Größenordnungen handelt, nicht in Betracht.

Setzen wir also in der im vorigen Paragraphen gefundenen Gleichung

[26]
$$\overline{\varepsilon^2} = \overline{E^2},$$

und nach dem Stefan-Boltzmannschen Gesetze

[27]
$$\bar{E} = c\,v\,T^4,$$

wobei v das Volumen in cm^3 und c die Konstante dieses Gesetzes bedeutet, so müssen wir für $\sqrt[3]{v}$ einen Wert von der Größenordnung der Wellenlänge maximaler Strahlungsenergie erhalten, welche der betreffenden Temperatur entspricht.

Man erhält:

$$\sqrt[3]{v} = \frac{2\sqrt[3]{\dfrac{\varkappa}{c}}}{T} = \frac{0{,}42}{T},$$

[28] wobei für $\varkappa$ der aus der kinetischen Gastheorie gefundene Wert und für c der Wert $7{,}06 \cdot 10^{-15}$ gesetzt ist.

362   *A. Einstein.   Allgemeine molekulare Theorie der Wärme.*

Ist λ_m die Wellenlänge des Energiemaximums der Strahlung, so liefert die Erfahrung:

$$\lambda_m = \frac{0{,}293}{T}.$$ [29]

Man sieht, daß sowohl die Art der Abhängigkeit von der Temperatur als auch die Größenordnung von λ_m mittels der allgemeinen molekularen Theorie der Wärme richtig bestimmt werden kann, und ich glaube, daß diese Übereinstimmung bei der großen Allgemeinheit unserer Voraussetzungen nicht dem Zufall zugeschrieben werden darf.   [30]

Bern, den 27. März 1904.

(Eingegangen 29. März 1904.)

Published in *Annalen der Physik* 14 (1904): 354–362. Dated Bern, 27 March 1904, received 29 March 1904, published 2 June 1904.

[1] *Einstein 1903* (Doc. 4).

[2] See, e.g., *Boltzmann 1877*, § V. This paper is regularly cited by Planck in this connection (see note 3). Boltzmann briefly discussed the topic of this paper in *Boltzmann 1898a*, pp. 40–42, referring the reader to the 1877 paper for more details. Einstein later referred to the expression for the entropy, with $\omega(E)$ construed as a probability, as Boltzmann's principle (see, e.g., *Einstein 1905i* [Doc. 14], p. 140).

[3] See, e.g., *Planck 1901a*, p. 556.

[4] As Einstein explained in *Einstein 1911c*, p. 175, this derivation of the second law is intended to remedy a defect in the derivation given in the previous paper. See *Einstein 1903* (Doc. 4), § 7, § 8, and § 9, and note 17.

[5] *Planck 1901b* is entitled "Ueber die Elementarquanta der Materie und der Elektricität." For additional evidence that Einstein had read this paper, see note 19.

[6] See *Einstein 1903* (Doc. 4), p. 176.

[7] See *Einstein 1903* (Doc. 4), p. 176, where $\omega(E)$ is defined, in the notation of the present paper, by:

$$\omega(E) = \int_E^{E + \delta E} dp_1 \ldots dp_n.$$

[8] "Isopyknische Zustandsänderungen" ("isopycnic changes of state") are defined in *Einstein 1903* (Doc. 4), p. 179.

[9] See *Einstein 1903* (Doc. 4), p. 176. Einstein assumed, as noted on p. 354, that $h = 1/4\kappa T$, where T is the temperature.

[10] Compare this model of a composite system of heat reservoirs with the slightly different model in *Einstein 1903* (Doc. 4), § 8 and § 9.

[11] The individual heat reservoirs are assumed to be thermodynamically and mechanically isolated from one another, so that the total probability factorizes.

[12] By "Gesamtzustand" ("total state") Einstein means the state of the composite system corresponding to the l component reservoirs' each having total energy E_1', E_2', . . . , E_l', respectively.

[13] Here Einstein assumes a monotonic increase in the probability of the "total state" ("Gesamtzustand") (see the preceding note), whereas the analogous assumption in *Einstein 1903* (Doc. 4), p. 184, concerns the probability

of a distribution of microstates over a hypersurface of constant energy.

[14] It is not clear just what Einstein meant by the term "kinetische Atomtheorie" ("kinetic theory of atoms"). The term suggests a kinetic theory, like the kinetic theory of gases, but applicable to any system with an atomistic structure, that is, possessing a finite number of degrees of freedom. See the editorial note, "Einstein on the Foundations of Statistical Physics," note 59.

[15] See *Boltzmann 1898a*, § 42. What was "geläufig" ("customary") was Boltzmann's proof of the equipartition theorem (see *Einstein 1902b* [Doc. 3], note 22); Boltzmann did not go on to estimate the value of κ (or, equivalently, the Boltzmann constant, *k*—see *Einstein 1902b* [Doc. 3], note 25), as Einstein did below.

[16] *Einstein 1903* (Doc. 4). The incompressibility condition, $\Sigma(\partial\varphi_\nu/\partial p_\nu) = 0$, is not stated explicitly in the cited paper; this is Einstein's first mention of it in print. See *Einstein 1903* (Doc. 4), note 7.

[17] The first exponent in both the numerator and the denominator on this line should be $-\Phi(x_1 \ldots x_n)/2\kappa T_0$; a minus sign should be added to the exponent inside the integrals in both the numerator and the denominator on this line and the next. The factor preceding the left-hand side of the next line should be 3/2.

[18] *Boltzmann 1898a*.

[19] This is the same value of *R* given in *Planck 1901b*, p. 564, which is probably Einstein's source (see note 5).

[20] Nowhere in *Meyer, O. E. 1877, 1895*, or *1899* does one find the value of *N* given here. However, exactly this value is given in *Planck 1901b*, p. 566, where Planck cited *Meyer, O. E. 1899*, p. 337. At the time of this paper, there was considerable uncertainty regarding the value of Avogadro's number, *N*. For a discussion of the problems involved in its determination, see the editorial note, "Einstein's Dissertation on the Determination of Molecular Dimensions," pp. 179–182.

[21] Einstein's value for κ yields a value of 1.30 x 10⁻¹⁶, for Boltzmann's constant. Planck calculated the value 1.346 x 10⁻¹⁶ with the help of his formula for the energy spectrum of blackbody radiation (see, e.g., *Planck 1901b*, p. 565). The agreement between the two values is surprising, given the uncertainty in the value of *N* (see note 20).

[22] In this and in three of the following four equations, ω*E* should be ω(*E*).

[23] In the bracketed expression, $\overline{E}^2$ should be E^2.

[24] The first term on the right-hand side should be $\overline{E^2}$.

[25] The first term on the left-hand side should be $\overline{E^2}$.

[26] The term on the right-hand side should be $\overline{E^2}$.

[27] See *Stefan 1879* and *Boltzmann 1884*.

[28] Einstein's value for *c* is the same as that given in *Planck 1901a*, p. 562, which, in turn, cites *Kurlbaum 1898*, p. 759. For κ, Einstein used the value 6.5 x 10⁻¹⁷ calculated above on p. 359.

[29] The equation $\lambda_m T$ = constant, a special case of Wien's displacement law (see *Wien 1893*), had been derived earlier by Einstein's ETH physics professor, H. F. Weber, from his semi-empirical radiation law (see *Weber, H. F. 1888*; for the early history of the displacement law, see *Kangro 1976*, chap. 3). The value of the constant used here appears to be the average of the value (0.294) given in *Lummer and Pringsheim 1899*, p. 218, and the value (0.292) given in *Paschen 1901b*, p. 657. See *Paschen 1901b* for a summary of the controversy over the value of this constant.

[30] In a letter to Conrad Habicht of 15 April 1904, Einstein wrote about this result: "I have now found in the simplest way the relation between the magnitude of the elementary quanta of matter and the wave lengths of radiation" ("Die Beziehung zwischen der Größe der Elementarquanta der Materie und den Strahlungswellenlängen habe ich nun in höchst simpler Weise gefunden").

EINSTEIN'S REVIEWS FOR THE
BEIBLÄTTER ZU DEN ANNALEN DER PHYSIK

Einstein began contributing reviews to the *Beiblätter zu den Annalen der Physik* in 1905, four years after the publication of his first paper in the *Annalen der Physik*. All but two of the twenty-three reviews Einstein wrote for the *Beiblätter* appeared in volume 29, published in 1905. Volume 30 (1906) contains one review, as does volume 31 (1907).[1]

The aim of the *Beiblätter* series, founded in 1877 by the longtime editor of the *Annalen*, Johann Christian Poggendorff, was to review "as completely as possible" ("mit möglichster Vollständigkeit") papers from journals in a variety of languages.[2] During the first quarter-century after 1877, the number of reviews published in the *Beiblätter* increased steadily. The number of reviews for the fifteen-year period from 1892 to 1906 is almost double that for the fifteen-year period from 1877 to 1891.[3] When Poggendorff initiated the supplementary series, only fifteen reviewers were needed; in 1905, Einstein was one of eighty-two reviewers.[4]

How Einstein came to be chosen as a reviewer (or if he volunteered to contribute) is uncertain. If the procedures followed in Einstein's case were the same as those followed in 1907, the editor of the *Beiblätter*, Walter König, invited Einstein to start writing reviews in a particular category, in Einstein's case the theory of heat ("Wärmelehre").[5] It is reasonable to assume that his activity as a reviewer of works on the theory of heat in the *Beiblätter* resulted from his earlier contributions on the same subject to the *Annalen*.[6] Einstein presumably received a list of journals, and had to indicate which ones were accessible to him. The decision about which papers to review was made by the editor. If a paper was not accessible to Einstein, an offprint was sent to him. At the end of the year Einstein received a modest payment for his contributions.[7]

[1] See *Klein and Needell 1977*, which first drew attention to these reviews.

[2] "An unsere Leser," dated 20 December 1876, appears in the front matter of the first volume of the *Beiblätter* (1877). Poggendorff aimed at reviewing German, French, English, and Italian papers, as well as "the less accessible" ("weniger verbreitete") Swedish, Danish, Dutch, and Russian literature.

[3] Approximately 18,000 reviews for the first period and over 30,000 for the second. Compare the forewords to the respective cumulative indexes in *Strobel 1893* (p. vi) and *Strobel 1909* (p. iv). A further indication of this expansion was the fate of the "Literature Survey" ("Literatur-Uebersicht") which appeared at the end of every volume of the *Beiblätter* up to 1904. In 1877, the first year of publication, this survey consisted of two pages; by 1900 it had grown to 138 pages; and in 1903, the last year in which such a survey was attempted, it ran to 178 pages and had to be abandoned.

[4] See vol. 29, p. [703], of the *Beiblätter* (1905).

[5] The description of reviewing procedures is based on a letter from König inviting Paul Ehrenfest to write reviews for the *Beiblätter* (see Walter König to Paul Ehrenfest, 28 June 1907, NeLR, Ehrenfest Scientific Correspondence 6–295).

[6] See *Einstein 1902b* (Doc. 3), *Einstein 1903* (Doc. 4), *Einstein 1904* (Doc. 5). His first two papers also employed thermodynamical arguments (see *Einstein 1901* [Doc. 1]), *Einstein 1902a* [Doc. 2]).

[7] In Ehrenfest's case, the payment was 30 Marks per "Druckbogen" (a sheet corresponding to sixteen printed pages), which is the same

Einstein probably was asked to list the foreign languages he could read. He presumably listed English, Italian, and French, since he reviewed papers in each of these languages. Einstein knew some Italian; his parents lived in Italy from 1894 to 1902, and he frequently spent time there between 1896 and 1901.[8] He took courses in Italian and French at the Aargau Kantonsschule.[9] It is uncertain how much English he knew. He managed to review four English papers, perhaps with the help of his wife.[10]

Conforming to the pattern of most other reviews in the *Beiblätter*, Einstein's are generally succinct and to the point. In most cases he limits himself to a summary of the contents of the paper reviewed; in a few cases, however, he adds a critical remark (*Einstein 1905m* [Doc. 18]), draws attention to a glaring error (*Einstein 1905z* [Doc. 31]), or praises the author for clarity of exposition (*Einstein 1905y* [Doc. 30]). He clearly found some papers more interesting than others. See, for example, *Einstein 1905g* (Doc. 12), which even reproduces tables of experimental values. *Einstein 1905n* (Doc. 19), on the other hand, suggests that Einstein did not take the trouble (or was unable) to consult earlier papers by the same author; accordingly, the review is somewhat vague.

Writing reviews for the *Beiblätter* gave Einstein the chance to become acquainted with papers that had an intrinsic research interest for him. An example is provided by the *Festschrift* published on the occasion of Boltzmann's sixtieth birthday in February 1904, three of whose 117 contributions were reviewed by Einstein.[11] It contains a paper by Planck comparing the definitions of entropy used by Gibbs and by Boltzmann.[12] Although Einstein did not review this particular paper, it is possible that this paper, or its review in the *Beiblätter* (printed immediately following *Einstein 1905p* [Doc. 21]), first drew his attention to Gibbs's work.

Another example is provided by *Einstein 1906f* (Doc. 37), which reviews the first edition of Max Planck's influential book *Vorlesungen über die Theorie der Wärmestrahlung*.[13] The review clearly demonstrates Einstein's familiarity with Planck's work on radiation theory; it may have been assigned to him because of his critical discussion of Planck's theory earlier that year, in a paper on the generation and absorption of light.[14]

After the review of Planck's book, only one item by Einstein appeared in the *Beiblätter*. *Einstein 1907i* (Doc. 46) reviews the second part of a two-volume work, the first part of

---

rate for contributors to similar review journals, *Die Fortschritte der Physik* and the *Jahrbuch über die Fortschritte der Mathematik* (see *Kirchner 1962*, pp. 461–463).

[8] See Vol. 1, "Chronology," pp. 372–376.

[9] See Final Grades, Aargau Kantonsschule, 5 September 1896 (Vol. 1, Doc. 19); *Matura* Examination (B) French: 'My Future Plans,' 18 September 1896 (Vol. 1, Doc. 22); and Aargau Kantonsschule, Curriculum (Vol. 1, Appendix D).

[10] Apparently, Mileva Einstein-Marić had studied English: the verso of a letter from Einstein contains English exercises in her hand (see

Einstein to Mileva Marić, after 28 November 1898, Vol. 1, Doc. 43, textual note).

[11] *Meyer, S. 1904*.

[12] *Planck 1904*.

[13] *Planck 1906c*. Einstein's review of this book is the longest of all his reviews, covering almost two and a half pages.

[14] See *Einstein 1906d* (Doc. 34). This review is the only one by Einstein not classified under "Wärmelehre," but under "Optik." For a discussion of Einstein's views on Planck's theory, see the editorial note, "Einstein's Early Work on the Quantum Hypothesis," pp. 137–138.

which he had reviewed in 1905. The rather abrupt reduction in the extent of Einstein's reviewing work beginning in 1906 may be connected with his promotion early that year to Technical Expert, Second Class, at the Swiss Patent Office. It is also possible that the reviewing work took up too much of his time.

The reviews have been dated by the approximate date of their publication. In 1904 and thereafter, each yearly volume consists of twenty-four undated issues.[15] For convenience in placing the reviews in chronological sequence, we have assumed that two issues appeared each month, and accordingly date each review to the first or second half of the appropriate month.

[15] Twenty-three issues were devoted to reviews, one to the annual index.

6. Review of Giuseppe Belluzzo, "Principles of Graphic Thermodynamics"

[*Einstein 1905a*]

First half of March 1905

PUBLISHED IN: *Beiblätter zu den Annalen der Physik* 29 (1905), no. 5: 235–236.

von etwa 0,5 Kal. Glaswolle, die von Wasser benetzt wird, zeigt den Effekt wohl in einer Wasserdampfatmosphäre, aber nicht beim Eintauchen in Wasser, ebenso verhält sich Baumwolle gegenüber absolutem Alkohol. F. K.

---

22. *M. W. Travers. Researches on the attainement of very low temperatures. Part. 1* (Smithsonian micellaneous collections part of vol. 46. 32 S. Washington, Smithsonian Institution, 1904). — Die Schrift gibt eine genaue Beschreibung der vom Verf. zur Verflüssigung von Wasserstoff benutzten Apparate sowie eine kurze Übersicht über die Theorie der bei der Verflüssigung auftretenden Vorgänge. Fch.

---

23. *M. Centnerszwer. Über eine Änderung der Methode von Cailletet und Mathias zur Bestimmung des kritischen Volumens* (ZS. f. phys. Chem. 49, S. 198—207. 1904). — Sind die kleinen Glasröhren, in denen die betreffenden Flüssigkeiten auf ihre kritischen Temperaturen erhitzt werden so gefüllt, daß bei dieser die kritische Dichte noch nicht erreicht ist, so verschwindet beim Erwärmen die *Flüssigkeit* als „letzter Tropfen"; ist die kritische Dichte überschritten, so verschwindet der *Dampf* als „letztes Gasbläschen". Man kann so bei einer größeren Anzahl von Versuchen neben der kritischen Temperatur auch die kritische Dichte erhalten, indem man die Röhren verschieden hoch füllt und aus den Dimensionen der Röhre und der Menge der Flüssigkeit die Dichte bei der kritischen Temperatur berechnet. Es wurde gefunden:

	Krit. Temp.	Krit. Dichte
Methylchlorid	143,0°	0,370
Äthyläther	194,4	0,258
Methylalkohol	240,2	0,275

Fch.

24. *G. Belluzzo. Prinzipien der graphischen Thermodynamik 1* (N. Cim. (5) 8, S. 196—222, 241—263. 1904). — Die offenbar für Techniker bestimmte Arbeit zerfällt in vier Abteilungen, von denen die erste Zustandsänderungen beliebiger Fluide graphisch behandelt. So wird in § 3 in der pv-Ebene die bekannte Flächenkonstruktion der vom Körper geleisteten Arbeit (L), der Energiezunahme (ΔE) und der aufgenommenen Wärme (G) gegeben, und in § 4 und 5 mit G und T (absolute

236   **Wärmelehre.**   **Beibl. 1905.**

Temperatur) bez. mit G und $1/T$ als Koordinaten die Entropiezunahme für eine beliebige Zustandsänderung als Fläche dargestellt. Es folgt die Theorie der Kreisprozesse und die
Definition der Reversibilität bez. Irreversibilität der Prozesse.
Es wird ein Prozeß als reversibel bez. irreversibel betrachtet,
je nachdem während des Prozesses der auf dem Fluidum
lastende Druck gleich dem inneren Druck des Fluidums ist oder
nicht; diese Festsetzung, welche übrigens für das Folgende
belanglos ist, hat keinen Sinn, da sonst bei allen irreversibeln
Prozessen das Prinzip der Gleichheit von Aktion und Reaktion
nicht erfüllt wäre. Die zweite Abteilung der Arbeit enthält
die Anwendung der Theorie auf ideale Gase; es werden die
Zustandsänderungen bei konstantem Volumen, konstantem Druck
und bei konstanter Temperatur, sowie die adiabatische und die
allgemeine polytropische Zustandsänderung untersucht. Der
letzte Paragraph beschäftigt sich mit dem Ausfluß der Gase
durch Röhren; die Hypothese von Saint-Venant und Wanzel
[1] wird durch (bereits bekannte) theoretische Erwägungen ersetzt.
Die dritte und vierte Abteilung der Arbeit enthalten die Lehre
von den gesättigten und überhitzten Wasserdämpfen in ganz
entsprechender Behandlungsweise, wobei die Theorie des Aus-
flusses des Wasserdampfes durch Röhren und die Theorie der
Verbesserung des Nutzeffektes . von Dampfmaschinen durch
Überhitzung besonders berücksichtigt sind. Als Zustandsgleichung für Wasserdämpfe wird nach Battelli und Tumlirz
[2] $p(v + \text{konst.}) = \text{konst. } T$ benutzt.   **A. E.**

Published in *Beiblätter zu den Annalen der Physik* 29 (1905): 235–236. Published in no. 5 [first
half of March].

[1] *Saint-Venant and Wantzel 1839* assumes
that the mean pressure at the orifice of a tube
from which a gas escapes is equal to the external
pressure, provided the latter is not too small. For
small values of the external pressure, the pressure at the orifice is assumed to be constant. See
Zeuner 1905, pp. 229–241, and *Weyrauch
1907*, pp. 305–306, which was reviewed by Einstein in *Einstein 1907i* (Doc. 46).

[2] Tumlirz used an equation of state of this
form, which is a simplified version of Battelli's
equation of state, to fit data on superheated
steam (*Battelli 1889, Tumlirz 1899*). See also
Weyrauch 1904, which was reviewed by Einstein in the same issue of the *Beiblätter, Einstein
1905f* (Doc. 11). For contemporary reviews of
equations of state, see *Zeuner 1906*, pp. 192–
257, and *Weyrauch 1907*, pp. 70–75, 91, which
was reviewed by Einstein in *Einstein 1907i*
(Doc. 46).

7. Review of Albert Fliegner, "On Clausius's Law of Entropy"

[Einstein 1905b]

First half of March 1905

PUBLISHED IN: *Beiblätter zu den Annalen der Physik* 29 (1905), no. 5: 236–237.

Temperatur) bez. mit G und $1/T$ als Koordinaten die Entropiezunahme für eine beliebige Zustandsänderung als Fläche dargestellt. Es folgt die Theorie der Kreisprozesse und die
Definition der Reversibilität bez. Irreversibilität der Prozesse.
Es wird ein Prozeß als reversibel bez. irreversibel betrachtet,
je nachdem während des Prozesses der auf dem Fluidum
lastende Druck gleich dem inneren Druck des Fluidums ist oder
nicht; diese Festsetzung, welche übrigens für das Folgende
belanglos ist, hat keinen Sinn, da sonst bei allen irreversibeln
Prozessen das Prinzip der Gleichheit von Aktion und Reaktion
nicht erfüllt wäre. Die zweite Abteilung der Arbeit enthält
die Anwendung der Theorie auf ideale Gase; es werden die
Zustandsänderungen bei konstantem Volumen, konstantem Druck
und bei konstanter Temperatur, sowie die adiabatische und die
allgemeine polytropische Zustandsänderung untersucht. Der
letzte Paragraph beschäftigt sich mit dem Ausfluß der Gase
durch Röhren; die Hypothese von Saint-Venant und Wanzel
wird durch (bereits bekannte) theoretische Erwägungen ersetzt.
Die dritte und vierte Abteilung der Arbeit enthalten die Lehre
von den gesättigten und überhitzten Wasserdämpfen in ganz
entsprechender Behandlungsweise, wobei die Theorie des Aus-
flusses des Wasserdampfes durch Röhren und die Theorie der
Verbesserung des Nutzeffektes . von Dampfmaschinen durch
Überhitzung besonders berücksichtigt sind. Als Zustandsgleichung für Wasserdämpfe wird nach Battelli und Tumlirz
$p(v + \text{konst.}) = \text{konst.}\ T$ benutzt.   A. E.

---

25. **A. Fliegner.** *Über den Clausiusschen Entropiesatz*
(Vierteljahresschr. d. naturf. Ges. Zürich S. 1—48. 1903). —
Der Verf. untersucht die Entropieänderung eines Systems
während eines als streng unstetig vorausgesetzten Prozesses
(unstetige Expansion eines Fluidums) und zieht aus seinen
Rechnungen den Schluß, daß am Anfange der plötzlichen
Expansion die Entropie des Systems abnehme. Betrachtungen
über nicht-umkehrbare chemische Prozesse führen den Verf.
zu dem Schlusse, daß die Gleichung $dQ/T \leqq dS$ nur für
exotherme, nicht aber für endotherme Prozesse gelte. Ebenso
soll die Gleichung für Kältemischungen nicht gelten. Es ist
daher begreiflich, wenn der Verf. mit dem Satze schließt:

Bd. 29. No. 5.   **Wärmelehre.**   **237**

„Die Frage, ob sich die Entropie des Weltalls überhaupt ändert, und wenn ja, in welchem Sinne, geht also gegenwärtig noch gar nicht zu beantworten, und sie wird wohl auch immer unentschieden bleiben".   **A. E.**   [1]

Published in *Beiblätter zu den Annalen der Physik* 29 (1905): 236–237. Published in no. 5 [first half of March].

[1] Fliegner's aim is to investigate Clausius's theorem that the entropy of the universe tends to a maximum. The statement Einstein quotes is the conclusion Fliegner draws from his result that the second law in the form $dQ/T \leq dS$ does not apply to all processes in nature.

8. Review of William McFadden Orr,
"On Clausius' Theorem for Irreversible Cycles, and on the Increase of Entropy"

[Einstein 1905c]

First half of March 1905

PUBLISHED IN: *Beiblätter zu den Annalen der Physik* 29 (1905), no. 5: 237.

Bd. 29. No. 5.   Wärmelehre.   **237**

„Die Frage, ob sich die Entropie des Weltalls überhaupt
ändert, und wenn ja, in welchem Sinne, geht also gegenwärtig
noch gar nicht zu beantworten, und sie wird wohl auch immer
unentschieden bleiben".   A. E.

---

26. **Mc. F. Orr.** *Über Clausius' Theorem der irreversibeln
Kreisprozesse und über das Wachsen der Entropie* (Phil. Mag.
(8) **46**, S. 509—527. 1904). — Der Verf. zeigt, daß Planck
in den „Vorlesungen über Thermodynamik" die Begriffe „rever-   [1]
sibel" und „irreversibel" in einem etwas anderen Sinne anwendet, als er sie definiert. Sodann führt er eine Reihe von   [2]
Einwänden an, welche gegen verschiedene Darstellungsweisen
der Grundlagen der Thermodynamik vorgebracht werden können;
unter diesen Einwänden ist der von Bertrand besonders be-   [3]
achtenswert, daß nämlich Druck, Temperatur und Entropie nur
für den Fall definiert sind, daß wenigstens genügend kleine
Teile eines Systems als im Gleichgewicht befindlich angesehen
werden können; ein ähnlicher Einwand wird bezüglich der
zugeführten Wärme erhoben.   A. E

Published in *Beiblätter zu den Annalen der Physik* 29 (1905): 237. Published in no. 5 [first half of March].

[1] *Planck 1897*; Orr quotes from the English translation (*Planck 1903b*).

[2] According to Planck's definition, a process is reversible if the initial state can be completely restored, i.e., without any change whatsoever, not only in the system under consideration, but also in the rest of the universe. Orr claimed that Planck actually uses a different definition, namely that a process is reversible if the initial state can be restored without any exchange of heat with its surroundings. See also Einstein's review of Planck's reply to Orr, *Einstein 1905o* (Doc. 20).

[3] See *Bertrand 1887*, pp. 265–273.

9. Review of George Hartley Bryan, "The Law of Degradation of Energy as the Fundamental Principle of Thermodynamics"

[*Einstein 1905d*]

First half of March 1905

PUBLISHED IN: *Beiblätter zu den Annalen der Physik* 29 (1905), no. 5: 237.

„Die Frage, ob sich die Entropie des Weltalls überhaupt ändert, und wenn ja, in welchem Sinne, geht also gegenwärtig noch gar nicht zu beantworten, und sie wird wohl auch immer unentschieden bleiben". A. E.

26. **Mc. F. Orr.** *Über Clausius' Theorem der irreversibeln Kreisprozesse und über das Wachsen der Entropie* (Phil. Mag. (8) **46**, S. 509—527. 1904). — Der Verf. zeigt, daß Planck in den „Vorlesungen über Thermodynamik" die Begriffe „reversibel" und „irreversibel" in einem etwas anderen Sinne anwendet, als er sie definiert. Sodann führt er eine Reihe von Einwänden an, welche gegen verschiedene Darstellungsweisen der Grundlagen der Thermodynamik vorgebracht werden können; unter diesen Einwänden ist der von Bertrand besonders beachtenswert, daß nämlich Druck, Temperatur und Entropie nur für den Fall definiert sind, daß wenigstens genügend kleine Teile eines Systems als im Gleichgewicht befindlich angesehen werden können; ein ähnlicher Einwand wird bezüglich der zugeführten Wärme erhoben. A. E

27. **G. H. Bryan.** *Das Gesetz von der Entwertung der Energie als Fundamentalprinzip der Thermodynamik* (Boltzmann-Festschrift, S. 123—136. 1904). — Der Verf. geht aus vom Energieprinzip sowie vom Prinzip der Abnahme der freien Energie. Die freie Energie (available Energy) eines Systems [1] wird definiert als die maximale mechanische Arbeit, welche das System bei mit den äußeren Bedingungen verträglichen Veränderungen leisten kann. Es folgt die Definition der dem System zugeführten Wärme. Dann werden der Begriff des thermischen Gleichgewichtes, der zweite Hauptsatz, der Begriff der absoluten Temperatur und der Energiebegriff aus den angegebenen Grundprinzipien in eleganter Weise entwickelt, und schließlich noch die Gleichungen des thermodynamischen Gleichgewichtes hergeleitet. A. E.

Published in *Beiblätter zu den Annalen der Physik* 29 (1905): 237. Published in no. 5 [first half of March].

[1] Bryan showed that the ''available energy'' is either $U - TS$, $U - TS + pV$, or, in one particular case, U.

10. Review of Nikolay Nikolayevich Schiller, "Some Concerns Regarding the Theory of Entropy Increase Due to the Diffusion of Gases Where the Initial Pressures of the Latter Are Equal"

[Einstein 1905e]

First half of March 1905

PUBLISHED IN: *Beiblätter zu den Annalen der Physik* 29 (1905), no. 5: 237–238.

„Die Frage, ob sich die Entropie des Weltalls überhaupt ändert, und wenn ja, in welchem Sinne, geht also gegenwärtig noch gar nicht zu beantworten, und sie wird wohl auch immer unentschieden bleiben". A. E.

---

26. *Mc. F. Orr.* *Über Clausius' Theorem der irreversibeln Kreisprozesse und über das Wachsen der Entropie* (Phil. Mag. (8) **46**, S. 509—527. 1904). — Der Verf. zeigt, daß Planck in den „Vorlesungen über Thermodynamik" die Begriffe „reversibel" und „irreversibel" in einem etwas anderen Sinne anwendet, als er sie definiert. Sodann führt er eine Reihe von Einwänden an, welche gegen verschiedene Darstellungsweisen der Grundlagen der Thermodynamik vorgebracht werden können; unter diesen Einwänden ist der von Bertrand besonders beachtenswert, daß nämlich Druck, Temperatur und Entropie nur für den Fall definiert sind, daß wenigstens genügend kleine Teile eines Systems als im Gleichgewicht befindlich angesehen werden können; ein ähnlicher Einwand wird bezüglich der zugeführten Wärme erhoben. A. E

---

27. *G. H. Bryan.* *Das Gesetz von der Entwertung der Energie als Fundamentalprinzip der Thermodynamik* (Boltzmann-Festschrift, S. 123—136. 1904). — Der Verf. geht aus vom Energieprinzip sowie vom Prinzip der Abnahme der freien Energie. Die freie Energie (available Energy) eines Systems wird definiert als die maximale mechanische Arbeit, welche das System bei mit den äußeren Bedingungen verträglichen Veränderungen leisten kann. Es folgt die Definition der dem System zugeführten Wärme. Dann werden der Begriff des thermischen Gleichgewichtes, der zweite Hauptsatz, der Begriff der absoluten Temperatur und der Energiebegriff aus den angegebenen Grundprinzipien in eleganter Weise entwickelt, und schließlich noch die Gleichungen des thermodynamischen Gleichgewichtes hergeleitet. A. E.

---

28. *N. Schiller.* *Einige Bedenken betreffend die Theorie der Entropievermehrung durch Diffusion der Gase bei einander gleichen Anfangsspannungen der letzteren* (Boltzmann-Festschrift, S. 350—366. 1904). — Es wird zunächst gezeigt, daß

238   **Wärmelehre.**   **Beibl. 1905.**

man ein homogenes Gas ohne Einfuhr von Arbeit und Wärme
isothermisch auf ein *n*-mal kleineres Volumen bringen kann,
unter der Annahme der Existenz von Wänden, welche für
einen Teil der Masse eines Gases durchlässig sind, für die
[1] übrige Masse des Gases dagegen nicht; in dieser Annahme
liegt nach der Ansicht des Verf. kein Widerspruch. Sodann
wird dargetan, daß der Ausdruck der Entropie eines aus räumlich getrennten Gasen von derselben Temperatur und demselben Druck bestehenden Systems die Form hat:

[2]
$$S = (\Sigma\, m_i\, R_i)\, \lg v + f(\Theta);$$

die Entropie des Systems nach erfolgter Diffusion kann durch
dieselbe Formel dargestellt werden. Daraus wird geschlossen,
[3] daß die Entropie vor und nach erfolgter Diffusion dieselbe sei.
Zu dem nämlichen Resultat gelangt der Verf. durch eine
Überlegung, welche hier nicht wiedergegeben werden kann.
Bei dieser Überlegung wird operiert mit einer Fläche, welche
ein chemisch homogenes Gas in zwei Teile trennt, derart, daß der
Gasdruck in beiden Teilen bei thermischem und mechanischem
Gleichgewicht verschieden ist. Dabei wird (implizite) an-
genommen, daß beim Hindurchtreten von Gas durch diese
Fläche von der letzteren auf das Gas keine Arbeit über-
tragen wird.   A. E.

Published in *Beiblätter zu den Annalen der Physik* 29 (1905): 237–238. Published in no. 5 [first
half of March].

[1] Schiller draws the same parallel between
semipermeable membranes and external forces
as Einstein did in *Einstein 1902a* (Doc. 2).

[2] m_i is the mass of gas i, $R_i = n_i\, R/m_i$, with

n_i the number of moles of gas i and R the gas
constant; v is the total volume of the gases, and
$f(\theta)$ is a function of the temperature θ.

[3] In contrast to the commonly accepted result
(first derived in *Boltzmann 1878b*) that diffusion
causes an increase in entropy, unless the diffusing gases are chemically identical.

11. Review of Jakob Johann Weyrauch, "On the Specific Heats of Superheated Water Vapor"

[Einstein 1905f]

First half of March 1905

PUBLISHED IN: *Beiblätter zu den Annalen der Physik* 29 (1905), no. 5: 240.

240   **Wärmelehre.**   **Beibl. 1905.**

seine Ideen auseinandersetzt, beschränken, und benutzen diese Gelegenheit, auf jene Diskussion nachträglich noch einmal hinzuweisen.]

30. *J. R. Benton. Thermodynamische Formeln für isotrope, in einer Richtung gedehnte Körper* (Phys. Rev. 16, S. 11 —16. 1903). — Wenn ein Metallstab in der Längsrichtung durch eine gleichförmige Spannung P (innerhalb der Elastizitätsgrenze) gedehnt wird, sonst aber keine äußeren Kräfte auf ihn wirken, so sind die Länge des Stabes, seine Energie und Entropie durch P und die Temperatur T völlig bestimmt. Hieraus sind drei Differentialgleichungen ableitbar, von denen die eine den von Lord Kelvin berechneten (negativen) Wert von dT/dP für eine adiabatische Spannungsänderung dP liefert. Die beiden anderen zeigen, daß der thermische Ausdehnungskoeffizient bei zunehmender Spannung wächst, wenn der Elastizitätsmodul bei einer Erwärmung abnimmt, und daß die spezifische Wärme (bei konstanter Spannung) bei einer Spannungszunahme wächst, wenn der Ausdehnungskoeffizient mit der Temperatur zunimmt. — Die Zahlenrechnung ergibt für diese Zunahme der spezifischen Wärme bei den meisten Metallen so kleine Werte, daß sie sich der Beobachtung entziehen.   Lck.

31. *J. J. Weyrauch. Über die spezifischen Wärmen des*
[1] *überhitzten Wasserdampfes* (S.-A. 9 S. ZS. d. Ver. D. Ing. 1904). — Bisherige Ermittelungen der spezifischen Wärme c_p werden angeführt und verglichen (I). Es werden für die Praxis ge-
[2] eignete Zustandsgleichungen für Wasserdampf angegeben und
[3] diskutiert (II) und vermittelst derjenigen von Zeuner c_p und c_v für gesättigten Dampf (III) und c_p und c_v für beliebig überhitzten Dampf auf thermodynamischem Wege hergeleitet. Hierauf werden Gesamtwärme und Dampfwärme bestimmt (V). Unter (VI) und (VII) folgen die wärmetheoretischen Hauptgleichungen für überhitzte Dämpfe, deren Anwendung auf besondere Fälle und einige Zahlenbeispiele.   A. E.

Published in *Beiblätter zu den Annalen der Physik* 29 (1905): 240. Published in no. 5 [first half of March].

[1] "S.-A." is an abbreviation for "Separat-Abdruck."

[2] One of the equations of state Weyrauch discusses is the one proposed by Tumlirz. See *Einstein 1905a* (Doc. 6), note 2.

[3] Zeuner's equation of state (*Zeuner 1867a, 1867b*) is $pV = RT - cp^\mu$ (c, μ constants). For another discussion of Zeuner's equation, see *Weyrauch 1907*, pp. 90–91, which was reviewed by Einstein in *Einstein 1907i* (Doc. 46).

12. Review of Jacobus Henricus Van 't Hoff, ''The Influence of the Change in Specific Heat on the Work of Conversion''

[*Einstein 1905g*]

First half of March 1905

PUBLISHED IN: *Beiblätter zu den Annalen der Physik* 29 (1905), no. 5: 240–242.

seine Ideen auseinandersetzt, beschränken, und benutzen diese Gelegenheit, auf jene Diskussion nachträglich noch einmal hinzuweisen.]

30. *J. R. Benton. Thermodynamische Formeln für isotrope, in einer Richtung gedehnte Körper* (Phys. Rev. 16, S. 11 —16. 1903). — Wenn ein Metallstab in der Längsrichtung durch eine gleichförmige Spannung P (innerhalb der Elastizitätsgrenze) gedehnt wird, sonst aber keine äußeren Kräfte auf ihn wirken, so sind die Länge des Stabes, seine Energie und Entropie durch P und die Temperatur T völlig bestimmt. Hieraus sind drei Differentialgleichungen ableitbar, von denen die eine den von Lord Kelvin berechneten (negativen) Wert von dT/dP für eine adiabatische Spannungsänderung dP liefert. Die beiden anderen zeigen, daß der thermische Ausdehnungskoeffizient bei zunehmender Spannung wächst, wenn der Elastizitätsmodul bei einer Erwärmung abnimmt, und daß die spezifische Wärme (bei konstanter Spannung) bei einer Spannungszunahme wächst, wenn der Ausdehnungskoeffizient mit der Temperatur zunimmt. — Die Zahlenrechnung ergibt für diese Zunahme der spezifischen Wärme bei den meisten Metallen so kleine Werte, daß sie sich der Beobachtung entziehen.   Lck.

31. *J. J. Weyrauch. Über die spezifischen Wärmen des überhitzten Wasserdampfes* (S.-A. 9 S. ZS. d. Ver. D. Ing. 1904). — Bisherige Ermittelungen der spezifischen Wärme c_p werden angeführt und verglichen (I). Es werden für die Praxis geeignete Zustandsgleichungen für Wasserdampf angegeben und diskutiert (II) und vermittelst derjenigen von Zeuner c_p und c_v für gesättigten Dampf (III) und c_p und c_v für beliebig überhitzten Dampf auf thermodynamischem Wege hergeleitet. Hierauf werden Gesamtwärme und Dampfwärme bestimmt (V). Unter (VI) und (VII) folgen die wärmetheoretischen Hauptgleichungen für überhitzte Dämpfe, deren Anwendung auf besondere Fälle und einige Zahlenbeispiele.   A. E.

32. *J. H. van't Hoff. Einfluß der Änderungen der spezifischen Wärme auf die Umwandlungsarbeit* (Boltzmann-Festschrift, S. 233—241. 1903). — Der Verf. zeigt auf thermodynamischem Wege, daß die (nach außen abgegebene) Um-

Bd. 29. No. 5.   Wärmelehre.   241

wandlungsarbeit E eines Systems A in ein System B (z. B. durch Schmelzung) bei isothermischer Umwandlung in der Form dargestellt werden kann:

$$E = E_0 + A T - S T \lg T.$$

(A ist eine Konstante, T die absolute Temperatur, $S = S_A - S_B$ die Differenz der spezifischen Wärmen, welche als unabhängig von T vorausgesetzt werden. $A T$ wird aus Analogiegründen $\left(\text{weil bei isothermischer Ausdehnung eines Gases } E = A T = 2 T \lg (v_B / v_A)\right)$ als durch Konzentrationsänderung [1] bedingt angesehen.

Die Gleichung wird auf Versuche von Richards angewendet, [2] welcher für Umwandlungen von der Art

$$\mathrm{Mg} + \mathrm{ZnSO_4 . Aq} = \mathrm{Zn} + \mathrm{MgSO_4 . Aq}$$

(wobei ursprünglich $\mathrm{ZnSO_4}$ und entstehendes $\mathrm{MgSO_4}$ gleich konzentriert) auf elektrischem Wege fand:

$$\frac{d E}{d T} = - \varkappa S,$$

wobei $\varkappa$ annähernd gleich für alle untersuchten Umwandlungen. Unter Fortlassung des Termes $A T$ erhält der Verf. aus obiger Gleichung

$$\frac{d E}{d T} = - S(1 + \lg T) = - 6{,}7\ S. \qquad [3]$$

Mittelwerte aus Beobachtungen ergaben: [4]

Reaktion	$\left(\dfrac{d E}{d T}\right):(-S)$	Reaktion	$\left(\dfrac{d E}{d T}\right):(-S)$
$\mathrm{Mg} + \mathrm{ZnSO_4}$	5	$\mathrm{Zn} + \mathrm{NiSO_4}$	8
$\mathrm{Mg} + \mathrm{CuSO_4}$	5,4	$\mathrm{Fe} + \mathrm{CuSO_4}$	7,5
$\mathrm{Mg} + \mathrm{NiSO_4}$	5,9	$\mathrm{Ni} + \mathrm{CuSO_4}$	7
$\mathrm{Mg} + \mathrm{FeSO_4}$	6,3	$\mathrm{Zn} + \mathrm{CuSO_4}$	7,4
$\mathrm{Zn} + \mathrm{FeSO_4}$	7,3	$\mathrm{Fe} + \mathrm{NiSO_4}$	7,1

Die Gleichung für E liefert ferner, auf Schmelzung sowie auf Umwandlung allotroper Elemente und polymorpher Verbindungen angewendet (wieder unter Vernachlässigung des Gliedes $A T$) den Satz: Die bei höherer Temperatur stabile Form (z. B. Flüssigkeit) hat die größere spezifische Wärme. Die Folgerung wird fast durchweg durch das Experiment bestätigt. Endlich wird aus der Gleichung gefolgert, daß die Thomson-Berthelotsche Regel bei tiefen Temperaturen gültig sein muß, [5]

242   **Wärmelehre.**   **Beibl. 1905.**

daß aber bei höheren Temperaturen durch das Glied $- S\,T\,\lg T$
Abweichungen veranlaßt werden können, wenn $S_A > S_B$.

————————

A. E.

Published in *Beiblätter zu den Annalen der Physik* 29 (1905): 240–242. Published in no. 5 [first half of March].

[1] For one mole, $E = RT \log (v_B/v_A)$; $R \approx 2$ cal/deg.

[2] *Richards 1902*.

[3] The second equality holds for $T = 18°C$.

[4] The experimental data are taken from *Richards 1902*.

[5] The rule of Thomsen and Berthelot states that chemical reactions take place in the direction of enthalpy or heat production. This rule is not valid at ordinary temperatures. For a contemporary discussion, see *Nernst 1898*, pp. 627–633.

13. Review of Arturo Giammarco, "A Case of Corresponding States in Thermodynamics"

[Einstein 1905h]

First half of March 1905

PUBLISHED IN: *Beiblätter zu den Annalen der Physik* 29 (1905), no. 5: 246–247.

des normalen Zustandes. Aus diesen Messungen wird der zweite Virialkoeffizient, d. h. B in der Formel $p = A + B/v + \ldots$ (Comm. No. 71; Beibl. 26, S. 261) berechnet und dann verglichen mit den Werten, die für B gefunden werden aus der allgemeinen reduzierten Zustandsgleichung von Comm. No. 71 mittels der kritischen Daten von Brinkman (Diss. Amsterdam 1904). Es ergibt sich, daß die zweiten Virialkoeffizienten für die Gemische mit genügender Annäherung aus dem Gesetze der korrespondierenden Zustände berechnet werden können. Nur Chlormethyl gibt einen etwas abweichenden Wert. Den von Leduc und Sacerdote für Chlormethyl gefundenen Wert (Beibl. 23, S. 19) finden die Verf. jedoch im Einklang mit dem genannten Gesetz.

Zweitens durch Bestimmung der Koexistenzbedingungen bei niedriger Temperatur. Mit dem Taupunktsapparat, beschrieben in der vorigen Mitteilung, wurde bestimmt: Anfangskondensationsdruck bei -25°C. für ein Gemisch vom Molekulargehalt 0,5042 an Chlormethyl. Mit dem ebenda beschriebenen Piezometer wurden gemessen: Kondensationsdruck des reinen Chlormethyls bei -25° und $-37,4^\circ$C., Endkondensationsdruck und -volum des genannten Gemisches bei $-38,5^\circ$C. Vergleichung mit dem Gesetz korrespondierender Zustände ergibt, daß aus den Koexistenzbedingungen Abweichungen dieses Gesetzes bei Isothermen der Gemische von Chlormethyl und Kohlensäure folgen, welche Abweichungen bei Flüssigkeitsdichten und niedrigen Temperaturen sehr deutlich werden.

Keesom.

---

39. A. Giammarco. *Ein Fall von übereinstimmenden Zuständen* (N. Cim. (5) 5, S. 377—391. 1904). — Hat man in einer geschlossenen zylindrischen Röhre eine Flüssigkeit (Volumen v) und darüber ihren gesättigten Dampf (Volumen v'), und trägt man v/v' als Funktion der absoluten Temperatur T in rechtwinkligen Koordinaten auf, so erhält man je nach der Menge der eingeschlossenen Substanz eine Kurve, welche ein Maximum $(v/v')_{max}$ besitzt, oder eine Kurve, welche gegen die Abszissenachse konvex ist, oder (als Grenzfall) eine Kurve, welche gegen die kritische Temperatur hin geradlinig verläuft. Der Verf. untersuchte so Äther, Alkohol und Chloroform und findet, daß die genannten Maxima $(v/v')_{max}$ auf einer Geraden liegen. Nach dem Gesetze der übereinstimmenden Zustände

[1] [2]

Bd. 29.   No. 5.   **Wärmelehre.**   **247**

müssen zwei Temperaturen T und T', in welchen zwei verschiedene Substanzen das nämliche $(v / v')_{max}$ besitzen, übereinstimmende Temperaturen sein (Methode zur Bestimmung
übereinstimmender Temperaturen), also $T / T_c = T' / T_c'$. Der
Verf. findet aus seinen Beobachtungen unter Benutzung der
(absoluten) kritischen Temperaturen von Äther (467°), Alkohol
(517°), Chloroform (541°), (Bureau des Longitudes 1902):   [3]

Korresp. abs. Temperaturen			$(V / V')_{max}$			$\dfrac{T}{T_c}$
Äther	Alkohol	Chloroform	Äther	Alkohol	Chloroform	
387°	428,07	447,09	0,320	0,320	0,330	0,828
391	432,8	452,8	0,340	0,340	0,350	0,837
394	435,8	456	0,355	0,356	0,360	0,843
404	447	467,9	0,395	0,400	0,409	0,865
414	456,5	478	0,440	0,440	0,448	0,883
423	468,2	489,6	0,490	0,490	0,495	0,905
427	472,7	494,4	0,510	0,510	0,511	0,914
437	485,3	505,8	0,556	0,556	0,556	0,935
458	506,6	530	0,655	0,652	0,652	0,981
467	517	541	0,695	0,698	0,698	1

[4]

Der Verf. findet bei Beobachtung der den Grenzfall bildenden Kurve, daß bei steigender bez. fallender Temperatur
das Verschwinden bez. Erscheinen des Meniskus bei der nämlichen Temperatur (der kritischen Temperatur) erfolgt.

                                        A. E.

Published in *Beiblätter zu den Annalen der Phy-sik* 29 (1905): 246–247. Published in no. 5 [first half of March].

[1] Van der Waals's theory of corresponding states is discussed in *Boltzmann 1898a*, pp. 23–27 (although the term is not mentioned).

[2] Giammarco's paper was published in 1903.
[3] *Annuaire 1902*, pp. 560–561.
[4] From this table the author verifies that $(V/V')_{max}$ has the same value for equal values of T/T_c.

EINSTEIN'S EARLY WORK ON THE
QUANTUM HYPOTHESIS

I

From 1905 through 1909 Einstein published five major papers on the hypothesis of energy quanta, its theoretical implications, and its use in the explanation of various phenomena: *Einstein 1905i* (Doc. 14), *Einstein 1906d* (Doc. 34), *Einstein 1907a* (Doc. 38) with a correction in *Einstein 1907d* (Doc. 42), *Einstein 1909b* (Doc. 56), and *Einstein 1909c* (Doc. 60). In addition, he reviewed Planck's lectures on thermal radiation[1] in *Einstein 1906f* (Doc. 37), and with Walter Ritz summarized their differences on the radiation problem in *Ritz and Einstein 1909* (Doc. 57).

In describing four of his 1905 papers, Einstein characterized only the one on the quantum hypothesis as revolutionary.[2] It is now regarded as revolutionary in challenging the unlimited validity of Maxwell's theory of light and suggesting the existence of light quanta. The paper shows that, at a sufficiently high frequency, the entropy of equilibrium thermal (or "black-body")[3] radiation behaves as if the radiation consists of a gas of independent "quanta of light energy" ("Lichtenergiequanten," or simply "light quanta," "Lichtquanten"), each with energy proportional to the frequency. Einstein showed how to explain several otherwise puzzling phenomena by assuming that the interaction of light with matter consists of the emission or absorption of such energy quanta.

In subsequent papers, Einstein examined the tacit assumptions underlying Planck's derivation of the energy distribution law for black-body radiation,[4] explored several further implications of the law itself, and showed that the concept of energy quantization is applicable not only to radiation but also to material oscillators. *Einstein 1906d* (Doc. 34) demonstrates that Planck's derivation presupposes quantization of the energy of charged oscillators in interaction with thermal radiation. *Einstein 1907a* (Doc. 38) explains the anomalous behavior of specific heats with decreasing temperature by treating the atoms or ions in a solid as a lattice of quantized oscillators.[5] *Einstein 1909b* (Doc. 56) shows that Planck's law gives rise to fluctuations in the energy and pressure of black-body radiation that appear to arise from two independent causes: light waves and light quanta.

In a masterful address to the Salzburg meeting of the Gesellschaft Deutscher Naturforscher und Ärzte, *Einstein 1909c* (Doc. 60), Einstein summarized his views on radia-

---

[1] *Planck 1906c.*

[2] See Einstein to Conrad Habicht, 18 May–8 June 1905. The paper on the quantum hypothesis is *Einstein 1905i* (Doc. 14); the other three papers mentioned are on: relativity theory, *Einstein 1905r* (Doc. 23); molecular dimensions, *Einstein 1905j* (Doc. 15); and Brownian motion, *Einstein 1905k* (Doc. 16).

[3] The term "black body" ("schwarzer Körper") was in general use by the turn of the century (see, e.g., *Rayleigh 1900*, p. 539, and *Planck 1900d*, p. 764). Einstein first used the term in *Einstein 1904* (Doc. 5), p. 354, but referred to the radiation itself as "temperature radiation" ("Temperaturstrahlung"), p. 361. *Einstein 1905i* (Doc. 14) refers to it as "black radiation" ("schwarze Strahlung").

[4] Planck presented this law in *Planck 1900c*. He gave derivations of it in *Planck 1900e, 1901a,* and *1906c.*

[5] *Einstein 1907d* (Doc. 42) considers both charged and uncharged lattice oscillators.

tion, and for the first time publicly linked his work on relativity and on the quantum hypothesis. He demonstrated the fundamental inability of classical mechanics and Maxwell's electrodynamics to account for a number of phenomena connected with electromagnetic radiation, again showed the utility of the quantum hypothesis in efforts to comprehend the structure of radiation, and speculated about modifications of existing theories that might lead to a consistent theory of radiation explaining the quantum hypothesis.[6]

II

Einstein started to study black-body radiation well before 1905. Mach's *Wärmelehre*, which Einstein read in 1897 or shortly thereafter, contains two chapters on thermal radiation, culminating in a discussion of Kirchhoff's work.[7] Kirchhoff showed that the energy emission spectrum of a perfectly black body (defined as one absorbing all incident radiation) at a given temperature is a universal function of the temperature and wave length. He inferred that equilibrium thermal radiation in a cavity with walls maintained at a certain temperature behaves like radiation emitted by a black body at the same temperature.[8]

H. F. Weber, Einstein's physics professor at the ETH, attempted to determine the universal black-body radiation function. He made measurements of the energy spectrum and proposed an empirical formula for the distribution function.[9] He showed that, as a consequence of his formula, $\lambda_m = $ constant$/T$ (where λ_m is the wavelength at the maximum intensity of the distribution), thus anticipating Wien's formulation of the displacement law for black-body radiation.[10] Weber described his work in a course at the ETH given during the winter semester of 1898–1899, for which Einstein registered.[11]

By March 1899, Einstein had started to think seriously about the problem of radiation.[12] In the spring of 1901, he was closely following Planck's work on black-body radiation.[13] Originally, Planck had hoped to explain irreversibility by studying electro-

[6] For further discussion of Einstein's papers on the quantum hypothesis, see *Hermann 1969*; *Jammer 1966*; *Jungnickel and McCormmach 1986b*, chaps. 25–26; *Klein 1977, 1979, 1980*; *Kuhn 1978*; *Mehra and Rechenberg 1982*; and *Pais 1982*.

[7] See *Mach 1896*, pp. 124–152; Kirchhoff's work is discussed on pp. 140–144. For an indication that Einstein may have read this work around 1897, see Einstein to Mileva Marić, 10 September 1899 (Vol. 1, Doc. 54), note 8.

[8] See *Kirchhoff 1860*. For surveys of research on black-body radiation up to and including Planck's work, see *Kangro 1976*; *Kuhn 1978*; and *Pais 1982*, pp. 364–372.

[9] See *Weber, H. F. 1887, 1888*. For a survey of Weber's work, see *Kangro 1976*, pp. 37–40.

[10] Einstein used this result in his first discussion of black-body radiation, *Einstein 1904*

(Doc. 5), pp. 360–362, without citing a source. Wien formulated the displacement law in *Wien 1893*, which cites Weber's work (see p. 62). For a history of the displacement law, see *Kangro 1976*, chap. 3.

[11] For the contents of the courses, see the notes of Emil Teucher on "Prinzipien, Apparate und Messmethoden der Elektrotechnik" (SzZE Bibliothek, Hs 32). For Einstein's registration, see ETH Record and Grade Transcript (Vol. 1, Doc. 28, p. 47).

[12] See Einstein to Mileva Marić, 13 or 20 March, 1899 (Vol. 1, Doc. 45).

[13] See Einstein to Mileva Marić, 4 April 1901 and 10 April 1901 (Vol. 1, Docs. 96 and 97). He had evidently read some of Planck's papers on black-body radiation, and indicated his intention to read *Planck 1901a*. It is possible that Einstein had already read *Drude 1900c*, a

magnetic radiation, but came to recognize that this could not be done without introducing statistical elements into the argument.[14] In a series of papers published between 1897 and 1900, Planck utilized Maxwell's electrodynamics to develop a theory of thermal radiation in interaction with one or more identical, charged harmonic oscillators within a cavity.[15] He was only able to account for the irreversible approach to thermal equilibrium by employing methods analogous to those Boltzmann used in kinetic theory. Planck introduced the notion of ''natural'' (that is, maximally disordered) radiation, which he defined in analogy with Boltzmann's definition of molecular chaos.[16] Using Maxwell's theory, Planck derived a relation between the average energy $\bar{E}_v$ of a charged oscillator of frequency v in equilibrium with thermal radiation and the energy density per unit frequency ρ_v of radiation at the same frequency:

$$\bar{E}_v = \frac{c^3}{8\pi v^2} \rho_v ,$$

(1)

where c is the speed of light.[17]

Planck calculated the average energy of an oscillator by making assumptions about the entropy of the oscillators that enabled him to derive Wien's law for the black-body spectrum, which originally seemed well supported by the experimental evidence.[18] But by the turn of the century new observations showed systematic deviations from Wien's law for large values of λT.[19]

Planck 1900b presents a new energy density distribution formula that agreed closely with observations over the entire spectrum:[20]

$$\rho_v = \frac{8\pi h v^3}{c^3} \frac{1}{e^{hv/kT} - 1} .$$

(2)

In this expression, now known as Planck's law or Planck's formula, $k = R/N$ is Boltzmann's constant, R is the gas constant, N is Avogadro's (or Loschmidt's) number, and h

---

copy of which is in his personal library. Section III includes a detailed discussion of black-body radiation, including Wien's law and references to one of Planck's papers on the subject (see pp. 482–484).

[14] See *Planck 1900a*, pp. 614–621.

[15] Planck summarized his work in *Planck 1900a*. Planck's work on black-body radiation has been extensively discussed. See, e.g., *Hermann 1969*, pp. 5–28; *Kangro 1976*, pp. 149–223; *Klein 1962, 1963a, 1966; Kuhn 1978*; and *Needell 1980*. For Planck, in contrast to Einstein, the oscillators and the cavity were ideal devices, introduced to facilitate the study of

thermal radiation.

[16] For a discussion of differences between Boltzmann's ''molecular chaos'' and Planck's ''natural radiation,'' see *Kuhn 1978*, pp. 120–125.

[17] See *Planck 1900e*, p. 241.

[18] See *Wien 1896*.

[19] This possibility, first suggested in *Lummer and Pringsheim 1899*, is confirmed in *Rubens and Kurlbaum 1901*, which summarizes earlier research.

[20] Empirical evidence confirming Planck's formula is reported, e.g., in *Rubens and Kurlbaum 1901* and in *Paschen 1901a*.

is a new constant (later called Planck's constant). To derive this formula,[21] Planck calculated the entropy of the oscillators, using what Einstein later called "the Boltzmann principle":[22] $S = k \cdot \log W$, where S is the entropy of a macroscopic state of the system, the probability of which is W. Following Boltzmann, Planck took W proportional to the number of "complexions," or possible microconfigurations of the system corresponding to its state. He calculated this number by dividing the total energy of the state into a finite number of elements of equal magnitude, and counting the number of possible ways of distributing these energy elements among the individual oscillators. If the size of the energy elements is set equal to $h\nu$, where ν is the frequency of the oscillators, an expression for the entropy of an oscillator results that leads to eq. (2).

Although Einstein expressed misgivings about Planck's approach in 1901,[23] he did not mention Planck or black-body radiation in his papers until 1904.[24] A study of the foundations of statistical physics, which he undertook between 1902 and 1904, provided Einstein with the tools he needed to analyze Planck's derivation and to explore its consequences.[25] At least three elements of Einstein's "general molecular theory of heat" ("allgemeine molekulare Theorie der Wärme")[26] were central to his subsequent work on the quantum hypothesis: the introduction of the canonical ensemble; the interpretation of probability in Boltzmann's principle; and the study of energy fluctuations in thermal equilibrium.

In an analysis of the canonical ensemble, Einstein proved that the equipartition theorem holds for any system in thermal equilibrium.[27] In 1905 he showed that, when applied to an ensemble of charged harmonic oscillators in equilbrium with thermal radiation, the equipartition theorem leads, via eq. (1), to a black-body distribution law now known as the Rayleigh-Jeans law:[28]

$$\rho_\nu = \frac{8\pi\nu^2}{c^3} kT. \tag{3}$$

Despite its rigorous foundation in classical physics, eq. (3) agrees with the observed en-

[21] See *Planck 1900e*, and *1901a*.

[22] *Einstein 1905i* (Doc. 14), p. 140.

[23] See Einstein to Mileva Marić, 4 April 1901 (Vol. 1, Doc. 96). In explanation of his misgivings, he cited his objection to Planck's use of one type of oscillator, with fixed period and damping constant (see Einstein to Marić, 10 April 1901, Vol. 1, Doc. 97).

[24] See *Einstein 1904* (Doc. 5), p. 354.

[25] See *Einstein 1902b* (Doc. 3), *Einstein 1903* (Doc. 4), and *Einstein 1904* (Doc. 5). For a discussion of this work, see the editorial note, "Einstein on the Foundations of Statistical Physics," pp. 41–55.

[26] See the title of *Einstein 1904* (Doc. 5).

[27] See *Einstein 1902b*, pp. 427–428.

[28] See *Einstein 1905i* (Doc. 14), § 1. Einstein rederived this result in *Einstein 1907a* (Doc. 38), and *Einstein 1909b* (Doc. 56). An expression equivalent to eq. (3) was obtained in 1905 by Rayleigh and Jeans by other methods (see *Rayleigh 1905a*, *1905b*, and *Jeans 1905a*, which appeared after the receipt of Einstein's paper). In 1900 Rayleigh obtained the proportionality between T and the energy of a vibration mode by applying the equipartition theorem to the normal modes of the radiation field (*Rayleigh 1900*).

ergy distribution only for small values of v/T; indeed, as Einstein noted, it implies an infinite total radiant energy.[29]

Einstein posed a question in 1906, which preoccupied him and others at the time: "How is it that Planck did not arrive at the same formula [eq. (3)], but at the expression . . . [eq. 2]?" ("Woher kommt es, daß Hr. Planck nicht zu der gleichen Formel, sondern zu dem Ausdruck . . . gelangt ist?")[30] One answer lies in Planck's definition of W in Boltzmann's principle, which, as Einstein repeatedly noted, differs fundamentally from his own definition of probabilities as time averages.[31] As noted above, Planck interpreted W as proportional to the number of complexions of a system. As Einstein pointed out in 1909, such a definition of W is equivalent to his definition only if all complexions corresponding to a given total energy are equally probable. However, if this is assumed to be the case for an ensemble of oscillators in thermal equilibrium with radiation, the Rayleigh-Jeans law results.[32] Hence, the validity of Planck's law implies that all complexions cannot be equally probable. Einstein showed that, if the energies available to a canonical ensemble of oscillators are arbitrarily restricted to multiples of the energy element hv, then all possible complexions are not equally probable, and Planck's law results.[33]

A third element of Einstein's work on statistical physics that is central to his work on the quantum hypothesis is his method for calculating mean square fluctuations in the state variables of a system in thermal equilibrium. He employed the canonical ensemble to calculate such fluctuations in the energy of mechanical systems, and then applied the result to a nonmechanical system—black-body radiation, deducing a relation that agrees with Wien's displacement law.[34] This agreement confirms the applicability of statistical concepts to radiation, and may have suggested to him the possibility that radiation could be treated as a system with a finite number of degrees of freedom, a possibility he raised at the outset of his first paper on the quantum hypothesis.[35]

In connection with his work on Brownian motion in 1905–1906, Einstein developed additional methods for calculating fluctuations, methods which he later applied to the analysis of black-body radiation. In particular, he developed a method based on the inversion of Boltzmann's principle, which may be used even in the absence of a microscopic model

[29] See *Einstein 1905i* (Doc. 14), p. 136. Ehrenfest later called this divergent behavior the "Rayleigh-Jeans catastrophe in the ultraviolet" ("Rayleigh-Jeans-Katastrophe im Ultravioletten") (*Ehrenfest 1911*, p. 92).

[30] *Einstein 1906d* (Doc. 34), p. 200. The problem is also discussed in *Einstein 1907a* (Doc. 38) and *Einstein 1909b* (Doc. 56); and, e.g., in *Ehrenfest 1906* and *Rayleigh 1905b*.

[31] The difference between their definitions is stated particularly clearly in *Einstein 1909b* (Doc. 56), sec. 4, pp. 187–188. Einstein first gave his definition in *Einstein 1903* (Doc. 4), pp. 171–172. Einstein's definition is discussed in the editorial note, "Einstein on the Founda-

tions of Statistical Physics," p. 52; *Klein 1974b* and *Pais 1982*, chap. 4.

[32] See *Einstein 1909b* (Doc. 56), p. 187.

[33] See *Einstein 1906d* (Doc. 34), pp. 201–203, and *Einstein 1907a* (Doc. 38), pp. 182–184.

[34] See *Einstein 1904* (Doc. 5), especially p. 362. For further discussion of his work on energy fluctuations, see *Klein 1967*, and the editorial notes, "Einstein on the Foundations of Statistical Physics," p. 54, and "Einstein on Brownian Motion," pp. 206–222.

[35] See *Einstein 1905i* (Doc. 14), pp. 132–133.

of the system.[36] If the entropy of a system is given as a function of its macroscopic state variables, then Boltzmann's principle, in the form $W = \exp(S/k)$, can be used to calculate the probability of a state, and hence of the fluctuations of any state variable. In 1909 Einstein used this method to calculate the fluctuations in the energy of black-body radiation in a given region of space.[37] The stochastic method, used in the same paper to calculate fluctuations in the pressure of radiation, is based on his work on Brownian motion. Pressure fluctuations maintain the Brownian motion of a small mirror moving through the radiation field, in the face of the retarding force exerted on the mirror by the average radiation pressure.[38] The results of these two fluctuation calculations are discussed in the next section.

Einstein's work on relativity also contributed to the development of his views on the nature of light. By eliminating the concept of the ether and showing that a flux of radiant energy transfers inertial mass, the theory of relativity demonstrated that light no longer need be treated as a disturbance in a hypothetical medium, but could be regarded as composed of independent structures, to which mass is attributed.[39]

III

Among Einstein's papers on the quantum hypothesis, the 1905 paper is unique in arguing for the notion of light quanta without using either the formal apparatus of his statistical papers or Planck's law.[40] As noted in the previous section, Einstein demonstrated that only the limiting form of Planck's formula for small values of v/T is consistent with the accepted foundations of statistical mechanics and electrodynamics. At the other extreme, at which Wien's distribution law holds, "the theoretical foundations we have utilized . . . fail . . . completely" ("die von uns benutzten theoretischen Grundlagen . . . versagen . . . vollständig").[41] As he explained later that year, this failure "seems to me to have its basis in a fundamental incompleteness of our physical concepts" ("scheint mir in einer elementaren Unvollkommenheit unserer physikalischen Anschauungen ihren Grund zu ha-

---

[36] The method is first discussed in *Einstein 1907b* (Doc. 39).

[37] See *Einstein 1909b* (Doc. 56), pp. 188–189. The equation for the energy fluctuations that results from Boltzmann's principle is equivalent to one that Einstein derived earlier from the canonical distribution in *Einstein 1904* (Doc. 5), pp. 359–360, as a simple thermodynamic argument shows. Einstein evidently preferred a formula not dependent on the canonical distribution for the application to black-body radiation.

[38] See *Einstein 1909b* (Doc. 56), pp. 189–190, and *Einstein 1909c* (Doc. 60), pp. 497–498. For a discussion of the stochastic method, see the editorial note, "Einstein on Brownian Motion," pp. 213–215.

[39] The conclusions are implicit in some of Einstein's earlier papers but are first made explicit in *Einstein 1909c* (Doc. 60). For discussion of the elimination of the ether and the inertia of energy, see the editorial note, "Einstein on the Theory of Relativity," pp. 253–274.

[40] A year later Einstein explained that initially he had regarded Planck's theory as forming "in a certain sense an antithesis" ("in gewisser Beziehung ein Gegenstück") to his own work (*Einstein 1906d* [Doc. 34], p. 199). While "Gegenstück" can also mean "complement," the context in which it is used here suggests "antithesis."

[41] *Einstein 1905i* (Doc. 14), p. 137.

ben'').[42] In order to suggest what new concepts might be needed, he focused on the problematic Wien region.

Using Wien's law,[43] Einstein showed that the expression for the volume dependence of the entropy of radiation at a given frequency is similar in form to that of the entropy of an ideal gas. He concluded that "monochromatic radiation of low density (within the range of validity of Wien's radiation formula) behaves thermodynamically as though it consisted of quanta of energy, which are independent of one another, and of magnitude $R\beta v/N$" ("[m]onochromatische Strahlung von geringer Dichte (innerhalb des Gültigkeits- bereiches der Wienschen Strahlungsformel) verhält sich in wärmetheoretischer Beziehung so, wie wenn sie aus voneinander unabhängigen Energiequanten von der Größe $R\beta v/N$ bestünde").[44] Einstein opened the paper by pointing out the "fundamental formal dis- tinction" ("tiefgreifender formaler Unterschied") between current theories of matter, in which the energy of a body is represented as a sum over a finite number of degrees of freedom, and Maxwell's theory, in which the energy is a continuous spatial function hav- ing an infinite number of degrees of freedom. He suggested that the inability of Maxwell's theory to give an adequate account of radiation might be remedied by a theory in which radiant energy is distributed discontinuously in space.[45] Einstein formulated "the light quantum hypothesis" ("die Lichtquantenhypothese")[46] that

> the energy of a light ray emitted from a point [is] not continuously distributed over an ever increasing space, but consists of a finite number of energy quanta which are localized at points in space, which move without dividing, and which can only be produced and absorbed as complete units.

> bei Ausbreitung eines von einem Punkte ausgehenden Lichtstrahles [ist] die Energie nicht kontinuierlich auf größer und größer werdende Räume verteilt, sondern es besteht dieselbe aus einer endlichen Zahl von in Raumpunkten lokalisierten Energiequanten, welche sich bewegen, ohne sich zu teilen und nur als Ganze absorbiert und erzeugt werden können.[47]

As noted in the previous section, Einstein first asserted that Planck's derivation implic- itly assumes quantization of the energies of charged oscillators in his second paper on

[42] *Einstein 1906b* (Doc. 32), p. 375.

[43] See *Wien 1896*.

[44] *Einstein 1905i* (Doc. 14), p. 143. $R\beta/N$ is equivalent to Planck's *h*. Except in reviewing *Planck 1906c* (see *Einstein 1906f* [Doc. 37]), Einstein did not use *h* before *Einstein 1909b* (Doc. 56). In 1901, Einstein had pointed out a parallelism between the kinetic energy of mole- cules and the thermal energy spectrum (see Ein- stein to Mileva Marić, 30 April 1901 [Vol. 1, Doc. 102]).

[45] See *Einstein 1905i* (Doc. 14), pp. 132– 133. Einstein reiterated that radiation in a finite region of space should be described by a finite

number of degrees of freedom in *Einstein 1907h* (Doc. 45), p. 372. For further discussion of this point, see *McCormmach 1970b*. For an analysis of *Einstein 1905i* (Doc. 14), see *Klein 1963b*.

[46] *Einstein 1906d* (Doc. 34), p. 203.

[47] *Einstein 1905i* (Doc. 14), p. 133. Four years later Einstein stated the light-quantum hy- pothesis more cautiously, as the assumption "that emission and absorption of radiation only take place in quanta of this magnitude of energy [*hv*]" ("daß Emission und Absorption von Strahlung überhaupt nur in Quanten von dieser Energiegröße stattfinde") (*Einstein 1909c* [Doc. 60], p. 495).

quanta, *Einstein 1906d* (Doc. 34). He returned to this question in *Einstein 1907a* (Doc. 38), showing that, if the structure function in phase space he had introduced earlier[48] is assumed to restrict the oscillators to orbits with energies that are integral multiples of $h\nu$, then the average oscillator energy in a canonical ensemble yields Planck's law, when substituted in eq. (1). This assumption about the oscillator energies is inconsistent with Planck's derivation of eq. (1), which assumes that the energy of a radiating oscillator varies continuously. So Einstein also had to assume that, in spite of this inconsistency, Maxwell's theory gives the correct average energy of an oscillator in a radiation field.

Except for his paper on specific heats, *Einstein 1907a* (Doc. 38), submitted in November 1906, Einstein did not write another major paper on the quantum hypothesis for over two years. Although his publications concentrate on relativity theory during this period, he did not cease to wrestle with the problem of quanta. He wrote Laub in April? 1909 about his intense struggle with the problem, especially as it pertains to light:

> I am ceaselessly occupied with the problem of the constitution of radiation. . . . This question of quanta is so extraordinarily important and difficult that every-one should be concerned with it.

> Ich beschäftige mich unablässig mit der Frage der Konstitution der Strahlung. . . . Diese Quantenfrage ist so ungemein wichtig und schwer, dass sich alle darum bemühen sollten.[49]

Einstein presented the results of those two years of reflection in *Einstein 1909b* (Doc. 56), and in his Salzburg lecture, *Einstein 1909c* (Doc. 60), which are discussed in sections V and VI below.

IV

In addition to their contributions to theory, each of Einstein's first three papers on the quantum hypothesis also provides ingenious explanations of observed phenomena or predictions of new ones. *Einstein 1905i* (Doc. 14) examines three interactions of light with matter, treated "as if light consisted of such energy quanta" ("wie wenn das Licht aus derartigen Energiequanten bestünde"):[50] Stokes's rule for fluorescence; the ionization of gases by ultraviolet light; and the photoelectric effect. In *Einstein 1906d* (Doc. 34), he deduced a relationship between the electromotive force series for metals and their photoelectric sensitivity. *Einstein 1907a* (Doc. 38) offers an explanation of the anomalous decrease of specific heats with decreasing temperature. Einstein's explanations of the photoelectric effect and of the behavior of specific heats proved to be especially significant.

Einstein 1905i (Doc. 14) proposes what later became known as Einstein's photoelectric equation,

---

[48] See, in particular, *Einstein 1904* (Doc. 5), p. 359.

[49] Einstein to Jakob Laub, April? 1909.

[50] *Einstein 1905i* (Doc. 14), p. 144.

$$E_{max} = (R/N)\beta\nu - P, \tag{4}$$

where E_{max} is the maximum kinetic energy of the photoelectrons, $R\beta/N$ is equivalent to Planck's h, ν is the frequency of the incident radiation, and P is the work function of the metal emitting the electrons. Although his derivation of this equation was later considered to be a leading achievement of that paper—it is cited in his 1922 Nobel Prize award[51]— for almost two decades the argument failed to persuade most physicists of the validity of the light quantum hypothesis. Lenard's experimental studies,[52] to which Einstein referred, only provide qualitative evidence for an increase of E_{max} with frequency. For almost a decade the quantitative relationship between electron energy and radiation frequency was in doubt.[53] By about 1914, there was a substantial body of evidence tending to support eq. (4).[54] Millikan's studies clinched the case for almost all physicists.[55] But even the confirmation of Einstein's photoelectric equation did not bring about widespread acceptance of the concept of light quanta. Alternative interpretations of the photoelectric effect still received general support for a number of years.[56]

The earliest widely accepted empirical evidence for the quantum hypothesis came not from radiation phenomena, but from data on specific heats. *Einstein 1907a* (Doc. 38) utilizes the model of a solid as a lattice of atoms harmonically bound to their equilibrium positions. When the oscillators are treated classically, the equipartition theorem leads to the Dulong-Petit rule, predicting a constant specific heat for solids at all temperatures.[57] Treating each atom as a quantized three-dimensional harmonic oscillator, Einstein was able to explain the well-known anomalous decrease of the specific heats of certain solids with decreasing temperature, and to obtain a relationship between the specific heat of a solid and its selective absorption of infrared radiation.[58]

Einstein had long suspected the existence of a connection between specific heats and

[51] Though given in 1922, the award was for 1921. For a discussion of Einstein's Nobel Prize, see *Pais 1982*, pp. 502–512.

[52] See *Lenard 1902*.

[53] One experimental study even suggested a quadratic relationship between energy and frequency (see *Ladenburg 1907*). For a survey of the difficulties in testing Einstein's predictions and of experimental results prior to 1916, see *Millikan 1916b*, § 1.

[54] In his review, *Jeans 1914*, pp. 58–65, Jeans concluded that the linear proportionality between energy and frequency in eq. (4) is valid, but that the proportionality constant varies from metal to metal.

[55] See *Millikan 1916a, 1916b*.

[56] For a discussion of such explanations, see *Stuewer 1970*. Foremost among them was the "triggering hypothesis," first enunciated in *Lenard 1902* to account for the independence of electron emission velocity from light intensity,

but argued most forcefully by J. J. Thomson (see, e.g., *Thomson, J. J. 1905*, pp. 588–589). For discussions of the triggering hypothesis, see *McCormmach 1967* and *Wheaton 1978a, 1978b*. Einstein noted this hypothesis in his address to the Salzburg meeting, but did not discuss it "because this hypothesis is once again just about abandoned" ("[w]eil diese Hypothese bereits wieder so ziemlich verlassen ist") (*Einstein 1909c* [Doc. 60], p. 491).

[57] This model was used by Boltzmann to explain the Dulong-Petit rule for specific heats (see *Boltzmann 1871b*). He interpreted deviations from the law as indications that the atoms of the lattice are not harmonically bound.

[58] H. F. Weber, Einstein's physics professor, performed pioneering research on the deviation of the specific heats of several solids from the Dulong-Petit rule at decreasing temperatures, notably on diamond, which does not obey the rule at room temperature (see *Weber, H. F.*

the selective absorption of radiation.[59] Perhaps inspired by Planck's work, in 1901 Einstein wondered whether the internal kinetic energy of solids and liquids could be conceived of as "the energy of electric resonators" ("elektrische Resonatorenenergie"). If it could, then the "specific heat and absorption spectrum of bodies must then be related" ("[e]s müßten dann spezifische Wärme und Absorptionsspektrum der Körper zusammenhängen").[60] He tried to connect this model with deviations from the Dulong-Petit rule.[61]

In 1907 Einstein avoided the implications of the equipartition theorem for the specific heats of solids as he had avoided them for radiation theory, by introducing energy quanta. From the average energy of a quantized oscillator, Einstein derived an expression for the specific heat of a monatomic solid as a function of $\beta T/\nu$. The expression approaches zero with the temperature, and approaches the Dulong-Petit value at high temperatures. Considering the simplified nature of the model, the expression is in fairly good agreement with Weber's data on diamond.

A connection could be made with absorption results from Drude's optical dispersion theory.[62] Drude showed that the infrared optical eigenfrequencies of a solid are due to vibrations of the lattice ions, while the electrons are responsible for the ultraviolet eigenfrequencies. At room temperature, the value of Einstein's expression for specific heat effectively vanishes at a frequency well within the infrared region for most solids, and increases to the Dulong-Petit value for even smaller frequencies. Einstein concluded that only the lattice ions and atoms contribute to the specific heats of solids. Moreover, if a solid displays infrared absorption resonances, the temperature dependence of its specific heat can be determined from these resonant frequencies.[63]

In 1910 Nernst and his assistant Frederick A. Lindemann obtained general agreement between Einstein's predictions and observations of the variation with temperature of the specific heat of a number of solids.[64] In 1911 Nernst reported the first confirmation of the quantum hypothesis outside the field of radiation: "That the observations in their totality provide a brilliant confirmation of the quantum theory of Planck and Einstein is obvious" ("Daß in ihrer Gesamtheit die Beobachtungen eine glänzende Bestätigung der Quantentheorie von Planck und Einstein erbringen, liegt auf der Hand").[65]

1875, 1887). Einstein 1907a (Doc. 38), p. 189, cites Weber's results for diamond, taking them from tables in *Landolt and Börnstein 1905*.

[59] For a discussion of his earlier interest in such a connection, see Vol. 1, the editorial note, "Einstein on Thermal, Electrical, and Radiation Phenomena," pp. 235–237.

[60] Einstein to Mileva Marić, 23 March 1901 (Vol. 1, Doc. 93).

[61] See Einstein to Mileva Marić, 23 March 1901 and 27 March 1901 (Vol. 1, Docs. 93 and 94).

[62] See *Drude 1904a, 1904b*. For a discussion of Einstein's earlier interest in Drude's electron theory of metals, see Vol. 1, the editorial note, "Einstein on Thermal, Electrical, and Radiation Phenomena," pp. 235–237.

[63] Einstein later noted that, if the lattice atoms of the solid are not ionized, they will not be optically active (see *Einstein 1907d* [Doc. 42]).

[64] See *Nernst 1911a, 1911b, 1911c*. Nernst personally informed Einstein of the confirmation (see Einstein to Jakob Laub, 16 March 1910, and Einstein to Arnold Sommerfeld, July 1910). Einstein also informed Sommerfeld (see ibid.) of a recent confirmation of the predicted infrared absorption maximum for diamond.

[65] *Nernst 1911c*, p. 310. For further discus-

V

Einstein maintained that his analysis of Planck's law showed the need to modify the foundations of both electrodynamics and mechanics. In early 1908 he wrote: "I believe moreover that we are still far from having satisfactory elementary foundations for electrical and mechanical processes" ("Ich glaube übrigens, dass wir noch weit davon entfernt sind, befriedigende elementare Grundlagen für die elektrischen und mechanischen Vorgänge zu besitzen").[66]

Most leading theoreticians, however, were still not even convinced of the need to introduce energy quanta.[67] Doubts about the validity of the equipartition theorem were still expressed.[68] In 1906 Planck attempted to invalidate the ergodic assumption underlying the equipartition theorem by introducing a lower limit to the size of phase space cells of equal probability, their size being fixed by the quantum of action h.[69] Another obstacle to acceptance of the quantum hypothesis was the attempt to defend the validity of the Rayleigh-Jeans law.[70] H. A. Lorentz, universally recognized as an authority on theoretical issues, originally hoped to prove that the equipartition theorem simply does not apply to matter-free radiation.[71] He demonstrated just the opposite, however, in *Lorentz 1908a*, arriving at the Rayleigh-Jeans law and calling for new experiments to decide between it and Planck's law. His paper prompted a vigorous discussion of the radiation problem in the pages of the *Physikalische Zeitschrift*. Lummer and Pringsheim emphatically reiterated that experiment had already decided in Planck's favor,[72] Jeans defended his position on equipartition,[73] and Walter Ritz asserted that the exclusive use of retarded potentials would restrict the equipartition of energy.[74] *Lorentz 1908b* accepts the criticism of the experimentalists, admitting "that a derivation of the radiation laws from the electron theory will hardly be possible without a profound modification in the foundations of the latter" ("daß eine Ableitung der Strahlungsgesetze aus der Elektronentheorie schwerlich ohne tiefgehende Änderung ihrer Grundlagen möglich sein wird").[75] Lorentz's statement

sion of Einstein's work on specific heats and its reception, see *Klein 1965*; *Kuhn 1978*, pp. 210–220; *Pais 1982*, chap. 20; and *Mehra and Rechenberg 1982*, pp. 113–136.

[66] Einstein to Arnold Sommerfeld, 14 January 1908. See the Introduction, pp. xxi–xxix, for further discussion of Einstein's search for "satisfactory elementary foundations."

[67] Ehrenfest is a noteworthy exception. See, e.g., *Ehrenfest 1906*.

[68] See, e.g., *Rayleigh 1905a, 1905b*. For a discussion of earlier doubts, see *Brush 1976*, vol. 2, pp. 356–363.

[69] See *Planck 1906c*, pp. 154–156, 178. Planck's attempt is discussed in the editorial note, "Einstein on the Foundations of Statistical Physics," p. 49. *Planck 1906c* is further discussed in *Kuhn 1978*, pp. 114–134, and *Jungnickel and McCormmach 1986b*, pp. 265–

268.

[70] For example, Jeans argued that the law does not hold for laboratory black-body radiation because such radiation has not yet reached thermal equilibrium (see *Jeans 1905c*, p. 293). Rayleigh had stated in 1900 that "although for some reason not yet explained the doctrine [of equipartition] fails in general, it seems possible that it may apply to the graver modes [long wavelengths]" (*Rayleigh 1900*, p. 540). For a discussion of attempts to account for the failure to obtain Planck's law from the equipartition theorem, see *Garber 1976*.

[71] Lorentz stated this hope in *Lorentz 1908a*, p. 19, and *Lorentz 1908b*, p. 562.

[72] See *Lummer and Pringsheim 1908*.

[73] See *Jeans 1908*.

[74] See *Ritz 1908b*.

[75] *Lorentz 1908b*, p. 562. See also Hendrik

was an important turning point leading to the general recognition that Planck's law implies fundamental changes in the foundations of physics.[76]

Even those who accepted the need for some form of energy quantization objected to the concept of light quanta, since numerous optical phenomena seem to require a wave theory for their explanation.[77] In 1907 Planck wrote Einstein: "I am not looking for the meaning of the elementary quantum of action (light quantum) in the vacuum, but at the places of absorption and emission, and assume that vacuum processes are *exactly* described by Maxwell's equations" ("ich suche die Bedeutung des elementaren Wirkungsquantums (Lichtquants) nicht im Vakuum, sondern an der Stelle der Absorption u. Emission, und nehme an, daß die Vorgänge im Vakuum durch die Maxwellsche Gleichungen *genau* dargestellt werden").[78]

Of Einstein's German-speaking colleagues, only the experimentalist Johannes Stark supported the idea of an atomistic structure of radiation in 1909.[79] He argued for the existence of localized energy quanta in X-rays, in opposition to the prevailing electromagnetic wave-pulse interpretation.[80] He observed that the absorption of X-rays by a metal results in the ejection of electrons of almost the same energy as the cathode rays producing the X-rays, no matter how great the distance traveled by the X-rays between production and absorption.[81]

Although Einstein was familiar with *Lorentz 1908b* and the discussion papers of Jeans and Ritz, he had not read *Lorentz 1908a* before publication of *Einstein 1909b* (Doc. 56), his contribution to the discussion initiated by Lorentz's paper.[82] Shortly afterwards, he sent Lorentz a copy of his paper, describing the work as "the insignificant result of years of reflection" ("das geringfügige Ergebnis von jahrelangem Nachdenken"), adding: "I have not been able to attain a real grasp of the matter" ("Zu einem wirklichen Erfassen der Sache habe ich nicht vordringen können").[83] The paper presents

---

Lorentz to Wilhelm Wien, 6 June 1908, quoted in part in *Kuhn 1978*, p. 302, note 19.

[76] For further discussion of *Lorentz 1908a* and its aftermath, see *Kuhn 1978*, pp. 189–205.

[77] Of particular concern were observations of interference effects with radiation of even the weakest intensity (see, e.g., *Taylor 1909*). For this argument against light quanta, see Hendrik Lorentz to Einstein, 6 May 1909. For a summary of the views of several German theorists on this question, see *Jungnickel and McCormmach 1986b*, pp. 305–306.

[78] Max Planck to Einstein, 6 July 1907. See also Planck's comment on *Einstein 1909c* (Doc. 60), in *Einstein et al. 1909c* (Doc. 61). Laue wrote Einstein in a similar vein a year earlier (see Max Laue to Einstein, 2 June 1906). Planck's struggle with the foundations of his radiation formula throughout this period are examined in *Needell 1980*.

[79] See *Stark 1909a*, and Johannes Stark to

Einstein, 8 April 1909. For a discussion of Stark's research on X-rays, including his support of the quantum hypothesis, see *Wheaton 1978a*, pp. 125–143.

[80] For a discussion of the "pulse theory," see *Wheaton 1978a*.

[81] Einstein referred to this phenomenon in support of the light quantum hypothesis during his Salzburg lecture, and Stark reviewed his findings during the discussion of Einstein's lecture (see *Einstein 1909c* [Doc. 60], p. 492, and Stark's comment in *Einstein et al. 1909c* [Doc. 61]). Einstein continued the discussion of X-rays in a letter to Arnold Sommerfeld of 19 January 1910.

[82] This paper was published on 15 March 1909. He wrote Lorentz on 13 April 1909 that he had just read *Lorentz 1909a*—a slight revision of *Lorentz 1908a*—in terms indicating that he had not previously read *Lorentz 1908a*.

[83] Einstein to Hendrik Lorentz, 30 March

several considerations, from which it follows for me that not only the mechanics of molecules, but also Maxwell-Lorentz electrodynamics cannot be brought into harmony with the [Planck] radiation formula.

einige Betrachtungen, aus denen für mich hervorgeht, dass nicht nur die Mole-kularmechanik sondern auch die Maxwell-Lorentz'sche Elektrodynamik mit der Strahlungsformel nicht in Einklang gebracht werden kann.[84]

After replying in turn to Ritz and Jeans,[85] Einstein stated, more clearly than he had pre-viously done so, what he regarded as the deviations from proper statistical foundations in Planck's derivation of the radiation law.[86] The major novelties of the paper, however, are two arguments for the existence of light quanta based upon the analysis of fluctuations in black-body radiation. As noted in section II, Einstein did not apply his canonical fluctua-tion formula, perhaps to forestall doubts about the applicability of this formula to radia-tion.[87] Instead, he applied an equivalent formula based on Boltzmann's principle, a prin-ciple he held to be universally valid. In addition to energy fluctuations, Einstein calculated fluctuations in the radiation pressure by studying the Brownian motion of a small, two-sided mirror in a radiation-filled cavity.[88] In both instances, Planck's law yields expres-sions for the fluctuations that are the sum of two terms, which "behave like fluctuations (errors) which arise from causes independent of one another" ("sich verhalten wie Schwankungen (Fehler), welche voneinander unabhängigen Ursachen entspringen").[89] One of the terms, which dominates at low frequencies, Einstein interpreted as due to the interference of independent, random waves; the other, which dominates at high frequen-cies, he interpreted as due to "complexes of energy hv of small extension, moving inde-pendently of one another" ("von einander unabhängig beweglichen, wenig ausgedehnten Komplexen von der Energie hv").[90] It is not enough to assume that the emission and absorption of light take place via energy quanta; the structure of the radiation itself shows evidence of such quanta. Although light quanta cannot be deduced from Planck's law, he concluded, "one can indeed assert that the quantum theory offers . . . the simplest inter-pretation" ("[m]an kann aber wohl behaupten, daß die Quantentheorie die einfachste Interpretation . . . liefert") of this law.[91]

1909. Presumably the "question" ("Sache") to which he refers is the "Strahlungsproblem" in the title of the paper.

[84] Einstein to Hendrik Lorentz, 30 March 1909.

[85] See *Einstein 1909b* (Doc. 56), pp. 185–187. Einstein's response to Ritz occasioned a re-joinder by Ritz, *Ritz 1909*, and a joint statement of their disagreement, *Ritz and Einstein 1909* (Doc. 57).

[86] See *Einstein 1909b* (Doc. 56), pp. 187–188.

[87] Einstein raised the question of the appli-

cability of statistical mechanics to thermal radia-tion in *Einstein 1909b* (Doc. 56), p. 186. *Ein-stein and Hopf 1910a*, *1910b* dispose of such doubts.

[88] Einstein later stated that in 1905 he knew Maxwell's theory leads to an incorrect expres-sion for the fluctuations of the radiation pressure on a mirror (see Einstein to Max von Laue, 17 January 1952).

[89] *Einstein 1909b* (Doc. 56), p. 190.

[90] *Einstein 1909b* (Doc. 56), p. 190.

[91] *Einstein 1909b* (Doc. 56), p. 191.

VI

The precise circumstances of Einstein's invitation to address the eighty-first meeting of the Gesellschaft Deutscher Naturforscher und Ärzte in Salzburg are unknown. The Gesellschaft, founded in 1822 to promote the unity of all natural sciences (including medicine) and all German-speaking scientists, had over three thousand members.[92] Einstein attended the Salzburg meeting, the first international conference at which he spoke, as an "honored guest" ("Ehrengast") of the Deutsche Physikalische Gesellschaft.[93] While in Salzburg, he met many of the leading figures of physics for the first time.[94] His talk was given to the Gesellschaft's "Section for Physics, including instruments and scientific photography" ("Abteilung für Physik, einschl. Instrumentenkunde und wissenschaftliche Photographie") on the afternoon of 21 September.[95]

According to one report, Einstein's "broadly conceived lecture" ("grossangelegten Vortrag") at Salzburg was "the main event of the physics section of this conference of scientists" ("das Hauptereignis der physikalischen Sektion dieses Naturforschertages").[96] Einstein opened the session with an address "On the Development of Our Views Concerning the Nature and Constitution of Radiation."[97] *Einstein 1909c* (Doc. 60), received shortly after the meeting, is the written version of his talk. It is ostensibly merely a review of his work on light over the preceding years.[98] Yet it is much more: it is the first synthetic account of the profound transformation in the concept of light brought about by the theory of relativity and the quantum hypothesis, and of the profound implications of this tranformation for the development of physics. Einstein maintained that the long-held ether hypothesis had, by 1909, been reduced to "a superseded position" ("einen überwundenen Standpunkt"); and that light, long regarded as completely described by a wave theory, possessed properties that were understandable only from the standpoint of an emission theory.[99] He elaborated on both claims for an audience that was favorably disposed toward the first assertion, but still skeptical of the second.

Einstein focused first upon the problems of the electrodynamics of moving bodies, and the resolution by the relativity principle of the previously "extremely unsatisfactory" ("höchst unbefriedigende")[100] situation in this field. The elimination of the ether implies

---

[92] See *Verhandlungen 1910*, part 2, p. 12.

[93] Recruitment Commission for the Chair of Mathematical Physics to Members of the Philosophical Faculty, University of Prague, 21 April 1910 (Cz-Ar, Einstein dossier).

[94] Einstein reported that he met for the first time Planck, Wien, Rubens, and Sommerfeld, of whom he was particularly fond (Einstein to Jakob Laub, 31 December 1909; see also Einstein to Arnold Sommerfeld, 29 September 1909).

[95] See *Versammlung 1909a*, part 2, pp. 26, 41.

[96] Recruitment Commission to Philosophical Faculty, University of Prague (see note 93).

[97] According to *Verhandlungen 1910*, part 2, p. 41, the title of Einstein's lecture was "On the More Recent Transformations Which Our Conceptions of the Nature of Light have Undergone" ("Über die neueren Umwandlungen, welche unsere Anschauungen über die Natur des Lichtes erfahren haben"). The text of *Einstein 1909c* (Doc. 60) was also printed in the *Physikalische Zeitschrift*.

[98] Einstein failed to send a copy to Laub "because it contains nothing new" ("weil sie nichts Neue[s] enthält") (Einstein to Jakob Laub, 31 December 1909).

[99] See *Einstein 1909c* (Doc. 60), p. 482.

[100] Ibid., p. 486.

that light propagating through empty space consists of electromagnetic fields behaving as "independent structures" ("selbständige Gebilde").[101] Moreover, "according to this theory [of relativity], light has the characteristic in common with corpuscular theory of transferring inertial mass from the emitting to the absorbing body" ("hat nach dieser Theorie mit einer Korpuskulartheorie des Lichtes das Merkmal gemeinsam, träge Masse vom emittierenden zum absorbierenden Körper zu übertragen").[102] In view of the presence of both wave and corpuscular terms in fluctuations of black-body radiation (see section V), he argued that a new "mathematical theory of radiation" ("mathematische Theorie der Strahlung") is needed, which "can be considered as a sort of fusion of the wave and the emission theory of light" ("sich als eine Art Verschmelzung von Undulations- und Emissionstheorie des Lichtes auffassen läßt").[103]

Einstein 1909b (Doc. 56) and *Einstein 1909c* (Doc. 60), together with Einstein's contemporary correspondence, report some of Einstein's intense, but unsuccessful, efforts to find an appropriate theory that would serve as such a "fusion."[104] Rather than maintaining his earlier inclination toward a corpuscular theory, Einstein now suggested that perhaps some nonlinear modification of the Maxwell field equations would yield not only the quantum of electric charge *e* as a consequence, rather than an assumption of the theory, but also the quantum structure of radiation. He suggested that the linear, homogeneous optical wave equation should be replaced by a nonlinear or inhomogeneous wave equation containing *e* as a coefficient.[105]

In a lengthy letter to Einstein, Lorentz criticized Einstein's search for a modification of electrodynamics that would encompass both electrons and quanta, remaining unconvinced of the existence of independent light quanta.[106] In his reply,[107] and in his Salzburg lecture, Einstein described further efforts to find a suitable modification. By the time of the Salzburg meeting he had hit upon an idea that allowed for both interference and individuality: in analogy to the electrostatic field surrounding an electron, a light quantum could be treated as a mathematical singularity surrounded by an extended vector field, all of the energy of the field being concentrated in such singularities.[108] He suggested that, if many singularities lie close together, their fields overlap to produce the effect of a continuous wave field. By this example Einstein attempted to show that the wave and quantum aspects of radiation, which he firmly believed to be inherent in Planck's law, are not necessarily incompatible. The problem of finding the correct relation between these two aspects of radiation dominated his subsequent work on the quantum hypothesis.

[101] Ibid., p. 487.

[102] Ibid., p. 490.

[103] Ibid., pp. 499 and 483.

[104] In addition to the papers cited, see, in particular: Einstein to Arnold Sommerfeld, 19 January 1910; Hendrik Lorentz to Einstein, 6 May 1909; Einstein to Jakob Laub, 19 May 1909; Einstein to Hendrik Lorentz, 23 May 1909; Einstein to Johannes Stark, 31 July 1909; Einstein to Michele Besso, 31 December 1909. For discussions of Einstein's efforts, see *McCormmach 1970a* and *Klein 1967*.

[105] See *Einstein 1909b* (Doc. 56), p. 192.

[106] See Hendrik Lorentz to Einstein, 6 May 1909.

[107] See Einstein to Hendrik Lorentz, 23 May 1909.

[108] See *Einstein 1909c* (Doc. 60), pp. 499–500, and *Einstein et al. 1909c* (Doc. 61). A year later, Lorentz still expressed doubts about light quanta, primarily because of the interference of very weak radiation (*Lorentz 1910*, pp. 1249–1250).

14. "On a Heuristic Point of View Concerning the Production and Transformation of Light"

[*Einstein 1905i*]

DATED Bern, 17 March 1905
RECEIVED 18 March 1905
PUBLISHED 9 June 1905

IN: *Annalen der Physik* 17 (1905): 132–148.

132

[1]

6. *Über einen die Erzeugung und Verwandlung des Lichtes betreffenden heuristischen Gesichtspunkt; von A. Einstein.*

Zwischen den theoretischen Vorstellungen, welche sich die Physiker über die Gase und andere ponderable Körper gebildet haben, und der Maxwellschen Theorie der elektromagnetischen Prozesse im sogenannten leeren Raume besteht ein tiefgreifender formaler Unterschied. Während wir uns nämlich den Zustand eines Körpers durch die Lagen und Geschwindigkeiten einer zwar sehr großen, jedoch endlichen Anzahl von Atomen und Elektronen für vollkommen bestimmt ansehen, bedienen wir uns zur Bestimmung des elektromagnetischen Zustandes eines Raumes kontinuierlicher räumlicher Funktionen, so daß also eine endliche Anzahl von Größen nicht als genügend anzusehen ist zur vollständigen Festlegung des elektromagnetischen Zustandes eines Raumes. Nach der Maxwellschen Theorie ist bei allen rein elektromagnetischen Erscheinungen, also auch beim Licht, die Energie als kontinuierliche Raumfunktion aufzufassen, während die Energie eines ponderabeln Körpers nach der gegenwärtigen Auffassung der Physiker als eine über die Atome und Elektronen erstreckte Summe darzustellen ist. Die Energie eines ponderabeln Körpers kann nicht in beliebig viele, beliebig kleine Teile zerfallen, während sich die Energie eines von einer punktförmigen Lichtquelle ausgesandten Lichtstrahles nach der Maxwellschen Theorie (oder allgemeiner nach jeder Undulationstheorie) des Lichtes auf ein stets wachsendes Volumen sich kontinuierlich verteilt.

[2]

Die mit kontinuierlichen Raumfunktionen operierende Undulationstheorie des Lichtes hat sich zur Darstellung der rein optischen Phänomene vortrefflich bewährt und wird wohl nie durch eine andere Theorie ersetzt werden. Es ist jedoch im Auge zu behalten, daß sich die optischen Beobachtungen auf zeitliche Mittelwerte, nicht aber auf Momentanwerte beziehen, und es ist trotz der vollständigen Bestätigung der Theorie der Beugung, Reflexion, Brechung, Dispersion etc. durch das

[3]

Erzeugung und Verwandlung des Lichtes. 133

Experiment wohl denkbar, daß die mit kontinuierlichen Raumfunktionen operierende Theorie des Lichtes zu Widersprüchen
mit der Erfahrung führt, wenn man sie auf die Erscheinungen
der Lichterzeugung und Lichtverwandlung anwendet.

Es scheint mir nun in der Tat, daß die Beobachtungen
über die „schwarze Strahlung", Photolumineszenz, die Er-   [4]
zeugung von Kathodenstrahlen durch ultraviolettes Licht und
andere die Erzeugung bez. Verwandlung des Lichtes betreffende
Erscheinungsgruppen besser verständlich erscheinen unter der   [5]
Annahme, daß die Energie des Lichtes diskontinuierlich im
Raume verteilt sei. Nach der hier ins Auge zu fassenden
Annahme ist bei Ausbreitung eines von einem Punkte ausgehenden Lichtstrahles die Energie nicht kontinuierlich auf
größer und größer werdende Räume verteilt, sondern es besteht dieselbe aus einer endlichen Zahl von in Raumpunkten
lokalisierten Energiequanten, welche sich bewegen, ohne sich
zu teilen und nur als Ganze absorbiert und erzeugt werden
können.

Im folgenden will ich den Gedankengang mitteilen und
die Tatsachen anführen, welche mich auf diesen Weg geführt
haben, in der Hoffnung, daß der darzulegende Gesichtspunkt
sich einigen Forschern bei ihren Untersuchungen als brauchbar erweisen möge.

§ 1. Über eine die Theorie der „schwarzen Strahlung" betreffende Schwierigkeit.

Wir stellen uns zunächst auf den Standpunkt der **Maxwell**schen Theorie und Elektronentheorie und betrachten folgenden Fall. In einem von vollkommen reflektierenden Wänden
eingeschlossenen Raumes befinde sich eine Anzahl Gasmoleküle und Elektronen, welche freibeweglich sind und aufeinander
konservative Kräfte ausüben, wenn sie einander sehr nahe
kommen, d. h. miteinander wie Gasmoleküle nach der kinetischen Gastheorie zusammenstoßen können.[1]) Eine Anzahl

---

1) Diese Annahme ist gleichbedeutend mit der Voraussetzung, daß
die mittleren kinetischen Energien von Gasmolekülen und Elektronen bei
Temperaturgleichgewicht einander gleich seien. Mit Hilfe letzterer Voraussetzung hat Hr. **Drude** bekanntlich das Verhältnis von thermischem und   [6]
elektrischem Leitungsvermögen der Metalle auf theoretischem Wege abgeleitet.

134   *A. Einstein.*

Elektronen sei ferner an voneinander weit entfernte Punkte
des Raumes gekettet durch nach diesen Punkten gerichtete,
den Elongationen proportionale Kräfte. Auch diese Elektronen
sollen mit den freien Molekülen und Elektronen in konservative
Wechselwirkung treten, wenn ihnen letztere sehr nahe kommen.
Wir nennen die an Raumpunkte geketteten Elektronen „Reso-
[7]   natoren"; sie senden elektromagnetische Wellen bestimmter
Periode aus und absorbieren solche.

Nach der gegenwärtigen Ansicht über die Entstehung des
Lichtes müßte die Strahlung im betrachteten Raume, welche
unter Zugrundelegung der Maxwellschen Theorie für den Fall
des dynamischen Gleichgewichtes gefunden wird, mit der
[8]   „schwarzen Strahlung" identisch sein — wenigstens wenn
Resonatoren aller in Betracht zu ziehenden Frequenzen als
[9]   vorhanden angesehen werden.

Wir sehen vorläufig von der von den Resonatoren emittierten und absorbierten Strahlung ab und fragen nach der
der Wechselwirkung (den Zusammenstößen) von Molekülen und
Elektronen entsprechenden Bedingung für das dynamische
Gleichgewicht. Die kinetische Gastheorie liefert für letzteres
die Bedingung, daß die mittlere lebendige Kraft eines Resonatorelektrons gleich der mittleren kinetischen Energie der fortschreitenden Bewegung eines Gasmoleküles sein muß. Zerlegen wir die Bewegung des Resonatorelektrons in drei aufeinander senkrechte Schwingungsbewegungen, so finden wir
für den Mittelwert $\bar{E}$ der Energie einer solchen geradlinigen
Schwingungsbewegung

$$\bar{E} = \frac{R}{N} T,$$

wobei R die absolute Gaskonstante, N die Anzahl der „wirk-
[10]   lichen Moleküle" in einem Grammäquivalent und T die absolute Temperatur bedeutet. Die Energie $\bar{E}$ ist nämlich wegen
der Gleichheit der zeitlichen Mittelwerte von kinetischer und
potentieller Energie des Resonators $^2/_3$ mal so groß wie die
lebendige Kraft eines freien, einatomigen Gasmoleküles. Würde
nun durch irgend eine Ursache — in unserem Falle durch
Strahlungsvorgänge — bewirkt, daß die Energie eines Resonators einen größeren oder kleineren zeitlichen Mittelwert als
$\bar{E}$ besitzt, so würden die Zusammenstöße mit den freien Elek-

tronen und Molekülen zu einer im Mittel von Null verschiedenen Energieabgabe an das Gas bez. Energieaufnahme von
dem Gas führen. Es ist also in dem von uns betrachteten
Falle dynamisches Gleichgewicht nur dann möglich, wenn jeder
Resonator die mittlere Energie $\bar{E}$ besitzt.

Eine ähnliche Überlegung machen wir jetzt bezüglich der
Wechselwirkung der Resonatoren und der im Raume vorhandenen Strahlung. Hr. Planck hat für diesen Fall die Bedingung des dynamischen Gleichgewichtes abgeleitet[1]) unter
der Voraussetzung, daß die Strahlung als ein denkbar ungeordnetster Prozeß[2]) betrachtet werden kann. Er fand:

$$\bar{E}_\nu = \frac{L^3}{8\pi\nu^2}\,\varrho_\nu.$$ [14]

$\bar{E}_\nu$ ist hierbei die mittlere Energie eines Resonators von der
Eigenfrequenz ν (pro Schwingungskomponente), L die Lichtgeschwindigkeit, ν die Frequenz und $\varrho_\nu\,d\nu$ die Energie pro
Volumeinheit desjenigen Teiles der Strahlung, dessen Schwingungszahl zwischen ν und $\nu + d\nu$ liegt.

---

1) M. Planck, Ann. d. Phys. **1.** p. 99. 1900.   [11]

2) Diese Voraussetzung läßt sich folgendermaßen formulieren. Wir   [12]
entwickeln die Z-Komponente der elektrischen Kraft (Z) in einem beliebigen Punkte des betreffenden Raumes zwischen den Zeitgrenzen $t = 0$
und $t = T$ (wobei T eine relativ zu allen in Betracht zu ziehenden
Schwingungsdauern sehr große Zeit bedeute) in eine Fouriersche Reihe

$$Z = \sum_{\nu=1}^{\nu=\infty} A_\nu \sin\left(2\pi\nu\,\frac{t}{T} + \alpha_\nu\right),$$

wobei $A_\nu \geqq 0$ und $0 \leqq \alpha_\nu \leqq 2\pi$. Denkt man sich in demselben Raumpunkte eine solche Entwickelung beliebig oft bei zufällig gewählten An-
fangspunkten der Zeit ausgeführt, so wird man für die Größen A_ν und α_ν
verschiedene Wertsysteme erhalten. Es existieren dann für die Häufigkeit der verschiedenen Wertekombinationen der Größen A_ν und α_ν
(statistische) Wahrscheinlichkeiten dW von der Form:

$$dW = f(A_1 A_2 \ldots \alpha_1 \alpha_2 \ldots)\,dA_1\,dA_2 \ldots d\alpha_1\,d\alpha_2 \ldots$$

Die Strahlung ist dann eine denkbar ungeordnetste, wenn

$$f(A_1, A_2 \ldots \alpha_1, \alpha_2 \ldots) = F_1(A_1)\,F_2(A_2) \ldots f_1(\alpha_1) \cdot f_2(\alpha_2) \ldots,$$

d. h. wenn die Wahrscheinlichkeit eines bestimmten Wertes einer der
Größen A bez. α von den Werten, welche die anderen Größen A bez. x [13]
besitzen, unabhängig ist. Mit je größerer Annäherung die Bedingung
erfüllt ist, daß die einzelnen Paare von Größen A_ν und α_ν von Emissions-
und Absorptionsprozessen *besonderer* Resonatorengruppen abhängen, mit
desto größerer Annäherung wird also in dem von uns betrachteten Falle
die Strahlung als eine „denkbar ungeordnetste" anzusehen sein.

136   *A. Einstein.*

Soll die Strahlungsenergie von der Frequenz ν nicht beständig im Ganzen weder vermindert noch vermehrt werden, so muß gelten:

$$\frac{R}{N} T = \bar{E} = E_\nu = \frac{L^3}{8\pi \nu^2} \varrho_\nu,$$

[15]
$$\varrho_\nu = \frac{R}{N} \frac{8\pi \nu^2}{L^3} T.$$

[16] Diese als Bedingung des dynamischen Gleichgewichtes gefundene Beziehung entbehrt nicht nur der Übereinstimmung mit der Erfahrung, sondern sie besagt auch, daß in unserem Bilde von einer bestimmten Energieverteilung zwischen Äther und Materie nicht die Rede sein kann. Je weiter nämlich der Schwingungszahlenbereich der Resonatoren gewählt wird, desto größer wird die Strahlungsenergie des Raumes, und wir erhalten in der Grenze

[17]
$$\int_0^\infty \varrho_\nu \, d\nu = \frac{R}{N} \frac{8\pi}{L^3} T \int_0^\infty \nu^2 \, d\nu = \infty.$$

[18]   § 2. **Über die Plancksche Bestimmung der Elementarquanta.**

Wir wollen im folgenden zeigen, daß die von Hrn. **Planck** gegebene Bestimmung der Elementarquanta von der von ihm aufgestellten Theorie der „schwarzen Strahlung" bis zu einem gewissen Grade unabhängig ist.

Die allen bisherigen Erfahrungen genügende **Planck**sche Formel[1]) für ϱ_ν lautet

[20]
$$\varrho_\nu = \frac{\alpha \nu^3}{e^{\frac{\beta \nu}{T}} - 1},$$

wobei

$$\alpha = 6{,}10 \; . \, 10^{-56},$$

$$\beta = 4{,}866 . \, 10^{-11}.$$

Für große Werte von T/ν, d. h. für große Wellenlängen und Strahlungsdichten geht diese Formel in der Grenze in folgende über:

$$\varrho_\nu = \frac{\alpha}{\beta} \nu^2 T.$$

---

[19]   1) M. **Planck**, Ann. d. Phys. 4. p. 561. 1901.

Man erkennt, daß diese Formel mit der in § 1 aus der Maxwellschen und der Elektronentheorie entwickelten übereinstimmt. Durch Gleichsetzung der Koeffizienten beider Formeln erhält man:

$$\frac{R}{N}\frac{8\pi}{L^3} = \frac{\alpha}{\beta}$$

oder

$$N = \frac{\beta}{\alpha}\frac{8\pi R}{L^3} = 6,17 \cdot 10^{23},$$

d. h. ein Atom Wasserstoff wiegt $1/N$ Gramm $= 1,62 \cdot 10^{-24}$ g. Dies ist genau der von Hrn. Planck gefundene Wert, welcher mit den auf anderen Wegen gefundenen Werten für diese Größe befriedigend übereinstimmt.   [21]

Wir gelangen daher zu dem Schlusse: Je größer die Energiedichte und die Wellenlänge einer Strahlung ist, als um so brauchbarer erweisen sich die von uns benutzten theoretischen Grundlagen; für kleine Wellenlängen und kleine Strahlungsdichten aber versagen dieselben vollständig.

Im folgenden soll die „schwarze Strahlung" im Anschluß an die Erfahrung ohne Zugrundelegung eines Bildes über die Erzeugung und Ausbreitung der Strahlung betrachtet werden.

§ 3. Über die Entropie der Strahlung.

Die folgende Betrachtung ist in einer berühmten Arbeit des Hrn. W. Wien enthalten und soll hier nur der Vollständigkeit halber Platz finden.   [22]

Es liege eine Strahlung vor, welche das Volumen v einnehme. Wir nehmen an, daß die wahrnehmbaren Eigenschaften der vorliegenden Strahlung vollkommen bestimmt seien, wenn die Strahlungsdichte $\varrho(\nu)$ für alle Frequenzen gegeben ist.[1)] Da Strahlungen von verschiedenen Frequenzen als ohne Arbeitsleistung und ohne Wärmezufuhr voneinander trennbar anzusehen sind, so ist die Entropie der Strahlung in der Form

$$S = v\int_0^\infty \varphi(\varrho, \nu)\, d\nu$$

darstellbar, wobei φ eine Funktion der Variabeln ϱ und ν

---

1) Diese Annahme ist eine willkürliche. Man wird naturgemäß an dieser einfachsten Annahme so lange festhalten, als nicht das Experiment dazu zwingt, sie zu verlassen.

bedeutet. Es kann φ auf eine Funktion von nur einer Variabeln reduziert werden durch Formulierung der Aussage, daß durch adiabatische Kompression einer Strahlung zwischen spiegelnden Wänden, deren Entropie nicht geändert wird. Wir wollen jedoch hierauf nicht eintreten, sondern sogleich untersuchen, wie die Funktion φ aus dem Strahlungsgesetz des schwarzen Körpers ermittelt werden kann.

Bei der „schwarzen Strahlung" ist ϱ eine solche Funktion von ν, daß die Entropie bei gegebener Energie ein Maximum ist, d. h. daß

$$\delta \int_0^\infty \varphi(\varrho, \nu)\, d\nu = 0\,,$$

wenn

$$\delta \int_0^\infty \varrho\, d\nu = 0\,.$$

Hieraus folgt, daß für jede Wahl des $\delta \varrho$ als Funktion von ν

$$\int_0^\infty \left(\frac{\partial \varphi}{\partial \varrho} - \lambda\right) \delta \varrho\, d\nu = 0\,,$$

wobei λ von ν unabhängig ist. Bei der schwarzen Strahlung ist also $\partial \varphi / \partial \varrho$ von ν unabhängig.

Für die Temperaturzunahme einer schwarzen Strahlung vom Volumen $\nu = 1$ um dT gilt die Gleichung:

$$dS = \int_{\nu=0}^{\nu=\infty} \frac{\partial \varphi}{\partial \varrho}\, d\varrho\, d\nu\,,$$

oder, da $\partial \varphi / \partial \varrho$ von ν unabhängig ist:

$$dS = \frac{\partial \varphi}{\partial \varrho}\, dE\,.$$

Da dE gleich der zugeführten Wärme und der Vorgang umkehrbar ist, so gilt auch:

$$dS = \frac{1}{T}\, dE\,.$$

Durch Vergleich erhält man:

$$\frac{\partial \varphi}{\partial \varrho} = \frac{1}{T}\,.$$

Dies ist das Gesetz der schwarzen Strahlung. Man kann also

aus der Funktion φ das Gesetz der schwarzen Strahlung und umgekehrt aus letzterem die Funktion φ durch Integration bestimmen mit Rücksicht darauf, daß φ für $\varrho = 0$ verschwindet.

§ 4. Grenzgesetz für die Entropie der monochromatischen Strahlung bei geringer Strahlungsdichte.

Aus den bisherigen Beobachtungen über die „schwarze Strahlung" geht zwar hervor, daß das ursprünglich von Hrn. W. Wien für die „schwarze Strahlung" aufgestellte Gesetz

$$\varrho = \alpha \, v^3 \, e^{-\beta \frac{v}{T}} \qquad [23]$$

nicht genau gültig ist. Dasselbe wurde aber für große Werte von v/T sehr vollkommen durch das Experiment bestätigt. [24] Wir legen diese Formel unseren Rechnungen zugrunde, behalten aber im Sinne, daß unsere Resultate nur innerhalb gewisser Grenzen gelten.

Aus dieser Formel ergibt sich zunächst:

$$\frac{1}{T} = -\frac{1}{\beta \, v} \lg \frac{\varrho}{\alpha \, v^3}$$

und weiter unter Benutzung der in dem vorigen Paragraphen gefundenen Beziehung:

$$\varphi \, (\varrho, v) = -\frac{\varrho}{\beta \, v} \left\{ \lg \frac{\varrho}{\alpha \, v^3} - 1 \right\}.$$

Es sei nun eine Strahlung von der Energie E gegeben, deren Frequenz zwischen v und $v + d\,v$ liegt. Die Strahlung nehme das Volumen v ein. Die Entropie dieser Strahlung ist:

$$S = v \, \varphi \, (\varrho, v) \, d\,v = -\frac{E}{\beta \, v} \left\{ \lg \frac{E}{v \, \alpha \, v^3 \, d\,v} - 1 \right\}. \qquad [25]$$

Beschränken wir uns darauf, die Abhängigkeit der Entropie von dem von der Strahlung eingenommenen Volumen zu untersuchen, und bezeichnen wir die Entropie der Strahlung mit S_0, falls dieselbe das Volumen v_0 besitzt, so erhalten wir:

$$S - S_0 = \frac{E}{\beta \, v} \lg \left(\frac{v}{v_0} \right).$$

Diese Gleichung zeigt, daß die Entropie einer monochromatischen Strahlung von genügend kleiner Dichte nach dem gleichen Gesetze mit dem Volumen variiert wie die Entropie eines idealen Gases oder die einer verdünnten Lösung. Die [26]

140   *A. Einstein.*

soeben gefundene Gleichung soll im folgenden interpretiert
werden unter Zugrundelegung des von Hrn. Boltzmann in
die Physik eingeführten Prinzips, nach welchem die Entropie
eines Systems eine Funktion der Wahrscheinlichkeit seines
[27] Zustandes ist.

§ 5. Molekulartheoretische Untersuchung der Abhängigkeit der Entropie von Gasen und verdünnten Lösungen vom Volumen.

Bei Berechnung der Entropie auf molekulartheoretischem
Wege wird häufig das Wort „Wahrscheinlichkeit" in einer
Bedeutung angewendet, die sich nicht mit der Definition der
[28] Wahrscheinlichkeit deckt, wie sie in der Wahrscheinlichkeitsrechnung gegeben wird. Insbesondere werden die „Fälle
gleicher Wahrscheinlichkeit" häufig hypothetisch festgesetzt
in Fällen, wo die angewendeten theoretischen Bilder bestimmt
genug sind, um statt jener hypothetischen Festsetzung eine
Deduktion zu geben. Ich will in einer besonderen Arbeit zeigen,
daß man bei Betrachtungen über thermische Vorgänge mit
der sogenannten „statistischen Wahrscheinlichkeit" vollkommen
[29] auskommt und hoffe dadurch eine logische Schwierigkeit zu
beseitigen, welche der Durchführung des Boltzmannschen
Prinzips noch im Wege steht. Hier aber soll nur dessen allgemeine Formulierung und dessen Anwendung auf ganz spezielle
Fälle gegeben werden.

Wenn es einen Sinn hat, von der Wahrscheinlichkeit eines
Zustandes eines Systems zu reden, wenn ferner jede Entropiezunahme als ein Übergang zu einem wahrscheinlicheren Zustande aufgefaßt werden kann, so ist die Entropie S_1 eines
Systems eine Funktion der Wahrscheinlichkeit W_1 seines
momentanen Zustandes. Liegen also zwei nicht miteinander
in Wechselwirkung stehende Systeme S_1 und S_2 vor, so kann
man setzen:

$$S_1 = \varphi_1 (W_1),$$
$$S_2 = \varphi_2 (W_2).$$

Betrachtet man diese beiden Systeme als ein einziges System
von der Entropie S und der Wahrscheinlichkeit W, so ist:

$$S = S_1 + S_2 = \varphi (W)$$

und

$$W = W_1 . W_2 .$$

Erzeugung und Verwandlung des Lichtes.   **141**

Die letztere Beziehung sagt aus, daß die Zustände der beiden Systeme voneinander unabhängige Ereignisse sind.

Aus diesen Gleichungen folgt:

$$\varphi(W_1 \cdot W_2) = \varphi_1(W_1) + \varphi_2(W_2)$$

und hieraus endlich

$$\varphi_1(W_1) = C \lg(W_1) + \text{konst.},$$
$$\varphi_2(W_2) = C \lg(W_2) + \text{konst.},$$
$$\varphi(W) = C \lg(W) + \text{konst.}$$

Die Größe C ist also eine universelle Konstante; sie hat, wie [30] aus der kinetischen Gastheorie folgt, den Wert R/N, wobei den Konstanten R und N dieselbe Bedeutung wie oben beizulegen ist. Bedeutet S_0 die Entropie bei einem gewissen Anfangszustande eines betrachteten Systems und W die relative Wahrscheinlichkeit eines Zustandes von der Entropie S, so erhalten wir also allgemein:

$$S - S_0 = \frac{R}{N} \lg W.$$

Wir behandeln zunächst folgenden Spezialfall. In einem Volumen v_0 sei eine Anzahl (n) beweglicher Punkte (z. B. Moleküle) vorhanden, auf welche sich unsere Überlegung beziehen soll. Außer diesen können in dem Raume noch beliebig viele andere bewegliche Punkte irgendwelcher Art vorhanden sein. Über das Gesetz, nach dem sich die betrachteten Punkte in dem Raume bewegen, sei nichts vorausgesetzt, als daß in bezug auf diese Bewegung kein Raumteil (und keine Richtung) von den anderen ausgezeichnet sei. Die Anzahl der betrachteten (ersterwähnten) beweglichen Punkte sei ferner so klein, daß von einer Wirkung der Punkte aufeinander abgesehen werden kann.

Dem betrachteten System, welches z. B. ein ideales Gas oder eine verdünnte Lösung sein kann, kommt eine gewisse Entropie S_0 zu. Wir denken uns einen Teil des Volumens v_0 von der Größe v und alle n beweglichen Punkte in das Volumen v versetzt, ohne daß an dem System sonst etwas geändert wird. Diesem Zustand kommt offenbar ein anderer Wert der Entropie (S) zu, und wir wollen nun die Entropiedifferenz mit Hilfe des B o l t z m a n n schen Prinzips bestimmen.

142   *A. Einstein.*

Wir fragen: Wie groß ist die Wahrscheinlichkeit des letzterwähnten Zustandes relativ zum ursprünglichen? Oder: Wie groß ist die Wahrscheinlichkeit dafür, daß sich in einem zufällig herausgegriffenen Zeitmoment alle n in einem gegebenen Volumen v_0 unabhängig voneinander beweglichen Punkte (zufällig) in dem Volumen v befinden?

Für diese Wahrscheinlichkeit, welche eine „statistische Wahrscheinlichkeit" ist, erhält man offenbar den Wert:

$$W = \left(\frac{v}{v_0}\right)^n;$$

man erhält hieraus durch Anwendung des Boltzmannschen Prinzipes:

$$S - S_0 = R\left(\frac{n}{N}\right) \lg\left(\frac{v}{v_0}\right).$$

Es ist bemerkenswert, daß man zur Herleitung dieser Gleichung, aus welcher das Boyle-Gay-Lussacsche Gesetz und das gleichlautende Gesetz des osmotischen Druckes leicht thermodynamisch ableiten kann[1]); keine Voraussetzung über das Gesetz zu machen braucht, nachdem sich die Moleküle bewegen.

§ 6. Interpretation des Ausdruckes für die Abhängigkeit der Entropie der monochromatischen Strahlung vom Volumen nach dem Boltzmannschen Prinzip.

Wir haben in § 4 für die Abhängigkeit der Entropie der monochromatischen Strahlung vom Volumen den Ausdruck gefunden:

$$S - S_0 = \frac{E}{\beta \nu} \lg\left(\frac{v}{v_0}\right).$$

Schreibt man diese Formel in der Gestalt:

$$S - S_0 = \frac{R}{N} \lg\left[\left(\frac{v}{v_0}\right)^{\frac{N}{R}\frac{E}{\beta\nu}}\right]$$

---

1) Ist E die Energie des Systems, so erhält man:

[31] $$-d(E - TS) = p\,dv = T\,dS = R\frac{n}{N}\frac{dv}{v};$$

also

$$pv = R\frac{n}{N}T.$$

Erzeugung und Verwandlung des Lichtes. **143**

und vergleicht man sie mit der allgemeinen, das Boltz-mannsche Prinzip ausdrückenden Formel

$$S - S_0 = \frac{R}{N} \lg W,$$

so gelangt man zu folgendem Schluß:

Ist monochromatische Strahlung von der Frequenz ν und der Energie E in das Volumen v_0 (durch spiegelnde Wände) eingeschlossen, so ist die Wahrscheinlichkeit dafür, daß sich in einem beliebig herausgegriffenen Zeitmoment die ganze Strahlungsenergie in dem Teilvolumen v des Volumens v_0 befindet:

$$W = \left(\frac{v}{v_0} \right)^{\frac{N}{R} \cdot \frac{E}{\beta \nu}}.$$

Hieraus schließen wir weiter:

Monochromatische Strahlung von geringer Dichte (innerhalb des Gültigkeitsbereiches der Wienschen Strahlungsformel) verhält sich in wärmetheoretischer Beziehung so, wie wenn sie aus voneinander unabhängigen Energiequanten von der Größe $R\beta\nu/N$ bestünde.

Wir wollen noch die mittlere Größe der Energiequanten der „schwarzen Strahlung" mit der mittleren lebendigen Kraft der Schwerpunktsbewegung eines Moleküls bei der nämlichen Temperatur vergleichen. Letztere ist $\frac{3}{2}(R/N)T$, während man für die mittlere Größe des Energiequantums unter Zugrunde-legung der Wienschen Formel erhält:

$$\frac{\displaystyle\int_0^\infty \alpha \nu^3 e^{-\frac{\beta \nu}{T}} \, d\nu}{\displaystyle\int_0^\infty \frac{N}{R\beta\nu} \alpha \nu^3 e^{-\frac{\beta \nu}{T}} \, d\nu} = 3 \frac{R}{N} T. \qquad [32]$$

Wenn sich nun monochromatische Strahlung (von hin-reichend kleiner Dichte) bezüglich der Abhängigkeit der Entropie vom Volumen wie ein diskontinuierliches Medium verhält, welches aus Energiequanten von der Größe $R\beta\nu/N$ besteht, so liegt es nahe, zu untersuchen, ob auch die Gesetze der

144   *A. Einstein.*

Erzeugung und Verwandlung des Lichtes so beschaffen sind,
wie wenn das Licht aus derartigen Energiequanten bestünde.
Mit dieser Frage wollen wir uns im folgenden beschäftigen.

[33]   **§ 7. Über die Stokessche Regel.**

Es werde monochromatisches Licht durch Photolumineszenz
in Licht anderer Frequenz verwandelt und gemäß dem eben
erlangten Resultat angenommen, daß sowohl das erzeugende
wie das erzeugte Licht aus Energiequanten von der Größe
$(R/N)\beta\nu$ bestehe, wobei ν die betreffende Frequenz bedeutet.
Der Verwandlungsprozeß wird dann folgendermaßen zu deuten
sein. Jedes erzeugende Energiequant von der Frequenz ν_1
wird absorbiert und gibt — wenigstens bei genügend kleiner
Verteilungsdichte der erzeugenden Energiequanten — für sich
allein Anlaß zur Entstehung eines Lichtquants von der
Frequenz ν_2; eventuell können bei der Absorption des erzeugenden Lichtquants auch gleichzeitig Lichtquanten von den
Frequenzen ν_3, ν_4 etc. sowie Energie anderer Art (z. B. Wärme)
entstehen. Unter Vermittelung von was für Zwischenprozessen
dies Endresultat zustande kommt, ist gleichgültig. Wenn die
photolumineszierende Substanz nicht als eine beständige Quelle
von Energie anzusehen ist, so kann nach dem Energieprinzip
die Energie eines erzeugten Energiequants nicht größer sein
als die eines erzeugenden Lichtquants; es muß also die Bezeichnung gelten:

$$\frac{R}{N}\beta\nu_2 \leqq \frac{R}{N}\beta\nu_1$$

oder

$$\nu_2 \leqq \nu_1 .$$

Dies ist die bekannte Stokessche Regel.

Besonders hervorzuheben ist, daß bei schwacher Belichtung
die erzeugte Lichtmenge der erregenden unter sonst gleichen
Umständen nach unserer Auffassung der erregenden Lichtstärke proportional sein muß, da jedes erregende Energiequant
einen Elementarprozeß von der oben angedeuteten Art verursachen wird, unabhängig von der Wirkung der anderen erregenden Energiequanten. Insbesondere wird es keine untere
Grenze für die Intensität des erregenden Lichtes geben, unter-
[34]   halb welcher das Licht unfähig wäre, lichterregend zu wirken.

Erzeugung und Verwandlung des Lichtes.   **145**

Abweichungen von der Stokesschen Regel sind nach der dargelegten Auffassung der Phänomene in folgenden Fällen denkbar: [35]

1. wenn die Anzahl der gleichzeitig in Umwandlung begriffenen Energiequanten pro Volumeneinheit so groß ist, daß ein Energiequant des erzeugten Lichtes seine Energie von mehreren erzeugenden Energiequanten erhalten kann;

2. wenn das erzeugende (oder erzeugte) Licht nicht von derjenigen energetischen Beschaffenheit ist, die einer „schwarzen Strahlung" aus dem Gültigkeitsbereich des Wienschen Gesetzes zukommt, wenn also z. B. das erregende Licht von einem Körper so hoher Temperatur erzeugt ist, daß für die in Betracht kommende Wellenlänge das Wiensche Gesetz nicht mehr gilt.

Die letztgenannte Möglichkeit verdient besonderes Interesse. Nach der entwickelten Auffassung ist es nämlich nicht ausgeschlossen, daß eine „nicht Wiensche Strahlung" auch in großer Verdünnung sich in energetischer Beziehung anders verhält als eine „schwarze Strahlung" aus dem Gültigkeitsbereich des Wienschen Gesetzes.

§ 8. Über die Erzeugung von Kathodenstrahlen durch Belichtung fester Körper.

Die übliche Auffassung, daß die Energie des Lichtes kontinuierlich über den durchstrahlten Raum verteilt sei, findet bei dem Versuch, die lichtelektrischen Erscheinungen zu erklären, besonders große Schwierigkeiten, welche in einer bahnbrechenden Arbeit von Hrn. Lenard dargelegt sind.[1)]   [36]

Nach der Auffassung, daß das erregende Licht aus Energiequanten von der Energie $(R/N)\beta\nu$ bestehe, läßt sich die Erzeugung von Kathodenstrahlen durch Licht folgendermaßen auffassen. In die oberflächliche Schicht des Körpers dringen Energiequanten ein, und deren Energie verwandelt sich wenigstens zum Teil in kinetische Energie von Elektronen. Die einfachste Vorstellung ist die, daß ein Lichtquant seine ganze Energie an ein einziges Elektron abgibt; wir wollen annehmen, daß dies vorkomme. Es soll jedoch nicht ausgeschlossen sein, daß Elektronen die Energie von Lichtquanten nur teilweise aufnehmen. Ein im Innern des Körpers mit kinetischer Energie

---

1) P. Lenard, Ann. d. Phys. **8**. p. 169 u. 170. 1902.   [37]

146   *A. Einstein.*

versehenes Elektron wird, wenn es die Oberfläche erreicht hat, einen Teil seiner kinetischen Energie eingebüßt haben. Außerdem wird anzunehmen sein, daß jedes Elektron beim Verlassen des Körpers eine (für den Körper charakteristische) Arbeit P [38] zu leisten hat, wenn es den Körper verläßt. Mit der größten Normalgeschwindigkeit werden die unmittelbar an der Oberfläche normal zu dieser erregten Elektronen den Körper verlassen. Die kinetische Energie solcher Elektronen ist

$$\frac{R}{N}\beta v - P.$$

Ist der Körper zum positiven Potential Π geladen und von Leitern vom Potential Null umgeben und ist Π eben imstande, einen Elektrizitätsverlust des Körpers zu verhindern, so muß sein:

$$\Pi \varepsilon = \frac{R}{N}\beta v - P,$$

[39] wobei ε die elektrische Masse des Elektrons bedeutet, oder

$$\Pi E = R\beta v - P',$$

wobei E die Ladung eines Grammäquivalentes eines einwertigen Ions und P' das Potential dieser Menge negativer Elektrizität in bezug auf den Körper bedeutet.[1]

[41]   Setzt man $E = 9,6 . 10^3$, so ist $\Pi . 10^{-8}$ das Potential in Volts, welches der Körper bei Bestrahlung im Vakuum annimmt.

Um zunächst zu sehen, ob die abgeleitete Beziehung der Größenordnung nach mit der Erfahrung übereinstimmt, setzen wir $P' = 0$, $v = 1,03 . 10^{15}$ (entsprechend der Grenze des Sonnenspektrums nach dem Ultraviolett hin) und $\beta = 4,866 . 10^{-11}$. [42] Wir erhalten $\Pi . 10^7 = 4,3$ Volt, welches Resultat der Größenordnung nach mit den Resultaten von Hrn. Lenard übereinstimmt.[2]

Ist die abgeleitete Formel richtig, so muß Π, als Funktion der Frequenz des erregenden Lichtes in kartesischen Koordinaten dargestellt, eine Gerade sein, deren Neigung von der [44] Natur der untersuchten Substanz unabhängig ist.

---

1) Nimmt man an, daß das einzelne Elektron durch das Licht aus einem neutralen Molekül unter Aufwand einer gewissen Arbeit losgelöst [40] werden muß, so hat man an der abgeleiteten Beziehung nichts zu ändern; nur ist dann P' als Summe von zwei Summanden aufzufassen.

[43]   2) P. Lenard, Ann. d. Phys. 8. p. 165 u. 184. Taf. I, Fig. 2. 1902.

Erzeugung und Verwandlung des Lichtes.   **147**

Mit den von Hrn. Lenard beobachteten Eigenschaften der lichtelektrischen Wirkung steht unsere Auffassung, soweit ich sehe, nicht im Widerspruch. Wenn jedes Energiequant des erregenden Lichtes unabhängig von allen übrigen seine Energie an Elektronen abgibt, so wird die Geschwindigkeitsverteilung der Elektronen, d. h. die Qualität der erzeugten Kathodenstrahlung von der Intensität des erregenden Lichtes unabhängig sein; andererseits wird die Anzahl der den Körper verlassenden Elektronen der Intensität des erregenden Lichtes unter sonst gleichen Umständen proportional sein.[1]

Über die mutmaßlichen Gültigkeitsgrenzen der erwähnten Gesetzmäßigkeiten wären ähnliche Bemerkungen zu machen wie bezüglich der mutmaßlichen Abweichungen von der Stokesschen Regel.

Im vorstehenden ist angenommen, daß die Energie wenigstens eines Teiles der Energiequanten des erzeugenden Lichtes je an ein einziges Elektron vollständig abgegeben werde. Macht man diese naheliegende Voraussetzung nicht, so erhält man statt obiger Gleichung die folgende:

$$\Pi E + P' \leqq R \beta v.$$

Für die Kathodenlumineszenz, welche den inversen Vorgang zu dem eben betrachteten bildet, erhält man durch eine der durchgeführten analoge Betrachtung: [45]

$$\Pi E + P' \geqq R \beta v.$$

Bei den von Hrn. Lenard untersuchten Substanzen ist $P E$ [46] stets bedeutend größer als $R \beta v$, da die Spannung, welche die Kathodenstrahlen durchlaufen haben müssen, um eben sichtbares Licht erzeugen zu können, in einigen Fällen einige Hundert, in anderen Tausende von Volts beträgt.[2] Es ist also anzunehmen, daß die kinetische Energie eines Elektrons zur Erzeugung vieler Lichtenergiequanten verwendet wird.

§ 9. **Über die Ionisierung der Gase durch ultraviolettes Licht.** [48]

Wir werden anzunehmen haben, daß bei der Ionisierung eines Gases durch ultraviolettes Licht je ein absorbiertes Licht-

1) P. Lenard, l. c. p. 150 und p. 166—168.
2) P. Lenard, Ann. d. Phys. 12. p. 469. 1903.   [47]

148   *A. Einstein.   Erzeugung und Verwandlung des Lichtes.*

energiequant zur Ionisierung je eines Gasmoleküles verwendet wird.   Hieraus folgt zunächst, daß die Ionisierungsarbeit (d. h. die zur Ionisierung theoretisch nötige Arbeit) eines Moleküles nicht größer sein kann als die Energie eines absorbierten wirksamen Lichtenergiequantes. Bezeichnet man mit J die (theoretische) Ionisierungsarbeit pro Grammäquivalent, so muß also sein:

$$R \beta v \geqq J.$$

[49] Nach Messungen Lenards ist aber die größte wirksame Wellenlänge für Luft ca. $1,9 \cdot 10^{-5}$ cm, also

$$R \beta v = 6,4 \cdot 10^{12} \text{ Erg} \geqq J.$$

Eine obere Grenze für die Ionisierungsarbeit gewinnt man auch aus den Ionisierungsspannungen in verdünnten Gasen. Nach J. Stark[1]) ist die kleinste gemessene Ionisierungsspannung (an Platinanoden) für Luft ca. 10 Volt.[2]) Es ergibt sich also für J die obere Grenze $9,6 \cdot 10^{12}$, welche nahezu gleich der eben gefundenen ist. Es ergibt sich noch eine andere Konsequenz, deren Prüfung durch das Experiment mir von großer Wichtigkeit zu sein scheint. Wenn jedes absorbierte Lichtenergiequant ein Molekül ionisiert, so muß zwischen der absorbierten Lichtmenge L und der Anzahl j der durch dieselbe ionisierten Grammoleküle die Beziehung bestehen:

[51]
$$j = \frac{L}{R \beta v}.$$

Diese Beziehung muß, wenn unsere Auffassung der Wirklichkeit entspricht, für jedes Gas gelten, welches (bei der betreffenden Frequenz) keine merkliche nicht von Ionisation begleitete Absorption aufweist.

Bern, den 17. März 1905.

[50]   1) J. Stark, Die Elektrizität in Gasen p. 57. Leipzig 1902.
   2) Im Gasinnern ist die Ionisierungsspannung für negative Ionen allerdings fünfmal größer.

(Eingegangen 18. März 1905.)

Published in *Annalen der Physik* 17 (1905): 132–148. Dated Bern, 17 March 1905, received 18 March 1905, published 9 June 1905.

[1] For a discussion of Einstein's use of the word "heuristisch," see *Klein 1982b*.

[2] Boltzmann noted a distinction between continuous and discrete energy distributions in *Boltzmann 1896*, p. 5.

[3] For a survey favorably comparing the optical predictions of electromagnetic theory with experiment, see *Wien 1909*, pp. 186–198.

[4] "Schwarze Strahlung" ("black radiation") is radiation emitted by a perfectly black body, i.e., with absorptivity of 1. The phrase was common at that time (see, e.g., *Graetz 1906*, p. 366).

[5] For treatments of similar phenomena, see *Einstein 1906d* (Doc. 34) and *Einstein 1907a* (Doc. 38).

[6] See *Drude 1900a* and *1900b*. In Drude's theory electrons are treated as freely moving charge carriers, similiar to molecules in the kinetic theory of gases. For a discussion of Einstein's early interest in the electron theory of metals and its relation to radiation theory, see Vol. 1, the editorial note, "Einstein on Thermal, Electrical, and Radiation Phenomena," pp. 235–237.

[7] In Planck's model, *Planck 1900a*, p. 70, black-body radiation was in equilibrium with bound electron "oscillators" ("Oszillatoren"), but not with free electrons and molecules.

[8] Kirchhoff inferred that thermal radiation at equilibrium in a cavity of perfectly reflecting walls is equivalent to black radiation of the same temperature (see *Kirchhoff 1860*, pp. 300–301). The equivalence of electromagnetic radiation in thermal equilibrium with oscillators to black radiation was the basis of Planck's work (see, e.g., *Planck 1900a*, pp. 69–70).

[9] Einstein was troubled earlier by Planck's seeming neglect of resonators of all frequencies (see Einstein to Mileva Marić, 10 April 1901 [Vol. 1, Doc. 97]).

[10] "Wirkliche Moleküle" ("real molecules") are presumably those that are not dissociated.

[11] *Planck 1900a*.

[12] Planck called this assumption "a special hypothesis" ("eine besondere Hypothese") (*Planck 1900a*, p. 71). Such "natural radiation" ("natürliche Strahlung") was originally defined in a somewhat different fashion from Einstein's

in *Planck 1898*, pp. 467–469, 473; *Planck 1899*, pp. 451–453; and *Planck 1900a*, pp. 88–91. In particular, Planck used perfect incoherence while Einstein employs "statistical probabilities" below.

[13] x should be α.

[14] This formula does not appear on p. 99 of *Planck 1900a*. It can be derived from equations on pp. 99 and 111. Planck first published it in *Planck 1900e*, p. 241.

[15] In 1900 Rayleigh obtained the proportionality between T and E_λ, the energy per vibration mode, by applying the equipartition theorem to the vibration modes of matter-free black-body radiation (see *Rayleigh 1900*). His result was $E_\lambda = cT/\lambda^4$, where c is a constant. Expressions equivalent to Einstein's equation were obtained in 1905 by Rayleigh and Jeans without the use of material resonators (see *Rayleigh 1905a, 1905b*, and *Jeans 1905a*, which appeared after the receipt of Einstein's paper).

[16] *Rubens and Kurlbaum 1901*, p. 666, states that Rayleigh's formula "fails in the region of shorter wavelengths. It also shows considerable systematic deviations from our observations" ("in dem Gebiet kurzer Wellenlängen versagt. Auch zeigt sie gegenüber unseren Beobachtungen erhebliche systematische Abweichungen"), but for large values of λT it agrees fairly well with the observed distribution. See also *Lummer and Pringsheim 1908*, p. 449.

[17] Rayleigh apparently noticed this difficulty in 1900 (*Rayleigh 1900*), asserting that the equipartition "doctrine" yields valid results only for "the graver modes." In May 1905 he stated: "According to [the equation for E_λ], if it were applicable to all wave-lengths, the total energy of radiation at a given temperature would be infinite" (*Rayleigh 1905a*, p. 55).

[18] Here "Elementarquanta" ("elementary quanta") refers to fundamental atomic constants. In *Planck 1901b*, Planck determined the mass of the hydrogen atom, Loschmidt's number (N), Boltzmann's constant, and the elementary electric charge.

[19] *Planck 1901a*.

[20] Planck's formula appears in *Planck 1901a*, p. 561, with the constants h and k, instead of α and β ($\alpha = 8\pi h/L^3$, $\beta = h/k$, and $k = R/N$). Planck's values for h and k, obtained in *Planck 1901a*, pp. 562–563, yield Einstein's values for α and β.

[21] See *Planck 1901b*, pp. 565–566, and

Meyer, O. E. 1899. For a discussion of Einstein's other methods for determining N, see the editorial note, "Einstein's Dissertation on the Determination of Molecular Dimensions," § IV, pp. 176–179.

[22] See *Wien 1894*.

[23] See *Wien 1896*, p. 667. Wien's formula was originally expressed as the density φ_λ of radiation of wavelength between λ and $\lambda + d\lambda$: $\varphi_\lambda = C/\lambda^5 e^{-\frac{c}{\lambda\theta}}$, where θ is the temperature, C is a constant, and c is the speed of light.

[24] The best contemporary experimental tests were reported in *Rubens and Kurlbaum 1901* and *Paschen 1901a*. This latter, p. 293, states: "For larger values of $1/\lambda T$ than about 0.0003 ($\lambda T <$ about 3000) the observations follow the Wien law within the margin of error. For smaller values of $1/\lambda T$ than about 0.0003 ($\lambda T >$ about 3000) deviations from the Wien law appear in all curves" ("Für grössere Werte von $1/\lambda T$ als ungefähr 0,0003 ($\lambda T < 3000$ ungefähr) folgen die Beobachtungen innerhalb der möglichen Fehler dem Wien'schen Gesetze. Für kleinere Werte von $1/\lambda T$ als ungefähr 0,0003 ($\lambda T > 3000$ ungefähr) treten in allen Curven Abweichungen vom Wien'schen Gesetze auf."

[25] S refers to radiation with frequencies between ν and $\nu + d\nu$, and $E = \rho\nu\, d\nu$.

[26] This comparison is further discussed in § 5.

[27] See *Boltzmann 1877*. The name "Boltzmannsches Prinzip" ("Boltzmann's principle"), used in the next section, is Einstein's terminology.

[28] For a discussion of Einstein's definition of probability, see the editorial note, "Einstein on the Foundations of Statistical Physics," p. 52.

[29] *Klein 1974b*, p. 190, suggests that the material Einstein presented in *Einstein 1909b* (Doc. 56), p. 187, is a partial fulfillment of this promise.

[30] See also *Einstein 1904* (Doc. 5), pp. 355 and 359. C is equivalent to Boltzmann's constant, now usually designated by k.

[31] The last term should be multiplied by T. For the extension of the law of Boyle and Gay-Lussac to dilute solutions, see *Van 't Hoff 1887*. See also *Einstein 1905q* (Doc. 22), § 1.

[32] In *Einstein 1906d* (Doc. 34), p. 203, Einstein used Planck's equation to show that the average energy of a Wien oscillator is much smaller than that of an energy quantum, $h\nu$.

[33] For a contemporary discussion of Stokes's rule, see *Winkelmann 1906b*, "Fluoreszenz," pp. 785–798.

[34] This conclusion is consistent with *Knoblauch 1895*, p. 198, which states that the intensities of fluorescent and incident light are proportional, "even when the intensity of the latter is varied in the ratio 1:6400" ("selbst wenn die Intensität des letzteren im Verhältniss 1:6400 verändert wird").

[35] For further discussion of such exceptions, see *Einstein 1909b* (Doc. 56), p. 191.

[36] For a contemporary survey of the photoelectric effect, with bibliography through 1904, see *Schweidler 1904*. In 1901 Einstein read *Lenard 1900b*, which discusses the photoelectric effect (see Einstein to Mileva Marić, 28 May 1901 [Vol. 1, Doc. 111]).

[37] *Lenard 1902*.

[38] For a discussion of the relationship between the photoelectric effect and the Volta effect, see *Einstein 1906d* (Doc. 34), pp. 203–206.

[39] "Elektrische Masse" ("electric mass") here refers to the charge of the electron.

[40] *Stark 1902*, which Einstein cites on p. 148, treats the ionization of a neutral gas molecule by ultraviolet light as "a separation of the negative electron from its positive remainder atom" ("eine Lostrennung des negativen Elektrons von seinem positiven Restatom") (p. 75). *Stark and Steubing 1908*, pp. 490–491, reports a "genetic" relationship between fluorescence and the photoelectric effect. The former is assumed to be a separation of an electron from a molecule, followed by the "reattachment of the electrons" ("Wiederanlagerung der Elektronen") with the emission of light; the latter is treated as the ejection of electrons from a metal.

[41] Using the value of N obtained on p. 137 and the value for the electric charge obtained in *Planck 1901b*, p. 566, converted from esu to coul, the correct value of E is 9.6×10^4 coul. Thus, Π should be multiplied by 10^{-7} to convert erg/coul to volts.

[42] "$\Pi \cdot 10^7$" should be "$\Pi \cdot 10^{-7}$".

[43] *Lenard 1902*. Lenard's values ranged between 1.5V to 3V, depending on the material studied. For a review of Lenard's work, see *Wheaton 1978b*.

[44] This and other predictions of Einstein's equation for ΠE were confirmed in experiments reported in *Millikan 1916a, 1916b*. Earlier tests

include *Hughes 1912* and *Richardson and Compton 1912*. Although the equation was generally accepted after Millikan's work, the light quantum concept was not accepted until the mid-1920s.

[45] For Lenard's studies of cathode luminescence, see *Lenard 1903*.

[46] "*PE*" should be "*ΠE*".

[47] *Lenard 1903*.

[48] For previous studies, see, e.g., *Lenard 1900a* and *1900c*; *Stark 1902*, pp. 74–79; and *Stark 1909b*, pp. 614–616.

[49] *Lenard 1900a*, p. 495.

[50] *Stark 1902*.

[51] This expression is equivalent to the law of photochemical equivalence, established in *Einstein 1912a*. This section may be viewed as initiating the field of quantum photochemistry.

EINSTEIN'S DISSERTATION
ON THE DETERMINATION OF
MOLECULAR DIMENSIONS

I

Einstein submitted a dissertation to the University of Zurich in 1901, about a year after graduation from the ETH, but withdrew it early in 1902.[1] In a successful second attempt three years later, he combined the techniques of classical hydrodynamics with those of the theory of diffusion to create a new method for the determination of molecular sizes and of Avogadro's number, a method he applied to solute sugar molecules.[2] The dissertation was completed on 30 April 1905 and submitted to the University of Zurich on 20 July.[3] On 19 August 1905, shortly after the thesis was accepted, the *Annalen der Physik* received a slightly different version for publication.[4]

Einstein 1906c (Doc. 33), published half a year later as a supplement to *Einstein 1906a*, utilizes experimental data not previously available to recalculate the size of sugar molecules. In 1911, after Jacques Bancelin found a discrepancy between the results of his experiments and Einstein's predictions, a calculational error in *Einstein 1905j* (Doc. 15) was discovered. Traces of an unsuccessful attempt by Einstein to locate the error, preserved as marginalia and interlineations in an offprint of the paper, are discussed in the annotations to Doc. 15. A correction of the error, which was found by Ludwig Hopf, then a collaborator of Einstein, is published in *Einstein 1911d*. The correction was reiterated in *Einstein 1920* and integrated into the republication of Einstein's dissertation in *Einstein 1922*.[5]

II

By 1905 several methods for the experimental determination of molecular dimensions were available.[6] Although estimates of upper bounds for the sizes of microscopic constituents of matter had been discussed for a long time, the first reliable methods for determining molecular sizes were developed in the second half of the nineteenth century, based

[1] For evidence of Einstein's submission of the dissertation, see the Receipt for the Return of Doctoral Fees, 1 February 1902 (Vol. 1, Doc. 132).

[2] See *Einstein 1905j* (Doc. 15). For a study of Einstein's dissertation, see *Pais 1982*, chap. 2, § 5.

[3] See Einstein to Rudolf Martin, 20 July 1905.

[4] The thesis was unanimously accepted by the Mathematics and Physics Faculty on 27 July 1905 (see Protokollbuch der Konferenz, Abteilung VI A, SzZE Bibliothek, Hs 1079:2). For the changes that Einstein made in the *Annalen*

version, *Einstein 1906a*, see the notes to *Einstein 1905j* (Doc. 15).

[5] For an English translation of *Einstein 1922*, see *Einstein 1926*. For the history of *Einstein 1922*, edited by Reinhold Fürth, see the editorial note, "Einstein on Brownian Motion," § I, p. 206. Some of Fürth's annotations for the 1922 edition (see *Fürth 1922*) have been utilized in these editorial notes.

[6] For a survey of the development of methods for the determination of molecular dimensions, see *Brush 1976*, pp. 75–78. For a contemporary survey, see *Meyer, O. E. 1899*, chap. 10.

on the kinetic theory of gases.[7] The study of phenomena as diverse as contact electricity in metals, the dispersion of light, and black-body radiation yielded new approaches to the problem of molecular dimensions.[8] Most of the methods available by the turn of the century gave values for the size of molecules and for Avogadro's number that are in more or less satisfactory agreement with each other.[9]

Although Einstein claimed that the method in his dissertation is the first to use phenomena in fluids in the determination of molecular dimensions,[10] the behavior of liquids plays a role in various earlier methods. For example, the comparison of densities in the liquid and gaseous states is an important part of Loschmidt's method, based on the kinetic theory of gases.[11] A method that depends entirely on the physics of liquids was known as early as 1816. Young's study of surface tension in liquids led to an estimate of the range of molecular forces,[12] and capillary phenomena were used later in several different ways to determine molecular sizes.[13]

A kinetic theory of liquids, comparable to the kinetic theory of gases, was not available, and the methods for deriving molecular volumes exclusively from the properties of liquids did not give very precise results.[14] Einstein's method, on the other hand, yields values comparable in precision to those provided by the kinetic theory of gases. While methods based on capillarity presuppose the existence of molecular forces, Einstein's central assumption is the validity of using classical hydrodynamics to calculate the effect of solute molecules, treated as rigid spheres, on the viscosity of the solvent in a dilute solution.[15]

Einstein's method is well suited to determine the size of solute molecules that are large compared to those of the solvent. In 1905 William Sutherland published a new method for determining the masses of large molecules that shares important elements with Einstein's.[16] Both methods make use of the molecular theory of diffusion that Nernst developed on the basis of Van 't Hoff's analogy between solutions and gases, and of Stokes's law of hydrodynamical friction.[17]

Sutherland was interested in the masses of large molecules because of the role they play in the chemical analysis of organic substances such as albumin.[18] In developing a new method for the determination of molecular dimensions, Einstein was concerned with several other problems on different levels of generality. An outstanding current problem of the theory of solutions was whether molecules of the solvent are attached to the molecules

[7] See, in particular, *Loschmidt 1865*.

[8] For methods based on contact electricity and the dispersion of light, see *Thomson, W. 1870*; for the determinations of Avogadro's number from black-body radiation, see *Planck 1901b* and *Einstein 1905i* (Doc. 14).

[9] See, e.g., *Meyer, O. E. 1899*, chap. 10.

[10] See *Einstein 1905j* (Doc. 15), p. 5.

[11] See *Loschmidt 1865*.

[12] See *Young 1816*.

[13] For a survey, see *Meyer, O. E. 1899*, chap. 10, § 122. For Einstein's knowledge of work on capillarity, see the editorial note, "Einstein on the Nature of Molecular Forces," pp. 3–4.

[14] For a discussion of the values given by these methods, see *Meyer, O. E. 1899*, chap. 10, § 122.

[15] For a discussion of his other assumptions, see § IV.

[16] See *Sutherland 1905*. Sutherland first outlined his method in 1904 (see ibid., p. 781). For a discussion of Sutherland's method, see § IV.

[17] See *Nernst 1888*, *Van 't Hoff 1887*, and *Stokes 1845*.

[18] See *Sutherland 1905*, p. 781.

or ions of the solute.[19] Einstein's dissertation contributed to the solution of this problem.[20] He recalled in 1909:

> At the time I used the viscosity of the solution to determine the volume of sugar dissolved in water because in this way I hoped to take into account the volume of any *attached* water molecules.

> Ich habe seinerzeit zur Bestimmung des Volumens des in Wasser aufgelösten Zuckers deswegen die Viskosität der Lösung benutzt, weil ich so das Volumen eventuel *angelagerter* Wassermoleküle mit zu berücksichtigen hoffte.[21]

The results obtained in his dissertation indicate that such an attachment does occur.[22]

Einstein's concerns extended beyond this particular question to more general problems of the foundations of the theory of radiation and the existence of atoms. He later emphasized:

> A precise determination of the size of molecules seems to me of the highest importance because Planck's radiation formula can be tested more precisely through such a determination than through measurements on radiation.

> Eine präzise Bestimmung der Grösse der Moleküle scheint mir deshalb von höchster Wichtigkeit, weil durch eine solche die Strahlungsformel von Planck schärfer geprüft werden kann als durch Strahlungsmessungen.[23]

The dissertation also marked the first major success in Einstein's effort to find further evidence for the atomic hypothesis, an effort that culminated in his explanation of Brownian motion.[24] By the end of 1905 Einstein had published three independent methods for determining molecular dimensions, and in the following years he found several more.[25]

[19] *Bousfield 1905b*, a study of the relationship between the sizes of ions and the electrical conductivity of electrolytes, calls this the most important open problem of the theory of aqueous solutions (p. 257). For contemporary reviews of research on hydrates, including a history of this problem, see *Washburn 1908* and *1909*, and *Dhar 1914*.

[20] *Einstein 1906a* is cited in *Washburn 1909*, p. 70, and in *Herzfeld 1921*, p. 1025, as providing evidence for the existence of an association between a solute molecule and molecules of the solvent.

[21] Einstein to Jean Perrin, 11 November 1909. The importance of this problem for Einstein is confirmed by a letter that Einstein wrote to Ludwig Hopf before 12 January 1911, emphasizing the significance of his equation for the coefficient of viscosity, "because from viscosity one can learn something about the volume of *dissolved* molecules" ("weil man aus der Vis-

kosität etwas erfahren kann über das Volumen *gelöster* Moleküle").

[22] See *Einstein 1905j* (Doc. 15), p. 18.

[23] Einstein to Jean Perrin, 11 November 1909.

[24] In his *Autobiographical Notes*, Einstein stated that his work on statistical mechanics, which preceded his dissertation, aimed at finding "facts . . . that would guarantee as much as possible the existence of atoms of definite finite size" ("Tatsachen . . . welche die Existenz von Atomen von bestimmter endlicher Grösse möglichst sicher stellten") (*Einstein 1979*, p. 44; translation, p. 45). For further discussion of Einstein's interest in the problem of molecular dimensions, see the editorial notes, "Einstein on the Foundations of Statistical Physics," p. 46, and "Einstein on Brownian Motion," pp. 206–222.

[25] In addition to the method published in the dissertation, other methods for the determina-

Of all these methods, the one in his dissertation is most closely related to his earlier studies of physical phenomena in liquids.[26]

III

Einstein's efforts to obtain a doctoral degree illuminate some of the institutional constraints on the development of his work on the problem of molecular dimensions. Einstein's choice of a theoretical topic for a dissertation at the University of Zurich was quite unusual, both because it was theoretical and because a dissertation theme was customarily assigned by the supervising professor.[27] By 1900, theoretical physics was slowly beginning to achieve recognition as an independent discipline in German-speaking countries, but it was not yet established at either the ETH or the University of Zurich. A beginning had been made at the ETH soon after its founding, with the appointment of the German mathematical physicist, Rudolf Clausius.[28] His departure a decade later may have been hastened by lack of official sympathy for a too-theoretical approach to the training of engineers and secondary-school teachers, the primary task of the school.[29]

Clausius's successor—after the position had been vacant for a number of years—was H. F. Weber, who occupied the chair for Mathematical and Technical Physics from 1875 until his death in 1912. During the last two decades of the nineteenth century, he did original research, mainly in experimental physics and electrotechnology, including work on a number of topics that were important for Einstein's later research,[30] such as blackbody radiation, the anomalous low-temperature behavior of specific heats, and the theory of diffusion; but his primary interests were never those of a theoretical physicist.[31]

The situation of theoretical physics at the University of Zurich at the turn of the century was hardly better. Four other major Swiss universities either had two full professorships in physics or one full and one nontenured position, while Zurich had only one physics chair, held by the experimentalist Alfred Kleiner.[32]

---

tion of molecular dimensions are presented in *Einstein 1905i* (Doc. 14) and *Einstein 1905k* (Doc. 16). For a discussion of Einstein's various methods, see *Pais 1982*, pp. 94–95.

[26] For a discussion of these studies, see Vol. 1, the editorial note, "Einstein on Molecular Forces," pp. 264–266, and the editorial note in this volume, "Einstein on the Nature of Molecular Forces," pp. 3–8.

[27] See the reports on dissertations in physics submitted between 1901 and 1905 to the University of Zurich (Promotionsgutachten, SzZSa, U 110 e 7, 8, and 9).

[28] For an account of the beginning of theoretical physics at the ETH, see *Jungnickel and McCormmach 1986a*, pp. 186–193; the first four of these pages discuss the Clausius appointment.

[29] According to the ETH's founding statute,

the "task of the polytechnic school consists of . . . educating engineers theoretically and as far as possible practically" ("Die Aufgabe der polytechnischen Schule besteht darin: Techniker . . . theoretisch und so weit tunlich praktisch auszubilden"). Mathematics and the natural sciences are assigned the role of "auxiliary sciences" ("Hilfswissenschaften"). See *Bundesgesez 1854*, article 2. For the preference for practical training by ETH students, see *Jungnickel and McCormmach 1986a*, p. 193.

[30] For additional information about Weber's activities at the ETH, see Vol. 1, the editorial note, "Einstein as a Student of Physics, and His Notes on H. F. Weber's Course," pp. 60–62.

[31] See Vol. 1, Biographies, pp. 387–388, and *Weiss 1912*.

[32] The four universities were Basel, Fri-

Since the ETH was not authorized to grant doctoral degrees until 1909,[33] a special arrangement enabled ETH students to obtain doctorates from the University of Zurich.[34] Most dissertations in physics by ETH students were prepared under Weber's supervision, with Kleiner as the second referee. As noted above, almost all physics dissertations prepared at the ETH and the University of Zurich between 1901 and 1905 were on experimental topics suggested to the students by their supervisor or at least closely related to the latter's research interests.[35] The range of topics was quite limited, and generally not at the forefront of experimental research. Thermal and electrical conductivity, and instruments for their measurement, were by far the most prominent subjects. General questions of theoretical physics, such as the properties of the ether or the kinetic theory of gases, occasionally found their way into examination papers (*Klausurarbeiten*),[36] but they were hardly touched upon in dissertations.

In the winter semester of 1900–1901, Einstein intended to work for a degree under Weber.[37] The topic may have been related to thermoelectricity, a field in which Einstein had shown an interest and in which several of Weber's doctoral students did experimental research.[38] After a falling out with Weber, Einstein turned to Kleiner for advice and comments on his work.[39]

Although Kleiner's research at this time focused on measuring instruments, he did have an interest in foundational questions of physics,[40] and Einstein's discussions with him covered a wide range of topics.[41] Einstein showed his first dissertation to Kleiner before submitting it to the university in November 1901.[42] This dissertation has not survived, and the evidence concerning its contents is somewhat ambiguous. In April 1901 Einstein wrote that he planned to summarize his work on molecular forces, up to that time mainly

bourg, Geneva, and Lausanne. For a sober assessment of his university's physics teaching, see Hans Schinz, Dekan, Philosophical Faculty II of the University of Zurich, to Erziehungsdirektion, Canton of Zurich, 10 September 1901 (SzZSa, U 110 b 1, Nr. 25). On Kleiner, see Vol. 1, Biographies, p. 383.

[33] See *Guggenbühl 1955*, pp. 133–135.
[34] See *Promotionsordnung 1899*, p. 2.
[35] See note 27 above.
[36] See the examination papers included with the reports on dissertations cited in note 27.
[37] For evidence of Einstein's intention to prepare a doctoral thesis under Weber, see the Questionnaire for Municipal Citizenship Applicants, 11–26 October 1900 (Vol. 1, Doc. 82).
[38] For the topics of dissertations submitted to the University of Zurich, see the Promotionsgutachten, SzZSa, U 110 e 7, 8, and 9. For Einstein's interest in thermoelectricity, see Einstein to Mileva Marić, 10 October 1899 (Vol. 1, Doc. 58).
[39] Einstein blamed Weber for the failure of

his attempt to obtain a position at the University of Göttingen in March 1901 (see Einstein to Mileva Marić, 23 March 1901 and 27 March 1901, Vol. 1, Docs. 93 and 94). Discussions with Kleiner are mentioned in Marić to Einstein, early November 1901 and 13 November 1901 (Vol. 1, Docs. 123 and 124), and in Einstein to Marić, 19 December 1901 (Vol. 1, Doc. 130).

[40] For surveys of Kleiner's research, see Vol. 1, Biographies, p. 383, and *Andenken/Kleiner 1916*. For evidence of Kleiner's interest in foundational questions, see, e.g., *Kleiner 1901*, pp. 21–23.
[41] For evidence of the range of Einstein's discussions with Kleiner, see Einstein to Mileva Marić, 19 December 1901 and 8 February 1902 (Vol. 1, Docs. 130 and 136).
[42] The dissertation was submitted to the University of Zurich on 23 November 1901 (see note 1). For evidence that Einstein had earlier submitted the dissertation to Kleiner, see Mileva Marić to Einstein, 13 November 1901 (Vol. 1, Doc. 124).

on liquids;[43] at the end of the year, Marić stated that he had submitted a work on molecular forces in gases.[44] Einstein himself wrote that it concerned "a topic in the kinetic theory of gases" ("ein Thema der kinetischen Theorie der Gase").[45] There are indications that the dissertation may have discussed Boltzmann's work on gas theory, as well as Drude's work on electron theory of metals.[46]

By February 1902 Einstein had withdrawn the dissertation, possibly at Kleiner's suggestion that he avoid a controversy with Boltzmann.[47] In view of the predominantly experimental character of the physics dissertations submitted to the University of Zurich at the time, lack of experimental confirmation for his theoretical results may have played a role in the decision to withdraw the thesis.[48] In January 1903 Einstein still expressed interest in molecular forces, but he stated that he was giving up his plan to obtain a doctorate, arguing that it would be of little help to him, and that "the whole comedy has become tiresome for me" ("mir die ganze Komödie langweilig geworden ist").[49]

Little is known about when Einstein started to work on the dissertation he completed in 1905.[50] By March 1903 some of the central ideas of the 1905 dissertation had already occurred to him.[51] Kleiner, one of the two faculty reviewers (*Gutachter*) of his disserta-

---

[43] See Einstein to Marcel Grossmann, 14 April 1901 (Vol. 1, Doc. 100).

[44] See Mileva Marić to Helene Savić, 23 November–mid-December 1901 (Vol. 1, Doc. 125).

[45] Einstein to the Swiss Patent Office, 18 December 1901 (Vol. 1, Doc. 129).

[46] Einstein to Mileva Marić, 17 December (Vol. 1, Doc. 128), states that, if Kleiner accepts the dissertation, "we'll see what stance the fine Mr. Drude takes" ("wollen wir sehen, wie sich der saubere Herr Drude dazu stellt"). Einstein to Marić, 8 February 1902 (Vol. 1, Doc. 136), mentions that part of one of two works previously submitted to Kleiner deals with Boltzmann's book. For a discussion of possible relations between Einstein's interests in kinetic theory and the electron theory of metals, see the editorial note, "Einstein on the Foundations of Statistical Physics," pp. 45–46.

[47] See the Receipt for the Return of Doctoral Fees, 1 February 1902 (Vol. 1, Doc. 132). According to a biography by Einstein's son-in-law, Kleiner rejected "an essay on the kinetic theory of gases" Einstein had given him in 1901 "out of consideration to his colleague Ludwig Boltzmann, whose train of reasoning Einstein had sharply criticized" (*Kayser 1930*, p. 69). See the preceding note for evidence that Einstein may have criticized Drude in his dissertation.

[48] This is suggested by Einstein's emphasis on his inability to provide such experimental confirmation in *Einstein 1902a* (Doc. 2); see p.

814.

[49] Einstein to Michele Besso, 22 January 1903.

[50] According to *Winteler-Einstein 1924*, p. 23, Einstein attempted to submit his work on the electrodynamics of moving bodies to the University of Zurich: "But the thing didn't seem quite right to the leading professors, as the wholly unknown author paid no heed to authority figures, even attacked them! So the work was simply rejected (irony of fate!) and the candidate saw himself compelled to write and submit another, more harmless work, on the basis of which he then obtained the title of Doctor Philosophiae" ("Allein die Sache schien den massgebenden Professoren nicht ganz geheuer, nahm doch der gänzlich unbekannte Verfasser keine Rücksicht auf die Meinung anerkannter Autoritäten, griff sie wohl gar noch an! So wurde die Arbeit schlechthin abgewiesen (Ironie des Schicksals!) u. der Kandidat sah sich gezwungen, eine andere harmlosere Abhandlung zu verfassen u. einzureichen, auf die hin er denn auch den Titel eines Doctor Philosophiae erhielt"). For a discussion of Einstein's contemporary work on the electrodynamics of moving bodies, see the editorial note, "Einstein on the Theory of Relativity," pp. 253–274.

[51] See Einstein to Michele Besso, 17 March 1903. The relationship of this letter to Einstein's dissertation is noted in *Holton 1980*, p. 54. The letter is discussed in detail in the following section.

tion, acknowledged that Einstein had chosen the topic himself and pointed out that "the arguments and calculations to be carried out are among the most difficult in hydrodynamics" ("die Ueberlegungen und Rechnungen, die durchzuführen sind, gehören zu den schwierigsten der Hydrodynamik"). The other reviewer, Heinrich Burkhardt, Professor of Mathematics at the University of Zurich, added: "the mode of treatment demonstrates *fundamental mastery of the relevant mathematical methods*" ("die Art der Behandlung zeugt von *gründlicher Beherrschung der in Frage kommenden mathematischen Methoden*").[52] Although Burkhardt checked Einstein's calculations, he overlooked a significant error in them.[53] The only reported criticism of Einstein's dissertation was for being too short.[54]

Compared to the other topics of his research at the time, his hydrodynamical method for determining molecular dimensions was a dissertation topic uniquely suited to the empirically oriented Zurich academic environment. In contrast to the Brownian motion work, for which the experimental techniques needed to extract information from observations were not yet available, Einstein's hydrodynamical method for determining the dimensions of solute molecules enabled him to derive new empirical results from data in standard tables.

IV

Like Loschmidt's method based on the kinetic theory of gases, Einstein's method depends on two equations for two unknowns, Avogadro's number N and the molecular radius P.[55] The first of Einstein's equations (the second equation on p. 21 of *Einstein 1905j* [Doc. 15]) follows from a relation between the coefficients of viscosity of a liquid with and without suspended molecules (k^* and k, respectively),[56]

$$k^* = k \left(1 + \varphi\right), \qquad\qquad (1)[57]$$

where φ is the fraction of the volume occupied by the solute molecules. This equation, in turn, is derived from a study of the dissipation of energy in the fluid.

[52] Both quotations are from the Gutachten über das Promotionsgesuch des Hrn. Einstein, 20–24 July 1905 (SzZSa, U 110 e 9).

[53] For a discussion of the discovery of this error and its correction, see § V.

[54] *Seelig 1960*, p. 112, reports: "Einstein later laughingly recounted that his dissertation was at first returned to him by Kleiner with the comment that it was too short. After he had added a single sentence, it was accepted without further comment") ("Lachend hat Einstein später erzählt, daß ihm seine Dissertation zuerst von Kleiner mit der Bemerkung zurückgeschickt wurde, sie sei zu kurz. Nachdem er noch einen einzigen Satz eingeschaltet hatte, sei sie still-schweigend angenommen worden").

[55] See *Loschmidt 1865*. In *Einstein 1905j* (Doc. 15), p. 5, the kinetic theory of gases is mentioned as the oldest source for the determination of molecular dimensions. Loschmidt's method is discussed in *Einstein 1915a*, p. 258.

[56] See § 1 and § 2 of *Einstein 1905j* (Doc. 15).

[57] This equation was later corrected; the error and the correct equation are discussed in § V. For later comments on the limitations of this equation, see Einstein to Hans Albert Einstein, before 13 December 1940; see also *Pais 1982*, p. 92.

Einstein's other fundamental equation (the third equation on p. 21 of *Einstein 1905j* [Doc. 15]) follows from an expression for the coefficient of diffusion D of the solute. This expression is obtained from Stokes's law for a sphere of radius P moving in a liquid, and Van 't Hoff's law for the osmotic pressure:

$$ D = \frac{R\,T}{6\,\pi\,k} \cdot \frac{1}{N\,P}, \tag{2} $$

where R is the gas constant, T the absolute temperature, and N Avogadro's number.

The derivation of eq. (1), technically the most complicated part of Einstein's thesis, presupposes that the motion of the fluid can be described by the hydrodynamical equations for stationary flow of an incompressible homogeneous liquid, even in the presence of solute molecules; that the inertia of these molecules can be neglected; that they do not affect each other's motions; and that they can be treated as rigid spheres moving in the fluid without slipping, under the sole influence of hydrodynamical stresses.[58] The hydrodynamic techniques needed are derived from *Kirchhoff 1897*, a book that Einstein first read during his student years.[59]

Eq. (2) follows from the conditions for the dynamical and thermodynamical equilibrium of the fluid. Its derivation requires the identification of the force on a single molecule, which appears in Stokes's law, with the apparent force due to the osmotic pressure (see *Einstein 1905j* [Doc. 15], p. 20). The key to handling this problem is the introduction of fictitious countervailing forces. Einstein had earlier introduced such fictitious forces: they are used in *Einstein 1902a* (Doc. 2) to counteract thermodynamical effects in proving the applicability to diffusion phenomena of a generalized form of the second law of thermodynamics;[60] they are also used in his papers on statistical physics.[61]

Einstein's derivation of eq. (2) does not involve the theoretical tools he developed in his work on the statistical foundations of thermodynamics; he reserved a more elaborate derivation, using these methods, for his first paper on Brownian motion.[62] Eq. (2) was derived independently, in somewhat more general form, by Sutherland in 1905.[63] To deal with the available empirical data, Sutherland had to allow for a varying coefficient of sliding friction between the diffusing molecule and the solution.

The basic elements of Einstein's method—the use of diffusion theory and the application of hydrodynamical techniques to phenomena involving the atomistic constitution of

[58] Einstein's derivation is only valid for Couette flow; a generalization to Poiseuille flow is given in *Simha 1936*. For a discussion of Einstein's assumptions, see *Pais 1982*, p. 90.

[59] See Einstein to Mileva Marić, 29 July 1900 and 1 August 1900 (Vol. 1, Docs. 68 and 69).

[60] For a discussion of this generalization, see the editorial note, "Einstein on the Nature of Molecular Forces," p. 8.

[61] See, in particular, *Einstein 1902b* (Doc. 3), § 10.

[62] This derivation, given in *Einstein 1905k* (Doc. 16), § 3, is cited in a footnote to *Einstein 1905j* (Doc. 15), p. 20, that was presumably added after the former paper had appeared. In the same paper, he also used eq. (2) to study the relation between diffusion and fluctuations.

[63] See *Sutherland 1905*, pp. 781–782.

matter or electricity—can be traced back to his earlier work.[64] Einstein's previous work had touched upon most aspects of the physics of liquids in which their molecular structure is assumed to play a role, such as Laplace's theory of capillarity, Van der Waals's theory of liquids, and Nernst's theory of diffusion and electrolytic conduction.[65]

Before Einstein's dissertation, the application of hydrodynamics to phenomena involving the atomic constitution of matter or electricity was restricted to consideration of the effects of hydrodynamical friction on the motion of ions. Stokes's law was employed in methods for the determination of the elementary charge[66] and played a role in studies of electrolytic conduction.[67] Einstein's interest in the theory of electrolytic conduction may have been decisive for the development of some of the main ideas in his dissertation.[68] This interest may have suggested a study of molecular aggregates in combination with water, as well as some of the techniques used in the dissertation.

In 1903 Einstein and Besso discussed a theory of dissociation that required the assumption of such aggregates, the "hypothesis of ionic hydrates" ("Ionenhydrathypothese"), as Besso called it,[69] claiming that this assumption resolves difficulties with Ostwald's law of dilution.[70] The assumption also opens the way to a simple calculation of the sizes of ions in solution, based on hydrodynamical considerations. In 1902 Sutherland had considered a calculation of the sizes of ions on the basis of Stokes's formula, but rejected it as in disagreement with experimental data.[71] Sutherland did not use the assumption of ionic hydrates, which can avoid such disagreement by permitting ionic sizes to vary with such physical conditions as temperature and concentration.[72] The idea of determining the sizes of ions by means of classical hydrodynamics occurred to Einstein in 1903, when he proposed to Besso what appears to be just the calculation that Sutherland had rejected:

> Have you already calculated the absolute magnitude of ions on the assumption that they are spheres and so large that the hydrodynamical equations for viscous fluids are applicable? With our knowledge of the absolute magnitude of the elec-

---

[64] Einstein presumably had acquired a basic knowledge of diffusion in liquids from his study of *Violle 1893* (see Vol. 1, "Albert Einstein—Beitrag für sein Lebensbild," p. lxiv). Chapter 4 contains an extensive treatment of diffusion and osmosis.

[65] See Vol. 1, the editorial note, "Einstein on Molecular Forces," pp. 264–266, and, in this volume, the editorial note, "Einstein on the Nature of Molecular Forces," pp. 3–8.

[66] See *Townsend 1920*, pp. 209–214, for a review of the use of Stokes's law in the interpretation of experiments on the determination of atomic charges.

[67] For a contemporary discussion of the application of Stokes's law to electrolytic conduction, see *Bousfield 1905b*. For a later survey, see

Herzfeld 1921, pp. 1011–1018.

[68] For evidence of this interest, see *Einstein 1902a* (Doc. 2) and Michele Besso to Einstein, 7–11 February 1903.

[69] See ibid.

[70] For a discussion of this law, its failure for strong electrolytes, and the resolution of this difficulty by the "hypothesis that the ions of an electrolyte consist of molecular aggregates in combination with water," see *Bousfield 1905a*, p. 563.

[71] In *Sutherland 1902*, Sutherland wrote: "Now this simple theory must have been written down by many a physicist and found to be wanting" (p. 167).

[72] This conclusion was drawn by Bousfield (see *Bousfield 1905b*, p. 264).

tron [charge] this would be a simple matter indeed. I would have done it myself but lack the reference material and the time; you could also bring in diffusion in order to obtain information about neutral salt molecules in solution.

Hast Du die absolute Größe der Ionen schon ausgerechnet unter der Voraussetzung, daß dieselben Kugeln und so groß sind, daß die Gleichungen der Hydrodynamik reibender Flüssigkeiten anwendbar sind. Bei unserer Kenntnis der absoluten Größe des Elektrons wäre dies ja eine einfache Sache. Ich hätte es selbst gethan, aber es fehlt mir an Litteratur und Zeit; auch die Diffusion könntest Du heranziehen, um über die neutralen Salzmoleküle in Lösung Aufschluss zu erhalten.[73]

This passage is remarkable, because both key elements of Einstein's method for the determination of molecular dimensions, the theories of hydrodynamics and diffusion, are already mentioned, although the reference to hydrodynamics probably covers only Stokes's law. While a program very similar to the first of Einstein's proposals to Besso is pursued in *Bousfield 1905a, 1905b*,[74] Einstein's dissertation can be seen to be an elaboration of the second proposal, regarding diffusion and neutral salt molecules. Einstein may thus have been proceeding similarly to Nernst, who first developed his theory of diffusion for the simpler case of nonelectrolytes.[75] The study of sugar solutions could draw upon extensive and relatively precise numerical data on viscosity and the coefficient of diffusion,[76] avoiding problems of dissociation and electrical interactions.[77]

V

The results obtained with Einstein's method for the determination of molecular dimensions differed from those obtained by other methods at the time, even when new data taken from *Landolt and Börnstein 1905* were used to recalculate them. In his papers on Brownian motion, Einstein cited either the value he obtained for Avogadro's number, or a more standard one.[78] Only once, in *Einstein 1908c* (Doc. 50), did he comment on the uncertainty in the determination of this number.[79] By 1909 Perrin's careful measurements of Brownian motion produced a new value for Avogadro's number, significantly different

---

[73] See Einstein to Michele Besso, 17 March 1903.

[74] For a critical evaluation of Bousfield's work, see *Dhar 1914*, p. 64.

[75] See *Nernst 1888*.

[76] The tables of data for sugar solutions in *Landolt and Börnstein 1894* and *1905* are extremely detailed.

[77] For a discussion of these problems, see *Sutherland 1902*, pp. 167ff, and, for a later review, *Dhar 1914*; for an account of the problem

of internal friction in electrolytes, see also *Herzfeld 1921*, pp. 1013–1018.

[78] In *Einstein 1905k* (Doc. 16), Einstein cited a value for Avogadro's number taken from the kinetic theory of gases; in *Einstein 1907c* (Doc. 40), he cited the value obtained in *Einstein 1906c* (Doc. 33); in *Einstein 1908c* (Doc. 50), he again cited a value close to that derived from gas theory.

[79] See *Einstein 1908c* (Doc. 50), p. 237, fn. 2.

from the values Einstein obtained from his hydrodynamical method and from Planck's black-body radiation law.[80] For Einstein, this discrepancy was particularly significant in view of what he regarded as the problematic nature of Planck's derivation of the radiation law.[81]

In 1909 Einstein drew Perrin's attention to his hydrodynamical method for determining the size of solute molecules. He emphasized that this method allows one to take into account the volume of any water molecules attached to the solute molecules, and suggested its application to the suspensions studied by Perrin.[82] In the following year, an experimental study of Einstein's formula for the viscosity coefficients (eq. [1] above) was performed in Perrin's laboratory by Jacques Bancelin.[83] Bancelin studied uniform aqueous emulsions of gamboge, prepared with the help of Perrin's method of fractional centrifugation. Bancelin confirmed that the increase in viscosity does not depend on the size of the suspended particles, but only on the fraction of the total volume that they occupy. However, he found a value for the increased viscosity that differs significantly from Einstein's prediction.[84] Bancelin sent a report of his experiments to Einstein, apparently citing a value of 3.9 for the coefficient of φ in eq. (1), instead of the predicted value of 1.[85]

After an unsuccessful attempt to find an error in his calculations,[86] Einstein wrote to his student and collaborator Ludwig Hopf:[87]

> I have checked my previous calculations and arguments and found no error in them. You would be doing a great service in this matter if you would carefully recheck my investigation. Either there is an error in the work, or the volume of

[80] For Einstein's derivation of Avogadro's number from the law of black-body radiation, see *Einstein 1905i* (Doc. 14), pp. 136–137.

[81] See Einstein to Jean Perrin, 11 November 1909, quoted in § II. For a discussion of Einstein's views on the foundations of Planck's theory, see the editorial note, "Einstein's Early Work on the Quantum Hypothesis," pp. 137–138.

[82] Einstein wrote: "It would perhaps not be uninteresting to apply to your suspensions the method for determining the volume of the suspended substance from the coefficients of viscosity and to make a comparison with the results of your methods" ("Es wäre vielleicht nicht uninteressant, die Methode zur Bestimmung des Volumens der suspendierten Substanz aus den Reibungskoeffizienten bei Ihren Suspensionen anzuwenden und mit den Resultaten Ihrer Methoden zu vergleichen") (Einstein to Jean Perrin, 11 November 1909).

[83] See Einstein to Jean Perrin, 12 January 1911 and *Bancelin 1911a* and *1911b*.

[84] For Einstein's value, see eq. (1).

[85] On 12 January 1911, Einstein wrote to Perrin: "You will in any case be familiar with Bancelin's report to me as well as with my reply" ("Der Bericht von Herrn Bancelin an mich sowie meine Antwort an ihn werden Ihnen jedenfalls bekannt sein"). This letter cites the value of 3.9 as Bancelin's result. In a letter to Hopf written shortly before, Einstein cited a value of 3.8 (see Einstein to Ludwig Hopf, before 12 January 1911). For further evidence of correspondence between Einstein and Bancelin, see also *Bancelin 1911a*, p. 1383.

[86] See *Einstein 1905j* (Doc. 15), note 14, for evidence of this attempt.

[87] Ludwig Hopf was Einstein's student at the University of Zurich. In the summer semester of 1910 he registered for Einstein's lectures on mechanics and kinetic theory of heat, and for his physics seminar (Student Files, SzZU, Kassa-Archiv). In the same year he published two joint papers with Einstein (*Einstein and Hopf 1910a*, and *1910b*).

Perrin's suspended substance in the suspended state is greater than Perrin believes.

Ich habe nun meine damaligen Rechnungen & Ueberlegungen geprüft und keinen Fehler darin gefunden. Sie würden sich sehr um die Sache verdient machen, wenn Sie meine Untersuchung seriös nachprüfen würden. Entweder ist ein Fehler in der Arbeit oder das Volumen von Perrins suspendierter Substanz ist in suspendiertem Zustande grösser als Perrin glaubt.[88]

Hopf found an error in the derivatives of the velocity components, which occur in the equations for the pressure components on p. 12 of Einstein's dissertation. After correction of this error, the coefficient of φ in eq. (1) becomes 2.5.[89]

By mid-January 1911 Einstein had informed Bancelin and Perrin of Hopf's discovery of the error in his calculations.[90] The remaining discrepancy between the corrected factor 2.5 in eq. (1) and Bancelin's experimental value of 3.9 led Einstein to suspect that there might also be an experimental error. He asked Perrin:

Wouldn't it be possible that your mastic particles, like colloids, are in a swollen state?[91] The influence of such a swelling 3.9/2.5 would be of rather slight influence on Brownian motion, so that it might possibly have escaped you.

Wäre es nicht möglich, dass Ihre Mastixteilchen nach Art von Kolloiden sich in gequollenem Zustand befinden? Der Einfluss einer derartigen Quellung 3,9/2,5 wäre ja auf die Brown'sche Bewegung von ziemlich geringem Einfluss, sodass er Ihnen möglicherweise entgangen sein könnte.[92]

On 21 January, Einstein submitted his correction (*Einstein 1911d*) for publication. He presented the corrected form of some of the equations in *Einstein 1905j* (Doc. 15), and

[88] Einstein to Ludwig Hopf, before 12 January 1911.

[89] In 1926, M. Kunitz claimed to have found another error in Einstein's derivation of the equation for the coefficients of viscosity (see John Northoff to Einstein, 5 April 1926, and the enclosed preprint of an article by Kunitz, published as *Kunitz 1926*). The supposed error is based on a misprint in *Einstein 1906a* (see *Kunitz 1926*). In his reply, Einstein proposed experiments on suspensions "for which φ is known, in order to test binding of H_2O in solutions" ("bei welchen φ bekannt ist, um dann bei Lösungen Bindung von H_2O zu prüfen") (Einstein to John Northoff, after 5 April 1926).

[90] Einstein to Jean Perrin, 12 January 1911. See this letter and *Bancelin 1911a*, p. 1383, for evidence of a letter by Einstein to Bancelin.

[91] Einstein later explained some peculiarities of viscosity in colloidal solutions: "A significantly greater increase in viscosity occurs in certain colloidal solutions of relatively small concentration, there being no sharply defined viscosity coefficient. Firm connections between particles then arise, forming chains throughout the volume, which, however, constantly re-form and dissolve over time according to statistical laws" ("Bei gewissen kolloidalen Lösungen von relativ kleiner Konzentration tritt oft eine bedeutend grössere Erhöhung der Viskosität ein, wobei es überhaupt keinen scharf definierten Viskositäts-Koeffizienten gibt. Es liegen dann feste Verbindungen der Teilchen vor, die Ketten durch das ganze Volumen bilden, die aber im Laufe der Zeit nach statistischen Gesetzen sich beständig neu bilden und wieder lösen") (Einstein to Hans Albert Einstein, before 13 December 1940).

[92] Einstein to Jean Perrin, 12 January 1911.

recalculated Avogadro's number. He obtained a value of 6.56×10^{23} per mole, a value that is close to those derived from kinetic theory and Planck's black-body radiation formula.

Bancelin continued his experiments, with results that brought experiment and theory into closer agreement. Four months later, he presented a paper on his viscosity measurements to the French Academy of Sciences,[93] giving a value of 2.9 as the coefficient of φ in eq. (1). Bancelin also recalculated Avogadro's number by extrapolating his results for emulsions to sugar solutions, and found a value of 7.0×10^{23} per mole.

Einstein's dissertation was at first overshadowed by his more spectacular work on Brownian motion, and it required an initiative by Einstein to bring it to the attention of his fellow scientists.[94] But the wide variety of applications of its results ultimately made the dissertation one of his most frequently cited papers.[95]

[93] See *Bancelin 1911a*. The paper was presented on 22 May 1911. Bancelin later published an article in German on the same results (see *Bancelin 1911b*).

[94] As is indicated by Einstein's letter of 11

November 1909 to Jean Perrin, he apparently drew Perrin's and thus Bancelin's attention to his work.

[95] See *Cawkell and Garfield 1980*.

15. A New Determination of Molecular Dimensions

[*Einstein 1905j*]

DATED Bern, 30 April 1905

PUBLISHED BY: Buchdruckerei K. J. Wyss, Bern (1906)

Slightly revised version published as "Eine neue Bestimmung der Moleküldimensionen," *Annalen der Physik* 19 (1906): 289–305 (*Einstein 1906a*). Dated Bern, 30 April 1905, received 19 August 1905, published 8 February 1906.

Revised version published in *Einstein 1922*, pp. 25–40.

The occurrence of page breaks in the slightly revised version published in the *Annalen der Physik* is noted on this document in the margins: the page indication is followed by the first word, part of a word, or part of an equation at the beginning of the page cited.

EINE NEUE BESTIMMUNG DER MOLEKÜLDIMENSIONEN

[1] ## INAUGURAL-DISSERTATION

ZUR

ERLANGUNG DER PHILOSOPHISCHEN DOKTORWÜRDE

DER

HOHEN PHILOSOPISCHEN FAKULTÄT
(MATHEMATISCH-NATURWISSENSCHAFTLICHE SEKTION)

DER

UNIVERSITÄT ZÜRICH

[2] VORGELEGT

VON

ALBERT EINSTEIN

AUS ZÜRICH

[3] Begutachtet von den Herren Prof. Dr. A. KLEINER
und
Prof. Dr. H. BURKHARDT

BERN

BUCHDRUCKEREI K. J. WYSS

1905

Meinem Freunde

Herrn Dr. Marcel Grossmann

GEWIDMET   [4]

Eine neue Bestimmung der Moleküldimensionen.

Die ältesten Bestimmungen der wahren Grösse der Moleküle hat die kinetische Theorie der Gase ermöglicht, während die an Flüssigkeiten beobachteten physikalischen Phänomene bis [5] jetzt zur Bestimmung der Molekülgrössen nicht gedient haben. Es liegt dies ohne Zweifel an den bisher unüberwindlichen Schwierigkeiten, welche der Entwickelung einer ins einzelne gehenden molekularkinetischen Theorie der Flüssigkeiten entgegenstehen. In dieser Arbeit soll nun gezeigt werden, dass man die Grösse der Moleküle des gelösten Stoffs in einer nicht dissoziierten verdünnten Lösung aus der inneren Reibung der Lösung und des reinen Lösungsmittels und aus der Diffusion des gelösten Stoffes im Lösungsmittel ermitteln kann, wenn das Volumen eines Moleküls des gelösten Stoffs gross ist gegen das Volumen eines Moleküls des Lösungsmittels. Ein derartiges gelöstes Molekül wird sich nämlich bezüglich seiner Beweglichkeit im Lösungsmittel und bezüglich seiner Beeinflussung der inneren Reibung des letzteren annähernd wie ein im Lösungsmittel suspendierter fester Körper verhalten, und es wird erlaubt sein, auf die Bewegung des Lösungsmittels in unmittelbarer Nähe eines Moleküls die hydrodynamischen Gleichungen anzuwenden, in welchen die Flüssigkeit als homogen betrachtet, eine molekulare Struktur derselben also nicht berücksichtigt wird. Als Form der festen Körper, welche die gelösten Moleküle darstellen sollen, wählen wir die Kugelform.

— 6 —

§ 1. Ueber die Beeinflussung der Bewegung einer Flüssigkeit durch eine sehr kleine in derselben suspendierte Kugel.

Es liege eine inkompressible homogene Flüssigkeit mit dem Reibungskoeffizienten k der Betrachtung zugrunde, deren Geschwindigkeitskomponenten u, v, w als Funktionen der Koordinaten x, y, z und der Zeit gegeben seien. Von einem beliebigen Punkt x_0, y_0, z_0 aus denken wir uns die Funktionen u, v, w als Funktionen von $x - x_0$, $y - y_0$, $z - z_0$ nach dem Taylorschen Satze entwickelt und um diesen Punkt ein so kleines Gebiet G abgegrenzt, dass innerhalb desselben nur die linearen Glieder dieser Entwickelung berücksichtigt werden müssen. Die Bewegung der in G enthaltenen Flüssigkeit kann dann bekanntlich als die Superposition dreier Bewegungen aufgefasst werden, nämlich   P.290[/dem]

1. einer Parallelverschiebung aller Flüssigkeitsteilchen ohne Aenderung von deren relativer Lage,

2. einer Drehung der Flüssigkeit ohne Aenderung der relativen Lage der Flüssigkeitsteilchen,

3. einer Dilatationsbewegung in drei aufeinander senkrechten Richtungen (den Hauptdilatationsrichtungen).   [6]

Wir denken uns nun im Gebiete G einen kugelförmigen starren Körper, dessen Mittelpunkt im Punkte x_0, y_0, z_0 liege und dessen Dimensionen gegen diejenigen des Gebietes G sehr klein seien. Wir nehmen ferner an, dass die betrachtete Bewegung eine so langsame sei, dass die kinetische Energie der Kugel sowie diejenige der Flüssigkeit vernachlässigt werden können. Es werde ferner angenommen, dass die Geschwindigkeitskomponenten eines Oberflächenelementes der Kugel mit den entsprechenden Geschwindigkeitskomponenten der unmittelbar benachbarten Flüssigkeitsteilchen übereinstimme, d. h., dass auch die (kontinuierlich gedachte) Trennungsschicht überall einen nicht unendlich kleinen Koeffizienten der inneren Reibung aufweise.   [7]

Es ist ohne weiteres klar, dass die Kugel die Teilbewegungen 1. und 2. einfach mitmacht, ohne die Bewegung der benachbarten Flüssigkeit zu modifizieren, da sich bei diesen Teilbewegungen die Flüssigkeit wie ein starrer Körper bewegt, und da wir die Wirkungen der Trägheit vernachlässigt haben.

— 7 —

Die Bewegung 3. aber wird durch das Vorhandensein der Kugel modifiziert, und es wird unsere nächste Aufgabe sein, den Einfluss der Kugel auf diese Flüssigkeitsbewegung zu untersuchen. Beziehen wir die Bewegung 3. auf ein Koordinatensystem, dessen Achsen den Hauptdilatationsrichtungen parallel sind, und setzen wir

$$x - x_0 = \xi,$$
$$y - y_0 = \eta,$$
$$z - z_0 = \zeta,$$

P.291[/so] so lässt sich jene Bewegung, falls die Kugel nicht vorhanden ist, durch die Gleichungen darstellen:

(1)
$$\begin{cases} u_0 = A\,\xi, \\ v_0 = B\,\eta, \\ w_0 = C\,\zeta\,; \end{cases}$$

A, B, C sind Konstanten, welche wegen der Inkompressibilität der Flüssigkeit die Bedingung erfüllen:

(2) $$A + B + C = 0.$$

Befindet sich nun im Punke x_0, y_0, z_0 die starre Kugel mit dem Radius P, so ändert sich in der Umgebung derselben die Flüssigkeitsbewegung. Im folgenden wollen wir der Bequemlichkeit wegen P als « endlich » bezeichnen, dagegen die Werte von ξ, η, ζ, für welche die Flüssigkeitsbewegung durch die Kugel nicht mehr merklich modifiziert wird, als „unendlich gross".

Zunächst ist wegen der Symmetrie der betrachteten Flüssigkeitsbewegung klar, dass die Kugel bei der betrachteten Bewegung weder eine Translation noch eine Drehung ausführen kann, [8] und wir erhalten die Grenzbedingungen:

$$u = v = w = 0 \text{ für } \rho = P,$$

wobei

$$\rho = \sqrt{\xi^2 + \eta^2 + \zeta^2} > 0$$

gesetzt ist. Hierbei bedeuten u, v, w die Geschwindigkeitskomponenten der nun betrachteten (durch die Kugel modifizierten) Bewegung. Setzt man

(3)
$$\begin{cases} u = A\,\xi + u_1, \\ v = B\,\eta + v_1, \\ w = C\,\zeta + w_1, \end{cases}$$

so müsste, da die in Gleichungen (3) dargestellte Bewegung im Unendlichen in die in Gleichungen (1) dargestellte über-

— 8 —

gehen soll, die Geschwindigkeiten u_1, v_1, w_1 im Unendlichen verschwinden.

Die Funktionen u, v, w haben den Gleichungen der Hydrodynamik zu genügen unter Berücksichtigung der inneren Reibung und unter Vernachlässigung der Trägheit. Es gelten also die Gleichungen [1])

P.292[/und]

[9]

$$(4) \quad \begin{cases} \dfrac{\delta p}{\delta \xi} = k \, \Delta u \quad \dfrac{\delta p}{\delta \eta} = k \, \Delta v \quad \dfrac{\delta p}{\delta \zeta} = \Delta w, \\[2mm] \dfrac{\delta u}{\delta \xi} + \dfrac{\delta v}{\delta \eta} + \dfrac{\delta w}{\delta \zeta} = 0, \end{cases}$$

[11]

wobei Δ den Operator

$$\frac{\delta^2}{\delta \xi^2} + \frac{\delta^2}{\delta \eta^2} + \frac{\delta^2}{\delta \zeta^2}$$

und p den hydrostatischen Druck bedeutet.

Da die Gleichungen (1) Lösungen der Gleichungen (4) und letztere linear sind, müssen nach (3) auch die Grössen u_1, v_1, w_1 den Gleichungen (4) genügen. Ich bestimmte u_1, v_1, w_1 und p nach einer im § 4 der erwähnten Kirchhoffschen Vorlesung angegebenen Methode [2]) und fand:

---

[1]) G. Kirchhoff, Vorlesungen über Mechanik. 26. Vorl. [10]

[2]) « Aus den Gleichungen (4) folgt $\Delta p = 0$. Ist p dieser Bedingung [12] gemäss angenommen und eine Funktion V bestimmt, die der Gleichung

$$\Delta V = \frac{1}{k} \, p$$

genügt, so erfüllt man die Gleichungen (4), wenn man

$$u = \frac{\delta V}{\delta \xi} + u', \quad v = \frac{\delta V}{\delta \eta} + v', \quad w = \frac{\delta V}{\delta \zeta} + w'$$

setzt und u', v' w' so wählt, dass $\Delta u' = 0$, $\Delta v' = 0$ und $\Delta w' = 0$ und

$$\frac{\delta u'}{\delta \xi} + \frac{\delta v'}{\delta \eta} + \frac{\delta w'}{\delta \zeta} = - \frac{1}{k} \, p$$

ist. »

Setzt man nun

$$\frac{p}{k} = 2 \, c \, \frac{\delta^2 \dfrac{1}{\rho}}{\delta \xi^3}$$

[13]

und im Einklang hiermit

$$V = c \, \frac{\delta^2 \rho}{\delta \xi^3} + b \, \frac{\delta^2 \dfrac{1}{\rho}}{\delta \xi^2} + \frac{a}{2} \Big(\xi^2 - \frac{\eta^2}{2} - \frac{\zeta^2}{2} \Big)$$

[14]

und

— 9 —

[16] P.293[/p =]

$$p = -\tfrac{5}{3} k\, P^3 \left\{ A\, \frac{\partial^2\left(\frac{1}{\rho}\right)}{\partial\,\xi^2} + B\, \frac{\partial^2\left(\frac{1}{\rho}\right)}{\partial\,\eta^2} + C\, \frac{\partial^2\left(\frac{1}{\delta}\right)}{\partial\,\zeta^2} \right\} + \text{konst.},$$

[17]

$$(5)\begin{cases} u = A\,\xi - \tfrac{5}{3} P^3\, A\, \dfrac{\xi}{\rho^3} - \dfrac{\partial D}{\partial\,\xi}, \\[2mm] v = B\,\eta - \tfrac{5}{3} P^3\, B\, \dfrac{\eta}{\rho^3} - \dfrac{\partial D}{\partial\,\eta}, \\[2mm] w = C\,\zeta - \tfrac{5}{3} P^3\, C\, \dfrac{\zeta}{\rho^3} - \dfrac{\partial D}{\partial\,\zeta}, \end{cases}$$

wobei

$$(5\,a)\quad \begin{cases} D = A\left\{ \tfrac{5}{6} P^3\, \dfrac{\partial^2 \rho}{\partial\,\xi^2} + \tfrac{1}{6} P^5\, \dfrac{\partial^2\left(\frac{1}{\rho}\right)}{\partial\,\xi^2} \right\} \\[4mm] \quad + B\left\{ \tfrac{5}{6} P^3\, \dfrac{\partial^2 \rho}{\partial\,\eta^2} + \tfrac{1}{6} P^5\, \dfrac{\partial^2\left(\frac{1}{\rho}\right)}{\partial\,\eta^2} \right\} \\[4mm] \quad + C\left\{ \tfrac{5}{6} P^3\, \dfrac{\partial^2 \rho}{\partial\,\zeta^2} + \tfrac{1}{6} P^5\, \dfrac{\partial^2\left(\frac{1}{\rho}\right)}{\partial\,\zeta^2} \right\}. \end{cases}$$

Es ist leicht zu beweisen, dass die Gleichungen (5) Lösungen der Gleichungen (4) sind. Denn da

$$\Delta\,\xi = 0,\ \ \Delta\,\frac{1}{\rho} = 0,\ \ \Delta\,\rho = \frac{2}{\rho}$$

und

$$\Delta\,\left(\frac{\xi}{\rho^3}\right) = -\,\frac{\partial}{\partial\,\xi}\left\{ \Delta\,\left(\frac{1}{\rho}\right) \right\} = 0,$$

erhält man

$$k\,\Delta\,u = -k\,\frac{\partial}{\partial\,\xi}\{\Delta\,D\} = -k\,\frac{\partial}{\partial\,\xi}\left\{ \tfrac{5}{3} P^3\, A\,\frac{\partial^2\frac{1}{\rho}}{\partial\,\xi^2} + \tfrac{5}{3} P^3\, B\,\frac{\partial^2\frac{1}{\rho}}{\partial\,\eta^2} + \ldots \right\}.$$

[15]

$$u' = -\,2\,c\,\frac{\partial\,\frac{1}{\delta}}{\partial\,\xi},\ \ v' = C.\ \ w' = 0,$$

so lassen sich die Konstanten a, b, c so bestimmen, dass für $\rho = P$ $u = v = w = 0$ ist. Durch Superposition dreier derartiger Lösungen erhält man die in den Gleichungen (5) und (5 a) angegebene Lösung.

— 10 —

Der zuletzt erhaltene Ausdruck ist aber nach der ersten der Gleichungen (5) mit $\frac{\delta n}{\delta \xi}$ identisch. Auf gleiche Weise zeigt man, [18] dass die zweite und dritte der Gleichungen (4) erfüllt ist. Ferner erhält man

$$\frac{\delta u}{\delta \xi} + \frac{\delta v}{\delta \eta} + \frac{\delta w}{\delta \xi} = (A + B + C)$$

$$+ \tfrac{5}{3} P^3 \left\{ A \frac{\delta^2 \left(\frac{1}{\rho} \right)}{\delta \xi^2} + B \frac{\delta^2 \left(\frac{1}{\rho} \right)}{\delta \eta^2} + C \frac{\delta^2 \left(\frac{1}{\rho} \right)}{\delta \zeta^2} \right\} - \Delta D.$$

Da aber nach Gleichung (5 a)   P.294[/Da]

$$\Delta D = \tfrac{5}{3} A P^3 \left\{ A \frac{\delta^2 \left(\frac{1}{\rho} \right)}{\delta \xi^2} + B \frac{\delta^2 \left(\frac{1}{\rho} \right)}{\delta \eta^2} + C \frac{\delta^2 \left(\frac{1}{\rho} \right)}{\delta \zeta^2} \right\},$$

so folgt, dass auch die letzte der Gleichungen (4) erfüllt ist. Was die Grenzbedingungen betrifft, so gehen zunächst für unendlich grosse ρ unsere Gleichungen für u, v, w in die Gleichungen (1) über. Durch Einsetzen des Wertes von D aus Gleichung (5 a) in die zweite der Gleichungen (5) erhält man:

(6)
$$\left\{ \begin{aligned} u &= A\xi - \tfrac{5}{2} \frac{P^3}{\rho^6} \xi \left(A\xi^2 + B\eta^2 + C\zeta^2 \right) \\ &\quad + \tfrac{5}{2} \frac{P^5}{\rho^7} \xi \left(A\xi^2 + B\eta^2 + C\zeta^2 \right) - \frac{P^5}{\rho^5} A\xi. \end{aligned} \right.$$
[19]

Man erkennt, dass u für $\rho = P$ verschwindet. Gleiches gilt aus Symmetriegründen für v und w. Es ist nun bewiesen, dass durch die Gleichungen (5) sowohl den Gleichungen (4) als auch den Grenzbedingungen der Aufgabe Genüge geleistet ist.

Es lässt sich auch beweisen, dass die Gleichungen (5) die einzige mit den Grenzbedingungen der Aufgabe verträgliche Lösung der Gleichungen (4) sind. Der Beweis soll hier nur angedeutet werden. Es mögen in einem endlichen Raume die Geschwindigkeitskomponenten u, v, w einer Flüssigkeit den Gleichungen (4) genügen. Existierte noch eine andere Lösung U, V, W der Gleichungen (4), bei welcher an den Grenzen des betrachteten Raumes $U = u$, $V = v$, $W = w$ ist, so ist $(U - u, V - v, W - w)$ eine Lösung der Geichungen (4), bei welcher die Geschwindigkeitskomponenten an der Grenze des Raumes verschwinden. Der in dem betrachteten Raume befindlichen Flüssigkeit wird also keine mechanische Arbeit zugeführt. Da wir die lebendige Kraft der Flüssigkeit vernachlässigt haben,

— 11 —

so folgt daraus, dass auch die im betrachteten Raume in Wärme verwandelte Arbeit gleich Null ist. Hieraus folgert man, dass [20] im ganzen Raume $u = u_1$, $v = v_1$, $w = w_1$ sein muss, falls der Raum wenigstens zum Teil durch ruhende Wände begrenzt ist. Durch Grenzübergang kann dies Resultat auch auf den Fall ausgedehnt werden, dass, wie in dem oben betrachteten Falle, der betrachtete Raum unendlich ist. Man kann so dartun, dass die oben gefundene Lösung die einzige Lösung der Aufgabe ist.

P.295[/Wir]   Wir legen nun um den Punkt x_0, y_0, z_0 eine Kugel vom Radius R, wobei R gegen P unendlich gross sei, und berechnen die Energie, welche in der innerhalb der Kugel befindlichen Flüssigkeit (in der Zeiteinheit) in Wärme verwandelt wird. Diese Energie W ist gleich der der Flüssigkeit mechanisch zugeführten Arbeit. Bezeichnet man die Komponenten des auf die Oberfläche der Kugel vom Radius R ausgeübten Druckes mit X_n, Y_n, Z_n, so ist:

$$W = \int (X_n u + Y_n v + Z_n w)\, ds,$$

wobei das Integral über die Oberfläche der Kugel vom Radius R zu erstrecken ist. Hierbei ist:

[21]
$$X_n = -\left(X\xi \frac{\xi}{\rho} + X\eta \frac{\eta}{\rho} + X\zeta \frac{\zeta}{\rho} \right),$$
$$Y_n = -\left(Y\xi \frac{\xi}{\rho} + Y\eta \frac{\eta}{\rho} + Y\zeta \frac{\zeta}{\rho} \right),$$
$$Z_n = -\left(Z\xi \frac{\xi}{\rho} + Z\eta \frac{\eta}{\rho} + Z\zeta \frac{\zeta}{\rho} \right),$$

[22] wobei

$$X_\xi = p - 2k\frac{\delta u}{\delta \xi}, \qquad Y_\zeta = Z_\eta = -k\left(\frac{\delta v}{\delta \zeta} + \frac{\delta w}{\delta \eta}\right),$$
$$Y_\eta = p - 2k\frac{\delta v}{\delta \eta}, \qquad Z_\xi = X_\zeta = -k\left(\frac{\delta w}{\delta \xi} + \frac{\delta u}{\delta \zeta}\right),$$
$$Z_\zeta = p - 2k\frac{\delta w}{\delta \zeta}, \qquad X_\eta = Y_\xi = -k\left(\frac{\delta u}{\delta \eta} + \frac{\delta v}{\delta \xi}\right).$$

Die Ausdrücke für u, v, w vereinfachen sich, wenn wir beachten, dass für $\rho = R$ die Glieder mit dem Faktor P^5/ρ^5 gegenüber denen mit dem Faktor P^3/ρ^3 verschwinden. Wir haben zu setzen:

— 12 —

$$(6\,a) \begin{cases} u = A\,\xi - \tfrac{5}{2}\,P^3\,\dfrac{\xi\,(A\,\xi^2 + \mathbf{B}\,\eta^2 + C\,\zeta^2)}{\rho^5}, \\[2mm] v = B\,\eta - \tfrac{5}{2}\,P^3\,\dfrac{\eta\,(A\,\xi^2 + B\,\eta^2 + C\,\zeta^2)}{\rho^5}. \\[2mm] w = C\,\zeta - \tfrac{5}{2}\,P^3\,\dfrac{\zeta\,(A\,\xi^2 + B\,\eta^2 + C\,\zeta^2)}{\rho^5}. \end{cases} \qquad [23]$$

Für p erhalten wir aus der ersten der Gleichungen (5) durch die entsprechenden Vernachlässigungen

$$p = -5\,k\,P^3\,\frac{A\,\xi^2 + B\,\eta^2 + C\,\zeta^2}{\rho^5} + \text{konst.} \qquad [24]$$

Wir erhalten zunächst: P.296[/Wir]

$$X_\xi = -2\,k\,A + 10\,k\,P^3\frac{A\,\xi^2}{\rho^5} - 25\,k\,P^3\frac{\xi^2\,(A\,\xi^2 + B\,\eta^2 + C\,\zeta^2)}{\rho^7}, \qquad [25]$$

$$X_\eta = \qquad\quad + 10\,k\,P^3\frac{A\,\xi\eta}{\rho^5} - 25\,k\,P^3\frac{\eta^2\,(A\,\xi^2 + B\,\eta^2 + C\,\zeta^2)}{\rho^7}, \qquad [26]$$

$$X_\zeta = \qquad\quad + 10\,k\,P^3\frac{A\,\xi\zeta}{\rho^5} + 25\,k\,P^3\frac{\zeta^2\,(A\,\xi^2 + B\,\eta^2 + C\,\zeta^2)}{\rho^7},$$

und hieraus

$$X_n = 2\,A\,k\,\frac{\xi}{\rho} - 10\,A\,k\,P^3\frac{\xi}{\rho^4} + 25\,k\,P^3\frac{\xi\,(A\,\xi^2 + B\,\eta^2 + C\,\zeta^2)}{\rho^6}. \qquad [27]$$

Mit Hilfe der durch zyklische Vertauschung abzuleitenden Ausdrücke für Y_n und Z_n erhält man unter Vernachlässigung aller Glieder, die das Verhältnis P/ρ in einer höheren als der dritten Potenz enthalten:

$$X_n\,u + Y_n\,v + Z_n\,w + \frac{2\,k}{\rho}\,(A^2\,\xi^2 + B^2\,\eta^2 + C^2\,\zeta^2) \qquad [28]$$

$$-\,10\,k\,\frac{P^3}{\rho^4}(A^2\,\xi^2 + . + .) + 20\,k\frac{P^3}{\rho^6}(A\,\xi^2 + . + .)^2.$$

Integriert man über die Kugel und berücksichtigt, dass

$$\int d\,s = 4\,R^2\,\pi,$$

$$\int \xi^2\,d\,s = \int \eta^2\,d\,s = \int \zeta^2\,d\,s = \tfrac{4}{3}\,\pi\,R^4,$$

$$\int \xi^4\,d\,s = \int \eta^4\,d\,s = \int \zeta^4\,d\,s = \tfrac{4}{5}\,\pi\,R^6,$$

$$\int \eta^2\,\zeta^2\,d\,s = \int \zeta^2\,\xi^2\,d\,s = \int \xi^2\,\eta^2\,d\,s = \tfrac{4}{15}\,\pi\,R^6, \qquad [29]$$

$$\int (A\,\xi^2 + B\,\eta^2 + C\,\zeta^2)^2\,d\,s = \tfrac{4}{15}\,\pi\,R^6\,(A^2 + B^2 + C^2). \qquad [30]$$

— 13 —

so erhält man:

[31]   (7) $W = \frac{8}{3} \pi R^3 k \delta^2 - \frac{8}{3} \pi P^3 k \delta^2 = 2 \delta^2 k (V - \Phi),$

wobei

[32]

$$\delta = A^2 + B^2 + C^2,$$
$$\frac{4}{3} \pi R^3 = V$$

und

$$\frac{4}{3} \pi P^3 = \Phi$$

gesetzt ist. Wäre die suspendierte Kugel nicht vorhanden ($\Phi = 0$), so erhielte man für die im Volumen V verzehrte Energie

(7 a) $W_0 = 2 \delta^2 k V.$

P.297[/Durch] Durch das Vorhandensein der Kugel wird also die verzehrte
[33]   Energie um $2 \delta^2 k \Phi$ verkleinert. Es ist bemerkenswert, dass der Einfluss der suspendierten Kugel auf die Grösse der verzehrten Energie gerade so gross ist, wie er wäre, wenn durch die Anwesenheit der Kugel die Bewegung der sie umgebenden Flüssigkeit gar nicht modifiziert würde.

§ 2. Berechnung des Reibungskoeffizienten einer Flüssigkeit, in welcher sehr viele kleine Kugeln in regelloser Verteilung suspendiert sind.

Wir haben im vorstehenden den Fall betrachtet, dass in einem Gebiete G von der oben definierten Grössenordnung eine relativ zu diesem Gebiete sehr kleine Kugel suspendiert ist und untersucht, wie dieselbe die Flüssigkeitsbewegung beeinflusst. Wir wollen nun annehmen, dass in dem Gebiete G unendlich viele Kugeln von gleichem, und zwar so kleinem Radius regellos verteilt sind, dass das Volumen aller Kugeln zusammen sehr klein sei gegen das Gebiet G. Die Zahl der auf die Volumeneinheit entfallenden Kugeln sei n, wobei n allenthalben in der Flüssigkeit bis auf Vernachlässigbares konstant sei.

Wir gehen nun wieder aus von einer Bewegung einer homogenen Flüssigkeit ohne suspendierte Kugeln und betrachten wieder die allgemeinste Dilatationsbewegung. Sind keine Kugeln vorhanden, so können wir bei passender Wahl des Koordinatensystems die Geschwindigkeitskomponenten u_0, v_0, w_0 in dem beliebigen Punkte x, y, z des Gebietes G darstellen durch die Gleichungen:

— 14 —

$$u_0 = A\,x,$$
$$v_0 = B\,y,$$
$$w_0 = C\,z,$$

wobei

$$A + B + C = 0.$$

Eine im Punkte x_v, y_v, z_v suspendierte Kugel beeinflusst nun [34]
diese Bewegung in der aus Gleichung (6) ersichtlichen Weise.
Da wir den mittleren Abstand benachbarter Kugeln als sehr
gross gegen deren Radius wählen, und folglich die von allen
suspendierten Kugeln zusammen herrührenden zusätzlichen P.298[/sus-
Geschwindigkeitskomponenten gegen u_0, v_0, w_0 sehr klein sind, pendierten]
so erhalten wir für die Geschwindigkeitskomponenten u, v, w
in der Flüssigkeit unter Berücksichtigung der suspendierten
Kugeln und unter Vernachlässigung von Gliedern höherer Ord-
nungen:

$$
(8)
\begin{cases}
u = A\,x - \sum \left\{ \frac{5}{2}\frac{P^3}{\rho_\nu^2}\frac{\xi_\nu\left(A\xi_\nu^2 + B\eta_\nu^2 + C\zeta_\nu^2\right)}{\rho_\nu^3} \right. \\
\qquad\qquad \left. - \frac{5}{2}\frac{P^5}{\rho_\nu^4}\frac{\xi_\nu\left(A\xi_\nu^2 + B\eta_\nu^2 + C\zeta_\nu^2\right)}{\rho_\nu^3} + \frac{P^5}{\rho_\nu^4}\frac{A\xi_\nu}{\rho_\nu} \right\}, \\[2ex]
v = B\,y - \sum \left\{ \frac{5}{2}\frac{P^3}{\rho_\nu^2}\frac{\eta_\nu\left(A\xi_\nu^2 + B\eta_\nu^2 + C\zeta_\nu^2\right)}{\rho_\nu^3} \right. \\
\qquad\qquad \left. - \frac{5}{2}\frac{P^5}{\rho_\nu^4}\frac{\eta_\nu\left(A\xi_\nu^2 + B\eta_\nu^2 + C\zeta_\nu^2\right)}{\rho_\nu^3} + \frac{P^5}{\rho_\nu^4}\frac{B\eta_\nu}{\rho_\nu} \right\}, \\[2ex]
w = C\,z - \sum \left\{ \frac{5}{2}\frac{P^3}{\rho_\nu^2}\frac{\zeta_\nu\left(A\xi_\nu^3 + B\eta_\nu^2 + C\zeta_\nu^2\right)}{\rho_\nu^3} \right. \\
\qquad\qquad \left. - \frac{5}{2}\frac{P^5}{\rho_\nu^4}\frac{\zeta_\nu\left(A\xi_\nu^2 + B\eta_\nu^2 + C\zeta_\nu^2\right)}{\rho_\nu^3} + \frac{P^5}{\rho_\nu^4}\frac{C\zeta_\nu}{\rho_\nu} \right\},
\end{cases}
$$

[35]

wobei die Summation über alle Kugeln des Gebietes G zu erstrecken ist und

$$\xi_\nu = x - x_\nu,$$
$$\eta_\nu = y - y_\nu, \quad \rho_\nu = \sqrt{\xi_\nu^2 + \eta_\nu^2 + \zeta_\nu^2};$$
$$\zeta_\nu = z - z_\nu,$$

gesetzt ist. x_ν, y_ν, z_ν sind die Koordinaten der Kugelmittel-
punkte. Aus den Gleichungen (7) und (7 a) schliessen wir ferner,

dass die Anwesenheit jeder der Kugeln bis auf unendlich Kleines
höherer Ordnung eine Verringerung der Wärmeproduktion pro
[36]   Zeiteinheit um $2\,\delta^2\,k\,\Phi$ zum Gefolge hat und dass im Gebiete
G die pro Volumeneinheit in Wärme verwandelte Energie den
Wert hat:

$$W = 2\,\delta^2\,k - 2\,n\,\delta^2\,k\,\Phi,$$

oder

(7 b) $$W = 2\,\delta^2\,k\,(1 - \varphi),$$

P.299[/wobei]   wobei φ den von den Kugeln eingenommenen Bruchteil des Volumens bedeutet.

[37]   Gleichung (7 b) erweckt den Anschein, als ob der Reibungskoeffizient der von uns betrachteten inhomogenen Mischung von
Flüssigkeit und suspendierten Kugeln (im folgenden kurz
„Mischung" genannt) kleiner sei als der Reibungskoeffizient k
der Flüssigkeit. Dies ist jedoch nicht der Fall, da A, B, C
nicht die Werte der Hauptdilatationen der in Gleichungen (8)
dargestellten Flüssigkeitsbewegung sind; wir wollen die Hauptdilatationen der Mischung A^*, B^*, C^* nennen. Aus Symmetriegründen folgt, dass die Hauptdilatationsrichtungen der Mischung
den Richtungen der Hauptdilatationen A, B, C, also den Ko-
ordinatenrichtungen parallel sind. Schreiben wir die Gleichungen
(8) in der Form:

$$u = A\,x + \sum u_\nu,$$

$$v = B\,y + \sum v_\nu,$$

$$w = C\,z + \sum w_\nu,$$

so erhalten wir:

$$A^* = \left(\frac{\delta u}{\delta x}\right)_{x=0} = A + \sum \left(\frac{\delta u_\nu}{\delta x}\right)_{x=0} = A - \sum \left(\frac{\delta u_\nu}{\delta x_\nu}\right)_{x=0}$$

Schliessen wir die unmittelbaren Umgebungen der einzelnen
Kugeln von der Betrachtung aus, so können wir die zweiten
und dritten Glieder der Ausdrücke von u, v, w weglassen und
erhalten für $x = y = z = 0$:

— 16 —

$$(9) \begin{cases} u_\nu = -\tfrac{5}{2} \dfrac{P^3}{r_\nu^2} \dfrac{x_\nu \left(A\,x_\nu^2 + B\,y_\nu^2 + C\,z_\nu^2 \right)}{r_\nu^3}, \\[2ex] v_\nu = -\tfrac{5}{2} \dfrac{P^3}{r_\nu^2} \dfrac{y_\nu \left(A\,x_\nu^2 + B\,y_\nu^2 + C\,z_\nu^2 \right)}{r_\nu^3}, \\[2ex] w_\nu = -\tfrac{5}{2} \dfrac{P^3}{r_\nu^2} \dfrac{x \left(A\,x_\nu^2 + B\,y_\nu^2 + C\,z_\nu^2 \right)}{r_\nu^3}, \end{cases} \qquad [38]$$

wobei

$$r_\nu = \sqrt{x_\nu^1 + y_\nu^2 + z_\nu^2} > 0$$

gesetzt ist. Die Summierung erstrecken wir über das Volumen einer Kugel K von sehr grossem Radius R, deren Mittelpunkt im Koordinatenursprung liegt. Betrachten wir ferner die *regellos* P.300[/*regellos*] verteilten Kugeln als *gleichmässig* verteilt und setzen an Stelle der Summe ein Integral, so erhalten wir:

$$A^* = A - n \int_K \frac{\delta u_\nu}{\delta x_\nu}\, d x_\nu\, d y_\nu\, d z_\nu,$$

$$= A - n \int \frac{u_\nu x_\nu}{r_\nu}\, d s,$$

wobei das letzte Integral über die Oberfläche der Kugel K zu erstrecken ist. Wir finden unter Berücksichtigung von (9):

$$A^* = A - \tfrac{5}{2} \frac{P^3}{R^6} n \int x_0^2 \left(A\,x_0^2 + B\,y_0^2 + C\,z_0^2 \right) d s, \qquad [39]$$

$$= A - n \left(\tfrac{4}{3} P^3 \pi \right) A = A \left(1 - \varphi \right).$$

Analog ist

$$B^* = B \left(1 - \varphi \right),$$
$$C^* = C \left(1 - \varphi \right).$$

Setzen wir

$$\delta^{*2} = A^{*2} + B^{*2} + C^{*2}, \qquad [40]$$

so ist bis auf unendlich Kleines höherer Ordnung:

$$\delta^{*2} = \delta^2 \left(1 - 2\,\varphi \right).$$

Wir haben für die Wärmentwicklung pro Zeit- und Volumeneinheit gefunden:

$$W^* = 2\,\delta^2\, k \left(1 - \varphi \right). \qquad [41]$$

— 17 —

Bezeichnen wir mit k^* den Reibungskoeffizienten des Gemisches, so ist:

$$W^* = 2\,\delta^{*2}\,k^*.$$

Aus den drei letzten Gleichungen erhält man unter Vernachlässigung von unendlich Kleinem höherer Ordnung:

[42]
$$k^* = k\,(1 + \varphi).$$

Wir erhalten also das Resultat:

[43] P.301
[/volumen]

Werden in einer Flüssigkeit sehr kleine starre Kugeln suspendiert, so wächst dadurch der Koeffizient der inneren Reibung um einen Bruchteil, der gleich ist dem Gesamtvolumen der in der Volumeneinheit suspendierten Kugeln, vorausgesetzt, dass dieses Gesamtvolumen sehr klein ist.

[44]
§ 3. Ueber das Volumen einer gelösten Substanz, deren Molekularvolumen gross ist gegenüber dem des Lösungsmittels.

Es liege eine verdünnte Lösung vor eines Stoffes, welcher in der Lösung nicht dissoziiert. Ein Molekül des gelösten Stoffes sei gross gegenüber einem Molekül des Lösungsmittels und werde als starre Kugel vom Radius P aufgefasst. Wir können dann das in § 2 gewonnene Resultat anwenden. Bedeutet k^* den Reibungskoeffizienten der Lösung, k denjenigen des reinen Lösungsmittels, so ist:

[45]
$$\frac{k^*}{k} = 1 + \varphi,$$

wobei φ das Gesamtvolumen der in Lösung befindlichen Moleküle pro Volumeneinheit ist.

Wir wollen φ für eine 1 proz. wässerige Zuckerlösung berechnen. Nach Beobachtungen von B u r k h a r d (Tabellen von L a n d o l t und B ö r n s t e i n) ist bei einer 1 proz. wässerigen Zuckerlösung $k^*/k = 1{,}0245$ (bei 20 ° C.), also $\varphi = 0{,}0245$ für (beinahe genau)

[46] 0,01 g Zucker. Ein Gramm in Wasser gelöster Zucker hat also auf den Reibungskoeffizienten denselben Einfluss wie kleine sus-

[47] pendierte starre Kugeln vom Gesamtvolumen 2,45 cm^3. Bei dieser Betrachtung ist der Einfluss des dem gelösten Zucker entsprechenden osmotischen Druckes auf die innere Reibung des Lösungsmittels vernachlässigt.

— 18 —

Es ist nun daran zu erinnern, dass 1 g festen Zuckers das Volumen 0,61 cm³ besitzt. Dasselbe Volumen findet man auch für das spezifische Volumen *s* des in Lösung befindlichen Zuckers, wenn man die Zuckerlösung als eine *Mischung* von Wasser und Zucker in gelöster Form auffasst. Die Dichte einer 1 proz. wässerigen Zuckerlösung (bezogen auf Wasser von derselben Temperatur) bei 17,5⁰ ist nämlich 1,00388. Man hat also (unter [48] Vernachlässigung des Dichteunterschiedes von Wasser von 4⁰ und Wasser von 17,5⁰):

$$\frac{1}{1,00388} = 0,99 + 0,01 \, s;$$

also $\qquad s = 0,61.$

Während also die Zuckerlösung, was ihre Dichte anbelangt, sich wie eine Mischung von Wasser und festem Zucker verhält, P.302[/hält,] ist der Einfluss auf die innere Reibung viermal grösser, als er [49] aus der Suspendierung der gleichen Zuckermenge resultieren würde. Es scheint mir dies Resultat im Sinne der Molekular- theorie kaum anders gedeutet werden zu können, als indem man annimmt, dass das in Lösung befindliche Zuckermolekül die Beweglichkeit des unmittelbar angrenzenden Wassers hemme, so dass ein Quantum Wasser, dessen Volumen ungefähr das Dreifache des Volumens des Zuckermoleküls ist, an das Zucker- [50] molekül gekettet ist.

Wir können also sagen, dass ein gelöstes Zuckermolekül (bezw. das Molekül samt dem durch dasselbe festgehaltene Wasser) in hydrodynamischer Beziehung sich verhält wie eine Kugel vom Volumen 2,45 . 342/N cm³, wobei 342 das Molekular- [51] gewicht des Zuckers und N die Anzahl der wirklichen Mole- küle in einem Grammmolekül ist.

§ 4. Ueber die Diffusion eines nicht dissoziierten Stoffes in flüssiger Lösung.

Es liege eine Lösung vor, wie sie in § 3 betrachtet wurde. Wirkt auf das Molekül, welches wir als eine Kugel vom Radius P betrachten, eine Kraft K, so bewegt sich das Molekül mit einer Geschwindigkeit ω, welche durch P und den Reibungskoeffi-

— 19 —

zienten k des Lösungsmittels bestimmt ist. Es besteht nämlich die Gleichung[1]):

$$(1) \qquad \omega = \frac{K}{6\,\pi\,k\,P}.$$

Diese Beziehung benutzen wir zur Berechnung des Diffusionskoeffizienten einer nicht dissoziierten Lösung. Bedeutet p den osmotischen Druck der gelösten Substanz, welcher bei der betrachteten verdünnten Lösung als die einzige bewegende Kraft anzusehen sei, so ist die auf die gelöste Substanz pro Volumeneinheit der Lösung in Richtung der X-Achse ausgeübte Kraft

[53] $= -\,\delta\,p/\delta\,x$. Befinden sich ρ Gramm in der Volumeneinheit und ist m das Molekulargewicht des gelösten Stoffes, N die Anzahl wirklicher Moleküle in einem Grammolekül, so ist

P.303[/lumen-einheit] $(\rho/m)\cdot N$ die Anzahl der (wirklichen) Moleküle in der Volumeneinheit und die auf ein Molekül infolge des Konzentrationsgefälles wirkende Kraft:

$$(2) \qquad K = -\,\frac{m}{\rho N}\,\frac{\delta\,p}{\delta\,x}.$$

Ist die Lösung genügend verdünnt, so ist der osmotische Druck durch die Gleichung gegeben:

[54] $$(3) \qquad p = \frac{R}{m}\,\rho\,T,$$

wobei T die absolute Temperatur und $R = 8{,}31\,.\,10^7$ ist. Aus den Gleichungen (1), (2) und (3) erhalten wir für die Geschwindigkeit der Wanderung der gelösten Substanz:

$$\omega = -\,\frac{R\,T}{6\,\pi k}\,\frac{1}{N\,P}\,\frac{1}{\rho}\,\frac{\delta\,\rho}{\delta\,x}.$$

Die pro Zeiteinheit durch die Einheit des Querschnittes in Richtung der X-Achse hindurchtretende Stoffmenge ist endlich

$$(4) \qquad \omega\,\rho = -\,\frac{R\,T}{6\,\pi k}\,\cdot\,\frac{1}{N\,P}\,\frac{\delta\,\rho}{\delta\,x}.$$

Wir erhalten also für den Diffusionskoeffizienten D:

[55] $$D = \frac{R\,T}{6\,n\,k}\,\cdot\,\frac{1}{N\,P}.$$

---

[52]   1) G. Kirchhoff, Vorlesungen über Mechanik. 26. Vorl., Gl. (22).

— 20 —

Man kann also aus dem Diffusionskoeffizienten und dem Koeffizienten der inneren Reibung des Lösungsmittels das Produkt aus der Anzahl N der wirklichen Moleküle in einem Grammolekül und dem hydrodynamisch wirksamen Molekularradius P berechnen.

In dieser Ableitung ist der osmotische Druck wie eine auf die einzelnen Moleküle wirkende Kraft behandelt worden, was offenbar der Auffassung der kinetischen Molekulartheorie nicht entspricht, da gemäss letzterer in dem vorliegenden Falle der osmotische Druck nur als eine scheinbare Kraft aufzufassen ist. Diese Schwierigkeit verschwindet jedoch, wenn man bedenkt, dass den (scheinbaren) osmotischen Kräften, welche den Konzentrationsverschiedenheiten der Lösung entsprechen, durch ihnen numerisch gleiche, entgegengesetzt gerichtete, auf die einzelnen Moleküle wirkende Kräfte das (dynamische) Gleichgewicht geleistet werden kann, wie auf thermodynamischem Wege leicht eingesehen werden kann. P.304[/gewicht]

Der auf die Masseneinheit wirkenden osmotischen Kraft $-\dfrac{1}{\rho}\dfrac{\delta p}{\delta x}$ kann durch die (an den einzelnen gelösten Molekülen angreifende) Kraft $-P_x$ das Gleichgewicht geleistet werden, wenn

$$-\frac{1}{\rho}\frac{\delta p}{\delta x} - P_x = 0.$$

Denkt man sich also an der gelösten Substanz (pro Masseneinheit) die zwei sich gegenseitig aufhebenden Kräftesysteme P_x und $-P_x$ angreifend, so leistet $-P_x$ dem osmotischen Drucke das Gleichgewicht, und es bleibt nur die dem osmotischen Drucke numerisch gleiche Kraft P_x als Bewegungsursache übrig. Damit ist die erwähnte Schwierigkeit beseitigt.[1]

§ 5. Bestimmung der Moleküldimensionen mit Hilfe der erlangten Relationen.

Wir haben in § 3 gefunden:

$$\frac{k^*}{k} = 1 + \varphi = 1 + \mathrm{n}.\tfrac{4}{3}\pi P^3, \qquad [57]$$

wobei n die Anzahl der gelösten Moleküle' pro Volumeneinheit und P den hydrodynamisch wirksamen Molekülradius bedeutet. Berücksichtigt man, dass

[1] Eine ausführliche Darlegung dieses Gedankenganges findet sich in Ann. d. Phys. 17. p. 549. 1905. [56]

— 21 —

$$\frac{n}{N} = \frac{\rho}{m},$$

wobei ρ die in der Volumeneinheit befindliche Masse des gelösten Stoffes und m dessen Molekulargewicht bedeutet, so erhält
man:

[58]
$$N\,P^3 = \frac{3}{4\,\pi}\,\frac{m}{\rho}\Big(\frac{k^*}{k} - 1\Big)$$

Andererseits wurde in § 4 gefunden:

$$N\,P = \frac{R\,T}{6\,\pi\,k}\,\frac{1}{D}.$$

Diese beiden Gleichungen setzen uns in den Stand, die Grössen
P.305[/ab P und N einzeln zu berechnen, von welchen sich N als unab-
hängig] hängig von der Natur des Lösungsmittels, der gelösten Substanz
und der Temperatur herausstellen muss, wenn unsere Theorie
den Tatsachen entspricht.

Wir wollen die Rechnung für wässerige Zuckerlösung durchführen. Nach den oben mitgeteilten Angaben über die innere
Reibung der Zuckerlösung folgt zunächst für 20° C:

[59]
$$N\,P^3 = 200.$$

Nach Versuchen von G r a h a m (berechnet von S t e f a n) ist
[60] der Diffusionskoeffizient von Zucker in Wasser bei 9,5° C. 0,384,
wenn der Tag als Zeiteinheit gewählt wird. Die Zähigkeit des
[61] Wassers bei 9,5° ist 0,0135. Wir wollen diese Daten in unsere
Formel für den Diffusionskoeffizienten einsetzen, trotzdem sie an
10 proz. Lösungen gewonnen sind und eine genaue Gültigkeit
unserer Formel bei so hohen Konzentrationen nicht zu erwarten
ist. Wir erhalten

$$N\,P = 2{,}08 \cdot 10^{16}.$$

Aus den für $N\,P^3$ und $N\,P$ gefundenen Werten folgt, wenn
wir die Verschiedenheit von P bei 9,5° und 20° vernachlässigen,

[62]
$$P = 9{,}9 \cdot 10^{-8}\ cm,$$
$$N = 2{,}1 \cdot 10^{23}.$$

Der für N gefundene Wert stimmt der Grössenordnung nach
mit den durch andere Methoden gefundenen Werten für diese
[63] Grösse befriedigend überein.

B e r n, den 30. April 1905.

Published by K. J. Wyss, Bern. Dated Bern, 30 April 1905.

[1] For Einstein's earlier attempts to obtain a doctorate, see the editorial note, "Einstein's Dissertation on the Determination of Molecular Dimensions," § III, pp. 173–176.

[2] The dissertation was formally submitted on 20 July 1905 (see Einstein to Rudolf Martin of this date).

[3] Alfred Kleiner was Professor of Physics at the University of Zurich. Einstein submitted the dissertation to him. At Kleiner's request, Heinrich Burkhardt, Professor of Mathematics at the University of Zurich, checked the calculations in Einstein's dissertation (see the Gutachten über das Promotionsgesuch des Hrn. Einstein, 20–24 July 1905, SzZSa, U 110 e 9).

[4] Einstein had stated earlier that he intended to dedicate his dissertation to Grossmann; see Einstein to Mileva Marić, 19 December 1901 (Vol. 1, Doc. 130).

[5] For a discussion of methods for determining molecular dimensions known at that time, see the editorial note, "Einstein's Dissertation on the Determination of Molecular Dimensions," § II, pp. 170–173.

[6] Einstein's argument closely follows *Kirchhoff 1897*, Lecture 10, pp. 95–108. For Einstein's reading of *Kirchhoff 1897*, see Einstein to Mileva Marić, 29 July 1900 and 1 August 1900 (Vol. 1, Docs. 68 and 69).

[7] The assumptions made for the liquid correspond to those introduced in *Kirchhoff 1897* on p. 374.

[8] For an argument in which similar boundary conditions for the hydrodynamical equations are used, see *Kirchhoff 1897*, pp. 378–379.

[9] The following equations are valid if the velocities are assumed to be infinitely small, and if the motion is stationary (see *Kirchhoff 1897*, p. 374). For a discussion of the Navier-Stokes equations, see *Brush 1976*, book 2, § 12.3, pp. 432–443.

[10] *Kirchhoff 1897*, Lecture 26.

[11] A factor k is missing on the right-hand side of the last equation on this line; this error is corrected in *Einstein 1906a*.

[12] *Kirchhoff 1897*, pp. 378–379. The Kirchhoff text begins: "From the equations 9) [corresponding to (4) in Einstein's text] follows $\Delta p = 0$; if one assumes p according to this condition . . ." ("Aus den Gleichungen 9) folgt $\Delta p = 0$; hat man dieser Bedingung gemäss p ange-

nommen . . .") and then continues as quoted by Einstein.

[13] The denominator on the right-hand side should be $\delta \xi^2$; this error is corrected in *Einstein 1906a*.

[14] The denominator of the first term on the right-hand side should be $\delta \xi^2$; this error is corrected in *Einstein 1906a*. A reprint of this article in the Einstein Archive shows marginalia and interlineations in Einstein's hand, the first of which refer to this and the following equation. The term "$+ g \frac{1}{\rho}$" was added to the right-hand side of the equation for V and then canceled. These marginalia and interlineations are presumably part of Einstein's unsuccessful attempt to find a calculational error; see note 26 below, and also the editorial note, "Einstein's Dissertation on the Determination of Molecular Dimensions," § V, pp. 179–182. On this assumption, they date from late 1910.

[15] The equation for u' should be (as corrected in *Einstein 1906a*): $u' = -2c \, \frac{\delta \frac{1}{\rho}}{\delta \xi}$. In the reprint mentioned in note 14, the first derivative with respect to ξ was changed to a second derivative and then changed back to a first derivative. At the bottom of the page, the following equations are written:

$b = -1/12 \, P^5 a$
$c = -5/12 \, P^3 a$
$g = 2/3 \, P^3 a$.

[16] The numerator of the last term in the curly parentheses should be "$\delta^2 (1/\rho)$", as corrected in *Einstein 1906a*.

[17] In Einstein's reprint (see note 14), a factor 2/3 was added to the first term on the right-hand side of this equation and then canceled.

[18] "$\frac{\delta n}{\delta \xi}$" should be "$\frac{\delta p}{\delta \xi}$" as corrected in *Einstein 1922*.

[19] The factor preceding the first parenthesis should be, as corrected in *Einstein 1906a*:
$$- 5/2 \, \frac{P^3}{\rho^5}.$$

[20] The equations should be: $u = U$, $v = V$, $w = W$.

[21] $X\xi$, $X\eta$, $X\zeta$, should be X_ξ, etc., as corrected in *Einstein 1906a*; analogous corrections apply to the subsequent two equations.

[22] For the following equations, see *Kirchhoff 1897*, p. 369.

[23] In Einstein's reprint (see note 14), the term

"$+ \dfrac{5}{6} P^3 \dfrac{A\xi}{\rho^3}$," is added to the right-hand side of this equation. After the last terms of the second and third equations, series of dots are added. These interlineations are presumably related to the marginal calculations indicated in note 14.

[24] In Einstein's reprint (see note 14), the term "$+ 5 k P^3 \dfrac{1}{\rho^3}$" is added to the right-hand side of this equation. This interlineation is presumably related to the marginal calculations indicated in note 14.

[25] In Einstein's reprint (see note 14), the term "$- \dfrac{5}{3} k P^3 A \left(\dfrac{1}{\rho^3} - 9 \dfrac{\xi}{\rho^5} \right)$" is added to the right-hand side of this equation. This addition is presumably related to the marginal calculations referred to in note 14.

[26] This equation and the subsequent one are incorrect. Apart from minor errors, they contain a calculational error bearing on the numerical factors. In *Einstein 1906a*, "$+25$" in front of the last term in the equation for X_ζ is changed to "-25". In Einstein's reprint (see note 14), the factor "ζ^2" in the last term on the right-hand side of this equation is corrected to "$\xi\zeta$", and the factor "η^2" in front of the parenthesis in the last term on the right-hand side of the equation for X_η is corrected to $\xi\eta$. The calculational error that is also contained in these equations, and some of its consequences, are corrected in *Einstein 1911d*. The corrections are integrated into the text of the reprint of this paper in *Einstein 1922*. For an account of the discovery of this error and of its correction, see the editorial note, "Einstein's Dissertation on the Determination of Molecular Dimensions," § V, pp. 179–182. The correct equations are:

$$X_\eta = + 5kP^3 \frac{(A + B)\xi\eta}{\rho^5}$$
$$- 25kP^3 \frac{\xi\eta(A\xi^2 + B\eta^2 + C\zeta^2)}{\rho^7}$$
$$X_\zeta = + 5kP^3 \frac{(A + C)\xi\zeta}{\rho^5}$$
$$- 25kP^3 \frac{\xi\zeta(A\xi^2 + B\eta^2 + C\zeta^2)}{\rho^7}$$

[27] "-10" should be replaced by "-5," and "25" by "20" (see previous note).

[28] The third "$+$" sign should be replaced by "$=$" as corrected in *Einstein 1922*. "-10"

should be replaced by "-5," and "20" by "15" (see note 26).

[29] In Einstein's reprint (see note 14), the factor "4/15" was changed to "8/15" and then changed back to "4/15".

[30] "4/15" should be replaced by "8/15" as corrected in Einstein's reprint (see note 14).

[31] This equation should be (see note 26):

$$W = 8/3 \, \pi \, R^3 \, k \, \delta^2 + 4/3 \, \pi \, P^3 \, k \, \delta^2$$
$$= 2 \, \delta^2 \, k \, (V + \Phi/2).$$

[32] "δ" should be "δ^2". This correction is made in Einstein's reprint (see note 14).

[33] It follows from the correction to eq. (7) that the dissipated energy is actually increased by half this amount. The statement in the text is only partially corrected in *Einstein 1922*; the amount is correctly given but still described as a diminution. The final sentence of this paragraph, which no longer applies to the corrected calculation, is omitted from *Einstein 1922*.

[34] The point should be denoted by $x_\nu, y_\nu, z_\nu,$ as corrected in *Einstein 1906a*.

[35] The second denominator should be ρ_ν^3 as corrected in *Einstein 1906a*.

[36] The heat production per unit time is actually increased by $\delta^2 \, k \, \Phi$. The correct equations are thus (see note 26): $W = 2 \, \delta^2 \, k + n \, \delta^2 \, k \, \Phi$, and $W = 2 \, \delta^2 \, k \, (1 + \varphi/2)$.

[37] The following two sentences are revised in *Einstein 1922*: "In order to calculate from equation (7b) the coefficient of friction of an inhomogeneous mixture of fluid and suspended spheres (in the following called "mixture" for short) which we are examining, we must further take into consideration that A, B, C are not values of the primary dilations of the motion of fluid represented in equation (8); we want to designate the primary dilations of the mixture as A^*, B^*, C^*" ("Um aus Gleichung (7b) den Reibungskoeffizienten der von uns betrachteten inhomogenen Mischung von Flüssigkeit und suspendierten Kugeln (im Folgenden kurz "Mischung" genannt) zu berechnen, müssen wir noch berücksichtigen, daß A, B, C nicht die Werte der Hauptdilatationen der in Gleichungen (8) dargestellten Flüssigkeitsbewegung sind; wir wollen die Hauptdilatationen der Mischung A^*, B^*, C^* nennen").

[38] In this and the following two equations, the sign after "$=$" should be "$+$." The third equation should have "z_ν" instead of "x" in the numerator; the latter correction is made in *Einstein 1906a*.

[39] The factor in front of the second term in the first equation is 5/2 (see *Einstein 1906a*). In deriving the second equation, Einstein used the equations at the bottom of p. 12 and the fact that $A + B + C = 0$.

[40] In Einstein's reprint (see note 14), "$= A^2 + B^2 + \delta^2 (1 - 2\varphi)$" is added to the right-hand side of this equation and then crossed out.

[41] The correct equation is (see note 26): $W^* = 2\,\delta^2\,k\,(1 + \varphi/2)$.

[42] The correct equation is (see note 26): $k^* = k\,(1 + 2.5\,\varphi)$.

[43] The fraction is actually 2.5 times the total volume of the suspended spheres (see note 26).

[44] The title of this paragraph in *Einstein 1906a* is: "On the Volume of a Dissolved Substance with a Molecular Volume Which is Large in Comparison to the Solvent" ("Über das Volumen einer gelösten Substanz von im Vergleich zum Lösungsmittel großem Molekularvolumen").

[45] The correct equation is (see note 26): $k^*/k = 1 + 2.5\,\varphi$.

[46] See *Landolt and Börnstein 1894*, p. 294. In *Einstein 1906c* (Doc. 33), Einstein uses more recent experimental data.

[47] The correct value is 0.98 cm³ (see note 26). The following sentence is omitted in *Einstein 1906a*.

[48] See *Landolt and Börnstein 1894*, p. 231.

[49] The viscosity is actually one and one-half times greater (see note 26).

[50] The quantity of water bound to a sugar molecule has a volume that is actually one-half that of the sugar molecule (see note 26). The existence of molecular aggregates in combination with water ("Hydrathüllen") was debated at that time (see the editorial note, "Einstein's Dissertation on the Determination of Molecular Dimensions," § II and IV, pp. 170–173, 176–179).

[51] The volume of the sphere is actually $0.98 \cdot 342/N$ *cm³* (see note 26).

[52] *Kirchhoff 1897*, p. 380. The formula was first derived in *Stokes 1845*. For a discussion of its applicability to objects of microscopic dimensions, see *Fürth 1922*, pp. 58–60, fn. 6.

[53] The connection between diffusion and osmotic pressure was first explained in *Nernst 1888* (see also *Nernst 1898*, p. 157; *Ostwald 1891*, p. 697).

[54] This equation was first derived in *Van 't Hoff 1887*; it is also mentioned in *Einstein 1905i* (Doc. 14), p. 142.

[55] The first denominator should be $6\,\pi\,k$, as corrected in *Einstein 1906a*. This equation was independently obtained in *Sutherland 1905* by a similar argument. The idea to use this formula for a determination of molecular dimension may have occurred to Einstein as early as 1903; see Einstein to Michele Besso, 17 March 1903, and the discussion of this formula in the editorial note, "Einstein's Dissertation on the Determination of Molecular Dimensions," § IV, pp. 176–179.

[56] *Einstein 1905k* (Doc. 16), § 3. This note was presumably added after the dissertation was submitted.

[57] The correct equation is (see note 26): $k^*/k = 1 + 2.5\,\varphi = 1 + 2.5\,n\,4/3\,\pi\,P^3$.

[58] The correct equation has an additional factor 2/5 on the right-hand side (see note 26).

[59] For the experimental data, see p. 17. The correct value is 80 (see note 26).

[60] The value given by Einstein is interpolated from the data in *Landolt and Börnstein 1894*, p. 306. In *Einstein 1906c* (Doc. 33), Einstein uses more recent experimental data.

[61] The value given by Einstein is an interpolation from the data in *Landolt and Börnstein 1894*, p. 288.

[62] The values obtained by using the correct equations (see *Einstein 1922*) are: $P = 6.2\ 10^{-8}$ cm; and $N = 3.3\ 10^{23}$ (per mole).

[63] For values of N obtained by other methods, see, for example, *Planck 1901b*.

EINSTEIN ON BROWNIAN MOTION

I

Einstein's study of Brownian motion constitutes one of the high points in the long tradition of research on the kinetic theory of heat and of his own contributions to this field.[1] Some of the consequences of his work were of great significance for the development of physics in the twentieth century. Einstein's derivation of the laws governing Brownian motion, and their subsequent experimental verification by Perrin and others, contributed significantly to the acknowledgment of the physical reality of atoms by the then still numerous skeptics. His papers on Brownian motion helped to establish the study of fluctuation phenomena as a new branch of physics. The methods he created in the course of his research prepared the way for statistical thermodynamics, later developed by Szilard and others, and for a general theory of stochastic processes.

Einstein published four major articles on Brownian motion in liquids between 1905 and 1908: *Einstein 1905k* (Doc. 16), *Einstein 1906b* (Doc. 32), *Einstein 1907c* (Doc. 40), and *Einstein 1908c* (Doc. 50). His own summary of a lecture to the Naturforschende Gesellschaft Bern, *Einstein 1907f*, is presented as Doc. 43. He also published three papers on related topics during this period: the first, originally published as Einstein's dissertation, *Einstein 1905j* (Doc. 15), deals with the determination of molecular dimensions;[2] the two others concern Brownian motion in condensers and its measurement, *Einstein 1907b* (Doc. 39), and *Einstein 1908a* (Doc. 48). Finally, a remark by Einstein on the measurement of Brownian motion made after a talk that was given at the 1909 Salzburg meeting of the Gesellschaft Deutscher Naturforscher und Ärzte is reproduced, together with other discussion contributions, as *Einstein et al. 1909a* (Doc. 58).

In 1920, Wolfgang Ostwald, editor of *Ostwald's Klassiker der exakten Wissenschaften*, proposed to Einstein an edition of his papers on Brownian motion and diffusion.[3] The resulting volume, *Einstein 1922*, reprints most of these papers.[4] An English translation appeared as *Einstein 1926*.

---

[1] For studies of Einstein's work on Brownian motion, see, e.g., *Brush 1968*, and *Pais 1982*, chap. 2, § 5.

[2] For a discussion of this paper, see the editorial note, "Einstein's Dissertation on the Determination of Molecular Dimensions," pp. 170–182.

[3] Wolfgang Ostwald to Einstein, 22 November 1920. Einstein immediately agreed, leaving to Ostwald the decision about which of his articles should be reprinted (Einstein to Wolfgang Ostwald, 22 November–20 December 1920). In his reply, Ostwald informed Einstein that he intended to appoint Reinhold Fürth as co-editor (Wolfgang Ostwald to Einstein, 20 December 1920). See *Fürth 1980*, for evidence that Fürth wrote Einstein, asking for his consent to the preparation of the edition.

[4] The reprinted papers are *Einstein 1905k* (Doc. 16); *Einstein 1906b* (Doc. 32); *Einstein 1906a*, incorporating the correction of a calculational error and *1906c* (Doc. 33); *Einstein 1907c* (Doc. 40); and *Einstein 1908c* (Doc. 50). Some of Fürth's annotations for the 1922 edition (*Fürth 1922*) have been utilized in these editorial notes.

II

Since at least the middle of the nineteenth century, a growing number of physicists and chemists had accepted the atomic hypothesis.[5] The assumption that matter consists of atoms and molecules suggested a number of relations between phenomena, both physical and chemical, that are unexpected from a purely macroscopic point of view. Various methods for the determination of molecular dimensions gave values that were often in surprisingly good agreement. The physical reality of atoms was not, however, universally accepted by the end of the century. There were still some fervent opponents of the atomic hypothesis, such as Wilhelm Ostwald and Georg Helm, who called themselves "energeticists" to indicate that they regarded the concept of energy as the most fundamental ontological concept of science.[6] Others, such as Ernst Mach, while adopting a hostile position with regard to the existence of entities not directly accessible to sense-experience, in particular, atoms, admitted that atomism may have a heuristic or didactic utility.[7] It was not uncommon, even among scientists who made explicit use of atomic assumptions in their work, to regard atomism as a mere working hypothesis.[8]

Although at the turn of the century the atomic hypothesis was proving its heuristic value in such new areas of research as the electron theory of metals[9] and stereochemistry,[10] some physicists had come to regard the theory of heat as an area in which the atomic hypothesis was no longer fruitful.[11] Einstein probably became aware of the controversy over the molecular theory of heat during his student years, when he read works by Mach, Ostwald, and Boltzmann.[12] In 1900 Einstein finished reading Boltzmann's *Gastheorie*,[13] in which Boltzmann, presumably reacting to a dispute with Ostwald and Helm, suggested that he was isolated in his support of the kinetic theory.[14] Although he criticized Boltz-

[5] For a survey of nineteenth-century debates on atomism and the kinetic theory of heat, see *Nye 1972*, chap. I, and *Brush 1976*, chap. I.

[6] For a comprehensive account of the energetics controversy in late nineteenth-century Germany, see *Deltete 1983*.

[7] See, e.g., *Mach 1896*, pp. 428–429. For a discussion of Mach's views on atomism and their historical background, see *Brush 1976*, pp. 274–299. For a comprehensive study of Mach's work, see *Blackmore 1972*. Ostwald believed in the heuristic value of kinetic-molecular hypotheses before he adopted a more radical position against atomism (see *Deltete 1983*, pp. 297ff.).

[8] See, e.g., the remarks in *Abraham 1905*, p. 2, and in *Bredig 1894*, p. 264.

[9] For an account of the early development of the electron theory of metals, see *Kaiser 1987*. Einstein's early interest in the subject is discussed in Vol. 1, the editorial note, "Einstein on

Thermal, Electrical, and Radiation Phenomena," pp. 235–237.

[10] For an account of the development of stereochemistry, see *Snelders 1976*.

[11] See *Nye 1972*, pp. 17–18, for a discussion of contemporary criticisms of the sterility of kinetic theory.

[12] For Einstein's reading of Mach, see Einstein to Mileva Marić, 10 September 1899 (Vol. 1, Doc. 54); for his reading of *Ostwald 1891, 1893*, see Einstein to Wilhelm Ostwald, 19 March 1901 (Vol. 1, Doc. 92); for his reading of *Boltzmann 1896, 1898a*, see Einstein to Marić, 10 September 1899, 13 September 1900, and 19 September 1900 (Vol. 1, Docs. 54, 75, and 76). Einstein's later favorable comments on the criticism of energetics in *Planck 1896* (*Einstein 1913*, p. 1077) suggest that he may have read this paper as a student.

[13] See the preceding note.

[14] See the preface to *Boltzmann 1898a*; for

mann for a lack of emphasis on the comparison of his theory with observation,[15] Einstein was firmly convinced of the principles of Boltzmann's theory.[16]

In his first published attempts at independent research,[17] Einstein took for granted the atomistic constitution of matter and of electricity. He developed a theory of molecular forces, on the basis of which he established a number of relations between observable phenomena.[18] Einstein's interest soon shifted from the details of molecular forces to the quest for facts, "which would guarantee as much as possible the existence of atoms of definite finite size" ("welche die Existenz von Atomen bestimmter endlicher Grösse möglichst sicher stellten"), as he later characterized this phase of his work.[19]

III

The irregular movement of microscopic particles suspended in a liquid had been noted long before the botanist Robert Brown published his careful observations in 1828,[20] but he was the first to emphasize its ubiquity, and to exclude its explanation as a vital phenomenon. Advances in observational technique and in theory served to eliminate a number of unsatisfactory explanations of Brownian motion by the end of the nineteenth century, if not to verify the correct one. Explanations of Brownian motion proposed after the exclusion of vital forces involved capillarity, convection currents, evaporation, interaction with light, and electrical forces. During the 1870s, the kinetic theory of heat was proposed as an explanation by several authors.[21] A powerful argument against this explanation was developed by the cytologist Karl von Nägeli in 1879.[22] He first used the equipartition theorem to calculate the average velocity of the molecules of the liquid, and then used the laws of elastic collision to obtain the velocity of a suspended particle. He concluded that the velocity of such a particle, because of its comparatively large mass, would be vanishingly small. William Ramsay and Louis-Georges Gouy independently tried to defend the molecular explanation of Brownian motion by assuming the existence of collective

accounts of the dispute, see *Brush 1976*, pp. 96–98, and *Deltete 1983*, pp. 416ff.

[15] On 30 April 1901, Einstein wrote to Mileva Marić: "I am now studying Boltzmann's *Gastheorie* once again. Everything is very fine, but too little value is placed on the comparison with reality" ("Ich studiere gegenwärtig wieder Boltzmanns Gastheorie. Alles ist sehr schön, aber zu wenig Wert gelegt auf den Vergleich mit der Wirklichkeit") (Vol. 1, Doc. 102).

[16] Einstein to Mileva Marić, 13 September 1900 (Vol. 1, Doc. 75).

[17] For these attempts, see *Einstein 1901* (Doc. 1) and *Einstein 1902a* (Doc. 2); for a discussion of the relevant letters, see Vol. 1, the editorial notes, "Einstein on Thermal, Electrical, and Radiation Phenomena," pp. 235–237, and "Einstein on Molecular Forces," pp. 264–266.

[18] See Vol. 1, the editorial note, "Einstein on Molecular Forces," pp. 264–266; and, in this volume, the editorial note, "Einstein on the Nature of Molecular Forces," pp. 3–8.

[19] See *Einstein 1979*, p. 44; translation, p. 45. For a discussion of this shift of interest, see the editorial notes, "Einstein on the Foundations of Statistical Physics," p. 46, and "Einstein's Dissertation on the Determination of Molecular Dimensions," pp. 174–176.

[20] See, e.g., *Brown 1828*. For contemporary reviews of Brownian motion, see *Smoluchowski 1906* and *De Haas-Lorentz 1913*. For historical accounts, see *Nye 1972* and *Brush 1968*.

[21] See *Brush 1968*, § 3.

[22] See *Nägeli 1879*.

motions of large numbers of atoms in liquids, an assumption suited to the refutation of arguments such as Nägeli's.[23]

In 1900 an entirely different way of applying the kinetic theory of heat to Brownian motion was investigated by Felix Exner, who assumed an equipartition of energy between the molecules of the liquid and the suspended particles.[24] He calculated the velocity of the molecules on the basis of observations that he interpreted as giving the mean velocities of the suspended particles, obtaining results that were not in agreement with contemporary estimates of molecular velocities. In Exner's work there is no fundamental difference between a solute molecule and a suspended particle. Einstein arrived at a similar conclusion, but instead of emphasizing the equipartition theorem, he took the osmotic pressure and its relation to the theory of diffusion and to the molecular theory of heat as the starting point of his analysis of Brownian motion:

> According to this theory, a solute molecule differs from a suspended body *solely* with regard to magnitude, and it is not apparent why the osmotic pressure of a number of suspended particles is not the same as that of the same number of solute molecules.

> Nach dieser Theorie unterscheidet sich eingelöstes Molekül von einem suspendierten Körper *lediglich* durch die Größe, und man sieht nicht ein, warum einer Anzahl suspendierter Körper nicht derselbe osmotische Druck entsprechen sollte, wie der nämlichen Anzahl gelöster Moleküle.[25]

On the other hand, Einstein pointed out, according to the "classical theory of thermodynamics" ("klassischen Theorie der Thermodynamik"),[26] suspended particles—as macroscopic objects—should *not* exert an osmotic pressure on a semipermeable wall. Before Einstein, no one seems to have recognized that this contrast provides a touchstone for the kinetic theory. His choice of a suspension to study the relations between the thermodynamic and atomic theories of heat amounted to a radical reversal of perspective. Usually the legitimacy of microscopic explanations of thermodynamic results was at issue. In this case, however, the question centered on the applicability of a thermodynamic concept— osmotic pressure—to the suspended particles.

In the course of studying colloidal solutions, the commonly made distinction between suspensions and solutions in nineteenth-century chemistry had gradually lost its absolute character.[27] The absence of any fundamental difference between solutions and suspensions was made strikingly clear in 1902, when observations performed with the newly

---

[23] See *Ramsay 1882* and *Gouy 1888*.

[24] See *Exner 1900*.

[25] *Einstein 1905k* (Doc. 16), p. 550. The attempt to use osmotic pressure as a distinguishing feature of solutions was also rejected by Zsigmondy (see the introduction to *Zsigmondy 1905*).

[26] *Einstein 1905k* (Doc. 16), p. 550. For a discussion of what Einstein meant by classical thermodynamics, see *Nye 1972*, p. 139, fn. 52.

[27] For a contemporary discussion of the distinction between solution and suspensions, see the introduction to *Zsigmondy 1905*. For a discussion of colloidal chemistry and its relation to the study of Brownian motion, see *Nye 1972*, pp. 98–102.

invented ultramicroscope[28] made it possible to resolve many colloidal solutions into their constituents.[29] The ultramicroscope not only demonstrated the physical reality of colloidal particles, but showed that irregular motion is one of their outstanding characteristics.[30]

Although the ultramicroscope brought closer the "distant reality" ("réalité lointaine")[31] of molecules, one of their fundamental properties, their velocities, remained inaccessible to measurement. The inconsistencies that result from presumed velocity measurements, such as Exner's, had hinted at this problem; but it was explicitly discussed for the first time in the theoretical studies of Brownian motion that Einstein and Smoluchowski independently published between 1905 and 1907.[32] Both introduced the mean square displacement of the suspended particles as the primary observable quantity in Brownian motion.[33] Einstein argued that dissipative forces change the direction and magnitude of the velocity of a suspended particle on such a short time scale that it cannot be measured.[34] This argument demonstrates the fundamental role of dissipation in Einstein's analysis of Brownian motion.

In summary, the study of previous explanations of Brownian motion shows that three elements of Einstein's approach are characteristic of his decisive progress: (1) he based his analysis on the osmotic pressure rather than on the equipartition theorem; (2) he identified the mean square displacements of suspended particles rather than their velocities as suitable observable quantities; and (3) he simultaneously applied the molecular theory of heat and the macroscopic theory of dissipation to the same phenomenon, rather than restricting each of these conceptual tools to a single scale, molecular or macroscopic.

IV

In his first paper on Brownian motion, *Einstein 1905k* (Doc. 16), Einstein proved:

> that, on the assumption of the molecular theory of heat, bodies of the order of magnitude of 1/1000 mm suspended in liquids must already carry out an observable random movement, which is generated by thermal motion.

> daß unter Voraussetzung der molekularen Theorie der Wärme in Flüssigkeiten suspendierte Körper von der Größenordnung 1/1000 mm bereits eine wahr-

[28] The ultramicroscope, developed by Siedentopf and Zsigmondy, is based on a new illumination technique that makes it possible to observe the diffraction discs of otherwise invisible objects; it increased the limit of visibility to approximately 5×10^{-3} micron. For a contemporary discussion of ultramicroscopes, see *Cotton and Mouton 1906*, chap. 3.

[29] See *Siedentopf and Zsigmondy 1903*.

[30] *Siedentopf and Zsigmondy 1903*. The identification of this irregular motion with Brownian motion was questioned (see, e.g.,

Ostwald 1907).

[31] *Perrin 1911*, p. 157.

[32] Smoluchowski was the first to discuss the impossibility of measuring the velocity of suspended particles (see *Smoluchowski 1906*). For Einstein's discussion of this problem, see *Einstein 1907c* (Doc. 40).

[33] For a more detailed discussion of Einstein's introduction of this quantity, see the following section.

[34] See *Einstein 1907c* (Doc. 40).

nehmbare ungeordnete Bewegung ausführen müssen, welche durch die Wärmebewegung erzeugt ist.[35]

Einstein wrote this paper "without knowing that observations concerning Brownian
motion were already long familiar" ("ohne zu wissen, dass Beobachtungen über die
'Brown'sche Bewegung' schon lange bekannt waren").[36] He did not mention Brownian
motion in the title of *Einstein 1905k* (Doc. 16), although in the text he conjectured that the
motion he predicted might be identical to Brownian motion.[37] Boltzmann's *Gastheorie*,
which Einstein carefully studied during his student years, explicitly denies that the thermal
motion of molecules in a gas leads to observable motions of suspended bodies.[38] (This
denial may be an instance of what Einstein referred to as Boltzmann's attaching too little
importance to a comparison of theory with observation.[39]) Some time between 1902 and
1905, Einstein read Poincaré's *Science et hypothèse*, which contains a brief discussion of
Gouy's work on Brownian motion, emphasizing Gouy's argument that Brownian motion
violates the second law of thermodynamics.[40] Einstein's second paper on Brownian motion, written after Siedentopf drew his attention to Gouy's work,[41] cites the observations
reported in *Gouy 1888* as qualitative confirmation of his results.

Einstein 1905k (Doc. 16) opens with the derivation of an expression for the coefficient
of diffusion in terms of the radius of the suspended particles, and the temperature and
viscosity of the liquid, an expression already obtained in Einstein's doctoral dissertation.[42] Unlike the previous derivation, however, the new one makes use of the methods
of statistical physics that Einstein developed. The new approach is different in two
respects:

(1) In his dissertation, which deals with solutions rather than suspensions,
Einstein simply assumed the validity of Van 't Hoff's law for the osmotic pressure. He now gave a derivation of this law from an expression for the free energy
of the suspension that follows from statistical mechanics.

(2) Rather than simply considering the equilibrium of forces acting on a single
molecule, Einstein derived the equilibrium between the osmotic pressure and a
friction force obeying Stokes's law by a thermodynamical argument.[43]

[35] Einstein to Conrad Habicht, 18 May–8
June 1905.

[36] *Einstein 1979*, p. 44; translation, p. 45.
See also Einstein to Michele Besso, 6 January
1948, and Einstein to Carl Seelig, 15 September
1952.

[37] In Einstein to Conrad Habicht, 18 May–8
June 1905, he wrote: "<Unexplained> movements of inanimate small suspended bodies have
in fact been observed by physiologists, movements which they call 'Brownian molecular motion' " ("[E]s sind <unerklärte> Bewegungen
lebloser kleiner suspendirter Körper in der That
beobachtet worden von den Physiologen,

welche Bewegungen von ihnen 'Brown-sche
Molekularbewegung' genannt wird").

[38] See *Boltzmann 1898a*, pp. 111–112. For
evidence that Einstein read the *Gastheorie*, see
note 12.

[39] See note 15 above.

[40] See *Poincaré 1902*, p. 209. For Einstein's
reading of Poincaré, see the Introduction, p.
xxv.

[41] See *Einstein 1906b* (Doc. 32), p. 371.

[42] See *Einstein 1905j* (Doc. 15), p. 19.

[43] A sketch of this derivation is found in *Einstein 1905j* (Doc. 15), § 4. Some of Einstein's
contemporaries considered the application of

The ensuing derivation of the diffusion equation is based on the introduction of a probability distribution for displacements. The introduction of such a distribution is presumably related to Einstein's previous use of probability distributions.[44] Einstein assumed the existence of a time interval, short with respect to the observation time, yet sufficiently long enough that the motions of a suspended particle in two successive time intervals can be treated as independent of each other. The displacement of the suspended particles can then be described by a probability distribution that determines the number of particles displaced by a certain distance in each time interval. Einstein derived the diffusion equation from an analysis of the time-dependence of the particle distribution, calculated from the probability distribution for displacements. This derivation is based on his crucial insight into the role of Brownian motion as the microscopic process responsible for diffusion on a macroscopic scale. Compared to such a derivation, one based on the analogy to the treatment of diffusion in the kinetic theory of gases may have appeared more problematic to Einstein because of the lack of a fully developed kinetic theory of liquids.[45]

The solution of the resulting diffusion equation, combined with his expression for the diffusion coefficient, yields an expression for the mean square displacement, λ_x, as a function of time (p. 559), an expression that Einstein suggested could be used experimentally to determine Avogadro's number N:

$$\lambda_x = \sqrt{t} \sqrt{\frac{RT}{N} \frac{1}{3\pi kP}} \quad , \tag{1}$$

where t is the time, R the gas constant, T the temperature, k the viscosity, and P the radius of the suspended particles.

Through his earlier work, Einstein was familiar with the theory of diffusion in both gases and liquids, as well as with other techniques needed for his analysis of Brownian motion.[46] In *Einstein 1902a* (Doc. 2) he suggested the replacement of semipermeable walls in thermodynamic arguments by external conservative forces, a method he stated to be particularly useful for treating arbitrary mixtures. In 1903 Einstein discussed the

---

Stokes's formula, which is derived for uniform motion, to Brownian motion as problematic (see *Smoluchowski 1906*, p. 775; *Perrin 1908b*; and *De Haas-Lorentz 1913*, pp. 55–57). For a discussion of this problem, including references, see *Fürth 1922*, pp. 58–60; fn. 6 to *Einstein 1905k* (Doc. 16).

[44] For Einstein's first use of probability distributions in his papers on statistical physics, see *Einstein 1902b* (Doc. 3), p. 422.

[45] For such a study of diffusion in liquids, based on the concept of mean free path, see *Riecke 1890*. In *Einstein 1905j* (Doc. 15), Einstein mentioned the "insuperable difficulties

confronting a detailed molecular-kinetic theory of fluids" ("unüberwindlichen Schwierigkeiten welche einer ins einzelne gehenden molekular-kinetischen Theorie der Flüssigkeiten entgegenstehen") (p. 5). For evidence of Einstein's skepticism with regard to an explanation of Brownian motion that follows the methods of the kinetic theory of gases, see his critical remarks on Smoluchowski's work in Einstein to Carl Seelig, 15 September 1952, quoted in § VI.

[46] For a discussion of Einstein's earlier interest in diffusion, see the editorial note, "Einstein's Dissertation on the Determination of Molecular Dimensions," pp. 177–179.

notions of semipermeable membrane and osmotic pressure in his correspondence with Michele Besso, showing interest in Sutherland's hypothesis on the mechanism of semipermeable membranes.[47] In his papers on statistical physics, Einstein generalized the idea of external conservative forces,[48] and noted the significant role of fluctuations in statistical physics. In *Einstein 1904* (Doc. 5) he derived an expression for mean square deviations from the average value of the energy of a system.[49]

His second paper on Brownian motion, *Einstein 1906b* (Doc. 32), shows "how Brownian motion is related to the foundations of the molecular theory of heat" ("wie die Brownsche Bewegung mit den Grundlagen der molekularen Theorie der Wärme zusammenhängt").[50] It includes two new fluctuation formulas, both of which are derived from the probability distribution for a canonical ensemble given in Einstein's papers on statistical physics.[51] The first formula (eq. [I] on his p. 373), which is closely related to the formula for energy fluctuations he had derived in 1904,[52] gives the probability of deviations from the equilibrium value, due to irregular molecular motions, of a suitable observable parameter α of a system subject to an external force with potential $\Phi(\alpha)$:[53]

$$dW = A' \, e^{-\frac{N}{RT}\Phi} d\alpha, \tag{I}$$

where dW is the probability that the value of the parameter lies between α and $\alpha + d\alpha$, and A' a constant. Einstein applied eq. (I) to a harmonic oscillator in equilibrium with a gas to derive the black-body radiation law in the limit of large wavelengths and high temperatures. An investigation of how small a particle must be in order to remain in suspension in a gravitational field provides another application.

Eq. (I) does not, however, allow the treatment of Brownian motion, a time-dependent process involving the interplay of fluctuations and dissipation. In order to derive a fluctuation formula that generalizes eq. (1), Einstein related a general dissipation mechanism, analogous to Stokes's law, to the condition for stability of the distribution, eq. (I), now interpreted as giving the number of systems in a certain state, rather than the probability of that state. The potential Φ refers to a fictitious force.[54] The resulting formula for the time dependence of the mean square fluctuation (eq. [II] on his p. 378) enabled Einstein to treat rotational as well as translational motions of suspended particles:

[47] See Michele Besso to Einstein, 7–11 February 1903, which indicates that there was additional correspondence on this subject. See *Sutherland 1897* for Sutherland's hypothesis.

[48] See, in particular, *Einstein 1902b* (Doc. 3), § 10.

[49] At this time, however, Einstein regarded black-body radiation as the only physical system for which experience suggests the existence of observable energy fluctuations (see *Einstein 1904* [Doc. 5], p. 361).

[50] *Einstein 1906b* (Doc. 32), p. 371.

[51] See *Einstein 1902b* (Doc. 3), § 3.

[52] See *Einstein 1904* (Doc. 5), § 4.

[53] The choice of the observable parameter, which is treated by analogy to the energy in the fluctuation formula given in *Einstein 1904* (Doc. 5), was the subject of correspondence between D.K.C. MacDonald and Einstein in 1953.

[54] For a discussion of Einstein's use of such fictitious forces, see the editorial note, "Einstein's Dissertation on the Determination of Molecular Dimensions," § IV, p. 177.

$$\sqrt{\overline{\Delta^2}} = \sqrt{\frac{2R}{N}} \sqrt{BTt}, \qquad\qquad\qquad \text{(II)}$$

where Δ is the change in the parameter α, and B is what Einstein calls the mobility of the system with respect to α. By considering the effect on the motion of dissipation alone as a function of time, Einstein was able to estimate the time interval below which his results are no longer valid, due to the breakdown of the assumed independence of events in successive time intervals.[55]

V

Einstein's further studies of fluctuation phenomena elaborate the two fundamental approaches in his first two papers on Brownian motion: the thermodynamical approach to fluctuations that emerged from his work on statistical mechanics, and the stochastic treatment of fluctuations that takes dissipation and time dependence into account.[56] By concentrating on a few basic features of statistical physics, Einstein further reduced the need for detailed microphysical assumptions and simplified the technical aspects of his analysis. All of these subsequent studies emphasize applications of his methods; by 1909, applications to black-body radiation were again of primary concern to Einstein.[57]

Einstein further developed the thermodynamical approach in a paper on voltage fluctuations in a condenser (*Einstein 1907b* [Doc. 39]). He gave a simple derivation of a formula for mean square fluctuations, a derivation that does not depend on dynamical premises, but is directly based on Boltzmann's principle relating probability and entropy, both conceived by Einstein as thermodynamical quantities. Consequently, he stated that his treatment of fluctuations does not require any ''definite stipulations concerning the molecular model to be applied'' (''bestimmte Festsetzungen in betreff des anzuwendenden molekularen Bildes'').[58] The new element in Einstein's argument is his definition of probability in Boltzmann's principle as the ''statistical probability of a state'' (''statistische Wahrscheinlichkeit eines Zustandes''), a concept introduced in *Einstein 1905i* (Doc. 14).[59] Einstein further elaborated this concept in *Einstein 1909b* (Doc. 56), where he applied the mean square fluctuation formula derived from Boltzmann's principle to energy fluctuations in black-body radiation.[60]

[55] For a discussion of subsequent work on the significance of such a time interval in the analysis of Brownian motion, see *Fürth 1922*, pp. 60–61, fn. 8.

[56] For studies of Einstein's thermodynamical approach to fluctuations, see *Klein 1967, 1974b, 1982a*. For a review of Einstein's work on the stochastic approach, see *Sciama 1979*.

[57] See *Einstein 1909b* (Doc. 56); for a discussion of Einstein's work on black-body radiation fluctuations, see the editorial note,

''Einstein's Early Work on the Quantum Hypothesis,'' p. 146.

[58] *Einstein 1907b* (Doc. 39), p. 569.

[59] For a discussion of the concept of probability developed in Doc. 14, see the editorial note, ''Einstein's Early Work on the Quantum Hypothesis,'' p. 138.

[60] For further discussion of this paper, see the editorial note, ''Einstein's Early Work on the Quantum Hypothesis,'' pp. 145–146.

In the same paper, Einstein also elaborated the stochastic approach. He calculated pressure fluctuations in black-body radiation from the condition that the momentum they impart to a small mirror moving through the radiation just compensates for the momentum lost due to the average radiation pressure on the mirror.[61] In his lectures on statistical physics, Einstein applied a similar technique to the derivation of eq. (II), his general formula for Brownian motion. Considering both the change, $- B a \alpha t$, in a parameter α due to dissipation in the presence of a harmonic force $- a \alpha$, and the changes Δ due to fluctuations, Einstein applied the stability condition:

$$\overline{(\alpha - B a \alpha t + \Delta)^2} = \overline{\alpha^2} \tag{2}$$

from which eq. (II) can be derived.[62] Einstein continued to apply the stochastic and the thermodynamical approaches to fluctuation phenomena in his further research, in particular to problems of radiation. The thermodynamical approach is discussed at length in the introductory paragraphs of Einstein's paper on critical opalescence (*Einstein 1910c*). The stochastic approach plays a major role in Einstein's effort to find acceptable methods for treating the statistical aspects of black-body radiation,[63] and in a method he developed for the statistical evaluation of observations.[64]

VI

Einstein's work on Brownian motion aroused widespread interest among physicists and chemists, as indicated by correspondence with Röntgen, Lorenz, Smoluchowski, and Svedberg.[65] It led to attempts at experimental verification of his results, discussed in section VII, as well as the publication of two important theoretical papers on Brownian motion by Marian Smoluchowski and Paul Langevin, which present results essentially identical to Einstein's, but with different derivations.

Shortly after the appearance of *Einstein 1906b* (Doc. 32), Smoluchowski submitted a paper on the kinetic theory of Brownian motion to the *Annalen* (*Smoluchowski 1906*). Although this publication was stimulated by Einstein's first two papers, Smoluchowski had been working on the subject for a number of years. In 1904 he published a paper on density fluctuations in gases (*Smoluchowski 1904*) that has several features in common

[61] A similar argument is given in more detail in *Einstein and Hopf 1910b*.

[62] A derivation along these lines is found in Einstein's notes for his lectures on the kinetic theory of heat at the University of Zurich, summer semester of 1910. For a similar derivation, see *Langevin 1908*.

[63] See *Einstein and Hopf 1910a, 1910b*.

[64] See *Einstein 1914a*, and an unpublished manuscript, "Eine Methode zur statistischen Verwertung von Beobachtungen scheinbar un-

regelmässig quasiperiodisch verlaufender Vorgänge," written in 1914, after Einstein's move to Berlin.

[65] See Wilhelm Röntgen to Einstein, 18 September 1906; Richard Lorenz to Einstein, 15 November 1907; and Einstein to Marian Smoluchowski, 11 June 1908. For evidence of early correspondence between Svedberg and Einstein, see The Svedberg to Einstein, 8 December 1919.

with his later work, as well as with Einstein's work on Brownian motion. Einstein may have read this paper, which appeared in *Meyer, S. 1904*.[66]

Smoluchowski apparently corresponded with Einstein regarding his first papers on Brownian motion.[67] Smoluchowski did not publish his results before he saw Einstein's papers, because of the lack of sufficient experimental evidence.[68] Now he wanted to contribute to a "clarification of views on this interesting subject . . . in particular because my method seems to me to be more direct, simpler, and thus perhaps also more convincing than that of Einstein" ("Klärung der Ansichten über diesen interessanten Gegenstand . . . insbesondere da mir meine Methode direkter, einfacher und darum vielleicht auch überzeugender zu sein scheint als jene Einsteins").[69]

Smoluchowski's analysis of Brownian motion embraced gases as well as liquids. In studying a sequence of collisions of a single suspended particle with the molecules of the surrounding medium, Smoluchowski followed the tradition of the kinetic theory of gases. Einstein later characterized Smoluchowski's treatment of Brownian motion as starting from the application of the equipartition theorem to the velocity of a suspended particle: "Smoluchowski arrives at the explanation of the phenomenon by quantitatively formulating [the condition] that this velocity is constantly destroyed by internal friction and constantly recreated by irregular molecular impulses" ("[I]ndem Smoluchowski quantitativ formuliert, daß diese Geschwindigkeit durch innere Reibung beständig vernichtet, durch unregelmäßige Molekularstöße immer wieder hergestellt wird, gelangt er zur Erklärung des Phänomens").[70] When much later he compared Smoluchowski's treatment to his own, Einstein emphasized the greater generality of his approach: "[Smoluchowski's paper is] based on mechanics, whereas my investigation essentially presupposes only the law of osmotic pressure" ("[Smoluchowskis Arbeit ist] auf die Mechanik gegründet, während

---

[66] See *Smoluchowski 1904*. Einstein reviewed several contributions in *Meyer, S. 1904*. For a discussion of Einstein's reviews in general, see the editorial note, "Einstein's Reviews for the *Beiblätter zu den Annalen der Physik*," pp. 109–111.

[67] Only one item has been found: Einstein's letter to Smoluchowski, 11 June 1908, in which Einstein sent some of his papers, "which might still be of interest" ("welche noch interessieren können") and asked for reprints of Smoluchowski's papers.

[68] See *Smoluchowski 1906*, p. 756.

[69] *Smoluchowski 1906*, p. 756. On p. 772, he developed his criticism of Einstein's methods in more detail: "Without entering into a discussion on the very sensible methods employed by Einstein, I would still like to remark that they rest on considerations of an indirect nature, which do not always seem convincing" ("Ohne in eine Diskussion der von Einstein befolgten, sehr

sinnreichen Methoden einzugehen, möchte ich doch bemerken, daß sie auf Überlegungen indirekter Art beruhen, welche nicht immer ganz überzeugend erscheinen"). In a footnote to this passage, Smoluchowski indicated the considerations he had in mind: "For example, application of the laws of osmotic pressure to those particles and calculation of their diffusion velocity or the application of Boltzmann's law (on the statistical distribution of the states of systems, which are subject to potential energies) to the frictional resistance experienced by a particle M" ("Zum Beispiel Übertragung der Gesetze des osmotischen Druckes auf jene Teilchen und Berechnung ihrer Diffusionsgeschwindigkeit oder die Anwendung des Boltzmannschen Satzes (über die statistische Verteilung der Zustände von Systemen, welche Potentialkräften unterworfen sind) auf den von einem Teilchen M erfahrenen Reibungswiderstand").

[70] *Einstein 1917b*, p. 737.

meine Untersuchung im Wesentlichen nur das Gesetz des osmotischen Druckes voraussetzt'').[71]

Smoluchowski obtained the same final result as Einstein for the mean square displacement of a particle suspended in a fluid, apart from a numerical factor. In 1908, this numerical discrepancy was the starting point of Langevin's analysis.[72] Langevin not only asserted that Smoluchowski's method leads to the same result as Einstein's if correctly applied, he also presented a new derivation of the mean square displacement formula.

Einstein's work on Brownian motion was also noticed by others outside the small circle of physicists with a specialized interest in the phenomenon. Between 1905 and the middle of 1906, it brought him into contact with Heinrich Zangger, Professor of Forensic Medicine at the University of Zurich, who had a strong interest in the physical properties of membranes,[73] and who apparently had made measurements of Brownian motion, which led to his acquaintance with Einstein.[74] Zangger continued to discuss the topic with Einstein.[75]

In a letter of 18 September 1906, Wilhelm Röntgen asked Einstein for his reaction to the ''objection already expressed by Gouy . . . that is directed against the assumption of molecular impulses as the source of motion of small bodies. This is quite difficult to bring into harmony with the second law of thermodynamics'' (''bereits von Gouy geäusserten Bedenken . . . dass sich gegen die Annahme von Molekularstössen als Ursache der Bewegungen kleiner Körperchen richtet. Dieselbe ist wohl schwer mit dem zweiten Hauptsatz der Thermodynamik in Einklang zu bringen''). In his papers, Einstein never directly addressed this and other objections raised to the kinetic interpretation of Brownian motion,[76] although they are indirectly addressed in attempts to make the fundamental

---

[71] Einstein to Carl Seelig, 15 September 1952. In this letter, Einstein recalled that Smoluchowski's work only concerned Brownian motion in gases and ''has not achieved the degree of acuity (precision) to be desired'' (''nicht den wünschbaren Grad von Schärfe (Genauigkeit) erreicht hat'').

[72] See *Langevin 1908*. For a critical discussion of various approaches to Brownian motion, see *Infeld 1940*. For a comparative discussion of Einstein's and Smoluchowski's work on Brownian motion, see *Teske 1969*.

[73] See *Zangger 1906, 1907*.

[74] In sketchy notes probably written in the 1950s, Zangger recalled how he met Einstein: ''Stodola [Professor of Mechanical Engineering at the ETH] told me he could say nothing concrete about my counting of Brownian m[olecular] motion. (1905/06 . . .) I should go to Einstein in Bern'' (''Stodola sagte mir, dass er über meine Zählung der Brown M Bewegung nichts Bestimmtes sagen könne. (1905/06 . . .)

Ich solle zu Einstein nach Bern'').

[75] A reference to their discussion of Brownian motion is found in Einstein to Heinrich Zangger, 7 November 1911. Zangger continued his research on Brownian motion in collaboration with a doctoral student, Paul Böhi (see *Zangger 1911* and *Böhi 1911*). For evidence of a letter from Einstein to Böhi, see *Böhi 1911*, p. 212.

[76] Einstein later used the example of Brownian motion in discussing an apparent contradiction between phenomenological thermodynamics and the kinetic theory of heat (*Einstein 1915a*, pp. 261–262). He emphasized that, from the phenomenon of Brownian motion, it is clear ''that the laws of phenomenological thermodynamics only possess approximate validity'' (''daß die Gesetze der phänomenologischen Wärmelehre nur angenäherte Gültigkeit besitzen''), and concluded: ''Thus, according to Boltzmann, the averaged laws of experience simulate the irreversibility of thermal processes

features of his theory accessible to mathematically less sophisticated readers.[77] In late 1907 or early 1908, the need for an elementary account of Brownian motion was suggested to Einstein by Richard Lorenz, Professor of Electrochemistry and Physical Chemistry at the ETH, who pointed out that such an account would be appreciated by chemists.[78] Misunderstandings of Einstein's theory among those trying to provide experimental support for it reinforced the need for such a theory.[79] *Einstein 1908c* (Doc. 50) was prepared in response to this.

In 1909, Einstein sent a reprint of at least one of his papers on Brownian motion to Ernst Mach. In the accompanying letter, as in later popular accounts, he emphasized the direct relationship between Brownian motion and "thermal motion" ("Wärmebewegung").[80] In 1915, Einstein wrote: "Under the microscope one, to some extent, immediately sees a part of thermal energy in the form of mechanical energy of moving particles" ("Man sieht gewissermaßen unter dem Mikroskop unmittelbar einen Teil der Wärmeenergie in Form von mechanischer Energie bewegter Teilchen").[81] Later, he stated his view of the influence his work on Brownian motion had on his contemporaries: "The agreement of these considerations [on Brownian motion] with experience together with Planck's determination of the true molecular size from the law of radiation (for high temperatures) convinced the skeptics, who were quite numerous at that time (Ostwald, Mach), of the reality of atoms" ("Die Übereinstimmung dieser Betrachtung mit der Erfahrung zusammen mit der Planck'schen Bestimmung der wahren Molekülgrösse aus dem Strahlungsgesetz (für hohe Temperaturen) überzeugte die damals zahlreichen Skeptiker (Ostwald, Mach) von der Realität der Atome").[82] The new results on Brownian motion indeed played a decisive role in convincing Ostwald of the existence of atoms.[83] Mach, on the other hand, remained skeptical on this point until at least 1910.[84]

---

for us" ("Die Durchschnittsgesetze der Erfahrung täuschen uns also nach Boltzmann die Nichtumkehrbarkeit der thermischen Prozesse vor").

[77] See especially *Einstein 1907c* (Doc. 40); *Einstein 1907f* (Doc. 43); *Einstein 1908c* (Doc. 50); and *Einstein 1915a.*

[78] For evidence of Lorenz's suggestion, see *Einstein 1908c* (Doc. 50), p. 235. Lorenz asked Einstein for reprints of his statistical papers, to which his attention was drawn by *Einstein 1905k* (Doc. 16) (Richard Lorenz to Einstein, 15 November 1907).

[79] For a discussion of these misunderstandings, see the following section.

[80] Einstein to Ernst Mach, 9 August 1909. Einstein forgot to include his papers, and actually sent them with his letter of 17 August 1909.

[81] *Einstein 1915a,* p. 261.

[82] *Einstein 1979,* p. 44; translation, p. 45.

[83] In 1906, Ostwald, reacting to the ultrami-

croscopic observations of colloidal particles reported in *Zsigmondy 1905,* admitted for the first time in print the existence of limitations to classical thermodynamics (see *Ostwald 1907* and the preface to *Ostwald 1909,* p. IV). For a discussion of Ostwald's reaction to the new work on Brownian motion, see *Brush 1968,* p. 35.

[84] In 1909, Mach reprinted his essay on the principle of conservation of energy (*Mach 1872*), which is critical of the kinetic theory of heat, and sent a copy to Einstein (see Einstein to Ernst Mach 9 August 1909). In *Mach 1910,* he reiterated his criticism of kinetic theory as "hypothetical-fictive physics" ("hypothetisch-fiktive Physik") (*Mach 1910,* p. 231). For discussions of the controversial question whether Mach ever accepted the reality of atoms, see *Blackmore 1972,* appendix, pp. 319–323, and *Wolters 1988,* § 2, pp. 172–175. For a discussion of the Einstein-Mach relationship and the problem of atomism, see *Klein 1986.*

VII

Several newly perfected techniques for the experimental investigation of Brownian motion, notably the ultramicroscope and new methods for preparing colloidal solutions, were available by the time Einstein published his first articles on the subject.[85] As one of the first applications of the ultramicroscope, Siedentopf and Zsigmondy observed Brownian motion in colloidal solutions, but did not perform precise measurements.[86] A book on the ultramicroscope and its applications by Aimé Cotton and Henri Mouton, published in 1906, helped stimulate interest in Brownian motion, and brought Einstein's theory to the attention of researchers in the field.[87] Using an ultramicroscope and a sophisticated observational technique, The Svedberg carried out careful measurements of Brownian motion with the aim of testing the interpretation of Brownian motion as caused by the thermal motions of molecules.[88] *Svedberg 1906b* reports on his attempts to test Einstein's theory, about which he had learned by reading *Cotton and Mouton 1906*.[89] Svedberg corresponded with Einstein on the subject of Brownian motion, sending one of his papers.[90]

Svedberg followed Zsigmondy in assuming two types of motions for colloidal particles, a translational motion and a "proper [Brownian] motion" ("Eigenbewegung").[91] Svedberg restricted his attention to the latter, and attempted to facilitate its measurement by superimposing a translatory motion. He described the resulting trajectories as "sinusoid-like" ("sinusoidähnlich"), but cautioned against concluding that the motion had an oscillatory character.[92] In the analysis of his results, however, Svedberg introduced a terminology that was adapted to the description of a simple oscillatory motion, relating the observed amplitudes to Einstein's root mean square displacement in *Svedberg 1906b*. Earlier, he had tried to estimate molecular velocities on the basis of the observed velocities of colloidal particles.[93] *Einstein 1907c* (Doc. 40), written mainly to correct the basic misunderstandings in Svedberg's work, showed that the velocities of ultramicroscopic particles, as calculated from the equipartition theorem, cannot be directly observed.[94] On 11 November 1909, Einstein wrote to Perrin:

[85] New methods to prepare metallic hydrosols, which are particularly suited for the ultramicroscopic observation of Brownian motion, were developed at this time. For contemporary accounts, see *Cotton and Mouton 1906*, pp. 90–94, and *Zsigmondy 1905*. As a starting point for his work on Brownian motion, Svedberg succeeded in preparing colloidal solutions containing particles of roughly the same size in different solvents (see *Svedberg 1906a*, p. 119).

[86] See *Siedentopf and Zsigmondy 1903*; *Siedentopf 1903*; *Zsigmondy 1905*, pp. 106–111; and the remark in *Einstein 1906b* (Doc. 32), p. 371. See also Siedentopf's discussion remarks in *Einstein et al. 1909a* (Doc. 58).

[87] Brownian motion is discussed in *Cotton and Mouton 1906*, pp. 96–106. A brief account

of *Einstein 1905k* (Doc. 16) and *Einstein 1906b* (Doc. 32) is given on pp. 103–104, including Einstein's formula for the mean square displacement of a suspended particle.

[88] See *Svedberg 1906a*. For a discussion of Svedberg's work on Brownian motion, see *Kerker 1976*.

[89] See *Svedberg 1910*, p. 574.

[90] For evidence that Einstein received a paper by Svedberg, see Einstein to The Svedberg, 14 December 1919.

[91] See *Svedberg 1906a*, p. 853. For Zsigmondy's discussion of the motion of colloidal particles, see *Zsigmondy 1905*, chap. 10.

[92] See *Svedberg 1906a*, p. 854.

[93] See *Svedberg 1906a*, pp. 856–859.

[94] Svedberg's errors are also mentioned in

The errors in Svedberg's method of observation and also in his theoretical treatment became clear to me at once. I wrote a minor correction at the time, which
only addressed the worst, as I couldn't bring myself to detract from Mr. S.'s
great pleasure in his work.

Die Fehler in der Beobachtungsweise und auch in der theoretischen Behandlung
bei Svedberg waren mir sogleich klar geworden. Ich schrieb damals eine kleine
Richtigstellung, welche nur das Schlimmste korrigierte, weil ich mich nicht
entschliessen konnte, Herrn S. das grosse Vergnügen an seiner Arbeit zu beein-
trächtigen.

In addition to the misconceptions underlying Svedberg's experimental work, his numerical results slightly disagreed with Einstein's predictions.[95] Other early experimental
work on Einstein's and Smoluchowski's theories, such as Felix Ehrenhaft's observations
of displacements of aerosol particles, Victor Henri's cinematographical measurements of
displacements of suspended particles,[96] or Max Seddig's study of the temperature dependence of Brownian motion,[97] provided qualitative confirmation of the theory; but the
work of Henri and Seddig also failed to yield quantitative agreement. As a consequence,
the kinetic interpretation of Brownian motion was not universally accepted in 1908 as the
exclusive explanation of the phenomenon.[98]

Einstein noted that the control of the temperature was the principal difficulty in obtaining satisfactory results from the photographic records of Seddig and Henri.[99] Before Jean
Perrin published his thorough experimental investigation of the phenomenon,[100] Einstein
was skeptical about the possibility of obtaining precise measurements of Brownian motion. In 1908, Einstein commented rather enthusiastically on Seddig's work in spite of its
shortcomings: "I have read Seddig's paper. He has done it very well. I cannot quite make
head or tail of his descriptions of the results" ("Die Arbeit von Seddig habe ich gelesen.
Er hat es sehr gut angestellt. Aus den Angaben über die Resultate werde ich nicht ganz
klug").[101] In 1909, Einstein wrote to Perrin: "I would have considered it impossible to
investigate Brownian motion so precisely; it is a stroke of luck for this subject that you
have taken it up" ("Ich hätte es für unmöglich gehalten, die Brown'sche Bewegung so

---

Langevin 1908, and by Siedentopf in a discussion remark (see *Einstein et al. 1909a* [Doc.
58]) following *Siedentopf 1909*. For Svedberg's
attempt to defend his experimental analysis, see
Svedberg 1910. For a review of criticisms of
Svedberg's work, see *Kerker 1976*, pp. 210–
212.

[95] See *Svedberg 1906b*, p. 910.

[96] See *Ehrenhaft 1907* and *Henri 1908*; for a
discussion of Henri's work, see *Nye 1972*, p.
126.

[97] See *Seddig 1907* and *1908*. For contemporary discussions of Seddig's work, see the dis-

cussion remarks by Einstein and Seddig in *Einstein et al. 1909a* (Doc. 58), and *Perrin 1911*,
p. 204; for a recent account, see *Nye 1972*, pp.
125–126.

[98] See *Nye 1972*, p. 126. For a contemporary
review of early experimental work on Brownian
motion, see *Cotton 1908*.

[99] See Einstein's remark in *Einstein et al.
1909a* (Doc. 58).

[100] For a discussion of Perrin's work on the
measurement of Brownian motion, see *Nye
1972*.

[101] Einstein to Jakob Laub, 30 July 1908.

präzis zu untersuchen; es ist ein Glück für diese Materie, dass Sie sich ihrer angenommen haben'').[102]

In a series of experiments, the first results of which were published in 1908, Perrin achieved an until then unmatched precision in the confirmation of almost all of Einstein's predictions. Like Einstein, Perrin recognized that the analogy established by Van 't Hoff between an ideal gas and a solution could be extended to colloidal solutions and suspensions, and that this analogy provides a unique means of obtaining evidence for atomism.[103] In his first experiments on Brownian motion, Perrin tested a formula for the vertical distribution of suspended particles under the influence of gravitation.[104] Although Perrin probably was aware of Einstein's theory through Langevin,[105] he was apparently unaware that Einstein had derived a similar formula.[106] Challenged by criticism, Perrin checked his assumption of the validity of Stokes's formula for the particles used in his experiments.[107] In two further papers published in 1908, Perrin applied his methods to a determination of Avogadro's number.[108]

In the same year, Perrin's doctoral student Chaudesaigues subjected Einstein's displacement formula to experimental tests. Contrary to Henri's results mentioned above, the results are in excellent agreement with theoretical predictions.[109] Perrin continued these successful experiments with the help of other students;[110] to Einstein's surprise, he was able to include rotational Brownian motion in his investigations.[111] On 11 November 1909, Einstein wrote to Perrin: ''I would not have considered a measurement of the rotations as feasible. In my eyes it was only a pretty trifle'' (''Eine Messung der Drehungen hätte ich nicht für ausführbar gehalten. Es war in meinen Augen nur eine hübsche Spielerei''). Perrin's success was based on the ingenious combination of several experimental techniques for preparing emulsions with precisely controllable particle sizes, and for measuring the particles' number and displacements. He summarized his results in various review articles and books that significantly furthered the general acceptance of atomism.[112]

Beginning in 1907, Einstein himself tried to contribute to the experimental study of fluctuation phenomena. His prediction of voltage fluctuations in condensers, published in *Einstein 1907b* (Doc. 39), stimulated him to explore the possibility of measuring small quantities of electricity in order to provide experimental support for ''a phenomenon in the field of electricity related to Brownian motion'' (''ein der Brownschen Bewegung verwandtes Phänomen auf dem Gebiete der Elektrizität'').[113] On 15 July 1907, he wrote

[102] Einstein to Jean Perrin, 11 November 1909.
[103] This conceptual background is discussed at length in *Perrin 1911*, pp. 166ff.
[104] See *Perrin 1908a*.
[105] See the discussion of this question in *Nye 1972*, p. 111.
[106] Although Einstein's name is mentioned in *Perrin 1908a* in connection with the validity of the equipartition theorem for suspended parti-

cles, none of his papers are cited.
[107] See *Perrin 1908b*, and the discussion in *Nye 1972*, pp. 108–109.
[108] See *Perrin 1908c* and *1908d*.
[109] See *Chaudesaigues 1908*.
[110] See, e.g., *Perrin and Dabrowski 1909*.
[111] See *Perrin 1909a*.
[112] See, e.g., *Perrin 1911* and *1914*.
[113] *Einstein 1907b* (Doc. 39), p. 572.

to his friends Conrad and Paul Habicht about his discovery of a method for the measurement of small quantities of electrical energy. Soon afterward the Habichts tried to build the device ("Maschinchen") proposed by Einstein.[114] At the end of 1907, Einstein dropped his idea of obtaining a patent for the device, "primarily because of the lack of interest by manufacturers" ("hauptsächlich wegen Interesselosigkeit der Fabrikanten").[115] Instead, he published a paper on the basic features of his method (*Einstein 1908a* [Doc. 48]), a paper which stimulated further work on the device proposed by Einstein.[116] While the use of the device for measuring fluctuation phenomena in conductors proved to be difficult,[117] experimental work done by others soon provided evidence for the atomistic constitution of matter and electricity that exceeded Einstein's initial expectations.[118]

[114] On 16 August 1907, Einstein wrote to Conrad and Paul Habicht: "I am not a little astonished at the breakneck speed with which you have made the 'Maschinchen' " ("Ich bin nicht wenig erstaunt über die rasende Schnelligkeit mit der Ihr das Maschinchen gemacht habt").

[115] Einstein to Conrad Habicht, 24 December 1907.

[116] See, e.g., *Habicht and Habicht 1910*; for further discussion, see Vol. 5, the editorial note,

"Einstein's 'Maschinchen' for the Measurement of Small Quantities of Electricity."

[117] For further discussion, see note 4 to *Einstein 1908a* (Doc. 48), p. 492.

[118] For Einstein's expectations, see Einstein to Jean Perrin, 11 November 1909, quoted earlier in this section. For a discussion of experimental studies of the atomistic constitution of electricity by Millikan and others, see *Holton 1978*.

16. ''On the Movement of Small Particles Suspended in Stationary Liquids Required by the Molecular-Kinetic Theory of Heat''

[Einstein 1905k]

DATED Bern, May 1905
RECEIVED 11 May 1905
PUBLISHED 18 July 1905

IN: *Annalen der Physik* 17 (1905): 549–560.

Republished in *Einstein 1922*, pp. 5–15.

549

5. *Über die von der molekularkinetischen Theorie der Wärme geforderte Bewegung von in ruhenden Flüssigkeiten suspendierten Teilchen;* von A. Einstein.

In dieser Arbeit soll gezeigt werden, daß nach der molekularkinetischen Theorie der Wärme in Flüssigkeiten suspendierte Körper von mikroskopisch sichtbarer Größe infolge der Molekularbewegung der Wärme Bewegungen von solcher Größe ausführen müssen, daß diese Bewegungen leicht mit dem Mikroskop nachgewiesen werden können. Es ist möglich, daß die hier zu behandelnden Bewegungen mit der sogenannten „Brownschen Molekularbewegung" identisch sind; die mir erreichbaren Angaben über letztere sind jedoch so ungenau, [1]   daß ich mir hierüber kein Urteil bilden konnte.

Wenn sich die hier zu behandelnde Bewegung samt den für sie zu erwartenden Gesetzmäßigkeiten wirklich beobachten läßt, so ist die klassische Thermodynamik schon für mikroskopisch unterscheidbare Räume nicht mehr als genau gültig anzusehen und es ist dann eine exakte Bestimmung der wahren Atomgröße möglich. Erwiese sich umgekehrt die Voraussage dieser Bewegung als unzutreffend, so wäre damit ein schwerwiegendes Argument gegen die molekularkinetische Auffassung [2]   der Wärme gegeben.

§ 1. Über den suspendierten Teilchen zuzuschreibenden osmotischen Druck.

Im Teilvolumen V^* einer Flüssigkeit vom Gesamtvolumen V seien z-Gramm-Moleküle eines Nichtelektrolyten gelöst. Ist das Volumen V^* durch eine für das Lösungsmittel, nicht aber für die gelöste Substanz durchlässige Wand vom reinen Lösungs-

A. Einstein.

mittel getrennt, so wirkt auf diese Wand der sogenannte osmotische Druck, welcher bei genügend großen Werten von V^*/z
der Gleichung genügt:

$$p\, V^* = R\, T\, z\,.$$ [3]

Sind hingegen statt der gelösten Substanz in dem Teilvolumen V^* der Flüssigkeit kleine suspendierte Körper vorhanden, welche ebenfalls nicht durch die für das Lösungsmittel durchlässige Wand hindurchtreten können, so hat man
nach der klassischen Theorie der Thermodynamik — wenigstens
bei Vernachlässigung der uns hier nicht interessierenden Schwerkraft — nicht zu erwarten, daß auf die Wand eine Kraft
wirke; denn die „freie Energie" des Systems scheint nach der
üblichen Auffassung nicht von der Lage der Wand und der
suspendierten Körper abzuhängen, sondern nur von den Ge-
samtmassen und Qualitäten der suspendierten Substanz, der
Flüssigkeit und der Wand, sowie von Druck und Temperatur.   [4]
Es kämen allerdings für die Berechnung der freien Energie
noch Energie und Entropie der Grenzflächen in Betracht
(Kapillarkräfte); hiervon können wir jedoch absehen, indem
bei den ins Auge zu fassenden Lagenänderungen der Wand
und der suspendierten Körper Änderungen der Größe und
Beschaffenheit der Berührungsflächen nicht eintreten mögen.

Vom Standpunkte der molekularkinetischen Wärmetheorie
aus kommt man aber zu einer anderen Auffassung. Nach
dieser Theorie unterscheidet sich eingelöstes Molekül von einem
suspendierten Körper *lediglich* durch die Größe, und man sieht
nicht ein, warum einer Anzahl suspendierter Körper nicht derselbe osmotische Druck entsprechen sollte, wie der nämlichen
Anzahl gelöster Moleküle. Man wird anzunehmen haben, daß
die suspendierten Körper infolge der Molekularbewegung der
Flüssigkeit eine wenn auch sehr langsame ungeordnete Bewegung in der Flüssigkeit ausführen; werden sie durch die
Wand verhindert, das Volumen V^* zu verlassen, so werden sie
auf die Wand Kräfte ausüben, ebenso wie gelöste Moleküle.
Sind also n suspendierte Körper im Volumen V^*, also $n/V^* = \nu$
in der Volumeneinheit vorhanden, und sind benachbarte unter
ihnen genügend weit voneinander entfernt, so wird ihnen ein
osmotischer Druck p entsprechen von der Größe:

Bewegung v. in ruhenden Flüssigkeiten suspendierten Teilchen.   551

$$p = \frac{RT}{V^*} \frac{n}{N} = \frac{RT}{N} \cdot v,$$

wobei N die Anzahl der in einem Gramm-Molekül enthaltenen wirklichen Moleküle bedeutet. Im nächsten Paragraph soll gezeigt werden, daß die molekularkinetische Theorie der Wärme wirklich zu dieser erweiterten Auffassung des osmotischen Druckes führt.

§ 2. Der osmotische Druck vom Standpunkte der molekular-kinetischen Theorie der Wärme.[1)]

Sind $p_1\, p_2 \ldots p_l$ Zustandsvariable eines physikalischen Systems, welche den momentanen Zustand desselben vollkommen bestimmen (z. B. die Koordinaten und Geschwindigkeitskomponenten aller Atome des Systems) und ist das vollständige System der Veränderungsgleichungen dieser Zustandsvariabeln von der Form

$$\frac{\partial p_\nu}{\partial t} = \varphi_\nu (p_1 \ldots p_l)(\nu = 1, 2 \ldots l)$$

[6]   gegeben, wobei $\Sigma \frac{\partial \varphi_\nu}{\partial p_\nu} = 0$, so ist die Entropie des Systems durch den Ausdruck gegeben:

[7]
$$S = \frac{\bar{E}}{T} + 2\varkappa \lg \int e^{-\frac{E}{2\varkappa T}} dp_1 \ldots dp_l.$$

Hierbei bedeutet T die absolute Temperatur, $\bar{E}$ die Energie des Systems, E die Energie als Funktion der p_ν. Das Integral ist über alle mit den Bedingungen des Problems vereinbaren Wertekombinationen der p_ν zu erstrecken. $\varkappa$ ist mit
[8]   der oben erwähnten Konstanten N durch die Relation $2\varkappa N = R$ verbunden. Für die freie Energie F erhalten wir daher:

$$F = -\frac{R}{N} T \lg \int e^{-\frac{EN}{RT}} dp_1 \ldots dp_l = -\frac{RT}{N} \lg B.$$

---

1) In diesem Paragraph sind die Arbeiten des Verfassers über die Grundlagen der Thermodynamik als bekannt vorausgesetzt (vgl. Ann. d.
[5]   Phys. **9**. p. 417. 1902; **11**. p. 170. 1903). Für das Verständnis der Resultate der vorliegenden Arbeit ist die Kenntnis jener Arbeiten sowie dieses Paragraphen der vorliegenden Arbeit entbehrlich.

Wir denken uns nun eine in dem Volumen V eingeschlossene Flüssigkeit; in dem Teilvolumen V^* von V mögen sich n gelöste Moleküle bez. suspendierte Körper befinden, welche im Volumen V^* durch eine semipermeabele Wand festgehalten seien; es werden hierdurch die Integrationsgrenzen des in den Ausdrücken für S und F auftretenden Integrales B beeinflußt. Das Gesamtvolumen der gelösten Moleküle bez. suspendierten Körper sei klein gegen V^*. Dies System werde im Sinne der erwähnten Theorie durch die Zustandsvariabeln $p_1 \ldots p_l$ vollständig dargestellt.

Wäre nun auch das molekulare Bild bis in alle Einzelheiten festgelegt, so böte doch die Ausrechnung des Integrales B solche Schwierigkeiten, daß an eine exakte Berechnung von F kaum gedacht werden könnte. Wir brauchen jedoch hier nur zu wissen, wie F von der Größe des Volumens V^* abhängt, in welchem alle gelösten Moleküle bez. suspendierten Körper (im folgenden kurz „Teilchen" genannt) enthalten sind.

Wir nennen x_1, y_1, z_1 die rechtwinkligen Koordinaten des Schwerpunktes des ersten Teilchens, x_2, y_2, z_2 die des zweiten etc., x_n, y_n, z_n die des letzten Teilchens und geben für die Schwerpunkte der Teilchen die unendlich kleinen parallelepipedförmigen Gebiete $dx_1\,dy_1\,dz_1,\ dx_2\,dy_2\,dz_2 \ldots dx_n\,dy_n\,dz_n$, welche alle in V^* gelegen seien. Gesucht sei der Wert des im Ausdruck für F auftretenden Integrales mit der Beschränkung, daß die Teilchenschwerpunkte in den ihnen soeben zugewiesenen Gebieten liegen. Dies Integral läßt sich jedenfalls auf die Form

$$dB = dx_1\,dy_1 \ldots dz_n \,.\, J$$

bringen, wobei J von $dx_1\,dy_1$ etc., sowie von V^*, d. h. von der Lage der semipermeabeln Wand, unabhängig ist. J ist aber auch unabhängig von der speziellen Wahl *der Lagen* der Schwerpunktsgebiete und von dem Werte von V^*, wie sogleich gezeigt werden soll. Sei nämlich ein zweites System von unendlich kleinen Gebieten für die Teilchenschwerpunkte gegeben und bezeichnet durch $dx_1'\,dy_1'\,dz_1',\ dx_2'\,dy_2'\,dz_2' \ldots dx_n'\,dy_n'\,dz_n'$, welche Gebiete sich von den ursprünglich gegebenen nur durch ihre Lage, nicht aber durch ihre Größe unterscheiden mögen und ebenfalls alle in V^* enthalten seien, so gilt analog:

$$dB' = dx_1'\,dy_1' \ldots dz_n' \,.\, J',$$

Bewegung v. in ruhenden Flüssigkeiten suspendierten Teilchen.   553

wobei

$$d x_1 \, d y_1 \ldots d z_n = d x'_1 \, d y'_1 \ldots d z'_n.$$

Es ist also:

$$\frac{d B}{d B'} = \frac{J}{J'}.$$

Aus der in den zitierten Arbeiten gegebenen molekularen Theorie der Wärme läßt sich aber leicht folgern[1]), daß $d B / B$ bez. $d B' / B$ gleich ist der Wahrscheinlichkeit dafür, daß sich in einem beliebig herausgegriffenen Zeitpunkte die Teilchenschwerpunkte in den Gebieten $(d x_1 \ldots d z_n)$ bez. in den Gebieten $(d x'_1 \ldots d z'_n)$ befinden. Sind nun die Bewegungen der einzelnen Teilchen (mit genügender Annäherung) voneinander unabhängig, ist die Flüssigkeit homogen und wirken auf die Teilchen keine Kräfte, so müssen bei gleicher Größe der Gebiete die den beiden Gebietssystemen zukommenden Wahrscheinlichkeiten einander gleich sein, so daß gilt:

$$\frac{d B}{B} = \frac{d B'}{B}.$$

Aus dieser und aus der zuletzt gefundenen Gleichung folgt aber

$$J = J'.$$

Es ist somit erwiesen, daß J weder von V^* noch von $x_1, y_1 \ldots z_n$ abhängig ist. Durch Integration erhält man

$$B = \int J \, d x_1 \ldots d z_n = J V^{*n}$$

und daraus

$$F = - \frac{R T}{N} \{ \lg J + n \lg V^* \}$$

und

[10]
$$p = - \frac{\partial F}{\partial V^*} = \frac{R T}{V^*} \frac{n}{N} = \frac{R T}{N} \nu.$$

Durch diese Betrachtung ist gezeigt, daß die Existenz des osmotischen Druckes eine Konsequenz der molekularkinetischen Theorie der Wärme ist, und daß nach dieser Theorie gelöste Moleküle und suspendierte Körper von gleicher Anzahl sich in bezug auf osmotischen Druck bei großer Verdünnung vollkommen gleich verhalten.

---

[9]   1) A. Einstein, Ann. d. Phys. **11**. p. 170. 1903.

554   *A. Einstein.*

§ 3. Theorie der Diffusion kleiner suspendierter Kugeln.   [11]

In einer Flüssigkeit seien suspendierte Teilchen regellos verteilt. Wir wollen den dynamischen Gleichgewichtszustand derselben untersuchen unter der Voraussetzung, daß auf die einzelnen Teilchen eine Kraft K wirkt, welche vom Orte, nicht aber von der Zeit abhängt. Der Einfachheit halber werde angenommen, daß die Kraft überall die Richtung der X-Achse habe.

Es sei ν die Anzahl der suspendierten Teilchen pro Volumeneinheit, so ist im Falle des thermodynamischen Gleichgewichtes ν eine solche Funktion von x, daß für eine beliebige virtuelle Verrückung δx der suspendierten Substanz die Variation der freien Energie verschwindet. Man hat also:

$$\delta F = \delta E - T \delta S = 0.$$ [12]

Es werde angenommen, daß die Flüssigkeit senkrecht zur X-Achse den Querschnitt 1 habe und durch die Ebenen $x = 0$ und $x = l$ begrenzt sei. Man hat dann:

$$\delta E = - \int_0^l K \nu \, \delta x \, dx$$

und

$$\delta S = \int_0^l R \frac{\nu}{N} \frac{\partial \delta x}{\partial x} \, dx = - \frac{R}{N} \int_0^l \frac{\partial \nu}{\partial x} \, \delta x \, dx.$$ [13]

Die gesuchte Gleichgewichtsbedingung ist also:

(1) $$- K \nu + \frac{R T}{N} \frac{\partial \nu}{\partial x} = 0$$

oder

$$K \nu - \frac{\partial p}{\partial x} = 0.$$

Die letzte Gleichung sagt aus, daß der Kraft K durch osmotische Druckkräfte das Gleichgewicht geleistet wird.

Die Gleichung (1) benutzen wir, um den Diffusionskoeffizienten der suspendierten Substanz zu ermitteln. Wir können den eben betrachteten dynamischen Gleichgewichtszustand als

die Superposition zweier in umgekehrtem Sinne verlaufender
[14] Prozesse auffassen, nämlich

1. einer Bewegung der suspendierten Substanz unter der
Wirkung der auf jedes einzelne suspendierte Teilchen wirkenden Kraft K,

2. eines Diffusionsvorganges, welcher als Folge der ungeordneten Bewegungen der Teilchen infolge der Molekularbewegung der Wärme aufzufassen ist.

Haben die suspendierten Teilchen Kugelform (Kugelradius P)
und besitzt die Flüssigkeit den Reibungskoeffizienten k, so
erteilt die Kraft K dem einzelnen Teilchen die Geschwindigkeit[1]

$$\frac{K}{6\,\pi\,k\,P},$$

und es treten durch die Querschnittseinheit pro Zeiteinheit

$$\frac{\nu\,K}{6\,\pi\,k\,P}$$

Teilchen hindurch.

Bezeichnet ferner D den Diffusionskoeffizienten der suspendierten Substanz und μ die Masse eines Teilchens, so treten
pro Zeiteinheit infolge der Diffusion

$$-\,D\,\frac{\partial\,(\mu\,\nu)}{\partial\,x}\ \text{Gramm}$$

oder

$$-\,D\,\frac{\partial\,\nu}{\partial\,x}$$

Teilchen durch die Querschnittseinheit.   Da dynamisches Gleichgewicht herrschen soll, so muß sein:

(2) $$\frac{\nu\,K}{6\,\pi\,k\,P}-D\,\frac{\partial\,\nu}{\partial\,x}=0.$$

Aus den beiden für das dynamische Gleichgewicht gefundenen Bedingungen (1) und (2) kann man den Diffusionskoeffizienten berechnen.   Man erhält:

$$D=\frac{R\,T}{N}\,\frac{1}{6\,\pi\,k\,P}.$$

Der Diffusionskoeffizient der suspendierten Substanz hängt also

---

1) Vgl. z. B. G. Kirchhoff, Vorlesungen über Mechanik, 26. Vor-
[15] lesung § 4.

556   *A. Einstein.*

außer von universellen Konstanten und der absoluten Temperatur nur vom Reibungskoeffizienten der Flüssigkeit und von
der Größe der suspendierten Teilchen ab.

§ 4. Über die ungeordnete Bewegung von in einer Flüssigkeit suspendierten Teilchen und deren Beziehung zur Diffusion.

Wir gehen nun dazu über, die ungeordneten Bewegungen
genauer zu untersuchen, welche, von der Molekularbewegung
der Wärme hervorgerufen, Anlaß zu der im letzten Paragraphen untersuchten Diffusion geben.

Es muß offenbar angenommen werden, daß jedes einzelne
Teilchen eine Bewegung ausführe, welche unabhängig ist von
der Bewegung aller anderen Teilchen; es werden auch die
Bewegungen eines und desselben Teilchens in verschiedenen
Zeitintervallen als voneinander unabhängige Vorgänge aufzufassen sein, solange wir diese Zeitintervalle nicht zu klein gewählt denken.

Wir führen ein Zeitintervall τ in die Betrachtung ein,
welches sehr klein sei gegen die beobachtbaren Zeitintervalle,
aber doch so groß, daß die in zwei aufeinanderfolgenden Zeitintervallen τ von einem Teilchen ausgeführten Bewegungen als
voneinander unabhängige Ereignisse aufzufassen sind.   [16]

Seien nun in einer Flüssigkeit im ganzen n suspendierte
Teilchen vorhanden. In einem Zeitintervall τ werden sich die
X-Koordinaten der einzelnen Teilchen um Δ vergrößern, wobei
Δ für jedes Teilchen einen anderen (positiven oder negativen)
Wert hat. Es wird für Δ ein gewisses Häufigkeitsgesetz gelten;
die Anzahl dn der Teilchen, welche in dem Zeitintervall τ
eine Verschiebung erfahren, welche zwischen Δ und $\Delta + d\Delta$
liegt, wird durch eine Gleichung von der Form

$$dn = n\,\varphi(\Delta)\,d\Delta$$

ausdrückbar sein, wobei

$$\int_{-\infty}^{+\infty} \varphi(\Delta)\,d\Delta = 1$$

und φ nur für sehr kleine Werte von Δ von Null verschieden
ist und die Bedingung

$$\varphi(\Delta) = \varphi(-\Delta)$$

erfüllt.

Bewegung v. in ruhenden Flüssigkeiten suspendierten Teilchen. 557

Wir untersuchen nun, wie der Diffusionskoeffizient von φ abhängt, wobei wir uns wieder auf den Fall beschränken, daß die Anzahl v der Teilchen pro Volumeneinheit nur von x und t abhängt.

Es sei $v = f(x, t)$ die Anzahl der Teilchen pro Volumeneinheit, wir berechnen die Verteilung der Teilchen zur Zeit $t + \tau$ aus deren Verteilung zur Zeit t. Aus der Definition der Funktion $\varphi(\varDelta)$ ergibt sich leicht die Anzahl der Teilchen, welche sich zur Zeit $t + \tau$ zwischen zwei zur X-Achse senkrechten Ebenen mit den Abszissen x und $x + dx$ befinden. Man erhält:

[17]
$$f(x, t + \tau)\, dx = dx . \int_{\varDelta = -\infty}^{\varDelta = +\infty} f(x + \varDelta)\, \varphi(\varDelta)\, d\varDelta .$$

Nun können wir aber, da τ sehr klein ist, setzen:

$$f(x, t + \tau) = f(x, t) + \tau\, \frac{\partial f}{\partial t} .$$

Ferner entwickeln wir $f(x + \varDelta, t)$ nach Potenzen von $\varDelta$:

$$f(x + \varDelta, t) = f(x, t) + \varDelta\, \frac{\partial f(x, t)}{\partial x} + \frac{\varDelta^2}{2!}\, \frac{\partial^2 f(x, t)}{\partial x^2} \cdots \text{in inf.}$$

Diese Entwicklung können wir unter dem Integral vornehmen, da zu letzterem nur sehr kleine Werte von $\varDelta$ etwas beitragen. Wir erhalten:

$$f + \frac{\partial f}{\partial t} \cdot \tau = f \cdot \int_{-\infty}^{+\infty} \varphi(\varDelta)\, d\varDelta + \frac{\partial f}{\partial x} \int_{-\infty}^{+\infty} \varDelta\, \varphi(\varDelta)\, d\varDelta$$

$$+ \frac{\partial^2 f}{\partial x^2} \int_{-\infty}^{+\infty} \frac{\varDelta^2}{2}\, \varphi(\varDelta)\, d\varDelta \cdots$$

Auf der rechten Seite verschwindet wegen $\varphi(x) = \varphi(-x)$ das zweite, vierte etc. Glied, während von dem ersten, dritten, fünften etc. Gliede jedes folgende gegen das vorhergehende sehr klein ist. Wir erhalten aus dieser Gleichung, indem wir berücksichtigen, daß

$$\int_{-\infty}^{+\infty} \varphi(\varDelta)\, d\varDelta = 1,$$

558   *A. Einstein.*

und indem wir

$$\frac{1}{\tau} \int\limits_{-\infty}^{+\infty} \frac{\Delta^2}{2} \, \varphi(\Delta) \, d\Delta = D$$

setzen und nur das erste und dritte Glied der rechten Seite berücksichtigen:

(1) $$\frac{\partial f}{\partial t} = D \frac{\partial^2 f}{\partial x^2}.$$ [18]

Dies ist die bekannte Differentialgleichung der Diffusion, und man erkennt, daß D der Diffusionskoeffizient ist.

An diese Entwicklung läßt sich noch eine wichtige Überlegung anknüpfen. Wir haben angenommen, daß die einzelnen Teilchen alle auf dasselbe Koordinatensystem bezogen seien. Dies ist jedoch nicht nötig, da die Bewegungen der einzelnen Teilchen voneinander unabhängig sind. Wir wollen nun die Bewegung jedes Teilchens auf ein Koordinatensystem beziehen, dessen Ursprung mit der Lage des Schwerpunktes des betreffenden Teilchens zur Zeit $t = 0$ zusammenfällt, mit dem Unterschiede, daß jetzt $f(x,t)\,dx$ die Anzahl der Teilchen bedeutet, deren X-Koordinaten von der Zeit $t = 0$ bis zur Zeit $t = t$ um eine Größe *gewachsen* ist, welche zwischen x und $x + dx$ liegt. Auch in diesem Falle ändert sich also die Funktion f gemäß Gleichung (1). Ferner muß offenbar für $x \gtrless 0$ und $t = 0$

$$f(x,t) = 0 \quad \text{und} \quad \int\limits_{-\infty}^{+\infty} f(x,t)\,dx = n$$

sein. Das Problem, welches mit dem Problem der Diffusion von einem Punkte aus (unter Vernachlässigung der Wechselwirkung der diffundierenden Teilchen) übereinstimmt, ist nun mathematisch vollkommen bestimmt; seine Lösung ist:

$$f(x,t) = \frac{n}{\sqrt{4\pi D}} \frac{e^{-\frac{x^2}{4Dt}}}{\sqrt{t}}.$$

Die Häufigkeitsverteilung der in einer beliebigen Zeit t erfolgten Lagenänderungen ist also dieselbe wie die der zu-

Bewegung v. in ruhenden Flüssigkeiten suspendierten Teilchen. 559

[19] fälligen Fehler, was zu vermuten war. Von Bedeutung aber ist, wie die Konstante im Exponenten mit dem Diffusionskoeffizienten zusammenhängt. Wir berechnen nun mit Hilfe dieser Gleichung die Verrückung λ_x in Richtung der *X*-Achse, welche ein Teilchen im Mittel erfährt, oder — genauer ausgedrückt — die Wurzel aus dem arithmetischen Mittel der Quadrate der Verrückungen in Richtung der *X*-Achse; es ist:

$$\lambda_x = \sqrt{\overline{x^2}} = \sqrt{2\,D\,t}\,.$$

Die mittlere Verschiebung ist also proportional der Quadratwurzel aus der Zeit. Man kann leicht zeigen, daß die Wurzel aus dem Mittelwert der Quadrate der *Gesamtverschiebungen* der Teilchen den Wert $\lambda_x \sqrt{3}$ besitzt.

§ 5. Formel für die mittlere Verschiebung suspendierter Teilchen. Eine neue Methode zur Bestimmung der wahren Größe der Atome.

In § 3 haben wir für den Diffusionskoeffizienten *D* eines in einer Flüssigkeit in Form von kleinen Kugeln vom Radius *P* suspendierten Stoffes den Wert gefunden:

$$D = \frac{RT}{N}\,\frac{1}{6\,\pi\,k\,P}\,.$$

Ferner fanden wir in § 4 für den Mittelwert der Verschiebungen der Teilchen in Richtung der *X*-Achse in der Zeit *t*:

$$\lambda_x = \sqrt{2\,D\,t}\,.$$

Durch Eliminieren von *D* erhalten wir:

$$\lambda_x = \sqrt{t}\,.\sqrt{\frac{RT}{N}\,\frac{1}{3\,\pi\,k\,P}}\,.$$

Diese Gleichung läßt erkennen, wie λ_x von *T*, *k* und *P* abhängen muß.

Wir wollen berechnen, wie groß λ_x für eine Sekunde ist, wenn *N* gemäß den Resultaten der kinetischen Gastheorie
[20] $6 . 10^{23}$ gesetzt wird; es sei als Flüssigkeit Wasser von 17^0 C.
[21] gewählt ($k = 1{,}35 . 10^{-2}$) und der Teilchendurchmesser sei
[22] 0,001 mm. Man erhält:

$$\lambda_x = 8 . 10^{-5}\,\text{cm} = 0{,}8\,\text{Mikron}\,.$$

Die mittlere Verschiebung in 1 Min. wäre also ca. 6 Mikron.

560   *A. Einstein. Bewegung etc.*

Umgekehrt läßt sich die gefundene Beziehung zur Bestimmung von N benutzen. Man erhält:

$$N = \frac{t}{\lambda_x^2} \cdot \frac{RT}{3\pi k P}.$$

Möge es bald einem Forscher gelingen, die hier auf-
geworfene, für die Theorie der Wärme wichtige Frage zu entscheiden!   [23]

Bern, Mai 1905.

(Eingegangen 11. Mai 1905.)

Published in *Annalen der Physik* 17 (1905):
549–560. Dated Bern, May 1905, received 11
May 1905, published 18 July 1905.

[1] For a discussion of Einstein's familiarity
with previous work on Brownian motion, see the
editorial note, "Einstein on Brownian Motion,"
§ IV, pp. 210–214.

[2] At this time, the kinetic theory of heat was
not universally accepted. For a brief discussion,
see the editorial note, "Einstein on Brownian
Motion," § II, pp. 207–208.

[3] This equation was first derived in *Van 't
Hoff 1887*; it is used in *Einstein 1905j* (Doc. 15),
p. 19, in a calculation of molecular dimensions.

[4] The concept of free energy is used in the
treatment of the osmotic pressure in *Helmholtz
1903*, part II, § 2; see, in particular, pp. 321–
326.

[5] *Einstein 1902b* (Doc. 3), *Einstein 1903*
(Doc. 4).

[6] For the role of this condition in Einstein's
statistical physics, see *Einstein 1903* (Doc. 4),
note 7.

[7] This expression was earlier derived in *Einstein 1902b* (Doc. 3), p. 432.

[8] For the derivation of this expression, see
Einstein 1904 (Doc. 5), § 3.

[9] *Einstein 1903* (Doc. 4); see, in particular,
§ 3.

[10] For a derivation of this equation based on
Boltzmann's principle, see *Einstein 1905i* (Doc.
14), § 5.

[11] The main results of this section are contained in *Einstein 1905j* (Doc. 15), § 4.

[12] For a discussion of thermodynamic equilibrium conditions involving the notion of free
energy, see, e.g., *Nernst 1898*, pp. 28–30.

[13] The first equation easily follows from an
expression for the entropy as the one derived in
Einstein 1905i (Doc. 14), p. 142, fn. 1.

[14] For a similar argument involving a diffusion process and motion of ions under the action
of an exterior force, see *Einstein 1902a* (Doc.
2).

[15] *Kirchhoff 1897*, p. 380. The formula was
first derived in *Stokes 1845*. For a discussion of
its applicability to objects of microscopic dimensions, see *Fürth 1922*, pp. 58–60, fn. 6.

[16] For a more detailed discussion of this
problem, see *Einstein 1906b* (Doc. 32), § 5; see
also *Fürth 1922*, pp. 60–61, fn. 8.

[17] f on the right-hand side is to be taken at the
time t.

[18] This equation was first derived in *Fick
1855*, in which the analogy with heat conduction
is used. For a brief discussion of various derivations, see *Herzfeld 1921*, § 22.

[19] A similar correspondence between the random error curve and the Maxwell distribution
was noted by Maxwell (*Maxwell 1860*).

[20] The value given for N is close to that in
Planck 1901b, but differs from those obtained in
Einstein 1905j (Doc. 15) and *Einstein 1906c*
(Doc. 33).

[21] The value for the viscosity of water is taken from *Einstein 1905j* (Doc. 15), p. 21. It actually refers to water at temperature 9.5°C.

[22] The size of the particles for which Einstein performed this calculation is also mentioned in his letter to Conrad Habicht of 18 May–8 June 1905; it approximately corresponds to the size of the largest particles studied by Felix Exner in his experiments on Brownian motion; see *Exner 1900*, p. 845.

[23] For a discussion of subsequent experimental verifications of Einstein's results, see the editorial note, "Einstein on Brownian Motion," § VII, pp. 219–222; see also *Fürth 1922*, pp. 63–64, fn. 13.

17. Review of Karl Fredrik Slotte, "On the Heat of Fusion"

[Einstein 1905l]

Second half of June 1905

PUBLISHED IN: *Beiblätter zu den Annalen der Physik* 29 (1905), no. 12: 623–624.

55. *G. Charpy* und *L. Grenet.* *Über die Umwandlungstemperaturen beim Stahl* (C. R. **138**, S. 567—568. 1904). — Zur Bestimmung der Umwandlungspunkte von Stahlproben wurden von den verschiedensten Forschern eine Reihe von Methoden verwandt. Die hauptsächlichsten sind, die Verfolgung des elektrischen Widerstandes, die Beobachtung der Volumänderung sowie die der Temperaturänderung beim Erhitzen. Die Verf. prüfen diese drei Methoden an einer Reihe von Proben und finden, daß die thermoelektrische und die dilatometrische außer bei sehr weichem Stahl wenig miteinander übereinstimmen. Qualitativ und meist auch quantitativ gut übereinstimmende Resultate liefert die Beobachtung der Leitfähigkeit und der Volumänderung.   G. J.

56. *A. Findlay.* *Gefrierpunktskurven dynamischer Isomeren. Ammoniumcyanat und Thioharnstoff* (J. chem. Soc. 85, S. 403—412. 1904). — Es wird gezeigt, daß die Gefrierpunktskurve für Ammoniumcyanat und Thioharnstoff die einfachste mögliche Form hat und einen eutektischen Punkt bei 104,3° ergibt. Nichts deutet auf eine beständige Verbindung zwischen den beiden Komponenten hin. Die Wärmetönung bei der Umwandlung des einen Körpers in den anderen scheint null oder jedenfalls sehr klein zu sein.   E. Bs.

57. *Ch. M. van Deventer.* *Über das Schmelzen von schwimmendem Eis* (Versl. K. Ak. van Wet. 13, S. 490—493. 1904). — Der folgende Satz wird bewiesen: Beim Schmelzen von einem Stück Eis, das in einem Gefäß mit Wasser schwimmt, ändert sich die Höhe des Wasserspiegels nicht. Wenn Eis von süßem Wasser in salzigem Wasser schwimmt und schmilzt, so wird der Wasserspiegel um einen berechneten Betrag erhöht.   Keesom.

58. *K. F. Slotte.* *Über die Schmelzwärme* (Öf. Finska Vet. Soc. Förh. 47, S. 1—8. 1904). — Der Verf. stützt sich auf eine von ihm früher durch elementare molekulartheoretische Betrachtungen abgeleitete Beziehung zwischen der Schmelzwärme l, der (absoluten) Schmelztemperatur T_1 und der spezifischen Wärme bei konstantem Druck c_p, welche angenähert [1] durch die Formel ausgedrückt wird: $l = 0,382 \, c_p \, T_1$. Diese

624   Wärmelehre.   **Beibl. 1905.**

Formel erweist sich in roher Annäherung als gültig sowohl
für Elemente wie für Verbindungen. Auch einige Stoffe, für
welche die Formel auch nicht annähernd stimmt, werden angegeben (Schwefel, Phosphor). Nebenbei sei bemerkt, daß die
angegebene Beziehung bis auf den Zahlenwert der Konstanten
eine Konsequenz des auf den festen Aggregatzustand ausgedehnten Gesetzes von den übereinstimmenden Zuständen ist.
Schließlich wird der Verf. durch eine hier nicht wiederzugebende
molekulartheoretische Betrachtung zu der Ansicht geführt, daß
man die beste Übereinstimmung der Theorie mit der Erfahrung
erziele, wenn man den Atomen' der einfachen festen Körper
geradlinige harmonische Schwingungen zuschreibt.   **A. E.**   [2]

Published in *Beiblätter zu den Annalen der Physik* 29 (1905): 623–624. Published in no. 12 [second half of June].

[1] See *Slotte 1900, 1902*.

[2] *Slotte 1900* gives a theory for the motion of molecules in the solid state that contains a constant characterizing the nature of the oscillations of the molecules. Comparison with the results for the heat of melting derived in *Slotte 1902* leads to the conclusion that Einstein cites. A similar model had already been considered in *Boltzmann 1876* in connection with specific heats of solids. See also Einstein's remarks on this subject in Doc. 38 (*Einstein 1907a*), p. 184.

18. Review of Karl Fredrik Slotte, "Conclusions Drawn from a Thermodynamic Equation"

[Einstein 1905m]

Second half of June 1905

PUBLISHED IN: *Beiblätter zu den Annalen der Physik* 29 (1905), no. 12: 629.

Bd. 29. No. 12.   Wärmelehre.   629

68. **K. F. Slotte.** *Folgerungen aus einer thermodynamischen Gleichung* (Öf. Finska Vet. Soc. Förh. 47, S. 1—3. 1904). — Aus der bekannten Gleichung

$$\left(\frac{dQ}{dv}\right)_T = T\left(\frac{dp}{dT}\right)_v$$ [1]

werden unter der durchaus unmotivierten Voraussetzung, daß $(dQ/dv)_T$ für unendlich kleine Werte von T endliche Werte besitze, einige Folgerungen betreffend das Verhalten der Körper in der Nähe des absoluten Nullpunktes der Temperatur gezogen. [2]

A. E.

Published in *Beiblätter zu den Annalen der Physik* 29 (1905): 629. Published in no. 12 [second half of June].

[1] This is one of the Maxwell relations, usually written as $(\partial S/\partial v)_T = (\partial p/\partial T)_v$.

[2] Slotte explores the consequences of this equation as $T \to 0$. He claims that for $T = 0$ $(\partial p/\partial T)_v$ cannot be finite, because this would imply $(dQ/dv)_T = 0$, in contradiction with "well-known laws of nature" ("bekannte Naturgesetze"). Actually, the third law of thermodynamics (which at the time had not yet been formulated) requires $(\partial S/\partial v)_T = 0$ for $T = 0$, so that $(\partial p/\partial T)_v$ also vanishes for $T = 0$.

19. Review of Emile Mathias,
"The Constant *a* of Rectilinear Diameters and the Laws of Corresponding States"

[Einstein 1905n]

Second half of June 1905

Published in: *Beiblätter zu den Annalen der Physik* 29 (1905), no. 12: 634–635.

634   Wärmelehre.   Beibl. 1905.

75. *H. W. Bakhuis Roozeboom und E. H. Büchner.* *Kritische Endpunkte in Dreiphasenkurven mit festen Phasen bei binären Gemischen, welche zwei Flüssigkeitsschichten zeigen* (Versl. K. Akad. van Wet. 13, S. 531—537. 1905). — Bei Systemen einer Komponente ist bis jetzt nur ein kritischer Endpunkt, und zwar beim Gleichgewicht Flüssigkeit–Dampf, gefunden worden. Für Systeme zweier Komponenten, die sich in allen Verhältnissen mischen, hat man statt eines kritischen Endpunktes eine kritische Kurve. Smits hat vor kurzem (zuletzt Beibl. 29, S. 248) gezeigt, wie für solche binäre Gemische die Dreiphasenkurve für das Gleichgewicht fest–flüssig–Gas zwei kritische Endpunkte bekommen kann. Jetzt wird in einer Reihe von p, t-Diagrammen gezeigt, wie kritische Endpunkte in den Dreiphasenkurven auftreten können, zunächst beim Gleichgewicht flüssig–flüssig–Gas, wenn also zwei sich nicht mischende Flüssigkeitsschichten vorkommen, sodann beim dreiphasigen Gleichgewicht, wo eine der Phasen fest ist. Ein solches Verhalten wurde von Hrn. Büchner experimentell festgestellt beim System: Diphenylamin–Kohlensäure.   Keesom.

---

76. *M. E. Mathias.* *Die Konstante a der rechtwinkligen Diameter und die Gesetze der übereinstimmenden Zustände* (J. de [1] Phys. (4) 4, S. 77—91. 1905). — Bezeichnet man mit y eine [2] vom Verf. bereits in früheren Arbeiten (J. de Phys. (3) 8, S. 407. 1899 und ebenda (3) 2, S. 5. 1893) untersuchte, von der Tem- [3] peratur linear abhängige Funktion der Dichte einer Flüssigkeit und der ihres gesättigten Dampfes, so gilt: $y = \Delta(1 + a[1 + m])$, wobei Δ die kritische Dichte, m die Temperatur bezogen auf die kritische als Einheit, und a eine Konstante bedeutet. Wenn das Gesetz der übereinstimmenden Zustände streng erfüllt wäre, [4] müßte a eine universelle Konstante sein. Es wird an Hand eines Erfahrungsmaterials von 37 Stoffen gezeigt, daß dies nicht der Fall ist. Während bei den meisten untersuchten Stoffen a von der Einheit wenig abweicht, hat diese Größe bei den schwer kondensierbaren Gasen bedeutend kleinere Werte, bei Wasserstoff den Wert 0,236. Der Verf. findet nun, daß die Größe $b = a / \sqrt{\Theta}$ (Θ = absolute kritische Temperatur) zwar auch keine universelle Konstante sei, jedoch bei Stoffen von ähnlicher chemischer Konstitution nahezu gleichen Wert

Bd. 29. No. 12.   Wärmelehre.   **635**

besitze; er schlägt vor, die Stoffe einzuteilen in „Serien" (Stoffe von nahezu gleichem b) und „Gruppen" (Stoffe von annähernd gleichem a).   **A. E.**

Published in *Beiblätter zu den Annalen der Physik* 29 (1905): 634–635. Published in no. 12 [second half of June].

[1] The proper translation of the French term "diamètre rectiligne" is "gerade Mittellinie," and not "rechtwinkliger Diameter." See also note 2.

[2] The function $y(T)$ represents the curve $\{\rho_{liq} + \rho_{vap}\}$ in the ρ-T diagram for vapor-liquid equilibrium (ρ is the density). This curve turns out to be approximately straight, which is in accord with Van der Waals's equation of state. For a review, see *Kamerlingh Onnes and Keesom 1912*, pp. 920–923.

[3] *Mathias 1893, 1899*.

[4] Van der Waals's theory of corresponding states is discussed in *Boltzmann 1898a*, pp. 23–27 (although the term is not mentioned).

20. Review of Max Planck, "On Clausius' Theorem for Irreversible Cycles, and on the Increase of Entropy"

[Einstein 1905o]

Second half of June 1905

PUBLISHED IN: *Beiblätter zu den Annalen der Physik* 29 (1905), no. 12: 635.

besitze; er schlägt vor, die Stoffe einzuteilen in „Serien" (Stoffe von nahezu gleichem *b*) und „Gruppen" (Stoffe von annähernd gleichem *a*).   A. E.

---

77. *J. Traube*. Beitrag zur Theorie von van der Waals (Physik. ZS. **4**, S 50—51. 1904). — Der Verf. findet eine Bestätigung seiner Anschauungen über Gasonen und Fluidonen in Messungen von Schüttarew über die Verdampfungswärme des Äthyläthers in der Nähe der kritischen Temperatur.

Ferner dehnt er die van-der-Waalssche Theorie auf feste Körper aus und findet für den Ausdehnungskoeffizienten des Kovolumens nahe $1/_{273}$ mit Ausnahme von Cl und Br. Ferner findet er Beziehungen des inneren Druckes zur Härte und Elastizität, und solche der van-der-Waalsschen Gleichung zum Schmelzpunkt, Siedepunkt und anderen Größen.   M. R.

---

78. *M. Planck*. Über Clausius' Theorem der irreversibeln Kreisprozesse und das Wachsen der Entropie (Phil. Mag. (9) **49**, S. 167—168. 1904). — In Entgegnung auf einige [1] von Hrn. Orr (Beibl. **29**, S. 237) gegen die vom Verf. gegebene Bearbeitung der Grundlagen der Thermodynamik erklärt der Verf., daß er die Ausdrücke „reversibel" und „irreversibel" [2] in demselben Sinne gebraucht habe wie Clausius. Er bestreitet, daß er die genannten Begriffe in anderem Sinne angewendet als definiert habe. Daß man nicht von der Temperatur und Dichte der Teilchen eines tumultuarisch bewegten Gases sprechen kann und ebensowenig von deren Entropie — sofern man nicht [3] die kinetische Gastheorie zu Hilfe nehmen will — gibt der Verf. zu. Er findet endlich, daß der von Hrn. Orr vorgeschlagene Beweisgang sich im Prinzip mit dem von Lord Kelvin ge- [4] gebenen decke und einen Zirkelschluß enthalte.   A. E.

Published in *Beiblätter zu den Annalen der Physik* 29 (1905): 635. Published in no. 12 [second half of June].

[1] *Orr 1904*. See also Einstein's review of Orr's paper, *Einstein 1905c* (Doc. 8).

[2] Planck actually claims that his statement that the process of heat conduction is irreversible has the same meaning as Clausius's theorem that heat cannot pass from a colder to a hotter body without some compensation.

[3] Planck explicitly refers to Boltzmann's definition of entropy for gases in terms of probability.

[4] According to Planck, Orr's proof of the "principle of increase of entropy" takes Kelvin's formulation of the second law as its starting point.

21. Review of Edgar Buckingham, "On Certain Difficulties Which Are Encountered in the Study of Thermodynamics"

[Einstein 1905p]

Second half of June 1905

PUBLISHED IN: *Beiblätter zu den Annalen der Physik* 29 (1905), no. 12: 635–636.

besitze; er schlägt vor, die Stoffe einzuteilen in „Serien" (Stoffe
von nahezu gleichem *b*) und „Gruppen" (Stoffe von annähernd
gleichem *a*).   A. E.

---

77. *J. Traube.* *Beitrag zur Theorie von van der Waals*
(Physik. ZS. 4, S 50—51. 1904). — Der Verf. findet eine Bestätigung seiner Anschauungen über Gasonen und Fluidonen
in Messungen von Schüttarew über die Verdampfungswärme
des Äthyläthers in der Nähe der kritischen Temperatur.

Ferner dehnt er die van-der-Waalssche Theorie auf feste
Körper aus und findet für den Ausdehnungskoeffizienten des
Kovolumens nahe $1/_{273}$ mit Ausnahme von Cl und Br. Ferner
findet er Beziehungen des inneren Druckes zur Härte und
Elastizität, und solche der van-der-Waalsschen Gleichung zum
Schmelzpunkt, Siedepunkt und anderen Größen.   M. R.

---

78. *M. Planck.* *Über Clausius' Theorem der irreversibeln Kreisprozesse und das Wachsen der Entropie* (Phil.
Mag. (9) 49, S. 167—168. 1904). — In Entgegnung auf einige
von Hrn. Orr (Beibl. 29, S. 237) gegen die vom Verf. gegebene
Bearbeitung der Grundlagen der Thermodynamik erklärt der
Verf., daß er die Ausdrücke „reversibel" und „irreversibel"
in demselben Sinne gebraucht habe wie Clausius. Er bestreitet,
daß er die genannten Begriffe in anderem Sinne angewendet
als definiert habe. Daß man nicht von der Temperatur und
Dichte der Teilchen eines tumultuarisch bewegten Gases sprechen
kann und ebensowenig von deren Entropie — sofern man nicht
die kinetische Gastheorie zu Hilfe nehmen will — gibt der Verf.
zu. Er findet endlich, daß der von Hrn. Orr vorgeschlagene
Beweisgang sich im Prinzip mit dem von Lord Kelvin gegebenen decke und einen Zirkelschluß enthalte.   A. E.

---

79. *E. Buckingham.* *Über gewisse Schwierigkeiten,
welchen man beim Studium der Thermodynamik begegnet* (Phil.
Mag. (9) 50, S. 208—214. 1904). — Der Verf. knüpft an eine
[1]   Arbeit von Hrn. Orr an (Beibl. 29, S. 237) und erklärt sich
mit den Resultaten der in jener Arbeit enthaltenen kritischen
Betrachtungen einverstanden. Er gibt ferner der Überzeugung
Ausdruck, daß es unmöglich sei, die Clausiussche Ungleichung

636   Wärmelehre.   Beibl. 1905.

$\int dQ \; T < 0$ ohne weitere Annahmen aus dem zweiten Hauptsatze — wie dieser von Lord Kelvin formuliert wurde — herzuleiten. Die Arbeit enthält noch einige kritische Bemerkungen zur Abhandlung von Hrn. Orr.   A. E.

Published in *Beiblätter zu den Annalen der Physik* 29 (1905): 635–636. Published in no. 12 [second half of June].

[1] *Orr 1904*. See Einstein's review of this paper, *Einstein 1905c* (Doc. 8). See also his review of *Planck 1905*, in *Einstein 1905o* (Doc. 20).

22. Review of Paul Langevin, "On a Fundamental Formula of the Kinetic Theory"

[Einstein 1905q]

Second half of June 1905

PUBLISHED IN: *Beiblätter zu den Annalen der Physik* 29 (1905), no. 12: 640–641.

93. *B. Weinstein. Entropie und innere Reibung* (Boltzmann-Festschrift, S. 510—517. 1904). — Mit Benutzung einer
sehr allgemeinen Beziehung zwischen der mittleren Bewegungs-
dauer τ und der mittleren Weglänge A eines Moleküls zwischen
zwei Zusammenstößen, die der Verf. im zweiten Bande seiner
Thermodynamik auf Grund der Theorie der festen Körper gegeben hat, wird durch Gleichsetzen der für ideale Gase geltenden bekannten Form der Entropieänderung und einer allgemeinen auf mechanischen Grundlagen beruhenden Form, in
welche unter anderem die Größe τ eingeht, ein neuer Wert der
mittleren molekularen Weglänge und mittleren molekularen Geschwindigkeit abgeleitet. Diese Werte werden in die bekannte
Formel des Reibungskoeffizienten der Gase eingeführt. Die neue
Form des Reibungskoeffizienten muß nach Ansicht des Verf. eine
noch bessere Übereinstimmung bez. der Abhängigkeit von der
Temperatur mit der Erfahrung zeigen, als die in seiner Thermodynamik in analoger aber nicht so allgemeiner Weise abgeleitete Form, und dieselbe gute Übereinstimmung in der Abhängigkeit von Druck, Dichte, Atomzahl und Molekulargewicht.
S. V.

94. *P. Langevin. Über eine fundamentale Formel der
kinetischen Gastheorie* (C. R. **140**, S. 35—38. 1905). — Der
Verf. teilt mit, daß er das Problem der Diffusion zweier Gase   [1]
unter Zugrundelegung beliebiger Wirkungsgesetze zwischen
den Molekülen unter der Annahme, daß auf die Moleküle
äußere Kräfte wirken, nach der Maxwell-Kirchhoffschen Methode   [2]
exakt gelöst hat, wobei er lediglich eine graphische Integration
nötig hat. Für den Fall, daß die Moleküle elastische, unend-   [3]
lich wenig deformierbare Kugeln sind und äußere Kräfte auf
die Moleküle nicht wirken, erhält der Verf. für die Diffusion
des einen Gases (Molekülmasse m_1) im anderen Gase (Molekülmasse m):

$$D = \frac{3}{16\, \sigma^2 M \sqrt{\dfrac{\pi\, h\, m\, m_1}{m + m_1}}}.$$ [4]

Hierbei bedeutet D die Diffusionskonstante, σ die Summe
der Molekülradien zweier ungleicher Moleküle, M die Anzahl
der Moleküle „m" pro Volumeneinheit, h drei Viertel des
reziproken Mittelwertes der Energie der fortschreitenden Be-

Bd. 29. No. 12.   Wärmelehre.   **641**

wegung eines Moleküls. Boltzmann fand nach der Clausius-

[5] schen Näherungsmethode

[6]
$$D = \frac{2}{3\pi\, \sigma^2\, M\, V \overline{\pi\, k\,(m + m_1)}}.$$

Beide Formel differieren besonders stark, wenn m und m_1
sehr verschieden sind. Es wird ferner mitgeteilt, daß der
Diffusionskoeffizient bei konstantem Druck wie $T^{3/2+2/n}$ variert,
wenn sich zwei ungleiche Moleküle mit einer der $n+1$-ten Potenz
des Abstandes der Molekülzentra umgekehrt proportionalen
Kraft abstoßen. Der Verf. hat die Theorie auch auf die
Ortsveränderungen elektrischer Ladungen in Gasen angewendet.
Er fand, daß die Annahme polarisierender, vom elektrischen
Teilchen auf die neutralen Moleküle wirkender Kräfte zur Erklärung von deren geringer Beweglichkeit nicht hinreiche,
sondern daß man in trockener Luft bei gewöhnlicher Temperatur
den negativen Ionen einen ungefähr zweimal, den positiven
einen ungefähr dreimal so großen Durchmesser zuschreiben
müsse als den neutralen Molekülen. Für Flammen findet der
Verf., daß aus den Erfahrungsresultaten zu schließen sei, daß
die Masse der negativen Elektrizitätsträger ungefähr tausendmal
kleiner sei als die der positiven, und daß die Masse der letzteren derjenigen des Wasserstoffatoms gleich sei; erstere entsprechen daher den Kathodenstrahlen, letztere den Goldstein-

[7] schen Strahlen.   **A. E.**

Published in *Beiblätter zu den Annalen der Physik* 29 (1905): 640–641. Published in no. 12
[second half of June].

[1] The paper gives no detailed calculations;
these were later given in *Langevin 1905b*.

[2] A method to calculate transport coefficients
was developed by Maxwell (*Maxwell 1867*). For
a detailed treatment, see *Kirchhoff 1894* and
Boltzmann 1896.

[3] In 1901 Einstein had tried to calculate the
diffusion coefficient (and other transport coefficients) for gases using his theory of molecular
forces (see Einstein to Mileva Marić, 15 April
1901 [Vol. 1, Doc. 101]).

[4] This expression was derived previously by

Stefan (*Stefan 1872*).

[5] Clausius assumed all molecules of the same
kind to have the same speed (*Clausius 1858*).

[6] Langevin derived this expression from a
more general one given by Boltzmann (*Boltzmann 1896*, p. 96) by assuming the number density of one of the two diffusing gases to be much
smaller than that of the other one, and by choos-
ing for the average speed of the molecules the
value found from the Maxwell-Boltzmann distribution (see also *Langevin 1905b*, p. 265).
Boltzmann himself pointed out difficulties with
his more general result.

[7] At the time, canal rays were also known as
Goldstein rays.

EINSTEIN ON THE THEORY OF RELATIVITY

I

Einstein was the first physicist to formulate clearly the new kinematical foundation for all of physics inherent in Lorentz's electron theory. This kinematics emerged in 1905 from his critical examination of the physical significance of the concepts of spatial and temporal intervals. The examination, based on a careful definition of the simultaneity of distant events, showed that the concept of a universal or absolute time, on which Newtonian kinematics is based, has to be abandoned; and that the Galilean transformations between the coordinates of two inertial frames of reference has to be replaced by a set of spatial and temporal transformations that agree formally with a set that Lorentz had introduced earlier, with a quite different interpretation. Through its interpretation of these transformations as elements of a space-time symmetry group corresponding to the new kinematics, the special theory of relativity, as it later came to be called, provided physicists with a powerful guide in the search for new dynamical theories of fields and particles, and gradually led to a deeper appreciation of the role of symmetry criteria in physics. The special theory of relativity also provided philosophers with abundant material for reflection on the new views of space and time. The special theory, like Newtonian mechanics, still assigns a privileged status to the class of inertial frames of reference. The attempt to generalize the theory to include gravitation led Einstein to formulate the equivalence principle in 1907. This was the first step in his search for a new theory of gravitation denying a privileged role to inertial frames, a theory that is now known as the general theory of relativity.

Einstein presented the special theory in *Einstein 1905r* (Doc. 23), a paper which is a landmark in the development of modern physics. In the first part of this paper Einstein presented the new kinematics, basing it on two postulates, the relativity principle and the principle of the constancy of the velocity of light. In the second part, he applied his kinematical results to the solution of a number of problems in the optics and electrodynamics of moving bodies. This volume includes a number of other papers on relativity. Three of these, *Einstein 1905s* (Doc. 24), *1906e* (Doc. 35), and *1907h* (Doc. 45), present arguments for one of the most important consequences of the theory, the equivalence of mass and energy. Two papers, *Einstein 1906g* (Doc. 36) and *1907e* (Doc. 41), suggest new experimental tests of the theory. In his reply to a paper by Ehrenfest, *Einstein 1907g* (Doc. 44), Einstein clarified the kinematical nature of the special theory. *Einstein 1907j* (Doc. 47) is the first major review of the foundations of the theory, as well as of its applications to date (corrections appeared in *Einstein 1908b* [Doc. 49]). The review also contains discussions of several topics Einstein had not previously treated. Particularly notable are Einstein's comments on the equivalence principle and its relationship to the problem of gravitation. A brief discussion comment, *Einstein et al. 1909b* (Doc. 59), concerns an objection to the theory. Two papers on the relativistic electrodynamics of moving media were written in collaboration with Jakob Laub, *Einstein and Laub 1908a* (Doc. 51), *1908b* (Doc. 52). These papers and corrections in *Einstein and Laub 1908c, 1909* (Docs. 53 and

54) are discussed in the editorial note, "Einstein and Laub on the Electrodynamics of Moving Media," pp. 503–507. Einstein's Salzburg talk, *Einstein 1909c* (Doc. 60), stating the relation between the theory of relativity and Einstein's views on the structure of radiation, is also discussed in the editorial note, "Einstein's Early Work on the Quantum Hypothesis," pp. 147–148, and in the Introduction, pp. xvii–xviii. In 1913, *Einstein 1905r* (Doc. 23) and *Einstein 1905s* (Doc. 24) were reprinted in *Blumenthal 1913*.[1]

Strictly speaking, it is anachronistic to use the term "the theory of relativity" in discussing Einstein's first papers on the subject.[2] In them he referred to the "principle of relativity" ("Prinzip der Relativität" or "Relativitätsprinzip").[3] Max Planck used the term "Relativtheorie" in 1906 to describe the Lorentz-Einstein equations of motion for the electron, and this expression continued to be used from time to time for several years.[4] Bucherer seems to have been the first person to use the term "Relativitätstheorie" in the discussion following Planck's lecture.[5] The term was used in an article by Ehrenfest[6] and adopted by Einstein in 1907, in his reply.[7] Although Einstein used the term from time to time thereafter, for several years he continued to employ "Relativitätsprinzip" in the titles of his articles.[8] In 1910 the mathematician Felix Klein suggested the name "Invariantentheorie," but this suggestion does not seem to have been adopted by any physicist.[9] In 1915 Einstein started to refer to his earlier work as "the special theory of relativity" ("die spezielle Relativitätstheorie") to contrast it with his later "general theory" ("allgemeine Theorie").[10] In *Einstein 1907j* (Doc. 47) he does refer to the need for generalizing the "principle of relativity" in order to include gravitation in the theory, but throughout the present volume the phrase "the theory of relativity" is used to denote the special theory.

II

In his 1905 paper, as well as in his 1907 and 1909 reviews of the theory, Einstein described the theory of relativity as arising from a specific problem: the apparent conflict between

[1] See Otto Blumenthal's foreword to *Blumenthal 1913*. Notes added to *Einstein 1905r* (Doc. 23) in this reprint are printed in editorial footnotes to that paper. For a discussion of the authorship of these notes, see *Einstein 1905r* (Doc. 23), note 8.

[2] See *Einstein 1905r* (Doc. 23), *1905s* (Doc. 24), *1906e* (Doc. 35), *1906g* (Doc. 36), and *1907e* (Doc. 41).

[3] He also referred to "die Relativitätselektrodynamik" in *Einstein 1907h* (Doc. 45) on pp. 372, 381, and to "das Relativitätssystem" in *Einstein 1907g* (Doc. 44) on p. 207.

[4] See *Planck 1906b*, p. 424. In a critical review of Kaufmann's experiments at the 1906 meeting of the Gesellschaft Deutscher Naturforscher und Ärzte, Planck compared the "Relativtheorie" of Lorentz and Einstein with the "Kugeltheorie" ("sphere theory") of Abraham. The term "Relativtheorie" was still in use in 1910 (see *Noether 1910*). Planck continued to use "Prinzip der Relativität" or "Relativitätsprinzip" to describe the approaches of Lorentz and Einstein, between which he did not distinguish. See *Planck 1906a, 1907a*.

[5] For Bucherer's comment, see *Discussion/Planck 1906*, p. 760.

[6] *Ehrenfest 1907*, p. 205.

[7] See *Einstein 1907g* (Doc. 44), pp. 206, 207; see also *Einstein 1907j* (Doc. 47), p. 439.

[8] He did not use "Relativitätstheorie" in a title until *Einstein 1911e*.

[9] See *Klein, F. 1910*, p. 287.

[10] See *Einstein 1915b*, p. 778.

the principle of relativity and the Maxwell-Lorentz theory of electrodynamics.[11] While the relativity principle asserts the physical equivalence of all inertial frames of reference, the Maxwell-Lorentz theory implies the existence of a privileged inertial frame.

The principle of relativity originated in classical mechanics.[12] Assuming Newton's laws of motion and central force interactions, it can be demonstrated that it is impossible to determine the state of motion of an inertial frame by means of mechanical experiments carried out within a closed system with center of mass at rest in this frame. This conclusion, well known and empirically well confirmed by the end of the nineteenth century, was sometimes called the principle of relative motion, or principle of relativity.[13]

The introduction of velocity-dependent forces between charged particles led to doubts about the validity of the relativity principle for magnetic interactions. The wave theory of light appeared to invalidate the principle for optical phenomena. The theory seemed to require an all-pervading medium, the so-called luminiferous ether, to explain the propagation of light in the absence of ordinary matter. The assumption that the ether moves together with matter seems excluded by the phenomenon of aberration and by Fizeau's results on the velocity of light in moving media. If the ether is not dragged with matter, it should be possible to detect motion relative to a reference frame fixed in the ether by means of optical experiments. However, all attempts to detect the motion of the earth through the ether by optical experiments failed.[14]

Maxwell's electromagnetic theory was intended to provide a unified explanation of electric, magnetic, and optical phenomena. With its advent, the question arose of the status of the principle of relativity for such phenomena. Does the principle follow from the fundamental equations of electrodynamics?[15] The answer to this question depends on the form of Maxwell's equations postulated for bodies in motion. Hertz developed an electrodynamics of moving bodies, based on the assumption that the ether moves with matter, in which the relativity principle holds.[16] In addition to its inability to account for the optical phenomena mentioned above, Hertz's theory was unable to explain several new electromagnetic phenomena, and it soon fell out of favor.[17]

[11] See *Einstein 1905r* (Doc. 23), pp. 891–892; *Einstein 1907j* (Doc. 47), pp. 411–413; and *Einstein 1909c* (Doc. 60), pp. 485–487. See *Miller 1981b* for a detailed study and an English translation of *Einstein 1905r* (Doc. 23).

[12] See *Torretti 1983* for a discussion of the Newtonian principle of relativity, with some historical references.

[13] See, e.g., *Violle 1892*, p. 90, and *Poincaré 1902*, pp. 135–137, 281, to name two sources that Einstein had read by 1905 (see Vol. 1, "Albert Einstein—Beitrag für sein Lebensbild," p. lxiv, for Violle; and *Solovine 1956*, p. viii, for Poincaré).

[14] For a survey of ether theories, centering on attempts to detect the motion of the earth through the ether, see *Hirosige 1976*; for a survey focusing on the Michelson-Morley experiment, see *Swenson 1972*. For a survey of work on the optics of moving bodies in the nineteenth century, see *Sesmat 1937*.

[15] For discussions of the problem of relative motion in current theories of electrodynamics, see *Cohn 1900, 1902*, and *Abraham/Föppl 1904*, pp. 430–436, which speaks of the "axiom of relative motion" ("Axiom der Relativbewegung").

[16] See *Hertz, H. 1890b*. For a review of Hertz's theory, see *Hirosige 1966*.

[17] It appears that Hertz was aware of the difficulties optical phenomena present for his theory. He explicitly stated that it was meant to explain electromagnetic phenomena in the narrower sense (see *Hertz, H. 1890b*). How-

By the turn of this century, when Einstein started working on the electrodynamics of moving bodies, Lorentz's very successful version of Maxwell's theory had gained wide acceptance.[18] Lorentz's electrodynamics is based on a microscopic theory that came to be known as the electron theory.[19] The theory makes a sharp distinction between ordinary, ponderable matter and the ether. Ordinary matter is composed of finite-sized material particles, at least some of which are electrically charged. All of space, even those regions occupied by material particles, is pervaded by the ether, a medium with no mechanical properties, such as mass. The ether is the seat of all electric and magnetic fields. Matter only influences the ether through charged particles, which create these fields. The ether only acts on matter through the electric and magnetic forces that the fields exert on charged particles. By assuming such atoms of electricity ("electrons"), the theory incorporates an important element of the pre-Maxwellian continental tradition into Maxwell's theory, from which it took the field equations.

The parts of the ether are assumed to be immobile relative to each other. Hence, Lorentz's ether defines a rigid reference frame, which is assumed to be inertial. It is in this frame that Maxwell's equations are valid; in other frames, the Galilei-transformed form of these equations hold. Hence it should be possible to detect the motion of the earth through the ether by suitably designed terrestrial electromagnetic or optical experiments. Lorentz was well aware of the failure of all attempts to detect the motion of the earth through the ether, in particular such sensitive optical attempts as the Michelson-Morley experiment,[20] and attempted to explain this failure on the basis of his theory.

His basic approach to this problem in 1895 was to use the theorem of "corresponding states" ("correspondierende Zustände") in combination with the well-known contraction hypothesis.[21] The theorem is essentially a calculational tool that sets up a correspondence between phenomena in moving systems and those in stationary sytems by introducing transformed coordinates and fields. On this basis, Lorentz was able to account for the failure of most electromagnetic experiments to detect the motion of the earth through the ether. In 1904 he showed how to explain the failure of all such experiments by a generalization of his theory. He introduced a set of transformations for the spatial and temporal coordinates (soon named the Lorentz transformations by Poincaré)[22] and for the electric and magnetic field components, such that under these transformations Maxwell's equations in the absence of charges take the same form in all inertial frames.[23] Lorentz's

ever, even here it yields incorrect predictions. For its rejection see, e.g., *Lorentz 1892c*, p. 6 (reprint edition); *Cohn 1900*, p. 518; *Cohn 1902*, p. 29; and *Abraham/Föppl 1904*, pp. 427–428, 435.

[18] Theories similar to Lorentz's electron theory were proposed by Wiechert (*Wiechert 1896*) and Larmor (*Larmor 1894, 1895, 1897*). Emil Cohn proposed a macroscopic electrodynamics of moving media (*Cohn 1900, 1902*), which received some attention. For a review of these theories, see *Hirosige 1966*. Einstein mentioned Cohn's theory in *Einstein 1907j* (Doc. 47), p.

413, and a copy of *Cohn 1904a* is in Einstein's reprint collection (now in IsReW).

[19] For Lorentz's program see, e.g., *Lorentz 1892c*, pp. 70–71. For reviews of his theory, see *Lorentz 1904c, 1909b*. For discussions of the theory, see *McCormmach 1970a, 1970b*.

[20] *Michelson 1881; Michelson and Morley 1887*. For Lorentz's first discussion of Michelson's experiment, see *Lorentz 1886*.

[21] See *Lorentz 1895*.

[22] See *Poincaré 1905b*.

[23] See *Lorentz 1904a*. Lorentz did not have transformation laws for the charge density and

approach to the explanation of failure of attempts to detect motion through the ether, thus, was to show that the basic equations of the electron theory, in spite of the fact that they single out the ether rest frame, can still be used to explain the failure of all optical and electromagnetic attempts to detect the earth's motion through the ether.

Einstein's work was based on a new outlook on the problem. Instead of regarding the failure of electromagnetic and optical experiments to detect the earth's motion through the ether as something to be deduced from the electrodynamical equations, he took this failure as empirical evidence for the validity of the principle of relativity in electrodynamics and optics. Indeed, he asserted the universal validity of the principle, making it a criterion for the acceptability of any physical law. In this respect he gave the principle of relativity a role similar to that of the principles of thermodynamics, an example that he later stated helped to guide him.[24] Rather than being deductions from other theories, such principles are taken as postulates for deductive chains of reasoning, resulting in a theory formulating general criteria that more specialized theories must satisfy.[25]

Einstein now confronted the problem of making Maxwell-Lorentz electrodynamics compatible with the principle of relativity. He did so by means of a principle drawn from electrodynamics, the principle of the constancy of the velocity of light. That the velocity of light is independent of that of its source, and has a constant value in the ether rest frame, can be deduced from the Maxwell-Lorentz theory. Einstein dropped the ether from consideration, and took the constancy of the velocity of light as a second postulate, supported by all the empirical evidence in favor of the Maxwell-Lorentz theory. When combined with the relativity principle, this leads to an apparently paradoxical conclusion: the velocity of light must be the same in all inertial frames. This result conflicts with the Newtonian law of addition of velocities, forcing a revision of the kinematical foundations of electrodynamics. Einstein showed that the simultaneity of distant events is only defined physically relative to a particular inertial frame, leading to kinematical transformations between the spatial and temporal coordinates of two inertial frames that agree formally with the transformations introduced by Lorentz in 1904.

Einstein next considered the implications of the new kinematics for electrodynamics and mechanics. By eliminating the concept of the ether, he in effect asserted that electromagnetic fields do not require an underlying substratum.[26] He showed that the Maxwell-Lorentz equations for empty space remain invariant in form under the new kinematical

---

velocity that would make the equations with sources fully invariant. These laws were supplied by Poincaré, who recognized that Lorentz's transformations form a group, under which the Maxwell-Lorentz equations remain invariant in form. Poincaré regarded these results as an explanation of the apparent universal validity of the principle of relativity (see *Poincaré 1905b, 1906*).

[24] Einstein first made the comparison between the relativity principle and the principles of thermodynamics in *Einstein 1907g* (Doc. 44). He stated that these principles had served as his "example" ("Vorbild") in *Einstein 1979*, p. 48.

[25] Einstein later formulated the distinction between "theories of principle" ("Prinziptheorien") and "constructive theories" ("konstruktive Theorien") to explain the relationship between the two types of theories (see *Einstein 1919*). See the Introduction, pp. xxi–xxii, for further discussion of Einstein's use of such principles.

[26] This concept of electromagnetic fields is first stated explicitly in *Einstein 1907j* (Doc. 47), p. 413; but it is implicit in *Einstein 1905i* (Doc. 14), as well as in *Einstein 1905r* (Doc. 23).

transformations when the transformation laws for the electric and magnetic fields are appropriately defined. He deduced appropriate transformation laws for charge densities and velocities from the requirement that Maxwell's equations remain invariant when convection currents are added. Finally, by assuming that Newton's equations hold for a charged particle at rest, he was able to use a kinematical transformation to deduce the equations of motion of a charged particle ("electron") with arbitrary velocity.

The problems connected with the formulation of an electrodynamics of moving bodies consistent with all the experimental evidence were discussed frequently during the years Einstein was working on his theory. Statements similar to many of the individual points made in *Einstein 1905r* (Doc. 23) occur in the contemporary literature, and Einstein may well have been familiar with some of the books and articles in which they do.[27] But his approach to the problem, leading to the peculiar combination of these ideas in his paper, is unique—particularly the recognition that a new kinematics of universal applicability is needed as the basis for a consistent approach to the electrodynamics of moving bodies.

III

Einstein's work on relativity grew out of his long-standing interest in the electrodynamics and optics of moving bodies.[28] His first scientific essay, written in 1895, discussed the propagation of light through the ether.[29] The next year, as he later recalled, the following problem started to puzzle him:

> If one were to pursue a light wave with the velocity of light, one would be confronted with a time-independent wave field. Such a thing doesn't seem to exist, however! This was the first childlike thought-experiment concerned with the special theory of relativity.

> Wenn man einer Lichtwelle mit Lichtgeschwindigkeit nachläuft, so würde man ein zeitunabhängiges Wellenfeld vor sich haben. So etwas scheint es aber doch nicht zu geben! Dies war das erste kindliche Gedanken-Experiment das mit der speziellen Relativitätstheorie zu tun hat.[30]

By this time Einstein presumably was familiar with the principle of relativity in classical mechanics. While preparing for the ETH entrance examination in 1895, he had studied the German edition of Violle's textbook.[31] Violle actually based his treatment of dynam-

[27] Some of these are cited in the editorial notes to *Einstein 1905r* (Doc. 23) (see notes 5, 6, 9, 10, 12). Two papers by Emil Cohn, *Cohn 1904a, 1904b*, published in response to *Lorentz 1904a*, deserve special mention. They are "remarkable . . . because in some respects they remind us of the point of view taken in the theory of relativity" (*Hirosige 1966*, p. 35). For evidence of Einstein's later familiarity with Cohn's work, see § V. However, there is still no evidence indicating when he first read Cohn.

[28] See Vol. 1, the editorial note, "Einstein on the Electrodynamics of Moving Bodies," pp. 223–225.

[29] See "On the Investigation of the State of the Ether in a Magnetic Field" (Vol. 1, Doc. 5).

[30] *Einstein 1955*, p. 146. See also *Einstein 1979*, pp. 48, 50.

[31] *Violle 1892, 1893*. For a reference to the evidence about when Einstein studied Violle's textbook, see note 13.

ics on the "principle of relative motions" ("Prinzip der relativen Bewegungen") together with the principle of inertia.[32]

About 1898, Einstein started to study Maxwell's electromagnetic theory, apparently with the help of Drude's textbook.[33] By 1899, after studying Hertz's papers on the subject, he was at work on the electrodynamics of moving bodies.[34] He discussed this topic a number of times in letters to Mileva Marić between 1899 and 1901,[35] once referring to "our work on relative motion" ("unsere Arbeit über die Relativbewegung").[36] In December 1901, Einstein also explained his ideas on the subject to Professor Alfred Kleiner of the University of Zurich, who encouraged him to publish them;[37] but there is no evidence that Kleiner played a further role in the development of these ideas.

Einstein's comments show that in 1899 his viewpoint on electrodynamics was similar to that of Lorentz;[38] but, aside from this similarity, there is no evidence that Einstein had yet read anything by Lorentz.[39] Shortly afterward, Einstein designed an experiment to test the effect of the motion of bodies relative to the ether on the propagation of light; in 1901 he designed a second such experiment, but was unable to carry out either.[40] Near the end of 1901, he was at work on what he described as "a capital paper" ("eine kapitale Abhandlung") on the electrodynamics of moving bodies, asserting his renewed conviction of the correctness of his "ideas on relative motion" ("Ideen über die Relativbewegung").[41] His words may indicate that he already doubted whether motion with respect to the ether is experimentally detectable.[42] Soon afterward he wrote that he intended to study Lorentz's theory in earnest.[43]

[32] See *Violle 1892*, pp. 90–91, for the statement of the two principles, and pp. 92–94 for the derivation of Newton's second law (see p. 90 for Violle's discussion of the role of such principles in physics). A copy of *Violle 1892* is in Einstein's personal library. A marginal annotation in Einstein's hand on p. 94 indicates that he read this section of the book.

[33] *Drude 1894*. For evidence that Einstein read Drude, see Einstein to Mileva Marić, 16 April–8 November 1898 and after 16 April 1898 (Vol. 1, Docs. 40 and 41), especially the descriptive notes to both letters.

[34] *Hertz, H. 1890a, 1890b*. For Einstein's discussion of Hertz's work, see Einstein to Mileva Marić, 10 August 1899 (Vol. 1, Doc. 52).

[35] See Vol. 1, the editorial note, "Einstein on the Electrodynamics of Moving Bodies," pp. 223–225, for a list and discussion of these letters.

[36] Einstein to Mileva Marić, 27 March 1901 (Vol. 1, Doc. 94). There is no indication in their letters of the nature of her collaboration, if any, in his research.

[37] See Einstein to Mileva Marić, 19 December 1901 (Vol. 1, Doc. 130).

[38] See Einstein to Mileva Marić, 10 August 1899 (Vol. 1, Doc. 52).

[39] Einstein may have already read *Wien 1898*, which includes a review of Lorentz's theory. For evidence that Einstein read this article, see Einstein to Mileva Marić, 28 September 1899 (Vol. 1, Doc. 57).

[40] See Einstein to Mileva Marić, 10 September 1899 (Vol. 1, Doc. 54), and Einstein to Marcel Grossmann, 6 September 1901 (Vol. 1, Doc. 122).

[41] Einstein to Mileva Marić, 17 December 1901 (Vol. 1, Doc. 128). *Earman et al. 1982* suggests that Einstein may have incorporated elements of his earlier work on electrodynamics into *Einstein 1905r* (Doc. 23).

[42] In 1899 he read *Wien 1898* (see note 39), which briefly describes a number of experimental attempts to detect the motion of the earth through the ether. The Michelson-Morley experiment is included among those "with negative result" ("mit negativem Ergebniss").

[43] See Einstein to Mileva Marić, 28 December 1901 (Vol. 1, Doc. 131). Einstein later stated that *Lorentz 1895* is the only work by Lorentz that he knew in 1905 (Einstein to Carl Seelig, 19 February 1955). It is possible, however, that he learned about subsequent developments

There is direct or strong indirect contemporary evidence that, by 1902, Einstein had read or was reading the following works on electrodynamics or optics: *Drude 1894*, *1900c*; *Helmholtz 1892*, *1897*; *Hertz, H. 1890a*, *1890b*; *Lorentz 1895*; *Voigt 1896*; and *Wien 1898*.[44] There is a later report that he read *Föppl 1894* as a student.[45] Comments in his letters on articles published in the *Annalen der Physik* between 1898 and 1901 indicate that during those years he looked at that journal regularly, and studied a number of articles in it.[46] It is reasonable to suppose that he continued to do so between 1902 and 1905.[47] During these years a number of significant articles on the electrodynamics and optics of moving bodies appeared in the *Annalen*.[48] He cited several works published before 1905 in his later articles on relativity, and it is possible that he read one or more of these before 1905.[49] Einstein's readings on the foundations of science during these years, as reported by his friend Maurice Solovine, are described in the Introduction. He later attributed great significance in the development of the theory of relativity to his reading of Hume, Mach, and Poincaré.[50]

Belief in the reality of the ether was widespread at the turn of the century.[51] However, Einstein was familiar with several works that questioned the certainty of its existence. Mill, in the course of a discussion of "the Hypothetical Method" in his *Logic*, gives a number of reasons for skepticism concerning "the prevailing hypothesis of a luminiferous

---

in Lorentz's work through the comments of others on Lorentz's widely discussed theory.

[44] For *Drude 1894*, see Einstein to Mileva Marić, after 16 April 1898 (Vol. 1, Doc. 41). For *Helmholtz 1892* and *Hertz, H. 1890a*, *1890b*, see Einstein to Marić, 10 August 1899 (Vol. 1, Doc. 52). For *Helmholtz 1897* and *Wien 1898*, see Einstein to Marić, 28 September 1899 (Vol. 1, Doc. 57). For *Voigt 1896*, see Einstein to Marić, 28 November 1901 (Vol. 1, Doc. 126). For *Drude 1900c* and *Lorentz 1895*, see Einstein to Marić, 28 December 1901 (Vol. 1, Doc. 131).

[45] See *Kayser 1930*, p. 49, and *Frank 1979*, p. 38. It is possible that Föppl's book was brought to his attention by Joseph Sauter, H. F. Weber's former *Assistent*, who was already at the Swiss Patent Office when Einstein started work there. Sauter cited *Föppl 1894* in an article attempting a mechanical explanation of Maxwell's equations (see *Sauter 1901*, p. 332).

[46] See Einstein to Mileva Marić, 28 May 1901 (Vol. 1, Doc. 111), for references to his reading of the *Annalen der Physik*, and to his study of papers in the *Annalen* by Lenard and Reinganum. See Einstein to Marić, 28 September 1899 (Vol. 1, Doc. 57), for a reference to a paper by Wien, and Einstein to Marić, 4 April 1901 (Vol. 1, Doc. 96), for references to papers

by Drude and Planck, all in the *Annalen*.

[47] In spite of his statement that the libraries were closed during his free time (Einstein to Johannes Stark, 25 September 1907), he must have found a way to follow some of the periodical literature. The same letter cites several recent journal articles, and his writings of this period refer to a number of others.

[48] See, e.g., *Abraham 1903*, *1904b*; *Bucherer 1903*; *Cohn 1902*; *Gans 1905*; *Hasenöhrl 1904*; *Nordmeyer 1903*; *Oppolzer 1902*; *Wien 1904*. For a discussion of Einstein's readings in electrodynamics before 1905, see *Miller 1981b*, pp. 87–92.

[49] For example, "the relevant works" ("die einschlägigen Arbeiten") of Emil Cohn are mentioned in *Einstein 1907j* (Doc. 47), p. 413. *Poincaré 1900* is cited in *Einstein 1906e* (Doc. 35).

[50] See the Introduction, pp. xxiii–xxiv.

[51] In 1909, Einstein cited a statement in *Khvolson 1902* as typical: "The probability of the hypothesis of the existence [of the ether] . . . approaches extremely near to certainty" ("Die Wahrscheinlichkeit der Hypothese von der Existenz . . . grenzt außerordentlich nahe an Gewißheit") (*Einstein 1909c* [Doc. 60], p. 482).

ether.''[52] Poincaré, in *La science et l'hypothèse*, raised the question of the existence of the ether, even if he offered no clear answer.[53] Ostwald, in his *Lehrbuch der allgemeinen Chemie*, suggested that the ether hypothesis could be replaced by a purely energetic treatment of radiation.[54]

Few contemporary documents throw any light on Einstein's work on electrodynamics between 1902 and 1905. On 22 January 1903, he wrote Michele Besso: "In the near future I want to deal with molecular forces in gases, and then make a comprehensive study of electron theory" ("In der nächsten Zeit will ich mich mit den Molekularkräften in Gasen abgeben, und dann umfassende Studien in Elektronentheorie machen").[55] On 5 December 1903, Einstein gave a talk to the Naturforschende Gesellschaft Bern on "The Theory of Electromagnetic Waves" ("die Theorie der elektromagnetischen Wellen").[56] By the time Einstein wrote his friend Conrad Habicht early in 1905, the theory was practically complete:

> The . . . paper exists only as a sketch and is an electrodynamics of moving bodies that utilizes a modification of the theory of space and time.

> Die . . . Arbeit liegt erst im Konzept vor und ist eine Elektrodynamik bewegter Körper unter Benützung einer Modifikation der Lehre von Raum und Zeit.[57]

Later statements by Einstein suggest several important elements in the development of his ideas on relativity before *Einstein 1905r* (Doc. 23) was written that are not recorded in any known contemporary documents. In 1932 he gave a general characterization of "the situation that led to setting up the theory of special relativity" ("die Situation, die zur Aufstellung der speziellen Relativitätstheorie geführt hat"):

> Mechanically all inertial systems are equivalent. In accordance with experience, this equivalence also extends to optics and electrodynamics. However, it did not appear that this equivalence could be attained in the theory of the latter. I soon reached the conviction that this had its basis in a deep incompleteness of the theoretical system. The desire to discover and overcome this generated a state of psychic tension in me that, after seven years of vain searching, was resolved by relativizing the concepts of time and length.

---

[52] See *Mill 1872*, vol. 2, pp. 12, 20 (see *Solovine 1956*, p. viii, for evidence that Einstein read Mill). Mill also noted several phenomena which appear to favor the emission theory (ibid., p. 23).

[53] See *Poincaré 1902*, pp. 199–202. For a reference to the evidence about when Einstein read this book, see note 13.

[54] See *Ostwald 1893*, part I, pp. 1014–1016. For evidence that Einstein read Ostwald's *Lehrbuch*, see Einstein to Wilhelm Ostwald, 19 March 1901 (Vol. 1, Doc. 92), and Einstein to Mileva Marić, 10 April 1901 (Vol. 1, Doc. 97).

[55] By "Elektronentheorie," Einstein presumably here meant Lorentz's "Elektronentheorie der Elektrodynamik" (*Abraham 1903*, p. 105). There is also an "Elektronentheorie der Metalle," in which Einstein had earlier shown great interest (see Vol. 1, the editorial note, "Einstein on Thermal, Electrical, and Radiation Phenomena," pp. 236–237).

[56] See the Minutes of the Society for that date in SzBe, and printed in *Verhandlungen 1904*, p. 328.

[57] Einstein to Conrad Habicht, 18 May–8 June 1905.

Mechanisch sind alle Inertialsysteme gleichwertig. Nach den Erfahrungen erstreckt sich diese Gleichwertigkeit auch auf die Optik, bezw. Elektrodynamik. In der Theorie der letzteren erschien aber diese Gleichwertigkeit unerreichbar. Ich gewann früh die Ueberzeugung, dass dies in einer tiefen Unvollkommenheit des theoretischen Systems seinen Grund habe. Der Wunsch diese aufzufinden und zu beheben, erzeugte einen Zustand psychischer Spannung in mir, der nach sieben Jahren vergeblichen Suchens durch Relativierung der Begriffe Zeit und Länge gelöst wurde.[58]

In 1952 he wrote:

My direct path to the special theory of relativity was mainly determined by the conviction that the electromotive force induced in a conductor moving in a magnetic field is nothing other than an electric field. But the result of Fizeau's experiment and the phenomenon of aberration also guided me.

Mein direkter Weg zur speziellen Relativitäts-Theorie wurde hauptsächlich durch die Überzeugung bestimmt, dass die in einem im Magnetfeld bewegten Leiter induzierte elektromotorische Kraft nichts anderes sei als ein elektrisches Feld. Aber auch das Ergebnis des Fizeau'schen Versuches und das Phänomen der Aberration führten mich.[59]

Beyond their well-known role as evidence against the assumption that the ether is completely carried along by moving matter, it is not clear what role the result of Fizeau's experiment and the phenomenon of aberration played in Einstein's thinking.[60] Possibly its role depended on the fact that, in both cases, the observed effect only depends on the motion of matter (water in the first case, star in second) relative to the earth, and not on the presumed motion of the earth with respect to the ether.

In the case of electromagnetic induction, Einstein earlier gave a more detailed account of its role. In 1920, Einstein wrote:

In setting up the special theory of relativity, the following . . . idea about Faraday's electromagnetic induction played a guiding role. According to Faraday, relative motion of a magnet and a closed electric circuit induces an electric current in the latter. Whether the magnet is moved or the conductor doesn't matter; only the relative motion is significant. . . . The phenomena of electromagnetic induction . . . compelled me to postulate the principle of (special) relativity.

Bei der Aufstellung der speziellen Relativitätstheorie hat für mich der folgende

[58] Einstein to Erika Oppenheimer, 13 September 1932.

[59] Message by Einstein, prepared for R. S. Shankland to read at a celebration of the centennial of Michelson's birth, 19 December 1952, at Case Institute.

[60] Einstein cited Fizeau's experiment in this connection in several early papers (see *Einstein 1909c*, *1910a*, *1911e*). Aberration was often similarly cited at the time (see, e.g., *Abraham 1905*, p. 342). *Einstein 1918* refutes the claim that aberration is compatible with an ether-drag theory.

. . . Gedanke über die Faraday'sche magnet-elektrische Induktion eine führende Rolle gespielt. Bei der Relativbewegung eines Magneten gegenüber einem elektrischen Stromkreise, wird nach Faraday in letzterem ein elektrischer Strom induziert. Ob der Magnet bewegt wird oder der Leiter, ist gleichgültig; es kommt nur auf die Relativ-Bewegung an. . . . Die Erscheinungen der magnet-elektrischen Induktion zwang mich dazu, das (spezielle) Relativitätsprinzip zu postulieren.[61]

In a footnote he added:

The difficulty to be overcome then lay in the constancy of the velocity of light in vacuum, which I first thought would have to be abandoned. Only after groping for years did I realize that the difficulty lay in the arbitrariness of the fundamental concepts of kinematics.

Die zu überwindende Schwierigkeit lag dann in der Konstanz der Vakuum-Lichtgeschwindigkeit, die ich zunächst aufgeben zu müssen glaubte. Erst nach jahrelangem Tasten bemerkte ich, dass die Schwierigkeit auf der Willkür der kinematischen Grundbegriffe beruhte.

His strong belief in the relativity principle and abandonment of "the constancy of the velocity of light in vacuum" led Einstein to explore the possibility of an emission theory of light. In such a theory the velocity of light is only fixed relative to that of its source, so it is clearly consistent with the relativity principle. Newton's corpuscular theory of light is an emission theory, and Einstein's search for such a theory may have been one source of his light quantum hypothesis. In 1912, commenting on Ritz's emission theory,[62] Einstein referred to "Ritz's conception, which before the theory of relativity was also mine" ("die Ritz'sche Auffassung, die . . . vor der Rel.-Theorie auch die meine war").[63] Elsewhere, he expanded on this theme:

I knew that the principle of the constancy of the velocity of light was something quite independent of the relativity postulate, and I weighed which was more probable, the principle of the constancy of c, as required by Maxwell's equations, or the constancy of c exclusively for an observer located at the light source. I decided in favor of the former . . .

Ich wusste wohl, dass das Prinzip von der Konstanz der Lichtgeschwindigkeit etwas von dem Relativitätspostulat ganz unabhängiges ist, und ich erwog, was wahrscheinlicher sei, das Prinzip von der Konstanz von c, wie es von den Maxwell'schen Gleichungen gefordert wird, oder die Konstanz von c, ausschliess-

[61] Unpublished manuscript, entitled "Fundamental Ideas and Methods of the Theory of Relativity, Presented as it Developed" ("Grundgedanken und Methoden der Relativitätstheorie, in ihrer Entwicklung dargestellt"). The undated manuscript was finished early in 1920 (see Einstein to Robert W. Lawson, 26 December 1919, and 22 January 1920).

[62] See, e.g., *Ritz 1908a, 1908b*.

[63] Einstein to Paul Ehrenfest, 25 April 1912. Ehrenfest had discussed Ritz's emission theory in *Ehrenfest 1912*.

lich für einen Beobachter, der bei der Lichtquelle sitzt. Ich entschied mich für das erstere . . .[64]

In 1924, Einstein described the sudden resolution of his dilemma:

> After seven years of reflection in vain (1898–1905), the solution came to me suddenly with the thought that our concepts and laws of space and time can only claim validity insofar as they stand in a clear relation to our experiences; and that experience could very well lead to the alteration of these concepts and laws. By a revision of the concept of simultaneity into a more malleable form, I thus arrived at the special theory of relativity.

> Nach siebenjährigem vergeblichem Nachdenken (1898–1905) kam mir plötzlich die Lösung mit dem Gedanken, daß unsere Begriffe und Gesetze über Raum und Zeit nur insofern Geltung beanspruchen dürfen, als sie mit den Erlebnissen in klaren Beziehungen stehen, und daß die Erfahrung sehr wohl dazu führen könne, daß wir diese Begriffe und Gesetze abändern. Durch eine Revision des Begriffes der Gleichzeitigkeit unter gestaltbarer Form gelangte ich so zur speziellen Relativitätstheorie.[65]

In a talk at Kyoto University in 1922, Einstein is reported to have said that, after a year of struggle with the problem of how to reconcile Lorentz's theory with his ideas on relativity, he visited a friend one day to discuss the problem in detail with him. The next day Einstein said to his friend: "Thanks to you, I have completely solved my problem."[66] The friend was presumably Michele Besso, then his colleague at the Swiss Patent Office and the only person whose help is acknowledged in Einstein's first paper on relativity.[67]

Work on this paper was apparently completed very rapidly after this. In 1952 Einstein wrote that "between the conception of the idea for the special theory of relativity and the completion of the relevant publication five or six weeks elapsed" ("Zwischen der Konzeption der Idee der speziellen Relativitätstheorie und der Beendigung der betreffenden Publikation sind fünf oder sechs Wochen vergangen").[68]

Einstein's comments suggest the following stages in his work on the theory of relativity:

(1) He became convinced that, as is the case for mechanical phenomena, only the

---

[64] Einstein to Paul Ehrenfest, 20 June 1912. The letter explains in some detail why he had rejected the emission theory. For additional accounts of his early interest in an emission theory, see Einstein to M. Viscardini, 28 April 1922; Einstein to C. O. Hines, February 1952; and Einstein to A. Rippenbein, after 25 August 1952.

[65] This recording is transcribed in *Herneck 1966a*; the quotation is from p. 134. *Herneck 1976*, p. 349, describes the provenance of the recording: "The discographic document of February 6, 1924, is registered in the catalogue of the former 'Institute for Sound Research' of the University of Berlin under the file number 'Autophon Nr. 56'. I have transcribed the text from a copy on magnetic tape" ("Das diskographische Dokument vom 6.2.1924 ist in der Kartei des ehemaligen 'Instituts für Lautforschung' der Berliner Universität unter der Signatur 'Autophon Nr. 56' registriert. Den Text habe ich nach einer Magnetbandkopie aufgezeichnet").

[66] See the report of Einstein's talk, given on 14 December 1922, in *Ishiwara 1971*, pp. 78–88.

[67] See *Einstein 1905r* (Doc. 23), p. 921.

[68] Einstein to Carl Seelig, 11 March 1952.

relative motions of ponderable bodies are significant in determining electromagnetic and optical phenomena; at some point, this conviction led him to abandon the concept of the ether.

(2) He temporarily abandoned Lorentz's theory of electrodynamics, which appears to attach physical significance to absolute motion (i.e., motion with respect to empty space or the ether).

(3) He explored the possibility of an alternative electrodynamical theory, which would justify the emission hypothesis about the velocity of light relative to its source.

(4) Abandoning such attempts, he reexamined Lorentz's theory, at some point focusing his concern on the conflict of his ideas on relative motion with a particular consequence of Lorentz's theory: the independence of velocity of light of the velocity of its source.

(5) He recognized that this conflict involves previously tacitly accepted kinematical assumptions about temporal and spatial intervals, leading him to examine the meaning of the concept of the simultaneity of distant events.

(6) He defined simultaneity physically, and constructed a new kinematical theory based on the relativity principle and the light principle, thus resolving the apparent conflict between them.

There have been a number of attempts at a detailed reconstruction of Einstein's development of the theory of relativity, attempts which often differ significantly in their conclusions.[69] Such a reconstruction has to take into account other strands in Einstein's work at this time. In particular, by the time he wrote the relativity paper, he no longer regarded Maxwell's electromagnetic theory as universally valid, and had already proposed his light quantum hypothesis.[70] He had also shown that the equipartition theorem, which his work on the foundations of thermodynamics convinced him is valid for the most general classical-mechanical system, combined with Maxwell's theory, leads to an incorrect law for black-body radiation. Thus, he already challenged the unlimited validity of both classical mechanics and of Maxwell's theory. For further discussion of his doubts about classical theories of matter and radiation, see the Introduction, pp. xvi–xxix, and the editorial note, "Einstein's Early Work on the Quantum Hypothesis," pp. 134–148.

Einstein later recalled that, uncertain how to proceed in the search for better theories of the structure of matter and radiation, he became convinced that "only the discovery of a universal formal principle could lead . . . to assured results" ("nur die Auffindung eines allgemeinen formalen Prinzips . . . zu gesicherten Ergebnissen führen könnte").[71] Such principles play a role analogous in this respect with the role played by the principles of thermodynamics. The theory of relativity is based on just such principles: even though suggested originally by specific mechanical and electromagnetic theories, the principles

[69] See, e.g., *Earman et al. 1982*; *Goldberg 1983*; *Hirosige 1976*; *Holton 1973*, part II; *Miller 1981b*; *Pais 1982*; *Schaffner 1982*.

[70] See *Einstein 1905i* (Doc. 14). In *Einstein 1907h* (Doc. 45), pp. 372–373, Einstein dis-

cussed why Maxwell's theory could nevertheless be used with confidence within the limits of its validity.

[71] *Einstein 1979*, p. 48; English translation, p. 49.

of relativity and of the constancy of the velocity of light are supported by empirical evidence that is independent of the validity of these theories. For further discussion of the role of such principles in Einstein's thought, see the Introduction, pp. xxi–xxii, xxvi.

IV

According to his sister's memoir, Einstein was anxious about whether his relativity paper would be accepted by the *Annalen der Physik*.[72] After it was accepted, he eagerly anticipated an immediate reaction to its publication, even though he expected it to be critical. He was greatly disappointed when his paper was not even mentioned in the following issues of the *Annalen*. Sometime afterward, she recounts, he received a letter from Planck, requesting explanations of a few obscure points in the work.[73]

> After the long period of waiting, this was the first sign that his paper was being read at all. The happiness of the young scholar was all the greater, since acknowledgment of his accomplishment came from one of the greatest contemporary physicists. . . . At that time Planck's interest signified infinitely much for the morale of the young physicist.
>
> Nach der langen Wartezeit war dies das erste Zeichen, dass seine Arbeit überhaupt gelesen worden war. Die Freude des jungen Gelehrten war um so grösser, da die Anerkennung seiner Leistung von einem der grössten Physiker der Gegenwart herrührte. . . . In jenem Zeitpunkt bedeutete das Interesse Plancks in moralischer Beziehung unendlich viel für den jungen Physiker.

Planck and Einstein continued to correspond, and Planck discussed Einstein's paper in the University of Berlin's physics colloquium during the fall of 1905.[74] During the next few years, Planck wrote several papers developing further consequences of the relativity principle,[75] and interested his assistant, Max Laue,[76] and one of his students, Kurd von Mosengeil,[77] in working on related problems. A few years later, Einstein paid tribute to Planck's role in promoting the theory of relativity:

[72] See *Winteler-Einstein 1924*, pp. 23–24. Winteler-Einstein also reports that Einstein originally intended to submit the work on special relativity as his doctoral dissertation (for her account of this claim, see the editorial note, "Einstein's Dissertation on the Determination of Molecular Dimensions," p. 175).

[73] It is not clear just when Planck first wrote to Einstein. His first known reference to Planck occurs in a letter of 3 May 1906 to Maurice Solovine: "My papers are well received and are giving rise to further research. Prof. Planck (Berlin) wrote me recently about this" ("Meine Arbeiten finden viel Würdigung und geben Anlass zu weiteren Untersuchungen. Prof. Planck

(Berlin) schrieb mir neulich darüber"). If Winteler-Einstein's account is correct, the letter from Planck referred to cannot be his first. The earliest surviving letter from Planck is dated 6 July 1907.

[74] According to Max Laue, who was then Planck's *Assistent*, Planck discussed Einstein's paper at one of the first, if not *the* first, of the physics colloquia during the winter semester. See *Laue 1952*.

[75] See *Planck 1906a*, *1907a*. He also analyzed Kaufmann's experiments during this period (see § V).

[76] See *Laue 1907*.

[77] See *Mosengeil 1907*.

The attention that this theory so quickly received from colleagues is surely to be ascribed in large part to the resoluteness and warmth with which he intervened for this theory.

Der Entschiedenheit und Wärme, mit der er für diese Theorie eingetreten ist, ist wohl zum großen Teil die Beachtung zuzuschreiben, die diese Theorie bei den Fachgenossen so schnell gefunden hat.[78]

Other physicists also started to discuss Einstein's work in 1905 and 1906. Two months after it appeared, Kaufmann cited it in a preliminary report of his recent experiments on the mass of electrons in β-rays.[79] The following year, in a fuller discussion of his results, while noting that the two theories yield the same equations of motion for the electron, he gave the first clear account of the basic theoretical difference between Lorentz's and Einstein's views.[80] Drude, the editor of the *Annalen*, cited Einstein's paper in the second edition of his standard text on optics,[81] as well as in his article on optics in the *Handbuch der Physik*.[82] Röntgen wrote to Einstein asking for copies of his papers on electrodynamics,[83] presumably in connection with a talk Röntgen was to give on the equations of motion of the electron.[84] Sommerfeld, who heard the talk, soon read Einstein's work and was so impressed that he decided to give a colloquium on it.[85] During 1907, Einstein was in correspondence about the theory with Planck, Laue, Wien,[86] and Minkowski.[87] In the same year, he was asked to write a review article on relativity, which appeared in Stark's *Jahrbuch der Radioaktivität* at the end of the year,[88] and a major publishing house inquired about the possibility of a book on his research.[89] A reference by Ehrenfest in 1907 to Einstein's theory as a "closed system" ("abgeschlossenes System")[90] led Einstein to clarify his view of the nature of the theory.[91] By 1908, the theory of relativity, though

[78] *Einstein 1913,* p. 1079.

[79] See *Kaufmann 1905*.

[80] See *Kaufmann 1906a*, pp. 491–493. See also *Kaufmann 1906b*, which contrasts Cohn's theory to those of Lorentz and Einstein.

[81] See *Drude 1906a*, p. 467.

[82] See *Drude 1906b*, p. 1387.

[83] See Wilhelm Röntgen to Einstein, 18 September 1906.

[84] The talk is discussed in a letter of Arnold Sommerfeld to Wilhelm Wien, 23 November 1906 (GyMDM, Wien Nachlaß, Mappe Sommerfeld).

[85] See Arnold Sommerfeld to Wilhelm Wien, 23 November 1906 (GyMDM, Wien Nachlaß, Mappe Sommerfeld). The letter indicates that Wien had previously commented to Sommerfeld on Einstein's paper. By early 1908, Sommerfeld was corresponding with Einstein (see *Eckert and Pricha 1984*). During the winter semester of 1908–1909 Sommerfeld gave a course on the theory of relativity at the University of Munich, which he believed to be the first

such course (see *Jungnickel and McCormmach 1986b*, p. 283).

[86] Einstein had a lengthy correspondence with Wien on the question of whether superluminal signal velocities are compatible with Maxwell's theory. See Vol. 5, the editorial note, "Einstein on Superluminal Signal Velocities."

[87] Minkowski wrote Einstein to request a copy of his paper, for discussion at a Göttingen seminar (see Hermann Minkowski to Einstein, 9 October 1907). The seminar on electrodynamics during the 1907 winter semester was conducted by Hilbert and Minkowski (see *Pyenson 1985*, p. 83).

[88] For correspondence with Stark about this review, see *Hermann 1966*. For Einstein's review paper, see *Einstein 1907j* (Doc. 47). The paper is discussed below in § V.

[89] See B. G. Teubner to Einstein, 3 October 1907.

[90] *Ehrenfest 1907*.

[91] See *Einstein 1907g* (Doc. 44). For a discussion of this paper, see § V below.

still controversial, and often not clearly distinguished from Lorentz's electron theory, was a major topic of discussion among leading German-speaking physicists.[92]

V

The topic "electrodynamics of moving bodies," as understood at the time Einstein wrote his 1905 paper, usually included not only the microscopic electron theory discussed in that paper, but also the macroscopic theory, involving conduction currents in polarizable and magnetizable moving media. The field equations of such a theory could either be postulated phenomenologically, as Cohn did;[93] or they could be derived from an underlying microscopic theory, as in Lorentz's electron theory.[94] Einstein did not turn to the macroscopic theory until 1908. His work with Laub on that subject is discussed in the editorial note, "Einstein and Laub on the Electrodynamics of Moving Media," pp. 503–507.

Since the theory of relativity grew out of Einstein's long-standing concern with electrodynamics, and his applications of the theory were primarily in this field, the theory was often looked upon as essentially another version of Lorentz's electron theory (see section II, pp. 256–257). Einstein soon felt the need to make clear the distinction between the kinematical results of the theory, deduced from the two principles of the theory,[95] and the application of such kinematical results to the solution of problems in the optics and electrodynamics of moving bodies, to the derivation of the equations of motion of a charged particle[96]—or indeed to any physical theory. He pointed out that the postulates of the theory do not constitute a "closed system" ("abgeschlossenes System"), but only a "heuristic principle, which considered by itself alone only contains assertions about rigid bodies, clocks, and light signals" ("heuristisches Prinzip, welches für sich allein betrachtet nur Aussagen über starre Körper, Uhren und Lichtsignale enthält"). Beyond such assertions, the theory could only establish "relations between otherwise apparently independent laws" ("Beziehungen zwischen sonst voneinander unabhängig erscheinenden Gesetzmäßigkeiten") of physics.[97]

A few months after first publishing the theory of relativity, Einstein discovered a relation that particularly intrigued him, the relation between inertial mass and energy. He wrote Conrad Habicht:

> One more consequence of the electrodynamical paper has also occurred to me. The principle of relativity, together with Maxwell's equations, requires that mass be a direct measure of the energy contained in a body; light transfers mass. A noticeable decrease of mass should occur in the case of radium. The argument is amusing and attractive; but I can't tell whether the Lord isn't laughing about it and playing a trick on me.

---

[92] For the reception of the theory of relativity in Germany and several other countries, see *Goldberg 1984*, part II; and *Glick 1987*.

[93] See *Cohn 1900, 1902, 1904a, 1904b*.

[94] See, e.g., *Lorentz 1904c*.

[95] See *Einstein 1905r* (Doc. 23), "Kinema-

tischer Teil," pp. 892–907.

[96] See *Einstein 1905r* (Doc. 23), "Elektrodynamischer Teil," pp. 907–921.

[97] *Einstein 1907g* (Doc. 44), pp. 206–207. The article is a reply to *Ehrenfest 1907*.

Eine Konsequenz der elektrodynamischen Arbeit ist mir noch in den Sinn gekommen. Das Relativitätsprinzip im Zusammenhang mit den Maxwellschen
Grundgleichungen verlangt nämlich, daß die Masse direkt ein Maß für die im
Körper enthaltene Energie ist; das Licht überträgt Masse. Eine merkliche Abnahme der Masse müßte beim Radium erfolgen. Die Überlegung ist lustig und
bestechend; aber ob der Herrgott nicht darüber lacht und mich an der Nase
herumgeführt hat, das kann ich nicht wissen.[98]

The idea that inertial mass is associated with electromagnetic energy was often discussed
before 1905. Around the turn of the century, it was suggested that all mechanical concepts
could be derived from those of electromagnetism.[99] In particular, there were attempts to
derive the entire inertial mass of the electron from the energy associated with its electromagnetic field.[100] It was also proved that a radiation-filled container manifests an apparent inertial mass, which (if the mass of the container is neglected) is proportional to the
energy of the enclosed radiation.[101]

In his first paper on this subject, Einstein argued that, as a consequence of the relativity
principle, inertial mass is associated with *all* forms of energy.[102] He was only able to
establish the association for a process involving the emission of electromagnetic radiation
by a system, but argued that the result is independent of the mechanism by which the
system loses energy. In addition, he was only able to show that changes in energy are
associated with changes in inertial mass equal to the changes in energy divided by c^2. His
argument was criticized in 1907 by Planck, who presented his own argument to show that
a transfer of heat is associated with a similarly related transfer of inertial mass.[103]

Soon afterward, Stark attributed the discovery of the relationship between mass and
energy to Planck.[104] Einstein wrote Stark:

> I was rather disturbed that you do not acknowledge my priority with regard to
> the connection between inertial mass and energy.

> Es hat mich etwas befremdet, daß Sie bezüglich des Zusammenhanges zwischen
> träger Masse u. Energie meine Priorität nicht anerkennen.[105]

After receiving a conciliatory reply from Stark, acknowledging his priority,[106] Einstein
replied, regretting his original testy reaction:

[98] Einstein to Conrad Habicht, end of June–
end of September 1905.

[99] *Wien 1900* explicitly states the program of
an electromagnetic foundation of mechanics.
For a survey of the so-called electromagnetic
world view, see *McCormmach 1970b*.

[100] See *Abraham 1902a, 1902b, 1903*. For a
contemporary review of such attempts, see
Bucherer 1904, pp. 51–68.

[101] See *Hasenöhrl 1904, 1905*, and *Mosengeil 1907*. For a contemporary review of work
on this topic that does not use the theory of rel-

ativity, see *Hasenöhrl 1909b*.

[102] See *Einstein 1905v* (Doc. 27).

[103] See *Planck 1907a*, § 17. The critical
comment on Einstein's argument is in a footnote
on p. 565. For a discussion of Planck's argument, and later criticisms of Einstein's derivation, see *Stachel and Torretti 1982*.

[104] See *Stark 1907*.

[105] Einstein to Johannes Stark, 17 February
1908.

[106] See Johannes Stark to Einstein, 19 February 1908.

People, to whom it is granted to contribute something to the progress of science, should not allow pleasure in the fruits of their common work to be clouded by such matters.

Die Leute, denen es vergönnt ist, zum Fortschritt der Wissenschaft etwas beizutragen, sollten sich die Freude über die Früchte gemeinsamer Arbeit nicht durch solche Dinge trüben lassen.[107]

Einstein returned to the relationshp between inertial mass and energy in 1906 and in 1907, giving more general arguments for their complete equivalence,[108] but he did not achieve the complete generality to which he aspired.[109] In his Salzburg talk, Einstein strongly emphasized that inertial mass is a property of all forms of energy, and therefore electromagnetic radiation must have mass. This conclusion strengthened Einstein's belief in the hypothesis that light quanta manifest particlelike properties.[110]

In 1905, Einstein proposed a number of other experimentally testable consequences of his theory, in particular the equations of motion of the electron.[111] His use of the terms "transverse mass" and "longitudinal mass" indicates that he was familiar with some of the earlier work on this topic.[112] The following year he suggested another experimental test of these equations employing cathode rays.[113]

In this paper, Einstein first mentioned Kaufmann's experimental investigations of the motion of electrons in β-rays.[114] Starting in 1901, Kaufmann had carried out a series of experiments on the deflection of β-rays by electric and magnetic fields. In 1905 he asserted that his recent experiments yielded data for the dependence of the transverse mass on velocity that were incompatible with the (identical) predictions of the Lorentz and Einstein theories, but were compatible with those of the Abraham and Bucherer electron models.[115] Kaufmann's work occasioned considerable discussion. Lorentz was disheartened by the apparent refutation of his theory.[116] Planck subjected the experiment to a

---

[107] Einstein to Johannes Stark, 22 February 1908.

[108] See *Einstein 1906e* (Doc. 35) and *Einstein 1907h* (Doc. 45).

[109] See *Einstein 1907h* (Doc. 45), pp. 371–372, for Einstein's dissatisfaction with arguments based on special cases.

[110] See *Einstein 1909c* (Doc. 60), p. 490.

[111] See *Einstein 1905r* (Doc. 23), § 10. Einstein restricted himself to slowly accelerated electrons.

[112] The concepts were introduced by Lorentz (see *Lorentz 1900*). The terms were apparently introduced by Abraham (see *Abraham 1902a*), and commonly employed thereafter in discussions of proposed equations of motion for the electron (see, e.g., *Abraham 1903, Bucherer 1904*).

[113] See *Einstein 1906g* (Doc 37).

[114] See *Kaufmann 1905, 1906a*, which include references to Kaufmann's earlier work. For citation of contemporary reviews of attempts to determine the variation of electron mass with velocity, see note 124. For recent accounts of Kaufmann's experiments and the subsequent discussion of them, see *Miller 1981b*, pp. 334–352; *Cushing 1981*.

[115] See *Kaufmann 1905*. For Lorentz's theory, see *Lorentz 1904a*. For Abraham's model, see *Abraham 1902a, 1902b, 1903*. For Bucherer's model see *Bucherer 1904*, pp. 57–58. Langevin had independently proposed the same hypothesis as Bucherer about the shape of a moving electron. See *Langevin 1905c*.

[116] See the letter of Hendrik Lorentz to Henri Poincaré, 8 March 1906, reproduced in *Miller 1980*, pp. 83–84; and Lorentz's comment in his 1906 lectures at Columbia University, printed in *Lorentz 1909b*, p. 213.

careful analysis, and concluded that Kaufmann's work could not be regarded as a definitive refutation of the Lorentz-Einstein predictions.[117] Röntgen, one of the leading German experimentalists, is reported also to have felt that Kaufmann's results were not decisive, because his observations were not that accurate.[118] In his 1907 review, Einstein discussed Kaufmann's results at some length, especially their apparent irreconcilability with the Lorentz-Einstein predictions.[119] Commenting on a figure showing Kaufmann's results and the relativistic predictions, Einstein wrote: "Considering the difficulty of the experiment, one might be inclined to regard the agreement as satisfactory" ("In Anbetracht der Schwierigkeit der Untersuchung möchte man geneigt sein, die Übereinstimmung als eine genügende anzusehen"). However, he noted that the deviations are systematic and well outside Kaufmann's error limits.

> Whether the systematic deviations are based upon a source of error not yet considered, or on lack of correspondence between the foundations of the theory of relativity and the facts, can only be decided with certainty when more manifold observational data are at hand.

> Ob die systematischen Abweichungen in einer noch nicht gewürdigten Fehlerquelle oder darin ihren Grund haben, daß die Grundlagen der Relativitätstheorie nicht den Tatsachen entsprechen, kann wohl erst dann mit Sicherheit entschieden werden, wenn ein mannigfaltigeres Beobachtungsmaterial vorliegen wird.

Although Einstein evidently accepted experiment as the ultimate arbiter of the fate of any theory, he was cautious about accepting Kaufmann's results as definitive, perhaps because of his familiarity with Planck's critical analyses of the experiments.[120] What he found even more difficult to accept were alternative equations of motion for the electron that are based on what he regarded as arbitrary dynamical assumptions about the shape of a moving electron. While conceding that Kaufmann's data seemed to favor the theories of Abraham and Bucherer,[121] Einstein concluded:

> In my opinion, however, a rather small probability should be ascribed to these theories, since their fundamental assumptions about the mass of a moving electron are not supported by theoretical systems that embrace wider complexes of phenomena.

> Jenen Theorien kommt aber nach meiner Meinung eine ziemlich geringe

[117] See *Planck 1906b*, and the discussion following his paper, reported in *Discussion/Planck 1906*; see also *Planck 1907b*.

[118] Röntgen's views were expressed in a talk to the Bavarian Academy of Sciences, reported in a letter of Arnold Sommerfeld to Wien, 23 November 1906 (GyMDM, Wien Nachlaß, Mappe Sommerfeld).

[119] See *Einstein 1907j* (Doc. 47), pp. 437–439.

[120] On 1 November 1907, Einstein thanked Stark for calling his attention to Planck's work on Kaufmann's experiments (*Planck 1906b*). Einstein evidently wrote to Planck requesting a copy, for Planck replied by postcard on 9 November, stating that he had sent copies of his two papers on the subject (*Planck 1906b, 1907b*) to Einstein, and adding further comments on the experiments.

[121] See note 115.

Wahrscheinlichkeit zu, weil ihre die Maße des bewegten Elektrons betreffenden Grundannahmen nicht nahe gelegt werden durch theoretische Systeme, welche größere Komplexe von Erscheinungen umfassen.[122]

This cautious attitude toward Kaufmann's results proved justified. During the following years, controversies over the interpretation of the experimental results prevented investigations of this topic from playing a decisive role in contemporary evaluations of the theory of relativity.[123] Bestelmeyer carried out β-ray experiments generally regarded as inconclusive, while Bucherer's results favoring the Lorentz-Einstein equations were seriously questioned.[124] Experiments using cathode rays, reported by several investigators starting in 1910, proved similarly inconclusive. Almost a decade elapsed until generally accepted data on the variation of electron mass with velocity were at hand; these data supported the relativistic predictions.[125]

In 1907, Einstein agreed to write a review article on relativity for Johannes Stark's *Jahrbuch der Radioaktivität und Elektronik*.[126] He indicated what he had read on the subject by this time:

> Besides my own, I know of a paper by H. A. Lorentz (1904),[127] one by E. Kohn,[128] one by Mosengeil,[129] as well as two by Planck.[130] I am not familiar with other relevant theoretical papers.[131]

> Ausser meinen eigenen Arbeiten ist mir eine Arbeit von H. A. Lorentz (1904), eine von E. Kohn, eine von Mosengeil sowie zwei von Planck bekannt. Andere die Sache betreffende theoretische Arbeiten sind mir nicht bekannt geworden.

Einstein indicated the care he had taken to make the review pedagogically useful:

> I have taken great pains to clarify the assumptions employed, by—as far as possible—introducing these assumptions one by one and pursuing their consequences in sequence.

> An die Klarlegung der benutzten Annahmen habe ich viel Sorgfalt verwendet,

[122] *Einstein 1907j* (Doc. 47), p. 439. For an earlier statement of Einstein's attitude toward the unifying power of theories, see Einstein to Marcel Grossmann, 14 April 1901 (Vol. 1, Doc. 100).

[123] For reviews of the early experiments on the variation of the mass of the electron with its velocity, see *Laub 1910*; *Laue 1911b*, pp. 16–18; *Guye and Lavanchy 1916*, pp. 288–292; *Pauli 1921*, p. 636; and *Lorentz 1922*, chap. VII.

[124] Einstein corresponded with Bucherer in 1908, shortly after the latter carried out his experiments.

[125] See *Guye and Lavanchy 1916*, and *Glitscher 1917*.

[126] Einstein to Johannes Stark, 25 September 1907. See also note 47.

[127] *Lorentz 1904a*.

[128] That is, Emil Cohn. See *Cohn 1900*, *1902*, *1904a*, and *1904b*. A copy of *Cohn 1904a* is in Einstein's collection of reprints, now in IsReW. See also note 27.

[129] *Mosengeil 1907*.

[130] Presumably a reference to *Planck 1906a*, *1907a*. Einstein did not yet know of *Planck 1906b*, *1907b* (see note 121).

[131] Einstein to Johannes Stark, 25 September 1907. In his reply (Stark to Einstein, 4 October 1907), Stark cited *Planck 1907a*, *Laue 1907*. Stark later told Einstein about Planck's studies of Kaufmann's experiments.

indem ich—soweit es anging—jene Annahmen einzeln einführte und der Reihe nach ihre Konsequenzen verfolgte.[132]

The review covers relativistic kinematics, optics, electromagnetic theory, the relativistic dynamics of a particle, and the relativistic dynamics and thermodynamics of an extended system. He summarized the results of a number of his earlier papers on the theory of relativity, sometimes simplifying his earlier proofs. He adopted Planck's approach to relativistic dynamics and thermodynamics, although often giving his own proofs of Planck's results.[133]

In his Salzburg lecture, Einstein reviewed the historical background of the theory of relativity in order to stress some consequences of the theory for the problem of the structure of the radiation field.[134] By discarding the concept of the ether and by showing light consists of "structures" ("Gebilde") that carry inertial mass from an emitter to an absorber, he stressed that the theory of relativity opens the way for a theory of light that includes both corpuscular and undulatory features. He went on to argue the necessity of such a theory. For further discussion of the Salzburg lecture, see the Introduction, pp. xvii–xviii, and the editorial note, "Einstein's Early Work on the Quantum Hypothesis," pp. 147–148.

Einstein wrote three additional reviews of relativity between 1910 and 1912.[135] *Einstein 1910a* and *Einstein 1911e*, written for nonspecialized physics audiences, discuss historical and foundational questions. The technically most detailed of all his reviews is an unpublished manuscript written about 1912.[136] After 1912, Einstein's active research interests no longer included what soon came to be known as the special theory of relativity. He occasionally wrote pedagogical articles on aspects of the special theory,[137] and continued to include reviews of the subject in many expositions of the theory of relativity as a whole.[138]

VI

The 1907 review concludes with Einstein's first published discussion of gravitation. The final section, entitled "Principle of Relativity and Gravitation" ("Relativitätsprinzip und Gravitation"),[139] takes the equality of gravitational and inertial mass as its starting

---

[132] Einstein to Johannes Stark, 1 November 1907.

[133] Although Einstein did not publish anything further on relativistic thermodynamics, it appears that later in his life he had doubts about the validity of Planck's approach (see, e.g., Einstein to Max von Laue, 29 January 1952).

[134] See *Einstein 1909c* (Doc. 60), pp. 482–490.

[135] After 1912, Einstein's reviews of relativity include discussions of the generalized theory, with primary emphasis on the problem of gravi-

tation.

[136] The untitled manuscript was written in 1912 for *Marx 1924*, publication of which was delayed due to the First World War (see Erich Marx to Einstein, 2 January 1922, and Einstein's undated reply to Marx's letter of 3 March 1922). See Vol. 4 for this manuscript.

[137] See, e.g., *Einstein 1935*.

[138] See *Einstein 1917a*, *Einstein 1921b*, for early examples.

[139] See *Einstein 1907j* (Doc. 47), § V, pp. 454–462.

point.[140] It follows from this equality that there is no observable difference between the behavior of a mechanical system in a gravitation-free uniformly accelerating reference frame and the behavior of the same system in an inertial frame in which there is a uniform gravitational field. Generalizing this mechanical equivalence, Einstein postulated "the complete physical equivalence of a gravitational field and a corresponding acceleration of the reference system" ("die völlige physikalische Gleichwertigkeit von Gravitationsfeld und entsprechender Beschleunigung des Bezugssystems"), which allows one "to replace a homogeneous gravitational field by a uniformly accelerated reference system" ("ein homogenes Gravitationsfeld durch ein gleichförmig beschleunigtes Bezugssystem zu ersetzen").[141] This equivalence enabled him to draw conclusions about the effects of a uniform gravitational field on physical processes from the analysis of such processes in a uniformly accelerated reference system. He thus deduced that spectral lines from the sun show a shift toward longer wave lengths when compared to the corresponding terrestrial lines (an effect now called the gravitational red shift) and that a light ray passing through a gravitational field is deflected from a straight-line path; but he noted that the effect is too small to be observed in the earth's gravitational field. Although his correspondence during the intervening years contains several references to the problem of gravitation, Einstein did not publish again on this topic until 1911 (see Volume 4, the editorial note, "Einstein on Gravitation and Relativity").[142]

[140] In *Einstein 1934* (first published in English in *Einstein 1933*), Einstein stated that he was convinced of this equality even though not yet aware of the experiments of Eötvös (see *Eöt-*

vös 1890).

[141] *Einstein 1907j* (Doc. 47), p. 454.

[142] See *Einstein 1911g*.

23. "On the Electrodynamics of Moving Bodies"

[Einstein 1905r]

DATED Bern, June 1905
RECEIVED 30 June 1905
PUBLISHED 26 September 1905

IN: *Annalen der Physik* 17 (1905): 891–921.

Republished in *Blumenthal 1913*, pp. 27–52.

3. *Zur Elektrodynamik bewegter Körper;*
von A. Einstein.

Daß die Elektrodynamik Maxwells — wie dieselbe gegenwärtig aufgefaßt zu werden pflegt — in ihrer Anwendung auf bewegte Körper zu Asymmetrien führt, welche den Phänomenen nicht anzuhaften scheinen, ist bekannt. Man denke z. B. an die elektrodynamische Wechselwirkung zwischen einem Magneten und einem Leiter. Das beobachtbare Phänomen hängt hier nur ab von der Relativbewegung von Leiter und Magnet, während nach der üblichen Auffassung die beiden Fälle, daß der eine oder der andere dieser Körper der bewegte sei, streng voneinander zu trennen sind. Bewegt sich nämlich der Magnet und ruht der Leiter, so entsteht in der Umgebung des Magneten ein elektrisches Feld von gewissem Energiewerte, welches an den Orten, wo sich Teile des Leiters befinden, einen Strom erzeugt. Ruht aber der Magnet und bewegt sich der Leiter, so entsteht in der Umgebung des Magneten kein elektrisches Feld, dagegen im Leiter eine elektromotorische Kraft, welcher an sich keine Energie entspricht, die aber — Gleichheit der Relativbewegung bei den beiden ins Auge gefaßten Fällen vorausgesetzt — zu elektrischen Strömen von derselben Größe und demselben Verlaufe Veranlassung gibt, wie im ersten Falle die elektrischen Kräfte.

Beispiele ähnlicher Art, sowie die mißlungenen Versuche, eine Bewegung der Erde relativ zum „Lichtmedium" zu konstatieren, führen zu der Vermutung, daß dem Begriffe der absoluten Ruhe nicht nur in der Mechanik, sondern auch in der Elektrodynamik keine Eigenschaften der Erscheinungen entsprechen, sondern daß vielmehr für alle Koordinatensysteme, für welche die mechanischen Gleichungen gelten, auch die gleichen elektrodynamischen und optischen Gesetze gelten, wie dies für die Größen erster Ordnung bereits erwiesen ist. Wir wollen diese Vermutung (deren Inhalt im folgenden „Prinzip der Relativität" genannt werden wird) zur Voraussetzung erheben und außerdem die mit ihm nur scheinbar unverträgliche

892   *A. Einstein.*

Voraussetzung einführen, daß sich das Licht im leeren Raume stets mit einer bestimmten, vom Bewegungszustande des emittierenden Körpers unabhängigen Geschwindigkeit V fortpflanze. [5] Diese beiden Voraussetzungen genügen, um zu einer einfachen und widerspruchsfreien Elektrodynamik bewegter Körper zu gelangen unter Zugrundelegung der Maxwellschen Theorie für ruhende Körper. Die Einführung eines „Lichtäthers" wird sich insofern als überflüssig erweisen, als nach der zu entwickelnden [6] Auffassung weder ein mit besonderen Eigenschaften ausgestatteter „absolut ruhender Raum" eingeführt, noch einem Punkte des leeren Raumes, in welchem elektromagnetische Prozesse stattfinden, ein Geschwindigkeitsvektor zugeordnet wird. [7]

Die zu entwickelnde Theorie stützt sich — wie jede andere Elektrodynamik — auf die Kinematik des starren Körpers, da die Aussagen einer jeden Theorie Beziehungen zwischen starren Körpern (Koordinatensystemen), Uhren und elektromagnetischen Prozessen betreffen. Die nicht genügende Berücksichtigung dieses Umstandes ist die Wurzel der Schwierigkeiten, mit denen die Elektrodynamik bewegter Körper gegenwärtig zu kämpfen hat.

I. Kinematischer Teil.

§ 1. Definition der Gleichzeitigkeit.

Es liege ein Koordinatensystem vor, in welchem die Newtonschen mechanischen Gleichungen gelten. Wir nennen [8] dies Koordinatensystem zur sprachlichen Unterscheidung von später einzuführenden Koordinatensystemen und zur Präzisierung der Vorstellung das „ruhende System".

Ruht ein materieller Punkt relativ zu diesem Koordinatensystem, so kann seine Lage relativ zu letzterem durch starre Maßstäbe unter Benutzung der Methoden der euklidischen Geometrie bestimmt und in kartesischen Koordinaten ausgedrückt werden.

Wollen wir die *Bewegung* eines materiellen Punktes beschreiben, so geben wir die Werte seiner Koordinaten in Funktion der Zeit. Es ist nun wohl im Auge zu behalten, daß eine derartige mathematische Beschreibung erst dann einen physikalischen Sinn hat, wenn man sich vorher darüber klar geworden ist, was hier unter „Zeit" verstanden wird.

Zur Elektrodynamik bewegter Körper.   893

Wir haben zu berücksichtigen, daß alle unsere Urteile, in welchen die Zeit eine Rolle spielt, immer Urteile über *gleichzeitige Ereignisse* sind. Wenn ich z. B. sage: „Jener Zug kommt hier um 7 Uhr an," so heißt dies etwa: „Das Zeigen des kleinen Zeigers meiner Uhr auf 7 und das Ankommen des Zuges sind gleichzeitige Ereignisse."[1])

Es könnte scheinen, daß alle die Definition der „Zeit" betreffenden Schwierigkeiten dadurch überwunden werden könnten, daß ich an Stelle der „Zeit" die „Stellung des kleinen Zeigers meiner Uhr" setze. Eine solche Definition genügt in der Tat, wenn es sich darum handelt, eine Zeit zu definieren ausschließlich für den Ort, an welchem sich die Uhr eben befindet; die Definition genügt aber nicht mehr, sobald es sich darum handelt, an verschiedenen Orten stattfindende Ereignisreihen miteinander zeitlich zu verknüpfen, oder — was auf dasselbe hinausläuft — Ereignisse zeitlich zu werten, welche in von der Uhr entfernten Orten stattfinden.

[9]

Wir könnten uns allerdings damit begnügen, die Ereignisse dadurch zeitlich zu werten, daß ein samt der Uhr im Koordinatenursprung befindlicher Beobachter jedem von einem zu wertenden Ereignis Zeugnis gebenden, durch den leeren Raum zu ihm gelangenden Lichtzeichen die entsprechende Uhrzeigerstellung zuordnet. Eine solche Zuordnung bringt aber den Übelstand mit sich, daß sie vom Standpunkte des mit der Uhr versehenen Beobachters nicht unabhängig ist, wie wir durch die Erfahrung wissen. Zu einer weit praktischeren Festsetzung gelangen wir durch folgende Betrachtung.

[10]

Befindet sich im Punkte *A* des Raumes eine Uhr, so kann ein in *A* befindlicher Beobachter die Ereignisse in der unmittelbaren Umgebung von *A* zeitlich werten durch Aufsuchen der mit diesen Ereignissen gleichzeitigen Uhrzeigerstellungen. Befindet sich auch im Punkte *B* des Raumes eine Uhr — wir wollen hinzufügen, „eine Uhr von genau derselben Beschaffenheit wie die in *A* befindliche" — so ist auch eine zeitliche Wertung der Ereignisse in der unmittelbaren Umgebung von

---

1) Die Ungenauigkeit, welche in dem Begriffe der Gleichzeitigkeit zweier Ereignisse an (annähernd) demselben Orte steckt und gleichfalls durch eine Abstraktion überbrückt werden muß, soll hier nicht erörtert werden.

894   *A. Einstein.*

B durch einen in B befindlichen Beobachter möglich. Es ist
aber ohne weitere Festsetzung nicht möglich, ein Ereignis in
A mit einem Ereignis in B zeitlich zu vergleichen; wir haben
bisher nur eine „A-Zeit" und eine „B-Zeit", aber keine für A
und B gemeinsame „Zeit" definiert. Die letztere Zeit kann
nun definiert werden, indem man *durch Definition* festsetzt, daß
die „Zeit", welche das Licht braucht, um von A nach B zu
gelangen, gleich ist der „Zeit", welche es braucht, um von B
nach A zu gelangen. Es gehe nämlich ein Lichtstrahl zur
„A-Zeit" t_A von A nach B ab, werde zur „B-Zeit" t_B in B
gegen A zu reflektiert und gelange zur „A-Zeit" t'_A nach A
zurück. Die beiden Uhren laufen definitionsgemäß synchron,
wenn

$$t_B - t_A = t'_A - t_B.$$

Wir nehmen an, daß diese Definition des Synchronismus
in widerspruchsfreier Weise möglich sei, und zwar für beliebig
viele Punkte, daß also allgemein die Beziehungen gelten:

1. Wenn die Uhr in B synchron mit der Uhr in A läuft,
so läuft die Uhr in A synchron mit der Uhr in B.

2. Wenn die Uhr in A sowohl mit der Uhr in B als auch
mit der Uhr in C synchron läuft, so laufen auch die Uhren in
B und C synchron relativ zueinander.

Wir haben so unter Zuhilfenahme gewisser (gedachter)
physikalischer Erfahrungen festgelegt, was unter synchron
laufenden, an verschiedenen Orten befindlichen, ruhenden
Uhren zu verstehen ist und damit offenbar eine Definition
von „gleichzeitig" und „Zeit" gewonnen. Die „Zeit" eines
Ereignisses ist die mit dem Ereignis gleichzeitige Angabe
einer am Orte des Ereignisses befindlichen, ruhenden Uhr,
welche mit einer bestimmten, ruhenden Uhr, und zwar für
alle Zeitbestimmungen mit der nämlichen Uhr, synchron läuft.

Wir setzen noch der Erfahrung gemäß fest, daß die
Größe

$$\frac{2\,\overline{A\,B}}{t'_A - t_A} = V$$

eine universelle Konstante (die Lichtgeschwindigkeit im leeren
Raume) sei.

Wesentlich ist, daß wir die Zeit mittels im ruhenden System

Zur Elektrodynamik bewegter Körper.   895

ruhender Uhren definiert haben; wir nennen die eben definierte Zeit wegen dieser Zugehörigkeit zum ruhenden System „die Zeit des ruhenden Systems".

§ 2. Über die Relativität von Längen und Zeiten.

Die folgenden Überlegungen stützen sich auf das Relativitätsprinzip und auf das Prinzip der Konstanz der Lichtgeschwindigkeit, welche beiden Prinzipien wir folgendermaßen definieren.

[11]

1. Die Gesetze, nach denen sich die Zustände der physikalischen Systeme ändern, sind unabhängig davon, auf welches von zwei relativ zueinander in gleichförmiger Translationsbewegung befindlichen Koordinatensystemen diese Zustandsänderungen bezogen werden.

[12]

2. Jeder Lichtstrahl bewegt sich im „ruhenden" Koordinatensystem mit der bestimmten Geschwindigkeit V, unabhängig davon, ob dieser Lichtstrahl von einem ruhenden oder bewegten Körper emittiert ist. Hierbei ist

[13]

$$\text{Geschwindigkeit} = \frac{\text{Lichtweg}}{\text{Zeitdauer}},$$

wobei „Zeitdauer" im Sinne der Definition des § 1 aufzufassen ist.

Es sei ein ruhender starrer Stab gegeben; derselbe besitze, mit einem ebenfalls ruhenden Maßstabe gemessen, die Länge l. Wir denken uns nun die Stabachse in die X-Achse des ruhenden Koordinatensystems gelegt und dem Stabe hierauf eine gleichförmige Paralleltranslationsbewegung (Geschwindigkeit v) längs der X-Achse im Sinne der wachsenden x erteilt. Wir fragen nun nach der Länge des *bewegten* Stabes, welche wir uns durch folgende zwei Operationen ermittelt denken:

a) Der Beobachter bewegt sich samt dem vorher genannten Maßstabe mit dem auszumessenden Stabe und mißt direkt durch Anlegen des Maßstabes die Länge des Stabes, ebenso, wie wenn sich auszumessender Stab, Beobachter und Maßstab in Ruhe befänden.

b) Der Beobachter ermittelt mittels im ruhenden Systeme aufgestellter, gemäß § 1 synchroner, ruhender Uhren, in welchen Punkten des ruhenden Systems sich Anfang und Ende des auszumessenden Stabes zu einer bestimmten Zeit t befinden.

58*

896   *A. Einstein.*

Die Entfernung dieser beiden Punkte, gemessen mit dem schon benutzten, in diesem Falle ruhenden Maßstabe ist ebenfalls eine Länge, welche man als „Länge des Stabes" bezeichnen kann.

Nach dem Relativitätsprinzip muß die bei der Operation a) zu findende Länge, welche wir „die Länge des Stabes im bewegten System" nennen wollen, gleich der Länge l des ruhenden Stabes sein.

Die bei der Operation b) zu findende Länge, welche wir „die Länge des (bewegten) Stabes im ruhenden System" nennen wollen, werden wir unter Zugrundelegung unserer beiden Prinzipien bestimmen und finden, daß sie von l verschieden ist.

Die allgemein gebrauchte Kinematik nimmt stillschweigend an, daß die durch die beiden erwähnten Operationen bestimmten Längen einander genau gleich seien, oder mit anderen Worten, daß ein bewegter starrer Körper in der Zeitepoche t in geometrischer Beziehung vollständig durch *denselben* Körper, wenn er in bestimmter Lage *ruht*, ersetzbar sei.

Wir denken uns ferner an den beiden Stabenden (A und B) Uhren angebracht, welche mit den Uhren des ruhenden Systems synchron sind, d. h. deren Angaben jeweilen der „Zeit des ruhenden Systems" an den Orten, an welchen sie sich gerade befinden, entsprechen; diese Uhren sind also „synchron im ruhenden System".

Wir denken uns ferner, daß sich bei jeder Uhr ein mit ihr bewegter Beobachter befinde, und daß diese Beobachter auf die beiden Uhren das im § 1 aufgestellte Kriterium für den synchronen Gang zweier Uhren anwenden. Zur Zeit[1] t_A gehe ein Lichtstrahl von A aus, werde zur Zeit t_B in B reflektiert und gelange zur Zeit t'_A nach A zurück. Unter Berücksichtigung des Prinzipes von der Konstanz der Lichtgeschwindigkeit finden wir:

$$t_B - t_A = \frac{r_{AB}}{V - v}$$

_____

1) „Zeit" bedeutet hier „Zeit des ruhenden Systems" und zugleich „Zeigerstellung der bewegten Uhr, welche sich an dem Orte, von dem die Rede ist, befindet".

und
$$t'_A - t_B = \frac{r_{AB}}{V + v},$$

wobei r_{AB} die Länge des bewegten Stabes — im ruhenden System gemessen — bedeutet. Mit dem bewegten Stabe bewegte Beobachter würden also die beiden Uhren nicht synchron gehend finden, während im ruhenden System befindliche Beobachter die Uhren als synchron laufend erklären würden.

Wir sehen also, daß wir dem Begriffe der Gleichzeitigkeit keine *absolute* Bedeutung beimessen dürfen, sondern daß zwei Ereignisse, welche, von einem Koordinatensystem aus betrachtet, gleichzeitig sind, von einem relativ zu diesem System bewegten System aus betrachtet, nicht mehr als gleichzeitige Ereignisse aufzufassen sind.

§ 3. Theorie der Koordinaten- und Zeittransformation von dem ruhenden auf ein relativ zu diesem in gleichförmiger Translationsbewegung befindliches System.

Seien im „ruhenden" Raume zwei Koordinatensysteme, d. h. zwei Systeme von je drei von einem Punkte ausgehenden, aufeinander senkrechten starren materiellen Linien, gegeben. Die X-Achsen beider Systeme mögen zusammenfallen, ihre Y- und Z-Achsen bezüglich parallel sein. Jedem Systeme sei ein starrer Maßstab und eine Anzahl Uhren beigegeben, und es seien beide Maßstäbe sowie alle Uhren beider Systeme einander genau gleich.

Es werde nun dem Anfangspunkte des einen der beiden Systeme (k) eine (konstante) Geschwindigkeit v in Richtung der wachsenden x des anderen, ruhenden Systems (K) erteilt, welche sich auch den Koordinatenachsen, dem betreffenden [14] Maßstabe sowie den Uhren mitteilen möge. Jeder Zeit t des ruhenden Systems K entspricht dann eine bestimmte Lage der Achsen des bewegten Systems und wir sind aus Symmetriegründen befugt anzunehmen, daß die Bewegung von k so beschaffen sein kann, daß die Achsen des bewegten Systems zur Zeit t (es ist mit „t" immer eine Zeit des ruhenden Systems bezeichnet) den Achsen des ruhenden Systems parallel seien.

Wir denken uns nun den Raum sowohl vom ruhenden System K aus mittels des ruhenden Maßstabes als auch vom bewegten System k mittels des mit ihm bewegten Maßstabes ausgemessen und so die Koordinaten x, y, z bez. ξ, η, ζ ermittelt. Es werde ferner mittels der im ruhenden System befindlichen ruhenden Uhren durch Lichtsignale in der in § 1 angegebenen Weise die Zeit t des ruhenden Systems für alle Punkte des letzteren bestimmt, in denen sich Uhren befinden; ebenso werde die Zeit τ des bewegten Systems für alle Punkte des bewegten Systems, in welchen sich relativ zu letzterem ruhende Uhren befinden, bestimmt durch Anwendung der in § 1 genannten Methode der Lichtsignale zwischen den Punkten, in denen sich die letzteren Uhren befinden.

Zu jedem Wertsystem x, y, z, t, welches Ort und Zeit eines Ereignisses im ruhenden System vollkommen bestimmt, gehört ein jenes Ereignis relativ zum System k festlegendes Wertsystem ξ, η, ζ, τ, und es ist nun die Aufgabe zu lösen, das diese Größen verknüpfende Gleichungssystem zu finden.

Zunächst ist klar, daß die Gleichungen *linear* sein müssen wegen der Homogenitätseigenschaften, welche wir Raum und Zeit beilegen.

Setzen wir $x' = x - vt$, so ist klar, daß einem im System k ruhenden Punkte ein bestimmtes, von der Zeit unabhängiges Wertsystem x', y, z zukommt. Wir bestimmen zuerst τ als Funktion von x', y, z und t. Zu diesem Zwecke haben wir in Gleichungen auszudrücken, daß τ nichts anderes ist als der Inbegriff der Angaben von im System k ruhenden Uhren, welche nach der im § 1 gegebenen Regel synchron gemacht worden sind.

Vom Anfangspunkt des Systems k aus werde ein Lichtstrahl zur Zeit τ_0 längs der X-Achse nach x' gesandt und von dort zur Zeit τ_1 nach dem Koordinatenursprung reflektiert, wo er zur Zeit τ_2 anlange; so muß dann sein:

$$\tfrac{1}{2}(\tau_0 + \tau_2) = \tau_1$$

oder, indem man die Argumente der Funktion τ beifügt und das Prinzip der Konstanz der Lichtgeschwindigkeit im ruhenden Systeme anwendet:

$$\tfrac{1}{2}\left[\tau(0,0,0,t) + \tau\left(0,0,0,\left\{t + \frac{x'}{V-v} + \frac{x'}{V+v}\right\}\right)\right]$$
$$= \tau\left(x',0,0,t + \frac{x'}{V-v}\right).$$

Hieraus folgt, wenn man x' unendlich klein wählt:

$$\tfrac{1}{2}\left(\frac{1}{V-v}+\frac{1}{V+v}\right)\frac{\partial\,\tau}{\partial\,t}=\frac{\partial\,\tau}{\partial\,x'}+\frac{1}{V-v}\frac{\partial\,\tau}{\partial\,t},$$

oder

$$\frac{\partial\,\tau}{\partial\,x'}+\frac{v}{V^2-v^2}\frac{\partial\,\tau}{\partial\,t}=0.$$

Es ist zu bemerken, daß wir statt des Koordinatenursprunges jeden anderen Punkt als Ausgangspunkt des Lichtstrahles hätten wählen können und es gilt deshalb die eben erhaltene Gleichung für alle Werte von x', y, z.

Eine analoge Überlegung — auf die H- und Z-Achse angewandt — liefert, wenn man beachtet, daß sich das Licht längs dieser Achsen vom ruhenden System aus betrachtet stets mit der Geschwindigkeit $\sqrt{V^2-v^2}$ fortpflanzt:

$$\frac{\partial\,\tau}{\partial\,y}=0$$

$$\frac{\partial\,\tau}{\partial\,z}=0.$$

Aus diesen Gleichungen folgt, da τ eine *lineare* Funktion ist:

$$\tau=a\left(t-\frac{v}{V^2-v^2}\,x'\right),$$

wobei a eine vorläufig unbekannte Funktion $\varphi(v)$ ist und der Kürze halber angenommen ist, daß im Anfangspunkte von k für $\tau=0$ $t=0$ sei.

Mit Hilfe dieses Resultates ist es leicht, die Größen ξ, η, ζ zu ermitteln, indem man durch Gleichungen ausdrückt, daß sich das Licht (wie das Prinzip der Konstanz der Lichtgeschwindigkeit in Verbindung mit dem Relativitätsprinzip verlangt) auch im bewegten System gemessen mit der Geschwindigkeit V fortpflanzt. Für einen zur Zeit $\tau=0$ in Richtung der wachsenden ξ ausgesandten Lichtstrahl gilt:

$$\xi=V\tau,$$

oder

$$\xi=a\,V\left(t-\frac{v}{V^2-v^2}\,x'\right)\cdot$$

Nun bewegt sich aber der Lichtstrahl relativ zum Anfangs-

900   *A. Einstein.*

punkt von k im ruhenden System gemessen mit der Geschwindigkeit $V - v$, so daß gilt:

$$\frac{x'}{V - v} = t.$$

Setzen wir diesen Wert von t in die Gleichung für ξ ein, so erhalten wir:

$$\xi = a \frac{V^2}{V^2 - v^2} x'.$$

Auf analoge Weise finden wir durch Betrachtung von längs den beiden anderen Achsen bewegte Lichtstrahlen:

$$\eta = V\tau = a V \left(t - \frac{v}{V^2 - v^2} x' \right),$$

wobei

$$\frac{y}{\sqrt{V^2 - v^2}} = t; \quad x' = 0;$$

also

$$\eta = a \frac{V}{\sqrt{V^2 - v^2}} y$$

und

$$\zeta = a \frac{V}{\sqrt{V^2 - v^2}} z.$$

Setzen wir für x' seinen Wert ein, so erhalten wir:

$$\tau = \varphi(v)\beta \left(t - \frac{v}{V^2} x \right),$$
$$\xi = \varphi(v)\beta (x - v t),$$
$$\eta = \varphi(v) y,$$
$$\zeta = \varphi(v) z,$$

wobei

$$\beta = \frac{1}{\sqrt{1 - \left(\frac{v}{V}\right)^2}}$$

und φ eine vorläufig unbekannte Funktion von v ist. Macht man über die Anfangslage des bewegten Systems und über den Nullpunkt von τ keinerlei Voraussetzung, so ist auf den rechten Seiten dieser Gleichungen je eine additive Konstante zuzufügen.

Wir haben nun zu beweisen, daß jeder Lichtstrahl sich, im bewegten System gemessen, mit der Geschwindigkeit V fortpflanzt, falls dies, wie wir angenommen haben, im ruhenden

System der Fall ist; denn wir haben den Beweis dafür noch
nicht geliefert, daß das Prinzip der Konstanz der Lichtgeschwindigkeit mit dem Relativitätsprinzip vereinbar sei.

Zur Zeit $t = \tau = 0$ werde von dem zu dieser Zeit gemeinsamen Koordinatenursprung beider Systeme aus eine Kugelwelle
ausgesandt, welche sich im System K mit der Geschwindigkeit V
ausbreitet. Ist (x, y, z) ein eben von dieser Welle ergriffener
Punkt, so ist also

$$x^2 + y^2 + z^2 = V^2 t^2.$$

Diese Gleichung transformieren wir mit Hilfe unserer Transformationsgleichungen und erhalten nach einfacher Rechnung:

$$\xi^2 + \eta^2 + \zeta^2 = V^2 \tau^2.$$

Die betrachtete Welle ist also auch im bewegten System
betrachtet eine Kugelwelle von der Ausbreitungsgeschwindigkeit V. Hiermit ist gezeigt, daß unsere beiden Grundprinzipien
[15] miteinander vereinbar sind.

In den entwickelten Transformationsgleichungen tritt noch
eine unbekannte Funktion φ von v auf, welche wir nun bestimmen wollen.

Wir führen zu diesem Zwecke noch ein drittes Koordinatensystem K' ein, welches relativ zum System k derart in Parallel-
translationsbewegung parallel zur Ξ-Achse begriffen sei, daß
sich dessen Koordinatenursprung mit der Geschwindigkeit $-v$
auf der Ξ-Achse bewege. Zur Zeit $t = 0$ mögen alle drei
Koordinatenanfangspunkte zusammenfallen und es sei für
$t = x = y = z = 0$ die Zeit t' des Systems K' gleich Null. Wir
nennen x', y', z' die Koordinaten, im System K' gemessen, und
erhalten durch zweimalige Anwendung unserer Transformationsgleichungen:

$$t' = \varphi(-v)\beta(-v)\left\{\tau + \frac{v}{V^2}\xi\right\} = \varphi(v)\varphi(-v)t,$$

$$x' = \varphi(-v)\beta(-v)\{\xi + v\tau\} \qquad = \varphi(v)\varphi(-v)x,$$

$$y' = \varphi(-v)\eta \qquad\qquad\qquad = \varphi(v)\varphi(-v)y,$$

$$z' = \varphi(-v)\zeta \qquad\qquad\qquad = \varphi(v)\varphi(-v)z.$$

Da die Beziehungen zwischen x', y', z' und x, y, z die Zeit t
nicht enthalten, so ruhen die Systeme K und K' gegeneinander,

und es ist klar, daß die Transformation von K auf K' die identische Transformation sein muß. Es ist also:

$$\varphi(v)\,\varphi(-v) = 1\,.$$

Wir fragen nun nach der Bedeutung von $\varphi(v)$. Wir fassen das Stück der H-Achse des Systems k ins Auge, das zwischen $\xi = 0$, $\eta = 0$, $\zeta = 0$ und $\xi = 0$, $\eta = l$, $\zeta = 0$ gelegen ist. Dieses Stück der H-Achse ist ein relativ zum System K mit der Geschwindigkeit v senkrecht zu seiner Achse bewegter Stab, dessen Enden in K die Koordinaten besitzen:

$$x_1 = v\,t, \quad y_1 = \frac{\cdot l}{\varphi(v)}, \quad z_1 = 0$$

und

$$x_2 = v\,t, \quad y_2 = 0, \qquad z_2 = 0\,.$$

Die Länge des Stabes, in K gemessen, ist also $l/\varphi(v)$; damit ist die Bedeutung der Funktion φ gegeben. Aus Symmetriegründen ist nun einleuchtend, daß die im ruhenden System gemessene Länge eines bestimmten Stabes, welcher senkrecht zu seiner Achse bewegt ist, nur von der Geschwindigkeit, nicht aber von der Richtung und dem Sinne der Bewegung abhängig sein kann. Es ändert sich also die im ruhenden System gemessene Länge des bewegten Stabes nicht, wenn v mit $-v$ vertauscht wird. Hieraus folgt:

$$\frac{l}{\varphi(v)} = \frac{l}{\varphi(-v)}\,,$$

oder

$$\varphi(v) = \varphi(-v)\,.$$

Aus dieser und der vorhin gefundenen Relation folgt, daß $\varphi(v) = 1$ sein muß, so daß die gefundenen Transformationsgleichungen übergehen in:

[16]

$$\tau = \beta\left(t - \frac{v}{V^2}\,x\right),$$
$$\xi = \beta\,(x - v\,t),$$
$$\eta = y\,,$$
$$\zeta = z\,,$$

wobei

$$\beta = \frac{1}{\sqrt{1 - \left(\dfrac{v}{V}\right)^2}}\,,$$

Zur Elektrodynamik bewegter Körper.   903

§ 4. Physikalische Bedeutung der erhaltenen Gleichungen, bewegte starre Körper und bewegte Uhren betreffend.

[17]   Wir betrachten eine starre Kugel[1]) vom Radius R, welche relativ zum bewegten System k ruht, und deren Mittelpunkt im Koordinatenursprung von k liegt. Die Gleichung der Oberfläche dieser relativ zum System K mit der Geschwindigkeit v bewegten Kugel ist:

$$\xi^2 + \eta^2 + \zeta^2 = R^2 .$$

Die Gleichung dieser Oberfläche ist in x, y, z ausgedrückt zur Zeit $t = 0$:

$$\frac{x^2}{\left(\sqrt{1 - \left(\frac{v}{V}\right)^2}\right)^2} + y^2 + z^2 = R^2 .$$

Ein starrer Körper, welcher in ruhendem Zustande ausgemessen die Gestalt einer Kugel hat, hat also in bewegtem Zustande — vom ruhenden System aus betrachtet — die Gestalt eines Rotationsellipsoides mit den Achsen

$$R\sqrt{1 - \left(\frac{v}{V}\right)^2}, \; R, \; R .$$

Während also die Y- und Z-Dimension der Kugel (also auch jedes starren Körpers von beliebiger Gestalt) durch die Bewegung nicht modifiziert erscheinen, erscheint die X-Dimension im Verhältnis $1 : \sqrt{1 - (v/V)^2}$ verkürzt, also um so stärker, je größer v ist. Für $v = V$ schrumpfen alle bewegten Objekte — vom „ruhenden" System aus betrachtet — in flächenhafte Gebilde zusammen. Für Überlichtgeschwindigkeiten werden unsere Überlegungen sinnlos; wir werden übrigens in den folgenden Betrachtungen finden, daß die Lichtgeschwindigkeit in unserer Theorie physikalisch die Rolle der unendlich großen Geschwindigkeiten spielt.

Es ist klar, daß die gleichen Resultate von im „ruhenden" System ruhenden Körpern gelten, welche von einem gleichförmig bewegten System aus betrachtet werden. —

Wir denken uns ferner eine der Uhren, welche relativ zum ruhenden System ruhend die Zeit t, relativ zum bewegten

---

1) Das heißt einen Körper, welcher ruhend untersucht Kugelgestalt besitzt.

904   *A. Einstein.*

System ruhend die Zeit τ anzugeben befähigt sind, im Koordi-
natenursprung von k gelegen und so gerichtet, daß sie die
Zeit τ angibt. Wie schnell geht diese Uhr, vom ruhenden
System aus betrachtet?

Zwischen die Größen x, t und τ, welche sich auf den Ort
dieser Uhr beziehen, gelten offenbar die Gleichungen:

$$\tau = \frac{1}{\sqrt{1-\left(\frac{v}{V}\right)^2}}\left(t-\frac{v}{V^2}x\right)$$

und

$$x = vt.$$

Es ist also

$$\tau = t\sqrt{1-\left(\frac{v}{V}\right)^2} = t-\left(1-\sqrt{1-\left(\frac{v}{V}\right)^2}\right)t,$$

woraus folgt, daß die Angabe der Uhr (im ruhenden System
betrachtet) pro Sekunde um $\left(1-\sqrt{1-(v/V)^2}\right)$ Sek. oder — bis
auf Größen vierter und höherer Ordnung um $\frac{1}{2}(v/V)^2$ Sek.
zurückbleibt.

Hieraus ergibt sich folgende eigentümliche Konsequenz.
Sind in den Punkten A und B von K ruhende, im ruhenden
System betrachtet, synchron gehende Uhren vorhanden, und
bewegt man die Uhr in A mit der Geschwindigkeit v auf der
Verbindungslinie nach B, so gehen nach Ankunft dieser Uhr
in B die beiden Uhren nicht mehr synchron, sondern die von A
nach B bewegte Uhr geht gegenüber der von Anfang an in B
befindlichen um $\frac{1}{2}tv^2/V^2$ Sek. (bis auf Größen vierter und
höherer Ordnung) nach, wenn t die Zeit ist, welche die Uhr
von A nach B braucht.

Man sieht sofort, daß dies Resultat auch dann noch gilt,
wenn die Uhr in einer beliebigen polygonalen Linie sich von A
nach B bewegt, und zwar auch dann, wenn die Punkte A
und B zusammenfallen.   [18]

Nimmt man an, daß das für eine polygonale Linie be-
wiesene Resultat auch für eine stetig gekrümmte Kurve gelte,
so erhält man den Satz: Befinden sich in A zwei synchron
gehende Uhren und bewegt man die eine derselben auf einer
geschlossenen Kurve mit konstanter Geschwindigkeit, bis sie
wieder nach A zurückkommt, was t Sek. dauern möge, so geht
die letztere Uhr bei ihrer Ankunft in A gegenüber der un-

bewegt gebliebenen um $\frac{1}{2} t (v/V)^2$ Sek. nach. Man schließt
[19] daraus, daß eine am Erdäquator befindliche Unruhuhr um einen
sehr kleinen Betrag langsamer laufen muß als eine genau
gleich beschaffene, sonst gleichen Bedingungen unterworfene,
an einem Erdpole befindliche Uhr.

§ 5. Additionstheorem der Geschwindigkeiten.

In dem längs der X-Achse des Systems K mit der Geschwindigkeit v bewegten System k bewege sich ein Punkt
gemäß den Gleichungen:

$$\xi = w_\xi \, \tau \,,$$

$$\eta = w_\eta \, \tau \,,$$

$$\zeta = 0 \,,$$

wobei w_ξ und w_η Konstanten bedeuten.

Gesucht ist die Bewegung des Punktes relativ zum System K.
Führt man in die Bewegungsgleichungen des Punktes mit Hilfe
der in § 3 entwickelten Transformationsgleichungen die Größen
x, y, z, t ein, so erhält man:

$$x = \frac{w_\xi + v}{1 + \dfrac{v\,w_\xi}{V^2}}\, t \,,$$

$$y = \frac{\sqrt{1 - \left(\dfrac{v}{V}\right)^2}}{1 + \dfrac{v\,w_\xi}{V^2}}\, w_\eta\, t \,,$$

$$z = 0 \,.$$

Das Gesetz vom Parallelogramm der Geschwindigkeiten gilt
also nach unserer Theorie nur in erster Annäherung. Wir
setzen:

$$U^2 = \left(\frac{d\,x}{d\,t}\right)^2 + \left(\frac{d\,y}{d\,t}\right)^2,$$

$$w^2 = w_\xi^2 + w_\eta^2$$

und

[20]

$$\alpha = \operatorname{arctg} \frac{w_y}{w_x} \,;$$

A. Einstein.

α ist dann als der Winkel zwischen den Geschwindigkeiten v und w anzusehen. Nach einfacher Rechnung ergibt sich:

$$U = \frac{\sqrt{(v^2 + w^2 + 2\,v\,w\cos\alpha) - \left(\dfrac{v\,w\sin\alpha}{V}\right)^2}}{1 + \dfrac{v\,w\cos\alpha}{V^2}}.$$

Es ist bemerkenswert, daß v und w in symmetrischer Weise in den Ausdruck für die resultierende Geschwindigkeit eingehen. Hat auch w die Richtung der X-Achse (Ξ-Achse), so erhalten wir:

$$U = \frac{v + w}{1 + \dfrac{v\,w}{V^2}}.$$

Aus dieser Gleichung folgt, daß aus der Zusammensetzung zweier Geschwindigkeiten, welche kleiner sind als V, stets eine Geschwindigkeit kleiner als V resultiert. Setzt man nämlich $v = V - \varkappa$, $w = V - \lambda$, wobei $\varkappa$ und λ positiv und kleiner als V seien, so ist:

$$U = V\,\frac{2\,V - \varkappa - \lambda}{2\,V - \varkappa - \lambda + \dfrac{\varkappa\,\lambda}{V}} < V.$$

Es folgt ferner, daß die Lichtgeschwindigkeit V durch Zusammensetzung mit einer „Unterlichtgeschwindigkeit" nicht geändert werden kann. Man erhält für diesen Fall:

$$U = \frac{V + w}{1 + \dfrac{w}{V}} = V.$$

Wir hätten die Formel für U für den Fall, daß v und w gleiche Richtung besitzen, auch durch Zusammensetzen zweier Transformationen gemäß § 3 erhalten können. Führen wir neben den in § 3 figurierenden Systemen K und k noch ein drittes, zu k in Parallelbewegung begriffenes Koordinatensystem k' ein, dessen Anfangspunkt sich auf der Ξ-Achse mit der Geschwindigkeit w bewegt, so erhalten wir zwischen den Größen x, y, z, t und den entsprechenden Größen von k' Gleichungen, welche sich von den in § 3 gefundenen nur dadurch unterscheiden, daß an Stelle von „v" die Größe

$$\frac{v + w}{1 + \dfrac{v\,w}{V^2}}$$

Zur Elektrodynamik bewegter Körper. 907

tritt; man sieht daraus, daß solche Paralleltransformationen —

[21] wie dies sein muß — eine Gruppe bilden.

Wir haben nun die für uns notwendigen Sätze der unseren zwei Prinzipien entsprechenden Kinematik hergeleitet und gehen dazu über, deren Anwendung in der Elektrodynamik zu zeigen.

II. Elektrodynamischer Teil.

§ 6. Transformation der Maxwell-Hertzschen Gleichungen für den leeren Raum. Über die Natur der bei Bewegung in einem Magnetfeld auftretenden elektromotorischen Kräfte.

Die Maxwell-Hertzschen Gleichungen für den leeren Raum mögen gültig sein für das ruhende System K, so daß

[22] gelten möge:

$$\frac{1}{V}\frac{\partial X}{\partial t} = \frac{\partial N}{\partial y} - \frac{\partial M}{\partial z}, \qquad \frac{1}{V}\frac{\partial L}{\partial t} = \frac{\partial Y}{\partial z} - \frac{\partial Z}{\partial y},$$

$$\frac{1}{V}\frac{\partial Y}{\partial t} = \frac{\partial L}{\partial z} - \frac{\partial N}{\partial x}, \qquad \frac{1}{V}\frac{\partial M}{\partial t} = \frac{\partial Z}{\partial x} - \frac{\partial X}{\partial z},$$

$$\frac{1}{V}\frac{\partial Z}{\partial t} = \frac{\partial M}{\partial x} - \frac{\partial L}{\partial y}, \qquad \frac{1}{V}\frac{\partial N}{\partial t} = \frac{\partial X}{\partial y} - \frac{\partial Y}{\partial x},$$

wobei (X, Y, Z) den **Vektor** der elektrischen, (L, M, N) den der magnetischen Kraft bedeutet.

Wenden wir auf diese Gleichungen die in § 3 entwickelte Transformation an, indem wir die elektromagnetischen Vorgänge auf das dort eingeführte, mit der Geschwindigkeit v bewegte Koordinatensystem beziehen, so erhalten wir die Gleichungen:

$$\frac{1}{V}\frac{\partial X}{\partial \tau} = \frac{\partial \beta\left(N - \frac{v}{V}\,Y\right)}{\partial \eta} - \frac{\partial \beta\left(M + \frac{v}{V}\,Z\right)}{\partial \zeta},$$

$$\frac{1}{V}\frac{\partial \beta\left(Y - \frac{v}{V}\,N\right)}{\partial \tau} = \frac{\partial L}{\partial \zeta} - \frac{\partial \beta\left(N - \frac{v}{V}\,Y\right)}{\partial \xi},$$

$$\frac{1}{V}\frac{\partial \beta\left(Z + \frac{v}{V}\,M\right)}{\partial \tau} = \frac{\partial \beta\left(M + \frac{v}{V}\,Z\right)}{\partial \xi} - \frac{\partial L}{\partial \eta},$$

$$\frac{1}{V}\frac{\partial L}{\partial \tau} = \frac{\partial \beta\left(Y - \frac{v}{V}\,N\right)}{\partial \zeta} - \frac{\partial \beta\left(Z + \frac{v}{V}\,M\right)}{\partial \eta},$$

908   *A. Einstein.*

$$\frac{1}{V}\frac{\partial\,\beta\left(M+\frac{v}{V}\,Z\right)}{\partial\,\tau} = \frac{\partial\,\beta\left(Z+\frac{v}{V}\,M\right)}{\partial\,\xi} - \frac{\partial\,X}{\partial\,\zeta}\,,$$

$$\frac{1}{V}\frac{\partial\,\beta\left(N-\frac{v}{V}\,Y\right)}{\partial\,\tau} = \frac{\partial\,X}{\partial\,\eta} - \frac{\partial\,\beta\left(Y-\frac{v}{V}\,N\right)}{\partial\,\xi}\,,$$

wobei

$$\beta = \frac{1}{\sqrt{1-\left(\frac{v}{V}\right)^2}}\,.$$

Das Relativitätsprinzip fordert nun, daß die **Maxwell-Hertz**schen Gleichungen für den leeren Raum auch im System *k* gelten, wenn sie im System *K* gelten, d. h. daß für die im bewegten System *k* durch ihre ponderomotorischen Wirkungen auf elektrische bez. magnetische Massen definierten Vektoren der elektrischen und magnetischen Kraft ($(X', Y'\, Z')$ und (L', M', N')) des bewegten Systems *k* die Gleichungen gelten:

$$\frac{1}{V}\frac{\partial\,X'}{\partial\,\tau} = \frac{\partial\,N'}{\partial\,\eta} - \frac{\partial\,M'}{\partial\,\zeta}\,,\qquad \frac{1}{V}\frac{\partial\,L'}{\partial\,\tau} = \frac{\partial\,Y'}{\partial\,\zeta} - \frac{\partial\,Z'}{\partial\,\eta}\,,$$

$$\frac{1}{V}\frac{\partial\,Y'}{\partial\,\tau} = \frac{\partial\,L'}{\partial\,\zeta} - \frac{\partial\,N'}{\partial\,\xi}\,,\qquad \frac{1}{V}\frac{\partial\,M'}{\partial\,\tau} = \frac{\partial\,Z'}{\partial\,\xi} - \frac{\partial\,X'}{\partial\,\zeta}\,,$$

$$\frac{1}{V}\frac{\partial\,Z'}{\partial\,\tau} = \frac{\partial\,M'}{\partial\,\xi} - \frac{\partial\,L'}{\partial\,\eta}\,,\qquad \frac{1}{V}\frac{\partial\,N'}{\partial\,\tau} = \frac{\partial\,X'}{\partial\,\eta} - \frac{\partial\,Y'}{\partial\,\xi}\,.$$

Offenbar müssen nun die beiden für das System *k* gefundenen Gleichungssysteme genau dasselbe ausdrücken, da beide Gleichungssysteme den **Maxwell-Hertz**schen Gleichungen für das System *K* äquivalent sind. Da die Gleichungen beider Systeme ferner bis auf die die Vektoren darstellenden Symbole übereinstimmen, so folgt, daß die in den Gleichungssystemen an entsprechenden Stellen auftretenden Funktionen bis auf einen für alle Funktionen des einen Gleichungssystems gemeinsamen, von ξ, η, ζ und τ unabhängigen, eventuell von v abhängigen Faktor $\psi(v)$ übereinstimmen müssen. Es gelten also die Beziehungen:

$$X' = \psi(v)\,X\,,\qquad\qquad L' = \psi(v)\,L\,,$$

$$Y' = \psi(v)\,\beta\left(Y-\frac{v}{V}\,N\right)\,,\qquad M' = \psi(v)\,\beta\left(M+\frac{v}{V}\,Z\right)\,,$$

$$Z' = \psi(v)\,\beta\left(Z+\frac{v}{V}\,M\right)\,,\qquad N' = \psi(v)\,\beta\left(N-\frac{v}{V}\,Y\right)\,.$$

Bildet man nun die Umkehrung dieses Gleichungssystems, erstens durch Auflösen der soeben erhaltenen Gleichungen, zweitens durch Anwendung der Gleichungen auf die inverse Transformation (von k auf K), welche durch die Geschwindigkeit $-v$ charakterisiert ist, so folgt, indem man berücksichtigt, daß die beiden so erhaltenen Gleichungssysteme identisch sein müssen:

$$\varphi(v) \cdot \varphi(-v) = 1.$$

Ferner folgt aus Symmetriegründen[1])

$$\varphi(v) = \varphi(-v);$$

es ist also

$$\varphi(v) = 1,$$

und unsere Gleichungen nehmen die Form an:

$$X' = X, \qquad\qquad L' = L,$$
$$Y' = \beta\left(Y - \frac{v}{V}N\right), \quad M' = \beta\left(M + \frac{v}{V}Z\right),$$
$$Z' = \beta\left(Z + \frac{v}{V}M\right), \quad N' = \beta\left(N - \frac{v}{V}Y\right).$$

Zur Interpretation dieser Gleichungen bemerken wir folgendes. Es liegt eine punktförmige Elektrizitätsmenge vor, welche im ruhenden System K gemessen von der Größe „eins" sei, d. h. im ruhenden System ruhend auf eine gleiche Elektrizitätsmenge im Abstand 1 cm die Kraft 1 Dyn ausübe. Nach dem Relativitätsprinzip ist diese elektrische Masse auch im bewegten System gemessen von der Größe „eins". Ruht diese Elektrizitätsmenge relativ zum ruhenden System, so ist definitionsgemäß der Vektor (X, Y, Z) gleich der auf sie wirkenden Kraft. Ruht die Elektrizitätsmenge gegenüber dem bewegten System (wenigstens in dem betreffenden Augenblick), so ist die auf sie wirkende, in dem bewegten System gemessene Kraft gleich dem Vektor (X', Y', Z'). Die ersten drei der obigen Gleichungen lassen sich mithin auf folgende zwei Weisen in Worte kleiden:

[23]

1. Ist ein punktförmiger elektrischer Einheitspol in einem elektromagnetischen Felde bewegt, so wirkt auf ihn außer der

---

1) Ist z. B. $X = Y = Z = L = M = 0$ und $N \neq 0$, so ist aus Symmetriegründen klar, daß bei Zeichenwechsel von v ohne Änderung des numerischen Wertes auch Y' sein Vorzeichen ändern muß, ohne seinen numerischen Wert zu ändern.

910   *A. Einstein.*

elektrischen Kraft eine „elektromotorische Kraft", welche unter   [24]
Vernachlässigung von mit der zweiten und höheren Potenzen
von v/V multiplizierten Gliedern gleich ist dem mit der
Lichtgeschwindigkeit dividierten Vektorprodukt der Bewegungs-
geschwindigkeit des Einheitspoles und der magnetischen Kraft.
(Alte Ausdrucksweise.)

2. Ist ein punktförmiger elektrischer Einheitspol in einem
elektromagnetischen Felde bewegt, so ist die auf ihn wirkende
Kraft gleich der an dem Orte des Einheitspoles vorhandenen
elektrischen Kraft, welche man durch Transformation des Feldes
auf ein relativ zum elektrischen Einheitspol ruhendes Koordinatensystem erhält. (Neue Ausdrucksweise.)

Analoges gilt über die „magnetomotorischen Kräfte". Man   [25]
sieht, daß in der entwickelten Theorie die elektromotorische
Kraft nur die Rolle eines Hilfsbegriffes spielt, welcher seine
Einführung dem Umstande verdankt, daß die elektrischen und
magnetischen Kräfte keine von dem Bewegungszustande des
Koordinatensystems unabhängige Existenz besitzen.

Es ist ferner klar, daß die in der Einleitung angeführte   [26]
Asymmetrie bei der Betrachtung der durch Relativbewegung
eines Magneten und eines Leiters erzeugten Ströme verschwindet.
Auch werden die Fragen nach dem „Sitz" der elektrodynamischen
elektromotorischen Kräfte (Unipolarmaschinen) gegenstandslos.   [27]

§ 7. **Theorie des Doppelerschen Prinzips und der Aberration.**

Im Systeme K befinde sich sehr ferne vom Koordinatenursprung eine Quelle elektrodynamischer Wellen, welche in
einem den Koordinatenursprung enthaltenden Raumteil mit
genügender Annäherung durch die Gleichungen dargestellt sei:

$$X = X_0 \sin \Phi, \quad L = L_0 \sin \Phi,$$
$$Y = Y_0 \sin \Phi, \quad M = M_0 \sin \Phi, \quad \Phi = \omega \left(t - \frac{a x + b y + c z}{V} \right).$$
$$Z = Z_0 \sin \Phi, \quad N = N_0 \sin \Phi,$$

Hierbei sind (X_0, Y_0, Z_0) und (L_0, M_0, N_0) die Vektoren, welche
die Amplitude des Wellenzuges bestimmen, a, b, c die Richtungskosinus der Wellennormalen.

Wir fragen nach der Beschaffenheit dieser Wellen, wenn
dieselben von einem in dem bewegten System k ruhenden

Beobachter untersucht werden. — Durch Anwendung der in
§ 6 gefundenen Transformationsgleichungen für die elektrischen
und magnetischen Kräfte und der in § 3 gefundenen Transformationsgleichungen für die Koordinaten und die Zeit erhalten wir unmittelbar:

$$X' = \qquad X_0 \sin \Phi', \qquad L' = \qquad L_0 \sin \Phi',$$

$$Y' = \beta \left(Y_0 - \frac{v}{V} N_0 \right) \sin \Phi', \qquad M' = \beta \left(M_0 + \frac{v}{V} Z_0 \right) \sin \Phi',$$

$$Z' = \beta \left(Z_0 + \frac{v}{V} M_0 \right) \sin \Phi', \qquad N' = \beta \left(N_0 - \frac{v}{V} Y_0 \right) \sin \Phi',$$

$$\Phi' = \omega' \left(\tau - \frac{a' \, \xi + b' \, \eta + c' \, \zeta}{V} \right),$$

wobei

$$\omega' = \omega \, \beta \left(1 - a \frac{v}{V} \right),$$

$$a' = \frac{a - \dfrac{v}{V}}{1 - a \dfrac{v}{V}},$$

$$b' = \frac{b}{\beta \left(1 - a \dfrac{v}{V} \right)},$$

$$c' = \frac{c}{\beta \left(1 - a \dfrac{v}{V} \right)}$$

gesetzt ist.

Aus der Gleichung für ω' folgt: Ist ein Beobachter relativ
zu einer unendlich fernen Lichtquelle von der Frequenz v mit
der Geschwindigkeit v derart bewegt, daß die Verbindungslinie „Lichtquelle–Beobachter" mit der auf ein relativ zur
Lichtquelle ruhendes Koordinatensystem bezogenen Geschwindigkeit des Beobachters den Winkel φ bildet, so ist die von
dem Beobachter wahrgenommene Frequenz v' des Lichtes
durch die Gleichung gegeben:

$$v' = v \frac{1 - \cos \varphi \, \dfrac{v}{V}}{\sqrt{1 - \left(\dfrac{v}{V} \right)^2}}.$$

Dies ist das **Doppelersche Prinzip** für beliebige Geschwindigkeiten. Für $\varphi = 0$ nimmt die Gleichung die übersichtliche Form an:

$$\nu' = \nu \sqrt{\frac{1 - \dfrac{v}{V}}{1 + \dfrac{v}{V}}} \,.$$

Man sieht, daß — im Gegensatz zu der üblichen Auffassung — für $v = -\infty$, $\nu = \infty$ ist.   [28]

Nennt man φ' den Winkel zwischen Wellennormale (Strahlrichtung) im bewegten System und der Verbindungslinie „Lichtquelle–Beobachter", so nimmt die Gleichung für a' die Form an: [29]

$$\cos \varphi' = \frac{\cos \varphi - \dfrac{v}{V}}{1 - \dfrac{v}{V} \cos \varphi} \,.$$

Diese Gleichung drückt das Aberrationsgesetz in seiner allgemeinsten Form aus. Ist $\varphi = \pi/2$, so nimmt die Gleichung die einfache Gestalt an:

$$\cos \varphi' = -\frac{v}{V} \,.$$

Wir haben nun noch die Amplitude der Wellen, wie dieselbe im bewegten System erscheint, zu suchen. Nennt man A bez. A' die Amplitude der elektrischen oder magnetischen Kraft im ruhenden bez. im bewegten System gemessen, so erhält man:

$$A'^2 = A^2 \frac{\left(1 - \dfrac{v}{V} \cos \varphi\right)^2}{1 - \left(\dfrac{v}{V}\right)^2} \,,$$ [30]

welche Gleichung für $\varphi = 0$ in die einfachere übergeht:

$$A'^2 = A^2 \frac{1 - \dfrac{v}{V}}{1 + \dfrac{v}{V}} \,.$$

Es folgt aus den entwickelten Gleichungen, daß für einen Beobachter, der sich mit der Geschwindigkeit V einer Lichtquelle näherte, diese Lichtquelle unendlich intensiv erscheinen müßte.

Zur Elektrodynamik bewegter Körper.   913

§ 8. Transformation der Energie der Lichtstrahlen. Theorie des auf vollkommene Spiegel ausgeübten Strahlungsdruckes.

Da $A^2/8\pi$ gleich der Lichtenergie pro Volumeneinheit ist, so haben wir nach dem Relativitätsprinzip $A'^2/8\pi$ als die Lichtenergie im bewegten System zu betrachten. Es wäre daher A'^2/A^2 das Verhältnis der „bewegt gemessenen" und [31] „ruhend gemessenen" Energie eines bestimmten Lichtkomplexes, wenn das Volumen eines Lichtkomplexes in K gemessen und in k gemessen das gleiche wäre. Dies ist jedoch nicht der Fall. Sind a, b, c die Richtungskosinus der Wellennormalen des Lichtes im ruhenden System, so wandert durch die Oberflächenelemente der mit Lichtgeschwindigkeit bewegten Kugelfläche

$$(x - Va\,t)^2 + (y - Vb\,t)^2 + (z - Vc\,t)^2 = R^2$$

keine Energie hindurch; wir können daher sagen, daß diese Fläche dauernd denselben Lichtkomplex umschließt. Wir fragen nach der Energiemenge, welche diese Fläche im System k betrachtet umschließt, d. h. nach der Energie des Lichtkomplexes relativ zum System k.

Die Kugelfläche ist — im bewegten System betrachtet — eine Ellipsoidfläche, welche zur Zeit $\tau = 0$ die Gleichung besitzt:

$$\left(\beta\,\xi - a\,\beta\,\frac{v}{V}\,\xi\right)^2 + \left(\eta - b\,\beta\,\frac{v}{V}\,\xi\right)^2 + \left(\zeta - c\,\beta\,\frac{v}{V}\,\xi\right)^2 = R^2.$$

Nennt man S das Volumen der Kugel, S' dasjenige dieses Ellipsoides, so ist, wie eine einfache Rechnung zeigt:

$$\frac{S'}{S} = \frac{\sqrt{1 - \left(\frac{v}{V}\right)^2}}{1 - \frac{v}{V}\cos\varphi}.$$

Nennt man also E die im ruhenden System gemessene, E' die im bewegten System gemessene Lichtenergie, welche von der betrachteten Fläche umschlossen wird, so erhält man:

$$\frac{E'}{E} = \frac{\dfrac{A'^2}{8\pi}S'}{\dfrac{A^2}{8\pi}S} = \frac{1 - \dfrac{v}{V}\cos\varphi}{\sqrt{1 - \left(\dfrac{v}{V}\right)^2}},$$

914   *A. Einstein.*

welche Formel für $\varphi = 0$ in die einfachere übergeht:

$$\frac{E'}{E} = \sqrt{\frac{1 - \frac{v}{V}}{1 + \frac{v}{V}}}\,.$$

Es ist bemerkenswert, daß die Energie und die Frequenz eines Lichtkomplexes sich nach demselben Gesetze mit dem Bewegungszustande des Beobachters ändern.

Es sei nun die Koordinatenebene $\xi = 0$ eine vollkommen spiegelnde Fläche, an welcher die im letzten Paragraph betrachteten ebenen Wellen reflektiert werden. Wir fragen nach dem auf die spiegelnde Fläche ausgeübten Lichtdruck und nach der Richtung, Frequenz und Intensität des Lichtes nach der Reflexion.   [32]

Das einfallende Licht sei durch die Größen A, $\cos\varphi$, ν (auf das System K bezogen) definiert. Von k aus betrachtet sind die entsprechenden Größen:

$$A' = A\,\frac{1 - \frac{v}{V}\cos\varphi}{\sqrt{1 - \left(\frac{v}{V}\right)^2}}\,,$$

$$\cos\varphi' = \frac{\cos\varphi - \frac{v}{V}}{1 - \frac{v}{V}\cos\varphi}\,,$$

$$\nu' = \nu\,\frac{1 - \frac{v}{V}\cos\varphi}{\sqrt{1 - \left(\frac{v}{V}\right)^2}}\,.$$

Für das reflektierte Licht erhalten wir, wenn wir den Vorgang auf das System k beziehen:

$$A'' = A'\,,$$

$$\cos\varphi'' = -\cos\varphi'\,,$$

$$\nu'' = \nu'\,.$$

Endlich erhält man durch Rücktransformieren aufs ruhende System K für das reflektierte Licht:

Zur Elektrodynamik bewegter Körper.   915

$$A''' = A'' \frac{1 + \frac{v}{V} \cos \varphi''}{\sqrt{1 - \left(\frac{v}{V}\right)^2}} = A \frac{1 - 2\frac{v}{V} \cos \varphi + \left(\frac{v}{V}\right)^2}{1 - \left(\frac{v}{V}\right)^2},$$

$$\cos \varphi''' = \frac{\cos \varphi'' + \frac{v}{V}}{1 + \frac{v}{V} \cos \varphi''} = - \frac{\left(1 + \left(\frac{v}{V}\right)^2\right) \cos \varphi - 2\frac{v}{V}}{1 - 2\frac{v}{V} \cos \varphi + \left(\frac{v}{V}\right)^2},$$

[33]
$$v''' = v'' \frac{1 + \frac{v}{V} \cos \varphi''}{\sqrt{1 - \left(\frac{v}{V}\right)^2}} = v \frac{1 - 2\frac{v}{V} \cos \varphi + \left(\frac{v}{V}\right)^2}{\left(1 - \frac{v}{V}\right)^2}.$$

Die auf die Flächeneinheit des Spiegels pro Zeiteinheit auftreffende (im ruhenden System gemessene) Energie ist offenbar $A^2/8\pi$ ($V \cos \varphi - v$). Die von der Flächeneinheit des Spiegels in der Zeiteinheit sich entfernende Energie ist $A'''^2/8\pi$ ($-V \cos \varphi''' + v$). Die Differenz dieser beiden Ausdrücke ist nach dem Energieprinzip die vom Lichtdrucke in der Zeiteinheit geleistete Arbeit. Setzt man die letztere gleich dem Produkt $P \cdot v$, wobei P der Lichtdruck ist, so erhält man:

[34]
$$P = 2\,\frac{A^2}{8\pi}\,\frac{\left(\cos \varphi - \frac{v}{V}\right)^2}{1 - \left(\frac{v}{V}\right)^2}.$$

In erster Annäherung erhält man in Übereinstimmung mit der
[35] Erfahrung und mit anderen Theorien

$$P = 2\,\frac{A^2}{8\pi}\,\cos^2 \varphi.$$

Nach der hier benutzten Methode können alle Probleme der Optik bewegter Körper gelöst werden. Das Wesentliche ist, daß die elektrische und magnetische Kraft des Lichtes, welches durch einen bewegten Körper beeinflußt wird, auf ein relativ zu dem Körper ruhendes Koordinatensystem transformiert werden. Dadurch wird jedes Problem der Optik bewegter Körper auf eine Reihe von Problemen der Optik ruhender Körper zurückgeführt.

916   *A. Einstein.*

§ 9. Transformation der **Maxwell-Hertz**schen Gleichungen mit Berücksichtigung der Konvektionsströme.

Wir gehen aus von den Gleichungen:

$$\frac{1}{V}\left\{u_x \varrho + \frac{\partial X}{\partial t}\right\} = \frac{\partial N}{\partial y} - \frac{\partial M}{\partial z}, \quad \frac{1}{V}\frac{\partial L}{\partial t} = \frac{\partial Y}{\partial z} - \frac{\partial Z}{\partial y},$$

$$\frac{1}{V}\left\{u_y \varrho + \frac{\partial Y}{\partial t}\right\} = \frac{\partial L}{\partial z} - \frac{\partial N}{\partial x}, \quad \frac{1}{V}\frac{\partial M}{\partial t} = \frac{\partial Z}{\partial x} - \frac{\partial X}{\partial z},$$

$$\frac{1}{V}\left\{u_z \varrho + \frac{\partial Z}{\partial t}\right\} = \frac{\partial M}{\partial x} - \frac{\partial L}{\partial y}, \quad \frac{1}{V}\frac{\partial N}{\partial t} = \frac{\partial X}{\partial y} - \frac{\partial Y}{\partial x},$$

wobei

$$\varrho = \frac{\partial X}{\partial x} + \frac{\partial Y}{\partial y} + \frac{\partial Z}{\partial z}$$

die 4π-fache Dichte der Elektrizität und (u_x, u_y, u_z) den Geschwindigkeitsvektor der Elektrizität bedeutet. Denkt man sich die elektrischen Massen unveränderlich an kleine, starre Körper (Ionen, Elektronen) gebunden, so sind diese Gleichungen die elektromagnetische Grundlage der **Lorentz**schen Elektrodynamik und Optik bewegter Körper.   [36]

Transformiert man diese Gleichungen, welche im System K gelten mögen, mit Hilfe der Transformationsgleichungen von § 3 und § 6 auf das System k, so erhält man die Gleichungen:

$$\frac{1}{V}\left\{u_\xi \varrho' + \frac{\partial X'}{\partial \tau}\right\} = \frac{\partial N'}{\partial \eta} - \frac{\partial M'}{\partial \zeta}, \quad \frac{\partial L'}{\partial \tau} = \frac{\partial Y'}{\partial \zeta} - \frac{\partial Z'}{\partial \eta},$$

$$\frac{1}{V}\left\{u_\eta \varrho' + \frac{\partial Y'}{\partial \tau}\right\} = \frac{\partial L'}{\partial \zeta} - \frac{\partial N'}{\partial \xi}, \quad \frac{\partial M'}{\partial \tau} = \frac{\partial Z'}{\partial \xi} - \frac{\partial X'}{\partial \zeta},$$

$$\frac{1}{V}\left\{u_\zeta \varrho' + \frac{\partial Z'}{\partial \tau}\right\} = \frac{\partial M'}{\partial \xi} - \frac{\partial L'}{\partial \eta}, \quad \frac{\partial N'}{\partial \tau} = \frac{\partial X'}{\partial \eta} - \frac{\partial Y'}{\partial \xi},$$

wobei

$$\frac{u_x - v}{1 - \dfrac{u_x v}{V^2}} = u_\xi,$$

$$\frac{u_y}{\beta\left(1 - \dfrac{u_x v}{V^2}\right)} = u_\eta, \quad \varrho' = \frac{\partial X'}{\partial \xi} + \frac{\partial Y'}{\partial \eta} + \frac{\partial Z'}{\partial \zeta} = \beta\left(1 - \frac{v\,u_x}{V^2}\right)\varrho.$$

$$\frac{u_z}{\beta\left(1 - \dfrac{u_x v}{V^2}\right)} = u_\zeta.$$

Da — wie aus dem Additionstheorem der Geschwindigkeiten
(§ 5) folgt — der Vektor (u_ξ, u_η, u_ζ) nichts anderes ist als
die Geschwindigkeit der elektrischen Massen im System k gemessen, so ist damit gezeigt, daß unter Zugrundelegung unserer
kinematischen Prinzipien die elektrodynamische Grundlage der
Lorentzschen Theorie der Elektrodynamik bewegter Körper
dem Relativitätsprinzip entspricht.

Es möge noch kurz bemerkt werden, daß aus den entwickelten Gleichungen leicht der folgende wichtige Satz gefolgert werden kann: Bewegt sich ein elektrisch geladener
Körper beliebig im Raume und ändert sich hierbei seine
Ladung nicht, von einem mit dem Körper bewegten Koordinatensystem aus betrachtet, so bleibt seine Ladung auch —
[37]   von dem „ruhenden“ System K aus betrachtet — konstant.

§ 10. Dynamik des (langsam beschleunigten) Elektrons.

In einem elektromagnetischen Felde bewege sich ein punkt-
förmiges, mit einer elektrischen Ladung ε versehenes Teilchen
(im folgenden „Elektron“ genannt), über dessen Bewegungsgesetz wir nur folgendes annehmen:

Ruht das Elektron in einer bestimmten Epoche, so erfolgt
in dem nächsten Zeitteilchen die Bewegung des Elektrons nach
den Gleichungen

$$\mu \frac{d^2 x}{d t^2} = \varepsilon X$$

$$\mu \frac{d^2 y}{d t^2} = \varepsilon Y$$

$$\mu \frac{d^2 z}{d t^2} = \varepsilon Z,$$

wobei x, y, z die Koordinaten des Elektrons, μ die Masse
des Elektrons bedeutet, sofern dasselbe langsam bewegt ist.

Es besitze nun zweitens das Elektron in einer gewissen
Zeitepoche die Geschwindigkeit v. Wir suchen das Gesetz,
nach welchem sich das Elektron im unmittelbar darauf folgenden Zeitteilchen bewegt.

Ohne die Allgemeinheit der Betrachtung zu beeinflussen,
können und wollen wir annehmen, daß das Elektron in dem
Momente, wo wir es ins Auge fassen, sich im Koordinaten-

918   *A. Einstein.*

sprung befinde und sich längs der X-Achse des Systems K mit [38]
der Geschwindigkeit v bewege. Es ist dann einleuchtend, daß
das Elektron im genannten Momente (t = 0) relativ zu einem
längs der X-Achse mit der konstanten Geschwindigkeit v
parallelbewegten Koordinatensystem k ruht.

Aus der oben gemachten Voraussetzung in Verbindung
mit dem Relativitätsprinzip ist klar, daß sich das Elektron in
der unmittelbar folgenden Zeit (für kleine Werte von t) vom
System k aus betrachtet nach den Gleichungen bewegt:

$$\mu \frac{d^2 \xi}{d\tau^2} = \varepsilon X',$$

$$\mu \frac{d^2 \eta}{d\tau^2} = \varepsilon Y',$$

$$\mu \frac{d^2 \zeta}{d\tau^2} = \varepsilon Z',$$

wobei die Zeichen ξ, η, ζ, τ, X', Y', Z' sich auf das System k
beziehen. Setzen wir noch fest, daß für $t = x = y = z = 0$
$\tau = \xi = \eta = \zeta = 0$ sein soll, so gelten die Transformationsgleichungen der §§ 3 und 6, so daß gilt:

$$\tau = \beta \left(t - \frac{v}{V^2} x \right),$$

$$\xi = \beta (x - vt), \qquad X' = X,$$

$$\eta = y, \qquad\qquad Y' = \beta \left(Y - \frac{v}{V} N \right),$$

$$\zeta = z, \qquad\qquad Z' = \beta \left(Z + \frac{v}{V} M \right).$$

Mit Hilfe dieser Gleichungen transformieren wir die obigen
Bewegungsgleichungen vom System k auf das System K und
erhalten: [39]

$$\textbf{(A)} \quad \begin{cases} \dfrac{d^2 x}{dt^2} = \dfrac{\varepsilon}{\mu} \dfrac{1}{\beta^3} X, \\[2mm] \dfrac{d^2 y}{dt^2} = \dfrac{\varepsilon}{\mu} \dfrac{1}{\beta} \left(Y - \dfrac{v}{V} N \right), \\[2mm] \dfrac{d^2 z}{dt^2} = \dfrac{\varepsilon}{\mu} \dfrac{1}{\beta} \left(Z + \dfrac{v}{V} M \right). \end{cases}$$

Wir fragen nun in Anlehnung an die übliche Betrachtungs-
weise nach der „longitudinalen" und „transversalen" Masse [40]

des bewegten Elektrons. Wir schreiben die Gleichungen (A) in der Form

$$\mu \beta^3 \frac{d^2 x}{d t^2} = \varepsilon X = \varepsilon X',$$

$$\mu \beta^2 \frac{d^2 y}{d t^2} = \varepsilon \beta \left(Y - \frac{v}{V} N \right) = \varepsilon Y',$$

$$\mu \beta^2 \frac{d^2 z}{d t^2} = \varepsilon \beta \left(Z + \frac{v}{V} M \right) = \varepsilon Z'$$

und bemerken zunächst, daß $\varepsilon X'$, $\varepsilon Y'$, $\varepsilon Z'$ die Komponenten der auf das Elektron wirkenden ponderomotorischen Kraft sind, und zwar in einem in diesem Moment mit dem Elektron mit gleicher Geschwindigkeit wie dieses bewegten System betrachtet. (Diese Kraft könnte beispielsweise mit einer im letzten System ruhenden Federwage gemessen werden.) Wenn wir nun diese Kraft schlechtweg „die auf das Elektron wirkende Kraft" [41] nennen und die Gleichung

Massenzahl $\times$ Beschleunigungszahl = Kraftzahl

aufrechterhalten, und wenn wir ferner festsetzen, daß die Beschleunigungen im ruhenden System K gemessen werden sollen, so erhalten wir aus obigen Gleichungen:

$$\text{Longitudinale Masse} = \frac{\mu}{\left(\sqrt{1 - \left(\frac{v}{V} \right)^2} \right)^3},$$

$$\text{Transversale Masse} = \frac{\mu}{1 - \left(\frac{v}{V} \right)^2}.$$

Natürlich würde man bei anderer Definition der Kraft und der Beschleunigung andere Zahlen für die Massen erhalten; man ersieht daraus, daß man bei der Vergleichung ver- [42] schiedener Theorien der Bewegung des Elektrons sehr vorsichtig verfahren muß.

Wir bemerken, daß diese Resultate über die Masse auch für die ponderabeln materiellen Punkte gilt; denn ein ponderabler materieller Punkt kann durch Zufügen einer *beliebig kleinen* elektrischen Ladung zu einem Elektron (in unserem Sinne) gemacht werden.

Wir bestimmen die kinetische Energie des Elektrons. Bewegt sich ein Elektron vom Koordinatenursprung des Systems K aus mit der Anfangsgeschwindigkeit 0 beständig auf der

A. Einstein.

X-Achse unter der Wirkung einer elektrostatischen Kraft *X*, so ist klar, daß die dem elektrostatischen Felde entzogene Energie den Wert $\int \varepsilon X \, dx$ hat. Da das Elektron langsam beschleunigt sein soll und infolgedessen keine Energie in Form von Strahlung abgeben möge, so muß die dem elektrostatischen Felde entzogene Energie gleich der Bewegungsenergie *W* des Elektrons gesetzt werden. Man erhält daher, indem man beachtet, daß während des ganzen betrachteten Bewegungsvorganges die erste der Gleichungen (A) gilt:

$$W = \int \varepsilon X \, dx = \int_{0}^{v} \beta^3 v \, dv = \mu V^2 \left\{ \frac{1}{\sqrt{1 - \left(\frac{v}{V}\right)^2}} - 1 \right\}.$$ [43]

W wird also für $v = V$ unendlich groß. Überlichtgeschwindigkeiten haben — wie bei unseren früheren Resultaten — keine Existenzmöglichkeit. [44]

Auch dieser Ausdruck für die kinetische Energie muß dem oben angeführten Argument zufolge ebenso für ponderable Massen gelten.

Wir wollen nun die aus dem Gleichungssystem (A) resultierenden, dem Experimente zugänglichen Eigenschaften der Bewegung des Elektrons aufzählen.

1. Aus der zweiten Gleichung des Systems (A) folgt, daß eine elektrische Kraft *Y* und eine magnetische Kraft *N* dann gleich stark ablenkend wirken auf ein mit der Geschwindigkeit *v* bewegtes Elektron, wenn $Y = N \cdot v/V$. Man ersieht also, daß die Ermittelung der Geschwindigkeit des Elektrons aus dem Verhältnis der magnetischen Ablenkbarkeit A_m und der elektrischen Ablenkbarkeit A_e nach unserer Theorie für beliebige Geschwindigkeiten möglich ist durch Anwendung des Gesetzes: [45]

$$\frac{A_m}{A_e} = \frac{v}{V}.$$

Diese Beziehung ist der Prüfung durch das Experiment zugänglich, da die Geschwindigkeit des Elektrons auch direkt, z. B. mittels rasch oszillierender elektrischer und magnetischer Felder, gemessen werden kann.

2. Aus der Ableitung für die kinetische Energie des Elektrons folgt, daß zwischen der durchlaufenen Potential-

Zur Elektrodynamik bewegter Körper.   921

differenz und der erlangten Geschwindigkeit v des Elektrons die Beziehung gelten muß:

$$P = \int X\,dx = \frac{\mu}{\varepsilon}\,V^2 \left\{ \frac{1}{\sqrt{1-\left(\frac{v}{V}\right)^2}} - 1 \right\}.$$

3. Wir berechnen den Krümmungsradius R der Bahn, wenn eine senkrecht zur Geschwindigkeit des Elektrons wirkende magnetische Kraft N (als einzige ablenkende Kraft) vorhanden ist. Aus der zweiten der Gleichungen (A) erhalten wir:

$$-\frac{d^2 y}{dt^2} = \frac{v^2}{R} = \frac{\varepsilon}{\mu}\,\frac{r}{V}\,N\cdot\sqrt{1-\left(\frac{v}{V}\right)^2}$$

oder

$$R = V^2\,\frac{\mu}{\varepsilon}\cdot\frac{\frac{v}{V}}{\sqrt{1-\left(\frac{v}{V}\right)^2}}\cdot\frac{1}{N}.$$

[46]   Diese drei Beziehungen sind ein vollständiger Ausdruck für die Gesetze, nach denen sich gemäß vorliegender Theorie das Elektron bewegen muß.

Zum Schlusse bemerke ich, daß mir beim Arbeiten an dem hier behandelten Probleme mein Freund und Kollege M. Besso treu zur Seite stand und daß ich demselben manche [47]   wertvolle Anregung verdanke.

Bern, Juni 1905.

(Eingegangen 30. Juni 1905.)

Published in *Annalen der Physik* 17 (1905): 891–921. Dated Bern, June 1905, received 30 June 1905, published 26 September 1905.

[1] As pointed out in *Holton 1967*, this example is discussed at length in *Föppl 1894* (see pp. 307–330, especially § 117 and § 119). For reports that Einstein read *Föppl 1894* while at the ETH, see *Kayser 1930*, p. 49, and *Frank 1979*, p. 38. A drastically modified discussion of electromagnetic induction and relative motion appears in Abraham's edition of Föppl's boo'

Abraham/Föppl 1904 (see § 4, chap. 2, especially § 85).

[2] For Einstein's earlier reading of *Wien 1898*, which discusses a number of such failed attempts, see Vol. 1, the editorial note, "Einstein on the Electrodynamics of Moving Bodies," p. 224. *Poincaré 1902*, which Einstein read and discussed with his friends of the Olympia Academy (see the Introduction, p. xxv), also refers to the failure of such attempts (see pp. 200–202).

[3] Einstein is presumably referring to the proof in *Lorentz 1895*, pp. 91–95. In the 1913 reprint of this paper, a note was added after "Ordnung": "The preceding paper by H. A. Lorentz [a German translation of *Lorentz 1904a*] was not yet known to the author" ("Die im Vorhergehenden abgedruckte Arbeit von H. A. Lorentz war dem Verfasser noch nicht bekannt") (*Blumenthal 1913*, p. 27). Einstein later stated that in 1905 he knew "only Lorentz's important 1895 work, but not Lorentz's later work" ("nur Lorentz bedeutendes Werk von 1895, aber nicht Lorentz' spätere Arbeit") (Einstein to Carl Seelig, 19 February 1955). In this paper, Einstein's notation for the velocity of light and many other quantities agrees with that used in *Lorentz 1895*.

[4] Einstein encountered the "principle of relative motions" ("Prinzip der relativen Bewegungen") a decade earlier in *Violle 1892* (see the editorial note, "Einstein on the Theory of Relativity," pp. 258–259). More recently (see note 2), he had read *Poincaré 1902*, which discusses "the principle of relative movement" ("le principe du mouvement relatif") on pp. 135–137 (see note 12 for Poincaré's definition of this principle). In discussing whether Lorentz's theory contradicts the principle, Poincaré used the phrase "le principe de relativité" (ibid., p. 281), translated into German as "das Prinzip der Relativität" (*Poincaré 1904a*, p. 243). There is a more extensive discussion of the principle and its relation to Lorentz's theory in *Poincaré 1904b*, pp. 310–312. *Abraham/Föppl 1904* mentions the "principle of relative motion" ("Prinzip der relativen Bewegung") on p. 398.

[5] See pp. 900–901 for Einstein's proof that this assumption is consistent with the relativity principle. In 1904, Wien and Abraham stressed the importance of the constancy of the velocity of light in electrodynamics. Wien rejected Cohn's electrodynamics of moving media (see *Cohn 1902*) because it violates "the constancy of the velocity of light and hence the foundations of Maxwell's theory. . . . For, once it has left the radiation source, it should not affect the velocity of propagation of radiation whether the source has moved or not" ("die Konstanz der Lichtgeschwindigkeit und damit die Grundlagen der Maxwellschen Theorie. . . . Denn es muß für die Fortpflanzungsgeschwindigkeit einer einmal von der Strahlungsquelle losgelösten Strahlung gleichgültig sein, ob sich diese bewegt hat oder nicht" (*Wien 1904*, p. 643). Abraham

stated: "The electromagnetic theory of light refers to an absolute motion of light, which proceeds in every direction with the same velocity (*c*)" ("Die elektromagnetische Lichttheorie spricht von einer absoluten Bewegung des Lichtes, die nach jeder Richtung hin mit derselben Geschwindigkeit (*c*) erfolgt") (*Abraham 1904b*, p. 238).

[6] See the editorial note, "Einstein on the Theory of Relativity," pp. 260–261, for Einstein's reading of books by Mill, Ostwald, and Poincaré that question the existence of the ether. In the years just prior to 1905, several physicists suggested the elimination of the ether from electrodynamics. Bucherer cited evidence that optical phenomena only depend on the relative motion of matter with respect to other matter (see *Bucherer 1903*, p. 282; *Bucherer 1904*, pp. 128, 130–131). Cohn noted there is no need to introduce an ether in his theory; it suffices to assume that electromagnetic energy can propagate in matter-free space (see *Cohn 1902*, p. 55).

[7] Einstein is evidently contrasting his approach to the electrodynamics of moving bodies with those of Lorentz and Hertz, respectively. For earlier comments on Hertz's electrodynamics, see Einstein to Mileva Marić, 10 August 1899 (Vol. 1, Doc. 52). For later comments on Lorentz and Hertz, see, e.g., *Einstein 1910a*, pp. 7–10, and *Einstein 1921a*, p. 782.

[8] In the 1913 reprint, a note was added after "gelten": "What is meant is, 'be valid in the first approximation' ") ("Gemeint ist: „in erster Annäherung gelten" ") (*Blumenthal 1913*, p. 28). The authorship of these unsigned additional notes is unclear (see *McCausland 1984*). The title page of *Blumenthal 1913* says "A Collection of Papers with Annotations by A. Sommerfeld" ("Eine Sammlung von Abhandlungen mit Anmerkungen von A. Sommerfeld"). However, the extensive notes by Sommerfeld on *Minkowski 1909*, printed after the reprint of that paper, are listed separately, under Sommerfeld's name, in the table of contents. Lorentz signed a note he added to the German translation of *Lorentz 1904a*. If Einstein did not write the additional notes to this paper, the contents of some suggest that he was consulted.

[9] *Poincaré 1902*, p. 111, states: "Not only do we not have a direct intuition of the equality of two time intervals, but we do not even have that of the simultaneity of two events that are produced at different localities; I have explained

this in an article entitled "The Measurement of Time [*Poincaré 1898*]" ("Non seulement nous n'avons pas l'intuition directe de l'égalité de deux durées, mais nous n'avons même pas celle de la simultanéité de deux événements qui se produisent sur des théâtres différents; c'est ce que j'ai expliqué dans un article intitulé la *Mesure du temps*"). In the German translation of Poincaré's book, the relevant passage of *Poincaré 1898* is translated in an editorial note to this paragraph, which includes a lengthy discussion of Poincaré's comments on simultaneity (see *Poincaré 1904a*, pp. 286–289).

[10] *Poincaré 1900*, p. 272, gives a physical interpretation of Lorentz's concept of local time (see *Lorentz 1895*, pp. 49–50; see also *Einstein 1907j* [Doc. 47], note 6). This interpretation is based on a clock-setting procedure, to which Einstein's is quite similar. *Poincaré 1900* is cited in *Einstein 1906e* (Doc. 35), p. 627; there is no evidence to indicate when Einstein first read it.

[11] For a discussion of the role of such principles in Einstein's research, see the Introduction, pp. xxi–xxii, xvi.

[12] *Poincaré 1902* formulates "le principe du mouvement relatif" in similar words: "The motion of an arbitrary system must obey the same laws, whether referred to fixed axes or to moving axes undergoing a uniform rectilinear motion" ("Le mouvement d'un système quelconque doit obéir aux mêmes lois, qu'on le rapporte à des axes fixes, ou à des axes mobiles entraînés dans un mouvement rectiligne et uniforme") (p. 135).

[13] See note 5.

[14] Einstein evidently was attempting to assure that initially identical measuring rods and clocks in the two inertial frames measure equal spatial and temporal intervals, respectively, once the two frames are in relative motion. He did not consider dynamical problems that arise in accelerating measuring rods and clocks in order to transfer them from one inertial frame to the other. The corresponding discussion in *Einstein 1907j* (Doc. 47), p. 418, is more cautious. For Einstein's first discussion of the relativistic dynamics of a rigid body, see *Einstein 1907h* (Doc. 45), § 3, pp. 379–382.

[15] In the 1913 reprint, the following note is appended to the end of this line: "The Lorentz transformation equations are more simply derivable directly from the condition that, as a consequence of these equations, the relation $\xi^2 +$ $\eta^2 + \zeta^2 - V^2\tau^2 = 0$ shall have the other $x^2 + y^2 + z^2 - V^2t^2 = 0$ as a consequence" ("Die Gleichungen der Lorentz-Transformation sind einfacher direkt aus der Bedingung abzuleiten, daß vermöge jener Gleichungen die Beziehung $\xi^2 + \eta^2 + \zeta^2 - V^2\tau^2 = 0$ die andere $x^2 + y^2 + z^2 - V^2t^2 = 0$ zur Folge haben soll") (*Blumenthal 1913*, p. 35). *Einstein 1907j* (Doc. 47), pp. 418–420, gives such a derivation.

[16] Mathematically similar transformation equations are introduced in *Larmor 1900* and *Lorentz 1904a*. *Voigt 1887* utilizes transformations differing only by a scale factor. *Wien 1904* introduces a different set of linear transformations on the x and t variables in Maxwell's equations in order to demonstrate that Lorentz's theory gives the same result whether an electron is taken to be at rest, with the ether flowing past it at a constant velocity; or the ether is taken to be at rest, with the electron moved through it at the same speed in the opposite direction.

[17] Einstein may have considered the kinematics of a rigid sphere because of Abraham's use of such a model of the electron. See, e.g., *Abraham 1903*, pp. 107–109.

[18] This result later became known as "the clock paradox." *Langevin 1911* seems to have first introduced human travelers, leading to the alternate name, "the twin paradox."

[19] In the 1913 reprint, the following note is appended to the word "Unruhuhr": "In contrast to the 'pendulum clock,' which—from the physical standpoint—is a system, to which the earth belongs; this had to be excluded" ("Im Gegensatz zu ,,Penduluhr'', welche—physikalisch betrachtet—ein System ist, zu welchem der Erdkörper gehört; dies mußte ausgeschlossen werden") (*Blumenthal 1913*, p. 38).

[20] The fraction should be "w_η/w_ξ."

[21] *Poincaré 1905b* notes that these transformations, together with the spatial rotations, form a group (see p. 1505).

[22] See *Hertz, H. 1890a*, as reprinted in *Hertz, H. 1892*, p. 214, for Hertz's form of Maxwell's equations (see Einstein to Mileva Marić, 10 August 1899, Vol. 1, Doc. 52, for evidence that Einstein read *Hertz, H. 1892*). Hertz first gave this formulation of Maxwell's theory in *Hertz, H. 1884*. Einstein used Hertz's notation for the field components, but Lorentz's notation for the speed of light.

[23] See *Einstein 1907j* (Doc. 47), p. 429, for a fuller discussion of the invariance of electric charge.

[24] *Hertz, H. 1890b*, as reprinted in *Hertz, H. 1892*, p. 264, defines this quantity (more precisely, the analogous quantity formed by using the magnetic polarization vector) as "that force which in the narrower sense is customarily designated as the electromotive force induced by motion" ("diejenige Kraft, welche in engerem Sinne als die durch Bewegung inducirte electromotorische Kraft bezeichnet zu werden pflegt").

[25] The term "motional magnetic force" was introduced by Heaviside (see *Heaviside 1892*, p. 446, for the concept, and p. 546 for the term). Einstein later defined the "magnetomotorische Kraft" as the force acting on a unit of magnetic charge moving through an electric field (see the unpublished 1912 review of the theory of relativity, described in the editorial note, "Einstein on the Theory of Relativity," p. 273, note 136). To the order of approximation used in the discussion of "elektromotorische Kraft," the magnetomotive force is given by $- 1/V$ [$\mathbf{v}$, $\mathbf{E}$], where $\mathbf{E} = (L,M,N)$, $\mathbf{v} = (v,0,0)$, and the bracket is the vector product.

[26] See *Einstein 1907j* (Doc. 47), p. 429, for a fuller discussion of the resolution of the problems of asymmetry and of unipolar induction.

[27] The phenomenon of unipolar induction was discovered by Faraday in 1831 (see *Faraday 1839*). Its theoretical explanation was the subject of considerable discussion around the turn of the century (for a review of this discussion, see *Miller 1981a*). *Föppl 1894* discusses the question on pp. 327–330 (see also the second edition, *Abraham/Föppl 1904*, pp. 405–409). Besso later suggested that he brought the subject to Einstein's attention (see Michele Besso to Einstein, 3 August 1952).

[28] In a reprint copy (in Gerald Holton's collection of Einstein's reprints), Einstein corrected this line: "für $v = -V$, $v = -\infty$".

[29] In a reprint copy (see the preceding note), Einstein canceled the phrase "Verbindungslinie ,,Lichtquelle-Beobachter" " and interlineated "Bewegungsrichtung."

[30] See *Einstein 1907j* (Doc. 47), p. 431, for a proof of this equation.

[31] Once the concept of the ether is eliminated (see p. 892 of this paper), light can no longer be treated as the state of a medium, but must be treated as an independent entity. See *Einstein 1907j* (Doc. 47), p. 413; for a fuller discussion of this question, see *Einstein 1909c* (Doc. 60). Einstein's use of the phrase "Lichtkomplex"

several times in this section may be connected with the discussion in *Einstein 1905i* (Doc. 14) of the inability of Maxwell's theory to treat the microstructure of light. In this paper, Einstein introduced the concept of "light quantum" ("Lichtquant"), with energy proportional to its frequency. The result proved later in this paragraph serves to demonstrate that this proportionality holds in any inertial frame.

[32] This problem was considered by Lorentz in *Lorentz 1892a*. Abraham considered it the fundamental problem of the thermodynamic theory of radiation (see *Abraham 1904b*, p. 240), devoting *Abraham 1904a*, and *1904b*, § 3, to its solution.

[33] In a reprint copy (see note 28), Einstein corrected the denominator in the final term to: "$1 - (v/V)^2$".

[34] Einstein's result agrees with one that Abraham had derived by a quite different argument (see *Abraham 1904a*, p. 91; *Abraham 1904b*, p. 257).

[35] Presumably, by "Erfahrung" Einstein meant the experimental verification (see *Lebedev 1901* and *Nichols and Hull 1903*) of the light pressure calculated by Maxwell (see, e.g., *Maxwell 1891*, pp. 440–441).

[36] Einstein accepted the electron theory quite soon after it was proposed (see Vol. 1, the editorial note, "Einstein on Thermal, Electrical, and Radiation Phenomena," pp. 235–237). In 1903 he undertook "comprehensive studies of electron theory" ("umfassende Studien in Elektronentheorie") (Einstein to Michele Besso, 22 January 1903).

[37] See *Einstein 1907j* (Doc. 47), pp. 429–430, for a proof of this assertion.

[38] In the 1913 reprint, "Koordinatensprung" is corrected to "Koordinatenursprung" (*Blumenthal 1913*, p. 49).

[39] For details of the derivation of the following equations, see *Einstein 1907j* (Doc. 47), p. 432.

[40] Abraham introduced these terms (see the editorial note, "Einstein on the Theory of Relativity," p. 270).

[41] In the 1913 reprint, the following note is appended to "nennen": "The definition of force given here is not advantageous as was first noted by M. Planck. It is instead appropriate to define force in such a way that the laws of momentum and conservation of energy take the simplest form" ("Die hier gegebene Definition der Kraft ist nicht vorteilhaftig, wie zuerst von M. Planck

dargetan wurde. Es ist vielmehr zweckmäßig, die Kraft so zu definieren, daß der Impulssatz und der Energiesatz die einfachste Form annehmen'') (*Blumenthal 1913*, p. 51; for Planck's definition, see *Planck 1906a*). Einstein adopted Planck's definition in *Einstein 1907j* (Doc. 47).

[42] Abraham (*Abraham 1902a*, *1902b*, *1903*); Lorentz (*Lorentz 1904a*); and Bucherer (*Bucherer 1904*, pp. 57–58; see also *Langevin 1905c*) had proposed models of the electron that resulted in differing equations of motion. For a contemporary review, including an account of relevant experimental results, see *Bucherer 1904*, pp. 51–68.

[43] The term after the second equality sign should include a factor μ.

[44] Einstein returned to the question of the impossibility of superluminal velocities in *Einstein 1907h* (Doc. 45), pp. 381–382. For a discussion of his correspondence on this question a few years later, see Vol. 5, the editorial note, ''Einstein on Superluminal Signal Velocities.''

[45] See *Einstein 1907j* (Doc. 47), p. 437, for precise definitions and further discussion of A_m and A_e.

[46] See *Einstein 1907j* (Doc. 47), pp. 436–437, for a similar discussion of these three relationships, but which avoids the concepts of transverse and longitudinal mass.

[47] In a lecture given in 1922 at Kyoto University, Einstein mentioned an important discussion with a friend, presumably Michele Besso, reviewing Einstein's difficulties with the electrodynamics of moving bodies (see *Ishiwara 1971* for an account of this lecture). For the excerpt alluding to Besso's role, see the editorial note, ''Einstein on the Theory of Relativity,'' p. 264. See note 27 for Besso's suggestion of another possible contribution he made to the paper.

24. "Does the Inertia of a Body Depend upon its Energy Content?"

[*Einstein 1905s*]

DATED Bern, September 1905
RECEIVED 27 September 1905
PUBLISHED 21 November 1905

IN: *Annalen der Physik* 18 (1905): 639–641.

Republished in *Blumenthal 1913*, pp. 53–55.

13. *Ist die Trägheit eines Körpers von seinem Energieinhalt abhängig?* *von A. Einstein.*

Die Resultate einer jüngst in diesen Annalen von mir publizierten elektrodynamischen Untersuchung[1]) führen zu einer sehr interessanten Folgerung, die hier abgeleitet werden soll.

Ich legte dort die **Maxwell-Hertzschen Gleichungen** für den leeren Raum nebst dem **Maxwellschen Ausdruck** für die elektromagnetische Energie des Raumes zugrunde und außerdem das Prinzip:

Die Gesetze, nach denen sich die Zustände der physikalischen Systeme ändern, sind unabhängig davon, auf welches von zwei relativ zueinander in gleichförmiger Parallel-Translationsbewegung befindlichen Koordinatensystemen diese Zustandsänderungen bezogen werden (Relativitätsprinzip).

Gestützt auf diese Grundlagen[2]) leitete ich unter anderem das nachfolgende Resultat ab (l. c. § 8):

Ein System von ebenen Lichtwellen besitze, auf das Koordinatensystem (x, y, z) bezogen, die Energie l; die Strahlrichtung (Wellennormale) bilde den Winkel φ mit der x-Achse des Systems. Führt man ein neues, gegen das System (x, y, z) in gleichförmiger Paralleltranslation begriffenes Koordinatensystem (ξ, η, ζ) ein, dessen Ursprung sich mit der Geschwindigkeit v längs der x-Achse bewegt, so besitzt die genannte Lichtmenge — im System (ξ, η, ζ) gemessen — die Energie:

$$l^* = l \frac{1 - \dfrac{v}{V} \cos \varphi}{\sqrt{1 - \left(\dfrac{v}{V}\right)^2}},$$

wobei V die Lichtgeschwindigkeit bedeutet. Von diesem Resultat machen wir im folgenden Gebrauch.

---

[1]
1) A. Einstein, Ann. d. Phys. 17. p. 891. 1905.

2) Das dort benutzte Prinzip der Konstanz der Lichtgeschwindigkeit ist natürlich in den Maxwellschen Gleichungen enthalten.

Es befinde sich nun im System (x, y, z) ein ruhender Körper, dessen Energie — auf das System (x, y, z) bezogen — E_0 sei. Relativ zu dem wie oben mit der Geschwindigkeit v bewegten System (ξ, η, ζ) sei die Energie des Körpers H_0.

Dieser Körper sende in einer mit der x-Achse den Winkel φ bildenden Richtung ebene Lichtwellen von der Energie $L/2$ (relativ zu (x, y, z) gemessen) und gleichzeitig eine gleich große Lichtmenge nach der entgegengesetzten Richtung. Hierbei bleibt der Körper in Ruhe in bezug auf das System (x, y, z). Für diesen Vorgang muß das Energieprinzip gelten und zwar (nach dem Prinzip der Relativität) in bezug auf beide Koordinatensysteme. Nennen wir E_1 bez. H_1 die Energie des Körpers nach der Lichtaussendung relativ zum System (x, y, z) bez. (ξ, η, ζ) gemessen, so erhalten wir mit Benutzung der oben angegebenen Relation:

$$E_0 = E_1 + \left[\frac{L}{2} + \frac{L}{2} \right],$$

$$H_0 = H_1 + \left[\frac{L}{2} \frac{1 - \frac{v}{V} \cos \varphi}{\sqrt{1 - \left(\frac{v}{V}\right)^2}} + \frac{L}{2} \frac{1 + \frac{v}{V} \cos \varphi}{\sqrt{1 - \left(\frac{v}{V}\right)^2}} \right]$$

$$= H_1 + \frac{L}{\sqrt{1 - \left(\frac{v}{V}\right)^2}} .$$

Durch Subtraktion erhält man aus diesen Gleichungen:

$$(H_0 - E_0) - (H_1 - E_1) = L \left\{ \frac{1}{\sqrt{1 - \left(\frac{v}{V}\right)^2}} - 1 \right\}.$$

Die beiden in diesem Ausdruck auftretenden Differenzen von der Form $H - E$ haben einfache physikalische Bedeutungen. H und E sind Energiewerte desselben Körpers, bezogen auf zwei relativ zueinander bewegte Koordinatensysteme, wobei der Körper in dem einen System (System (x, y, z)) ruht. Es ist also klar, daß die Differenz $H - E$ sich von der kinetischen Energie K des Körpers in bezug auf das andere System (System (ξ, η, ζ)) nur durch eine additive Konstante C unterscheiden kann, welche von der Wahl der willkürlichen addi-

Trägheit eines Körpers von seinem Energieinhalt abhängig?   641

tiven Konstanten der Energien H und E abhängt. Wir können also setzen:

$$H_0 - E_0 = K_0 + C,$$
$$H_1 - E_1 = K_1 + C,$$

da C sich während der Lichtaussendung nicht ändert. Wir erhalten also:

$$K_0 - K_1 = L \left\{ \frac{1}{\sqrt{1 - \left(\frac{v}{V}\right)^2}} - 1 \right\}.$$

Die kinetische Energie des Körpers in bezug auf (ξ, η, ζ) nimmt infolge der Lichtaussendung ab, und zwar um einen von den Qualitäten des Körpers unabhängigen Betrag. Die Differenz $K_0 - K_1$ hängt ferner von der Geschwindigkeit ebenso ab wie die kinetische Energie des Elektrons (l. c. § 10).

Unter Vernachlässigung von Größen vierter und höherer Ordnung können wir setzen:

[2]
$$K_0 - K_1 = \frac{L}{V^2} \frac{v^2}{2}.$$

Aus dieser Gleichung folgt unmittelbar:

Gibt ein Körper die Energie L in Form von Strahlung ab, so verkleinert sich seine Masse um L/V^2. Hierbei ist es offenbar unwesentlich, daß die dem Körper entzogene Energie gerade in Energie der Strahlung übergeht, so daß wir zu der allgemeineren Folgerung geführt werden:

Die Masse eines Körpers ist ein Maß für dessen Energieinhalt; ändert sich die Energie um L, so ändert sich die Masse in demselben Sinne um $L/9 . 10^{20}$, wenn die Energie in Erg und die Masse in Grammen gemessen wird.

Es ist nicht ausgeschlossen, daß bei Körpern, deren Energieinhalt in hohem Maße veränderlich ist (z. B. bei den
[3] Radiumsalzen), eine Prüfung der Theorie gelingen wird.

Wenn die Theorie den Tatsachen entspricht, so überträgt die Strahlung Trägheit zwischen den emittierenden und absor-
[4] bierenden Körpern.

Bern, September 1905.

(Eingegangen 27. September 1905.)

Published in *Annalen der Physik* 18 (1905): 639–641. Dated Bern, September 1905, received 27 September 1905, published 21 November 1905.

[1] *Einstein 1905r* (Doc. 23).

[2] Einstein used the Newtonian limit of the body's kinetic energy in order to evaluate the change in its rest mass.

[3] See *Einstein 1907j* (Doc. 47), p. 442, for further discussion of the possible use of radioactive materials to test the dependence of inertial mass on energy content.

[4] See *Einstein 1909c* (Doc. 60), p. 490, for further discussion of the significance of the transfer of inertial mass by radiation.

25. Review of Heinrich Birven,
Fundamentals of the Mechanical Theory of Heat

[*Einstein 1905t*]

Second half of September 1905

PUBLISHED IN: *Beiblätter zu den Annalen der Physik* 29 (1905), no. 18: 950.

950   Wärmelehre.   Beibl. 1905.

73. **H. Birven.** *Grundzüge der mechanischen Wärme-theorie* (128 S. ℳ 2.80. Stuttgart u. Berlin, F. Grub, 1905). — Den Inhalt des Büchleins bildet eine gedrängte, elementare Darlegung der Thermodynamik der Gase und Dämpfe sowie deren Anwendung in der Theorie der Dampfmaschine und der Kältemaschinen. Wenn das Büchlein auch in bezug auf die grundlegenden Definitionen und Darlegungen manche Ungenauig-keit aufweist (vgl. z. B. die Definition der Entropie S. 50), so [1] dürfte dasselbe doch manchem vor dem Examen stehenden, mit lückenhaften Kollegienheften versehenen Polytechniker einen Dienst leisten. A. E. [2]

Published in *Beiblätter zu den Annalen der Physik* 29 (1905): 950. Published in no. 18 [second half of September].

[1] Birven defines the entropy as $S = \Sigma \ Q/T$, where the sum extends over a given process or cycle and Q is the amount of heat given off or absorbed.

[2] In the preface Birven states that his book is primarily intended for students at higher technical institutes. Einstein himself was not very diligent at taking notes during his studies at the ETH. The detailed notes taken by his classmate Marcel Grossmann were his "saving anchor" ("Rettungsanker") while he was preparing for his ETH exams (*Einstein 1955*, p. 147).

26. Review of Auguste Ponsot, "Heat in the Displacement of the Equilibrium of a Capillary System"

[*Einstein 1905u*]

Second half of September 1905

PUBLISHED IN: *Beiblätter zu den Annalen der Physik* 29 (1905), no. 18: 952.

952   Wärmelehre.   Beibl. 1905.

sich die Zahlenwerte einiger in der Theorie auftretender Konstanten; im besonderen ergibt sich das Verhältnis der spezifischen Wärmen:

$$k = \frac{5}{6\,n} + 1,$$

wenn n die Anzahl der Atome im Molekül bezeichnet. Der Verf. zeigt an vielen Beispielen, daß diese Formel in der Erfahrung bestätigt wird.   A. K.

---

77. **M. Ponsot.** *Die Wärmezufuhr bei einer Gleichgewichtsänderung eines kapillaren Systems* (C. R. **140**, S. 1176 —1179. 1905). — Der Verf. untersucht die Grundlagen der thermodynamischen Theorie der Kapillarität und findet in denselben eine nicht exakt richtige Voraussetzung; über die Größenordnung der Ungenauigkeiten, welche aus dieser Voraussetzung erwachsen, ist nichts gesagt.   A. E. [1]

Published in *Beiblätter zu den Annalen der Physik* 29 (1905): 952. Published in no. 18 [second half of September].

[1] Ponsot claims that it is not permissible to treat the temperature T and the surface area s as completely independent thermodynamic variables.

27. Review of Karl Bohlin, "On Impact Considered as the Basis of Kinetic Theories of Gas Pressure and of Universal Gravitation"

[*Einstein 1905v*]

Second half of September 1905

PUBLISHED IN: *Beiblätter zu den Annalen der Physik* 29 (1905), no. 18: 952–953.

952   Wärmelehre.   Beibl. 1905.

sich die Zahlenwerte einiger in der Theorie auftretender Konstanten; im besonderen ergibt sich das Verhältnis der spezifischen Wärmen:

$$k = \frac{5}{6\,n} + 1,$$

wenn n die Anzahl der Atome im Molekül bezeichnet. Der Verf. zeigt an vielen Beispielen, daß diese Formel in der Erfahrung bestätigt wird.   A. K.

---

77. **M. Ponsot.** *Die Wärmezufuhr bei einer Gleichgewichtsänderung eines kapillaren Systems* (C. R. 140, S. 1176—1179. 1905). — Der Verf. untersucht die Grundlagen der thermodynamischen Theorie der Kapillarität und findet in denselben eine nicht exakt richtige Voraussetzung; über die Größenordnung der Ungenauigkeiten, welche aus dieser Voraussetzung erwachsen, ist nichts gesagt.   A. E.

---

78. **G. Jäger.** *Zur kinetischen Theorie der Abhängigkeit der Gasdichte von den äußeren Kräften* (Wien. Ber. 113, IIa, S. 1289—1302. 1904). — Nachdem zunächst an einem Beispiel dargetan wird, wie notwendig es ist, für eine richtige Bestimmung der Anzahl der Moleküle in der Volumeneinheit die Bedingung des molekular-ungeordneten Zustandes eines Gases nicht außer Acht zu lassen, zeigt der Verf. im folgenden in Erweiterung seiner Ausführungen in Drudes Ann. 11, S 1071. 1903 über den gleichen Gegenstand, daß das Maxwell-Boltzmannsche Gesetz für die Verteilung der Moleküle eines Gases unter Zulassung eines beliebigen Kraftfeldes von solcher Art, daß alle vorkommenden Kräfte ein Potential besitzen, stets seine Gültigkeit behält, auch wenn man der Betrachtung so kleine Räume zugrunde legt, daß sie selbst im Vergleich zu den Entfernungen der Moleküle nicht groß sind.   S. V.

---

79. **K. Bohlin.** *Über den Stoß als Fundament der Theorien des Gasdruckes und der Gravitation* (Arch. för Math. Ast. 1, S. 522—540. 1904). — Ausgehend von der Bemerkung, daß sowohl in der kinetischen Theorie der Gase als auch in der dynamischen Theorie der Gravitation abstoßende Kräfte zwischen [1]

Bd. 29. No. 18.   **Wärmelehre.**   **953**

den Teilchen lediglich zur Erklärung der Zusammenstöße eingeführt werden, sucht der Verf. die Einführung abstoßender Kräfte
[2] überhaupt zu vermeiden. Er versucht, den Stoß lediglich auf die
Wirkung von anziehenden Kräften, zwischen den die zusammen-
stoßenden Körper konstituierenden Korpuskeln zurückzuführen.
Er vertritt dabei den Standpunkt, daß jede Anziehungskraft
aus der Stoßwirkung relativ unendlich kleiner Korpuskeln und
jeder Stoß durch die Anziehung relativ unendlich kleiner Korpuskeln (kinetisch) zu erklären sei. Es werden also Korpuskeln
von unendlich vielen Größenordnungen zur Darstellung der
elementaren Eigenschaften der Materie eingeführt.   A. E.

Published in *Beiblätter zu den Annalen der Physik* 29 (1905): 952–953. Published in no. 18 [second half of September].

[1] For a review of kinetic theories of gravitation, see *Zenneck 1903*, pp. 57–64.

[2] Bohlin is in particular concerned with the explanation of the elasticity of molecules in terms of attractive forces.

28. Review of Georges Meslin, "On the Constant in Mariotte and Gay-Lussac's Law"

[Einstein 1905w]

First half of November 1905

PUBLISHED IN: *Beiblätter zu den Annalen der Physik* 29 (1905), no. 21: 1114.

1114   Wärmelehre.   Beibl. 1905.

Sie bemerken nämlich in einer Nattererschen Röhre, die mit einer Lösung von Alizarin in Alkohol (bez. Kaliumbichromat in Wasser) gefüllt ist, beim kritischen Punkt der Lösung eine Färbung in der ganzen Ausdehnung der Röhre; wäre nur Dampf in der Röhre vorhanden, so müßte die in der Flüssigkeit gelöste Substanz zu Boden sinken, da die Substanz in dem Dampf der Flüssigkeit bei Temperaturen, die wenig unter der kritischen liegen, nicht merklich gelöst wird.

Raveau weist darauf hin, daß der Schluß unter der Voraussetzung gezogen ist, daß sich die Eigenschaft des Dampfes, bez. des Lösungsvermögens in der unmittelbaren Nähe des kritischen Punktes nicht ändert. Die Erfahrung lehrt im allgemeinen eine starke Änderung der Eigenschaften der Substanzen in der Nähe des kritischen Punktes, die man dann als durch die Existenz einer solchen Mischung hervorgerufen ansehen muß.   S. V.

---

37. *L. Levi-Bianchini.* *Über den kritischen Punkt von verdünnten Salzlösungen* (Gaz. chim. (1) 35, S. 160—163. 1905). — Eine vorläufige Mitteilung über die kritischen Erscheinungen von verdünnten Salzlösungen. Für Lösungen von LiCl, LiBr, LiJ, NaBr, KBr in Methylalkohol nimmt die kritische Temperatur mit der Salzkonztrationen zu. Bei vielen anderen Salzlösungen tritt Zersetzung vor der Erreichnng der kritischen Temperatur ein. Ausführlichere und quantitative Angaben über seine Versuche wird der Verf. in einer nächsten Abhandlung mitteilen.   Chilesotti.

---

38. *G. Meslin.* *Über die Konstante des Mariotte-Gay-Lussacschen Gesetzes* (J. de Phys. (4) 4, S. 252—256. 1905). —
[1] **Es wird gezeigt, daß der Quotient der genannten Konstanten und des mechanischen Wärmeäquivalentes, der einen von 2 wenig abweichenden Wert besitzt, von der Wahl der Einheiten für Masse, Länge und Zeit unabhängig ist, daß aber dem Zahlenwert dieses Quotienten doch keine physikalische Bedeutung beizumessen ist, da er von der Wahl der Einheit des**
[2] **Molekulargewichtes abhängt.**   A. E.

Published in *Beiblätter zu den Annalen der Physik* 29 (1905): 1114. Published in no. 21 [first half of November].
[1] The constant R in the law $pV = RT$.

[2] Meslin uses a thermodynamical argument; within the framework of the kinetic theory of gases his conclusion is trivially true.

29. Review of Albert Fliegner, ''The Efflux of Hot Water from Container Orifices''

[*Einstein 1905x*]

First half of November 1905

PUBLISHED IN: *Beiblätter zu den Annalen der Physik* 29 (1905), no. 21: 1115.

Bd. 29. No. 21.   Wärmelehre.   1115

39. *W. Holtz. Einfache Vorlesungsthermoskope für die Verdichtungswärme der Gase* (CZtg. f. Opt. u. Mech. 26, S. 17. 1905). — Eine Breguetsche Spirale wird, am Stöpsel hängend, in eine verschlossene Flasche gebracht. Durch einen Gummiballon läßt sich die Luft in der Flasche komprimieren; dabei zeigt die Spirale die Erwärmung durch einen Ausschlag an, der allmählich zurückgeht und sich dann bei nachfolgender Wiederausdehnung umkehrt.

Die Arbeit enthält ferner nützliche Angaben zur Selbstanfertigung von Breguetschen Spiralen aus Messingstreifen, die geeignet zusammengedreht und dann einseitig versilbert oder vernickelt werden, oder, noch einfacher, nur einen einseitigen Paraffinüberzug erhalten.   Bdkr.

---

40. *A. Fliegner. Das Ausströmen heißen Wassers aus Gefäßmündungen* (Schweiz. Bauztg. 45, S. 282, 306. 1904). —
[1] Nach Zeuner und Lorenz ist die Abweichung der Versuchs-
[2] resultate von Pulin und Bonnin betreffend die Ausflußgeschwindigkeit von unter dem Druck seines Dampfes stehendem Wasser auf eine Art Verdampfungsverzug zurückzuführen. Im Gegensatz hierzu vertritt der Verf. die Auffassung, es sei die Nichtübereinstimmung der Versuche mit der Theorie darauf zurückzuführen, daß bei den genannten Versuchen das zur Ausströmung gelangende Wasser infolge von Temperaturverschiedenheiten im Wassergefäß von etwas tieferer Temperatur gewesen sei, als dem im Gefäß vorhandenen Dampfdruck entspricht. Es wird außerdem ein Versuch angeführt, bei dem die genannte Fehlerquelle vermieden war und dessen Ergebnis im Einklang mit der Theorie ist.   A. E.

Published in *Beiblätter zu den Annalen der Physik* 29 (1905): 1115. Published in no. 21 [first half of November].

[1] See *Zeuner 1906*, pp. 156–158; *Lorenz 1904*, pp. 245–247.

[2] The experimental results of Pulin and Bonnin were published by Sauvage (*Sauvage 1892*).

30. Review of Jakob Johann Weyrauch, *An Outline of the Theory of Heat. With Numerous Examples and Applications.* Part 1

[*Einstein 1905y*]

Second half of November 1905

PUBLISHED IN: *Beiblätter zu den Annalen der Physik* 29 (1905), no. 22: 1152–1153.

normal ca. 2,1; vgl. van't Hoffs Vorles. III, S. 72), übrigens das benutzte Temperaturintervall nur 20° umfaßt, scheint die Bedeutung dieser Berechnung fraglich.   Bdkr.

26. *J. Bolle und Ph. A. Guye. Oberflächenspannungen einiger organischer Flüssigkeiten* (J. Chim. Phys. 3, S. 38. 1905). — Die Arbeit enthält die Resultate der Messung der Oberflächenspannung von Butyl- und Phenylisosulfocyanat, Ortho- und Metakresol, Chinolin und Phenol bei verschiedenen Temperaturen. Für die drei letztgenannten Substanzen zeigt sich der Temperaturkoeffizient der molekularen Oberflächenspannung zum Teil merklich vom Normalwert (2—2,4) abweichend, woraus auf Polymerisation geschlossen wird. In der gleichen Arbeit werden Bestimmungen der Dichte und Wärmeausdehnung für dieselben Substanzen und einige andere veröffentlicht.   Bdkr.

Ph. A. Guye. Neue Untersuchungen über das Atomgewicht des Stickstoffs (Arch. de Genève 20, S. 231—258. 1905).

S. Valentiner und R. Schmidt. Über eine neue Methode der Darstellung von Neon, Krypton und Xenon (Drudes Ann. 18, S. 187—197. 1905; Berl. Ber. 1905, 816—820).

A. Skrabal. Zur Kinetik der Oxydationsvorgänge (Die Permanganat-Oxalsäurereaktion) (ZS. f. Elektrochem. 11, S. 653—657. 1905).

C. Forch. Die Oberflächenspannung von anorganischen Salzlösungen (Drudes Ann. 17, S. 744—762. 1905).

Wärmelehre.

27. *J. J. Weyrauch. Grundriß der Wärmetheorie. Mit zahlreichen Beispielen und Anwendungen. Erste Hälfte* (313 S. Stuttgart, K. Wittwer, 1905). — Das Werk ist aus Vorlesungen entstanden, die der Verf. an der Stuttgarter technischen Hochschule gehalten hat, und enthält in der Hauptsache die Theorie der thermischen Hauptsätze und im Anschluß an diese in übersichtlicher und ausführlicher Darlegung die Theorien der

Bd. 29. No. 22.   Wärmelehre.   1153

verschiedenen Wärmemotoren.   Das Werk ist zum Selbst-
studium sehr geeignet, indem auf die Vollendung desselben in
didaktischer Beziehung große Sorgfalt verwendet ist.   Um
einerseits die Theorie in geschlossener, übersichtlicher Form
darstellen zu können und andererseits dem Leser die entwickelten abstrakten Resultate möglichst lebendig zum Bewußt-
sein zu bringen, sind zwischen die theoretischen Darlegungen
zahlreiche Beispiele und Aufgaben eingeschaltet, die sich keineswegs auf für den Techniker wichtige Anwendungen beschränken.
Viele Beispiele sind der Geschichte der Wärmetheorie entnommen, wobei besonders die Gedankengänge Robert Mayers
ausführlich dargelegt sind; auch die Arbeit und Wirkungsweise [1]
„menschlicher Motoren" ist einer eingehenden Betrachtung
unterzogen und mit künstlichen Wärmemotoren verglichen.   Die
sieben Abschnitte der vorliegenden ersten Hälfte des Werkes
sind folgendermaßen betitelt: I. Erhaltung der Energie. Erster
Hauptsatz.   II. Wärme und Arbeit.   Zweiter Hauptsatz.
III. Über Wärmemotoren im allgemeinen. IV. Von den Gasen.
V. Über Luftmaschinen. VI. Aus der Chemie und kinetischen
Gastheorie.   VII. Über Verbrennungsmotoren. — Technische
und physikalische Vorkenntnisse werden nicht vorausgesetzt,
wohl aber die Elemente der Infinitesimalrechnung. Auf konstruktive Einzelheiten der kalorischen Maschinen wird nur
insoweit eingegangen, als zur Darlegung der Theorien der
einzelnen Maschinengattungen und zur Untersuchung ihrer
Wirkungsgrade erforderlich ist. Dem Bande ist eine Tabelle
der gebrauchten Buchstabenbezeichnungen, ferner ein Namen-
und Sachregister beigegeben, so daß das Werk auch als Nach-
schlagewerk mit Vorteil benutzt werden und jeder der vielen
vorkommenden Erfahrungswerte (z. B. Heizwerte verschiedener
Brennstoffe, bisher praktisch erzielte Wirkungsgrade der einzelnen Wärmemotoren etc.) leicht aufgefunden werden kann.
Die zweite Hälfte des Werkes soll — wie im Vorwort mitgeteilt wird — im Laufe des nächsten Jahres erscheinen. Sie [2]
wird die Lehre von den gesättigten und überhitzten Dämpfen,
die Abschnitte über Aerostatik, Aerodynamik und feste Körper,
sowie die entsprechenden Anwendungen enthalten, in ganz
ähnlicher Darstellung wie die erste Hälfte.   A. E.

Published in *Beiblätter zu den Annalen der Physik* 29 (1905): 1152–1153. Published in no. 22 [second half of November].

[1] Weyrauch edited a collection of Mayer's major scientific works (*Mayer 1893a*) and a collection of documents related to Mayer's life (*Mayer 1893b*); he also authored a biographical sketch of Mayer (*Weyrauch 1890*). In addition, he wrote on the history of thermodynamics (see, e.g., *Weyrauch 1885*).

[2] The preface is dated summer 1904; the second part did not appear until 1907 (*Weyrauch 1907*). It was reviewed by Einstein in *Einstein 1907i* (Doc. 46).

31. Review of Albert Fliegner, "On the Thermal Value of Chemical Processes"

[*Einstein 1905z*]

Second half of November 1905

PUBLISHED IN: *Beiblätter zu den Annalen der Physik* 29 (1905), no. 22: 1158.

1158   **Wärmelehre.**   **Beibl. 1905.**

34. *A. Fliegner. Über den Wärmewert chemischer Vorgänge* (Vierteljschr. d. naturf. Ges. Zürich 50, S. 201—212. 1905). — Der Verf. übersieht, daß die Definitionsgleichung der Entropie $dS = dQ/T$ nur für *reversible* Vorgänge gültig ist und gelangt infolgedessen zu dem Resultat, daß man von der Entropieänderung eines Systems durch einen chemischen Vorgang nicht sprechen könne.   A. E.

Published in *Beiblätter zu den Annalen der Physik* 29 (1905): 1158. Published in no. 22 [second half of November].

32. ''On the Theory of Brownian Motion''

[*Einstein 1906b*]

DATED Bern, December 1905
RECEIVED 19 December 1905
PUBLISHED 8 February 1906

IN: *Annalen der Physik* 19 (1906): 371–381.

Republished in *Einstein 1922*, pp. 15–25.

7. *Zur Theorie der Brownschen Bewegung;*
von A. Einstein.

Kurz nach dem Erscheinen meiner Arbeit über die durch die Molekulartheorie der Wärme geforderte Bewegung von in Flüssigkeiten suspendierten Teilchen[1]) teilte mir Hr. Sieden-
[2] topf (Jena) mit, daß er und andere Physiker — zuerst wohl
[3] Hr. Prof. Gouy (Lyon) — durch direkte Beobachtung zu der Überzeugung gelangt seien, daß die sogenannte Brownsche Bewegung durch die ungeordnete Wärmebewegung der Flüssig-keitsmoleküle verursacht sei.[2]) Nicht nur die qualitativen Eigen-schaften der Brownschen Bewegung, sondern auch die Größen-ordnung der von den Teilchen zurückgelegten Wege entspricht durchaus den Resultaten der Theorie. Ich will hier nicht eine Vergleichung des mir zur Verfügung stehenden dürftigen Erfahrungsmaterials mit den Resultaten der Theorie anstellen, sondern diese Vergleichung denjenigen überlassen, welche das
[5] Thema experimentell behandeln.

Die nachfolgende Arbeit soll meine oben genannte Arbeit in einigen Punkten ergänzen. Wir leiten hier nicht nur die fortschreitende, sondern auch die Rotationsbewegung suspen-dierter Teilchen ab für den einfachsten Spezialfall, daß die
[6] Teilchen Kugelgestalt besitzen. Wir zeigen ferner bis zu wie kurzen Beobachtungszeiten das in jener Abhandlung gegebene Resultat gilt.

Für die Herleitung wollen wir uns hier einer allgemeineren Methode bedienen, teils um zu zeigen, wie die Brownsche Bewegung mit den Grundlagen der molekularen Theorie der Wärme zusammenhängt, teils um die Formeln für die fort-schreitende und für die rotierende Bewegung durch eine ein-heitliche Untersuchung entwickeln zu können. Es sei näm-lich α ein beobachtbarer Parameter eines im Temperatur-

---

[1]   1) A. Einstein, Ann. d. Phys. **17**. p. 549. 1905.
[4]   2) M. Gouy, Journ. de Phys. (2) **7**. p. 561. 1888.

gleichgewicht befindlichen physikalischen Systems und es sei angenommen, daß das System bei jedem (möglichen) Wert von α im sogenannten indifferenten Gleichgewicht sich befinde. Nach der klassischen Thermodynamik, die zwischen Wärme und anderen Energiearten *prinzipiell* unterscheidet, finden spontane Änderungen von α nicht statt, wohl aber nach der molekularen Theorie der Wärme. Wir wollen im nachfolgenden untersuchen, nach welchen Gesetzen jene Änderungen gemäß der letzteren Theorie stattfinden müssen. Wir haben dann jene Gesetze auf folgende Spezialfälle anzuwenden:

1. α ist die x-Koordinate des Schwerpunktes eines in einer (der Schwerkraft nicht unterworfenen) homogenen Flüssigkeit suspendierten Teilchens von Kugelgestalt.

2. α ist der Drehwinkel, welcher die Lage eines in einer Flüssigkeit suspendierten, um einen Durchmesser drehbaren Teilchens von Kugelgestalt bestimmt.

§ 1. Über einen Fall thermodynamischen Gleichgewichtes.

In einer Umgebung von der absoluten Temperatur T befinde sich ein physikalisches System, das mit dieser Umgebung in thermischer Wechselwirkung stehe und im Zustand des Temperaturgleichgewichtes sei. Dies System, das also ebenfalls die absolute Temperatur T besitzt, sei im Sinne der molekularen Theorie der Wärme vollständig bestimmt[1]) durch die Zustandsvariabeln $p_1 \ldots p_n$. Als Zustandsvariable $p_1 \ldots p_n$ können in den zu behandelnden Spezialfällen die Koordinaten und Geschwindigkeitskomponenten aller das betrachtete System bildenden Atome gewählt werden.

Es gilt für die Wahrscheinlichkeit dafür, daß in einem zufällig herausgegriffenen Zeitpunkt die Zustandsvariabeln $p_1 \ldots p_n$ in dem n-fach unendlich kleinen Gebiete $(dp_1 \ldots dp_n)$ liegen, die Gleichung[2]):

$$(1) \qquad dw = C e^{-\frac{N}{RT}E} dp_1 \ldots dp_n,$$

wobei C eine Konstante, R die universelle Konstante der Gasgleichung, N die Anzahl der wirklichen Moleküle in einem Grammmolekül und E die Energie bedeutet.

---

1) Vgl. Ann. d. Phys. **17.** p. 549. 1905.   [7]

2) l. c. § 3 und 4.   [8]

Es sei α ein beobachtbarer Parameter des Systems und es entspreche jedem Wertsystem $p_1 \ldots p_n$ ein bestimmter Wert α. Wir bezeichnen mit $A\, d\alpha$ die Wahrscheinlichkeit dafür, daß in einem zufällig herausgegriffenen Zeitpunkt der Wert des Parameters α zwischen α und $\alpha + d\alpha$ liege. Es ist dann

$$(2) \qquad A\, d\alpha = \int\limits_{d\alpha} C e^{-\frac{N}{RT} E}\, dp_1 \ldots dp_n,$$

wenn das Integral der rechten Seite über alle Wertkombinationen der Zustandsvariabeln erstreckt wird, deren α-Wert zwischen α und $\alpha + d\alpha$ liegt.

Wir beschränken uns auf den Fall, daß aus der Natur des Problems ohne weiteres klar ist, daß allen (möglichen) Werten von α dieselbe Wahrscheinlichkeit (Häufigkeit) zukommt, daß also die Größe A von α unabhängig ist.

[9]

Es liege nun ein zweites physikalisches System vor, das sich von dem soeben betrachteten einzig darin unterscheide, daß auf das System eine nur von α abhängige Kraft vom Potential $\Phi(\alpha)$ wirke. Ist E die Energie des vorhin betrachteten Systems, so ist $E + \Phi$ die Energie des jetzt betrachteten, so daß wir die der Gleichung (1) analoge Beziehung erhalten:

$$dw' = C' e^{-\frac{N}{RT}(E + \Phi)}\, dp_1 \ldots dp_n.$$

Hieraus folgt für die Wahrscheinlichkeit dW dafür, daß in einem beliebig herausgegriffenen Zeitpunkt der Wert von α zwischen α und $\alpha + d\alpha$ liegt, die der Gleichung (2) analoge Beziehung:

$$(I) \quad \begin{cases} dW = \int C' e^{-\frac{N}{RT}(E + \Phi)}\, dp_1 \ldots dp_n = \dfrac{C'}{C} e^{-\frac{N}{RT}\Phi} A\, d\alpha \\[2mm] \qquad\qquad\qquad\qquad\qquad\qquad = A' e^{-\frac{N}{RT}\Phi}\, d\alpha, \end{cases}$$

wobei A' von α unabhängig ist.

Diese Beziehung, welche dem von **Bolzmann** in seinen gastheoretischen Untersuchungen vielfach benutzten Exponentialgesetz genau entspricht, ist für die molekulare Theorie der Wärme charakteristisch. Sie gibt Aufschluß darüber, wieviel sich ein einer konstanten äußeren Kraft unterworfener Parameter eines Systems infolge der ungeordneten Molekularbewegung

[10]

374   *A. Einstein.*

von dem Werte entfernt, **welcher** dem stabilen Gleichgewicht
entspricht.

§ 2. Anwendungsbeispiele für die in § 1 abgeleitete Gleichung.

Wir betrachten einen Körper, dessen Schwerpunkt sich
längs einer Geraden (*X*-Achse eines Koordinatensystems) bewegen kann. Der Körper sei von einem Gase umgeben und
es herrsche thermisches und mechanisches Gleichgewicht. Nach
der Molekulartheorie wird sich der Körper infolge der Un-
gleichheit der Molekularstöße längs der Geraden in unregelmäßiger Weise hin und her bewegen, derart, daß bei dieser
Bewegung kein Punkt der Geraden bevorzugt ist — vorausgesetzt, daß auf den Körper in Richtung der Geraden keine
anderen Kräfte wirken als die Stoßkräfte der Moleküle. Die
Abszisse x des Schwerpunktes ist also ein Parameter des
Systems, welcher die oben für den Parameter α vorausgesetzten Eigenschaften besitzt.

Wir wollen nun eine auf den Körper in Richtung der
Geraden wirkende Kraft $K = -Mx$ einführen. Dann wird
der Schwerpunkt des Körpers nach der Molekulartheorie
ebenfalls ungeordnete Bewegungen ausführen, ohne sich jedoch
viel vom Punkte $x = 0$ zu entfernen, während er nach der
klassischen Thermodynamik im Punkte $x = 0$ ruhen müßte.
Nach der Molekulartheorie ist (Formel I)

$$d\,W = A'\, e^{-\frac{N}{R\,T} M \frac{x^2}{2}}\, d\,x,$$

gleich der Wahrscheinlichkeit dafür, daß in einem zufällig gewählten Zeitpunkt der Wert der Abszisse x zwischen x und
$x + dx$ liegt. Hieraus findet man den mittleren Abstand des
Schwerpunktes vom Punkte $x = 0$:

$$\sqrt{x^2} = \frac{\displaystyle\int_{-\infty}^{+\infty} x^2\, A'\, e^{-\frac{N}{R\,T}\frac{Mx^2}{2}}\, d\,x}{\displaystyle\int_{-\infty}^{+\infty} A'\, e^{-\frac{N}{R\,T}\frac{Mx^2}{2}}\, d\,x} = \sqrt{\frac{R\,T}{N\,M}}\cdot \qquad [11]$$

Damit $\sqrt{x^2}$ genügend groß sei, um der Beobachtung zugänglich zu sein, muß die die Gleichgewichtslage des Körpers

bestimmende Kraft sehr klein sein. Setzen wir als untere
[12] Grenze des Beobachtbaren $\sqrt{\overline{x^2}} = 10^{-4}$ cm, so erhalten wir
für $T = 300$ $M =$ ca. $5 . 10^{-6}$. Damit der Körper mit dem
Mikroskop beobachtbare Schwankungen ausführe, darf also die
auf ihn wirkende Kraft bei einer Elongation von 1 cm nicht
[13] mehr als 5 milliontel Dyn betragen.

  Wir wollen noch eine theoretische Bemerkung an die ab-
[14] geleitete Gleichung anknüpfen. Der betrachtete Körper trage
eine über einen sehr kleinen Raum verteilte elektrische Ladung
und es sei das den Körper umgebende Gas so verdünnt, daß
der Körper eine durch das umgebende Gas nur schwach modi-
fizierte Sinusschwingung ausführe. Der Körper strahlt dann
elektrische Wellen in den Raum aus und empfängt Energie
aus der Strahlung des umliegenden Raumes; er vermittelt also
einen Energieaustausch zwischen Strahlung und Gas. Wir gelangen zu einer Ableitung des Grenzgesetzes der Temperaturstrahlung, welches für große Wellenlängen und für hohe
Temperaturen zu gelten scheint, indem wir die Bedingung
dafür aufstellen, daß der betrachtete Körper im Durchschnitt
ebensoviel Strahlung emittiert als absorbiert. Man gelangt
so[1]) zu der folgenden Formel für die der Schwingungszahl ν
entsprechende Strahlungsdichte ϱ_ν:

$$\varrho_\nu = \frac{R}{N} \frac{8 \pi \nu^2}{L^3} T,$$

wobei L die Lichtgeschwindigkeit bedeutet.

  Die von Hrn. Planck gegebene Strahlungsformel[2]) geht
für kleine Periodenzahlen und hohe Temperaturen in diese
Formel über. Aus dem Koeffizienten des Grenzgesetzes läßt
sich die Größe N bestimmen, und man erhält so die Planck-
[17] sche Bestimmung der Elementarquanta. Die Tatsache, daß man
auf dem angedeuteten Wege nicht zu dem wahren Gesetz der
Strahlung, sondern nur zu einem Grenzgesetz gelangt, scheint
mir in einer elementaren Unvollkommenheit unserer physi-
[18] kalischen Anschauungen ihren Grund zu haben.

  Wir wollen nun die Formel (I) noch dazu verwenden, zu
entscheiden, wie klein ein suspendiertes Teilchen sein muß,

---

[15] 1) Vgl. Ann. d. Phys. **17.** p. 549. 1905. § 1 und 2.
[16] 2) M. Planck, Ann. d. Phys. **1.** p. 99. 1900.

376   *A. Einstein.*

damit es trotz der Wirkung der Schwere dauernd suspendiert
bleibe. Wir können uns dabei auf den Fall beschränken, daß
das Teilchen spezifisch schwerer ist als die Flüssigkeit, da der
entgegengesetzte Fall vollkommen analog ist.

Ist v das Volumen des Teilchens, ϱ dessen Dichte, ϱ_0 die
Dichte der Flüssigkeit, g die Beschleunigung der Schwere und
x der vertikale Abstand eines Punktes vom Boden des Ge-
fäßes, so ergibt Gleichung (I)

$$d\,W = \text{konst.}\, e^{-\frac{N}{R\,T}\,v\,(\varrho - \varrho_0)\,g\,x}\,d\,x.$$

Man wird also dann finden, daß suspendierte Teilchen in
einer Flüssigkeit zu schweben vermögen, wenn für Werte
von x, die nicht wegen ihrer Kleinheit sich der Beobachtung
entziehen, die Größe

$$\frac{N}{R\,T}\,v\,(\varrho - \varrho_0)\,g\,x$$

keinen allzu großen Wert besitzt — vorausgesetzt, daß an den
Gefäßboden gelangende Teilchen nicht durch irgendwelche Umstände an demselben festgehalten werden.   [19]

§ 3. Über die von der Wärmebewegung verursachten Veränderungen des Parameters α.   [20]

Wir kehren wieder zu dem in § 1 behandelten allgemeinen
Falle zurück, für den wir Gleichung (I) abgeleitet haben. Der
einfacheren Ausdrucksweise und Vorstellung halber wollen wir
aber nun annehmen, daß eine sehr große Zahl (n) identischer
Systeme von der dort charakterisierten Art vorliege; wir haben
es dann mit Anzahlen statt mit Wahrscheinlichkeiten zu tun.
Gleichung (I) sagt dann aus:
Von N Systemen liegt bei

(I a) $d\,n = \varphi\, e^{-\frac{N}{R\,T}\,\Phi}\,d\,\alpha = F(\alpha)\,d\,\alpha$

Systemen der Wert des Parameters α in einem zufällig herausgegriffenen Zeitpunkt zwischen α und $\alpha + d\,\alpha$.

Diese Beziehung wollen wir dazu benutzen, die Größe der
durch die ungeordneten Wärmevorgänge erzeugten unregelmäßigen Veränderungen des Parameters α zu ermitteln. Zu
diesem Zweck drücken wir in Zeichen aus, daß die Funktion $F(\alpha)$

sich unter der vereinten Wirkung der dem Potential Φ entsprechenden Kraft und des ungeordneten Wärmeprozesses sich
innerhalb der Zeitspanne t nicht ändert; t bedeute hierbei eine
so kleine Zeit, daß die zugehörigen Änderungen der Größen α
der einzelnen Systeme als unendlich kleine Argumentänderungen
der Funktion $F(\alpha)$ betrachtet werden können.

Trägt man auf einer Geraden von einem bestimmten Null-
punkte aus den Größen α numerisch gleiche Strecken ab, so
entspricht jedem System ein Punkt (α) auf dieser Geraden.
$F(\alpha)$ ist die Lagerungsdichte der Systempunkte (α) auf der
Geraden. Durch einen beliebigen Punkt (α_0) der Geraden
müssen nun während der Zeit t genau soviele Systempunkte
in dem einen Sinne hindurchwandern, wie in dem anderen
Sinne.

Die dem Potential Φ entsprechende Kraft bewirke eine
Änderung von α von der Größe

$$\Delta_1 = - B \frac{\partial \Phi}{\partial \alpha} t,$$

wobei B von α unabhängig sei, d. h. die Änderungsgeschwindigkeit von α sei proportional der wirkenden Kraft und unabhängig
vom Werte des Parameters. Den Faktor B nennen wir die
,,Beweglichkeit des Systems in bezug auf α".

Würde also die äußere Kraft wirken, ohne daß der unregelmäßige molekulare Wärmeprozeß die Größen α änderte,
so gingen durch den Punkt (α_0) während der Zeit t

$$n_1 = B \left(\frac{\partial \Phi}{\partial \alpha} \right)_{\alpha = \alpha_0} . t . F(\alpha_0)$$

Systempunkte nach der negativen Seite hindurch.

Es sei ferner die Wahrscheinlichkeit dafür, daß der Parameter α eines Systems infolge des ungeordneten Wärmeprozesses
innerhalb der Zeit t eine Änderung erfahre, deren Wert zwischen Δ
und $\Delta + d\Delta$ liegt, gleich $\psi(\Delta)$, wobei $\psi(\Delta) = \psi(-\Delta)$ und ψ
von' α unabhängig sei. Die Anzahl der infolge des ungeord-
neten Wärmeprozesses durch den Punkt (α_0) während der
Zeit t nach der positiven Seite hin wandernden Systempunkte
ist dann:

$$n_2 = \int_{\Delta = 0}^{\Delta = \infty} F(\alpha_0 - \Delta) \chi(\Delta) d\Delta,$$

378   *A. Einstein.*

wenn

$$\int\limits_{\Delta}^{\infty} \psi(\Delta)\, d\Delta = \chi(\Delta)$$

gesetzt wird. Die Anzahl der nach der negativen Seite infolge des ungeordneten Wärmeprozesses wandernden Systempunkte ist:

$$n_3 = \int\limits_{\Delta}^{\infty} F(\alpha_0 + \Delta)\, \chi(\Delta)\, d\Delta.$$ [21]

Der mathematische Ausdruck für die Unveränderlichkeit der Funktion F ist also:

$$- n_1 + n_2 - n_3 = 0.$$

Setzt man die für n_1, n_2, n_3 gefundenen Ausdrücke ein und berücksichtigt, daß Δ unendlich klein ist bez. daß $\psi(\Delta)$ nur für unendlich kleine Werte von Δ von 0 verschieden ist, so erhält man hieraus nach einfacher Rechnung:

$$B \left(\frac{\partial \Phi}{\partial \alpha} \right)_{\alpha = \alpha_0} F(\alpha_0)\, t + \tfrac{1}{2} F'(\alpha_0)\, \overline{\Delta^2} = 0.$$

Hierbei bedeutet

$$\overline{\Delta^2} = \int\limits_{-\infty}^{+\infty} \Delta^2\, \psi(\Delta)\, d\Delta$$

den Mittelwert der Quadrate der durch den unregelmäßigen Wärmeprozeß während der Zeit t hervorgerufenen Änderungen der Größen α. Aus dieser Beziehung erhält man unter Berücksichtigung von Gleichung (Ia):

II) $$\sqrt{\overline{\Delta^2}} = \sqrt{\frac{2R}{N}} \cdot \sqrt{B\,T\,t}.$$

Hierbei bedeutet R die Konstante der Gasgleichung $(8,31 \cdot 10^7)$, N die Anzahl der wirklichen Moleküle in einem Grammolekül (ca. $4 \cdot 10^{23}$), B die „Beweglichkeit des Systems in bezug auf   [22] den Parameter α", T die absolute Temperatur, t die Zeit, innerhalb welcher die durch den ungeordneten Wärmeprozeß hervorgerufenen Änderungen von α stattfinden.

§ 4. Anwendung der abgeleiteten Gleichung auf die Brownsche Bewegung.

Wir berechnen nun mit Hilfe der Gleichungen (II) zunächst die mittlere Verschiebung, die ein kugelförmiger, in einer Flüssigkeit suspendierter Körper während der Zeit t in einer bestimmten Richtung (X-Richtung eines Koordinatensystems) erleidet. Zu diesem Zweck haben wir in jene Gleichung den entsprechenden Wert für B einzusetzen.

Wirkt auf eine Kugel vom Radius P, die in einer Flüssigkeit vom Reibungskoeffizienten k suspendiert ist, eine Kraft K, so bewegt sie sich mit der Geschwindigkeit[1]) $K/6\,\pi\,k\,P$. Es ist also zu setzen

$$B = \frac{1}{6\,\pi\,k\,P},$$

so daß man — in Übereinstimmung mit der oben zitierten Arbeit — für die mittlere Verschiebung der suspendierten Kugel in Richtung der X-Achse den Wert erhält:

$$\sqrt{\overline{\varDelta_x^2}} = \sqrt{t}\,\sqrt{\frac{R\,T}{N}\,\frac{1}{3\,\pi\,k\,P}}.$$

Wir behandeln zweitens den Fall, daß die betrachtete Kugel in der Flüssigkeit um einen ihrer Durchmesser (ohne Lagerreibung) frei drehbar gelagert sei und fragen nach der mittleren Drehung $\sqrt{\overline{\varDelta_r^2}}$ der Kugel während der Zeit t infolge des ungeordneten Wärmeprozesses.

Wirkt auf eine Kugel vom Radius P, die in einer Flüssigkeit vom Reibungskoeffizienten k drehbar gelagert ist, das Drehmoment D, so dreht sie sich mit der Winkelgeschwindigkeit[1])

[24]
$$\psi = \frac{D}{8\,\pi\,k\,P^3}.$$

Es ist also zu setzen:

$$B = \frac{1}{8\,\pi\,k\,P^3}.$$

Man erhält also:

$$\sqrt{\overline{\varDelta_r^2}} = \sqrt{t}\,\sqrt{\frac{R\,T}{N}\,\frac{1}{4\,\pi\,k\,P^3}}.$$

Die durch die Molekularbewegung erzeugte Drehbewegung sinkt also mit wachsendem P viel rascher als die fortschreitende Bewegung.

Für $P = 0,5$ mm und Wasser von 17^0 liefert die Formel für den im Mittel in einer Sekunde zurückgelegten Winkel etwa 11 Bogensekunden, in der Stunde ca. 11 Bogenminuten.

---

[23]   1) Vgl. G. Kirchhoff, Vorles. über Mechanik. 26. Vorl.

380   *A. Einstein.*

Für $P = 0{,}5$ Mikron und Wasser von 17^0 erhält man für
$t = 1$ Sekunde ca. 100 Winkelgrade.   [25]

    Bei einem frei schwebenden suspendierten Teilchen finden
drei voneinander unabhängige derartige Drehbewegungen statt.

    Die für $\sqrt{\overline{\varDelta^2}}$ entwickelte Formel ließe sich noch auf andere
Fälle anwenden. Setzt man z. B. für B den reziproken elek-
trischen Widerstand eines geschlossenen Stromkreises ein, so
gibt sie an, wieviel Elektrizität im Durchschnitt während der
Zeit t durch irgend einen Leiterquerschnitt geht, welche Beziehung abermals mit dem Grenzgesetz der Strahlung des
schwarzen Körpers für große Wellenlängen und hohe Temperaturen zusammenhängt. Da ich jedoch keine durch das   [26]
Experiment kontrollierbare Konsequenz mehr habe auffinden
können, scheint mir die Behandlung weiterer Spezialfälle unnütz.

§ 5. Über die Gültigkeitsgrenze der Formel für $\sqrt{\overline{\varDelta^2}}$. [27]

    Es ist klar, daß die Formel (II) nicht für beliebig kleine
Zeiten gültig sein kann. Die mittlere Veränderungsgeschwindig-
keit von α infolge des Wärmeprozesses

$$\frac{\sqrt{\overline{\varDelta^2}}}{t} = \sqrt{\frac{2\,R\,T\,B}{N}} \cdot \frac{1}{\sqrt{t}}$$

wird nämlich für unendlich kleine Zeitdauer t unendlich groß,
was offenbar unmöglich ist, denn es müßte sich ja sonst jeder
suspendierte Körper mit unendlich großer Momentangeschwindigkeit bewegen. Der Grund liegt daran, daß wir in unserer
Entwickelung implizite angenommen haben, daß der Vorgang
während der Zeit t als von dem Vorgange in den unmittelbar
vorangehenden Zeiten unabhängiges Ereignis aufzufassen sei.
Diese Annahme trifft aber um so weniger zu, je kleiner die
Zeiten t gewählt werden. Wäre nämlich zur Zeit $z = 0$

$$\frac{d\,\alpha}{d\,t} = \beta_0$$ [28]

der Momentanwert der Änderungsgeschwindigkeit, und würde
die Änderungsgeschwindigkeit β in einem gewissen darauf
folgenden Zeitintervall durch den ungeordneten thermischen
Prozeß nicht beeinflußt, sondern die Änderung von β lediglich

Theorie der Brownschen Bewegung.   **381**

durch den passiven Widerstand $(1/B)$ bestimmt, so würde für $d\beta/dz$ die Beziehung gelten:

$$- \mu \frac{d\beta}{dz} = \frac{\beta}{B}.$$

μ ist hierbei durch die Festsetzung definiert, daß $\mu(\beta^2/2)$ die der Änderungsgeschwindigkeit β entsprechende Energie sein soll. In dem Falle der Translationsbewegung der suspendierten Kugel wäre also z. B. $\mu(\beta^2/2)$ die kinetische Energie der Kugel samt der kinetischen Energie der mitbewegten Flüssigkeit. Durch Integration folgt:

$$\beta = \beta_0\, e^{-\frac{z}{\mu B}}.$$

Aus diesem Resultat folgert man, daß die Formel (II) nur [29] für Zeitintervalle gilt, welche groß sind gegen μB.

Für Körperchen von 1 Mikron Durchmesser und von der Dichte $\varrho = 1$ in Wasser von Zimmertemperatur ist die untere Grenze der Gültigkeit der Formel (II) ca. 10^{-7} Sekunden; diese untere Grenze für die Zeitintervalle wächst proportional dem Quadrat des Radius des Körperchens. Beides gilt sowohl für die fortschreitende wie für die Rotationsbewegung der Teilchen.

Bern, Dezember 1905.

(Eingegangen 19. Dezember 1905.)

Published in *Annalen der Physik* 19 (1906): 371–381. Dated Bern, December 1905, received 19 December 1905, published 8 February 1906.

[1] *Einstein 1905k* (Doc. 16).

[2] Henry Friedrich Wilhelm Siedentopf, one of the inventors of the ultramicroscope, worked at the Zeiss optical works in Jena.

[3] Louis-Georges Gouy, Professor of Physics at the University of Lyons, worked mainly in optics.

[4] *Gouy 1888*.

[5] For references to the contemporary experimental research, see the editorial note, "Einstein on Brownian Motion," § III and § VII, pp. 208–210, 219–222.

[6] Rotational Brownian motion is discussed heuristically in *Gouy 1888*. It was studied experimentally in 1909 by Perrin (*Perrin 1909a*).

[7] *Einstein 1905k* (Doc. 16). Einstein presumably meant to refer to *Einstein 1903* (Doc. 4). A reference to this paper was added to the republished version in *Einstein 1922*.

[8] Formula (1) is derived in § 3 of *Einstein 1903* (Doc. 4). In § 4, the factor of the energy in the exponential is related, using the equipartition theorem, to Boltzmann's constant, and thus to N/RT.

[9] In 1953, a criticism of this assumption (in D.K.C. MacDonald to Einstein, 20 February 1953) led to a discussion of the interpretation of entropy and probability (see Einstein to D.K.C. MacDonald, 2 March 1953 and 30 March 1953).

[10] Boltzmann generalized Maxwell's expo-

nential distribution law to include the case of external forces acting on the gas molecules (*Boltzmann 1868*; see also *Boltzmann 1896*, § 19).

[11] A square root sign over the second term of the equation is missing.

[12] The development of the ultramicroscope by Siedentopf and Zsigmondy had, in fact, shifted the lower limit of observability to ca. 10^{-6} cm (*Siedentopf and Zsigmondy 1903*). For a contemporary discussion of microscopic observability, see *Cotton and Mouton 1906*, chap. 1.

[13] For an elaboration of Einstein's study of Brownian motion under the influence of an elastic force, including a proposed experimental verification, see, e.g., *Smoluchowski 1913*; see also *Fürth 1922*, p. 65, fn. 15.

[14] The following argument is presented in greater detail in *Einstein 1905i* (Doc. 14), § 1 and § 2.

[15] *Einstein 1905k* (Doc. 16). Einstein presumably meant to refer to *Einstein 1905i* (Doc. 14), § 1 and § 2, as corrected in *Einstein 1922*, and, in particular, to p. 136.

[16] *Planck 1900a*. This paper does not, however, give the formula for black-body radiation to which Einstein referred. This formula is given, e.g., in *Planck 1901a*, which is cited by Einstein elsewhere for this formula (see, e.g., *Einstein 1905i* [Doc. 14], p. 136).

[17] See *Planck 1901b*.

[18] For a discussion of Einstein's views on this fundamental imperfection, see the editorial note, "Einstein's Early Work on the Quantum Hypothesis," pp. 139–141.

[19] In 1908, Perrin reported on experiments showing that the vertical distribution of granules of gamboge in a liquid is exponential (*Perrin 1908a*). Perrin mentioned Einstein's name, but gave a derivation of the exponential distribution that differs from Einstein's.

[20] The basic results of this section are generalizations of those derived in *Einstein 1905k* (Doc. 16), § 4.

[21] The integral should extend from 0 to ∞.

[22] This value for N is close to the value de-

rived in *Einstein 1906c* (Doc. 33); in *Einstein 1922*, the more accurate value 6×10^{23} per mole is given instead. For a discussion of the discrepancy between these values, and its origin, see the editorial note, "Einstein's Dissertation on the Determination of Molecular Dimensions," § V, pp. 179–182.

[23] *Kirchhoff 1897* (see, in particular, p. 380); for Einstein's previous use of Stokes's formula for the derivation of the mean square displacement in Brownian motion, see *Einstein 1905k* (Doc. 16), § 3 and § 5.

[24] See *Kirchhoff 1897*, in particular, pp. 375–376.

[25] Using the values for R and N given on p. 378, and the value of k given in *Einstein 1905j* (Doc. 15), p. 21, one obtains the result given by Einstein. In 1909, Perrin performed an experimental test of Einstein's formula (see *Perrin 1909a*). He worked with granules of resin, having a diameter of ca. 13 μ, which contained small inclusions that enabled him to follow their rotational motion. The experimental values he found are in good agreement with those predicted by Einstein's formula. See also note 6.

[26] See the previous discussion on p. 375. For an account of this and other closely related problems, see *De Haas-Lorentz 1913*, chap. 7 (in particular, pp. 87–88). Fluctuations of the potential difference between the plates of a condenser are treated in *Einstein 1907b* (Doc. 39). Einstein's study of charge and potential fluctuations was the starting point of his attempt to develop methods for the measurements of small quantities of electricity (see *Einstein 1908a* [Doc. 48] and Vol. 5, the editorial note, "Einstein's 'Maschinchen' for the Measurement of Small Quantities of Electricity").

[27] For a closely related discussion, see *Einstein 1907c* (Doc. 40).

[28] The equation should be "$d\alpha/dz = \beta_o$".

[29] This restriction was removed later by including the threshold value μB of the time in a description of Brownian motion valid for all time intervals (see the discussion in *Fürth 1922*, pp. 60–61, fn. 8).

33. "Supplement" to "A New Determination of Molecular Dimensions"

[*Einstein 1906c*]

DATED Bern, January 1906
PUBLISHED 8 February 1906

IN: *Annalen der Physik* 19 (1906): 305–306.

This supplement is appended to the version of Document 15 which appears in *Annalen der Physik* 19 (1906): 289–305. The journal version, also dated 30 April 1905, was received 19 August 1905.

Revised version published in *Einstein 1922*, p. 40.

Neue Bestimmung der Moleküldimensionen.   305

abhängig von der Natur des Lösungsmittels, der gelösten Substanz und der Temperatur herausstellen muß, wenn unsere Theorie den Tatsachen entspricht.

Wir wollen die Rechnung für wässerige Zuckerlösung durchführen. Nach den oben mitgeteilten Angaben über die innere Reibung der Zuckerlösung folgt zunächst für 20° C.:

$$N P^3 = 200.$$

Nach Versuchen von Graham (berechnet von Stefan) ist der Diffusionskoeffizient von Zucker in Wasser bei 9,5° C. 0,384, wenn der Tag als Zeiteinheit gewählt wird. Die Zähigkeit des Wassers bei 9,5° ist 0,0135. Wir wollen diese Daten in unsere Formel für den Diffusionskoeffizienten einsetzen, trotzdem sie an 10 proz. Lösungen gewonnen sind und eine genaue Gültigkeit unserer Formel bei so hohen Konzentrationen nicht zu erwarten ist. Wir erhalten

$$N P = 2,08 . 10^{16}.$$

Aus den für $N P^3$ und $N P$ gefundenen Werten folgt, wenn wir die Verschiedenheit von P bei 9,5° und 20° vernachlässigen,

$$P = 9,9 . 10^{-8} \text{ cm},$$

$$N = 2,1 . 10^{23}.$$

Der für N gefundene Wert stimmt der Größenordnung nach mit den durch andere Methoden gefundenen Werten für diese Größe befriedigend überein.

Bern, den 30. April 1905.

(Eingegangen 19. August 1905.)

Nachtrag.

In der neuen Auflage der physikalisch-chemischen Tabellen von Landolt und Börnstein finden sich weit brauchbarere [1] Angaben zur Berechnung der Größe des Zuckermoleküls und der Anzahl N der wirklichen Moleküle in einem Grammmolekül.

Thovert fand (Tab. p. 372) für den Diffusionskoeffizienten [2] von Zucker in Wasser bei 18,5° C. und der Konzentration

306   *A. Einstein.   Neue Bestimmung der Moleküldimensionen.*

0,005 Mol./Liter den Wert 0,33 cm²/Tage. Aus einer Tabelle

[3] mit Beobachtungsresultaten von H o s k i n g (Tab. p. 81) findet man ferner durch Interpolation, daß bei verdünnter Zucker-lösung einer Zunahme des Zuckergehaltes um 1 Proz. bei 18,5° C. eine Zunahme des Viskositätskoeffizienten um 0,00025 entspricht.

[4]   Unter Zugrundelegung dieser Angaben findet man
$$P = 0{,}78 \cdot 10^{-6}\,\text{mm}$$
und
$$N = 4{,}15 \cdot 10^{23}.$$
B e r n, Januar 1906.

Published in *Annalen der Physik* 19 (1906): 305–306. Dated Bern, January 1906, published 8 February 1906.

[1] *Landolt and Börnstein 1905*.

[2] *Landolt and Börnstein 1905*, p. 372.

[3] *Landolt and Börnstein 1905*, p. 81.

[4] See *Einstein 1905j* (Doc. 15), § 5. For later corrections of these values, based on the discovery of a calculational error, see *Einstein 1911d* and *1922*, where *N* was recalculated to be 6.56 × 10²³. For a discussion of the error, see the editorial note, ''Einstein's Dissertation on the Determination of Molecular Dimensions,'' § V, pp. 179–182.

34. "On the Theory of Light Production and Light Absorption"

[*Einstein 1906d*]

DATED Bern, March 1906
RECEIVED 13 March 1906
PUBLISHED 11 May 1906

IN: *Annalen der Physik* 20 (1906): 199–206.

12. *Zur Theorie der Lichterzeugung und Lichtabsorption; von A. Einstein.*

In einer letztes Jahr erschienenen Arbeit[1]) habe ich gezeigt, daß die Maxwellsche Theorie der Elektrizität in Verbindung mit der Elektronentheorie zu Ergebnissen führt, die mit den Erfahrungen über die Strahlung des schwarzen Körpers im Widerspruch sind. Auf einem dort dargelegten Wege wurde ich zu der Ansicht geführt, daß Licht von der Frequenz ν lediglich in Quanten von der Energie $(R/N)\beta\nu$ absorbiert und emittiert werden könne, wobei R die absolute Konstante der auf das Grammolekül angewendeten Gasgleichung, N die Anzahl der wirklichen Moleküle in einem Grammolekül, β den Exponentialkoeffizienten der Wienschen (bez. der Planckschen) Strahlungsformel und ν die Frequenz des betreffenden Lichtes bedeutet. Diese Beziehung wurde entwickelt für einen Bereich, der dem Bereich der Gültigkeit der Wienschen Strahlungsformel entspricht.

Damals schien es mir, als ob die Plancksche Theorie der Strahlung[2]) in gewisser Beziehung ein Gegenstück bildete zu meiner Arbeit. Neue Überlegungen, welche im § 1 dieser Arbeit mitgeteilt sind, zeigten mir aber, daß die theoretische Grundlage, auf welcher die Strahlungstheorie von Hrn. Planck ruht, sich von der Grundlage, die sich aus der Maxwellschen Theorie und Elektronentheorie ergeben würde, unterscheidet, und zwar gerade dadurch, daß die Plancksche Theorie implizite von der eben erwähnten Lichtquantenhypothese Gebrauch macht.

In § 2 der vorliegenden Arbeit wird mit Hilfe der Lichtquantenhypothese eine Beziehung zwischen Voltaeffekt und lichtelektrischer Zerstreuung hergeleitet.

---

1) A. Einstein, Ann. d. Phys. 17. p. 132. 1905.
2) M. Planck, Ann. d. Phys. 4. p. 561. 1901.

200   *A. Einstein.*

§ 1. Die **Plancksche Theorie der Strahlung** und die Lichtquanten.

In § 1 meiner oben zitierten Arbeit habe ich gezeigt, daß die Molekulartheorie der Wärme zusammen mit der Maxwellschen Theorie der Elektrizität und Elektronentheorie zu der mit der Erfahrung im Widerspruch stehenden Formel für die Strahlung des schwarzen Körpers führt:

$$(1) \qquad \varrho_\nu = \frac{R}{N}\,\frac{8\pi\nu^2}{L^3}\,T.$$

Hierbei bedeutet ϱ_ν die Dichte der Strahlung bei der Temperatur T, deren Frequenz zwischen ν und $\nu + 1$ liegt.

Woher kommt es, daß Hr. Planck nicht zu der gleichen Formel, sondern zu dem Ausdruck

$$(2) \qquad \varrho_\nu = \frac{\alpha\,\nu^3}{e^{\frac{\beta\nu}{T}} - 1} \qquad\qquad [5]$$

gelangt ist?

Hr. Planck hat abgeleitet[1]), daß die mittlere Energie $\bar{E}_\nu$ eines Resonators von der Eigenfrequenz ν, der sich in einem mit ungeordneter Strahlung erfüllten Raume befindet, durch die Gleichung [7]

$$(3) \qquad \bar{E}_\nu = \frac{L^3}{8\pi\nu^2}\,\varrho_\nu \qquad\qquad [8]$$

gegeben ist. Damit war das Problem der Strahlung des schwarzen Körpers reduziert auf die Aufgabe, $\bar{E}_\nu$ als Funktion der Temperatur zu bestimmen. Die letztere Aufgabe aber ist gelöst, wenn es gelingt, die Entropie eines aus einer großen Anzahl im dynamischen Gleichgewicht sich befindender, miteinander in Wechselwirkung stehender, gleich beschaffener Resonatoren von der Eigenfrequenz ν zu berechnen.

Die Resonatoren denken wir uns als Ionen, welche um eine Gleichgewichtslage geradlinige Sinusschwingungen auszuführen vermögen. Bei der Berechnung dieser Entropie spielt die Tatsache, daß die Ionen elektrische Ladungen besitzen, keine Rolle; wir haben diese Ionen einfach als Massenpunkte (Atome) aufzufassen, deren Momentanzustand durch ihre momentane Abweichung x von der Gleichgewichtslage und

---

1) M. Planck, Ann. d. Phys. **1**. p. 99. 1900.   [6]

durch ihre Momentangeschwindigkeit $dx/dt = \xi$ vollkommen bestimmt ist.

Damit bei thermodynamischem Gleichgewicht die Zustandsverteilung dieser Resonatoren eine eindeutig bestimmte sei, hat man anzunehmen, daß außer den Resonatoren frei bewegliche Moleküle in beliebig kleiner Zahl vorhanden seien, welche dadurch, daß sie mit den Ionen zusammenstoßen, Energie von [9] Resonator zu Resonator übertragen können; die letzteren Moleküle werden wir bei Berechnung der Entropie nicht berücksichtigen.

Wir könnten E_r als Funktion der Temperatur aus dem **Maxwell-Boltzmann**schen Verteilungsgesetz ermitteln und [10] würden dadurch zu der ungültigen Strahlungsformel (1) gelangen. Zu dem von Hrn. **Planck** eingeschlagenen Wege wird man in folgender Weise geführt.

Es seien $p_1 \ldots p_n$ geeignet gewählte Zustandsvariable [1]), welche den Zustand eines physikalischen Systems vollkommen bestimmen (z. B. in unserem Falle die Größen x und ξ sämtlicher Resonatoren). Die Entropie S dieses Systems bei der absoluten Temperatur T ist dargestellt durch die Gleichung [2]):

[12] (4) $$ S = \frac{\bar{H}}{T} + \frac{R}{N} \lg \int e^{-\frac{N}{RT}H} \, dp_1 \ldots dp_n, $$

wobei $\bar{H}$ die Energie des Systems bei der Temperatur T, H die Energie als Funktion der $p_1 \ldots p_n$ bedeutet, und das Integral über alle möglichen Wertkombinationen der $p_1 \ldots p_n$ zu erstrecken ist.

Besteht das System aus sehr vielen molekularen Gebilden — und nur in diesem Falle hat die Formel Bedeutung und Gültigkeit, so tragen nur solche Wertkombinationen der $p_1 \ldots p_n$ merklich zu dem Werte des in S auftretenden Integrales bei, [13] deren H sehr wenig von $\bar{H}$ abweicht.[3]) Berücksichtigt man dies, so ersieht man leicht, daß bis auf Vernachlässigbares gesetzt werden kann:

[14] $$ S = \frac{R}{N} \lg \int_{\bar{H}}^{\bar{H}+\varDelta \bar{H}} dp_1 \ldots dp_n, $$

[11]   1) A. Einstein, Ann. d. Phys. **11**. p. 170. 1903.
          2) l. c. § 6.
          3) Folgt aus § 3 und § 4 l. c.

202   *A. Einstein.*

wobei ΔH zwar sehr klein, aber doch so groß gewählt sei,
daß $R \lg (\Delta H)/N$ eine vernachlässigbare Größe ist. S ist dann
von der Größe von ΔH unabhängig.

Setzt man nun die Variabeln x_a und ξ_a der Resonatoren
an Stelle der $d p_1 \ldots d p_n$ in die Gleichung ein und berücksichtigt man, daß für den α^{ten} Resonator die Gleichung

$$\int_{E_a}^{E_a + d E_a} d x_a \, d \xi_a = \text{konst. } d E_a$$

gilt (da E_a eine quadratische, homogene Funktion von x_a und ξ_a
ist), so erhält man für S den Ausdruck:

(5) $$S = \frac{R}{N} \lg W,$$

wobei

(5 a) $$W = \int_{H}^{H + \Delta H} d E_1 \ldots d E_n$$

gesetzt ist.

Würde man S nach dieser Formel berechnen, so würde
man wieder zu der ungültigen Strahlungsformel (1) gelangen.   [15]
Zur Planckschen Formel aber gelangt man, indem man
voraussetzt, daß die Energie E_a eines Resonators nicht jeden
beliebigen Wert annehmen kann, sondern nur Werte, welche
ganzzahlige Vielfache von ε sind, wobei

$$\varepsilon = \frac{R}{N} \beta \nu.$$

Setzt man nämlich $\Delta H = \varepsilon$, so ersieht man sofort aus
Gleichung (5 a), daß nun W bis auf einen belanglosen Faktor
gerade in diejenige Größe übergeht, welche Hr. Planck „Anzahl der Komplexionen" genannt hat.   [16]

Wir müssen daher folgenden Satz als der Planckschen
Theorie der Strahlung zugrunde liegend ansehen:

Die Energie eines Elementarresonators kann nur Werte
annehmen, die ganzzahlige Vielfache von $(R/N) \beta \nu$ sind; die
Energie eines Resonators ändert sich durch Absorption und
Emission sprungweise, und zwar um ein ganzzahliges Vielfache von $(R/N) \beta \nu$.

Diese Voraussetzung involviert aber noch eine zweite, indem sie im Widerspruch steht mit der theoretischen Grundlage, aus der heraus Gleichung (3) entwickelt ist. Wenn die Energie eines Resonators sich nur sprungweise ändern kann, so kann nämlich zur Ermittelung der mittleren Energie eines in einem Strahlungsraum befindlichen Resonators die übliche Theorie der Elektrizität nicht Anwendung finden, da diese keine *ausgezeichneten* Energiewerte eines Resonators kennt. Es liegt also der Planckschen Theorie die Annahme zugrunde:

Obwohl die Maxwellsche Theorie auf Elementarresonatoren nicht anwendbar ist, so ist doch die *mittlere* Energie eines in einem Strahlungsraume befindlichen Elementarresonators gleich derjenigen, welche man mittels der Maxwellschen Theorie [17] der Elektrizität berechnet.

Der letztere Satz wäre ohne weiteres plausibel, wenn in allen Teilen des Spektrums, die für die Beobachtung in Betracht kommen, $\varepsilon = (R/N)\beta\nu$ klein wäre gegen die mittlere Energie E_ν eines Resonators; dies ist aber durchaus nicht der Fall. Innerhalb des Gültigkeitsbereiches der Wienschen Strahlungsformel ist nämlich $e^{\beta\nu/T}$ groß gegen 1. Man beweist nun leicht, daß nach der Planckschen Strahlungstheorie E_ν/ε innerhalb des Gültigkeitsbereiches der Wienschen Strahlungsformel den Wert $e^{-\beta\nu/T}$ hat; E_ν ist also weit kleiner als ε. Es kommt also überhaupt nur wenigen Resonatoren ein von Null verschiedener Wert der Energie zu.

Die vorstehenden Überlegungen widerlegen nach meiner Meinung durchaus nicht die Plancksche Theorie der Strahlung; sie scheinen mir vielmehr zu zeigen, daß Hr. Planck in seiner Strahlungstheorie ein neues hypothetisches Element — die Lichtquantenhypothese — in die Physik eingeführt hat.

§ 2. Eine zu erwartende quantitative Beziehung zwischen lichtelektrischer Zerstreuung und Voltaeffekt.

Ordnet man die Metalle nach ihrer lichtelektrischen Empfindlichkeit in eine Reihe, so erhält man bekanntlich die Voltasche Spannungsreihe, wobei die Metalle desto lichtempfindlicher sind, je näher sie dem elektropositiven Ende [18] [19] der Spannungsreihe liegen.

204   *A. Einstein.*

Man begreift diese Tatsache bis zu einem gewissen Grade unter alleiniger Zugrundelegung der Annahme, daß die die wirksamen Doppelschichten erzeugenden, hier nicht zu untersuchenden Kräfte nicht an der Berührungsfläche zwischen Metall und Metall, sondern an der Berührungsfläche zwischen Metall und Gas ihren Sitz haben.

Jene Kräfte mögen an der Oberfläche eines an ein Gas angrenzenden Metallstückes M eine elektrische Doppelschicht erzeugen, welcher eine Potentialdifferenz V zwischen Metall und Gas entspreche — positiv gerechnet, wenn das Metall das höhere Potential besitzt.

Es seien V_1 und V_2 die Spannungsdifferenzen zweier Metalle M_1 und M_2 bei elektrostatischem Gleichgewichte, falls die Metalle gegeneinander isoliert sind. Bringt man die beiden Metalle zur Berührung, so wird das elektrische Gleichgewicht gestört und es findet ein vollständiger[1]) Spannungsausgleich zwischen den Metallen statt. Dabei werden sich über die vorerwähnten Doppelschichten an den Grenzflächen Metall–Gas einfache Schichten superponieren; diesen entspricht ein elektrostatisches Feld im Luftraume, dessen Linienintegral gleich der Voltadifferenz ist.

Nennt man V_{l_1} bez. V_{l_2} die elektrischen Potentiale in Punkten des Gasraumes, welche den einander berührenden Metallen unmittelbar benachbart sind, und V' das Potential im Innern der Metalle, so ist

$$V' - V_{l_1} = V_1,$$
$$V' - V_{l_2} = V_2,$$

also

$$V_{l_2} - V_{l_1} = V_1 - V_2.$$

Die elektrostatisch meßbare Voltadifferenz ist also numerisch gleich der Differenz der Potentiale, welche die Metalle im Gase annehmen, falls sie voneinander isoliert sind.

Ionisiert man das Gas, so findet im Gasraum eine durch die daselbst vorhandenen elektrischen Kräfte hervorgerufene Wanderung der Ionen statt, welcher Wanderung in den Metallen ein Strom entspricht, der an der Berührungsstelle der Metalle

---

1) Von der Wirkung der thermoelektrischen Kräfte sehen wir ab.

[20]   vom Metall mit größerem V (schwächer elektropositiv) nach dem
[21]   Metall mit kleinerem V (stärker elektropositiv) gerichtet ist.

Es befinde sich nun ein Metall M isoliert in einem Gase.
Seine der Doppelschicht entsprechende Potentialdifferenz gegen
das Gas sei V. Um die Einheit negativer Elektrizität aus
dem Metall in das Gas zu befördern, muß eine dem Potential V
numerisch gleiche Arbeit geleistet werden. Je größer V, d. h.

[22]   je weniger elektropositiv das Metall ist, desto mehr Energie
ist also für die lichtelektrische Zerstreuung nötig, desto weniger
lichtelektrisch empfindlich wird also das Metall sein.

Soweit übersieht man die Tatsachen, ohne über die Natur
der lichtelektrischen Zerstreuung Annahmen zu machen. Die
Lichtquantenhypothese liefert aber außerdem eine quantitative
Beziehung zwischen Voltaeffekt und lichtelektrischer Zerstreuung.
Es wird nämlich einem negativen Elementarquantum (Ladung ε)
mindestens die Energie $V\varepsilon$ zugeführt werden müssen, um es
aus dem Metall in das Gas zu bewegen. Es wird also eine
Lichtart nur dann negative Elektrizität aus dem Metall entfernen können, wenn das „Lichtquant" der betreffenden Lichtart mindestens den Wert $V\varepsilon$ besitzt. Wir erhalten also:

$$V\varepsilon \leqq \frac{R}{N}\beta\nu,$$

oder

$$V \leqq \frac{R}{A}\beta\nu,$$

wobei A die Ladung eines Grammoleküls eines einwertigen
Ions ist.

Nehmen wir nun an, daß ein Teil der absorbierenden
Elektronen das Metall zu verlassen befähigt ist, sobald die
Energie der Lichtquanten $V\varepsilon$ übertrifft[1]) — welche Annahme
sehr plausibel ist —, so erhalten wir

$$V = \frac{R}{A}\beta\nu,$$

wobei ν die kleinste lichtelektrisch wirksame Frequenz bedeutet.

Sind also ν_1 und ν_2 die kleinsten Lichtfrequenzen, welche
auf die Metalle M_1 und M_2 wirken, so soll für die Voltasche

---

1) Von der thermischen Energie der Elektronen ist dabei abgesehen.

206   *A. Einstein.   Theorie der Lichterzeugung u. Lichtabsorption.*

Spannungsdifferenz V_{12} der beiden Metalle die Gleichung gelten:

$$-V_{12} = V_1 - V_2 = \frac{R}{A}\,\beta\,(v_1 - v_2),$$

oder, wenn V_{12} in Volt gemessen wird:

$$V_{12} = 4{,}2 \cdot 10^{-15}\,(v_2 - v_1).$$

In dieser Formel ist folgender, im großen ganzen jedenfalls gültige Satz enthalten: Je stärker elektropositiv ein Metall ist, desto kleiner ist die unterste wirksame Lichtfrequenz für   [23] das betreffende Metall. Es wäre von hohem Interesse zu wissen, ob die Formel auch in quantitativer Beziehung als Ausdruck der Tatsachen zu betrachten ist.   [24]

Bern, März 1906.

(Eingegangen 13. März 1906.)

Published in *Annalen der Physik* 20 (1906): 199–206. Dated Bern March 1906, received 13 March 1906, published 11 May 1906.

A set of page proofs, with several corrections in Einstein's hand, was given to Einstein's Bern acquaintance, Adolf Gasser (Gasser Nachlaß, Basel). Einstein's corrections are indicated in the relevant notes below.

[1] *Einstein 1905i* (Doc. 14).

[2] See *Einstein 1905i* (Doc. 14), pp. 133–134, for Einstein's derivation of the energy distribution of black-body radiation.

[3] That is, for large values of v/T. See *Einstein 1905i* (Doc. 14), p. 139.

[4] *Planck 1901a*. On the proof of this page "A. Einstein" appears in fn. 2 instead of "M. Planck." The error is not corrected in the proof. Planck's formula is eq. (12) of the page cited.

[5] Comparison between this expression and *Planck 1901a*, p. 561, eq. (12), shows that $\alpha = 8\pi h/c^3$, where c is the speed of light (L in Einstein's notation) and $\beta = h/k = hN/R$.

[6] *Planck 1900a*.

[7] Einstein is referring to Planck's "natural radiation" ("natürliche Strahlung"). For Einstein's definition, see *Einstein 1905i* (Doc. 14), p. 135, fn. 2.

[8] This equation does not appear on p. 99 of *Planck 1900a*. It can be derived from equations on pp. 99 and 111. Planck first published the equation in *Planck 1900e*, p. 241.

[9] This is substantially Einstein's model in *Einstein 1905i* (Doc. 14), § 1.

[10] See *Einstein 1905i* (Doc. 14), p. 136.

[11] *Einstein 1903* (Doc. 4).

[12] An equation equivalent to eq. (4) is derived in *Einstein 1903* (Doc. 4), § 6, p. 182.

[13] See *Einstein 1903* (Doc. 4), pp. 174–178.

[14] The limits on the integral here and in eq. (5a) should be $\overline{H}$ and $\overline{H} + \Delta\overline{H}$, and ΔH in the next lines of text should be $\Delta\overline{H}$.

[15] For details of the calculation, see, e.g., *Kuhn 1978*, pp. 183–184.

[16] In *Planck 1900d*, p. 240, and *Planck 1901a*, p. 557, Planck refers to W as "the number of all possible complexions" ("die Anzahl

aller möglichen Complexionen''). For details of the calculation referred to in the text, see, e.g., *Kuhn 1978*, p. 184.

[17] In *Einstein 1907h* (Doc. 45), p. 372, Einstein stated conditions under which Maxwell's theory of radiation can be utilized.

[18] On the page proof, ''elektronegativen'' appeared in place of ''elektropositiven.'' Einstein corrected it to ''elektropositiven.''

[19] The phenomenon described in the text was first reported in *Elster and Geitel 1891* for alkali metals and their amalgams. Prior to *Millikan 1916b*, observations involving the nonalkaline metals seemed to contradict their result. *Millikan 1916b* confirms it, as well as Einstein's predictions in this section. Contact electricity is discussed in Einstein's ETH notes, H. F. Weber's Lectures on Physics, ca. December 1897–ca. June 1898 (Vol. 1, Doc. 37), pp. 178–181. For a review of the Voltaic effect and the electro-motive force series, see *Ostwald 1893*, chap. 8. For a review of contact electricity, including the electromotive force series, see *Auerbach 1893*.

[20] On the page proof ''stärker'' appeared in place of ''schwächer.'' Einstein corrected it to ''schwächer.''

[21] On the page proof ''schwächer'' appeared in place of ''stärker.'' Einstein corrected it to ''stärker.''

[22] On the page proof ''stärker'' appeared in place of ''weniger.'' Einstein corrected it to ''weniger.''

[23] On the page proof ''höher'' appeared in place of ''kleiner.'' Einstein corrected it to ''kleiner.''

[24] Einstein's equation for V_{12} and its generalization are discussed in *Millikan 1916a*; experimental tests of the generalized equation are reported in *Millikan 1916b*.

35. "The Principle of Conservation of Motion of the Center of Gravity and the Inertia of Energy"

[Einstein 1906e]

DATED Bern, May 1906
RECEIVED 17 May 1906
PUBLISHED 26 June 1906

IN: *Annalen der Physik* 20 (1906): 627–633.

13. *Das Prinzip von der Erhaltung der Schwerpunktsbewegung und die Trägheit der Energie; von A. Einstein.*

In einer voriges Jahr publizierten Arbeit[1]) habe ich gezeigt, daß die Maxwellschen elektromagnetischen Gleichungen in Verbindung mit dem Relativitätsprinzip und Energieprinzip zu der Folgerung führen, daß die Masse eines Körpers bei Änderung von dessen Energieinhalt sich ändere, welcher Art auch jene Energieänderung sein möge. Es zeigte sich, daß einer Energieänderung von der Größe ΔE eine gleichsinnige Änderung der Masse von der Größe $\Delta E/V^2$ entsprechen müsse, wobei V die Lichtgeschwindigkeit bedeutet.

In dieser Arbeit will ich nun zeigen, daß jener Satz die notwendige und hinreichende Bedingung dafür ist, daß das Gesetz von der Erhaltung der Bewegung des Schwerpunktes (wenigstens in erster Annäherung) auch für Systeme gelte, in welchen außer mechanische auch elektromagnetische Prozesse vorkommen. Trotzdem die einfachen formalen Betrachtungen, die zum Nachweis dieser Behauptung durchgeführt werden müssen, in der Hauptsache bereits in einer Arbeit von H. Poincaré enthalten sind[2]), werde ich mich doch der Übersichtlichkeit halber nicht auf jene Arbeit stützen.

§ 1. Ein Spezialfall.

K sei ein im Raume frei schwebender, ruhender starrer Hohlzylinder. In A sei eine Einrichtung, um eine bestimmte Menge S strahlender Energie durch den Hohlraum nach B zu senden. Während der Aussendung jener Strahlungsmenge wirkt ein Strahlungsdruck auf die linke Innenwand des Hohlzylinders K, der letzterem eine gewisse nach links gerichtete Geschwindigkeit verleiht. Besitzt der Hohlzylinder die Masse M,

---

[1]   1) A. Einstein, Ann. d. Phys. 18. p. 639. 1905.
[2]   2) H. Poincaré, Lorentz-Festschrift p. 252. 1900.

628 *A. Einstein.*

so ist diese Geschwindigkeit, wie aus den Gesetzen des Strahlungsdruckes leicht zu beweisen, gleich $\dfrac{1}{V}\dfrac{S}{M}$, wobei V die Lichtgeschwindigkeit bedeutet. Diese Geschwindigkeit behält K so lange, bis der Strahlenkomplex, dessen räumliche Ausdehnung im Verhältnis zu der des Hohlraumes von K sehr klein sei, in B absorbiert ist. Die Dauer der Bewegung des Hohlzylinders ist (bis auf Glieder höherer Ordnung) gleich α/V, wenn α die Entfernung zwischen A und B bedeutet. Nach Absorption des Strahlenkomplexes in B ruht der Körper K wieder. Bei dem betrachteten Strahlungsvorgang hat sich K um die Strecke

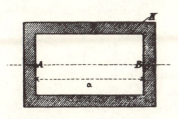

$$\delta = \frac{1}{V}\frac{S}{M}\cdot\frac{\alpha}{V}$$

nach links verschoben.

Im Hohlraum von K sei ein der Einfachheit halber masselos gedachter Körper k vorhanden nebst einem (ebenfalls masselosen) Mechanismus, um den Körper k, der sich zunächst in B befinden möge, zwischen B und A hin und her zu bewegen. Nachdem die Strahlungsmenge S in B aufgenommen ist, werde diese Energiemenge auf k übertragen, und hierauf k nach A bewegt. Endlich werde die Energiemenge S in A wieder vom Hohlzylinder K aufgenommen und k wieder nach B zurückbewegt. Das ganze System hat nun einen vollständigen Kreisprozeß durchgemacht, den man sich beliebig oft wiederholt denken kann.

Nimmt man an, daß der Transportkörper k auch dann masselos ist, wenn er die Energiemenge S aufgenommen hat, so muß man auch annehmen, daß der Rücktransport der Energiemenge S nicht mit einer Lagenänderung des Hohlzylinders K verbunden sei. Der Erfolg des ganzen geschilderten Kreisprozesses besteht also einzig in einer Verschiebung δ des ganzen Systems nach links, welche Verschiebung durch Wiederholung des Kreisprozesses beliebig groß gemacht werden kann. Wir erhalten also das Resultat, daß ein ursprünglich ruhendes System, ohne daß äußere Kräfte auf dasselbe wirken, die Lage

Prinzip von der Erhaltung der Schwerpunktsbewegung etc.   **629**

seines Schwerpunktes beliebig viel verändern kann, und zwar ohne daß das System irgend eine dauernde Veränderung erlitte.

Es ist klar, daß das erlangte Resultat keinen inneren Widerspruch enthält; wohl aber widerstreitet es den Grundgesetzen der Mechanik, nach denen ein ursprünglich ruhender Körper, auf welchen andere Körper nicht einwirken, keine Translationsbewegung ausführen kann.

Setzt man jedoch voraus, daß jeglicher Energie E die Trägheit E/V^2 zukomme, so verschwindet der Widerspruch mit den Elementen der Mechanik. Nach dieser Annahme besitzt nämlich der Transportkörper, während er die Energiemenge S von B nach A transportiert, die Masse S/V^2; und da der Schwerpunkt *des ganzen Systems* während dieses Vorganges nach dem Schwerpunktssatz ruhen muß, so erfährt der Hohlzylinder K während desselben im ganzen eine Verschiebung S' nach rechts von der Größe

$$\delta' = \alpha \cdot \frac{S}{V^2} \cdot \frac{1}{M}.$$

Ein Vergleich mit dem oben gefundenen Resultat zeigt, daß (wenigstens in erster Annäherung) $\delta = \delta'$ ist, daß also die Lage des Systems vor und nach dem Kreisprozeß dieselbe ist. Damit ist der Widerspruch mit den Elementen der Mechanik beseitigt.

§ 2. Über den Satz von der Erhaltung der Bewegung des Schwerpunktes.

Wir betrachten ein System von n diskreten materiellen Punkten mit den Massen $m_1, m_2 \ldots m_n$ und den Schwerpunktskoordinaten $x_1 \ldots z_n$. Diese materiellen Punkte seien in thermischer und elektrischer Beziehung nicht als Elementargebilde (Atome, Moleküle), sondern als Körper im gewöhnlichen Sinne von geringen Dimensionen aufzufassen, deren Energie durch die Schwerpunktsgeschwindigkeit nicht bestimmt sei. Diese Massen mögen sowohl durch elektromagnetische Vorgänge als auch durch konservative Kräfte (z. B. Schwerkraft, starre Verbindungen) aufeinander einwirken; wir wollen jedoch annehmen, daß sowohl die potenzielle Energie der konservativen Kräfte als auch die

630   *A. Einstein.*

kinetische Energie der Schwerpunktsbewegung der Massen stets als unendlich klein relativ zu der „inneren" Energie der Massen $m_1 \ldots m_n$ aufzufassen seien.

Es mögen im ganzen Raume die Maxwell-Lorenzschen Gleichungen

(1) $$\begin{cases} \dfrac{u}{V}\varrho + \dfrac{1}{V}\dfrac{dX}{dt} = \dfrac{\partial N}{\partial y} - \dfrac{\partial M}{\partial z}, \\[2ex] \dfrac{u}{V}\varrho + \dfrac{1}{V}\dfrac{dY}{dt} = \dfrac{\partial L}{\partial z} - \dfrac{\partial N}{\partial x}, \\[2ex] \dfrac{u}{V}\varrho + \dfrac{1}{V}\dfrac{dZ}{dt} = \dfrac{\partial M}{\partial x} - \dfrac{\partial L}{\partial y}, \\[2ex] \dfrac{1}{V}\dfrac{dL}{dt} = \dfrac{\partial Y}{\partial z} - \dfrac{\partial Z}{\partial y}, \\[2ex] \dfrac{1}{V}\dfrac{dM}{dt} = \dfrac{\partial Z}{\partial x} - \dfrac{\partial X}{\partial z}, \\[2ex] \dfrac{1}{V}\dfrac{dN}{dt} = \dfrac{dX}{\partial y} - \dfrac{\partial Y}{\partial x} \end{cases}$$

[3]

gelten, wobei

$$\varrho = \frac{\partial X}{\partial x} + \frac{\partial Y}{\partial y} + \frac{\partial Z}{\partial z}$$

die 4π-fache Dichte der Elektrizität bedeutet.

Addiert man die der Reihe nach mit

$$\frac{V}{4\pi} X x, \quad \frac{V}{4\pi} Y x \ldots \frac{V}{4\pi} N x$$

multiplizierten Gleichungen (1) und integriert man dieselben über den ganzen Raum, so erhält man nach einigen partiellen Integrationen die Gleichung

(2) $$\begin{cases} \displaystyle\int \frac{\varrho}{4\pi} x(uX + vY + wZ)d\tau \\[2ex] + \dfrac{d}{dt}\left\{\displaystyle\int x \cdot \frac{1}{8\pi}(X^2 + Y^2 \ldots + N^2)d\tau\right\} \\[2ex] - \dfrac{V}{8\pi}\displaystyle\int (YN - ZM)d\tau = 0. \end{cases}$$

[4]

Das erste Glied dieser Gleichung stellt die von dem elektromagnetischen Felde den Körpern $m_1 \ldots m_n$ zugeführte Energie dar. Nach unserer Hypothese von der Abhängigkeit der Massen von der Energie hat man daher das erste Glied der Summe dem Ausdruck

[5]

$$V^2 \sum x_\nu \frac{dm_\nu}{dt}$$

Prinzip von der Erhaltung der Schwerpunktsbewegung etc.   631

gleichzusetzen, da wir nach dem Obigen annehmen, daß die einzelnen materiellen Punkte m_ν ihre Energie und daher auch ihre Masse *nur* durch Aufnahme von elektromagnetischer Energie ändern.

Schreiben wir ferner auch dem elektromagnetischen Felde eine Massendichte (ϱ_e) zu, die sich von der Energiedichte durch den Faktor $1/V^2$ unterscheidet, so nimmt das zweite Glied der Gleichung die Form an:

$$V^2 \frac{d}{dt}\left\{\int x\,\varrho_e\,d\tau\right\}.$$

Bezeichnet man mit J das im dritten Gliede der Gleichung (2) auftretende Integral, so geht letztere über in:

$$(2a)\qquad \sum\left(x_\nu \frac{dm_\nu}{dt}\right) + \frac{d}{dt}\left\{\int x\,\varrho_e\,d\tau\right\} - \frac{1}{4\pi V}J = 0.$$

Wir haben nun die Bedeutung des Integrales J aufzusuchen. Multipliziert man die zweite, dritte, fünfte und sechste der Gleichungen (1) der Reihe nach mit den Faktoren NV, $-MV$, $-ZV$, YV, addiert und integriert über den Raum, so erhält man nach einigen partiellen Integrationen

$$[6]\quad (3)\qquad \frac{dJ}{dt} = -4\pi V\int\frac{\varrho}{4\pi}\left(X + \frac{v}{V}N - \frac{w}{V}M\right)d\tau = -4VR_x,$$

wobei R_x die algebraische Summe der X-Komponenten aller vom elektromagnetischen Felde auf die Massen $m_1 \ldots m_n$ ausgeübten Kräfte bedeutet. Da die entsprechende Summe aller von den konservativen Wechselwirkungen herrührenden Kräfte verschwindet, so ist R_x gleichzeitig die Summe der X-Komponenten *aller* auf die Massen m_ν ausgeübten Kräfte.

Wir wollen uns nun zunächst mit Gleichung (3) befassen, welche von der Hypothese, daß die Masse von der Energie abhängig sei, unabhängig ist. Sehen wir zunächst von der Abhängigkeit der Massen von der Energie ab und bezeichnen wir mit $\mathfrak{X}_\nu$ die Resultierende aller X-Komponenten der auf m_ν wirkenden Kräfte, so haben wir für die Masse m_ν die Bewegungsgleichung aufzustellen:

$$(4)\qquad m_\nu\frac{d^2 x_\nu}{dt^2} = \frac{d}{dt}\left\{m_\nu\frac{dx_\nu}{dt}\right\} = \mathfrak{X}_\nu.$$

632   *A. Einstein.*

folglich erhalten wir auch:

(5)
$$\frac{d}{dt} \sum \left(m_\nu \frac{dx_\nu}{dt} \right) = \sum \mathfrak{X}_\nu = R_x.$$

Aus Gleichung (5) und Gleichung (3) erhält man

(6)
$$\frac{J}{4\pi V} + \sum m_\nu \frac{dx_\nu}{dt} = \text{konst.}$$

Führen wir nun die Hypothese wieder ein, daß die Größen m_ν von der Energie also auch von der Zeit abhängen, so stellt sich uns die Schwierigkeit entgegen, daß für diesen Fall die mechanischen Gleichungen nicht mehr bekannt sind; das erste Gleichheitszeichen der Gleichung (4) gilt nun nicht mehr. Es ist jedoch zu beachten, daß die Differenz

$$\frac{d}{dt} \left\{ m_\nu \frac{dx_\nu}{dt} \right\} - m_\nu \frac{d^2 x_\nu}{dt^2} = \frac{dm_\nu}{dt} \frac{dx_\nu}{dt}$$

$$= \frac{1}{V^2} \int \frac{\varrho}{4\pi} \frac{dx_\nu}{dt} (u X + v Y + w Z) d\tau$$

in den Geschwindigkeiten vom zweiten Grade ist. Sind daher alle Geschwindigkeiten so klein, daß Glieder zweiten Grades vernachlässigt werden dürfen, so gilt auch bei Veränderlichkeit der Masse m_ν die Gleichung

$$\frac{d}{dt} \left(m_\nu \frac{dx_\nu}{dt} \right) = \mathfrak{X}_\nu$$

sicher mit der in Betracht kommenden Genauigkeit. Es gelten dann auch die Gleichungen (5) und (6), und man erhält aus den Gleichungen (6) und (2a):

(2b)
$$\frac{d}{dt} \left[\sum (m_\nu x_\nu) + \int x \varrho_e d\tau \right] = \text{konst.}$$

Bezeichnet ξ die X-Koordinate des Schwerpunktes der ponderabelen Massen und der Energiemasse des elektromagnetischen Feldes, so ist

$$\xi = \frac{\sum (m_\nu x_\nu) + \int x \varrho_e d\tau}{\sum m_\nu + \int \varrho_e d\tau},$$

wobei nach dem Energieprinzip der Wert des Nenners der

Prinzip von der Erhaltung der Schwerpunktsbewegung etc.   633

rechten Seite von der Zeit unabhängig ist.[1]) Wir können daher
Gleichung (2b) auch in der Form schreiben:

$$(2\,\mathrm{c}) \qquad\qquad \frac{d\,\xi}{d\,t} = \text{konst.}$$

Schreibt man also jeglicher Energie E die träge Masse E/V^2
zu, so gilt — wenigstens in erster Annäherung — das Prinzip
von der Erhaltung der Bewegung des Schwerpunktes auch für
Systeme, in denen elektromagnetische Prozesse vorkommen.

Aus der vorstehenden Untersuchung folgt, daß man entweder auf den Grundsatz der Mechanik, nach welchem ein
ursprünglich ruhender, äußeren Kräften nicht unterworfener
Körper keine Translationsbewegung ausführen kann, verzichten
oder annehmen muß, daß die Trägheit eines Körpers nach
dem angegebenen Gesetze von dessen Energieinhalt abhänge.

Bern, Mai 1906.

----

1) Nach der in dieser Arbeit entwickelten Auffassung ist der Satz
von der Konstanz der Masse ein Spezialfall des Energieprinzipes.

(Eingegangen 17. Mai 1906.)

Published in *Annalen der Physik* 20 (1906):
627–633. Dated Bern, May 1906, received 17
May 1906, published 26 June 1906.

[1] *Einstein 1905s* (Doc. 24).

[2] *Poincaré 1900*.

[3] The first term in this equation should be
"$\frac{v}{V}\rho$," and the first term in the following equation should be "$\frac{w}{V}\rho$."

[4] The coefficient of the integral should be
"$-V/4\pi$."

[5] The first term actually represents the x-component of time derivative of the center of energy.

[6] The last term should be "$-4\pi VR_x$."

36. ''On a Method for the Determination of the Ratio of the Transverse and the Longitudinal Mass of the Electron''

[*Einstein 1906g*]

DATED Bern, August 1906
RECEIVED 4 August 1906
PUBLISHED 20 November 1906

IN: *Annalen der Physik* 21 (1906): 583–586.

9. *Über eine Methode zur Bestimmung des Verhältnisses der transversalen und longitudinalen Masse des Elektrons; von A. Einstein.*

Drei die Kathodenstrahlen betreffende Größen gibt es, welche einer präzisen Beobachtung zugänglich sind, nämlich die Spannung, welche den Strahlen ihre Geschwindigkeit verleiht (Erzeugungsspannung), die elektrostatische Ablenkbarkeit [1] und die magnetische Ablenkbarkeit. Zwischen diesen drei Größen gibt es zwei voneinander unabhängige Beziehungen, deren Kenntnis für bedeutende Strahlengeschwindigkeiten von hervorragendem theoretischen Interesse ist. Eine dieser Be- [2] ziehungen wurde für β-Strahlen von Hrn. Kaufmann untersucht, nämlich der Zusammenhang zwischen magnetischer und elektrostatischer Ablenkbarkeit.

Im folgenden soll darauf aufmerksam gemacht werden, daß eine zweite Beziehung zwischen diesen Größen mit hinreichender Genauigkeit bestimmt werden kann, nämlich die Beziehung zwischen Erzeugungsspannung und elektrostatischer Ablenkbarkeit der Kathodenstrahlen oder — was dasselbe bedeutet — das Verhältnis der transversalen zur longitudinalen [3] Masse des Elektrons in Funktion der Erzeugungsspannung.

Wenn das Quadrat der Geschwindigkeit der Elektronen sehr klein ist gegenüber dem Quadrat der Lichtgeschwindigkeit, so gelten für die Bewegung des Elektrons die Gleichungen

$$\frac{d^2 x}{d t^2} = - \frac{\varepsilon}{\mu_0} X \text{ etc.,}$$

wobei ε/μ_0 das Verhältnis der Ladung zur Masse des Elektrons, x, y, z die Koordinaten des Elektrons und X, Y, Z die Komponenten der elektrischen Kraft des Feldes bedeuten, falls andere Kräfte als elektrostatische nicht auf das Elektron wirken. Wir nehmen an, die Elektronen bewegen sich mit der Anfangsgeschwindigkeit Null von einem gewissen Punkte x_0, y_0, z_0 (Kathode) aus. Die Bewegung ist dann eindeutig

584   *A. Einstein.*

bestimmt durch obige Gleichungen; sie sei gegeben durch die Gleichungen

$$x = \varphi_1(t),$$
$$y = \varphi_2(t),$$
$$z = \varphi_3(t).$$

Denkt man sich alle elektrostatischen Kraftkomponenten überall mit n^2 multipliziert, so bewegt sich nunmehr — wie leicht aus den obigen Bewegungsgleichungen zu ersehen ist — das Elektron gemäß den Gleichungen

$$x = \varphi_1(nt),$$
$$y = \varphi_2(nt),$$
$$z = \varphi_3(nt).$$

Hieraus folgt, daß bei Proportionaländerung des Feldes wohl die Geschwindigkeit, nicht aber die Bahn der Elektronen sich ändert.

Eine Änderung der Bahn tritt bei Proportionaländerung des Feldes offenbar erst bei solchen Elektrongeschwindigkeiten ein, bei welchen das Verhältnis von transversaler und longitudinaler Masse merklich von der Einheit abweicht. Wählt man das elektrostatische Feld derart, daß die Kathodenstrahlen eine stark gekrümmte Bahn durchlaufen, so werden bereits geringe Verschiedenheiten der transversalen und longitudinalen

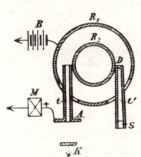

Masse einen beobachtbaren Einfluß auf die Bahnkurve haben. Nebenstehende schematische Skizze zeigt eine Anordnung, mittels welcher man das Verhältnis der transversalen zur longitudinalen Masse des Elektrons nach dem angedeuteten Prinzip bestimmen könnte. Die Kathodenstrahlen erlangen zwischen der geerdeten Kathode K und der an die positive Klemme der Stromquelle M angeschlossenen, zugleich als Blende dienenden Anode A ihre Geschwindigkeit, werden hierauf durch das mit A verbundene Röhrchen t in den Raum zwischen den Metallzylindern R_1 und R_2 eingeführt. R_1 ist geerdet, R_2 mit t, also mit dem positiven Pol der Stromquelle leitend verbunden, deren negativer Pol geerdet ist. Die Dimensionen seien so gewählt, daß sich langsame Kathodenstrahlen annähernd in einem Kreise bewegen, und zwar in geringer Entfernung von R_2. Die Strahlen gelangen hierauf in die mit R_2 metallisch verbundene, etwas konische Metallröhre t', in welcher sich der phosphoreszierende Schirm S befindet. Auf letzteren falle der Schatten des am inneren Ende von t' angeordneten vertikalen Drahtes D.

Bei Anwendung langsamer Kathodenstrahlen erhält der Schatten von D auf S eine ganz bestimmte Lage (Nullage). Erhöht man die Erzeugungsspannung der Strahlen, so wandert der Drahtschatten. Durch Einschalten einer Batterie B in die Erdungsleitung von R_1 werde jedoch der Schatten wieder in die Nullage zurückgeführt.

Bezeichnet man mit Π das Potential, bei welchem die Ablenkung der schattenbildenden Strahlen erfolgt, so ist Π auch diejenige Spannung, welcher die in Ablenkung begriffenen Strahlen ihre kinetische Energie verdanken. Bezeichnet ferner ϱ den Krümmungsradius der schattenbildenden Strahlen, so ist

[4]
$$\frac{\mu_t}{\mu_l} = \frac{\varrho}{2} \frac{X}{\Pi} .$$

Hierbei bedeutet μ_t die „transversale Masse" des Elektrons, μ_l diejenige longitudinale Masse, welche durch die Gleichung

$$\text{Kinetische Energie} = \mu_l \frac{v^2}{2}$$

definiert ist und X die ablenkende elektrische Kraft.

Nennt man P das Potential von R_2 (Potential des positiven Poles der Stromquelle M), p das Potential von R_1, bei welchem sich der Schatten in der Nullage befindet, so ist

$$\Pi = P - \alpha(P - p),$$

wobei α eine von den Apparatdimensionen abhängige, gegen 1 [5] kleine Konstante bedeutet. Ferner ist die Größe X der Spannung $P - p$ proportional. Man erhält also aus obiger Gleichung

$$\frac{\mu_t}{\mu_l} = \text{konst.} \frac{P - p}{P - \alpha(P - p)},$$

oder (mit einigen erlaubten Vernachlässigungen)

$$\frac{\mu_t}{\mu_l} = \text{konst.} \left[1 - (1 + \alpha)\frac{p}{P}\right].$$

586   *A. Einstein.   Methode zur Bestimmung etc.*

Da α offenbar mit genügender Genauigkeit ermittelt werden kann und P und p bis auf wenige Prozent genau meßbar sind, so ist die Genauigkeit, mit welcher die Abweichung der Größe μ_t/μ_l von der Einheit ermittelt werden kann, im wesentlichen bestimmt durch die Genauigkeit, mit welcher auf die Nullage des Drahtschattens eingestellt werden kann. Man überzeugt sich leicht, daß letztere Genauigkeit so groß gemacht werden kann, daß eine Abweichung der Größe μ_t/μ_l von der Einheit um 0,3 Proz. (entsprechend einer Schattenverschiebung von ca. 1 mm, wenn $\overline{DS} = 10$ cm) noch bemerkt werden kann. Zu erwähnen ist insbesondere, daß die unvermeidlichen Schwankungen, denen beim Experiment das Potential P unterworfen ist, nur von unbedeutendem Einfluß auf die Genauigkeit der Messung sein können. [6]

Wir wollen noch die Beziehung zwischen μ_t/μ_l und Π in erster Annäherung angeben, wie sie sich aus den verschiedenen Theorien ergibt. Wird Π in Volt ausgedrückt, so gilt

nach der Theorie von **Bucherer:**   [7]

$$\frac{\mu_t}{\mu_l} = 1 - 0{,}0070 \cdot \frac{\Pi}{10\,000},$$

nach der Theorie von **Abraham:**   [8]

$$\frac{\mu_t}{\mu_l} = 1 - 0{,}0084 \cdot \frac{\Pi}{10\,000},$$

nach der Theorie von **Lorentz** und **Einstein:**   [9]

$$\frac{\mu_t}{\mu_l} = 1 - 0{,}0104 \cdot \frac{\Pi}{10\,000}.$$

Da ich nicht in der Lage bin, selbst experimentell zu arbeiten, würde es mich freuen, wenn sich ein Physiker für die dargelegte Methode interessierte. [10]

Bern, August 1906.

(Eingegangen 4. August 1906.)

Published in *Annalen der Physik* 21 (1906): 583–586. Dated Bern, August 1906, received 4 August 1906, published 20 November 1906.

[1] For the concepts of electrostatic and magnetic power of deflection, see *Einstein 1905r* (Doc. 23), p. 920, and *Einstein 1907j* (Doc. 47), p. 437.

[2] See *Kaufmann 1906a*, which also cites Kaufmann's earlier papers.

[3] For Einstein's original definitions of the transverse and longitudinal mass of an electron, see *Einstein 1905r* (Doc. 23), p. 919. For his definition of the longitudinal mass in this paper, see p. 585.

[4] This formula is easily derived from the definition of μ_l given below.

[5] α is small because the electrons move in circular trajectories close to R_2.

[6] See *Einstein 1907b* (Doc. 39), in particular p. 571, for a discussion of these fluctuations.

[7] See *Bucherer 1904*, p. 58.

[8] See *Abraham 1902a, 1902b, 1903*.

[9] See *Lorentz 1904a, Einstein 1905r* (Doc. 23). Einstein apparently was the first person to use the phrase, "Theorie von Lorentz und Einstein," which soon was adopted by others (see, e.g., *Planck 1906b*, p. 424). Kaufmann had earlier referred to the "Lorentz-Einsteinschen Grundannahme" (*Kaufmann 1906a*, p. 495), but was careful to distinguish between the two theories (see ibid., pp. 491–493).

[10] In 1907 Einstein noted that sufficiently fast cathode rays for such tests were still not available. See *Einstein 1907j* (Doc. 47), p. 437.

37. Review of Max Planck, *Lectures on the Theory of Thermal Radiation*

[*Einstein 1906f*]

First half of August 1906

PUBLISHED IN: *Beiblätter zu den Annalen der Physik* 30 (1906), no. 15: 764–766.

bestätigen vollkommen das obige Gesetz. Der Wert 17 für den Diffusionskoeffizienten von Wasser unter Atmosphärendruck bei 17,7° ist in guter Übereinstimmung mit den nach anderen Methoden gefundenen Werten. Voraussetzung für die Genauigkeit der Methode ist die Abwesenheit von Verunreinigungen, welche die Kapillarkonstante des Wassers in der Umgebung des Bläschens beeinflussen.   W. H.

A. Einstein. Das Prinzip von der Erhaltung der Schwerpunktsbewegung und die Trägheit der Energie (Drudes Ann. 20, S. 627—633. 1906).

L. Fejér. Gleichgewicht im widerstehenden Mittel (Verhandl. d. Naturf. zu Meran 1905, II, 1, S. 223. 1906).

S. Berliner. Über das Verhalten des Gußeisens bei langsamen Belastungswechseln (Drudes Ann. 20, S. 527—562. 1906).

Th. Lohnstein. Zur Theorie des Abtropfens. Nachtrag und weitere Belege (Drudes Ann. 20, S. 606—618. 1906).

Optik.

19. *M. Planck. Vorlesungen über die Theorie der Wärmestrahlung* (222 S. ℳ 7,80. Leipzig, J. A. Barth, 1906). — In dem vorliegenden Buche sind die grundlegenden Arbeiten Kirchhoffs, W. Wiens und des Verf. zu einem Ganzen von wunderbarer Klarheit und Einheitlichkeit vereinigt worden, so daß das Buch vortrefflich dazu geeignet ist, den Leser — auch wenn demselben das behandelte Gebiet noch ganz fremd sein sollte — vollständig mit der Materie vertraut zu machen.

Im ersten Abschnitt (S. 1—23) werden zunächst die grundlegenden Begriffe und Bezeichnungen (wie „Emissionskoeffizient", „Zerstreuungskoeffizient", „spiegelnde Fläche", „glatte" und „rauhe Oberfläche", „schwarze Fläche", „schwarzer Körper", „Absorptionskoeffizient", „Strahlenbündel", „Intensität", „Strahlungsdichte") festgelegt und — soweit sie definitionsgemäß zusammenhängen — mathematisch verknüpft. Hierauf (S. 23—48) wird die Clausiussche Beziehung über das Verhältnis der Strahlungsdichten in Medien von verschiedenen Brechungs-

[1]

[2]

[3]

Bd. 30. No. 15. Optik. 765

exponenten, sowie die Kirchhoffsche Beziehung zwischen Emissions- und Absorptionsvermögen hergeleitet. [4]

Während im bisherigen nur die Sätze der Strahlenoptik benutzt wurden, wird im zweiten Abschnitt (S. 49—99) die Maxwellsche Theorie benutzt, jedoch ausschließlich zur Herleitung des Strahlendruckes. Die Größe des letzteren kann, wie der Verf. hervorhebt, nicht aus energetischen Betrachtungen gewonnen werden. Mit Hilfe des erlangten Ausdruckes für [5] den Strahlungsdruck werden das Stephan-Boltzmannsche Gesetz und das Wiensche Verschiebungsgesetz hergeleitet und die Begriffe „Temperatur der monochromatischen Strahlung", „Temperatur eines monochromatischen Elementarstrahlenbündels" festgelegt.

Das Wiensche Verschiebungsgesetz liefert für die Dichte u der Energie im Normalspektrum die Gleichung $u = \nu^3 \varphi(T/\nu)$, wobei T die absolute Temperatur und ν die Schwingungszahl bedeutet. Die Abschnitte drei und vier des Buches (S. 100—179) enthalten eine Darlegung der grundlegenden Untersuchungen des Verf., welche die Bestimmung der im Wienschen Verschiebungsgesetz auftretenden Funktion φ zum Ziele haben. Obwohl es nun nicht gelungen ist, unter alleiniger Benutzung von durch Erfahrung hinreichend gestützten theoretischen Hilfsmitteln auf rein deduktivem Wege dies Ziel zu erreichen, indem [6] sich der Verf. einer lediglich durch Analogie gestützten Hypothese bedient, so wird doch jeder unbefangene Leser finden, daß [7] dem erlangten Resultat eine große Wahrscheinlichkeit zukommt.

Der Gang der Untersuchung ist folgender: Es wird zunächst unter Zugrundelegung der Maxwellschen Gleichungen die Schwingungsgleichung eines in einem Strahlungsfelde befindlichen Resonators von kleinen Dimensionen und kleiner Dämpfung aufgestellt. Hierauf wird mit Hilfe der Schwingungsgleichung die mittlere Energie eines in einem stationären Strahlungsfelde befindlichen Resonators, sowie unter Benutzung des zweiten Hauptsatzes die „Temperatur des Resonators" in Abhängigkeit von der genannten universellen Funktion bestimmt. Damit ist das Problem der Energieverteilung im Normalspektrum [8] reduziert auf die Aufgabe, die Entropie eines aus einer großen Anzahl Strahlungsresonatoren gleicher Frequenz bestehenden Systems zu bestimmen.

Zur Lösung der letztgenannten Aufgabe wird zunächst in
[9] Anlehnung an Bolzmanns Arbeiten dargelegt, daß man in der
Gastheorie zu einer richtigen Bestimmung der Entrophie S
geführt wird, wenn man setzt $S = k \log W$, wobei k eine (universelle) Konstante und W die Anzahl der „Komplexionen"
bedeutet. Letztere Größe gibt die Manigfaltigkeit aller derjenigen möglichen Verteilungen der Elementarvariabeln an,
welche zum Komplex beobachtbarer Größen, denen die Entropie
S entspricht, gehören.

Um die Größe W durch Abzählung ermitteln zu können,
muß das ganze verfügbare Gebiet der Zustandsvariabeln in
diskrete Elementargebiete zerlegt werden. Das Resultat hängt
im allgemeinen sowohl von der absoluten Größe wie vom
[10] Größenverhältnis dieser Elementargebiete ab. Während nun
zur Bestimmung der Größe W eines Resonatorensystems das
Größen*verhältnis* der Elementargebiete gewählt wird, wie bei
einem sinusartig schwingenden Gebilde in der Gastheorie,
werden — im Gegensatz zu der bisher in der Gastheorie allgemein benutzten Annahme unendlich kleiner Elementargebiete
[11] — die Elementargebiete von endlicher Größe ($= h\nu$) gewählt,
wobei ν die Frequenz und h eine universelle Konstante bedeutet; $h\nu$ hat die Dimension einer Energie. Der Verf. weist
wiederholt auf die Notwendigkeit der Einführung dieser universellen Konstanten h hin und betont die Wichtigkeit einer
(in dem Buche nicht gegebenen) physikalischen Deutung
derselben.

Aus dem auf dem angedeuteten Wege gewonnenen Ausdruck für die Entropie S wird dann die bekannte Plancksche
Strahlungsformel

$$u = \frac{8\pi k \nu^3}{C^3} \cdot \frac{1}{e^{h\nu/k.T} - 1}$$

abgeleitet. Der vierte Abschnitt enthält ferner die Plancksche
Bestimmung der Elementarquanta, sowie Besprechungen strah-
[12] lungstheoretischer Arbeiten verschiedener Autoren.

Der letzte Abschnitt des Buches (S. 180—222), welcher
die irreversibeln Strahlungsvorgänge behandelt, gewährt einen
tiefen Einblick in das Wesen der Irreversibilität thermischer
Vorgänge. A. E.

Published in *Beiblätter zu den Annalen der Physik* 30 (1906): 764–766. Published in no. 15 [first half of August].

[1] According to the foreword, dated Easter 1906, Planck delivered the lectures at the University of Berlin during the winter semester of 1905–1906 (*Planck 1906c*, p. v).

[2] Works of the authors cited by Einstein and treated in *Planck 1906c* include *Kirchhoff 1860*, *Wien 1894, 1896*, and *Planck 1900a, 1901a*.

[3] *Planck 1906c*, p. 38, fn. 1, cites p. 594 of *Kirchhoff 1882*, a reprinting of *Kirchhoff 1860*, and *Clausius 1864*.

[4] *Planck 1906c*, p. 42, cites p. 574 of *Kirchhoff 1882*, a reprinting of *Kirchhoff 1860*.

[5] See *Planck 1906c*, p. 58.

[6] Einstein showed in *Einstein 1905i* (Doc. 14) that the molecular theory of heat and Maxwell's theory yield what is now called the Rayleigh-Jeans formula. Einstein reiterated in *Einstein 1906d* (Doc. 34), p. 200, that this formula contradicts "experience" ("Erfahrung").

[7] Einstein may be referring to Planck's treatment of "complexions" in radiation theory in analogy with Boltzmann's treatment of complexions in kinetic theory of gases (see *Planck 1906c*, § 136–138). For Einstein's criticism of Planck's treatment, see the editorial note, "Einstein's Early Work on the Quantum Hypothesis," p. 138.

[8] The word "Normalspektrum" ("normal spectrum"), used by Planck, but not otherwise by Einstein, refers to the black-body energy distribution.

[9] See *Planck 1906c*, pp. 129–140. In connection with Boltzmann's notion of "molecular chaos" ("molekulare Unordnung"), p. 134, fn. 1, refers to *Boltzmann 1896*, p. 21, to the conclusion of *Boltzmann 1878a*, and to *Burbury 1894*.

[10] The remainder of this paragraph and the next refer to material in *Planck 1906c*, pp. 148–179.

[11] Planck emphasizes (*Planck 1906c*, pp. 156 and 178–179) that if *h* is taken to be arbitrarily small, then the Rayleigh-Jeans distribution results. See the following note.

[12] *Einstein 1905i* (Doc. 14) is among the works discussed in *Planck 1906c*, p. 160, fn. 1. Planck argued that the Rayleigh-Jeans formula results from the incorrect assumption that the equipartition theorem is valid for all values of λT, referring the reader to pp. 177–179, where he states that the above assumption requires the probability of any system lying in a small region of phase space to be proportional to the size of the region, no matter how small. The last condition is not fulfilled, however, for thermal radiation, since no system can lie in a cell smaller than the finite size represented by *h*. For further discussion, see the editorial note, "Einstein on the Foundations of Statistical Physics," p. 49.

38. ''Planck's Theory of Radiation and the Theory of Specific Heat''

[Einstein 1907a]

DATED Bern, November 1906
RECEIVED 9 November 1906
PUBLISHED 28 December 1906 (for January 1907)

IN: *Annalen der Physik* 22 (1907): 180–190.

180

9. *Die Plancksche Theorie der Strahlung und die Theorie der spezifischen Wärme; von A. Einstein.*

In zwei früheren Arbeiten[1]) habe ich gezeigt, daß die Interpretation des Energieverteilungsgesetzes der schwarzen Strahlung im Sinne der Boltzmannschen Theorie des zweiten Hauptsatzes uns zu einer neuen Auffassung der Phänomene der Lichtemission und Lichtabsorption führt, die zwar noch keineswegs den Charakter einer vollständigen Theorie besitzt, die aber insofern bemerkenswert ist, als sie das Verständnis einer Reihe von Gesetzmäßigkeiten erleichtert. In der vorliegenden Arbeit soll nun dargetan werden, daß die Theorie der Strahlung — und zwar speziell die Plancksche Theorie — zu einer Modifikation der molekular-kinetischen Theorie der Wärme führt, durch welche einige Schwierigkeiten beseitigt werden, die bisher der Durchführung jener Theorie im Wege standen. Auch wird sich ein gewisser Zusammenhang zwischen dem thermischen und optischen Verhalten fester Körper ergeben. [2]

Wir wollen zuerst eine Herleitung der mittleren Energie des Planckschen Resonators geben, die dessen Beziehung zur Molekularmechanik klar erkennen läßt.

Wir benutzen hierzu einige Resultate der allgemeinen molekularen Theorie der Wärme.[1]) Es sei der Zustand eines [3] Systems im Sinne der molekularen Theorie vollkommen bestimmt durch die (sehr vielen) Variabeln $P_1, P_2 \ldots P_n$. Der Verlauf der molekularen Prozesse geschehe nach den Gleichungen

$$\frac{dP_\nu}{dt} = \Phi_\nu(P_1, P_2 \ldots P_n), \quad (\nu = 1, 2 \ldots n)$$

und es gelte für alle Werte der P_ν die Beziehung

(1)
$$\sum \frac{\partial \Phi_\nu}{\partial P_\nu} = 0 .$$
[4]

---

1) A. Einstein, Ann. d. Phys. **17**. p. 132. 1905 u. **20**. p. 199. 1905. [1]

Plancksche Theorie der Strahlung etc.   181

Es sei ferner ein Teilsystem des Systemes der P_ν bestimmt durch die Variabeln $p_1 \dots p_m$ (welche zu den P_ν gehören), und es sei angenommen, daß sich die Energie des ganzen Systems mit großer Annäherung aus zwei Teilen zusammengesetzt denken lasse, von denen einer (E) *nur* von den $p_1 \dots p_m$ abhänge, während der andere von $p_1 \dots p_m$ unabhängig sei. E sei ferner unendlich klein gegen die Gesamtenergie des Systems.

Die Wahrscheinlichkeit dW dafür, daß die p_ν in einem zufällig herausgegriffenen Zeitpunkt in einem unendlich kleinen Gebiete $(dp_1, dp_2 \dots dp_m)$ liegen, ist dann durch die Gleichung gegeben[1)]

[6]   (2)
$$dW = Ce^{-\frac{N}{RT}E}dp_1 \dots dp_m.$$

Hierbei ist C eine Funktion der absoluten Temperatur (T), N die Anzahl der Moleküle in einem Grammäquivalent, R die Konstante der auf das Grammolekül bezogenen Gasgleichung.

Setzt man

$$\int_{dE} dp_1 \dots dp_m = \omega(E)dE,$$

wobei das Integral über alle Kombinationen der p_ν zu erstrecken ist, welchen Energiewerte zwischen E und $E + dE$ entsprechen, so erhält man

(3)
$$dW = Ce^{-\frac{N}{RT}E}\omega(E)dE.$$

Setzt man als Variable P_ν die Schwerpunktskoordinaten und Geschwindigkeitskomponenten von Massenpunkten (Atomen, Elektronen), und nimmt man an, daß die Beschleunigungen nur von den Koordinaten, nicht aber von den Geschwindigkeiten abhängen, so gelangt man zur molekular-kinetischen Theorie der Wärme. Die Relation (1) ist hier erfüllt, so daß auch Gleichung (2) gilt.

Denkt man sich speziell als System der p_ν ein elementares Massenteilchen gewählt, welches längs einer Geraden Sinusschwingungen auszuführen vermag, und bezeichnet man mit x bez. ξ momentane Distanz von der Gleichgewichtslage bez. Geschwindigkeit desselben, so erhält man

(2 a)
$$dW = Ce^{-\frac{N}{RT}E}dx\,d\xi$$

[5]   1) A. Einstein, Ann. d. Phys. **11.** p. 170 u. f. 1903.

182   *A. Einstein.*

und, da $\int d\,x\,d\,\xi = \text{konst.}\,d\,E$, also $\omega = \text{konst.}$ zu setzen ist[1]):

(3 a) $d\,W = \text{konst.}\,e^{-\frac{N}{R\,T}\,E}\,d\,E.$

Der Mittelwert der Energie des Massenteilchens ist also:

(4) $\bar{E} = \dfrac{\int E\,e^{-\frac{N}{R\,T}\,E}\,d\,E}{\int e^{-\frac{N}{R\,T}\,E}\,d\,E} = \dfrac{R\,T}{N}.$

Formel (4) kann offenbar auch auf ein geradlinig schwingendes Ion angewendet werden. Tut man dies, und berücksichtigt man, daß zwischen dessen mittlerer Energie $\bar{E}$ und der Dichte ϱ_ν der schwarzen Strahlung für die betreffende Frequenz nach einer Planckschen Untersuchung[2]) die Beziehung

(5) $\bar{E}_\nu = \dfrac{L^3}{8\,\pi\,\nu^2}\,\varrho_\nu$ [8]

gelten muß, so gelangt man durch Elimination von $\bar{E}$ aus (4) und (5) zu der Reileighschen Formel

(6) $\varrho_\nu = \dfrac{R}{N}\dfrac{8\,\pi\,\nu^2}{L^3}\,T,$ [9]

welcher bekanntlich nur die Bedeutung eines Grenzgesetzes für große Werte von T/ν zukommt.   [10]

Um zur Planckschen Theorie der schwarzen Strahlung zu gelangen, kann man wie folgt verfahren.[3]) Man behält [12] Gleichung (5) bei, nimmt also an, daß durch die Maxwellsche Theorie der Elektrizität der Zusammenhang zwischen Strahlungsdichte und $\bar{E}$ richtig ermittelt sei. Dagegen verläßt man Gleichung (4), d. h. man nimmt an, daß die Anwendung der molekular-kinetischen Theorie den Widerspruch mit der Erfahrung bedinge. Hingegen halten wir an den Formeln (2) und (3) der allgemeinen molekularen Theorie der Wärme fest. Statt daß wir indessen gemäß der molekular-kinetischen Theorie

$\omega = \text{konst.}$

setzen, setzen wir $\omega = 0$ für alle Werte von E, welche den Werten 0, ε, 2 ε, 3 ε etc. nicht außerordentlich nahe liegen. Nur

---

1) Weil $E = a\,x^2 + b\,\xi^2$ zu setzen ist.

2) M. Planck, Ann. d. Phys. 1. p. 99. 1900.   [7]

3) Vgl. M. Planck, Vorlesungen über die Theorie der Wärmestrahlung. J. Ambr. Barth. 1906. §§ 149, 150, 154, 160, 166.   [11]

zwischen 0 und $0 + \alpha$, ε und $\varepsilon + \alpha$, $2\,\varepsilon$ und $2\,\varepsilon + \alpha$ etc. (wobei α unendlich klein sei gegen ε) sei ω von Null verschieden, derart, daß

$$\int_0^a \omega \, dE = \int_\varepsilon^{\varepsilon + a} \omega \, dE = \int_{2\varepsilon}^{2\varepsilon + a} \omega \, dE = \ldots = A$$

sei. Diese Festsetzung involviert, wie man aus Gleichung (3) sieht, die Annahme, daß die Energie des betrachteten Elementargebildes lediglich solche Werte annehme, die den Werten 0, ε, $2\,\varepsilon$ etc. unendlich nahe liegen.

Unter Benutzung der eben dargelegten Festsetzung für ω erhält man mit Hilfe von (3):

[13] $$\bar{E} = \frac{\int E e^{-\frac{N}{RT}E} \omega(E)\, dE}{\int e^{-\frac{N}{RT}} \omega(E)\, dE} = \frac{0 + A\,\varepsilon\, e^{-\frac{N}{RT}\varepsilon} + A\cdot 2\,\varepsilon\, e^{-\frac{N}{RT}2\varepsilon} \ldots}{A + A\, e^{-\frac{N}{RT}\varepsilon} + A e^{-\frac{N}{RT}2\varepsilon} + \ldots}$$

$$= \frac{\varepsilon}{e^{\frac{N}{RT}\varepsilon} - 1} \cdot$$

[14] Setzt man noch $\varepsilon = (R/N)\,\beta\,\nu$ (gemäß der Quantenhypothese), so erhält man hieraus:

[15] (7) $$\bar{E} = \frac{\dfrac{R}{N}\beta\,\nu}{e^{\frac{\beta\nu}{T}} - 1},$$

sowie mit Hilfe von (5) die **Plancksche Strahlungsformel**:

$$\varrho_\nu = \frac{8\,\pi}{L^3} \cdot \frac{R\,\beta}{N} \frac{\nu^3}{e^{\frac{\beta\nu}{T}} - 1} \cdot$$

Gleichung (7) gibt die Abhängigkeit der mittleren Energie des Planckschen Resonators von der Temperatur an.

———————

Aus dem Vorhergehenden geht klar hervor, in welchem Sinne die molekular-kinetische Theorie der Wärme modifiziert werden muß, um mit dem Verteilungsgesetz der schwarzen Strahlung in Einklang gebracht zu werden. Während man sich nämlich bisher die molekularen Bewegungen genau denselben Gesetzmäßigkeiten unterworfen dachte, welche für die Bewegungen der Körper unserer Sinnenwelt gelten (wir fügen

wesentlich nur das Postulat vollständiger Umkehrbarkeit hinzu),
sind wir nun genötigt, für schwingungsfähige Ionen bestimmter
Frequenz, die einen Energieaustausch zwischen Materie und
Strahlung vermitteln können, die Annahme zu machen, daß
die Mannigfaltigkeit der Zustände, welche sie anzunehmen vermögen, eine geringere sei als bei den Körpern unserer Erfahrung. Wir mußten ja annehmen, daß der Mechanismus   [16]
der Energieübertragung ein solcher sei, daß die Energie des
Elementargebildes ausschließlich die Werte 0, $(R/N)\beta v$,
$2(R/N)\beta v$ etc. annehmen könne.[1])

Ich glaube nun, daß wir uns mit diesem Resultat nicht
zufrieden geben dürfen. Es drängt sich nämlich die Frage
auf: Wenn sich die in der Theorie des Energieaustausches
zwischen Strahlung und Materie anzunehmenden Elementargebilde nicht im Sinne der gegenwärtigen molekular-kinetischen
Theorie auffassen lassen, müssen wir dann nicht auch die
Theorie modifizieren für die anderen periodisch schwingenden
Gebilde, welche die molekulare Theorie der Wärme heran-
zieht? Die Antwort ist nach meiner Meinung nicht zweifelhaft. Wenn die Plancksche Theorie der Strahlung den Kern
der Sache trifft, so müssen wir erwarten, auch auf anderen
Gebieten der Wärmetheorie Widersprüche zwischen der gegenwärtigen molekular-kinetischen Theorie und der Erfahrung zu
finden, die sich auf dem eingeschlagenen Wege heben lassen.
Nach meiner Meinung trifft dies tatsächlich zu, wie ich im
folgenden zu zeigen versuche.

Die einfachste Vorstellung, die man sich über die Wärmebewegung in festen Körpern bilden kann, ist die, daß die
einzelnen in denselben enthaltenen Atome Sinusschwingungen
um Gleichgewichtslagen ausführen. Unter dieser Voraus-   [17]
setzung erhält man durch Anwendung der molekular-kinetischen
Theorie (Gleichung (4)) unter Berücksichtigung des Umstandes,
daß jedem Atom drei Bewegungsfreiheiten zuzuschreiben sind,

----

1) Es ist übrigens klar, daß diese Voraussetzung auch auf schwingungsfähige Körper auszudehnen ist, die aus beliebig vielen Elementargebilden bestehen.

für die auf das Grammäquivalent bezogene spezifische Wärme des Stoffes

$$c = 3\,R\,n$$

oder — in Grammkalorien ausgedrückt —

$$c = 5{,}94\,n,$$

wenn n die Anzahl der Atome im Molekül bedeutet. Es ist bekannt, daß diese Beziehung für die meisten Elemente und für viele Verbindungen im festen Aggregatzustand mit bemerkenswerter Annäherung erfüllt ist (Doulong-Petitsches [18] Gesetz, Regel von F. Neumann und Kopp).

Betrachtet man jedoch die Tatsachen etwas genauer, so begegnet man zwei Schwierigkeiten, die der Anwendbarkeit der Molekulartheorie enge Grenzen zu ziehen scheinen.

1. Es gibt Elemente (Kohlenstoff, Bor und Silizium), welche im festen Zustande bei gewöhnlicher Temperatur eine [19] bedeutend kleinere spezifische Atomwärme besitzen als 5,94. Es haben ferner alle festen Verbindungen, in denen Sauerstoff, Wasserstoff oder mindestens eines der eben genannten Elemente vorkommen, eine kleinere spezifische Wärme pro Gramm- [20] molekül als $n \cdot 5{,}94$.

2. Hr. Drude hat gezeigt[1]), daß die optischen Erscheinungen (Dispersion) dazu führen, jedem Atom einer Verbindung mehrere unabhängig voneinander bewegliche Elementarmassen zuzuschreiben, indem er mit Erfolg die ultraroten Eigenfrequenzen auf Schwingungen der Atome (Atomionen), die ultravioletten Eigenfrequenzen auf Schwingungen von Elek- [22] tronen zurückführte. Hieraus ergibt sich für die molekularkinetische Theorie der Wärme eine zweite bedeutende Schwierigkeit, indem die spezifische Wärme — da die Zahl der beweglichen Massenpunkte pro Molekül größer ist als dessen Atomzahl — den Wert $5{,}94\,n$ beträchtlich übersteigen müßte.

Nach dem Obigen ist hierzu folgendes zu bemerken. Wenn wir die Träger der Wärme in festen Körpern als periodisch schwingende Gebilde ansehen, deren Frequenz von ihrer Schwingungsenergie unabhängig ist, dürfen wir nach der Planckschen Theorie der Strahlung nicht erwarten, daß die

----

[21]   1) P. Drude, Ann. d. Phys. **14**. p. 677. 1904.

186   *A. Einstein.*

spezifische Wärme stets den Wert 5,94 n besitze. Wir haben vielmehr zu setzen (7)

$$\bar{E} = \frac{3R}{N} \frac{\beta\nu}{e^{\frac{\beta\nu}{T}} - 1} \cdot$$

Die Energie von N solchen Elementargebilden, in Grammkalorien gemessen, hat daher den Wert

$$5,94 \frac{\beta\nu}{e^{\frac{\beta\nu}{T}} - 1},$$

so daß jedes derartige schwingende Elementargebilde zur spezifischen Wärme pro Grammäquivalent den Wert

$$(8) \qquad 5,94 \frac{e^{\frac{\beta\nu}{T}} \cdot \left(\frac{\beta\nu}{T}\right)^2}{\left(e^{\frac{\beta\nu}{T}} - 1\right)^2} \cdot$$

beiträgt. Wir bekommen also, indem wir über alle Gattungen von schwingenden Elementargebilden summieren, welche in dem

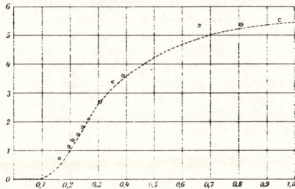

betreffenden festen Stoffe vorkommen, für die spezifische Wärme pro Grammäquivalent den Ausdruck[1]

$$(8a) \qquad c = 5,94 \sum \frac{e^{\frac{\beta\nu}{T}} \left(\frac{\beta\nu}{T}\right)^2}{\left(e^{\frac{\beta\nu}{T}} - 1\right)^2} \cdot$$

Die vorstehende Figur[2] zeigt den Wert des Ausdruckes (8) in Funktion von $x = (T/\beta\nu)$. Wenn $(T/\beta\nu) > 0,9$, unterscheidet

---

1) Die Betrachtung läßt sich leicht auf anisotrope Körper ausdehnen.
2) Vgl. deren gestrichelte Kurve.

sich der Beitrag des Gebildes zur molekularen spezifischen Wärme nicht beträchtlich vom Werte 5,94, der auch aus der bisher akzeptierten molekular-kinetischen Theorie sich ergibt; je kleiner ν ist, bei um so tieferen Temperaturen wird dies bereits der Fall sein. Wenn dagegen $(T/\beta\nu) < 0,1$, so trägt das betreffende Elementargebilde nicht merklich zur spezifischen Wärme bei. Dazwischen findet ein anfänglich rascheres, dann langsameres Wachsen des Ausdruckes (8) statt.

Aus dem Gesagten folgt zunächst, daß die zur Erklärung der ultravioletten Eigenfrequenzen anzunehmenden schwingungsfähige Elektronen bei gewöhnlicher Temperatur ($T = 300$) zur spezifischen Wärme nicht merklich beitragen können; denn die Ungleichung $(T/\beta\nu) < 0,1$ geht für $T = 300$ über in die

[23] Ungleichung $\lambda < 4,8\,\mu$. Wenn dagegen ein Elementargebilde die Bedingung $\lambda > 48\,\mu$ erfüllt, so muß es nach dem Obigen bei gewöhnlicher Temperatur zur spezifischen Wärme pro Grammäquivalent nahezu den Beitrag 5,94 liefern.

Da für die ultraroten Eigenfrequenzen im allgemeinen

[24] $\lambda > 4,8\,\mu$ ist, so müssen nach unserer Auffassung jene Eigenschwingungen einen Beitrag zur spezifischen Wärme liefern, und zwar einen um so bedeutenderen, je größer das betreffende λ

[25] ist. Nach Drudes Untersuchungen sind es die ponderablen Atome (Atomionen) selbst, welchen diese Eigenfrequenzen zuzuschreiben sind. Es liegt also am nächsten, als Träger der Wärme in festen Körpern (Isolatoren) ausschließlich die positiven Atomionen zu betrachten.

Wenn die ultraroten Eigenschwingungsfrequenzen ν eines festen Körpers bekannt sind, so wäre also nach dem Gesagten dessen spezifische Wärme sowie deren Abhängigkeit von der

[26] Temperatur durch Gleichung (8a) vollkommen bestimmt. Deutliche Abweichungen von der Beziehung $c = 5,94\,n$ wären bei gewöhnlicher Temperatur zu erwarten, wenn der betreffende Stoff eine optische ultrarote Eigenfrequenz aufweist, für welche $\lambda < 48\,\mu$; bei genügend tiefen Temperaturen sollen die spezifischen Wärmen aller festen Körper mit sinkender Temperatur bedeutend

[27] abnehmen. Ferner muß das Doulong-Petitsche Gesetz sowie das allgemeinere Gesetz $c = 5,94\,n$ für alle Körper bei genügend hohen Temperaturen gelten, falls sich bei letzteren keine neuen

[28] Bewegungsfreiheiten (Elektronionen) bemerkbar machen.

188   *A. Einstein.*

Die beiden oben genannten Schwierigkeiten werden durch
die neue Auffassung beseitigt, und ich halte es für wahrscheinlich, daß letztere sich im Prinzip bewähren wird. Daran,
daß sie den Tatsachen exakt entspreche, ist natürlich nicht
zu denken. Die festen Körper erfahren beim Erwärmen [29]
Änderungen der molekularen Anordnung (z. B. Volumände-
rungen), die mit Änderungen des Energieinhaltes verbunden
sind; alle festen Körper, die elektrisch leiten, enthalten frei
bewegliche Elementarmassen, die zur spezifischen Wärme einen
Beitrag liefern; die ungeordneten Wärmeschwingungen sind
vielleicht von etwas anderer Frequenz als die Eigenschwingungen der nämlichen Elementargebilde bei optischen Prozessen.
Endlich aber ist die Annahme, daß die in Betracht kommenden Elementargebilde eine von der Energie (Temperatur) unabhängige Schwingungsfrequenz besitzen, ohne Zweifel unzulässig.

Immerhin ist es interessant, unsere Konsequenzen mit
der Erfahrung zu vergleichen. Da es sich nur um rohe Annäherung handelt, nehmen wir gemäß der F. Neumann-Kopp-
schen Regel an, daß jedes Element, auch wenn dasselbe abnorm
kleine spezifische Wärme besitzt, in allen seinen festen Verbindungen den gleichen Beitrag zur molekularen spezifischen
Wärme liefere. Die in nachstehender Tabelle angegebenen
Zahlen sind dem Lehrbuche der Chemie von Roskoe ent- [30]
nommen. Wir bemerken, daß alle Elemente von abnorm kleiner
Atomwärme kleines Atomgewicht besitzen; dies ist nach unserer

Element	Spezifische Atomwärme	$\lambda_{ber.}$
S und P	5,4	42
Fl	5	33
O	4	21
Si	3,8	20
B	2,7	15
H	2,3	13
C	1,8	12

Auffassung zu erwarten, da ceteris paribus kleinen Atom-
gewichten große Schwingungsfrequenzen entsprechen. In der
letzten Spalte der Tabelle sind die Werte von λ in Mikron
angegeben, wie sie sich aus diesen Zahlen unter der Annahme,

daß letztere für $T = 300$ gelten, mit Hilfe der dargestellten Beziehung zwischen x und c ergeben.

Wir entnehmen ferner den Tabellen von Landolt und Börnstein einige Angaben über ultrarote Eigenschwingungen (metallische Reflexion, Reststrahlen) einiger durchsichtiger fester Körper; die beobachteten λ sind in nachstehender Tabelle [31] unter „$\lambda_{\text{beob.}}$" angegeben; die Zahlen unter „$\lambda_{\text{ber.}}$" sind obiger Tabelle entnommen, soweit sie sich auf Atome von abnorm kleiner spezifischer Wärme beziehen; für die übrigen soll $\lambda > 48\,\mu$ sein.

Körper	$\lambda_{\text{beob.}}$	$\lambda_{\text{ber.}}$
CaFl	24; 31,6	33; > 48
NaCl	51,2	> 48
KCl	61,2	> 48
CaCO$_3$	6,7; 11,4; 29,4	12; 21; > 48
SiO$_2$	8,5; 9,0; 20,7	20; 21

In der Tabelle enthalten NaCl und KCl nur Atome von normaler spezifischer Wärme; in der Tat sind die Wellenlängen ihrer ultraroten Eigenschwingungen größer als $48\,\mu$. Die übrigen Stoffe enthalten lauter Atome mit abnorm kleiner spezifischer Wärme (ausgenommen Ca); in der Tat liegen die Eigenfrequenzen dieser Stoffe zwischen $4,8\,\mu$ und $48\,\mu$. Im allgemeinen sind die aus den spezifischen Wärmen theoretisch ermittelten λ erheblich größer als die beobachteten. Diese Abweichungen können vielleicht in einer starken Veränderlichkeit der Frequenz des Elementargebildes mit der Energie desselben ihre Erklärung finden. Wie dem auch sein mag, jedenfalls ist die Übereinstimmung der beobachteten und berechneten λ hinsichtlich der Reihenfolge, sowie hinsichtlich der Größenordnung sehr bemerkenswert.

Wir wollen nun die Theorie noch auf den Diamanten anwenden. Die ultrarote Eigenfrequenz desselben ist nicht bekannt, läßt sich jedoch unter Zugrundelegung der dargelegten Theorie berechnen, wenn für einen Wert von T die molekulare spezifische Wärme c bekannt ist; das zu c gehörige x läßt sich aus der Kurve unmittelbar entnehmen, und man bestimmt hieraus λ nach der Beziehung $(TL/\beta\lambda) = x$.

190   *A. Einstein. Plancksche Theorie der Strahlung etc.*

Ich benutze die Beobachtungsresultate von H. F. Weber, die ich den Tabellen von Landolt und Börnstein entnahm [32] (vgl. nachstehende Tabelle). Für $T = 331,3$ ist $c = 1,838$; hieraus folgt nach der angegebenen Methode $\lambda = 11,0 \, \mu$. Unter Zugrundelegung dieses Wertes sind die in der dritten Spalte der Tabelle nach der Formel $x = (TL/\beta\lambda)$ berechnet $(\beta = 4,86 \cdot 10^{-11})$.

T	c	x
222,4	0,762	0,1679
262,4	1,146	0,1980
283,7	1,354	0,2141
306,4	1,582	0,2312
331,3	1,838	0,2500
358,5	2,118	0,2705
413,0	2,661	0,3117
479,2	3,280	0,3615
520,0	3,631	0,3924
879,7	5,290	9,6638
1079,7	5,387	0,8147
1258,0	5,507	0,9493

[33]

Die Punkte, deren Abszissen diese Werte von x, deren Ordinaten die in der Tabelle angegebenen, aus Beobachtungen Webers ermittelten Werte von c sind, sollen auf der oben dargestellten x, c-Kurve liegen. Wir haben diese Punkte — mit Ringen bezeichnet — in die obige Figur eingetragen; sie liegen tatsächlich nahezu auf der Kurve. Wir haben also anzunehmen, daß die elementaren Träger der Wärme beim Diamanten nahezu monochromatische Gebilde sind.

Es ist also nach der Theorie zu erwarten, daß der Diamant bei $\lambda = 11 \, \mu$ ein Absorptionsmaximum aufweist.   [34]

Bern, November 1906.

(Eingegangen 9. November 1906.)

Published in *Annalen der Physik* 22 (1907): 180–190. Dated Bern, November 1906, received 9 November 1906, published 28 December 1906 (for January 1907).

[1] *Einstein 1905i* (Doc. 14) and *Einstein 1906d* (Doc. 34). The second work was published in 1906.

[2] Einstein had looked for such a link in 1901. See Einstein to Mileva Marić, 23 March 1901 (Vol. 1, Doc. 93).

[3] Evidently, this note is a reference to p. 181, fn. 1.

[4] This condition assures the validity of Liouville's theorem for this system. See the editorial note, "Einstein on the Foundations of Statistical Physics," p. 52.

[5] *Einstein 1903* (Doc. 4).

[6] This equation is equivalent to *Einstein 1903* (Doc. 4), p. 176, eq. (3).

[7] *Planck 1900a*.

[8] This formula does not appear on p. 99 of *Planck 1900a*. It can be derived from equations on pp. 99 and 111. Planck first published the equation in *Planck 1900e*, p. 241.

[9] When expressed as ρ_λ, the energy density per unit wavelength, this is now called the Rayleigh-Jeans law (see *Rayleigh 1900*, *1905a*, *1905b*; *Jeans 1905a*, *1905b*). *Planck 1906c*, p. 159, calls it the "Rayleigh radiation law" ("Rayleighsches Strahlungsgesetz").

[10] Einstein noted this in *Einstein 1905i* (Doc. 14), p. 136.

[11] These sections of *Planck 1906c* include the derivation of the distribution law. A common feature is Planck's division of energies distributed over individual oscillators into integral multiples of $h\nu$. For Einstein's review of *Planck 1906c*, see *Einstein 1906f* (Doc. 37).

[12] The procedure outlined in the next few sentences is similar to Einstein's procedure in *Einstein 1906d* (Doc. 34).

[13] A factor of E is missing from the exponential in the denominator of the first equality.

[14] As in *Einstein 1905i* (Doc. 14) and *Einstein 1906d* (Doc. 34), β corresponds to $Nh/R = h/k$, where h and k are Planck's and Boltzmann's constants, respectively.

[15] Planck had obtained an equivalent result in *Planck 1906c*, p. 157, eq. (231), from his expression for the radiation's entropy S, eq. (227), and the relation $dS/dU = 1/T$, where U is the average oscillator energy.

[16] For further discussion of this assertion, see *Einstein 1907h* (Doc. 45), pp. 372–373.

[17] A model of this type is used in *Boltzmann 1871b* and in *Boltzmann 1876*, pp. 556–559, for calculating the specific heat of a monatomic solid. The results are equivalent to the expressions given here near the top of p. 185.

[18] According to the Dulong-Petit rule, for a number of solid monatomic elements the product of the atomic weight and the atomic specific heat is a constant. Neumann extended the Dulong-Petit rule to compounds (*Neumann, F. E. 1831*): for chemically similar compounds, the product of the molecular specific heat and the molecular weight is a constant. According to the Kopp rule (*Kopp 1864*), each element contributes the same amount to the specific heat of a compound as its specific heat as a free element. Using this rule, one can calculate the molecular specific heat of a substance from the atomic specific heats of its constituents. For a brief statement of the Kopp rule, see p. 188, where it is called the F. Neumann-Kopp rule. For a contemporary discussion of these rules, see *Winkelmann 1906a*. For an earlier reference to the rules, see Einstein to Mileva Marić, 23 March 1901 (Vol. 1, Doc. 93).

[19] H. F. Weber observed this phenomenon for the three elements in question in *Weber, H. F. 1875*. Einstein may have learned of Weber's results while at the ETH. See Vol. 1, the editorial note, "Einstein as a Student of Physics, and His Notes on H. F. Weber's Course," pp. 60–62.

[20] These assertions are supported by data given in *Landolt and Börnstein 1905*, pp. 387–392.

[21] *Drude 1904a*.

[22] These results are stated on p. 682 of *Drude 1904a*.

[23] According to *Landolt and Börnstein 1905*, p. 611, the wavelengths of ultraviolet light range from 0.1μ to 0.36μ.

[24] According to *Landolt and Börnstein 1905*, p. 611, the wavelengths of infrared light range from 0.81μ to 61.1μ.

[25] *Drude 1904a* and *1904b*. In *Einstein 1907d* (Doc. 42) Einstein corrected this and the following sentence.

[26] For later attempts to obtain the infrared proper frequencies from the physical properties of a solid, see *Lindemann 1910* and *Einstein 1911b*.

[27] This important prediction was tested and confirmed by Nernst (see *Nernst 1911a*, *1911b*, *1911c*). He showed not only that the specific

heats decrease with temperature, but that they all approach zero at small absolute temperatures (small $T/\beta\nu$), as Einstein's formula implies.

[28] This assertion is generally confirmed by the tables in *Landolt and Börnstein 1905*, pp. 385–406.

[29] *Nernst 1911c*, pp. 308–309, indicates that discrepancies do occur between Einstein's predictions and experimental data, especially at lower absolute temperatures (small $T/\beta\nu$) where Einstein's predicted curve is too low. For Einstein's response, see *Einstein 1911f*.

[30] The work in question may be *Roscoe et al. 1898*, a copy of which is in Einstein's personal library.

[31] The data for $\lambda_{beob.}$ in the table are taken from *Landolt and Börnstein 1905*, p. 620.

[32] The data in the table below are derived from data in *Landolt and Börnstein 1905*, p. 384, which in turn are derived from *Weber, H. F. 1875*. In the table below the temperatures given in *Landolt and Börnstein 1905*, p. 384, have been converted from Celsius to absolute and the specific heats multiplied by 12, the atomic weight of carbon, to obtain the atomic specific heat, c.

[33] In the third column of the table, 9,6638 should be 0,6638.

[34] This prediction is modified in *Einstein 1907d* (Doc. 42).

39. "On the Limit of Validity of the Law of Thermodynamic Equilibrium and on the Possibility of a New Determination of the Elementary Quanta"

[Einstein 1907b]

DATED Bern, December 1906
RECEIVED 12 December 1906
PUBLISHED 5 March 1907

IN: *Annalen der Physik* 22 (1907): 569–572.

569

10. *Über die Gültigkeitsgrenze des Satzes vom thermodynamischen Gleichgewicht und über die Möglichkeit einer neuen Bestimmung der Elementarquanta; von A. Einstein.* [1]

Der Zustand eines physikalischen Systems sei im Sinne der Thermodynamik bestimmt durch die Parameter λ, μ etc. (z. B. Anzeige eines Thermometers, Länge oder Volumen eines Körpers, Substanzmenge einer gewissen Art in einer Phase). Ist das System mit anderen Systemen nicht in Wechselwirkung, was wir annehmen, so wird nach der Tkermodynamik Gleichgewicht bei bestimmten Werten λ_0, μ_0 etc. der Parameter statthaben, für welche Werte die Entropie S des Systems ein Maximum ist. Nach der molekularen Theorie der Wärme jedoch ist dies nicht genau, sondern nur angenähert richtig; nach dieser Theorie besitzt der Parameter λ auch bei Temperaturgleichgewicht keinen konstanten Wert, sondern einen unregelmäßig schwankenden, der sich von λ_0 allerdings nur äußerst selten beträchtlich entfernt.

Die theoretische Untersuchung des statistischen Gesetzes, welchem diese Schwankungen unterworfen sind, scheint auf den ersten Blick bestimmte Festsetzungen in betreff des anzuwendenden molekularen Bildes zu erfordern. Dies ist jedoch [2] nicht der Fall. Es genügt vielmehr im wesentlichen, die bekannte Boltzmannsche Beziehung anzuwenden, welche die Entropie S mit der statistischen Wahrscheinlichkeit eines Zustandes verbindet. Diese Beziehung lautet bekanntlich

$$S = \frac{R}{N} \lg W,$$

wobei R die Konstante der Gasgleichung und N die Anzahl der Moleküle in einem Grammäquivalent bedeutet.

Wir fassen einen Zustand des Systems ins Auge, in welchem der Parameter λ den von λ_0 sehr wenig abweichenden Wert $\lambda_0 + \varepsilon$ besitzt. Um den Parameter λ auf umkehrbarem Wege vom Werte λ_0 zum Werte λ bei konstanter Energie E

570   A. Einstein.

zu bringen, wird man eine Arbeit A dem System zuführen und die entsprechende Wärmemenge dem System entziehen müssen. Nach thermodynamischen Beziehungen ist:

$$A = \int dE - \int T dS,$$

oder, da die betrachtete Änderung unendlich klein und $\int dE = 0$ ist:

$$A = -T(S - S_0).$$

Andererseits ist aber nach dem Zusammenhang zwischen Entropie und Zustandswahrscheinlichkeit:

$$S - S_0 = \frac{R}{N} \lg \left(\frac{W}{W_0}\right).$$

Aus den beiden letzten Gleichungen folgt:

$$A = -\frac{RT}{N} \lg \frac{W}{W_0}$$

oder

$$W = W_0\, e^{-\frac{N}{RT} A}.$$

Dies Resultat insolviert eine gewisse Ungenauigkeit, indem man ja eigentlich nicht von der Wahrscheinlichkeit eines *Zustandes*, sondern nur von der Wahrscheinlichkeit eines Zustands-

[3] *gebietes* reden kann. Schreiben wir statt der gefundenen Gleichung

$$dW = \text{konst.}\, e^{-\frac{N}{RT} A}\, d\lambda,$$

so ist das letztere Gesetz ein exaktes. Die Willkür, welche darin liegt, daß wir das Differential von λ und nicht das Differential irgendeiner Funktion von λ in die Gleichung ein-

[4] gesetzt haben, wird auf unser Resultat nicht von Einfluß sein.

Wir setzen nun $\lambda = \lambda_0 + \varepsilon$ und beschränken uns auf den Fall, daß A nach positiven Potenzen von ε entwickelbar ist, und daß nur das erste nicht verschwindende Glied dieser Ent-wickelung zum Werte des Exponenten merklich beiträgt bei solchen Werten von ε, für welche die Exponentialfunktion noch merklich von Null verschieden ist. Wir setzen also $A = a\varepsilon^2$ und erhalten:

$$dW = \text{konst.}\, e^{-\frac{N}{RT} a\varepsilon^2}\, d\varepsilon.$$

Thermodynamisches Gleichgewicht.   571

Es gilt also in diesem Falle für die Abweichungen ε das Gesetz der zufälligen Fehler. Für den Mittelwert der Arbeit A erhält man den Wert:

$$\bar{A} = \frac{1}{2} \frac{R}{N} T.$$

Das Quadrat der Schwankung ε eines Parameters λ ist also im Mittel so groß, daß die äußere Arbeit A, welche man bei strenger Gültigkeit der Thermodynamik anwenden müßte, um den Parameter λ bei konstanter Energie des Systems von λ_0 auf $\lambda^0 + \sqrt{\overline{\varepsilon^2}}$ zu verändern, gleich $\frac{1}{2} \frac{R}{N} T$ ist (also gleich dem dritten Teil der mittleren kinetischen Energie eines Atoms).

Führt man für R und N die Zahlenwerte ein, so erhält man angenähert:

$$\bar{A} = 10^{-16} T.$$

Wir wollen nun das gefundene Resultat auf einen kurz geschlossenen Kondensator von der (elektrostatisch gemessenen) Kapazität c anwenden. Ist $\sqrt{\overline{p^2}}$ die Spannung (elektrostatisch), welche der Kondensator im Mittel infolge der molekularen Unordnung annimmt, so ist

$$\bar{A} = \tfrac{1}{2} c \, \overline{p^2} = 10^{-16} T.$$

Wir nehmen an, der Kondensator sei ein Luftkondensator und er bestehe aus zwei ineinandergeschobenen Plattensystemen von je 30 Platten. Jede Platte habe von den benachbarten des anderen Systems im Mittel den Abstand 1 mm. Die Größe der Platten sei 100 cm². Die Kapazität c ist dann ca. 5000. [5] Für gewöhnliche Temperatur erhält man dann

$$\sqrt{\overline{p^2_{\text{stat.}}}} = 3,4 \cdot 10^{-9}.$$ [6]

In Volt gemessen erhält man

$$\sqrt{\overline{p^2_{\text{Volt}}}} = 10^{-6}.$$ [7]

Denkt man sich die beiden Plattensysteme relativ zueinander beweglich, so daß sie vollständig auseinander geschoben werden können, so kann man erzielen, daß die Kapazität nach dem Auseinanderschieben von der Größenordnung 10 ist.

Nennt man π die Potentialdifferenz, welche durch das Auseinanderschieben aus p entsteht, so hat man

$$\sqrt{\pi^2} = 10^{-6} \cdot \frac{5000}{10} = 0,0005 \text{ Volt}.$$

Schließt man also den Kondensator bei zusammengeschobenen Plattensystemen kurz, und schiebt man dann, nachdem die Verbindung unterbrochen ist, die Plattensysteme auseinander, so erhält man zwischen den Plattensystemen Spannungsdifferenzen von der Größenordnung eines halben Millivolt.

Es scheint mir nicht ausgeschlossen zu sein, daß diese
[8] Spannungsdifferenzen der Messung zugänglich sind. Falls man nämlich Metallteile elektrisch verbinden und trennen kann, ohne daß hierbei noch andere *unregelmäßige* Potentialdifferenzen von gleicher Größenordnung wie die soeben berechneten auftreten, so muß man durch Kombination des obigen Plattenkondensators mit einem Multiplikator zum Ziele gelangen
[9] können. Es wäre dann ein der Brownschen Bewegung verwandtes Phänomen auf dem Gebiete der Elektrizität gegeben, daß zur Ermittelung der Größe N benutzt werden könnte.

Bern, Dezember 1906.

(Eingegangen 12. Dezember 1906.)

Published in *Annalen der Physik* 22 (1907): 569–572. Dated Bern, December 1906, received 12 December 1906, published 5 March 1907.

[1] "Elementarquanta" is used to designate fundamental constants such as the true mass of atoms and thus Avogadro's number (see *Planck 1901b* and *Einstein 1905i*).

[2] Explicit molecular-kinetic assumptions were used, e.g., in Smoluchowski's work on fluctuation phenomena (see *Smoluchowski 1906*). For a discussion of Einstein's efforts to reduce the role of such assumptions in the study of fluctuation phenomena, see the editorial note, "Einstein on Brownian Motion," p. 214.

[3] This question is discussed in more detail in *Einstein 1910c*, as well as in Einstein's later correspondence; see, e.g., Einstein to Michele Besso, 11 September 1911 and 20–30 September, 1911; Einstein to D.K.C. MacDonald, 2 March 1953 and 30 March 1953 (the latter makes explicit reference to the problem of charge fluctuations in a condenser treated later in the text).

[4] Einstein discussed this question in his correspondence; see the letters listed in the preceding note. In Einstein to D.K.C. MacDonald, 30 March 1953, he points out that, given the realizability of the thermodynamic state in question and knowing "to which region of states the probability refers" ("auf was für ein Zustandsgebiet sich die Wahrscheinlichkeit bezieht"), the freedom in choice of a parameter does not cause serious difficulty.

[5] The value of the capacitance is given in statfarads (electrostatic units). The formula applied is: $c = (2n - 1) F/4\pi a$, where F is the area of a plate, a the distance between two plates, and $2n$ their total number. This standard formula occurs in Einstein's ETH notes (H. F. Weber's Lectures on Physics, ca. December 1897–ca. June 1898, Vol. 1, Doc. 37, p. 168).

[6] In his calculation, Einstein assumed a temperature of 289°K.

[7] 1 statvolt (electrostatic units) = 300 volt.

[8] Electrometers, such as the quadrant electrometer developed by Dolezalek (*Dolezalek 1901*) or that developed by Kleiner (*Kleiner 1906*), did in fact allow for the measurement of potential differences of the order of magnitude indicated by Einstein. For a survey of instruments then current, see *Cermak 1918*, pp. 141–142. The quadrant electrometer is also discussed in Einstein's ETH notes (H. F. Weber's Lectures on Physics, ca. December 1897–ca. June 1898, Vol. 1, Doc. 37, pp. 156–158).

[9] Following this line of reasoning, Einstein proposed an instrument for the measurement of small quantities of electricity in *Einstein 1908a* (Doc. 48). Irregular potential differences, which arise from surface effects between the metal parts in the measuring device described, prevent the effect from being measured as proposed by Einstein (see, e.g., *Schmidt 1918*, p. 35). For a discussion of attempts at experimental verification of this effect, see Vol. 5, the editorial note, "Einstein's 'Maschinchen' for the Measurement of Small Quantities of Electricity."

40. "Theoretical Remarks on Brownian Motion"

[*Einstein 1907c*]

DATED Bern, January 1907
RECEIVED 22 January 1907
PUBLISHED 8 February 1907

IN: *Zeitschrift für Elektrochemie und angewandte physikalische Chemie* 13 (1907): 41–42.

Republished in *Einstein 1922*, pp. 40–43.

Z. f. Elektroch. Bd. **13.** 8. Februar 1907. Nr. 6. (S. 41—48.)

THEORETISCHE BEMERKUNGEN ÜBER DIE BROWNSCHE BEWEGUNG.

Von *A. Einstein*.

[1] eranlasst durch die neulich in der Z. f. Elektroch. erschienene Svedbergsche Untersuchung über die Bewegung suspendierter Teilchen, halte ich es für angezeigt, auf einige durch die molekulare Theorie der Wärme geforderten Eigenschaften dieser Bewegung hinzuweisen. Ich hoffe, durch das Nachfolgende den Physikern, welche sich experimentell mit dieser Sache abgeben, die Deutung ihres Beobachtungsmaterials sowie die Vergleichung desselben mit [2] der Theorie etwas zu erleichtern.

1. Die molekulare Theorie der Wärme erlaubt, den Mittelwert der Momentangeschwindigkeit zu berechnen, welche ein Teilchen bei der absoluten Temperatur T besitzt. Die kinetische Energie der Schwerpunktsbewegung eines Teilchens ist nämlich unabhängig von der Grösse und Beschaffenheit des Teilchens und unabhängig von der Beschaffenheit seiner Umgebung, z. B. der Flüssigkeit, in der das Teilchen suspendiert ist; diese kinetische Energie ist gleich der eines einatomigen Gasmoleküls. Die mittlere Geschwindigkeit $\sqrt{\overline{v^2}}$ des Teilchens von der Masse m ist also bestimmt durch die Gleichung

[3]
$$m\frac{\overline{v^2}}{2} = \frac{3}{2}\frac{RT}{N},$$

wobei $R = 8,3 \cdot 10^7$, T die absolute Temperatur und N die Anzahl der wirklichen Moleküle in [4] einem Gramm-Molekül (etwa $4 \cdot 10^{23}$) bedeutet. Wir wollen $\sqrt{\overline{v^2}}$, sowie die im folgenden noch in Betracht kommenden Grössen berechnen für Teilchen kolloïdaler Platinlösungen, wie sie Herr [5] Svedberg untersucht hat. Für diese Teilchen haben wir $m = 2,5 \cdot 10^{-15}$ zu setzen, so dass wir für $T = 292$ erhalten:

[6]
$$\sqrt{\overline{v^2}} = \sqrt{\frac{3RT}{mN}} = 8,6 \text{ cm/Sek.}$$

2. Wir wollen nun nachsehen, ob Aussicht vorhanden ist, diese enorme Geschwindigkeit an einem suspendierten Teilchen wirklich zu beobachten.

Wüssten wir nichts von molekularer Wärmetheorie, so würden wir folgendes erwarten. Angenommen, man erteilt einem in einer Flüssigkeit suspendierten Teilchen durch einen von aussen auf dasselbe ausgeübten Kraftimpuls eine Geschwindigkeit, so wird diese Geschwindigkeit rasch aufgezehrt durch die Reibung der Flüssigkeit. Wir vernachlässigen die Trägheit der letzteren und beachten, dass der Widerstand, den das mit der Geschwindigkeit v bewegte Teilchen erfährt, $6\pi k P v$ ist, wobei k den Viskositätskoëffizienten der Flüssigkeit und P den Radius des Teilchens bedeutet. Wir erhalten die Gleichung:

[7]
$$m\frac{dv}{dt} = -6\pi k P v.$$

Für die Zeit ϑ, in welcher die Geschwindigkeit auf ein Zehntel ihres anfänglichen Wertes sinkt, ergibt sich hieraus

$$\vartheta = \frac{m}{0,434 \cdot 6\pi k P}.$$

Für die oben erwähnten Platinteilchen (in Wasser) haben wir $P = 2,5 \cdot 10^{-6}$ cm und $\eta = 0,01$ [8] zu setzen, so dass man erhält[1]:

$$\vartheta = 3,3 \cdot 10^{-7} \text{ Sekunden.}$$

Kehren wir zur Molekulartheorie der Wärme zurück, so haben wir diese Betrachtung zu modifizieren. Wir müssen zwar auch jetzt annehmen, dass das Teilchen eine anfängliche Bewegung während der sehr kurzen Zeit ϑ durch Reibung nahezu verliere. Gleichzeitig müssen wir aber annehmen, dass das Teilchen während dieser Zeit durch einen der inneren Reibung inversen Vorgang neue Bewegungsimpulse erhalte, so dass es eine Geschwindigkeit beibehält, die im Mittel gleich $\sqrt{\overline{v^2}}$ ist. Da wir aber Richtung und Grösse dieser Bewegungsimpulse uns als (nahezu) unabhängig von der anfänglichen Bewegungsrichtung und Geschwindigkeit des Teilchens vorstellen müssen, so müssen wir schliessen, dass Geschwindigkeit und Bewegungsrichtung des Teilchens sich schon in der ausserordentlich kurzen Zeit ϑ sehr stark ändere, und zwar in ganz unregelmässiger Weise.

Es ist daher — wenigstens für ultramikroskopische Teilchen — unmöglich, $\sqrt{\overline{v^2}}$ durch Beobachtungen zu ermitteln.

3. Beschränkt man sich auf die Untersuchung der Wege oder — präziser ausgedrückt — der Lagenänderungen in Zeiten τ, welche wesentlich grösser sind als ϑ, so gilt nach der Molekulartheorie der Wärme [9]

$$\sqrt{\overline{\lambda_x^2}} = \sqrt{\tau}\sqrt{\frac{RT}{N}\frac{1}{3\pi k P}},$$

falls man mit λ_x die während τ stattfindende Aenderung der x-Koordinate des Teilchens be-

---

[1] Für „mikroskopische" Teilchen ist ϑ bedeutend grösser, indem ϑ unter sonst gleichen Umständen dem Quadrat des Teilchenradius proportional ist.

42   ZEITSCHRIFT FÜR ELEKTROCHEMIE.   [Nr. 6.

zeichnet. Man kann als mittlere Geschwindigkeit im Zeitintervall τ die Grösse

$$\frac{\sqrt{\overline{\lambda_x{}^2}}}{\tau} = \frac{w}{\sqrt{\tau}}$$

definieren, wobei zur Abkürzung

$$\sqrt{\frac{RT}{N}\frac{1}{3\pi kP}} = w$$

gesetzt ist. Diese mittlere Geschwindigkeit ist aber desto grösser, je kleiner τ ist; so lange τ gross ist gegen ϑ, nähert sich dieselbe mit abnehmendem τ keinem Grenzwert.

Da ein mit bestimmten Beobachtungsmitteln in bestimmter Weise operierender Beobachter niemals die in beliebig kleinen Zeiten zurückgelegten Wege wahrnehmen kann, so wird stets eine gewisse mittlere Geschwindigkeit ihm als [10] Momentangeschwindigkeit erscheinen. Es ist aber klar, dass der so ermittelten Geschwindigkeit keine objektive Eigenschaft der untersuchten Bewegung entspricht — wenigstens falls die Theorie den Tatsachen entspricht. [11]

Bern, Januar 1907.

(Eingegangen: 22. Januar.)

Published in *Zeitschrift für Elektrochemie und angewandte physikalische Chemie* 13 (1907): 41–42. Dated Bern, January 1907, received 22 January 1907, published 8 February 1907.

[1] Einstein was presumably referring to *Svedberg 1906a*, *1906b*. The latter paper is a review of the experimental results reported on in the first paper in the light of Einstein's work on Brownian motion. For evidence of the fact that Svedberg sent one of his papers to Einstein, see Einstein to The Svedberg, 14 December 1919.

[2] Svedberg's interpretation of his observations was based on some erroneous assumptions concerning the nature of Brownian motion. On 11 November 1909, Einstein commented on Svedberg's assumptions in a letter to Jean Perrin (see the editorial note, "Einstein on Brownian Motion," § VII, pp. 219–222.

[3] Svedberg related the apparent velocities of the Brownian particles he observed to the velocity of molecules calculated from the equipartition theorem (see *Svedberg 1906a*). Felix Exner had earlier used a similar argument for the evaluation of his observations of Brownian motion (see *Exner 1900*, pp. 845–847).

[4] The value given for N is approximately the same as that given in *Einstein 1906c* (Doc. 33). In *Einstein 1922*, the more accurate value 6 × 10^{23} per mole is given instead.

[5] Svedberg's work on Brownian motion was based on the preparation of organosols of metals such as platinum by a new method, which allowed one to produce colloids consisting of par-

ticles of the same substance in different liquids, obtaining colloidal particles of approximately constant size (see *Svedberg 1905* and *1906a*).

[6] The value for m (in gm) is that given in *Svedberg 1906a*, p. 859. According to this paper, the temperatures at which the experiments were performed varied between 18° and 20°C. The observed "velocity" of the platinum particles is stated to be approximately 0.03 cm/sec.

[7] For Einstein's previous use of Stokes's law in his study of Brownian motion, see *Einstein 1905k* (Doc. 16), § 3. For an argument related to the present use of this law, see *Einstein 1906b* (Doc. 32), § 5.

[8] η should be k, as corrected in *Einstein 1922*. The viscosity of water and the particle radius correspond to the values given in *Svedberg 1906b*, p. 910. Water is one of several solvents that Svedberg used for his experiments.

[9] The following argument is based on results derived in *Einstein 1905k* (Doc. 16).

[10] Svedberg referred to the apparent velocity of colloidal particles as their mean speed (*Svedberg 1906a*, p. 858). He superimposed a translatory motion of the solution as a whole on the Brownian motion of the particles, thus obtaining "sinusoidlike" ("sinusoidähnliche") trajectories that are accessible to measurement (*Svedberg 1906a*, p. 854).

[11] For references to later experimental verifications of Einstein's theory, see the editorial note, "Einstein on Brownian Motion," § VII, pp. 219–222.

41. "On the Possibility of a New Test of the Relativity Principle"

[*Einstein 1907e*]

DATED Bern, March 1907
RECEIVED 17 March 1907
PUBLISHED 28 May 1907

IN: *Annalen der Physik* 23 (1907): 197–198.

12. *Über die Möglichkeit*
einer neuen Prüfung des Relativitätsprinzips;
von A. Einstein.

In einer letztes Jahr erschienenen wichtigen Arbeit[1]) hat
Hr. J. Stark dargetan, daß die bewegten positiven Ionen der
Kanalstrahlen Linienspektra emittieren, indem er den Doppler-
Effekt nachwies und messend verfolgte. Er stellte auch Untersuchungen an in der Absicht, einen Effekt zweiter Ordnung
(proportional $(v/V)^2$) nachzuweisen und zu messen; die nicht
speziell für diesen Zweck eingerichtete Versuchsanordnung ge-
nügte jedoch nicht zur Erlangung eines sicheren Resultates.

Ich will im nachfolgenden kurz zeigen, daß das Relativitätsprinzip in Verbindung mit dem Prinzip der Konstanz der Geschwindigkeit des Lichtes jenen Effekt vorauszubestimmen
gestattet. Wie ich in einer früheren Arbeit[2]) gezeigt habe,
geht aus jenen Prinzipien hervor, daß eine gleichförmig bewegte Uhr, vom „ruhenden" System aus beurteilt, langsamer
läuft als von einem mitbewegten Beobachter aus beurteilt.
Bezeichnet v die Anzahl der Schläge der Uhr pro Zeiteinheit
für den ruhenden, v_0 die entsprechende Anzahl für den mitbewegten Beobachter, so ist

[3]
$$\frac{v}{v_0} = \sqrt{1 - \left(\frac{v}{V}\right)^2}$$

oder in erster Annäherung

$$\frac{v - v_0}{v_0} = -\frac{1}{2}\left(\frac{v}{V}\right)^2.$$

Das Strahlung von bestimmten Frequenzen aussendende und
absorbierende Atomion der Kanalstrahlen ist nun als eine
rasch bewegte Uhr aufzufassen, und es ist daher die soeben
angegebene Beziehung auf dasselbe anwendbar.

[1]   1) J. Stark, Ann. d. Phys. **21.** p. 401. 1906.
[2]   2) A. Einstein, Ann. d. Phys. **17.** p. 903. 1905.

198   *A. Einstein.   Prüfung des Relativitätsprinzips.*

Es ist aber zu beachten, daß die Frequenz ν_0 (für den mitbewegten Beobachter) unbekannt ist, so daß die obige Beziehung der experimentellen Prüfung nicht direkt zugänglich ist. Es ist aber anzunehmen, daß ν_0 auch gleich ist der Frequenz, welche dasselbe Ion im ruhenden Zustand emittiert bez. absorbiert, und zwar aus folgendem Grunde. Aus der Tatsache, daß dasselbe Linienspektrum unter sehr verschiedenen Bedingungen entsteht, entnehmen wir, daß die Frequenz ν_0 nicht abhängig ist von Wechselwirkungen zwischen bewegten Ionen und ruhendem Gas, sondern daß sie dem Ion allein eigentümlich ist; hieraus folgert man direkt mit Hilfe des Relativitätsprinzips, daß ν_0 gleich sein muß der Frequenz der von einem ruhenden Ion emittierten bez. absorbierten Strahlung.

Die Gleichung

$$\frac{\nu - \nu_0}{\nu_0} = -\frac{1}{2}\left(\frac{v}{V}\right)^2$$

gibt also direkt den gesuchten Effekt zweiter Ordnung.

Die von Hrn. Stark für den Effekt angegebenen Zahlenwerte sind mehr als zehnmal so groß als die aus der angegebenen Formel hervorgehenden. Es erscheint mir wahr- [4]
scheinlich, daß sichere Resultate in der vorliegenden Frage erst dann zu erwarten sind, wenn es gelungen ist, (nicht-leuchtende?) Kanalstrahlen im völlig gasfreien Raume zu erzielen.   [5]

Bern, März 1907.

(Eingegangen 17. März 1907.)

Published in *Annalen der Physik* 23 (1907): 197–198. Dated Bern, March 1907, received 17 March 1907, published 28 May 1907.

[1] *Stark 1906*.

[2] *Einstein 1905r* (Doc. 23).

[3] See *Einstein 1905r* (Doc. 23), p. 904.

[4] See Einstein to Johannes Stark, 13 April 1907, for an indication that Stark may have intended to further investigate the effect discussed in this paper.

[5] See *Ives and Stilwell 1938* for the first reported observations of the transverse Doppler effect.

42. ''Correction to My Paper: 'Planck's Theory of Radiation, etc.' ''

[Einstein 1907d]

RECEIVED 3 March 1907
PUBLISHED 4 April 1907

IN: *Annalen der Physik* 22 (1907): 800.

800

14. *Berichtigung zu meiner Arbeit:*. *„Die Plancksche Theorie der Strahlung etc.";* *von A. Einstein.*

In der genannten, im Januarheft dieses Jahres erschienenen Arbeit habe ich geschrieben: „Nach D r u d e s Untersuchungen sind es die ponderabeln Atome (Atomionen) selbst, welchen diese Eigenfrequenzen zuzuschreiben sind. Es liegt also am nächsten, als Träger der Wärme in festen Körpern (Isolatoren) ausschließlich die positiven Atomionen zu betrachten."   [1]

Dieser Satz ist in zwei Beziehungen nicht aufrecht zu erhalten. Erstens sind nicht nur positiv, sondern auch negativ geladene Atomionen anzunehmen. Zweitens aber — und dies ist das Wesentliche — wird durch D r u d e s Untersuchungen nicht die Annahme gerechtfertigt, daß jedes schwingungsfähige Elementargebilde, welches als Träger von Wärme auftritt, stets eine elektrische Ladung besitze. Man kann also wohl aus der Existenz eines Absorptionsgebietes (unter den angegebenen Einschränkungen) auf die Existenz einer Gattung von Elementar-   [2] gebilden schließen, welche zur spezifischen Wärme einen Beitrag von charakteristischer Temperaturabhängigkeit liefert; der   [3] umgekehrte Schluß ist aber nicht statthaft, da es sehr wohl ungeladene Wärmeträger geben kann, d. h. solche, die sich optisch nicht bemerkbar machen. Letzteres ist besonders zu erwarten bei chemisch nicht gebundenen Atomen.

Der im letzten Satz der Abhandlung aus den Eigenschaften der spezifischen Wärme des Diamanten gezogene Schluß ist daher ebenfalls unstatthaft. Es sollte heißen:

„Es ist also nach der Theorie zu erwarten, daß der Diamant entweder bei $\lambda = 11\,\mu$ ein Absorptionsmaximum aufweist, oder daß derselbe überhaupt keine optisch nachweisbare ultrarote Eigenfrequenz besitzt."   [4]

(Eingegangen 3. März 1907.)

Berichtigung.

Bd. 22, p. 287 ist Zeile 4 von unten in Gleichung (2) der Buchstabe π zu streichen.

Druck von Metzger & Wittig in Leipzig.

Published in *Annalen der Physik* 22 (1907): 800. Received 3 March 1907, published 4 April 1907.

[1] *Einstein 1907a* (Doc. 38), p. 187. The frequencies in question are in the infrared region. The reference is to *Drude 1904a*, especially p. 682.

[2] Presumably Einstein is referring to absorption of waves in the infrared region, specifically wavelengths greater than 4.8μ. See *Einstein 1907a* (Doc. 38), p. 187.

[3] That is, one that shows the temperature dependence of specific heats derived in *Einstein 1907a* (Doc. 38).

[4] Einstein told Sommerfeld in 1910 about the approximate confirmation of the first part of this statement (see Einstein to Arnold Sommerfeld, July 1910).

43. Author's abstract of lecture: "On the Nature of the Movements of Microscopically Small Particles Suspended in Liquids"

[Einstein 1907f]

Lecture held in Bern, 23 March 1907

REPORT PUBLISHED IN: *Naturforschende Gesellschaft Bern. Mitteilungen* (1907), no. 1038.

— VII —

[1] **1038. Sitzung vom 23. März 1907.**
Abends 8 Uhr im Storchen.

[2] Vorsitzender: Herr Ed. Fischer. Anwesend: 20 Mitglieder und Gäste.

1. Herr **A. Einstein** spricht „**Ueber die Natur der Bewegungen mikroskopisch kleiner, in Flüssigkeiten suspendierter Teilchen.**"

[3] Mikroskopisch kleine, in Flüssigkeiten suspendierte leblose Teilchen (z. B. von der Grössenordnung 0,001 mm Durchmesser) führen unregelmässige Bewegungen aus, welche desto lebhafter sind, je kleiner der Teilchendurchmesser und die Viskosität der Flüssigkeit und je höher die Temperatur ist (Brown'sche Bewegung). Nach [4] kurzer Darlegung verschiedener Erklärungsversuche wird vom Vortragenden mit Hilfe der kinetischen Theorie der Wärme auf elemen- [5] tarem Wege eine einfache Formel für die von den Teilchen zurück- [6] gelegten Wegstrecken abgeleitet. (Autoreferat.)

[7] Weiteres darüber siehe: Ann. d. Physik 4. 17 1905, pag. 549.
Ann. d. Physik 4. 19. 1906, pag. 371.

2. Herr **W. Rytz** spricht über „**Beiträge zur Kientaler-Pilzflora.**" (Siehe die Abhandlungen dieses Bandes.)

Published in *Naturforschende Gesellschaft Bern. Mitteilungen* (1907), no. 1038.

[1] Einstein gave the lecture which the following report summarizes at a meeting of the Naturforschende Gesellschaft Bern on 23 March. He had been a member of the Society since 2 May 1903 (see the Minutes for that date, SzBe).

[2] Eduard Fischer was Professor of Botany at the University of Bern.

[3] Einstein had earlier characterized Brownian motion as "movements of inorganic small suspended bodies" ("Bewegungen lebloser kleiner suspendirter Körper") in Einstein to Conrad Habicht, 18 May–8 June 1905. The same size of the suspended particles is mentioned in this letter, as well as in *Einstein 1905k* (Doc. 16), p. 559.

[4] For a contemporary critical analysis of such attempted explanations of Brownian motion, see *Smoluchowski 1906.*

[5] Einstein's expression for the mean square displacement was first derived in *Einstein 1905k* (Doc. 16), p. 559. For a more elementary derivation, which does not use the concept of "mean square displacement," see *Einstein 1908c* (Doc. 50), pp. 237 238.

[6] Following the lecture, a discussion took place in which Hermann Sahli, Professor of Medicine at the University of Bern, Michele Besso, and Einstein participated (see the Minutes of the Society for that date, SzBe).

[7] *Einstein 1905k* (Doc. 16); *Einstein 1906b* (Doc. 32).

44. ''Comments on the Note of Mr. Paul Ehrenfest: 'The Translatory Motion of Deformable Electrons and the Area Law' ''

[*Einstein 1907g*]

DATED Bern, 14 April 1907
RECEIVED 16 April 1907
PUBLISHED 28 May 1907

IN: *Annalen der Physik* 23 (1907): 206–208.

206

15. *Bemerkungen zu der Notiz von Hrn. Paul Ehrenfest: „Die Translation deformierbarer Elektronen und der Flächensatz"; von A. Einstein.*

[1]

In der genannten Abhandlung sind folgende Bemerkungen enthalten:

[2] „Die Lorentzsche Relativitätselektrodynamik wird in der Formulierung, in der sie Hr. Einstein publiziert hat, ziemlich allgemein als abgeschlossenes System angesehen. Dementsprechend muß sich aus ihr rein deduktiv eine Antwort auf die Frage ergeben, die man durch Übertragung des Abraham-

[3] schen Problems vom starren auf das deformierbare Elektron erhält: Angenommen, es existiere ein deformierbares Elektron,

[4] das in der Ruhe irgend eine nicht-kugelförmige und nicht ellipsoidische Gestalt besitzt. Bei gleichförmiger Translation erfährt dieses Elektron nach Hrn. Einstein die bekannte Lorentz-Kontraktion. Ist nun für dieses Elektron gleichförmige Translation nach jeder Richtung hin kräftefrei möglich

[5] oder nicht?"

Hierzu habe ich folgendes zu bemerken:

1. Das Relativitätsprinzip oder — genauer ausgedrückt — das Relativitätsprinzip zusammen mit dem Prinzip von der Konstanz der Lichtgeschwindigkeit ist nicht als ein „abgeschlossenes System", ja überhaupt nicht als System aufzufassen, sondern lediglich als ein heuristisches Prinzip, welches für sich allein betrachtet nur Aussagen über starre Körper, Uhren und Lichtsignale enthält. Weiteres liefert die Relativitätstheorie nur dadurch, daß sie Beziehungen zwischen

Bemerkungen zu der Notiz von P. Ehrenfest.   207

sonst voneinander unabhängig erscheinenden Gesetzmäßigkeiten
fordert.   [6]

Die Theorie der Bewegung des Elektrons beispielsweise
kommt folgendermaßen zustande. Man setzt die Maxwellschen Gleichungen für das Vakuum für ein Koordinatenzeit-
system voraus. Durch Anwendung der vermittelst des Rela-
tivitätssystems hergeleiten Ort-Zeit-Transformation findet man
die Transformationsgleichungen für die elektrischen und magnetischen Kräfte. Unter Benutzung der letzteren findet man
durch abermalige Anwendung der Ort-Zeit-Transformation aus
dem Gesetz für die Beschleunigung des langsam bewegten
Elektrons (welches angenommen bez. der Erfahrung entnommen wurde) das Gesetz für die Beschleunigung des beliebig rasch bewegten Elektrons. Es handelt sich hier also   [7]
keineswegs um ein „System", in welchem implizite die einzelnen
Gesetze enthalten wären, und nur durch Deduktion daraus
gefunden werden könnten, sondern nur um ein Prinzip, das
(ähnlich wie der zweite Hauptsatz der Wärmetheorie) gewisse
Gesetze auf andere zurückzuführen gestattet.   [8]

2. Als man sich noch nicht auf das Relativitätsprinzip
stützte, sondern die Bewegungsgesetze des Elektrons auf
elektrodynamischem Wege zu ermitteln strebte, sah man sich
genötigt, über die Verteilung der Elektrizität bestimmtere
Annahmen zu machen, damit das Problem kein unbestimmtes
sei. Man dachte sich dabei die Elektrizität auf einem
(starren) Gerüst verteilt. Es ist wohl zu beachten, daß die   [9]
Gesetze, nach welchen ein solches Gebilde sich bewegt, nicht
aus der Elektrodynamik allein hergeleitet werden können.
Das Gerüst ist ja nichts anderes als die Einführung von
Kräften, welche den elektrodynamischen das Gleichgewicht
leisten. Wenn wir das Gerüst als einen starren (d. h. durch
äußere Kräfte nicht deformierbaren) Körper ansehen, so
kann das Problem der Bewegung des Elektrons dann und
nur dann auf deduktivem Wege ohne Willkür gelöst werden,
wenn die Dynamik des starren Körpers hinreichend genau bekannt ist.   [10]

Falls die Relativitätstheorie zutrifft, sind wir von letzterem Ziele noch weit entfernt. Wir besitzen erst eine Kinematik der Paralleltranslation und einen Ausdruck für die

208   *A. Einstein.   Bemerkungen zu der Notiz von P. Ehrenfest.*

kinetische Energie eines in Paralleltranslation begriffenen Körpers, falls letzterer mit anderen Körpern nicht in Wechselwirkung steht¹); im übrigen ist sowohl die Dynamik als auch die Kinematik des starren Körpers für den vorliegenden Fall noch als unbekannt zu betrachten.

Bern, den 14. April 1907.

1) Daß letztere Einschränkung wesentlich ist, werde ich demnächst [11]   in einer Arbeit zeigen.

(Eingegangen 16. April 1907.)

Published in *Annalen der Physik* 23 (1907): 206–208. Dated Bern, 14 April 1907, received 16 April 1907, published 28 May 1907.

[1] *Ehrenfest 1907*.

[2] In Ehrenfest's text, "Relativitäts-Elektrodynamik."

[3] Abraham's problem is stated in Ehrenfest's paper: "Abraham has drawn attention to the fact that, for a (rigid) *non*-spherical electron, *force-free* uniform translation in every direction cannot take place" ("Hr. Abraham hat darauf hingewiesen, daß für ein (starres) *nicht*-kugelförmiges Elektron gleichförmige Translation nicht nach jeder Richtung hin *kräftefrei* stattfinden kann") (*Ehrenfest 1907*, p. 204). In Ehrenfest's text, "deformierbare" is in italics.

[4] In Ehrenfest's text, "nicht" is in italics both times; it is also followed by a hyphen the

second time.

[5] In Ehrenfest's text, the final sentence is in italics.

[6] For a similar discussion of the significance of the principle of relativity, see Einstein to Arnold Sommerfeld, 14 January 1908.

[7] See *Einstein 1905r* (Doc. 23), § 10, pp. 917–919.

[8] This is the first time that Einstein compared the relativity principle to the laws of thermodynamics. For a discussion of this comparison, see the Introduction, pp. xxi–xxii.

[9] See *Abraham 1902a, 1902b, 1903*.

[10] For Einstein's first discussion of the problem of rigid-body dynamics, see *Einstein 1907h* (Doc. 45), § 3, pp. 379–382.

[11] See *Einstein 1907h* (Doc. 45), § 1, pp. 373–377.

45. "On the Inertia of Energy Required by the Relativity Principle"

[Einstein 1907h]

DATED Bern, May 1907
RECEIVED 14 May 1907
PUBLISHED 13 June 1907

IN: *Annalen der Physik* 23 (1907): 371–384.

12. *Über die vom Relativitätsprinzip geforderte Trägheit der Energie; von A. Einstein.*

Das Relativitätsprinzip führt in Verbindung mit den Maxwellschen Gleichungen zu der Folgerung, daß die Trägheit eines Körpers mit dessen Energieinhalt in ganz bestimmter Weise wachse bez. abnehme. Betrachtet man nämlich einen Körper, der gleichzeitig nach zwei entgegengesetzten Richtungen eine bestimmte Strahlungsenergie aussendet, und untersucht man diesen Vorgang von zwei relativ zueinander gleichförmig bewegten Koordinatensystemen aus [1]), von denen das eine relativ zu dem Körper ruht, und wendet man auf den Vorgang — von beiden Koordinatensystemen aus — das Energieprinzip an, so gelangt man zu dem Resultat, daß einem Energiezuwachs ΔE des betrachteten Körpers stets ein Massenzuwachs $\Delta E / V^2$ entsprechen müsse, wobei V die Lichtgeschwindigkeit bedeutet.

Der Umstand, daß der dort behandelte spezielle Fall eine Annahme von so außerordentlicher Allgemeinheit (über die Abhängigkeit der Trägheit von der Energie) notwendig macht, fordert dazu auf, in allgemeinerer Weise die Notwendigkeit bez. Berechtigung der genannten Annahme zu prüfen. Insbesondere erhebt sich die Frage: Führen nicht andere spezielle Fälle zu mit der genannten Annahme unvereinbaren Folgerungen? Einen ersten Schritt in dieser Hinsicht habe ich letztes Jahr unternommen [2]), indem ich zeigte, daß jene Annahme den Widerspruch der Elektrodynamik mit dem Prinzip von der Konstanz der Schwerpunktsbewegung (mindestens was die Glieder erster Ordnung anbelangt) aufhebt.

Die *allgemeine* Beantwortung der aufgeworfenen Frage ist darum vorläufig nicht möglich, weil wir ein vollständiges, dem

[1]   1) A. Einstein, Ann. d. Phys. **18.** p. 639. 1905.
[2]   2) A. Einstein, Ann. d. Phys. **20.** p. 627. 1906.

372 *A. Einstein.*

Relativitätsprinzip entsprechendes Weltbild einstweilen nicht besitzen. Wir müssen uns vielmehr auf die speziellen Fälle beschränken, welche wir ohne Willkür vom Standpunkt der Relativitätselektrodynamik gegenwärtig behandeln können. Zwei solche Fälle werden wir im folgenden betrachten; bei dem ersten derselben besteht das System, dessen träge Masse untersucht werden soll, in einem starren, starr elektrisierten Körper, bei dem zweiten Fall aus einer Anzahl gleichförmig bewegter Massenpunkte, welche aufeinander keine Kräfte ausüben.

Bevor ich mit der Untersuchung beginne, muß ich hier noch eine Bemerkung über den mutmaßlichen Gültigkeitsbereich der Maxwellschen Gleichungen für den leeren Raum einschieben, um einem naheliegenden Einwand zu begegnen. In früheren Arbeiten habe ich gezeigt, daß unser heutiges elektromechanisches Weltbild nicht geeignet ist, die Entropieeigenschaften der Strahlung sowie die Gesetzmäßigkeiten der Emission und Absorption der Strahlung und die der spezifischen Wärme zu erklären; es ist vielmehr nach meiner Meinung nötig anzunehmen, daß die Beschaffenheit eines jeglichen periodischen Prozesses eine derartige ist, daß eine Umsetzung der Energie nur in bestimmten Quanten von endlicher Größe (Lichtquanten) vor sich gehen kann, daß also die Mannigfaltigkeit der in Wirklichkeit möglichen Prozesse eine kleinere ist als die Mannigfaltigkeit der im Sinne unserer heutigen theoretischen Anschauungen möglichen Prozesse.[1]) Einen Strahlungsvorgang im besonderen hätten wir uns so zu denken, daß der momentane elektromagnetische Zustand in einem Raumteile durch eine *endliche* Zahl von Größen vollständig bestimmt sei — im Gegensatze zur Vektorentheorie der Strahlung. Solange wir jedoch nicht im Besitz eines Bildes sind, welches den genannten Forderungen entspricht, werden wir uns naturgemäß in allen Fragen, welche nicht Entropieverhältnisse sowie Umwandlungen elementar kleiner Energiemengen betreffen, der gegenwärtigen Theorie bedienen, ohne fürchten zu müssen, dadurch zu unrichtigen Resultaten zu gelangen. Wie ich mir die heutige Sachlage in diesen Fragen denke, kann

---

1) A. Einstein, Ann. d. Phys. 17. p. 132. 1905; 20. p. 199. 1906 [3] und 22. p. 180. 1907.

ich am anschaulichsten durch folgenden fingierten Fall illustrieren.

Man denke sich, daß die molekularkinetische Theorie der Wärme noch nicht aufgestellt, daß aber mit voller Sicherheit nachgewiesen sei, daß die Brownsche Bewegung (Bewegung von in Flüssigkeiten suspendierten Teilchen) nicht auf äußerer Energiezufuhr beruhe, sondern daß klar erkannt sei, daß jene Bewegungen mit Hilfe der Mecbanik und Thermodynamik nicht erklärt werden können. Man würde bei dieser Sachlage mit Recht zu dem Schlusse geführt, daß eine tiefgreifende Änderung der theoretischen Grundlagen Platz greifen müsse. Trotzdem würde sich aber niemand scheuen, bei Behandlung aller Fragen, welche sich nicht auf Momentanzustände in kleinen Raumteilen beziehen, die Grundgleichungen der Mechanik und Thermodynamik anzuwenden. In diesem Sinne können wir nach meiner Meinung mit Zuversicht unsere Betrachtungen auf die Maxwellschen Gleichungen stützen.

Es scheint mir in der Natur der Sache zu liegen, daß das Nachfolgende zum Teil bereits von anderen Autoren klargestellt sein dürfte. Mit Rücksicht darauf jedoch, daß hier die betreffenden Fragen von einem neuen Gesichtspunkt aus behandelt sind, glaubte ich, von einer für mich sehr umständlichen Durchmusterung der Literatur absehen zu dürfen, zumal zu hoffen ist, daß diese Lücke von anderen Autoren noch ausgefüllt werden wird, wie dies in dankenswerter Weise bei meiner ersten Arbeit über das Relativitätsprinzip durch Hrn. [4] Planck und Hrn. Kaufmann bereits geschehen ist.

§ 1. Über die kinetische Energie eines in gleichförmiger Translation begriffenen, äußeren Kräften unterworfenen starren Körpers.

Wir betrachten einen in gleichförmiger Translationsbewegung (Geschwindigkeit v) in Richtung der wachsenden x-Koordinate eines ruhend gedachten Koordinatensystems (x, y, z) befindlichen starren Körper. Wirken äußere Kräfte nicht auf ihn, so ist nach der Relativitätstheorie seine kinetische Energie K_0 gegeben durch die Gleichung[1])

[5]   1) A. Einstein, Ann. d. Phys. 17. p. 917 ff. 1905.

374   *A. Einstein.*

$$K_0 = \mu V^2 \left\{ \frac{1}{\sqrt{1 - \left(\frac{v}{V}\right)^2}} - 1 \right\},$$

wobei μ seine Masse (im gewöhnlichen Sinne) und V die Lichtgeschwindigkeit im Vakuum bedeutet. Wir wollen nun zeigen, daß nach der Relativitätstheorie dieser Ausdruck nicht mehr gilt, falls äußere Kräfte auf den Körper wirken, welche einander das Gleichgewicht halten. Um den Fall behandeln zu können, müssen wir voraussetzen, daß jene Kräfte elektrodynamische seien. Wir denken uns daher den Körper starr elektrisiert (mit kontinuierlich verteilter Elektrizität), und es wirke auf ihn ein elektromagnetisches Kraftfeld. Die elektrische Dichte denken wir uns allenthalben als sehr gering und das Kraftfeld als intensiv, derart, daß die den Wechselwirkungen zwischen den elektrischen Massen des Körpers entsprechenden Kräfte gegenüber den vom äußeren Kraftfelde auf die elektrischen Ladungen des Körpers ausgeübten Kräfte vernachlässigt werden können.[1]) Die von dem Kraftfeld auf den Körper zwischen den Zeiten t_0 und t_1 übertragene Energie ΔE ist gegeben durch den Ausdruck:

$$\Delta E = \int_{t_0}^{t_1} dt \int v X \frac{\varrho}{4\pi} \, dx \, dy \, dz,$$

wobei das Raumintegral über den Körper zu erstrecken und

$$\varrho = \frac{\partial X}{\partial x} + \frac{\partial Y}{\partial y} + \frac{\partial Z}{\partial z}$$

gesetzt ist. Diesen Ausdruck transformieren wir nach den in der oben zitierten Abhandlung angegebenen Transformationsgleichungen[2]) auf dasjenige Ort-Zeitsystem (ξ, η, ζ, τ), welches einem relativ zu dem Körper ruhenden, zu (x, y, z) parallelachsigen Koordinatensystem entspricht. Man erhält so in einer Bezeichnung, welche der in jener Abhandlung benutzten genau entspricht, nach einfacher Rechnung

$$\Delta E = \iint \beta v X' \frac{\varrho'}{4\pi} \, d\xi \, d\eta \, d\zeta \, d\tau,$$

---

1) Wir führen diese Annahme ein, um annehmen zu können, daß die wirkenden Kräfte vermöge der Art, wie sie erzeugt sind, keinen beschränkenden Bedingungen unterworfen seien.

2) A. Einstein, Ann. d. Phys. 17. §§ 3 u. 6. 1905.   [6]

wobei β wie dort den Ausdruck

$$\frac{1}{\sqrt{1-\left(\frac{v}{V}\right)^2}}$$

bedeutet. Es ist zu beachten, daß gemäß unseren Voraussetzungen die Kräfte X' keine beliebigen sein dürfen. Sie
müssen vielmehr zu jeder Zeit so beschaffen sein, daß der
betrachtete Körper keine Beschleunigung erfährt. Hierfür erhält man nach einem Satze der Statik die notwendige (aber
nicht hinreichende) Bedingung, daß von einem mit dem Körper
bewegten Koordinatensystem aus betrachtet die Summe der
X-Komponenten der auf den Körper wirkenden Kräfte stets
verschwindet. Man hat also für jedes τ:

$$\int X' \varrho' d\xi \, d\eta \, d\zeta = 0 \, .$$

Wären also die Grenzen für τ in dem obigen Integralausdruck
für ΔE von ξ, η, ζ unabhängig, so wäre $\Delta E = 0$. Dies ist
jedoch nicht der Fall. Aus der Transformationsgleichung

$$t = \beta \left(\tau + \frac{v}{V^2} \xi \right)$$

folgt nämlich unmittelbar, daß die Zeitgrenzen im bewegten
System sind:

$$\tau = \frac{t_0}{\beta} - \frac{v}{V^2} \xi \quad \text{und} \quad \tau = \frac{t_1}{\beta} - \frac{v}{V^2} \xi \, .$$

Wir denken uns das Integral im Ausdruck für ΔE in drei
Teile zerlegt.

Der erste Teil umfasse die Zeiten τ zwischen

$$\frac{t_0}{\beta} - \frac{v}{V^2} \xi \quad \text{und} \quad \frac{t_0}{\beta},$$

der zweite Teil zwischen

$$\frac{t_0}{\beta} \quad \text{und} \quad \cdot \frac{t_1}{\beta},$$

der dritte zwischen

$$\frac{t_1}{\beta} \quad \text{und} \quad \frac{t_1}{\beta} - \frac{v}{V^2} \xi \, .$$

Der zweite Teil verschwindet, weil er von ξ, η, ζ unabhängige Zeitgrenzen hat. Der erste und dritte Teil hat überhaupt nur dann einen bestimmten Wert, wenn die Annahme

gemacht wird, daß in der Nähe der Zeiten $t = t_0$ und $t = t_1$ die auf den Körper wirkenden Kräfte von der Zeit unabhängig seien, derart, daß für alle Punkte des starren Körpers zwischen den Zeiten

$$\tau = \frac{t_0}{\beta} - \frac{v}{V^2}\xi \quad \text{und} \quad \tau = \frac{t_0}{\beta}$$

bez. zwischen

$$\tau = \frac{t_1}{\beta} \quad \text{und} \quad \tau = \frac{t_1}{\beta} - \frac{v}{V^2}\xi$$

die elektrische Kraft X' von der Zeit unabhängig ist. Nennt man X_0' bez. X_1' die in diesen beiden Zeiträumen vorhandenen X', so erhält man:

$$\varDelta E = -\int \frac{v^2}{V^2}\beta \frac{\xi X_1' \varrho'}{4\pi} d\xi\, d\eta\, d\zeta + \int \frac{v^2}{V^2}\beta \frac{\xi X_0' \varrho'}{4\pi} d\xi\, d\eta\, d\zeta.$$

Nimmt man ferner an, daß am Anfang $(t = t_0)$ keine Kräfte auf den Körper wirken, so verschwindet das zweite dieser Integrale. Mit Rücksicht darauf, daß

$$\frac{X_1' \varrho'}{4\pi} d\xi\, d\eta\, d\zeta$$

die ξ-Komponente K_ξ der auf das Raumelement wirkenden ponderomotorischen Kraft ist, erhält man

$$\varDelta E = -\frac{\left(\dfrac{v}{V}\right)^2}{\sqrt{1 - \left(\dfrac{v}{V}\right)^2}}\sum(\xi\, K_\xi),$$

wobei die Summe über alle Massenelemente des Körpers zu erstrecken ist.

Wir haben also folgendes merkwürdige Resultat erhalten. Setzt man einen starren Körper, auf den ursprünglich keine Kräfte wirken, dem Einflusse von Kräften aus, welche dem Körper keine Beschleunigung erteilen, so leisten diese Kräfte — von einem relativ zu dem Körper bewegten Koordinatensystem aus betrachtet — eine Arbeit $\varDelta E$ auf den Körper, welche lediglich abhängt von der endgültigen Kräfteverteilung und der Translationsgeschwindigkeit. Nach dem Energieprinzip folgt hieraus unmittelbar, daß die kinetische Energie eines Kräften unterworfenen starren Körpers um $\varDelta E$ größer ist als die kinetische Energie desselben, ebenso rasch bewegten Körpers, falls keine Kräfte auf denselben wirken.

§ 2. Über die Trägheit eines elektrisch geladenen starren Körpers.

Wir betrachten abermals einen starren, starr elektrisierten Körper, welcher eine gleichförmige Translationsbewegung im Sinne der wachsenden x-Koordinaten eines „ruhenden" Koordinatensystems ausführt (Geschwindigkeit v). Ein äußeres elektromagnetisches Kraftfeld sei nicht vorhanden. Wir wollen indessen jetzt das von den elektrischen Massen des Körpers erzeugte elektromagnetische Feld berücksichtigen. Wir berechnen zunächst die elektromagnetische Energie

$$E_e = \frac{1}{8\pi} \int (X^2 + Y^2 + Z^2 + L^2 + M^2 + N^2)\, dx\, dy\, dz.$$

Zu diesem Zweck transformieren wir diesen Ausdruck unter Benutzung der in der mehrfach zitierten Abhandlung enthaltenen Transformationsgleichungen, indem wir unter dem Integral Größen einführen, welche sich auf ein mit dem Körper bewegtes Koordinatensystem beziehen. Wir erhalten so:

$$E_e = \frac{1}{8\pi} \int \frac{1}{\beta} \left[X'^2 + \frac{1 + \left(\frac{v}{V}\right)^2}{1 - \left(\frac{v}{V}\right)^2}(Y'^2 + Z'^2) \right] d\xi\, d\eta\, d\zeta.$$

Es ist zu beachten, daß der Wert dieses Ausdruckes abhängt von der Orientierung des starren Körpers relativ zur Bewegungsrichtung. Wenn sich daher die gesamte kinetische Energie des elektrisierten Körpers ausschließlich zusammensetzte aus der kinetischen Energie K_0, welche dem Körper wegen seiner ponderabeln Masse zukommt, und dem Überschuß der elektromagnetischen Energie des bewegten Körpers über die elektrostatische Energie des Körpers für den Fall der Ruhe, so wären wir damit zu einem Widerspruche gelangt, wie leicht aus folgendem zu ersehen ist.

Wir denken uns, der betrachtete Körper sei relativ zu dem mitbewegten Koordinatensystem in unendlich langsamer Drehung begriffen, ohne daß äußere Einwirkungen während dieser Bewegung auf ihn stattfinden. Es ist klar, daß diese

378   *A. Einstein.*

Bewegung kräftefrei möglich sein muß, da ja nach dem Relativitätsprinzip die Bewegungsgesetze des Körpers relativ zu dem mitbewegten System dieselben sind wie die Bewegungsgesetze in bezug auf ein „ruhendes" System. Wir betrachten nun den gleichförmig bewegten und unendlich langsam sich drehenden Körper vom „ruhenden" System aus. Da die Drehung unendlich langsam sein soll, trägt sie zur kinetischen Energie nichts bei. Der Ausdruck der kinetischen Energie ist daher in dem betrachteten Fall derselbe wie wenn keine Drehung, sondern ausschließlich gleichförmige Paralleltranslation stattfände. Da nun der Körper relativ zur Bewegungsrichtung im Laufe der Bewegung verschiedene (beliebige) Lagen annimmt, und während der ganzen Bewegung das Energieprinzip gelten muß, so ist klar, daß eine Abhängigkeit der kinetischen Energie eines in Translationsbewegung begriffenen elektrisierten Körpers von der Orientierung unmöglich ist.

Dieser Widerspruch wird durch die Resultate des vorigen Paragraphen beseitigt. Die kinetische Energie des betrachteten Körpers kann nämlich nicht berechnet werden wie die eines starren Körpers, auf den keine Kräfte wirken. Wir haben vielmehr gemäß § 1 zu berücksichtigen, daß unser starrer Körper Kräften unterworfen ist, welche ihre Ursache in der Wechselwirkung zwischen den elektrischen Massen haben. Bezeichnen wir also mit K_0 die kinetische Energie für den Fall, daß keine elektrischen Ladungen vorhanden sind, so erhalten wir für die gesamte kinetische Energie K des Körpers den Ausdruck

$$K = K_0 + \Delta E + (E_e - E_a),$$

wobei E_a die elektrostatische Energie des betrachteten Körpers im Zustand der Ruhe bedeutet. In unserem Falle hat man

$$\Delta E = - \frac{v^2}{V^2}\,\beta\,\frac{1}{4\pi}\int \xi\,X'\left(\frac{\partial X'}{\partial \xi} + \frac{\partial Y'}{\partial \eta} + \frac{\partial Z'}{\partial \zeta}\right) d\xi\,d\eta\,d\zeta,$$

woraus man durch partielle Integration mit Berücksichtigung des Umstandes, daß X', Y', Z' von einem Potential ableitbar sind, erhält

$$\Delta E = \frac{v^2}{V^2}\,\beta\,\frac{1}{8\pi}\int (X'^2 - Y'^2 - Z'^2)\,d\xi\,d\eta\,d\zeta.$$

Berücksichtigt man die im § 1 angegebenen Ausdrücke für K_0 und β, so erhält man für die kinetische Energie des elektrisierten starren Körpers den Ausdruck

$$K = \left(\mu + \frac{E_s}{V^2}\right) \cdot V^2 \left(\frac{1}{\sqrt{1 - \left(\frac{v}{V}\right)^2}} - 1\right).$$

Dieser Ausdruck ist, wie es sein muß, von der Orientierung des Körpers relativ zur Translationsrichtung unabhängig. Vergleicht man den Ausdruck für K mit dem für die Energie K_0 eines nicht elektrisch geladenen Körpers

$$K_0 = \mu\, V^2 \left(\frac{1}{\sqrt{1 - \left(\frac{v}{V}\right)^2}} - 1\right),$$

so erkennt man, daß der elektrostatisch geladene Körper eine träge Masse besitzt, welche die des nicht geladenen Körpers um die durch das Quadrat der Lichtgeschwindigkeit dividierte elektrostatische Energie übertrifft. Der Satz von der Trägheit der Energie wird also durch unser Resultat in dem behandelten speziellen Fall bestätigt.

§ 3. Bemerkungen betreffend die Dynamik des starren Körpers.

Nach dem Vorangehenden könnte es scheinen, als ob wir von dem Ziele, eine dem Relativitätsprinzip entsprechende Dynamik der Paralleltranslation des starren Körpers zu schaffen, [7] nicht mehr weit entfernt wären. Man muß sich indessen daran erinnern, daß die im § 1 ausgeführte Untersuchung die Energie des Kräften unterworfenen starren Körpers nur für den Fall lieferte, daß jene Kräfte zeitlich konstant sind. Wenn zur Zeit t_1 die Kräfte X' von der Zeit abhängen, so erweist sich die Arbeit ΔE, also auch die Energie des starren Körpers, nicht nur als abhängig von denjenigen Kräften, welche zu *einer* bestimmten Zeit herrschen.

Um die hier vorliegende Schwierigkeit möglichst drastisch zu beleuchten, denken wir uns folgenden einfachen Spezialfall. Wir betrachten einen starren Stab AB, welcher relativ zu einem Koordinatensystem (ξ, η, ζ) ruhe, wobei die Stabachse in der ζ-Achse ruhe. Zu einer bestimmten Zeit τ_0 mögen

380   *A. Einstein.*

auf die Stabenden für ganz kurze Zeit entgegengesetzt gleiche
Kräfte P wirken, während der Stab in allen übrigen Zeiten
Kräften nicht unterworfen sei. Es ist klar, daß die genannte,
zur Zeit τ_0 auf den Stab ausgeübte Wirkung eine Bewegung

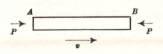

des Stabes *nicht* erzeugt. Wir betrachten nun genau denselben Vorgang von einem zum vorher benutzten parallelachsigen Koordinatensystem aus, relativ zu welchem
sich unser Stab in der Richtung A—B mit der Geschwindigkeit v
bewegt. Von dem letztgenannten Koordinatensystem aus beurteilt,
wirken nun aber die Kraftimpulse in A und B nicht gleichzeitig;
der Impuls in B ist vielmehr gegen den Impuls in A verspätet um
$l\beta(v/V^2)$ Zeiteinheiten, wobei l die (ruhend gemessene) Stab-
länge bedeutet. Wir sind also zu dem folgenden sonderbar
aussehenden Resultat gekommen. Auf den bewegten Stab AB
wirkt zuerst in A ein Kraftimpuls und darauf nach einiger
Zeit ein entgegengesetzter in B. Diese beiden Kraftimpulse
kompensieren einander derart, daß die Bewegung des Stabes
durch sie nicht modifiziert wird. Noch merkwürdiger erscheint
der Fall, wenn wir nach der Energie des Stabes fragen zu
einer Zeit, in welcher der Impuls in A bereits vorbei ist,
während der Impuls in B noch nicht zu wirken begonnen hat.
Der Impuls in A hat auf den Stab Arbeit übertragen (weil
der Stab bewegt ist); um diese Arbeit muß sich also die
Energie des Stabes vermehrt haben. Gleichwohl hat sich weder
die Geschwindigkeit des Stabes noch sonst eine auf ihn Bezug
habende Größe, von der wir die Energiefunktion des Stabes
abhängen lassen könnten, geändert. Es scheint also eine Verletzung des Energieprinzipes vorzuliegen.

Die *prinzipielle* Lösung dieser Schwierigkeit liegt auf der
Hand. Indem wir implizite annehmen, durch die auf den
Stab wirkenden Kräfte und durch die in demselben Augenblick
herrschende Stabgeschwindigkeit den Momentanzustand des
Stabes vollständig bestimmen zu können, nehmen wir an, daß
ein Geschwindigkeitszuwachs des Körpers durch die ihn erzeugende, irgendwo am Körper angreifende Kraft *momentan*
erzeugt werde, daß also die Ausbreitung der auf einen Punkt
des Körpers ausgeübten Kraft über den ganzen Körper keine

Zeit erfordere. Eine derartige Annahme ist, wie nachher gezeigt wird, mit dem Relativitätsprinzip nicht vereinbar. Wir sind also in unserem Falle offenbar genötigt, bei Einwirkung des Impulses in A eine Zustandsänderung unbekannter Qualität im Körper anzunehmen, welche sich mit endlicher Geschwindigkeit in demselben ausbreitet und in kurzer Zeit eine Beschleunigung des Körpers bewirkt, falls innerhalb dieser Zeit nicht noch andere Kräfte auf den Körper wirken, deren Wirkungen die der erstgenannten kompensieren. Wenn also die Relativitätselektrodynamik richtig ist, sind wir noch weit davon entfernt, eine [8] Dynamik der Paralleltranslation des starren Körpers zu besitzen.

Wir wollen nun zeigen, daß nicht nur die Annahme *momentaner* Ausbreitung irgend einer Wirkung, sondern allgemeiner jede Annahme von der Ausbreitung einer Wirkung mit Überlichtgeschwindigkeit mit der Relativitätstheorie nicht [9] vereinbar ist.

Längs der x-Achse eines Koordinatensystems (x, y, z) erstrecke sich ein Materialstreifen, relativ zu welchem sich eine gewisse Wirkung mit der Geschwindigkeit W fortzupflanzen vermöge, und es möge sowohl in $x = 0$ (Punkt A) als auch in $x = + l$ (Punkt B) sich je ein relativ zum Koordinatensystem (x, y, z) ruhender Beobachter befinden. Der Beobachter in A sende vermittelst der oben genannten Wirkung Zeichen zu dem Beobachter in B durch den Materialstreifen, welch letzterer nicht ruhe, sondern sich mit der Geschwindigkeit $v (< V)$ in der negativen x-Richtung bewege. Das Zeichen wird dann, wie aus § 5 (l. c.) hervorgeht, mit der Geschwindigkeit

$$\frac{W - v}{1 - \dfrac{W v}{V^2}}$$

von A nach B übertragen. Die Zeit T, welche zwischen Zeichengebung in A und Zeichenempfang in B verstreicht, ist also

$$T = l \frac{1 - \dfrac{W v}{V^2}}{W - v}.$$

Die Geschwindigkeit v kann jeglichen Wert annehmen, der kleiner ist als V. Wenn also $W > V$ ist, wie wir angenommen haben, so kann man v stets so wählen, daß $T < 0$ ist. Dies Resultat besagt, daß wir einen Übertragungsmechanismus für möglich halten müßten, bei dessen Benutzung die erzielte Wirkung der (etwa von einem Willensakt begleiteten) Ursache *vorangeht.* Wenn dies Resultat auch, meiner Meinung nach, rein logisch genommen keinen Widerspruch enthält, so widerstreitet es doch so unbedingt dem Charakter unserer gesamten Erfahrung, daß durch dasselbe die Unmöglichkeit der Annahme $W > V$ zur Genüge erwiesen ist. [10]

§ 4. Über die Energie eines Systems, welches aus einer Anzahl kräftefrei bewegter Massenpunkte besteht.

Betrachtet man den Ausdruck für die kinetische Energie k eines mit der Geschwindigkeit v bewegten Massenpunktes (μ)

$$k = \mu V^2 \left\{ \frac{1}{\sqrt{1 - \left(\frac{v}{V}\right)^2}} - 1 \right\},$$

so fällt auf, daß dieser Ausdruck die Gestalt einer Differenz besitzt. Es ist nämlich

$$k = \left| \mu V^2 \frac{1}{\sqrt{1 - \left(\frac{v}{V}\right)^2}} \right._{v=0}^{v=v}.$$

Frägt man nicht speziell nach der kinetischen Energie, sondern nach der Energie ε des bewegten Massenpunktes schlechtweg, so ist $\varepsilon = k + \text{konst.}$ Während man nun in der klassischen Mechanik die willkürliche Konstante in dieser Gleichung am bequemsten verschwinden läßt, erhält man in der Relativitäts-mechanik den einfachsten Ausdruck für ε, indem man den Nullpunkt der Energie so wählt, daß die Energie ε_0 für den ruhenden Massenpunkt μV^2 gesetzt wird.[1]) Man erhält dann

$$\varepsilon = \mu V^2 \frac{1}{\sqrt{1 - \left(\frac{v}{V}\right)^2}}.$$

An dieser Wahl des Nullpunktes der Energie werden wir im folgenden festhalten.

----

1) Es ist zu beachten, daß die vereinfachende Festsetzung $\mu V^2 = \varepsilon_0$ zugleich der Ausdruck des Prinzipes der Äquivalenz von Masse und Energie ist, und daß im Falle des masselosen elektrisierten Körpers ε_0 [11] nichts anderes ist als seine elektrostatische Energie.

Wir führen nun wieder die zwei stets relativ zueinander bewegten Koordinatensysteme (x, y, z) und (ξ, η, ζ) ein. Relativ zu (ξ, η, ζ) sei ein Massenpunkt μ mit der Geschwindigkeit w bewegt in einer Richtung, welche mit der positiven ξ-Achse den Winkel φ bilde. Unter Benutzung der in § 5 (l. c.) hergeleiteten Beziehungen läßt sich leicht die Energie ε des Massenpunktes, bezogen auf das System (x, y, z) bestimmen. Man erhält

$$\varepsilon = \mu V^2 \frac{1 + \dfrac{v\, w \cos \varphi}{V^2}}{\sqrt{1 - \dfrac{v^2}{V^2}}\sqrt{1 - \dfrac{w^2}{V^2}}}.$$

Sind mehrere Massenpunkte vorhanden, denen verschiedene Massen, Geschwindigkeiten und Bewegungsrichtungen zukommen, so erhalten wir für deren Gesamtenergie E den Ausdruck

$$E = \frac{1}{\sqrt{1 - \left(\dfrac{v}{V}\right)^2}}\left\{ \sum \mu\, V^2 \cdot \frac{1}{\sqrt{1 - \left(\dfrac{w}{V}\right)^2}} \right\}$$
$$+ \frac{v}{\sqrt{1 - \left(\dfrac{v}{V}\right)^2}}\left\{ \sum \frac{\mu\, w \cos \varphi}{\sqrt{1 - \left(\dfrac{w}{V}\right)^2}} \right\}.$$

Bis jetzt haben wir über den Bewegungszustand des Systems (ξ, η, ζ) relativ zu den bewegten Massen nichts festgesetzt. Wir können und wollen hierüber nun folgende, den Bewegungszustand von (ξ, η, ζ) eindeutig bestimmende Bedingungen festsetzen:

$$\sum \frac{\mu\, w_\xi}{\sqrt{1 - \left(\dfrac{w}{V}\right)^2}} = 0, \quad \sum \frac{\mu\, w_\eta}{\sqrt{1 - \left(\dfrac{w}{V}\right)^2}} = 0,$$

$$\sum \frac{\mu\, w_\zeta}{\sqrt{1 - \left(\dfrac{w}{V}\right)^2}} = 0,$$

wobei w_ξ, w_η, w_ζ die Komponenten von w bezeichnen. Dieser Festsetzung entspricht in der klassischen Mechanik die Bedingung, daß das Bewegungsmoment des Massensystems in bezug auf (ξ, η, ζ) verschwinde. Dann erhalten wir

$$E = \left(\sum \mu\, V^2 \cdot \frac{1}{\sqrt{1 - \left(\dfrac{w}{V}\right)^2}} \right) \cdot \frac{1}{\sqrt{1 - \left(\dfrac{v}{V}\right)^2}},$$

384   *A. Einstein. Trägheit der Energie.*

oder, indem man die Energie E_0 des Systems relativ zum System (ξ, η, ζ) einführt:

$$E = \frac{E_0}{V^2} \cdot V^2 \frac{1}{\sqrt{1 - \left(\frac{v}{V}\right)^2}}.$$

Vergleicht man diesen Ausdruck mit dem für die Energie eines mit der Geschwindigkeit v bewegten Massenpunktes

$$\varepsilon = \mu V^2 \frac{1}{\sqrt{1 - \left(\frac{v}{V}\right)^2}},$$

so erhält man folgendes Resultat: In bezug auf die Abhängigkeit der Energie vom Bewegungszustand des Koordinatensystems, auf welches die Vorgänge bezogen werden, läßt sich ein System gleichförmig bewegter Massenpunkte ersetzen durch einen einzigen Massenpunkt von der Masse $\mu = E_0/V^2$.

Ein System bewegter Massenpunkte besitzt also — als Ganzes genommen — desto mehr Trägheit, je rascher die Massenpunkte relativ zueinander bewegt sind. Die Abhängigkeit ist wieder gegeben durch das in der Einleitung angegebene Gesetz.

Bern, Mai 1907.

(Eingegangen 14. Mai 1907.)

Published in *Annalen der Physik* 23 (1907): 371–384. Dated Bern, May 1907, received 14 May 1907, published 13 June 1907.

[1] *Einstein 1905v* (Doc. 27).

[2] *Einstein 1906e* (Doc. 35).

[3] *Einstein 1905i* (Doc. 14), *Einstein 1906d* (Doc. 34), and *Einstein 1907a* (Doc. 38).

[4] See *Planck 1906a, 1906b*, and *Kaufmann 1906a*.

[5] *Einstein 1905r* (Doc. 23).

[6] *Einstein 1905r* (Doc. 23).

[7] See *Einstein 1907g* (Doc. 44), p. 207, for an earlier comment on the problem of rigid-body dynamics.

[8] A paper by Born proposing a Lorentz-invariant definition of a rigid body (*Born 1909*) initiated an intensive discussion of the relativistic kinematics and dynamics of rigid motions in the course of the next few years. Finally, using arguments similar to those of this section of Einstein's paper, Laue showed that an extended relativistic body must have an infinite number of degrees of freedom (*Laue 1911a*). This does not preclude it from undergoing relativistic rigid motions, which had been shown to possess only three degrees of freedom (see *Herglotz 1910, Noether 1910*). For a review of the discussion on rigidity, see *Pauli 1921*, pp. 689–691.

[9] About the time of this paper, Einstein had an extensive correspondence with Wien and Sommerfeld on the possibility of transmitting signals at superluminal velocities. For a discussion of this correspondence, see Vol. 5, the editorial note, "Einstein on Superluminal Signal Velocities."

[10] Two arguments against the possibility of superluminal velocities for an electron are given in *Einstein 1905r* (Doc. 23), pp. 906, 920.

[11] This is Einstein's first use of the phrase "the equivalence of mass and energy." Earlier, he had stated that the inertia or mass of a body depends on, or is a measure of, its energy content (see *Einstein 1905v* [Doc. 27], *Einstein 1906e* [Doc. 35]).

46. Review of Jakob Johann Weyrauch, *An Outline of the Theory of Heat. With Numerous Examples and Applications.* Part 2

[*Einstein 1907i*]

Second half of August 1907

PUBLISHED IN: *Beiblätter zu den Annalen der Physik* 31 (1907), no. 16: 777–778.

13. *A. Marshall. Die Dampfdrücke binärer Gemische. I. Die möglichen Typen von Dampfdruckkurven* (J. chem. Soc. 89, S. 1350—1386. 1906). — Der Verf. unterscheidet die Kurven, welche die Änderungen des Teildruckes mit dem molekularen Gehalt vorstellen, nach vier Typen. Von diesen beziehen sich drei auf Stoffe, die sich in allen Verhältnissen mischen; diese unterscheiden sich dadurch voneinander, daß p/x, Partialdruck dividiert durch Molekulargehalt, gleich dem Dampfdruck der reinen Substanz, kleiner oder größer ist. Die Totaldruckkurven werden erhalten durch Addierung von zwei Partialdruckkurven, welche demselben Typus angehören. Der Verf. leitet daraus ab, daß in der Totaldruckkurve von in allen Verhältnissen sich mischenden Stoffen nur *ein* Maximum oder Minimum auftreten kann. Die Totaldruckkurven werden in zwölf Typen klassifiziert; von jedem dieser Typen sind Beispiele bekannt. Die von Duhem „Regnaultsches Gesetz" genannte Regel, daß für teilweise mischbare Flüssigkeiten der Totaldruck des heterogenen Gemisches gleich sei dem Dampfdrucke der flüchtigeren Komponenten im ungemischten Zustande, gilt approximativ für Methylacetat und Wasser und für Äther und Wasser; die Regel ist aber eine rohe Annäherung, und nur in einigen Fällen anwendbar. Der Verf. hat die Dampfdrucke folgender Gemische experimentell untersucht: Nitroglycerol und Aceton, Diäthylamin und Aceton, Äthylalkohol und Methyläthylketon, Wasser und Methyläthylketon, Wasser und Methylacetat, Wasser und Äther, Wasser und Amylalkohol. Alle diese Erörterungen sind abhängig von der Voraussetzung, daß für den Dampf die idealen Gasgesetze gültig sind, was bei den Beobachtungstemperaturen auch wohl der Fall war.   Keesom.

_____

14. *J. J. Weyrauch. Grundriß der Wärmetheorie.*
[1] *2. Band* (412 S. $\mathscr{M}$ 16,00. Stuttgart, K. Wittwer, 1907). — Der vorliegende 2. Band handelt über die gesättigten und überhitzten Dämpfe, über Dampfmaschinen, Aërostatik, Aërodynamik (Bewegung in Kanälen, Ausfluß aus Gefäßmündungen) und über feste Körper; derselbe ist zusammen mit dem 1. Band vortrefflich geeignet, den mit den Elementen der Differential- und Integralrechnung Vertrauten derart in die

778   **Wärmelehre.**   **Beibl. 1907.**

Theorie der Wärme einzuführen, wie es für die Behandlung technischer Aufgaben (besonders Kraftmaschinen betreffende) nützlich ist. 150 Aufgaben und 250 meist den Verhältnissen der Praxis entsprechend gewählte Zahlenbeispiele, welche in den Text eingeflochten sind, geben dem Studierenden Gelegenheit zu reichlicher Übung, so daß er sich verhältnismäßig leicht die zur Lektüre technischer Abhandlungen und zur Berechnung spezieller Aufgaben erforderliche Gewandtheit anzueignen vermag. Außerdem gibt ihm das Werk die nötigen empirischen Daten an die Hand, sowie zahlreiche Angaben von meist technischer Literatur. Die Redaktion des Buches ist eine einfache und klare, die Anwendung der mathematischen Zeichen eine konsequente, das Inhaltsverzeichnis übersichtlich und vollständig, so daß man sich ohne viel Zeitverlust über besondere Fragen zu informieren vermag.   A. E.

[2]

Published in *Beiblätter zu den Annalen der Physik* 31 (1907): 777–778. Published in no. 16 [second half of August].

[1] For Einstein's review of the first part of this work, see *Einstein 1905y* (Doc. 30).

[2] The two volumes of this work contain 150 numbered sections, 46 numbered problems, and 204 numbered numerical examples.

47. ''On the Relativity Principle and the Conclusions Drawn from It''

[*Einstein 1907j*]

RECEIVED 4 December 1907
PUBLISHED 22 January 1908

IN: *Jahrbuch der Radioaktivität und Elektronik* 4 (1907): 411–462.

Einstein, Relativitätsprinzip u. die aus demselben gezog. Folgerungen. 411

Durch Siedepunktserhöhungsbestimmungen an den wäßrigen Lösungen kommt Benrath zu dem Schluß, daß nur diejenigen Substanzen, z. B. $CdCl_2$, welche die Blaufärbung verhindern, Komplexe bilden, während diejenigen, z. B. $CaCl_2$, welche die Blaufärbung bewirken, keine Komplexe bilden. Und wenn dies auch aus seinen Resultaten hervorzugehen scheint, so ist immer nicht zu leugnen, daß die Überführungsversuche die Existenz von Komplexen im letzteren Fall unzweideutig beweisen. Benrath gibt ja zu, daß solche Komplexe bei höheren Konzentrationen existieren, aber wenn bei höheren Konzentrationen, warum nicht auch bei etwas niedrigeren Konzentrationen? Es handelt sich nur um die relativen Mengen, und wenn Benrath noch behauptet, daß die Bildung solcher Komplexe sich „thermodynamisch wohl kaum begründen läßt", so kann man hierzu nur bemerken, daß die Thermodynamik wohl kaum etwas damit zu tun hat.

Aus diesen neueren Arbeiten geht deutlich hervor, daß die Frage nach der Ursache der Farbenänderungen bei Kobaltsalzen noch nicht erledigt ist und daß deren Erledigung noch weiterer exakter Versuche bedarf.

Nachtrag III.

Zu dem Kapitel über den Einfluß des Aggregatzustandes. K. Arndt hat auf die Einwendungen von Lorenz kurz geantwortet (Ber. d. d. chem. Ges. **40**, 3612—3614, 1907) und beabsichtigt in der Zeitschrift für Elektrochemie ausführlich dieselben zu widerlegen.

(Eingegangen 1. Oktober 1907.)

Über das Relativitätsprinzip und die aus demselben gezogenen Folgerungen.

Von A. Einstein.

Die Newtonschen Bewegungsgleichungen behalten ihre Form, wenn man auf ein neues, relativ zu dem ursprünglich benutzten in gleichförmiger Translationsbewegung begriffenes Koordinatensystem transformiert nach den Gleichungen

28*

412 Einstein, Relativitätsprinzip u. die aus demselben gezog. Folgerungen.

$$x' = x - vt$$
$$y' = y$$
$$z' = z$$

Solange man an der Meinung festhielt, daß auf die Newtonschen Bewegungsgleichungen die ganze Physik aufgebaut werden könne, konnte man also nicht daran zweifeln, daß die Naturgesetze gleich ausfallen, auf welches von relativ zueinander gleichförmig bewegten (beschleunigungsfreien) Koordinatensystemen sie auch bezogen werden mögen. Jene Unabhängigkeit vom Bewegungszustande des benutzten Koordinatensystems, im folgenden „Relativitätsprinzip" genannt, schien aber mit einem Male in Frage gestellt durch die glänzenden Bestätigungen, welche die H. A. Lorentzsche Elektrodynamik bewegter Körper erfahren hat.[1]) Jene Theorie ist nämlich auf die Voraussetzung eines ruhenden, unbeweglichen Lichtäthers gegründet; ihre Grundgleichungen sind nicht so beschaffen, daß sie bei Anwendung der obigen Transformationsgleichungen in Gleichungen von der gleichen Form übergehen.

Seit dem Durchdringen jener Theorie mußte man erwarten, daß es gelingen werde, einen Einfluß der Bewegung der Erde relativ zum Lichtäther auf die optischen Erscheinungen nachzuweisen. Lorentz [2] bewies allerdings bekanntlich in jener Arbeit, daß nach seinen Grundannahmen eine Beeinflussung des Strahlenganges bei optischen Versuchen durch jene Relativbewegung nicht zu erwarten sei, sofern man sich bei der Rechnung auf die Glieder beschränkt, in denen das Verhältnis $\frac{v}{c}$ jener Relativgeschwindigkeit zur Lichtgeschwindigkeit im Vakuum in der ersten Potenz auftritt. Aber das negative Resultat des Experimentes von Michelson und Morley[2]) zeigte, daß in einem bestimmten Falle auch ein Effekt zweiter Ordnung $\left(\text{proportional } \frac{v^2}{c^2}\right)$ nicht vorhanden war, trotzdem er nach den Grundlagen der Lorentzschen Theorie bei dem Versuche sich hätte bemerkbar machen müssen.

Es ist bekannt, daß jener Widerspruch zwischen Theorie und [4] Experiment durch die Annahme von H. A. Lorentz und Fitzgerald, nach welcher bewegte Körper in der Richtung ihrer Bewegung eine bestimmte Kontraktion erfahren, formell beseitigt wurde. Diese ad

---

[1]   1) H. A. Lorentz, Versuch einer Theorie der elektrischen und optischen Erscheinungen in bewegten Körpern. Leiden 1895. Neudruck Leipzig 1906.

[3]   2) A. A. Michelson und E. W. Morley, Amer. Journ. of Science (3) **34**, S. 333, 1887.

Einstein, Relativitätsprinzip u. die aus demselben gezog. Folgerungen. 413

hoc eingeführte Annahme erschien aber doch nur als ein künstliches Mittel, um die Theorie zu retten; der Versuch von Michelson und Morley hatte eben gezeigt, daß Erscheinungen auch da dem Relativitätsprinzip entsprechen, wo dies nach der Lorentzschen Theorie nicht einzusehen war. Es hatte daher den Anschein, als ob die Lorentzsche Theorie wieder verlassen und durch eine Theorie ersetzt werden müsse, deren Grundlagen dem Relativitätsprinzip entsprechen, denn eine solche Theorie ließe das negative Ergebnis des Versuches von Michelson und Morley ohne weiteres voraussehen. [5]

Es zeigte sich aber überraschenderweise, daß es nur nötig war, den Begriff der Zeit genügend scharf zu fassen, um über die soeben dargelegte Schwierigkeit hinweg zu kommen. Es bedurfte nur der Erkenntnis, daß man eine von H. A. Lorentz eingeführte Hilfsgröße, welche er „Ortszeit" nannte, als „Zeit" schlechthin definieren kann. [6] Hält man an der angedeuteten Definition der Zeit fest, so entsprechen die Grundgleichungen der Lorentzschen Theorie dem Relativitätsprinzip, wenn man nur die obigen Transformationsgleichungen durch solche ersetzt, welche dem neuen Zeitbegriff entsprechen. Die Hypothese von H. A. Lorentz und Fitzgerald erscheint dann als eine zwingende Konsequenz der Theorie. Nur die Vorstellung eines Lichtäthers als des Trägers der elektrischen und magnetischen Kräfte paßt nicht in die hier dargelegte Theorie hinein; elektromagnetische Felder erscheinen nämlich hier nicht als Zustände irgendeiner Materie, sondern als selbständig existierende Dinge, die der ponderabeln Materie gleichartig sind und mit ihr das Merkmal der Trägheit gemeinsam haben. [7]

Im folgenden ist nun der Versuch gemacht, die Arbeiten zu einem Ganzen zusammenzufassen, welche bisher aus der Vereinigung von H. A. Lorentzscher Theorie und Relativitätsprinzip hervorgegangen sind.

In den ersten beiden Teilen der Arbeit sind die kinematischen Grundlagen sowie deren Anwendung auf die Grundgleichungen der Maxwell-Lorentzschen Theorie behandelt; dabei hielt ich mich an Arbeiten[1]) von H. A. Lorentz (Versl. Kon. Akad. v. Wet., Amsterdam 1904) und A. Einstein (Ann. d. Phys. **16**, 1905). [9]

In dem ersten Abschnitt, in dem ausschließlich die kinematischen Grundlagen der Theorie angewendet worden sind, habe ich auch einige optische Probleme (Dopplersches Prinzip, Aberration, Mitführung des

---

1) Es kommen auch noch die einschlägigen Arbeiten von E. Cohn [8] in Betracht, von welchen ich aber hier keinen Gebrauch gemacht habe.

414 Einstein, Relativitätsprinzip u. die aus demselben gezog. Folgerungen.

Lichtes durch bewegte Körper) behandelt; auf die Möglichkeit einer
derartigen Behandlungsweise wurde ich durch eine mündliche Mitteilung
[10]   und eine Arbeit (Ann. d. Phys. **23**, 989, 1907) von Herrn M. Laue,
und durch eine (allerdings korrekturbedürftige) Arbeit von Herrn J. Laub
[11]   (Ann. d. Phys. **32**, 1907) aufmerksam.

Im dritten Teil ist die Dynamik des materiellen Punktes (Elektrons)
entwickelt. Zur Ableitung der Bewegungsgleichungen benutzte ich die-
[12]   selbe Methode wie in meiner oben genannten Arbeit. Die Kraft ist
[13]   definiert wie in der Planckschen Arbeit. Auch die Umformungen der
Bewegungsgleichungen des materiellen Punktes, welche die Analogie der
Bewegungsgleichungen mit denen der klassischen Mechanik so deutlich
hervortreten lassen, sind dieser Arbeit entnommen.

Der vierte Teil befaßt sich mit den allgemeinen Folgerungen, betreffend die Energie und die Bewegungsgröße physikalischer Systeme,
zu welchen die Relativitätstheorie führt. Dieselben sind in den
Originalabhandlungen:

[14]   A. Einstein, Ann. d. Phys. **18**, 639, 1905 und Ann. d. Phys.
23, 371, 1907, sowie M. Planck, Sitzungsber. d. Kgl. Preuß.
Akad. d. Wissensch. XXIX, 1907

entwickelt worden, hier aber auf einem neuen Wege abgeleitet, der
— wie mir scheint — den Zusammenhang jener Anwendungen mit
den Grundlagen der Theorie besonders klar erkennen läßt. Auch die
Abhängigkeit der Entropie und Temperatur vom Bewegungszustande
ist hier behandelt; bezüglich der Entropie hielt ich mich ganz an
die zuletzt zitierte Plancksche Abhandlung, die Temperatur bewegter
Körper definierte ich wie Herr Mosengeil in seiner Arbeit über die
bewegte Hohlraumstrahlung.[1])

Das wichtigste Ergebnis des vierten Teiles ist das von der trägen
Masse der Energie. Dies Resultat legt die Frage nahe, ob die Energie
auch schwere (gravitierende) Masse besitze. Ferner drängt sich die
Frage auf, ob das Relativitätsprinzip auf beschleunigungfrei bewegte Systeme beschränkt sei. Um diese Fragen nicht ganz unerörtert
zu lassen, habe ich dieser Abhandlung einen fünften Teil hinzugefügt,
welcher eine neue relativitätstheoretische Betrachtung über Beschleunigung und Gravitation enthält.

---

[15]   1) Kurd von Mosengeil, Ann. d. Phys. **22**, 867, 1907.

Einstein, Relativitätsprinzip u. die aus demselben gezog. Folgerungen. 415

I. Kinematischer Teil.

§ 1. Prinzip von der Konstanz der Lichtgeschwindigkeit. Definition der Zeit. Relativitätsprinzip.

Um irgendeinen physikalischen Vorgang beschreiben zu können, müssen wir imstande sein, die in den einzelnen Punkten des Raumes stattfindenden Veränderungen örtlich und zeitlich zu werten.

Zur örtlichen Wertung eines in einem Raumelement stattfindenden Vorganges von unendlich kurzer Dauer (Punktereignis) bedürfen wir eines Cartesischen Koordinatensystems, d. h. dreier aufeinander senkrecht stehender, starr miteinander verbundener, starrer Stäbe, sowie eines starren Einheitsmaßstabes.[1] Die Geometrie gestattet, die Lage eines Punktes bezw. den Ort eines Punktereignisses durch drei Maßzahlen (Koordinaten x, y, z) zu bestimmen.[2] Für die zeitliche Wertung eines Punktereignisses bedienen wir uns einer Uhr, die relativ zum Koordinatensystem ruht und in deren unmittelbarer Nähe das Punktereignis stattfindet. Die Zeit des Punktereignisses ist definiert durch die gleichzeitige Angabe der Uhr.

Wir denken uns in vielen Punkten relativ zum Koordinatensystem ruhende Uhren angeordnet. Dieselben seien sämtlich gleichwertig, d. h. die Differenz der Angaben zweier solcher Uhren soll ungeändert bleiben, falls sie nebeneinander angeordnet werden. Denkt man sich diese Uhren irgendwie eingestellt, so erlaubt die Gesamtheit der Uhren, falls letztere in genügend kleinen Abständen angeordnet sind, ein beliebiges Punktereignis — etwa mittels der nächstgelegenen Uhr — zeitlich zu werten.

Der Inbegriff dieser Uhrangaben liefert uns aber gleichwohl noch keine „Zeit", wie wir sie für physikalische Zwecke nötig haben. Wir bedürfen vielmehr hierzu noch einer Vorschrift, nach welcher diese Uhren relativ zueinander eingestellt werden sollen.

Wir nehmen nun an, die Uhren können so gerichtet werden, daß die Fortpflanzungsgeschwindigkeit eines jeden Lichtstrahles im Vakuum — mit Hilfe dieser Uhren gemessen — allenthalben gleich einer universellen Konstante c wird, vorausgesetzt, daß das Koordinatensystem nicht beschleunigt ist. Sind A und B zwei relativ zum Koordinatensystem ruhende, mit Uhren ausgestattete Punkte, deren Entfernung r beträgt, und ist t_A die An-

---

1) Statt von „starren" Körpern, könnte hier sowie im folgenden ebenso gut von deformierenden Kräften nicht unterworfenen festen Körpern gesprochen werden.

2) Hierzu braucht man noch Hilfsstäbe (Lineale, Zirkel).

[16]

416 Einstein, Relativitätsprinzip u. die aus demselben gezog. Folgerungen.

gabe der Uhr in A, wenn ein durch das Vakuum in der Richtung AB sich fortpflanzender Lichtstrahl den Punkt A erreicht, t_B die Angabe der Uhr in B beim Eintreffen des Lichtstrahles in B, so soll also, wie auch die den Lichtstrahl emittierende Lichtquelle, sowie andere Körper bewegt sein mögen, stets

$$\frac{r}{t_B - t_A} = c$$

sein.

Daß die hier gemachte Annahme, welche wir „Prinzip von der Konstanz der Lichtgeschwindigkeit" nennen wollen, in der Natur wirklich erfüllt sei, ist keineswegs selbstverständlich, doch wird dies — wenigstens für ein Koordinatensystem von bestimmtem Bewegungszustande — wahrscheinlich gemacht durch die Bestätigungen, welche die, auf die Voraussetzung eines absolut ruhenden Äthers gegründete Lorentzsche Theorie [1]) durch das Experiment erfahren hat.[2])

Den Inbegriff der Angaben aller gemäß dem vorhergehenden gerichteter Uhren, welche man sich in den einzelnen Raumpunkten relativ zum Koordinatensystem ruhend angeordnet denken kann, nennen wir die zu dem benutzten Koordinatensystem gehörige Zeit oder kurz die Zeit dieses Systems.

Das benutzte Koordinatensystem samt Einheitsmaßstab und den zur Ermittlung der Zeit des Systems dienenden Uhren, nennen wir „Bezugssystem S". Wir denken uns die Naturgesetze in bezug auf das Bezugssystem S ermittelt, welches etwa zunächst relativ zur Sonne ruhe. Hierauf werde das Bezugssystem S durch irgendeine äußere Ursache eine Zeitlang beschleunigt und gelange schließlich wieder in einen beschleunigungsfreien Zustand. Wie werden die Naturgesetze ausfallen, wenn man die Vorgänge auf das nunmehr in einem anderen Bewegungszustande befindliche Bezugssystem S bezieht?

[19] In bezug hierauf machen wir nun die denkbar einfachste und durch das Experiment von Michelson und Morley nahe gelegte Annahme: Die Naturgesetze sind unabhängig vom Bewegungszustande des Bezugssystems, wenigstens falls letzterer ein beschleunigungsfreier ist.

Auf diese Annahme, welche wir „Relativitätsprinzip" nennen,

[17]   1) H. A. Lorentz, Versuch einer Theorie der elektrischen und optischen Erscheinungen in bewegten Körpern. Leiden 1895.
[18]   2) Insbesondere kommt in Betracht, daß diese Theorie den Mitführungskoeffizienten (Fizeauscher Versuch) im Einklang mit der Erfahrung lieferte.

Einstein, Relativitätsprinzip u. die aus demselben gezog. Folgerungen. 417

sowie auf das oben angegebene Prinzip von der Konstanz der Lichtgeschwindigkeit werden wir uns im folgenden stützen.

§ 2. Allgemeine Bemerkungen, Raum und Zeit betreffend.

1. Wir betrachten eine Anzahl beschleunigungsfrei und gleich
bewegter (d. h. relativ zueinander ruhender) starrer Körper. Nach
dem Relativitätsprinzip schließen wir, daß die Gesetze, nach denen
sich diese Körper relativ zueinander räumlich gruppieren lassen, bei
Änderung des gemeinsamen Bewegungszustandes dieser Körper sich
nicht ändern. Daraus folgt, daß die Gesetze der Geometrie die
Lagerungsmöglichkeiten starrer Körper stets in der gleichen Weise
bestimmen, unabhängig von deren gemeinsamem Bewegungszustande.
Aussagen über die Gestalt eines beschleunigungsfrei bewegten Körpers
haben daher unmittelbar einen Sinn. Wir wollen die Gestalt eines
Körpers im dargelegten Sinn, die „Geometrische Gestalt" desselben [20]
nennen. Letztere ist offenbar nicht vom Bewegungszustande eines
Bezugssystems abhängig.

2. Eine Zeitangabe hat gemäß der in § 1 gegebenen Definition
der Zeit nur mit Bezug auf ein Bezugssystem von bestimmtem Bewegungszustande einen Sinn. Es ist daher zu vermuten (und wird
sich im folgenden zeigen), daß zwei räumlich distante Punktereignisse,
welche in bezug auf ein Bezugssystem S gleichzeitig sind, in bezug
auf ein Bezugssystem S' von anderem Bewegungszustande im allgemeinen nicht gleichzeitig sind.

3. Ein aus den materiellen Punkten P bestehender Körper
bewege sich irgendwie relativ zu einem Bezugssystem S. Zur Zeit t
von S besitzt jeder materielle Punkt P eine bestimmte Lage in S,
d. h. er koinzidiert mit einem bestimmten, relativ zu S ruhendem
Punkte Π. Den Inbegriff der Lagen der Punkte Π relativ zum
Koordinatensystem von S nennen wir die Lage, den Inbegriff der
Lagenbeziehungen der Punkte Π untereinander die kinematische Gestalt
des Körpers in bezug auf S für die Zeit t. Ruht der Körper relativ
zu S, so ist seine kinematische Gestalt in bezug auf S mit seiner
geometrischen Gestalt identisch.

Es ist klar, daß relativ zu einem Bezugssystem S ruhende Beobachter nur die auf S bezogene kinematische Gestalt eines relativ
zu S bewegten Körpers zu ermitteln vermögen, nicht aber dessen
geometrische Gestalt.

Im folgenden werden wir gewöhnlich nicht explizite zwischen

418 Einstein, Relativitätsprinzip u. die aus demselben gezog. Folgerungen.

geometrischer und kinematischer Gestalt unterscheiden; eine Aussage geometrischen Inhaltes betrifft die kinematische bezw. geometrische Gestalt, je nachdem dieselbe auf ein Bezugssystem S bezogen ist oder nicht.

[21]   § 3. Koordinaten-Zeit-Transformation.

S und S' seien gleichwertige Bezugssysteme, d. h. diese Systeme mögen gleichlange Einheitsmaßstäbe und gleichlaufende Uhren besitzen, falls diese Gegenstände im Zustande relativer Ruhe miteinander verglichen werden. Es ist dann einleuchtend, daß jedes Naturgesetz, das in bezug auf S gilt, in genau gleicher Form auch in bezug auf S' gilt, falls S und S' relativ zueinander ruhen. Das Relativitätsprinzip verlangt jene vollkommene Übereinstimmung auch für den Fall, daß S' relativ zu S in gleichförmiger Translationsbewegung begriffen ist. Im speziellen muß sich also für die Lichtgeschwindigkeit im Vakuum in bezug auf beide Bezugssysteme dieselbe Zahl ergeben.

Ein Punktereignis sei relativ zu S durch die Variabeln x, y, z, t relativ zu S' durch die Variabeln x', y', z', t', bestimmt, wobei S und S' beschleunigungsfrei und relativ zueinander bewegt seien. Wir fragen nach den Gleichungen, welche zwischen den erstgenannten und den letztgenannten Variabeln bestehen.

Von diesen Gleichungen können wir sofort aussagen, daß sie in bezug auf die genannten Variabeln linear sein müssen, weil die Homogenitätseigenschaften des Raumes und der Zeit dies erfordern. Daraus folgt im speziellen, daß die Koordinatenebenen von S' — auf das Bezugssystem S bezogen — gleichförmig bewegte Ebenen sind; doch werden diese Ebenen im allgemeinen nicht aufeinander senkrecht stehen. Wählen wir jedoch die Lage der x'-Achse so, daß letztere — auf S bezogen — die gleiche Richtung hat, wie die auf S bezogene Translationsbewegung von S', so folgt aus Symmetriegründen, daß die auf S bezogenen Koordinatenebenen von S' aufeinander senkrecht stehen müssen. Wir können und wollen die Lagen der beiden Koordinatensysteme im speziellen so wählen, daß die x-Achse von S und die x'-Achse von S' dauernd zusammenfallen und daß die auf S bezogene y'-Achse von S' parallel der y-Achse von S ist. Ferner wollen wir als Anfangspunkt der Zeit in beiden Systemen den Augenblick wählen, in welchem die Koordinatenanfangspunkte koinzidieren; dann sind die gesuchten linearen Transformationsgleichungen homogen.

Aus der nun bekannten Lage der Koordinatenebenen von S' relativ

Einstein, Relativitätsprinzip u. die aus demselben gezog. Folgerungen. 419

zu S schließen wir unmittelbar, daß je zwei der folgenden Gleichungen gleichbedeutend sind:

$$x' = o \quad \text{und} \quad x - vt = o$$
$$y' = o \quad \text{und} \quad y = o$$
$$z' = o \quad \text{und} \quad z = o$$

Drei der gesuchten Transformationsgleichungen sind also von der Form:

$$x' = a(x - vt)$$
$$y' = by$$
$$z' = cz.$$

Da die Ausbreitungsgeschwindigkeit des Lichtes im leeren Raume in bezug auf beide Bezugssysteme gleich c ist, so müssen die beiden Gleichungen:

$$x^2 + y^2 + z^2 = c^2 t^2$$

und

$$x'^2 + y'^2 + z'^2 = c^2 t'^2$$

gleichbedeutend sein. Hieraus und aus den soeben für x', y', z' gefundenen Ausdrücken schließt man nach einfacher Rechnung, daß die gesuchten Transformationsgleichungen von der Form sein müssen:

$$t' = \varphi(v) \cdot \beta \cdot \left(t - \frac{v}{c^2} x \right)$$
$$x' = \varphi(v) \cdot \beta \cdot (x - vt)$$
$$y' = \varphi(v) \cdot y$$
$$z' = \varphi(v) \cdot z.$$

Dabei ist

$$\beta = \frac{1}{\sqrt{1 - \dfrac{v^2}{c^2}}}$$

gesetzt.

Die noch unbestimmt gebliebene Funktion von v wollen wir nun bestimmen. Führen wir ein drittes mit S und S' gleichwertiges Bezugssystem S'' ein, welches relativ zu S' mit der Geschwindigkeit $-v$ bewegt und ebenso relativ zu S' orientiert ist, wie S' relativ zu S, so erhalten wir durch zweimalige Anwendung der eben erlangten Gleichungen

$$t'' = \varphi(v) \cdot \varphi(-v) \cdot t$$
$$x'' = \varphi(v) \cdot \varphi(-v) \cdot x$$
$$y'' = \varphi(v) \cdot \varphi(-v) \cdot y$$
$$z'' = \varphi(v) \cdot \varphi(-v) \cdot z.$$

Da die Koordinatenanfangspunkte von S und S'' dauernd zu-

420   Einstein, Relativitätsprinzip u. die aus demselben gezog. Folgerungen.

sammenfallen, die Achsen gleich orientiert und die Systeme „gleichwertige" sind, so ist diese Substitution die identische [1]), so daß

$$\varphi(v) \cdot \varphi(-v) = 1 .$$

Da ferner die Beziehung zwischen y und y' vom Vorzeichen von v nicht abhängen kann, ist,

$$\varphi(v) = \varphi(-v) .$$

Es ist also [2]) $\varphi(v) = 1$, und die Transformationsgleichungen lauten

$$\left. \begin{aligned} t' &= \beta\left(t - \frac{v}{c^2}x\right) \\ x' &= \beta(x - vt) \\ y' &= y \\ z' &= z \end{aligned} \right\} \quad \ldots \ldots (1)$$

wobei

$$\beta = \frac{1}{\sqrt{1 - \dfrac{v^2}{c^2}}} .$$

Löst man die Gleichungen (1) nach x, y, z, t auf, so erhält man die nämlichen Gleichungen, nur daß die „gestrichenen" durch die gleichnamigen „ungestrichenen" Größen und umgekehrt ersetzt sind, und v durch $-v$ ersetzt ist. Es folgt dies auch unmittelbar aus dem Relativitätsprinzip und aus der Erwägung, daß S relativ zu S' eine Paralleltranslation in Richtung der X'-Achse mit der Geschwindigkeit $-v$ ausführt.

[22]

Allgemein erhält man gemäß dem Relativitätsprinzip aus jeder richtigen Beziehung zwischen „gestrichenen" (mit Bezug auf S' definierten) und „ungestrichenen" (mit Bezug auf S definierten) Größen oder zwischen Größen nur einer dieser Gattungen wieder eine richtige Beziehung, wenn man die ungestrichenen durch die entsprechenden gestrichenen Zeichen und umgekehrt sowie v durch $-v$ ersetzt.

§ 4.   Folgerungen aus den Transformationsgleichungen, starre Körper und Uhren betreffend.

1. Relativ zu S' ruhe ein Körper. x_1', y_1', z_1' und x_2' y_2' z_2' seien die auf S' bezogenen Koordinaten zweier materieller Punkte desselben. Zwischen den Koordinaten x_1, y_1, z_1 und x_2, y_2, z_2 dieser Punkte in

---

1) Dieser Schluß ist auf die physikalische Voraussetzung gegründet, daß die Länge eines Maßstabes, sowie die Ganggeschwindigkeit einer Uhr dadurch keine dauernde Änderung erleiden, daß diese Gegenstände in Bewegung gesetzt und wieder zur Ruhe gebracht werden.

2) $\varphi(v) = -1$ kommt offenbar nicht in Betracht.

Einstein, Relativitätsprinzip u. die aus demselben gezog. Folgerungen. 421

bezug auf das Bezugssystem S bestehen zu jeder Zeit t von S nach den soeben abgeleiteten Transformationsgleichungen die Beziehungen

$$\left.\begin{aligned} x_2 - x_1 &= \sqrt{1 - \frac{v^2}{c^2}}(x_2{}' - x_1{}') \\ y_2 - y_1 &= y_2{}' - y_1{}' \\ z_2 - z_1 &= z_2{}' - z_1{}' \end{aligned}\right\} \quad \ldots \quad (2)$$

Die kinematische Gestalt eines in gleichförmiger Translations-bewegung begriffenen Körpers hängt also ab von dessen Geschwindig-keit relativ zum Bezugssystem, und zwar unterscheidet sich die kine-matische Gestalt des Körpers von seiner geometrischen Gestalt lediglich durch eine Verkürzung in Richtung der Relativbewegung im Verhältnis $1 : \sqrt{1 - \frac{v^2}{c^2}}$. Eine Relativbewegung von Bezugssystemen mit Über-lichtgeschwindigkeit ist mit unseren Prinzipien nicht vereinbar.

2. Im Koordinatenanfangspunkt von S' sei eine Uhr ruhend an-geordnet, welche v_0 mal schneller laufe als die zur Zeitmessung in den Systemen S und S' benutzten Uhren. d. h. diese Uhr führe v_0-Perioden aus in einer Zeit, in welcher die Angabe einer relativ zu ihr ruhenden Uhr von der Art der in S und S' zur Zeitmessung be-nutzten Uhren um eine Einheit zunimmt. Wie schnell geht die erst-genannte Uhr vom System S aus betrachtet?

Die betrachtete Uhr beendet jeweilen eine Periode in den Zeit-epochen $t_n{}' = \frac{n}{v_o}$, wobei n die ganzen Zahlen durchläuft, und für die Uhr dauernd $x' = 0$ ist. Hieraus erhält man mit Hilfe der beiden ersten Transformationsgleichungen für die Zeitepochen t_n, in denen die Uhr, von S aus betrachtet, jeweilen eine Periode beendet

$$t_n = \beta\, t_n{}' = \frac{\beta}{v_o} n .$$

Vom System S aus betrachtet führt die Uhr also, pro Zeiteinheit $v = \frac{v_o}{\beta} = v_o \sqrt{1 - \frac{v^2}{c^2}}$ Perioden aus; oder: eine relativ zu einem Bezugssystem mit der Geschwindigkeit v gleichförmig bewegte Uhr geht von diesem Bezugssystem aus beurteilt im Verhältnis $1 : \sqrt{1 - \frac{v^2}{c^2}}$ langsamer als die nämliche Uhr, falls sie relativ zu jenem Bezugs-system ruht.

Die Formel $v = v_o \sqrt{1 - \frac{v^2}{c^2}}$ gestattet eine sehr interessante

422 Einstein, Relativitätsprinzip u. die aus demselben gezog. Folgerungen.

Anwendung. Herr J. Stark hat im vorigen Jahre gezeigt[1]), daß die die Kanalstrahlen bildenden Ionen Linienspektra emittieren, indem er eine als Dopplereffekt zu deutende Verschiebung von Spektrallinien beobachtete.

Da der einer Spektrallinie entsprechende Schwingungsvorgang wohl als ein intraatomischer Vorgang zu betrachten ist, dessen Frequenz durch das Ion allein bestimmt ist, so können wir ein solches Ion als eine Uhr von bestimmter Frequenzzahl ν_0 ansehen, welch letztere man z. B. erhält, wenn man das von gleich beschaffenen, relativ zum Beobachter ruhenden Ionen ausgesandte Licht untersucht. Die obige Betrachtung zeigt nun, daß der Einfluß der Bewegung auf die von dem Beobachter zu ermittelnde Lichtfrequenz durch den Dopplereffekt noch nicht vollständig gegeben ist. Die Bewegung verringert vielmehr außerdem die (scheinbare) Eigenfrequenz der emittierenden Ionen [24] gemäß obiger Beziehung.[2])

§ 5. Additionstheorem der Geschwindigkeiten.

Relativ zum System S' bewege sich ein Punkt gleichförmig gemäß den Gleichungen

$$x' = u_x' t$$
$$y' = u_y' t$$
$$z' = u_z' t.$$

Ersetzt man x', y', z', t' durch ihre Ausdrücke in x, y, z, t vermittels der Transformationsgleichungen (1), so erhält man x, y, z in Funktion von t, also auch die Geschwindigkeitskomponenten w_x, w_y, w_z des Punktes in bezug auf S. Es ergibt sich so

$$
\left.
\begin{aligned}
u_x &= \frac{u_x' + v}{1 + \frac{v u_x'}{c^2}} \\[2em]
u_y &= \frac{\sqrt{1 - \frac{v^2}{c^2}}}{1 + \frac{v u_x'}{c^2}} u_y' \\[2em]
u_x &= \frac{\sqrt{1 - \frac{v^2}{c^2}}}{1 + \frac{v n_x'}{c^2}} u_z'
\end{aligned}
\right\} \quad \cdots \cdots \quad (3)
$$

[25]

[23]   1) J. Stark, Ann. d. Phys. **21**, 401, 1906.
2) Vgl. hierzu § 6 Gleich. (4a).

Einstein, Relativitätsprinzip u. die aus demselben gezog. Folgerungen. 423

Das Gesetz vom Parallelogramm der Geschwindigkeiten gilt also nur in erster Annäherung. Setzen wir

$$u^2 = u_x{}^2 + u_y{}^2 + u_z{}^2$$
$$u'^2 = u_x'{}^2 + u_y'{}^2 + u_z'{}^2$$

und bezeichnen wir mit α den Winkel zwischen der x'-Achse (v) und der Bewegungsrichtung des Punktes in bezug auf S' (w'), so ist

$$u = \frac{\sqrt{(v^2 + u'^2 + 2vu'\cos\alpha) - \left(\dfrac{vu'\sin\alpha}{c^2}\right)^2}}{1 + \dfrac{vu'\cos\alpha}{c^2}}. \qquad [26]$$

Sind beide Geschwindigkeiten (v und u') gleichgerichtet, so hat man:

$$u = \frac{v + u'}{1 + \dfrac{vu'}{c^2}}.$$

Aus dieser Gleichung folgt, daß aus der Zusammensetzung zweier Geschwindigkeiten, welche kleiner sind als c, stets eine Geschwindigkeit resultiert, die kleiner als c ist. Setzt man nämlich $v = c - k$, $u' = c - \lambda$, wobei k und λ positiv und kleiner als c seien, so ist:

$$u = c - \frac{2c - k - \lambda}{2c - k - \lambda + \dfrac{k\lambda}{c}} < c.$$

Es folgt ferner, daß die Zusammensetzung der Lichtgeschwindigkeit c und einer „Unterlichtgeschwindigkeit" wieder die Lichtgeschwindigkeit c ergibt.

Aus dem Additionstheorem der Geschwindigkeiten ergibt sich ferner noch die interessante Folgerung, daß es keine Wirkung geben kann, welche zur willkürlichen Signalgebung verwendet werden kann, und die sich schneller fortpflanzt als das Licht im Vakuum. Es [27] erstrecke sich nämlich längs der x-Achse von S ein Materialstreifen, relativ zu welchem sich eine gewisse Wirkung (vom Materialstreifen aus beurteilt) mit der Geschwindigkeit W fortzupflanzen vermöge, und es befinde sich sowohl im Punkte $x = 0$ (Punkt A) als auch im Punkte $x = \lambda$ (Punkt B) der x-Achse ein relativ zu S ruhender Beobachter. [28] Der Beobachter in A sende vermittels der oben genannten Wirkung Zeichen zu dem Beobachter in B durch den Materialstreifen, welch letzterer nicht ruhe, sondern mit der Geschwindigkeit $v(<c)$ sich in der negativen x-Richtung bewege. Das Zeichen wird dann, wie aus der ersten der Gleichungen (3) hervorgeht, mit der Geschwindigkeit

424 Einstein, Relativitätsprinzip u. die aus demselben gezog. Folgerungen.

$\dfrac{W-v}{1-\dfrac{Wv}{c^2}}$ von A nach B übertragen. Die hierzu nötige Zeit T ist also

$$T = l\,\frac{1-\dfrac{Wv}{c^2}}{W-v}.$$

Die Geschwindigkeit v kann jeglichen Wert unter c annehmen. Wenn also $W > c$ ist, wie wir angenommen haben, so kann man v stets so wählen, daß $T < O$. Dies Resultat besagt, daß wir einen Übertragungsmechanismus für möglich halten müßten, bei dessen Benutzung die erzielte Wirkung der Ursache vorangeht. Wenn dies Resultat auch, meiner Ansicht nach, rein logisch genommen, keinen Widerspruch enthält, so widerstreitet es doch derart dem Charakter unserer gesamten Erfahrung, daß durch dasselbe die Unmöglichkeit der Annahme $W > c$ zur Genüge erwiesen erscheint.

[29]   § 6. Anwendungen der Transformationsgleichungen auf einige Probleme der Optik.

Der Lichtvektor einer im Vakuum sich fortpflanzenden ebenen Lichtwelle sei, auf das System S bezogen, proportional zu

$$\sin \omega\left(t - \frac{lx + my + nz}{c}\right),$$

auf S' bezogen sei der Lichtvektor des nämlichen Vorganges proportional zu

[30]
$$\sin \omega'\left(t' - \frac{l'x + m'y + n'z}{c}\right).$$

Die im § 3 entwickelten Transformationsgleichungen verlangen, daß zwischen den Größen ω, l, m, n und ω', l', m', n' die folgenden Beziehungen bestehen:

$$\left.\begin{aligned} \omega' &= \omega\beta\left(1 - l\frac{v}{c}\right) \\[2mm] l' &= \frac{l - \dfrac{v}{c}}{1 - l\dfrac{v}{c}} \\[2mm] m' &= \frac{m}{\beta\left(1 - l\dfrac{v}{c}\right)} \\[2mm] n' &= \frac{n}{\beta\left(1 - l\dfrac{v}{c}\right)} \end{aligned}\right\} \quad \ldots \ldots \quad (4)$$

Einstein, Relativitätsprinzip u. die aus demselben gezog. Folgerungen. **425**

Die Formel 'für ω' wollen wir in zwei verschiedenen Weisen deuten, je nachdem wir uns den Beobachter als bewegt und die (unendlich ferne) Lichtquelle als ruhend, oder umgekehrt ersteren als ruhend und letztere als bewegt betrachten.

    1. Ist ein Beobachter relativ zu einer unendlich fernen Licht- [31] quelle von der Frequenz ν mit der Geschwindigkeit v derart bewegt, daß die Verbindungslinie „Lichtquelle-Beobachter" mit der auf ein relativ zur Lichtquelle ruhendes Koordinatensystem bezogenen Geschwindigkeit des Beobachters den Winkel φ bildet, so ist die von dem Beobachter wahrgenommene Frequenz ν' des Lichtes gegeben durch die Gleichung

$$\nu' = \nu \,\frac{1 - \cos\varphi\,\dfrac{v}{c}}{\sqrt{1 - \dfrac{v^2}{c^2}}}$$

    2. Ist eine Lichtquelle, welche bezogen auf ein mit ihr bewegtes System die Frequenz ν_0 besitzt, derart bewegt, daß die Verbindungslinie „Lichtquelle-Beobachter" mit der auf ein relativ zum Beobachter ruhendes System bezogenen Geschwindigkeit der Lichtquelle den Winkel φ bildet, so ist die vom Beobachter wahrgenommene Frequenz ν durch die Gleichung gegeben

$$\nu = \nu_0 \,\frac{\sqrt{1 - \dfrac{v^2}{c^2}}}{1 - \cos\varphi\,\dfrac{v}{c}} \quad \ldots \ldots \quad (4\,a)$$

Die beiden letzten Gleichungen drücken das **Doppler**sche Prinzip in seiner allgemeinen Fassung aus; die letzte Gleichung läßt erkennen, wie die beobachtbare Frequenz des von Kanalstrahlen emittierten (bezw. absorbierten) Lichtes von der Bewegungsgeschwindigkeit der die Strahlen bildenden Ionen und von der Richtung des Visierens abhängt.

    Nennt man ferner φ bezw. φ' den Winkel zwischen der Wellen- [32] normale (Strahlrichtung) und der Richtung der Relativbewegung von S' gegen S (d. h. mit der x- bezw. x'-Achse), so nimmt die Gleichung für l' die Form an

$$\cos\varphi' = \frac{\cos\varphi - \dfrac{v}{c}}{1 - \cos\varphi\,\dfrac{v}{c}}\cdot$$

Diese Gleichung zeigt den Einfluß der Relativbewegung des Beobachters auf den scheinbaren Ort einer unendlich fernen Lichtquelle (Aberration).

426 Einstein, Relativitätsprinzip u. die aus demselben gezog. Folgerungen.

Wir wollen noch untersuchen, wie rasch sich das Licht in einem in Richtung des Lichtstrahles bewegten Medium fortpflanzt. Das Medium ruhe relativ zum System S', und der Lichtvektor sei proportional zu

$$\sin \omega' \left(t' - \frac{x'}{V'} \right)$$

bezw. zu

$$\sin \omega \left(t - \frac{x}{V} \right),$$

je nachdem der Vorgang auf S' oder auf S bezogen wird.

Die Transformationsgleichungen ergeben

$$\omega = \beta \omega' \left(1 + \frac{v}{V'} \right)$$
$$\frac{\omega}{V} = \beta \frac{\omega'}{V'} \left(1 + \frac{V'v}{c^2} \right).$$

Hierbei ist V' als aus der Optik ruhender Körper bekannte Funktion von ω' zu betrachten. Durch Division dieser Gleichungen erhält man

$$V = \frac{V' + v}{1 + \frac{V'v}{c^2}},$$

welche Gleichung man auch unmittelbar durch Anwendung des Additionstheorems der Geschwindigkeiten hätte erhalten können.[1] Falls V' als bekannt anzusehen ist, löst die letzte Gleichung die Aufgabe vollständig. Falls aber nur die auf das „ruhende" System S bezogene Frequenz (ω) als bekannt anzusehen ist, wie z. B. bei dem bekannten Experiment von Fizeau, sind die beiden obigen Gleichungen in Verbindung mit der Beziehung zwischen ω' und V' zu verwenden zur Bestimmung der drei Unbekannten ω', V' und V.

Ist ferner G bezw. G' die auf S bezw. S' bezogene Gruppengeschwindigkeit, so ist nach dem Additionstheorem der Geschwindigkeiten

$$G = \frac{G' + v}{1 + \frac{G'v}{c^2}}.$$

Da die Beziehung zwischen G' und ω' aus der Optik ruhender Körper zu entnehmen ist[2]), und ω' nach dem Obigen aus ω berechenbar ist, so ist die Gruppengeschwindigkeit G auch dann berechenbar,

[33]   1) Vgl. M. Laue, Ann. d. Phys. **23**, 989, 1907.

[34]   2) Es ist nämlich $G' = \dfrac{V'}{1 + \dfrac{1}{V'} \dfrac{dV'}{d\omega'}}$

Einstein, Relativitätsprinzip u. die aus demselben gezog. Folgerungen. 427

wenn lediglich die auf S bezogene Frequenz des Lichtes sowie die Natur und die Bewegungsgeschwindigkeit des Körpers gegeben ist.

II. Elektrodynamischer Teil.

§ 7. Transformation der Maxwell-Lorentzschen Gleichungen.   [35]

Wir gehen aus von den Gleichungen

$$\frac{1}{c}\left\{u_x\varrho + \frac{\partial X}{\partial t}\right\} = \frac{\partial N}{\partial y} - \frac{\partial M}{\partial z}$$

$$\frac{1}{c}\left\{u_y\varrho + \frac{\partial Y}{\partial t}\right\} = \frac{\partial L}{\partial z} - \frac{\partial N}{\partial x} \qquad \cdots \cdots (5)$$

$$\frac{1}{c}\left\{u_z\varrho + \frac{\partial Z}{\partial t}\right\} = \frac{\partial M}{\partial x} - \frac{\partial L}{\partial y}$$

$$\frac{1}{c}\frac{\partial L}{\partial t} = \frac{\partial Y}{\partial z} - \frac{\partial Z}{\partial y}$$

$$\frac{1}{c}\frac{\partial M}{\partial t} = \frac{\partial Z}{\partial x} - \frac{\partial X}{\partial z} \qquad \cdots \cdots (6)$$

$$\frac{1}{c}\frac{\partial N}{\partial t} = \frac{\partial X}{\partial y} - \frac{\partial Y}{\partial x}$$

In diesen Gleichungen bedeutet

(X, Y, Z) den Vektor der elektrischen Feldstärke,

(L, M, N) den Vektor der magnetischen Feldstärke,

$\varrho = \dfrac{\partial X}{\partial x} + \dfrac{\partial Y}{\partial y} + \dfrac{\partial Z}{\partial z}$ die 4π-fache Dichte der Elektrizität,

$(u_x,\ u_y,\ u_z)$ den Geschwindigkeitsvektor der Elektrizität.

Diese Gleichungen in Verbindung mit der Annahme, daß die elektrischen Massen unveränderlich an kleine starre Körper (Ionen, Elektronen) gebunden seien, bilden die Grundlage der Lorentzschen Elektrodynamik und Optik bewegter Körper.

Transformiert man diese Gleichungen, welche in bezug auf das System S gelten mögen, mit Hilfe der Transformationsgleichungen (1) auf das relativ zu S wie bei den bisherigen Betrachtungen bewegte System S', so erhält man die Gleichungen

$$\frac{1}{c}\left\{u_x'\varrho' + \frac{\partial X'}{\partial t'}\right\} = \frac{\partial N'}{\partial y'} - \frac{\partial M'}{\partial z'}$$

$$\frac{1}{c}\left\{u_y'\varrho' + \frac{\partial Y'}{\partial t'}\right\} = \frac{\partial L'}{\partial z'} - \frac{\partial N'}{\partial x'} \qquad \cdots \cdots (5')$$

$$\frac{1}{c}\left\{u_z'\varrho' + \frac{\partial Z'}{\partial t'}\right\} = \frac{\partial M'}{\partial x'} - \frac{\partial L'}{\partial z'}$$

428 Einstein, Relativitätsprinzip u. die aus demselben gezog. Folgerungen.

$$\frac{1}{c}\frac{\partial L'}{\partial t'} = \frac{\partial Y'}{\partial z'} - \frac{\partial Z'}{\partial y'} \left.\vphantom{\begin{array}{c}1\\1\\1\end{array}}\right\}$$

$$\frac{1}{c}\frac{\partial M'}{\partial t'} = \frac{\partial Z'}{\partial x'} - \frac{\partial X'}{\partial z'} \quad\bigg\} \quad \ldots \ldots \quad (6')$$

$$\frac{1}{c}\frac{\partial N'}{\partial t'} = \frac{\partial X'}{\partial y'} - \frac{\partial Y'}{\partial x'}$$

wobei gesetzt ist:

$$X' = X$$

$$Y' = \beta\left(Y - \frac{v}{c}N\right) \Bigg\} \quad \ldots \ldots \quad (7\,\text{a})$$

$$Z' = \beta\left(Z + \frac{v}{c}M\right)$$

$$L' = L$$

$$M' = \beta\left(M + \frac{v}{c}Z\right) \Bigg\} \quad \ldots \ldots \quad (7\,\text{b})$$

$$N' = \beta\left(N - \frac{v}{c}Y\right)$$

$$\varrho' = \frac{\partial X'}{\partial x'} + \frac{\partial Y'}{\partial y'} + \frac{\partial Z'}{\partial z'} = \beta\left(1 - \frac{v\,u_x}{c^2}\right)\varrho \quad \ldots \ldots \quad (8)$$

$$u_x' = \frac{u_x - v}{1 - \dfrac{u_x v}{c^2}}$$

$$u_y' = \frac{u_y}{\beta\left(1 - \dfrac{u_x v}{c^2}\right)} \quad\Bigg\} \quad \ldots \ldots \quad (9)$$

$$u_z' = \frac{u_x}{\beta\left(1 - \dfrac{u_x v}{c^2}\right)}$$

Die erlangten Gleichungen sind von derselben Gestalt wie die Gleichungen (5) und (6). Aus dem Relativitätsprinzip folgt andererseits, daß die elektrodynamischen Vorgänge, auf S' bezogen, nach den gleichen Gesetzen verlaufen wie die auf S bezogenen. Wir schließen hieraus zunächst, daß X', Y', Z' bezw. L', M', N' nichts anderes sind als die Komponenten der auf S' bezogenen elektrischen bezw. magnetischen Feldstärke.[1]) Da ferner gemäß den Umkehrungen der

---

1) Die Übereinstimmung der gefundenen Gleichungen mit den Gleichungen (5) und (6) läßt zwar die Möglichkeit offen, daß sich die Größen X' usw. von den auf S' bezogenen Feldstärken um einen konstanten **Faktor** unterscheiden. Daß dieser Faktor gleich 1 sein muß, läßt sich aber leicht [36] auf ganz ähnliche Weise zeigen wie in § 3 bei der Funktion $\varphi(v)$.

Einstein, Relativitätsprinzip u. die aus demselben gezog. Folgerungen. 429

Gleichungen (3) die in den Gleichungen (9) auftretenden Größen u_x', u_y', u_z' gleich sind den Geschwindigkeitskomponenten der Elektrizität in bezug auf S', so ist ϱ' die auf S' bezogene Dichte der Elektrizität. Die elektrodynamische Grundlage der Maxwell-Lorentzschen Theorie entspricht also dem Prinzip der Relativität.

Zur Interpretation der Gleichungen (7a) bemerken wir folgendes.   [37] Es liege eine punktförmige Elektrizitätsmenge vor, welche relativ zu S ruhend in bezug auf S von der Größe „eins" sei, d. h. auf eine gleiche, ebenfalls in bezug auf S ruhende Elektrizitätsmenge im Abstand 1 cm die Kraft 1 Dyn ausübe. Nach dem Relativitätsprinzip ist diese elektrische Masse auch dann gleich „eins", wenn sie relativ zu S' ruht und von S' aus untersucht wird.[1] Ruht diese Elektrizitätsmenge relativ zu S, so ist (X, Y, Z) definitionsgemäß gleich der auf sie wirkenden Kraft, wie sie z. B. mittels einer relativ zu S ruhenden Federwage gemessen werden könnte. Die analoge Bedeutung hat der Vektor (X', Y', Z') mit Bezug auf S'.

Gemäß den Gleichungen (7a) und (7b) kommt einer elektrischen bezw. magnetischen Feldstärke an und für sich keine Existenz zu, indem es von der Wahl des Koordinatensystems abhängen kann, ob an einer Stelle (genauer: in der örtlich-zeitlichen Umgebung eines Punktereignisses) eine elektrische bezw. magnetische Feldstärke vorhanden ist oder nicht. Man ersieht ferner, daß die bisher eingeführten „elektromotorischen" Kräfte, welche auf eine in einem Magnetfelde bewegte elektrische Masse wirken, nichts anderes sind als „elektrische" Kräfte, falls man ein zu der betrachteten elektrischen Masse ruhendes Bezugssystem einführt. Die Fragen über den Sitz jener elektromotorischen Kräfte (bei Unipolarmaschinen) werden daher gegenstandslos; die Antwort fällt nämlich verschieden aus, je nach der Wahl des Bewegungszustandes des benutzten Bezugssystems.

Die Bedeutung der Gleichung (8) erkennt man aus folgendem.   [38] Ein elektrisch geladener Körper ruhe relativ zu S'. Seine auf S' bezogene Gesamtladung ε' ist dann $\int \dfrac{\varrho'}{4\pi}\, dx'\, dy'\, dz'$. Wie groß ist seine Gesamtladung ε zu einer bestimmten Zeit t von S?

Aus den drei letzten der Gleichungen (1) folgt, daß für konstantes t die Beziehung gilt:

$$dx'\, dy'\, dz' = \beta\, dx\, dy\, dz .$$

---

1) Dieser Schluß gründet sich ferner auf die Annahme, daß die Größe einer elektrischen Masse von deren Bewegungsvorgeschichte unabhängig ist.

430 **Einstein**, Relativitätsprinzip u. die aus demselben gezog. Folgerungen.

Gleichung (8) lautet in unserem Falle:

$$\varrho' = \frac{1}{\beta}\,\varrho.$$

Aus diesen beiden Gleichungen folgt, daß

$$\varepsilon' = \varepsilon$$

sein muß. Gleichung (8) sagt also aus, daß die elektrische Masse eine vom Bewegungszustand des Bezugssystems unabhängige Größe ist. Bleibt also die Ladung eines beliebig bewegten Körpers vom Standpunkt eines mitbewegten Bezugssystems konstant, so bleibt sie auch in bezug auf jedes andere Bezugssystem konstant.

Mit Hilfe der Gleichungen (1), (7), (8) und (9) läßt sich jedes Problem der Elektrodynamik und Optik bewegter Körper, in welchem nur Geschwindigkeiten, nicht aber Beschleunigungen eine wesentliche Rolle spielen, auf eine Reihe von Problemen der Elektrodynamik bezw. Optik ruhender Körper zurückführen.

Wir behandeln noch ein einfaches Anwendungsbeispiel für die hier entwickelten Beziehungen. Eine ebene, im Vakuum sich fortpflanzende Lichtwelle sei relativ zu S dargestellt durch die Gleichungen

[39]
$$X = X_0 \sin \Phi \qquad\qquad L = L_0 \sin \Phi$$
$$Y = Y_0 \sin \Phi \qquad\qquad M = M_0 \sin \Phi \qquad\qquad \Phi = w\left(t - \frac{lx + my + nz}{c}\right)$$
$$Z = Z_0 \sin \Phi \qquad\qquad N = N_0 \sin \Phi$$

Wir fragen nach der Beschaffenheit dieser Welle, wenn dieselbe auf das System S' bezogen wird.

Durch Anwendung der Transformationsgleichungen (1) und (7) erhält man

$$X' = X_0 \sin \Phi' \qquad\qquad\qquad L' = L_0 \sin \Phi'$$
$$Y' = \beta\left(Y_0 - \frac{v}{c}\,N_0\right)\sin \Phi' \qquad\qquad M' = \beta\left(M_0 + \frac{v}{c}\,Z_0\right)\sin \Phi'$$
$$Z' = \beta\left(Z_0 + \frac{v}{c}\,M_0\right)\sin \Phi' \qquad\qquad N' = \beta\left(N_0 - \frac{v}{c}\,Y_0\right)\sin \Phi'$$

[40]
$$\Phi' = w'\left(t - \frac{l'x' + m'y' + n'z'}{c}\right).$$

Daraus, daß die Funktionen X' usw. den Gleichungen (5') und (6') genügen müssen, folgt, daß auch in bezug auf S' Wellennormale, elektrische Kraft und magnetische Kraft aufeinander senkrecht stehen, und daß die beiden letzteren einander gleich sind. Die Beziehungen, die aus der Identität $\Phi = \Phi'$ fließen, haben wir schon in § 6 behandelt; wir haben hier nur noch Amplitude und Polarisationszustand der Welle in bezug auf S' zu ermitteln.

Einstein, Relativitätsprinzip u. die aus demselben gezog. Folgerungen. 431

Wir wählen die X-Y-Ebene parallel zur Wellennormale und behandeln zunächst den Fall, daß die elektrische Schwingung parallel zur Z-Achse erfolgt. Dann haben wir zu setzen:

$$X_0 = 0 \qquad\qquad L_0 = -A \sin \varphi$$
$$Y_0 = 0 \qquad\qquad M_0 = -A \cos \varphi$$
$$Z_0 = A \qquad\qquad N_0 = 0,$$

wobei φ den Winkel zwischen Wellennormale und X-Achse bezeichnet. Es folgt nach dem Obigen

$$X' = 0 \qquad\qquad\qquad L' = -A \sin \varphi \sin \Phi'$$

$$Y' = 0 \qquad\qquad\qquad M' = \beta \left(-\cos \varphi + \frac{v}{c} \right) A \sin \Phi'$$

$$Z' = \beta \left(1 - \frac{v}{c} \cos \varphi \right) A \sin \varphi' \qquad N' = 0.$$

Bedeutet also A' die Amplitude der Welle in bezug auf S', so ist

$$A' = A \frac{1 - \dfrac{v}{c} \cos \varphi}{\sqrt{1 - \dfrac{v^2}{c^2}}} \quad \ldots \ldots \ldots \quad (10)$$

Für den Spezialfall, daß die magnetische Kraft senkrecht auf der Richtung der Relativbewegung und der Wellennormale steht, gilt offenbar die gleiche Beziehung. Da man aus diesen beiden Spezialfällen den allgemeinen Fall durch Superposition konstruieren kann, so folgt, daß bei der Einführung eines neuen Bezugssystems S' die Beziehung (10) allgemein gilt, und daß der Winkel zwischen der Polarisationsebene und einer zur Wellennormale und zur Richtung der Relativbewegung parallelen Ebene in den beiden Bezugssystemen derselbe ist.

III. Mechanik des materiellen Punktes (Elektrons).

§ 8. Ableitung der Bewegungsgleichungen des (langsam beschleunigten) materiellen Punktes bezw. Elektrons.   [41]

In einem elektromagnetischen Felde bewege sich ein mit einer elektrischen Ladung ε versehenes Teilchen (im folgenden „Elektron" genannt), über dessen Bewegungsgesetz wir folgendes annehmen:

Ruht das Elektron in einem bestimmten Zeitpunkt in bezug auf ein (beschleunigungsfreies) System S', so erfolgt dessen Bewegung im nächsten Zeitteilchen in bezug auf S' nach den Gleichungen

432 Einstein, Relativitätsprinzip u. die aus demselben gezog. Folgerungen.

$$\mu \frac{d^2 x_0'}{dt'^2} = \varepsilon X'$$

$$\mu \frac{d^2 y_0'}{dt'^2} = \varepsilon Y'$$

$$\mu \frac{d^2 z_0'}{dt'^2} = \varepsilon Z',$$

wobei x_0', y_0', z_0' die Koordinaten des Elektrons in bezug auf S' bezeichnen, und μ eine Konstante bedeutet, welche wir die Masse des Elektrons nennen.

Wir führen ein System S ein, relativ zu welchem S' wie bei unseren bisherigen Untersuchungen bewegt sei, und transformieren unsere Bewegungsgleichungen mittels der Transformationsgleichungen (1) und (7a).

Erstere lauten in unserem Falle

$$t' = \beta \left(t - \frac{v}{c^2} x_0 \right)$$
$$x_0' = \beta (x_0 - v t)$$
$$y^{0'} = y_0$$
$$z_0' = z_0 .$$

Aus diesen Gleichungen erhalten wir, indem wir $\frac{dx_0}{dt} = \dot{x}_0$ usw. setzen:

$$\frac{dx_0'}{dt'} = \frac{\beta(\dot{x}_0 - v)}{\beta \left(1 - \frac{v \dot{x}_0}{c^2} \right)} \quad \text{usw.}$$

[42]

$$\frac{d^2 x_0'}{dt'^2} = \frac{\frac{d}{dt}\left\{ \frac{dx_0'}{dt'} \right\}}{\beta \left(1 - \frac{v x_0'}{c^2} \right)} = \frac{1}{\beta} \frac{\left(1 - \frac{v \dot{x}_0}{c^2} \right)\ddot{x}_0 + (\dot{x}_0 - v)\frac{v \ddot{x}_0}{c^2}}{\left(1 - \frac{v \dot{x}_0}{c^2} \right)} \quad \text{usw.}$$

Setzt man diese Ausdrücke, nachdem man in ihnen $\dot{x}_0 = v$, $\dot{y}_0 = 0$, $\dot{z}_0 = 0$ gesetzt hat, in die obigen Gleichungen ein, so erhält man, indem man gleichzeitig X', Y', Z' mittels der Gleichungen (7a) ersetzt

$$\mu \beta^3 \ddot{x}_0 = \varepsilon X$$

$$\mu \beta \ddot{y}_0 = \varepsilon \left(Y - \frac{v}{c} N \right)$$

$$\mu \beta \ddot{z}_0 = \varepsilon \left(Z + \frac{v}{c} M \right) .$$

Diese Gleichungen sind die Bewegungsgleichungen des Elektrons für den Fall, daß in dem betreffenden Augenblick $\dot{x}_0 = v$, $\dot{y}_0 = o$, $\dot{z}_0 = o$ ist. Man kann also auf den linken Seiten statt v die durch die Gleichung

Einstein, Relativitätsprinzip u. die aus demselben gezog. Folgerungen. 433

$$q = \sqrt{\dot{x}_0{}^2 + \dot{y}_0{}^2 + \dot{z}_0{}^2}$$

definierte Geschwindigkeit q einsetzen und auf den rechten Seiten v durch $\dot{x}_0$ ersetzen. Außerdem fügen wir die durch zyklische Vertauschung aus $\dfrac{\dot{x}_0}{c} M$ und $-\dfrac{\dot{x}_0}{c} N$ zu gewinnenden Glieder, welche in dem betrachteten Spezialfalle verschwinden, an den entsprechenden Stellen hinzu. Indem wir den Index bei x_0 usw. weglassen, erhalten wir so die für den betrachteten Spezialfall mit den obigen gleichbedeutenden Gleichungen: [43]

$$\left. \begin{aligned} \frac{d}{dt}\left\{ \frac{\mu \dot{x}}{\sqrt{1 - \dfrac{q^2}{c^2}}} \right\} &= K_x \\[2em] \frac{d}{dt}\left\{ \frac{\mu \dot{y}}{\sqrt{1 - \dfrac{q^2}{c^2}}} \right\} &= K_y \\[2em] \frac{d}{dt}\left\{ \frac{\mu \dot{z}}{\sqrt{1 - \dfrac{q^2}{c^2}}} \right\} &= K_z, \end{aligned} \right\} \quad \cdots \cdots \quad (11)$$

wobei gesetzt ist:

$$\left. \begin{aligned} K_x &= \varepsilon \left\{ X + \frac{\dot{y}}{c} N - \frac{\dot{z}}{c} M \right\} \\[1em] K_y &= \varepsilon \left\{ Y + \frac{\dot{z}}{c} L - \frac{\dot{x}}{c} N \right\} \\[1em] K_z &= \varepsilon \left\{ Z + \frac{\dot{x}}{c} M - \frac{\dot{y}}{c} L \right\} \end{aligned} \right\} \quad \cdots \cdots \quad (12)$$

Diese Gleichungen ändern ihre Form nicht, wenn man ein neues, relativ ruhendes Koordinatensystem mit anders gerichteten Achsen einführt. Sie gelten daher allgemein, nicht nur, wenn $\dot{x} = \dot{z} = 0$ ist.

Den Vektor $(K_x,\ K_y,\ K_z)$ nennen wir die auf den materiellen Punkt wirkende Kraft. In dem Falle, daß q^2 gegen c^2 verschwindet, gehen $K_x,\ K_y,\ K_z$ nach Gleichungen (11) in die Kraftkomponenten gemäß Newtons Definition über. Im nächsten Paragraphen ist ferner dargelegt, daß in der Relativitätsmechanik jener Vektor auch im übrigen dieselbe Rolle spielt wie die Kraft in der klassischen Mechanik.

Wir wollen an den Gleichungen (11) auch in dem Falle festhalten, daß die auf den Massenpunkt ausgeübte Kraftwirkung nicht elektromagnetischer Natur ist. In diesem Falle haben die Gleichungen

434 Einstein, Relativitätsprinzip u. die aus demselben gezog. Folgerungen.

(11) keinen physikalischen Inhalt, sondern sie sind dann als Definitionsgleichungen der Kraft aufzufassen.

§ 9. Bewegung des Massenpunktes und mechanische Prinzipien.

Multipliziert man die Gleichungen (5) und (6) der Reihe nach mit $\dfrac{X}{4\pi}, \dfrac{Y}{4\pi} \ldots \dfrac{N}{4\pi}$ und integriert über einen Raum, an dessen Grenzen die Feldstärken verschwinden, so erhält man

$$\int \frac{\varrho}{4\pi}(u_x X + u_y Y + u_z Z)d\omega + \frac{dE_e}{dt} = 0, \quad \ldots \quad (13)$$

wobei

$$E_e = \int\left[\frac{1}{8\pi}(X^2 + Y^2 + Z^2) + \frac{1}{8\pi}(L^2 + M^2 + N^2)\right]d\omega$$

die elektromagnetische Energie des betrachteten Raumes ist. Das erste Glied der Gleichung (13) ist nach dem Energieprinzip gleich der Energie, welche vom elektromagnetischen Felde pro Zeiteinheit an die Träger der elektrischen Massen abgegeben wird. Sind elektrische Massen mit einem materiellen Punkte starr verbunden (Elektron), so ist der auf sie entfallende Anteil jenes Gliedes gleich dem Ausdruck

$$\varepsilon(X\dot{x} + Y\dot{y} + Z\dot{z}),$$

wenn (X, Y, Z) die äußere elektrische Feldstärke bezeichnet, d. h. die Feldstärke abzüglich derjenigen, welche von der Ladung des Elektrons selbst herrührt. Dieser Ausdruck geht vermöge der Gleichungen (12) über in

$$K_x\dot{x} + K_y\dot{y} + K_z\dot{z}.$$

Der im vorigen Paragraph als „Kraft" bezeichnete Vektor (K_x, K_y, K_z) steht also zu der geleisteten Arbeit in derselben Beziehung wie bei der Newtonschen Mechanik.

[44] Multipliziert man also die Gleichungen (11) der Reihe nach mit x, y, z, addiert und integriert über die Zeit, so muß sich die kinetische Energie des materiellen Punktes (Elektrons) ergeben. Man erhält

$$\int(K_x\dot{x} + K_y\dot{y} + K_z\dot{z})dt = \frac{\mu c^2}{\sqrt{1 - \dfrac{q^2}{c^2}}} + \text{const.} \quad \ldots \quad (14)$$

Daß die Bewegungsgleichungen (11) mit dem Energieprinzip im Einklang sind, ist damit gezeigt. Wir wollen nun dartun, daß sie auch dem Prinzip von der Erhaltung der Bewegungsgröße entsprechen.

Multipliziert man die zweite und dritte der Gleichungen (5) und

Einstein, Relativitätsprinzip u. die aus demselben gezog. Folgerungen. 435

die zweite und dritte der Gleichungen (6) der Reihe nach mit $\dfrac{N}{4\pi}$, $\dfrac{-M}{4\pi}$, $\dfrac{-Z}{4\pi}$, $\dfrac{Y}{4\pi}$, addiert und integriert über einen Raum, an dessen Grenzen die Feldstärken verschwinden, so erhält man

$$\frac{d}{dt}\left[\int \frac{1}{4\pi c}(YN-ZM)\,d\omega\right] + \int \frac{\varrho}{4\pi}\left(X + \frac{u_y}{c}N - \frac{u_z}{c}M\right)d\omega = 0 \quad (15)$$

oder gemäß den Gleichungen (12)

$$\frac{d}{dt}\left[\int \frac{1}{4\pi c}(YN-ZM)\,d\omega\right] + \Sigma K_x = 0 \quad . \quad . \quad . \quad (15a)$$

Sind die elektrischen Massen an frei bewegliche materielle Punkte (Elektronen) gebunden, so geht diese Gleichung vermöge (11) über in

$$\frac{d}{dt}\left[\int \frac{1}{4\pi c}\,YN-ZM) + \Sigma\right]\frac{\mu\dot{x}}{\sqrt{1-\dfrac{q^2}{c^2}}} = 0 \quad . \quad . \quad (15b) \quad [45]$$

Diese Gleichung drückt in Verbindung mit den durch zyklische Vertauschung zu gewinnenden den Satz von der Erhaltung der Bewegungsgröße in dem hier betrachteten Falle aus. Die Größe $\xi = \dfrac{\mu\dot{x}}{\sqrt{1-\dfrac{v^2}{c^2}}}$ [46]

spielt also die Rolle der Bewegungsgröße des materiellen Punktes, und es ist gemäß Gleichungen (11) wie in der klassischen Mechanik

$$\frac{d\xi}{dt} = K_x.$$

Die Möglichkeit, eine Bewegungsgröße des materiellen Punktes einzuführen, beruht darauf, daß in den Bewegungsgleichungen die Kraft bezw. das zweite Glied der Gleichung (15) als Differentialquotient nach der Zeit dargestellt werden kann.

Man sieht ferner unmittelbar, daß unseren Bewegungsgleichungen des materiellen Punktes die Form der Bewegungsgleichungen von Lagrange gegeben werden kann; denn es ist gemäß Gleichungen (11)

$$\frac{d}{dt}\left[\frac{\partial H}{\partial \dot{x}}\right] = K_x \text{ usw.},$$

wobei

$$H = -\mu c^2\sqrt{1-\frac{q^2}{c^2}} + \text{const} \quad [47]$$

gesetzt ist. Die Bewegungsgleichungen lassen sich auch darstellen in der Form des Hamiltonschen Prinzips

436 Einstein, Relativitätsprinzip u. die aus demselben gezog. Folgerungen.

$$\int_{t_0}^{t_1} (dH + A)dt = 0,$$

wobei die Zeit t sowie die Anfangs- und Endlage unvariiert bleibt und A die virtuelle Arbeit bezeichnet:

$$A = K_x \partial x + K_y \partial y + K_z \partial z.$$

Endlich stellen wir noch die **Hamilton**schen kanonischen Bewegungsgleichungen auf. Hierzu dient die Einführung der „Impulskoordinaten" (Komponenten der Bewegungsgröße) ξ, η, ζ, wobei wie oben gesetzt ist

[48]

$$\xi = \frac{\partial H}{\partial \dot{x}} = \frac{\mu x}{\sqrt{1 - \dfrac{q^2}{c^2}}} \text{ usw.}$$

Betrachtet man die kinetische Energie L als Funktion von ξ, η, ζ und setzt $\xi^2 + \eta^2 + \zeta^2 = \varrho^2$, so ergibt sich

$$L = \mu c^2 \sqrt{1 + \frac{\varrho^2}{\mu^2 c^2}} + \text{const}$$

und die **Hamilton**schen Bewegungsgleichungen werden:

$$\frac{d\xi}{dt} = K_x \qquad \frac{d\eta}{dt} = K_y \qquad \frac{d\zeta}{dt} = K_z$$

$$\frac{dx}{dt} = \frac{\partial L}{\partial \xi} \qquad \frac{dy}{dt} = \frac{\partial L}{\partial \eta} \qquad \frac{dz}{dt} = \frac{\partial L}{\partial \zeta}.$$

§ 10. Über die Möglichkeit einer experimentellen Prüfung [49] der Theorie der Bewegung des materiellen Punktes. Kaufmannsche Untersuchung.

Eine Aussicht auf Vergleichung der im letzten Paragraphen abgeleiteten Resultate mit der Erfahrung ist nur da vorhanden, wo bewegte, mit einer elektrischen Ladung versehene Massenpunkte Geschwindigkeiten besitzen, deren Quadrat gegenüber c^2 nicht zu vernachlässigen ist. Diese Bedingung ist bei den rascheren Kathodenstrahlen und bei den von radioaktiven Substanzen ausgesandten Elektronenstrahlen (β-Strahlen) erfüllt.

Es gibt drei Größen bei Elektronenstrahlen, deren gegenseitige Beziehungen Gegenstand einer genaueren experimentellen Untersuchung sein können, nämlich das Erzeugungspotential bezw. die kinetische Energie der Strahlen, die Ablenkbarkeit durch ein elektrisches Feld und die Ablenkbarkeit durch ein magnetisches Feld.

Das Erzeugungspotential Π ist gemäß (14) gegeben durch die Formel

Einstein, Relativitätsprinzip u. die aus demselben gezog. Folgerungen. 437

$$\Pi \varepsilon = \mu \left\{ \frac{c^2}{\sqrt{1 - \dfrac{q^2}{c^2}}} - 1 \right\} \qquad [50]$$

Zur Berechnung der andern beiden Größen schreiben wir die letzte der Gleichungen (11) hin für den Fall, daß die Bewegung momentan parallel zur X-Achse ist; man erhält, falls man mit ε den absoluten Betrag der Ladung des Elektrons bezeichnet,

$$-\frac{d^2 z}{d t^2} = \frac{\varepsilon}{\mu} \sqrt{1 - \frac{q^2}{c^2}} \left(Z + \frac{q}{c} M \right) \cdot$$

Falls Z und M die einzigen ablenkenden Feldkomponenten sind, die Krümmung also in der XZ-Ebene erfolgt, ist der Krümmungsradius R der Bahn gegeben durch $\dfrac{q^2}{R} = \left[\dfrac{d^2 z}{d t^2} \right]$. Definiert man als elektrische bezw. magnetische Ablenkbarkeit die Größe $A_e = \dfrac{1}{R} : Z$ bzw. $A_m = \dfrac{1}{R} : M$ für den Fall, daß nur eine elektrische bezw. nur eine magnetische ablenkende Feldkomponente vorhanden ist, so hat man also

$$A_e = \frac{\varepsilon}{\mu} \frac{\sqrt{1 - \dfrac{q^2}{c^2}}}{q^2}$$

$$A_m = \frac{\varepsilon}{\mu} \frac{\sqrt{1 - \dfrac{q^2}{c^2}}}{cq} \cdot$$

Bei Kathodenstrahlen kommen alle drei Größen, Π, A_e und A für die Messung in Betracht; es liegen jedoch noch keine Untersuchungen bei genügend raschen Kathodenstrahlen vor. Bei β-Strahlen [51] sind (praktisch) nur die Größen A_e und A_m der Beobachtung zugänglich. Herr W. Kaufmann hat mit bewunderungswürdiger Sorgfalt die Beziehung zwischen A_m und A_e für die von einem Radium-Bromid-Körnchen ausgesandten β-Strahlen ermittelt.[1]

Sein Apparat, dessen hauptsächliche Teile in Fig. 1 in natürlicher [53] Größe dargestellt sind, bestand im wesentlichen in einem lichtdichten, im Innern eines evakuierten Glasgefäßes befindlichen Messinggehäuse H, auf dessen Boden A in einer kleinen Vertiefung O sich das Radiumkörnchen befand. Die von ihm ausgehenden β-Strahlen durchlaufen

---

1) W. Kaufmann, Über die Konstitution des Elektrons. Ann. d. [52] Phys. 19, 1906. Die beiden Figuren sind der Kaufmannschen Arbeit entnommen.

438 Einstein, Relativitätsprinzip u. die aus demselben gezog. Folgerungen.

den Zwischenraum zwischen zwei Kondensatorplatten P_1 und P_2, treten durch das Diaphragma D von 0,2 mm Durchmesser und fallen dann auf die photographische Platte. Die Strahlen wurden durch ein zwischen den Kondensatorplatten P_1 und P_2 gebildetes elektrisches Feld sowie

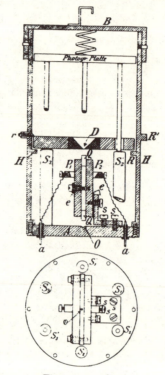

Fig. 1 (nat. Gr.).

durch ein von einem großen permanenten Magneten erzeugtes, in gleicher Richtung verlaufendes magnetisches Feld senkrecht dazu abgelenkt, so daß durch die Wirkung der Strahlen einer bestimmten Geschwindigkeit ein Punkt, durch die Wirkung der Teilchen von den verschiedenen Geschwindigkeiten zusammen eine Kurve auf der Platte markiert wurde.

[54]   Fig. 2 zeigt diese Kurve[1]), welche bis auf den Maßstab für Ab-

          1) Die in der Figur angegebenen Maßzahlen bedeuten Millimeter auf der photographischen Platte. Die gezeichnete Kurve ist nicht genau die
[55] beobachtete, sondern die „auf unendlich kleine Ablenkung reduzierte" Kurve.

Einstein, Relativitätsprinzip u. die aus demselben gezog. Folgerungen. 439

szisse und Ordinate die Beziehung zwischen A_m (Abszisse) und A_e (Ordinate) darstellt. Über der Kurve sind durch Kreuzchen der nach der Relativitätstheorie berechneten Kurve angegeben, wobei für $\frac{\varepsilon}{\mu}$ der Wert $1,878 \cdot 10^7$ angenommen ist.   [56]

In Anbetracht der Schwierigkeit der Untersuchung möchte man geneigt sein, die Übereinstimmung als eine genügende anzusehen. Die vorhandenen Abweichungen sind jedoch systematisch und erheblich

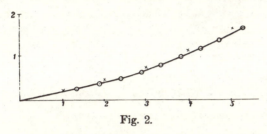

Fig. 2.

außerhalb der Fehlergrenze der Kaufmannschen Untersuchung. Daß die Berechnungen von Herrn Kaufmann fehlerfrei sind, geht daraus hervor, daß Herr Planck bei Benutzung einer anderen Berechnungsmethode zu Resultaten geführt wurde, die mit denen von Herrn Kaufmann durchaus übereinstimmen. [1])

Ob die systematischen Abweichungen in einer noch nicht gewürdigten Fehlerquelle oder darin ihren Grund haben, daß die Grundlagen der Relativitätstheorie nicht den Tatsachen entsprechen, kann wohl erst dann mit Sicherheit entschieden werden, wenn ein mannigfaltigeres Beobachtungsmaterial vorliegen wird.   [58]

Es ist noch zu erwähnen, daß die Theorien der Elektronenbewegung von Abraham [2]) und von Bucherer [3]) Kurven liefern, die sich der beobachteten Kurve erheblich besser anschließen als die aus der Relativitätstheorie ermittelte Kurve. Jenen Theorien kommt aber nach meiner Meinung eine ziemlich geringe Wahrscheinlichkeit zu, weil ihre die Maße des bewegten Elektrons betreffenden Grundannahmen nicht nahe gelegt werden durch theoretische Systeme, welche größere Komplexe von Erscheinungen umfassen.

1) Vergl. M. Planck, Verhandl. d. Deutschen Phys. Ges. VIII. Jahrg. [57] Nr. 20, 1906; IX. Jahrg. Nr. 14, 1907.

2) M. Abraham, Gött. Nachr. 1902.   [59]

3) A. H. Bucherer, Math. Einführung in die Elektronentheorie, S. 58, [60] Leipzig 1904.

440 Einstein, Relativitätsprinzip u. die aus demselben gezog. Folgerungen.

IV. Zur Mechanik und Thermodynamik der Systeme.

[61] **§ 11. Über die Abhängigkeit der Masse von der Energie.**

Wir betrachten ein von einer für Strahlung nicht durchlässigen Hülle umgebenes physikalisches System. Dies System schwebe frei im Raume und sei keinen andern Kräften unterworfen, als den Einwirkungen elektrischer und magnetischer Kräfte des umgebenden Raumes. Durch letztere kann auf das System Energie in Form von Arbeit und Wärme übertragen werden, welche Energie im Innern des Systems irgendwelche Verwandlungen erfahren kann. Die von dem System aufgenommene Energie ist, auf das System S bezogen, gemäß (13) gegeben durch den Ausdruck

$$\int dE = \int dt \int \frac{\varrho}{4\pi} (X_a\, u_x + Y_a\, u_y + Z_a\, u_z)\, d\omega,$$

wobei (X_a, Y_a, Z_a) den Feldvektor des äußern, nicht zum System gerechneten Feldes und $\frac{\varrho}{4\pi}$ die Elektrizitätsdichte in der Hülle bedeutet.

Diesen Ausdruck transformieren wir mittels der Umkehrungen der Gleichungen (7a), (8) und (9), indem wir berücksichtigen, daß gemäß den Gleichungen (1) die Funktionaldeterminante

$$\frac{D\,(x',\, y',\, z',\, t')}{D\,(x, y, z, t)}$$

gleich eins ist. Wir erhalten so

$$\int dE = \beta \iint \frac{\varrho'}{4\pi} (u_{x'}\, X_a' + u_{y'}\, Y_a' + u_{z'}\, Z_a')\, d\omega'\, dt'$$

$$+ \beta v \iint \frac{\varrho'}{4\pi} \Big(X_a' + \frac{u_{y'}}{c}\, N_a' - \frac{u_{z'}}{c}\, M_a' \Big) d\omega'\, dt',$$

oder, da auch in bezug auf S' das Energieprinzip gelten muß, in leicht verständlicher Schreibweise

$$dE = \beta\, dE' + \beta v \int [\Sigma K_x']\, dt'. \tag{16}$$

Wir wollen diese Gleichung auf den Fall anwenden, daß sich das betrachtete System derart gleichförmig bewegt, daß es als Ganzes relativ zu dem Bezugsystem S' ruht. Dann dürfen wir, falls die Teile des Systems relativ zu S' so langsam bewegt sind, daß die Quadrate der Geschwindigkeiten relativ zu S' gegenüber c^2 zu vernachlässigen sind, in bezug auf S' die Sätze der Newtonschen Mechanik anwenden. Es kann also nach dem Schwerpunktsatz das betrachtete System (genauer gesagt, dessen Schwerpunkt) nur dann dauernd in Ruhe bleiben, wenn für jedes t'

$$\Sigma K_x' = 0$$

Einstein, Relativitätsprinzip u. die aus demselben gezog. Folgerungen. 441

ist. Trotzdem braucht das zweite Glied auf der rechten Seite der Gleichung (16) nicht zu verschwinden, weil die zeitliche Integration nicht zwischen zwei bestimmten Werten von t', sondern zwischen zwei bestimmten Werten von t auszuführen ist.

Wenn aber am Anfang und am Ende der betrachteten Zeitspanne keine äußeren Kräfte auf das Körpersystem wirken, so verschwindet jenes Glied, so daß wir einfach erhalten

$$d\,E = \beta \cdot d\,E'.$$

Aus dieser Gleichung schließen wir zunächst, daß die Energie eines (gleichförmig) bewegten Systems, das nicht unter dem Einfluß äußerer Kräfte steht, eine Funktion zweier Variabeln ist, nämlich der Energie E_0 des Systems relativ zu einem mitbewegten Bezugssystem[1]), und der Translationsgeschwindigkeit q des Systems, und wir erhalten

$$\frac{\partial\,E}{\partial\,E_0} = \frac{1}{\sqrt{1 - \dfrac{q^2}{c^2}}}.$$

Daraus folgt

$$E = \frac{1}{\sqrt{1 - \dfrac{q^2}{c^2}}}\,E + \varphi(q), \qquad\qquad [62]$$

wobei $\varphi(q)$ eine vorläufig unbekannte Funktion von q ist. Den Fall, daß E_0 gleich 0 ist, d. h. daß die Energie des bewegten Sytems Funktion der Geschwindigkeit q allein ist, haben wir bereits in den § 8 und 9 untersucht. Aus Gleichung (14) folgt unmittelbar, daß wir zu setzen haben

$$\varphi(q) = \frac{\mu\,c^2}{\sqrt{1 - \dfrac{q^2}{c^2}}} + \text{const.}$$

Wir erhalten also

$$E = \left(\mu + \frac{E_0}{c^2}\right)\frac{c^2}{\sqrt{1 - \dfrac{q^2}{c^2}}}, \qquad\qquad (16\,\mathrm{a})$$

wobei die Integrationskonstante weggelassen ist. Vergleicht man diesen Ausdruck für E mit dem in Gleichung (14) enthaltenen Ausdruck für die kinetische Energie des materiellen Punktes, so erkennt man, daß

---

1) Hier so wie im folgenden versehen wir ein Zeichen mit dem unteren Index „0", um anzudeuten, daß die betreffende Größe sich auf ein relativ zu dem betrachteten physikalischen System ruhendes Bezugssystem bezieht. Da das betrachtete System relativ zu S' ruht, können wir also hier E' durch E_0 ersetzen.

442 Einstein, Relativitätsprinzip u. die aus demselben gezog. Folgerungen.

beide Ausdrücke von derselben Form sind; bezüglich der Abhängigkeit der Energie von der Translationsgeschwindigkeit verhält sich das betrachtete physikalische System wie ein materieller Punkt von der Masse M, wobei M von dem Energieinhalt E_0 des Systems abhängt nach der Formel

$$M = \mu + \frac{E_0}{c^2}. \qquad (17)$$

Dies Resultat ist von außerordentlicher theoretischer Wichtigkeit, weil in demselben die träge Masse und die Energie eines physikalischen Systems als gleichartige Dinge auftreten. Eine Masse μ ist in bezug auf Trägheit äquivalent mit einem Energieinhalt von der Größe μc^2. Da wir über den Nullpunkt von E_0 willkürlich verfügen können, sind wir nicht einmal imstande, ohne Willkür zwischen einer „wahren" und einer „scheinbaren" Masse des Systems zu unterscheiden. Weit natürlicher erscheint es, jegliche träge Masse als einen Vorrat von Energie aufzufassen.

Der Satz von der Konstanz der Masse ist nach unserem Resultat für ein einzelnes physikalisches System nur dann zutreffend, wenn dessen Energie konstant bleibt; er ist dann gleichbedeutend mit dem Energieprinzip. Allerdings sind die Änderungen, welche die Masse physikalischer Systeme bei den bekannten physikalischen Vorgängen erfährt, stets unmeßbar klein. Die Abnahme der Masse eines Systems, welches 1000 Gramm-Kalorien abgibt, beträgt z. B. $4{,}6 \cdot 10^{-11}$ gr.

Beim radioaktiven Zerfall eines Stoffes werden ungeheure Energiemengen frei; ist die bei einem derartigen Prozeß auftretende Verminderung der Masse nicht groß genug, um konstatiert zu werden?

[63]   Herr Planck schreibt hierüber: „Nach J. Precht[1]) entwickelt ein Grammatom Radium, wenn es von einer hinreichend dicken Bleischicht umgeben ist, pro Stunde $134{,}4 \times 225 = 30240$ gr-cal. Dies
[65] ergibt nach (17) für die Stunde eine Verminderung der Masse um

$$\frac{30240 \cdot 419 \cdot 10^5}{9 \cdot 10^{20}} \text{ gr} = 1{,}41 \cdot 10^{-6} \text{ mgr}$$

oder in einem Jahre eine Verminderung der Masse um 0,012 mgr. Dieser Betrag ist allerdings, besonders mit Rücksicht auf das hohe Atomgewicht des Radiums, immer noch so winzig, daß er wohl zunächst außer dem Bereich der möglichen Erfahrung liegt". Es liegt nahe, sich zu fragen, ob man nicht durch Anwendung einer indirekten Methode zum Ziele kommen könnte. Es sei M das Atomgewicht des

[64]   1) J. Precht, Ann. d. Phys. 21, 599, 1906.

Einstein, Relativitätsprinzip u. die aus demselben gezog. Folgerungen **443**

zerfallenden Atoms, m_1, m_2 etc. seien die Atomgewichte der Endprodukte des radioaktiven Zerfalls, dann muß sein

$$M - \Sigma m = \frac{E}{c^2},$$

wobei E die beim Zerfall eines Grammatoms entwickelte Energie bedeutet; diese kann berechnet werden, wenn man die bei stationärem Zerfall pro Zeiteinheit entwickelte Energie und die mittlere Zerfalldauer des Atoms kennt. Ob die Methode mit Erfolg angewendet werden kann, hängt in erster Linie davon ab, ob es radioaktive Reaktionen gibt, für welche $\dfrac{M - \Sigma m}{M}$ nicht allzu klein gegen 1 ist. Für den oben erwähnten Fall des Radiums ist — wenn man die Lebensdauer desselben zu 2600 Jahren annimmt — ungefähr

$$\frac{M - \Sigma m}{M} = \frac{12 \cdot 10^{-6} \cdot 2600}{250} = 0,00012 \,.$$

Wenn also die Lebensdauer des Radiums einigermaßen richtig bestimmt ist, müßte man die in Betracht kommenden Atomgewichte auf fünf Stellen genau kennen, um unsere Beziehung prüfen zu können. Dies ist natürlich ausgeschlossen. Es ist indessen möglich, daß radioaktive Vorgänge bekannt werden, bei welchen ein bedeutend größerer Prozentsatz der Masse des ursprünglichen Atoms sich in Energie diverser Strahlungen verwandelt als beim Radium. Es liegt wenigstens nahe, sich vorzustellen, daß die Energieentwickelung beim Zerfall eines Atoms bei verschiedenen Stoffen nicht minder verschieden sei als die Raschheit des Zerfalls.   [66]

Im vorhergehenden ist stillschweigend vorausgesetzt, daß eine derartige Massenänderung mit dem zur Messung von Massen gewöhnlich benutzten Instrument, der Wage, gemessen werden könne, daß also die Beziehung

$$M = \mu + \frac{E_0}{c^2}$$

nicht nur für die träge Masse, sondern auch für die gravitierende Masse gelte, oder mit anderen Worten, daß Trägheit und Schwere eines Systems unter allen Umständen genau proportional seien. Wir   [67] hätten also auch z. B. anzunehmen, daß in einem Hohlraum eingeschlossene Strahlung nicht nur Trägheit, sondern auch Gewicht besitze. Jene Proportionalität zwischen träger und schwerer Masse gilt aber ausnahmslos für alle Körper mit der bisher erreichten Genauigkeit,   [68] so daß wir bis zum Beweise des Gegenteils die Allgemeingültigkeit

30*

444 Einstein, Relativitätsprinzip u. die aus demselben gezog. Folgerungen.

annehmen müssen. Wir werden ferner im letzten Abschnitt dieser Abhandlung ein neues, die Annahme stützendes Argument finden.

§ 12. Energie und Bewegungsgröße eines bewegten Systems.

Wir betrachten wieder wie im vorigen Paragraphen ein frei im Raume schwebendes System, welches von einer für Strahlung nicht durchlässigen Hülle umgeben ist. Mit X_a, Y_a, Z_a etc. bezeichnen wir wieder die Feldstärken des äußeren elektromagnetischen Feldes, welches den Energieaustausch des Systems mit anderen Systemen vermittle. Auf dies äußere Feld können wir die Betrachtungen anwenden, welche uns zu Formel (15) geführt haben, so daß wir erhalten

$$\frac{d}{dt}\left[\int \frac{1}{4\pi c}\left(Y_a N_a - Z_a M_a\right) d\omega\right]$$
$$+ \int \frac{\varrho}{4\pi}\left(X_a + \frac{u_y}{c} N_a - \frac{u_z}{c} M_a\right) d\omega = 0.$$

Wir wollen nun annehmen, daß der Satz von der Erhaltung der Bewegungsgröße allgemein gelte. Dann muß der über die Systemhülle erstreckte Teil des zweiten Gliedes dieser Gleichung, als Differential-quotient nach der Zeit einer durch den Momentanzustand des Systems vollkommen bestimmten Größe G_x darstellbar sein, welche wir als die X-Komponente der Bewegungsgröße des Systems bezeichnen. Wir wollen nun das Transformationsgesetz der Größe G_x aufsuchen. Durch Anwendung der Transformationsgleichungen (1), (7), (8) und (9) erhalten wir auf ganz analogem Wege wie im vorigen Paragraphen die Beziehung

$$\int d G_x = \beta \int\int \frac{\varrho'}{4\pi}\left(X_a' + \frac{u_y'}{c} N_a' - \frac{u_z'}{c} M_a'\right) d\omega' \cdot dt'$$
$$+ \frac{\beta v}{c^2}\int\int \frac{\varrho'}{4\pi}\left(X_a u_x' + Y_a' u_y' + Z_a' u_x'\right) d\omega \cdot dt'$$

[69]

oder

$$d G_x = \beta \frac{v}{c^2} d E' + \beta \int\left\{\Sigma K_x'\right\} dt'. \tag{18}$$

Der Körper bewege sich wieder beschleunigungsfrei, derart, daß er dauernd in bezug auf S' ruht, dann ist wieder
$$\Sigma K_x' = 0.$$

Trotzdem die Grenzen der Zeitintegration von x' abhängen, verschwindet wieder das zweite Glied auf der rechten Seite der Gleichung, wenn der Körper vor und nach der betrachteten Veränderung äußeren Kräften nicht ausgesetzt ist; es ist dann

Einstein, Relativitätsprinzip u. die aus demselben gezog. Folgerungen. 445

$$d\,G_x = \beta\,\frac{v}{c^2}\,d\,E'\,.$$

Hieraus folgt, daß die Bewegungsgröße eines äußeren Kräften nicht ausgesetzten Systems eine Funktion nur zweier Variabeln ist, nämlich der Energie E_0 des Systems in bezug auf ein mitbewegtes Bezugssystem und der Translationsgeschwindigkeit q desselben. Es ist

$$\frac{\partial\,G}{\partial\,E_0} = \frac{\frac{q}{c^2}}{\sqrt{1-\frac{q^2}{c^2}}}\,.$$

Hieraus folgt

$$G = \frac{q}{\sqrt{1-\frac{q^2}{c^2}}}\cdot\left(\frac{E_0}{c^2}+\psi(q)\right),$$

wobei $\psi(q)$ eine vorläufig unbekannte Funktion von q ist. Da $\psi(q)$ nichts anderes ist als die Bewegungsgröße für den Fall, daß letztere durch die Geschwindigkeit allein bestimmt ist, schließen wir aus Formel (15b), daß

$$\psi(q) = \frac{\mu\,q}{\sqrt{1-\frac{q^2}{c^2}}}$$

ist. Wir erhalten also

$$G = \frac{q}{\sqrt{1-\frac{q^2}{c^2}}}\left\{\mu+\frac{E_0}{e^2}\right\} \qquad (18\,\mathrm{a}) \quad [70]$$

Dieser Ausdruck unterscheidet sich von dem für die Bewegungsgröße des materiellen Punktes nur dadurch, daß an Stelle von μ die Größe $\left(\mu+\frac{E_0}{c^2}\right)$ tritt, im Einklang mit dem Resultat des vorigen Paragraphen.

[71] Wir wollen nun Energie und Bewegungsgröße eines in bezug auf S ruhenden Körpers aufsuchen für den Fall, daß der Körper dauernden äußeren Kräften unterworfen ist. In diesem Falle ist zwar auch für jedes t'

$$\Sigma K_x' = 0,$$

aber das in den Gleichungen (16) und (18) auftretende Integral

$$\int[\Sigma K_x']\,dt'$$

verschwindet nicht, weil dasselbe nicht zwischen zwei bestimmten Werten von t', sondern von zwei bestimmten Werten von t zu er-

446 Einstein, Relativitätsprinzip u. die aus demselben gezog. Folgerungen.

strecken ist. Da nach der Umkehrung der ersten der Gleichungen (1)

$$t = \beta\left(t' + \frac{v}{c^2}\,x'\right),$$

so sind die Grenzen für die Integration nach t' gegeben durch

$$\frac{t_1}{\beta} - \frac{v}{c^2}\,x' \quad \text{und} \quad \frac{t_2}{\beta} - \frac{v}{c^2}\,x',$$

wobei t_1 und t_2 von x', y', z' unabhängig sind. Die Grenzen der Zeitintegration in bezug auf S' sind also von der Lage der Angriffspunkte der Kräfte abhängig. Wir zerlegen das obige Integral in drei Integrale:

$$\int [\Sigma K_x']\,dt' = \int\limits_{\frac{t_1}{\beta} - \frac{v}{c^2}x'}^{\frac{t_1}{\beta}} + \int\limits_{\frac{t_1}{\beta}}^{\frac{t_2}{\beta}} + \int\limits_{\frac{t_2}{\beta}}^{\frac{t_2}{\beta} - \frac{vx'}{c^2}}.$$

Das zweite dieser Integrale verschwindet, weil es konstante Zeitgrenzen hat. Wenn ferner die Kräfte K_x' beliebig rasch veränderlich sind, können wir die beiden anderen Integrale nicht auswerten; dann können wir bei Anwendung der hier benutzten Grundlagen von einer Energie bzw. Bewegungsgröße des Systems überhaupt nicht reden.[1]) Falls sich aber jene Kräfte in Zeiten von der Größenordnung $\frac{vx'}{c^2}$ sehr wenig ändern; so können wir setzen:

$$\int\limits_{t_1 - \frac{vx'}{c^2}}^{\frac{t_1}{\beta}} (\Sigma K_x')\,dt' = \Sigma K_x' \int\limits_{\frac{t_1}{\beta} - \frac{vx'}{c^2}}^{\frac{t_1}{\beta}} dt' = \frac{v}{c^2}\,\Sigma x'\,K_x'.$$

Nachdem das dritte Integral entsprechend ausgewertet ist, erhält man

$$\int (\Sigma K_x')\,dt' = -\,d\left\{\frac{v}{c^2}\,\Sigma x'\,K_x'\right\}.$$

Nun ist die Berechnung der Energie und der Bewegungsgröße aus den Gleichungen (16) und (18) ohne Schwierigkeit auszuführen. Man erhält

$$E = \left(\mu + \frac{E_0}{c^2}\right)\frac{c^2}{\sqrt{1 - \frac{q^2}{c^2}}} - \frac{\frac{q^2}{c^2}}{\sqrt{1 - \frac{q^2}{c^2}}}\,\Sigma(\delta_0\,K_{0\delta}) \qquad (16\mathrm{b})$$

[72] 1) Vergl A. Einstein, Ann. d. Phys. **23**, § 2, 1907.

Einstein, Relativitätsprinzip u. die aus demselben gezog. Folgerungen. 447

$$q = \frac{q}{\sqrt{1 - \dfrac{q^2}{c^2}}} \left(\mu + \frac{E_0 - \Sigma\,(\delta_0\,K_{0\delta})}{c^2} \right), \tag{18b}$$ [73]

wobei $K_{0\delta}$ die in die Bewegungsrichtung fallende Komponente einer auf ein mitbewegtes Bezugssystem bezogenen Kraft, δ_0 den in demselben System gemessenen Abstand des Angriffspunktes jener Kraft von einer zur Bewegungsrichtung senkrechten Ebene bedeutet.

Besteht, wie wir im folgenden annehmen wollen, die äußere Kraft in einem von der Richtung unabhängigen, überall auf die Oberfläche des Systems senkrecht wirkenden Druck p_0, so ist im speziellen

$$\Sigma\,(\delta_0\,K_{0\delta}) = -\,p_0\,V_0, \tag{19}$$

wobei V_0 das auf ein mitbewegtes Bezugssystem bezogene Volumen des Systems ist. Die Gleichungen (16b) und (18b) nehmen dann die Form an [74]

$$E = \left(\mu + \frac{E_0}{c^2} \right) \frac{c^2}{\sqrt{1 - \dfrac{q^2}{c^2}}} + \frac{\dfrac{q^2}{c^2}}{\sqrt{1 - \dfrac{q^2}{c^2}}}\, p_0\,V_0 \tag{16c}$$

$$G = \frac{q}{\sqrt{1 - \dfrac{q^2}{c^2}}} \left(\mu + \frac{E_0 + p_0\,V_0}{c^2} \right). \tag{18c}$$

§ 13. Volumen und Druck eines bewegten Systems. [75]
Bewegungsgleichungen.

Wir haben uns zur Bestimmung des Zustandes des betrachteten Systems der Größen E_0, p_0, V_0 bedient, welche mit Bezug auf ein mit dem physikalischen System bewegtes Bezugssystem definiert sind. Wir können uns aber statt der genannten auch der entsprechenden Größen bedienen, welche mit Bezug auf dasselbe Bezugssystem definiert sind, wie die Bewegungsgröße G. Zu diesem Zweck müssen wir untersuchen, wie sich Volumen und Druck bei Einführung eines neuen Bezugssystems ändern.

Ein Körper ruhe in bezug auf das Bezugssystem S'. V' sei sein Volumen in bezug auf S', V sein Volumen in bezug auf S. Aus Gleichungen (2) folgt unmittelbar

$$\int dx \cdot dy \cdot dz = \sqrt{1 - \frac{v^2}{c^2}} \int dx' \cdot dy' \cdot dz'$$

oder

$$V = \sqrt{1 - \frac{v^2}{c^2}} \cdot V'.$$

Ersetzt man gemäß der von uns benutzten Bezeichnungsweise V' durch V^0 und v durch q, so hat man

$$V = \sqrt{1 - \frac{q^2}{c^2}} \cdot V_0. \qquad (20)$$

Um ferner die Transformationsgleichung für die Druckkräfte zu ermitteln, müssen wir von den Transformationsgleichungen ausgehen, welche für Kräfte überhaupt gelten. Da wir ferner in § 8 die bewegenden Kräfte so definiert haben, daß sie durch die Kraftwirkungen elektromagnetischer Felder auf elektrische Massen ersetzt werden können, können wir uns hier darauf beschränken, die Transformationsgleichungen für letztere aufzusuchen.[1]

Die Elektrizitätsmenge ε ruhe in bezug auf S'. Die auf dieselbe wirkende Kraft ist gemäß den Gleichungen (12) gegeben durch die Gleichungen:

$$K_x = \varepsilon X \qquad\qquad K_x' = \varepsilon X'$$
$$K_y = \varepsilon\left(Y - \frac{v}{c}N\right) \qquad K_y' = \varepsilon Y'$$
$$K_z = \varepsilon\left(Z + \frac{v}{c}M\right) \qquad K_z' = \varepsilon Z'.$$

Aus diesen Gleichungen und den Gleichungen (7a) folgt:

$$\left.\begin{aligned} K_x' &= K_x \\ K_y' &= \beta \cdot K_y \\ K_z' &= \beta \cdot K_z \end{aligned}\right\} \qquad (21)$$

Nach diesen Gleichungen lassen sich Kräfte berechnen, wenn sie in bezug auf ein mitbewegtes Bezugssystem bekannt sind.

Wir betrachten nun eine auf das relativ zu S' ruhende Flächenelement s' wirkende Druckkraft

$$K_x' = p' \, s' \cdot \cos l' = p' \cdot s_x'$$
$$K_y' = p' \, s' \cdot \cos m' = p' \cdot s_y'$$
$$K_z' = p' \, s' \cdot \cos n' = p' \cdot s_z',$$

wobei l', m', n' die Richtungscosinus der (nach dem Innern des Körpers gerichteten) Normale, s_y', s_y', s_z' die Projektionen von s' bedeuten. Aus den Gleichungen (2) folgt, daß

---

1) Durch diesen Umstand wird auch das in den vorhergehenden Untersuchungen benutzte Verfahren gerechtfertigt, welches darin bestand, daß wir einzig Wechselwirkung rein elektromagnetischer Art zwischen dem betrachteten System und seiner Umgebung einführten. Die Resultate gelten ganz allgemein.

Einstein, Relativitätsprinzip u. die aus demselben gezog. Folgerungen. 449

$$s_x{}' = s_x$$
$$s_y{}' = \beta \cdot s_y$$
$$s_z{}' = \beta \cdot s_z \,,$$

wobei s_x, s_y, s_z die Projektionen des Flächenelements in bezug auf S sind. Für die Komponenten K_x, K_y, K_z der betrachteten Druckkraft in bezug auf S erhält man also aus den letzten drei Gleichungssystemen

$$K_x = K_x{}' = p' \cdot s_x{}' = p' \cdot s_x = p' \cdot s \cos l$$
$$K_y = \frac{1}{\beta} K_y{}' = \frac{1}{\beta} p' s_y{}' = p' \cdot s_y = p' \cdot s \cdot \cos m$$
$$K_z = \frac{1}{\beta} K_z{}' = \frac{1}{\beta} p' s_z{}' = p' \cdot s_z = p' \cdot s \cdot \cos n \,,$$

wobei s die Größe des Flächenelements, l, m, n die Richtungscosinus von dessen Normale in bezug auf S bezeichnen. Wir erhalten also das Resultat, daß der Druck p' in bezug auf das mitbewegte System sich in bezug auf ein anderes Bezugssystem durch einen ebenfalls senkrecht auf das Flächenelement wirkenden Druck von gleicher Größe ersetzen läßt. In der von uns benutzten Bezeichnungweise ist also

$$p = p_0 \,. \tag{22}$$

Die Gleichungen (16c), (20) und (22) setzen uns in den Stand, den Zustand eines physikalischen Systems statt durch die in bezug auf ein mitbewegtes Bezugssystem definierten Größen E_0, V_0, p_0 durch die Größen E, V, p zu bestimmen, welche in bezug auf dasselbe System definiert sind wie die Bewegungsgröße G und die Geschwindigkeit q des Systems. Falls z. B. der Zustand des betrachteten Systems für einen mitbewegten Beobachter durch zwei Variable (V_0 und E_0) vollkommen bestimmt ist, dessen Zustandsgleichung also als eine Beziehung zwischen p_0, V_0 und E_0 aufgefaßt werden kann, kann man mittels der genannten Gleichungen die Zustandsgleichung auf die Form

$$\varphi(q, p, V, E) = 0$$

bringen.

Formt man die Gleichung (18c) in entsprechender Weise um, so erhält man

$$G = q \left\{ \mu + \frac{E + pV}{c^2} \right\}, \tag{18d}$$ [76]

welche Gleichung in Verbindung mit den das Prinzip von der Erhaltung der Bewegungsgröße ausdrückenden Gleichungen

$$\frac{d G_x}{dt} = \Sigma K_x \text{ etc.}$$

die Translationsbewegung des Systems als Ganzes vollkommen be-

450   Einstein, Relativitätsprinzip u. die aus demselben gezog. Folgerungen.

stimmen, wenn außer den Größen ΣK_x etc. auch E, p und V als Funktionen der Zeit bekannt sind, oder wenn statt der letzten drei Funktionen drei ihnen äquivalente Angaben über die Bedingungen vorliegen, unter denen die Bewegung des Systems vor sich gehen soll.

§ 14. Beispiele.

[77]   Das betrachtete System bestehe in elektromagnetischer Strahlung, welche in einen masselosen Hohlkörper eingeschlossen sei, dessen Wandung dem Strahlungsdruck das Gleichgewicht leiste. Wenn keine äußeren Kräfte auf den Hohlkörper wirken, so können wir auf das ganze System (den Hohlkörper inbegriffen) die Gleichungen (16a) und (18a) anwenden. Es ist also:

$$E = \frac{E_0}{\sqrt{1 - \dfrac{q^2}{c^2}}}$$

[78]

$$G = \frac{q}{\sqrt{1 - \dfrac{q^2}{c^2}}} \, E_0 = q \, \frac{E}{c^2} \, ,$$

wobei E_0 die Energie der Strahlung in bezug auf ein mitbewegtes Bezugssystem bedeutet.

Sind dagegen die Wandungen des Hohlkörpers vollkommen biegsam und dehnbar, so daß dem auf sie von innen ausgeübten Strahlungsdruck durch äußere Kräfte, welche von nicht zu dem betrachteten System gehörigen Körpern ausgehen, das Gleichgewicht geleistet werden muß, so sind die Gleichungen (16c) und (18c) anzuwenden, in welche der bekannte Wert des Strahlungsdruckes

[79]

$$p_0 = \frac{1}{3} \, \frac{E_0}{c^2}$$

einzusetzen ist, so daß man erhält:

$$E = \frac{E_0 \left(1 + \dfrac{1}{3} \dfrac{q^2}{c^2} \right)}{\sqrt{1 - \dfrac{q^2}{c^2}}}$$

$$G = \frac{q}{\sqrt{1 - \dfrac{q^2}{c^2}}} \, \frac{\dfrac{4}{3} E_0}{c^2} \, .$$

Wir betrachten ferner den Fall eines elektrisch geladenen masselosen Körpers. Falls äußere Kräfte auf denselben nicht wirken, können

Einstein, Relativitätsprinzip u. die aus demselben gezog. Folgerungen. 451

wir wieder die Formeln (16a) und (18a) anwenden. Bezeichnet E_0 die elektrische Energie in bezug auf ein mitbewegtes Bezugssystem, so hat man

$$E = \frac{E_0}{\sqrt{1 - \dfrac{q^2}{c^2}}}$$

$$G = \frac{q}{\sqrt{1 - \dfrac{q^2}{c^2}}} \frac{\frac{4}{3} E_0}{c^2}.$$

[80]

Von diesen Werten entfällt ein Teil auf das elektromagnetische Feld, der Rest auf den masselosen, von seiten seiner Ladung Kräften unterworfenen Körper.[1])

§ 15. Entropie und Temperatur bewegter Systeme.

Wir haben bisher von den Variabeln, welche den Zustand eines physikalischen Systems bestimmen, nur Druck, Volumen, Energie, Geschwindigkeit und Bewegungsgröße benutzt, von den thermischen Größen aber noch nicht gesprochen. Es geschah dies deshalb, weil es für die Bewegung eines Systems gleichgültig ist, welcher Art die ihm zugeführte Energie ist, so daß wir bisher keine Ursache hatten, zwischen Wärme und mechanischer Arbeit zu unterscheiden. Nun aber wollen wir noch die thermischen Größen einführen.

Der Zustand eines bewegten Systems sei durch die Größen q, V, E vollkommen bestimmt. Für ein solches System haben wir offenbar als zugeführte Wärme dQ die gesamte Energiezunahme zu betrachten abzüglich der vom Drucke geleisteten und der auf Vergrößerung der Bewegungsgröße verwendeten Arbeit, so daß man hat

$$dQ = dE + p\,dV - q\,dQ. \tag{23}$$ [82]

Nachdem so die zugeführte Wärme für ein bewegtes System definiert ist, kann man durch Betrachtung von umkehrbaren Kreisprozessen die absolute Temperatur T und Entropie η des bewegten Systems in derselben Weise einführen, wie dies in den Lehrbüchern der Thermodynamik geschieht. Für umkehrbare Prozesse gilt auch hier die Gleichung

$$dQ = T\,d\eta. \tag{24}$$

Wir haben nun die Gleichungen abzuleiten, die zwischen den Größen dQ, η, T und den auf ein mitbewegtes Bezugssystem bezogenen entsprechenden Größen dQ_0, η_0, T_0 bestehen. Bezüglich der Entropie

---

1) Vgl. A. Einstein, Ann. d. Phys. (4) **23**, 373—379, 1907.   [81]

452   Einstein, Relativitätsprinzip u. die aus demselben gezog. Folgerungen.

wiederhole ich hier eine von Herrn Planck angegebene Überlegung[1]), indem ich bemerke, daß unter dem „gestrichenen" bezw. „ungestrichenen" Bezugssystem das Bezugssystem S' bezw. S zu verstehen ist.

„Wir denken uns den Körper aus einem Zustand, in welchem er für das ungestrichene Bezugssystem ruht, durch irgendeinen reversiblen, adiabatischen Prozeß in einen zweiten Zustand gebracht, in welchem er für das gestrichene Bezugssystem ruht. Bezeichnet man die Entropie des Körpers für das ungestrichene System im Anfangszustand mit η_1, im Endzustand mit η_2, so ist wegen der Reversibilität und Adiabasie $\eta_1 = \eta_2$. Aber auch für das gestrichene Bezugssystem ist der Vorgang reversibel und adiabatisch, also haben wir ebenso $\eta_1' = \eta_2'$."

„Wäre nun η_1' nicht gleich η_1, sondern etwa $\eta_1' > \eta_1$, so würde
[84]   das heißen: Die Entropie eines Körpers ist für das Bezugssystem, für welches er in Bewegung begriffen ist, größer als für dasjenige Bezugssystem, für welches er sich in Ruhe befindet. Dann müßte nach diesem Satze auch $\eta_2 > \eta_2'$ sein; denn im zweiten Zustand ruht der Körper
[85]   für das gestrichene Bezugssystem, während er für das ungestrichene in Bewegung begriffen ist. Diese beiden Ungleichungen widersprechen aber den oben aufgestellten beiden Gleichungen. Ebensowenig kann
[86] $\eta_1' > \eta_1$ sein; folglich ist $\eta_1' = \eta_1$, und allgemein $\eta' = \eta$, d. h. die Entropie des Körpers hängt nicht von der Wahl des Bezugssystems ab."

Bei Anwendung der von uns benutzten Bezeichnungsweise haben wir also zu setzen:

$$\eta = \eta_0 . \tag{25}$$

Führen wir ferner auf der rechten Seite der Gleichung (23) mittels der Gleichungen (16c), (18c), (20) und (22) die Größen E_0, p_0 und V_0 ein, so erhalten wir

$$dQ = \sqrt{1 - \frac{q^2}{c^2}}\,(dE_0 + p_0\,dV_0)$$

oder

$$dQ = dQ_0 \cdot \sqrt{1 - \frac{q^2}{c^2}} . \tag{26}$$

Da ferner gemäß (24) die beiden Gleichungen

$$dQ = T\,d\eta$$
[87] $$dQ_0 = T\,d\eta_0$$

gelten, so erhält man endlich mit Rücksicht auf (25) und (26)

[83]   1) M. Planck, Zur Dynamik bewegter Systeme. Sitzungsber. d. kgl. Preuß. Akad. d. Wissensch. 1907.

Einstein, Relativitätsprinzip u. die aus demselben gezog. Folgerungen. 453

$$\frac{T}{T_0} = \sqrt{1 - \frac{q^2}{c^2}}.$$ (27) [88]

Die Temperatur eines bewegten Systems ist also in bezug auf ein relativ zu ihm bewegtes Bezugssystem stets kleiner als in bezug auf ein relativ zu ihm ruhendes Bezugssystem.

§ 16. Dynamik der Systeme und Prinzip der kleinsten Wirkung.

Herr Planck geht in seiner Abhandlung „Zur Dynamik bewegter Systeme" vom Prinzip der kleinsten Wirkung (und von den Transformationsgleichungen für Druck und Temperatur der Hohlraumstrahlung) aus[1]) und gelangt zu Resultaten, mit welchen die hier entwickelten übereinstimmen. Es erhebt sich daher die Frage, wie die Grundlagen seiner und der vorliegenden Untersuchung zusammenhängen.

Wir sind ausgegangen vom Energieprinzip und vom Prinzip von der Erhaltung der Bewegungsgröße. Nennen wir F_x, F_y, F_z die Komponenten der Resultierenden der auf das System wirkenden Kräfte, so können wir die von uns benutzten Prinzipien für umkehrbare Prozesse und ein System, dessen Zustand durch die Variabeln q, V, T bestimmt ist, so formulieren:

$$d E = F_x dx + F_y dy + F_z dz - p dV + T dS$$ (28) [90]

$$F_x = \frac{d G_x}{d t} \text{ etc.}$$ (29)

Aus diesen Gleichungen erhält man, wenn man beachtet, daß

$$F_x dx = F_x \dot{x} dt = \dot{x} dG = d(\dot{x} G_x) - G_x d\dot{x} \text{ etc.}$$ [91]

und

$$T d\eta = d(T\eta) - \eta dT,$$

die Beziehung

$$d(-E + T\eta + q G) = G_x d\dot{x} + G_y d\dot{y} + G_z d\dot{z} + p dV + \eta dT.$$

Da auch die rechte Seite dieser Gleichung ein vollständiges Differential sein muß, so folgt unter Berücksichtigung von (29):

$$\frac{d}{dt}\left(\frac{\partial H}{\partial \dot{x}}\right) = F_x \qquad \frac{d}{dt}\left(\frac{\partial H}{\partial \dot{y}}\right) = F_y \qquad \frac{d}{dt}\left(\frac{\partial H}{\partial \dot{z}}\right) = F_z$$

$$\frac{\partial H}{\partial V} = p \qquad \frac{\partial H}{\partial T} = \eta.$$

Dies sind aber die mittels des Prinzips der kleinsten Wirkung ableitbaren Gleichungen, von denen Herr Planck ausgegangen ist. [92]

---

1) M. Planck, Zur Dynamik bewegter Systeme. Sitzungsber d. kgl. [89] Preuß. Akad. d. Wissensch. 1907.

454   Einstein, Relativitätsprinzip u. die aus demselben gezog. Folgerungen.

V. Relativitätsprinzip und Gravitation.

§ 17. Beschleunigtes Bezugssystem und Gravitationsfeld.

Bisher haben wir das Prinzip der Relativität, d. h. die Voraussetzung der Unabhängigkeit der Naturgesetze vom Bewegungszustande des Bezugssystems, nur auf beschleunigungsfreie Bezugssysteme angewendet. Ist es denkbar, daß das Prinzip der Relativität auch für Systeme gilt, welche relativ zueinander beschleunigt sind?

Es ist zwar hier nicht der Ort für die eingehende Behandlung dieser Frage. Da sich diese aber jedem aufdrängen muß, der die bisherigen Anwendungen des Relativitätsprinzips verfolgt hat, will ich es nicht unterlassen, zu der Frage hier Stellung zu nehmen.

Wir betrachten zwei Bewegungssysteme Σ_1 und Σ_2. Σ_1 sei in Richtung seiner X-Achse beschleunigt, und es sei γ die (zeitlich kon-
[93]   stante) Größe dieser Beschleunigung. Σ_2 sei ruhend; es befinde sich aber in einem homogenen Gravitationsfelde, das allen Gegenständen die Beschleunigung $-\gamma$ in Richtung der X-Achse erteilt.

[94]   Soweit wir wissen, unterscheiden sich die physikalischen Gesetze in bezug auf Σ_1 nicht von denjenigen in bezug auf Σ_2; es liegt dies daran, daß alle Körper im Gravitationsfelde gleich beschleunigt werden. Wir haben daher bei dem gegenwärtigen Stande unserer Erfahrung keinen Anlaß zu der Annahme, daß sich die Systeme Σ_1 und Σ_2 in irgendeiner Beziehung voneinander unterscheiden, und wollen daher im folgenden die völlige physikalische Gleichwertigkeit von Gravitationsfeld und entsprechender Beschleunigung des Bezugssystems annehmen.

Diese Annahme erweitert das Prinzip der Relativität auf den Fall der gleichförmig beschleunigten Translationsbewegung des Bezugssystems. Der heuristische Wert der Annahme liegt darin, daß sie ein homogenes Gravitationsfeld durch ein gleichförmig beschleunigtes Bezugssystem zu ersetzen gestattet, welch letzterer Fall bis zu einem gewissen Grade der theoretischen Behandlung zugänglich ist.

§ 18. Raum und Zeit in einem gleichförmig beschleunigten Bezugssystem.

Wir betrachten zunächst einen Körper, dessen einzelne materielle Punkte zu einer bestimmten Zeit t des beschleunigungsfreien Bezugssystems S, relativ zu S keine Geschwindigkeit, jedoch eine gewisse Beschleunigung besitzen. Was für einen Einfluß hat diese Beschleunigung γ auf die Gestalt des Körpers in bezug auf S?

Falls ein derartiger Einfluß vorhanden ist, wird er in einer Dila-

Einstein, Relativitätsprinzip u. die aus demselben gezog. Folgerungen.　455

tation nach konstantem Verhältnis in der Beschleunigungsrichtung sowie eventuell in den beiden dazu senkrechten Richtungen bestehen;
denn ein Einfluß anderer Art ist aus Symmetriegründen ausgeschlossen.
Jene von der Beschleunigung herrührenden Dilatationen müssen (falls
solche überhaupt existieren) gerade Funktionen von γ sein; sie können
also vernachlässigt werden, wenn man sich auf den Fall beschränkt, daß
γ so klein ist, daß Glieder zweiten und höheren Grades in γ vernachlässigt werden dürfen. Da wir uns im folgenden auf diesen Fall beschränken wollen, haben wir also einen Einfluß der Beschleunigung auf
die Gestalt eines Körpers nicht anzunehmen.

Wir betrachten nun ein relativ zu dem beschleunigungsfreien Bezugssystem S in Richtung von dessen X-Achse gleichförmig beschleunigtes Bezugssystem Σ. Uhren besw. Maßstab von Σ seien, ruhend
untersucht, gleich den Uhren bezw. dem Maßstab von S. Der Koordinatenanfang von Σ bewege sich auf der X-Achse von S, und die
Achsen von Σ seien denen von S dauernd parallel. Es existiert in
jedem Augenblick ein unbeschleunigtes Bezugssystem S', dessen Koordinatenachsen in dem betreffenden Augenblick (zu einer bestimmten
Zeit t' von S' mit den Koordinatenachsen von Σ zusammen fallen. Besitzt ein Punktereignis, welches zu dieser Zeit t' stattfindet, in bezug
auf Σ die Koordinaten ξ, η, ζ, so ist

$$\left.\begin{array}{l} x' = \xi \\ y' = \eta \\ z' = \zeta \end{array}\right\},$$

weil ein Einfluß der Beschleunigung auf die Gestalt der zur Messung
von ξ, η, ζ benutzten Meßkörper nach dem Obigen nicht anzunehmen
ist. Wir wollen uns ferner vorstellen, daß die Uhren von Σ zu dieser
Zeit t' von S' so gerichtet werden, daß ihre Angabe in diesem
Augenblick gleich t' ist. Wie steht es mit dem Gang der Uhren in
dem nächsten Zeitteilchen τ?

Zunächst haben wir zu berücksichtigen, daß ein spezifischer Einfluß der Beschleunigung auf den Gang der Uhren von Σ nicht in
Betracht fällt, da dieser von der Ordnung γ^2 sein müßte. Da ferner
der Einfluß der während τ erlangten Geschwindigkeit auf den Gang
der Uhren zu vernachlässigen ist, und ebenso die während der Zeit τ
von den Uhren relativ zu denen von S' zurückgelegten Wege von der
Ordnung τ^2, also zu vernachlässigen sind, so sind für das Zeitelement τ
die Angaben der Uhren von Σ durch die Angaben der Uhren von S'
vollkommen nutzbar.　　　　　　　　　　　　　　　　　　　　　　[95]

Aus dem Vorangehenden folgt, daß sich das Licht im Vakuum

456   **Einstein**, Relativitätsprinzip u. die aus demselben gezog. Folgerungen.

relativ zu Σ im Zeitelement τ mit der universellen Geschwindigkeit c fortpflanzt, falls wir die Gleichzeitigkeit in dem relativ zu Σ momentan ruhenden System S' definieren, und zur Zeit- bzw. Längenmessung Uhren bzw. Maßstäbe verwenden, welche jenen gleich sind, die in unbeschleunigten Systemen zur Ausmessung von Zeit und Raum benutzt werden. Das Prinzip von der Konstanz der Lichtgeschwindigkeit läßt sich also auch hier zur Definition der Gleichzeitigkeit verwenden, falls man sich auf sehr kleine Lichtwege beschränkt.

Wir denken uns nun die Uhren von Σ in der angegebenen Weise zu derjenigen Zeit $t = 0$ von S gerichtet, in welcher Σ relativ zu S momentan ruht. Der Inbegriff der Angaben der so gerichteten Uhren von [96] Σ werde die „Ortszeit" σ des Systems Σ genannt. Die physikalische Bedeutung der Ortszeit σ ist, wie man unmittelbar erkennt, die folgende. Bedient man sich zur zeitlichen Wertung der in den einzelnen Raumelementen von Σ stattfindenden Vorgänge jener Ortszeit σ, so können die Gesetze, denen jene Vorgänge gehorchen, nicht von der Lage des betreffenden Raumelementes, d. h. von dessen Koordinaten, abhängen, [97] falls man sich in den verschiedenen Raumelementen nicht nur gleichen Uhren, sondern auch sonst gleicher Meßmittel bedient.

Dagegen dürfen wir nicht die Lokalzeit σ als die „Zeit" von Σ schlechthin bezeichnen, und zwar deshalb, weil zwei in verschiedenen Punkten von Σ stattfindende Punktereignisse nicht dann im Sinne unserer obigen Definition gleichzeitig sind, wenn ihre Lokalzeiten σ einander gleich sind. Da nämlich irgend zwei Uhren von Σ zur Zeit $t = 0$ in bezug auf S synchron sind und den nämlichen Bewegungen unterworfen werden, so bleiben sie dauernd in bezug auf S synchron. Aus diesem Grunde laufen sie aber gemäß § 4 in bezug auf ein momentan relativ zu Σ ruhendes, in bezug auf S bewegtes Bezugssystem S' nicht synchron, also gemäß unserer Definition auch nicht in bezug auf Σ.

Wir definieren nun die „Zeit" τ des Systems Σ als den Inbegriff derjenigen Angaben der im Koordinatenanfangspunkt von Σ befindlichen Uhr, welche mit den zeitlich zu wertenden Ereignissen im Sinne der obigen Definition gleichzeitig sind.[1]

Wir wollen jetzt die Beziehung aufsuchen, welche zwischen der Zeit τ und der Ortszeit σ eines Punktereignisses besteht. Aus der ersten der Gleichungen (1) folgt, daß zwei Ereignisse in bezug auf S', also auch in bezug auf Σ gleichzeitig sind, wenn

---

1) Das Zeichen „τ" ist also hier in einem anderen Sinne verwendet als oben.

Einstein, Relativitätsprinzip u. die aus demselben gezog. Folgerungen. 457

$$t_1 - \frac{v}{c^2} x_1 = t_2 - \frac{v}{c^2} x_2,$$

wobei die Indizes die Zugehörigkeit zu dem einen bzw. andern Punkt-ereignis andeuten soll. Wir beschränken uns nun zunächst auf die Betrachtung so kurzer Zeiten[1]), daß alle Glieder, welche die zweite oder eine höhere Potenz von τ oder v enthalten, weggelassen werden dürfen; dann haben wir mit Rücksicht auf (1) und (29) zu setzen: [98]

$$x_2 - x_1 = x_2' - x_1' = \xi_2 - \xi_1$$
$$t_1 = \sigma_1 \qquad t_2 = \sigma_2$$
$$v = \gamma t = \gamma \tau,$$

[99]

so daß wir aus obiger Gleichung erhalten:

$$\sigma_2 - \sigma_1 = \frac{\gamma \tau}{c^2} (\xi_2 - \xi_1).$$

Verlegen wir das erste Punktereignis in den Koordinatenanfang, so daß $\sigma_1 = \tau$ und $\xi_1 = 0$, so erhalten wir unter Weglassung des Index für das zweite Punktereignis

$$\sigma = \tau \left(1 + \frac{\gamma \xi}{c^2} \right). \tag{30}$$

Diese Gleichung gilt zunächst, wenn τ und ξ unterhalb gewisser Grenzen liegen. Sie gilt offenbar für beliebig große τ, falls die Beschleunigung γ mit Bezug auf Σ konstant ist, weil die Beziehung zwischen σ und τ dann linear sein muß. Für beliebig große ξ gilt Gleichung (30) nicht. Daraus, daß die Wahl des Koordinatenanfangs-punktes auf die Relation nicht von Einfluß sein darf, schließt man nämlich, daß die Gleichung (30) genau genommen durch die Gleichung

$$\sigma = \tau\, e^{\frac{\gamma \xi}{c^2}}$$

ersetzt werden müßte. Wir wollen jedoch an der Formel (30) fest-halten.

Gleichung (30) ist nach § 17 auch auf ein Koordinatensytem an-zuwenden, in dem ein homogenes Schwerfeld wirkt. In diesem Falle haben wir $\Phi = \gamma \xi$ zu setzen, wobei Φ das Potential der Schwerkraft bedeutet, so daß wir erhalten ·

$$\sigma = \tau \left(1 + \frac{\Phi}{c^2} \right) \tag{30a}$$

Wir haben zweierlei Zeiten für Σ definiert. Welcher von beiden Definitionen haben wir uns für die verschiedenen Fälle zu bedienen? Nehmen wir an, es existiere an zwei Orten verschiedenen Gravitations-

---

1) Hierdurch wird gemäß (1) auch eine gewisse Beschränkung in bezug auf die Werte von $\xi = x'$ angenommen.

458 **Einstein, Relativitätsprinzip u. die aus demselben gezog. Folgerungen.**

potentials $(\gamma\ \xi)$ je ein physikalisches System, und wir wollen ihre physikalischen Größen vergleichen. Zu diesem Zwecke werden wir wohl am natürlichsten folgendermaßen vorgehen: Wir begeben uns mit unseren Meßmitteln zuerst zu dem ersten physikalischen System und führen dort unsere Messungen aus; hierauf begeben wir uns samt unsern Meßmitteln nach dem zweiten System, um hier die gleichen Messungen auszuführen. Ergeben die Messungen da und dort die gleichen Resultate, so werden wir die beiden physikalischen Systeme als „gleich" bezeichnen. Unter den genannten Meßmitteln befindet sich eine Uhr, mit welcher wir Lokalzeiten σ messen. Daraus folgt, daß wir uns zum Definieren der physikalischen Größen an einem Orte des Schwerfeldes naturgemäß der Zeit σ bedienen.

Handelt es sich aber um ein Phänomen, bei welchem an Orten verschiedenen Gravitationspotentials befindliche Gegenstände gleichzeitig berücksichtigt werden müssen, so haben wir uns bei den Gliedern, in welchen die Zeit explizite (d. h. nicht nur bei der Definition physikalischer Größen) vorkommt, der Zeit τ zu bedienen, da sonst die Gleichzeitigkeit der Ereignisse nicht durch die Gleichheit der Zeitwerte beider Ereignisse ausgedrückt würde. Da bei der Definition der Zeit τ nicht ein willkürlich gewählter Zeitpunkt, wohl aber eine an einem willkürlich gewählten Orte befindliche Uhr benutzt ist, so können bei Benutzung der Zeit τ die Naturgesetze nicht mit der Zeit, wohl aber mit dem Orte variieren.

§ 19. Einfluß des Gravitationsfeldes auf Uhren.

Befindet sich in einem Punkte P vom Gravitationspotential Φ eine Uhr, welche die Ortszeit angibt, so sind gemäß (30a) ihre Angaben $\left(1+\dfrac{\Phi}{c^2}\right)$ mal größer als die Zeit τ, d. h. sie läuft $\left(1+\dfrac{\Phi}{c^2}\right)$ mal schneller als eine gleich beschaffene, im Koordinatenanfangspunkt befindliche Uhr. Ein irgendwo im Raume befindlicher Beobachter nehme die Angaben dieser beiden Uhren irgendwie, z. B. auf optischem Wege, wahr. Da die Zeit $\Delta\tau$, welche zwischen dem Zeitpunkt einer Angabe einer der Uhren und der Wahrnehmung dieser Angabe durch den Beobachter verstreicht, von τ unabhängig ist, so läuft die Uhr in P für einen irgendwo im Raume befindlichen Beobachter $\left(1+\dfrac{\Phi}{c^2}\right)$ mal schneller als die Uhr im Koordinatenanfangspunkt. In diesem Sinne können wir sagen, daß der in der Uhr sich abspielende Vorgang — und allgemeiner jeder physikalische Prozeß — desto schneller abläuft,

Einstein, Relativitätsprinzip u. die aus demselben gezog. Folgerungen. 459

je größer das Gravitationspotential des Ortes ist, an dem er sich abspielt.

Es gibt nun „Uhren", welche an Orten verschiedenen Gravitationspotentials vorhanden sind und deren Ganggeschwindigkeit sehr genau kontrolliert werden kann; es sind dies die Erzeuger der Spektrallinien. Aus dem Obigen schließt man[1]), daß von der Sonnenoberfläche kommendes Licht, welches von einem solchen Erzeuger herrührt, eine um etwa zwei Millionstel größere Wellenlänge besitzt, als das von gleichen Stoffen auf der Erde erzeugte Licht.

[100]

§ 20. Einfluß der Schwere auf die elektromagnetischen Vorgänge.

Beziehen wir einen elektromagnetischen Vorgang in einem Zeitpunkt auf ein beschleunigungsfreies Bezugssystem S', das momentan relativ zu dem wie oben beschleunigten Bezugssystem Σ ruht, so gelten gemäß (5) und (6) die Gleichungen

$$\frac{1}{c}\left(\varrho' u_x' + \frac{\partial X'}{\partial t'}\right) = \frac{\partial N'}{\partial y'} - \frac{\partial M'}{\partial z'} \quad \text{etc.}$$

und

$$\frac{1}{c}\frac{\partial L'}{\partial t'} = \frac{\partial Y'}{\partial z'} - \frac{\partial Z'}{\partial y'} \quad \text{etc.}$$

Nach dem Obigen können wir die auf S' bezogenen Größen ϱ', u', X', L', x' etc. den entsprechenden auf Σ bezogenen Größen ϱ, u, X, L, ξ etc. ohne weiteres gleichsetzen, falls wir uns auf eine unendlich kurze Zeit beschränken[2]), welche der Zeit der relativen Ruhe von S' und Σ unendlich nahe liegt. Ferner haben wir t' durch die Lokalzeit σ zu ersetzen. Dagegen dürfen wir nicht einfach

$$\frac{\partial}{\partial t'} = \frac{\partial}{\partial \sigma}$$

setzen, und zwar deshalb, weil ein in bezug auf Σ ruhender Punkt, auf den sich die auf Σ transformierten Gleichungen beziehen sollen, relativ zu S' während des Zeitteilchens $dt' = d\sigma$ seine Geschwindigkeit ändert, welcher Änderung gemäß den Gleichungen (7a) und (7b) eine zeitliche Änderung der auf Σ bezogenen Feldkomponenten entspricht. Wir haben daher zu setzen:

---

1) Indem man voraussetzt, daß Gleichung (30a) auch für ein nichthomogenes Gravitationsfeld gelte.

2) Diese Beschränkung beeinträchtigt den Gültigkeitsbereich unserer Resultate nicht, da die abzuleitenden Gesetze der Natur der Sache nach von der Zeit nicht abhängen können.

31*

460 Einstein, Relativitätsprinzip u. die aus demselben gezog. Folgerungen.

$$\frac{\partial X'}{\partial t'} = \frac{\partial X}{\partial \sigma} \qquad\qquad \frac{\partial L'}{\partial t'} = \frac{\partial L}{\partial \sigma}$$

$$\frac{\partial Y'}{\partial t'} = \frac{\partial Y}{\partial \sigma} + \frac{\gamma}{c} N \qquad\qquad \frac{\partial M'}{\partial t'} = \frac{\partial M}{\partial \sigma} - \frac{\gamma}{c} Z$$

$$\frac{\partial Z'}{\partial t'} = \frac{\partial Z}{\partial \sigma} - \frac{\gamma}{c} M \qquad\qquad \frac{\partial N'}{\partial t'} = \frac{\partial N}{\partial \sigma} + \frac{\gamma}{c} Y.$$

Die auf Σ bezogenen elektromagnetischen Gleichungen lauten also zunächst

$$\frac{1}{c}\left(\varrho\, u_\xi + \frac{\partial X}{\partial \sigma}\right) = \frac{\partial N}{\partial \eta} - \frac{\partial M}{\partial \zeta}$$

$$\frac{1}{c}\left(\varrho\, u_\eta + \frac{\partial Y}{\partial \sigma} + \frac{\gamma}{c} N\right) = \frac{\partial L}{\partial \zeta} - \frac{\partial N}{\partial \xi}$$

$$\frac{1}{c}\left(\varrho\, \mathbf{u}_\xi + \frac{\partial Z}{\partial \sigma} - \frac{\gamma}{c} M\right) = \frac{\partial M}{\partial \xi} - \frac{\partial L}{\partial \eta}$$

$$\frac{1}{c}\frac{\partial L}{\partial \sigma} = \frac{\partial Y}{\partial \zeta} - \frac{\partial Z}{\partial \eta}$$

$$\frac{1}{c}\left(\frac{\partial M}{\partial \sigma} - \frac{\gamma}{c} Z\right) = \frac{\partial Z}{\partial \xi} - \frac{\partial X}{\partial \zeta}$$

$$\frac{1}{c}\left(\frac{\partial N}{\partial \sigma} + \frac{\gamma}{c} Y\right) = \frac{\partial X}{\partial \eta} - \frac{\partial Y}{\partial \xi}$$

Diese Gleichungen multiplizieren wir mit $\left(1 + \dfrac{\gamma\xi}{c^2}\right)$ und setzen zur Abkürzung

$$X^* = X\left(1 + \frac{\gamma\xi}{c^2}\right), \quad Y^* = Y\left(1 + \frac{\gamma\xi}{c^2}\right) \text{ etc.}$$

$$\varrho^* = \varrho\left(1 + \frac{\gamma\xi}{c^2}\right)$$

Wir erhalten dann, indem wir Glieder zweiten Grades in γ vernachlässigen, die Gleichungen:

$$\left.\begin{aligned} \frac{1}{c}\left(\varrho^* u_\xi + \frac{\partial X^*}{\partial \sigma}\right) &= \frac{\partial N^*}{\partial \eta} - \frac{\partial M^*}{\partial \zeta} \\[1ex] \frac{1}{c}\left(\varrho^* u_\eta + \frac{\partial Y^*}{\partial \sigma}\right) &= \frac{\partial L^*}{\partial \zeta} - \frac{\partial N^*}{\partial \xi} \\[1ex] \frac{1}{c}\left(\varrho^* u_\zeta + \frac{\partial Z^*}{\partial \sigma}\right) &= \frac{\partial M^*}{\partial \xi} - \frac{\partial L^*}{\partial \eta} \end{aligned}\right\} \tag{31 a}$$

$$\left.\begin{aligned} \frac{1}{c}\frac{\partial L^*}{\partial \sigma} &= \frac{\partial Y^*}{\partial \zeta} - \frac{\partial Z^*}{\partial \eta} \\[1ex] \frac{1}{c}\frac{\partial M^*}{\partial \sigma} &= \frac{\partial Z^*}{\partial \xi} - \frac{\partial X^*}{\partial \zeta} \\[1ex] \frac{1}{c}\frac{\partial N^*}{\partial \sigma} &= \frac{\partial X^*}{\partial \eta} - \frac{\partial Y^*}{\partial \xi} \end{aligned}\right\} \tag{32 a}$$

Einstein, Relativitätsprinzip u. die aus demselben gezog. Folgerungen. 461

Aus diesen Gleichungen ersieht man zunächst, wie das Gravitationsfeld die statischen und stationären Erscheinungen beeinflußt. Die geltenden Gesetzmäßigkeiten sind dieselben wie im gravitationsfreien Felde; nur sind die Feldkomponenten X etc. durch $X\left(1 + \dfrac{\gamma\xi}{c^2}\right)$ etc. und ϱ durch $\varrho\left(1 + \dfrac{\gamma\xi}{c^2}\right)$ ersetzt.

Um ferner den Verlauf nichtstationärer Zustände zu übersehen, bedienen wir uns der Zeit τ sowohl bei den nach der Zeit differenzierten Gliedern als auch für die Definition der Geschwindigkeit der Elektrizität, d. h. wir setzen gemäß (30)

$$\frac{\partial}{\partial\tau} = \left(1 + \frac{\gamma\xi}{e^2}\right)\frac{\partial}{\partial\tau} \qquad [101]$$

und

$$w_\xi = \left(1 + \frac{\gamma\xi}{c^2}\right). \qquad [102]$$

Wir erhalten so

$$\frac{1}{c\left(1 + \dfrac{\gamma\xi}{c^2}\right)}\left(\varrho^* w_\xi + \frac{\partial X^*}{\partial\tau}\right) = \frac{\partial N^*}{\partial\eta} - \frac{\partial M^*}{\partial\zeta} \text{ etc.} \qquad \textbf{(31 b)}$$

und

$$\frac{1}{c\left(1 + \dfrac{\gamma\xi}{c^2}\right)}\frac{\partial L^*}{\partial\tau} = \frac{\partial Y^*}{\partial\zeta} = \frac{\partial Z^*}{\partial\eta} \text{ etc.} \qquad \textbf{(32 b)} \quad [103]$$

Auch diese Gleichungen sind von derselben Form wie die entsprechenden des beschleunigungs- bzw. gravitationsfreien Raumes; hier tritt aber an die Stelle von c der Wert

$$c\left(1 + \frac{\gamma\xi}{c^2}\right) = c\left(1 + \frac{\Phi}{c^2}\right).$$

Es folgt hieraus, daß die Lichtstrahlen, welche nicht in der ξ-Achse verlaufen, durch das Gravitationsfeld gekrümmt werden; die Richtungsänderung beträgt, wie leicht zu ersehen, pro Zentimeter Lichtweg $\dfrac{\gamma}{c^2}\sin\varphi$, wobei φ den Winkel zwischen der Richtung der Schwerkraft und der des Lichtstrahles bedeutet. [104]

Mittels dieser Gleichungen und den aus der Optik ruhender Körper bekannten Gleichungen zwischen Feldstärke und elektrischer Strömung an einem Orte läßt sich der Einfluß des Gravitationsfeldes auf die optischen Erscheinungen bei ruhenden Körpern ermitteln. Es ist hierbei zu berücksichtigen, daß jene Gleichungen aus der Optik ruhender Körper für die Lokalzeit σ gelten. Leider ist der Einfluß des irdischen Schwerefeldes nach unserer Theorie ein so geringer (wegen der Klein-

462   Einstein, Relativitätsprinzip u. die aus demselben gezog. Folgerungen.

[105] heit von $\frac{\gamma x}{c^2}$), daß eine Aussicht auf Vergleichung der Resultate der Theorie mit der Erfahrung nicht besteht.

Multiplizieren wir die Gleichungen (31a) und (32a) der Reihe nach mit $\frac{X^*}{4\pi}\cdots\cdots\frac{N^*}{4\pi}$ und integrieren über den unendlichen Raum, so erhalten wir bei Benutzung unserer früheren Bezeichnungsweise:

[106]
$$\int \left(1 + \frac{\gamma\xi}{c^2}\right)^2 \frac{\varrho}{4\pi}(u\ X + u_\eta\ Y + u\ Z)\,d\omega$$
$$+ \int \left(1 + \frac{\gamma\xi}{c^2}\right)^2 \cdot \frac{1}{8\pi}\frac{\partial}{\partial\sigma}(X^2 + Y^2 \cdots + N^2)\,d\omega = 0.$$

[107] $\frac{\varrho}{4\pi}(u\ X + u_\eta\ Y + u\ Z)$ ist die der Materie pro Volumeneinheit und Einheit der Lokalzeit σ zugeführte Energie η_σ, falls diese Energie mittels an der betreffenden Stelle befindlicher Meßmittel gemessen

[108] wird. Folglich ist gemäß (30) $\eta_\tau = \eta^\sigma\left(1 - \frac{\gamma\xi}{c^2}\right)$ die der Materie pro Volumeneinheit und Einheit der Zeit τ zugeführte (ebenso gemessene) Energie. $\frac{1}{8\pi}(X^2 + Y^2 \cdots + N^2)$ ist die elektromagnetische Energie ε pro Volumeneinheit — ebenso gemessen. Berücksichtigen wir ferner, daß gemäß (30) $\frac{\partial}{\partial\sigma} = \left(1 - \frac{\gamma\xi}{c^2}\right)\frac{\partial}{\partial\tau}$ zu setzen ist, so erhalten wir

$$\int \left(1 + \frac{\gamma\xi}{c^2}\right)\eta_\tau\,d\omega + \frac{d}{d\tau}\left\{\int\left(1 + \frac{\gamma\xi}{c^2}\right)\varepsilon\,d\omega\right\} = 0.$$

Diese Gleichung drückt das Prinzip von der Erhaltung der Energie aus und enthält ein sehr bemerkenswertes Resultat. Eine Energie bzw. eine Energiezufuhr, welche — an Ort und Stelle gemessen — den Wert $E = \varepsilon\,d\omega$ bzw. $E = \eta\,d\omega\,d\tau$ hat, liefert zum Energieintegral außer dem ihrer Größe entsprechenden Wert E noch einen ihrer **Lage** entsprechenden Wert $\frac{E}{c^2}\gamma\xi = \frac{E}{c^2}\Phi$. Jeglicher Energie E kommt also im Gravitationsfelde eine Energie der Lage zu, die ebenso groß ist, wie die Energie der Lage einer „ponderabeln" Masse von der Größe $\frac{E}{c^2}$.

Der im § 11 abgeleitete Satz, daß einer Energiemenge E eine Masse von der Größe $\frac{E}{c^2}$ zukomme, gilt also, falls die im § 17 eingeführte Voraussetzung zutrifft, nicht nur für die **träge**, sondern **auch für die gravitierende Masse**.

(Eingegangen 4. Dezember 1907.)

Published in *Jahrbuch der Radioaktivität und Elektronik* 4 (1907): 411–462. Received 4 December 1907, published 22 January 1908.

[1] *Lorentz 1895*.

[2] See *Lorentz 1895*, § V.

[3] *Michelson and Morley 1887*, which gives the results of a more precise repetition of the experiment reported in *Michelson 1881*. This is Einstein's first known reference to the Michelson-Morley experiment.

[4] See *FitzGerald 1889* and *Lorentz 1892b*.

[5] This may be an allusion to attempts to set up an emission theory of light. For evidence of an attempt by Einstein before 1905, see the editorial note, "Einstein on the Theory of Relativity," pp. 263–264.

[6] *Lorentz 1895*, pp. 49–50, defines the local time t' for a frame of reference moving through the ether by the equation:

$$t' = t - p_x/V^2\, x - p_y/V^2\, y - p_z/V^2\, z,$$

where t is the (universal) time, $\mathbf{p}$ is the velocity vector of the moving frame with respect to the ether frame, and $\mathbf{r} = (x,y,z)$ is the position vector with respect to the ether frame of the point at which the local time is calculated. *Lorentz 1904a* defines the local time by an expression that is formally equivalent to Einstein's expression for the time relative to the moving frame of reference. *Cohn 1904b*, p. 1408, expresses a view similar to Einstein's on the role of the local time.

[7] See *Einstein 1905s* (Doc. 24), *Einstein 1906e* (Doc. 35), and *Einstein 1907h* (Doc. 45) for discussions of the inertia of electromagnetic energy.

[8] See *Cohn 1900, 1902, 1904a, 1904b*. A copy of *Cohn 1904a* is in Einstein's collection of reprints, now in IsReW.

[9] *Lorentz 1904a* and *Einstein 1905r* (Doc. 23) (the reference should be to vol. 17).

[10] *Laue 1907*. A reprint copy, dedicated "To the discoverer of the relativity principle" ("Dem Entdecker des Relativitätsprinzips"), containing a short calculation in Einstein's hand, is in Einstein's collection of reprints, now in IsReW.

[11] *Laub 1907* (the reference should be to vol. 23). In a letter of 4 September 1907, Max Laue informed Einstein that he had discovered Laub's paper after completing work on *Laue 1907*. Since Laub's paper contains several errors, Laue decided not to withdraw his paper, but to inform Laub of his objections.

[12] See *Einstein 1905r* (Doc. 23), pp. 917–920.

[13] *Planck 1906a. Planck 1907a*, cited below by Einstein in connection with this topic, is Planck's second paper on the topic.

[14] *Einstein 1905s* (Doc. 24), *1907h* (Doc. 45), and *Planck 1907a*.

[15] *Mosengeil 1907*. The article, based on Mosengeil's 1906 Berlin doctoral thesis, was prepared for publication by Planck, his thesis supervisor, after Mosengeil's death (see ibid., p. 867).

[16] The question of rigid bodies in the theory of relativity was an unsolved problem at this time (see *Einstein 1907h* [Doc. 45], § 3). Einstein here noted that only rigid motions of solid bodies are required.

[17] *Lorentz 1895*.

[18] See *Fizeau 1851*. For Lorentz's explanation of the dragging coefficient, see *Lorentz 1895*, pp. 96–99.

[19] See note 3.

[20] Einstein here introduced the terms "geometrische Gestalt" and "kinematische Gestalt" to generalize what he had earlier called "the length of the rod" ("die Länge des Stabes") and "the length of the (moving) rod in a system at rest" ("die Länge des (bewegten) Stabes im ruhenden System"), respectively (see *Einstein 1905r* [Doc. 23], p. 896).

[21] The derivation of the transformation equations given in this section differs substantially from that given in *Einstein 1905r* (Doc. 23), § 3.

[22] This rule had been given in *Planck 1907a*, p. 551.

[23] *Stark 1906*.

[24] Einstein discussed this question in more detail in *Einstein 1907e* (Doc. 41).

[25] The "n'_x" in the denominator should be "u'_x."

[26] The c in the denominator of the last term should not be squared.

[27] The discussion in the remainder of this paragraph follows *Einstein 1907h* (Doc. 45), pp. 381–382.

[28] The "λ" should be "l."

[29] The topics that are treated kinematically in this section are not discussed in *Einstein 1905r* (Doc. 23) until after the introduction of Maxwell's equations (see ibid., pp. 910–911).

[30] The "x" in the numerator should be "x'."

[31] The wording of this paragraph is almost

identical to that of the last paragraph on p. 911 of *Einstein 1905r* (Doc. 23).

[32] See the corresponding paragraph on aberration in *Einstein 1905r* (Doc. 23), p. 912, and note 29 to that paper.

[33] *Laue 1907*.

[34] In a reprint copy in the Holton collection of Einstein's reprints (see the Acknowledgments, p. xxxiv), Einstein changed this formula to:

$$G' = \frac{V'}{1 + \frac{\omega'}{V'} \frac{dV'}{d\omega'}}$$

The plus sign in the denominator should be a minus sign.

[35] Einstein here combined the treatment of what, in *Einstein 1905r* (Doc. 23), he called the Maxwell-Hertz equations for empty space (see ibid., § 6) and the Maxwell-Hertz equations with convection currents (see ibid., § 9). This section also includes some material treated in § 7 and § 8 of the 1905 paper.

[36] See *Einstein 1905r* (Doc. 23), pp. 908–909, for the proof.

[37] The discussion in the following two paragraphs corresponds to that in *Einstein 1905r* (Doc. 23), last portion of § 6, pp. 909–910.

[38] The following paragraph proves an assertion in *Einstein 1905r* (Doc. 23), final paragraph of § 9, p. 917.

[39] In the equation for Φ, "w" should be "ω".

[40] The "w'" should be "ω'."

[41] This section presents a modified treatment of the material in § 10 of *Einstein 1905r* (Doc. 23), largely based on the work of Planck (see *Planck 1906a*).

[42] In the first denominator, "x_0'" should be "$\dot{x}_0$." In the second denominator, the parenthesis should be raised to the third power.

[43] This form of the the equations of motion and the corresponding definition of the relativistic force differ from those given in *Einstein 1905r* (Doc. 23), § 10. Einstein followed Planck here (see *Planck 1906a*).

[44] The "x, y, z" should be "$\dot{x}, \dot{y}, \dot{z}$."

[45] In a reprint of this paper (see note 34), Einstein indicated that the square bracket after "Σ" should be deleted, and a square bracket placed after "$-ZM)$." An open parenthesis should also be inserted before "YN." See also *Einstein 1908b* (Doc. 49).

[46] Einstein here followed Planck in the definition of the relativistic momentum of a particle (see *Planck 1906a*).

[47] In this expression for the relativistic Lagrangian, as in the rest of this paragraph, Einstein followed Planck's treatment (see *Planck 1906a*).

[48] In the final term, "x" should be "$\dot{x}$."

[49] Einstein did not introduce the concepts of longitudinal and transverse mass, which he had used in earlier discussions of this topic. See *Einstein 1905r* (Doc. 23), pp. 918–921, and *Einstein 1906g* (Doc. 36).

[50] The c^2 inside the brackets should be placed outside.

[51] *Starke 1903*, e.g., indicates that it was difficult to exceed discharge potentials of 38,000 volts, corresponding to velocities of 1.17×10^{10} cm/sec for cathode rays.

[52] *Kaufmann 1906a*. This paper cites Kaufmann's earlier papers on his experiments, starting with *Kaufmann 1901*.

[53] The drawing of Kaufmann's apparatus on p. 438 is reproduced from *Kaufmann 1906a*, p. 496.

[54] The figure on p. 439 is based on fig. 11, table IV of *Kaufmann 1906a*.

[55] That is, the curve was calculated on the assumption that the deviations are negligibly small compared to the dimensions of the apparatus (see *Kaufmann 1906a*, p. 524).

[56] This is the value calculated by Kaufmann (see *Kaufmann 1906a*, p. 551).

[57] *Planck 1906b, 1907b*.

[58] See *Laub 1910* for a contemporary review of experiments by Kaufmann and others on the dependence of electron mass on velocity. For a discussion of these experiments, see the editorial note, "Einstein on the Theory of Relativity," pp. 270–272.

[59] *Abraham 1902a*.

[60] *Bucherer 1904*. See also *Langevin 1905c*.

[61] Einstein had previously discussed this topic in *Einstein 1905v* (Doc. 27) and *Einstein 1906e* (Doc. 35). His treatment here is similar to that in *Einstein 1907h* (Doc. 45), § 1 and § 2.

[62] The "E" on the right-hand side of this equation should be "E_0."

[63] The quotation is from *Planck 1907a*, § 18, p. 568. Planck's text has "1 gr Atom" where Einstein's text has "ein Grammatom."

[64] *Precht 1906*.

[65] In Planck's text (see note 63), the reference is to eq. (48) on p. 564 of *Planck 1907a*, which differs from Einstein's eq. (17). Planck

defines the mass M of a body as the limit of its momentum divided by its velocity as the velocity approaches zero (see ibid., p. 564). Applied to black-body radiation, this gives $M = (E_0 + pV_0)/c^2$, which is Planck's eq. (48).

[66] See *Cockcroft and Walton 1932* for the first reported verification of the mass-energy relationship in a nuclear reaction.

[67] See *Planck 1907a*, § 18, pp. 568–569, for a discussion of the relationship of inertial and gravitational mass.

[68] See *Eötvös 1890* for a report of the most accurate tests of this proportionality then available.

[69] The "X_a" should be "X_a'."

[70] The "e^2" should be "c^2."

[71] The following discussion is based on *Einstein 1907h* (Doc. 45), § 1.

[72] *Einstein 1907h* (Doc. 45); the reference should be to § 1.

[73] The "q" on the left-hand side of this equation should be "G."

[74] Planck derived results equivalent to the following two equations from a variational principle in *Planck 1907a*, § 11 and § 12.

[75] Planck derived transformation equations for volume and pressure by different methods in *Planck 1907a*, § 5 and § 6.

[76] The μ should be omitted from this equation.

[77] See *Mosengeil 1907* for a discussion of this system not involving the theory of relativity. *Planck 1907a*, § 1, summarizes Mosengeil's results; § 15 shows how to obtain them with the help of the relativistic transformation laws.

[78] The "E_0" should be "E_0/c^2."

[79] The "c^2" should be "V_0."

[80] In a reprint copy (see note 34), Einstein crossed out the factor "4/3" in this equation. See also *Einstein 1908b* (Doc. 49).

[81] *Einstein 1907h* (Doc. 45).

[82] In a reprint copy (see note 34), Einstein corrected the term "$-q\,dQ$" to "$-q\,dG$."

[83] *Planck 1907a*. The quotation is from p. 552. Planck used S as a symbol for the entropy throughout his text.

[84] Planck's text (see note 81) has "dasjenige" where Einstein's quotation has "das."

[85] Planck's text (see note 81) has "Bezugssystem" following "ungestrichene."

[86] Planck's text (see note 81) has "$S_1' < S_1$," where Einstein's quotation has "$\eta_1' > \eta_1$."

[87] The "T" in this equation should be "T_0."

[88] Planck gave a different derivation of this result in *Planck 1907a*, § 6. Einstein returned to the question of the transformation law for temperature many years later, and arrived at a different result:

$$T/T_0 = \frac{1}{\sqrt{1 - v^2/c^2}}$$

(see Einstein to Max von Laue, 27 January 1952).

[89] *Planck 1907a*.

[90] In a reprint copy (see note 34), Einstein corrected the coefficient of dz to F_z, and the coefficient of T to $d\eta$. See also *Einstein 1908b* (Doc. 49).

[91] In a reprint copy (see note 34), Einstein corrected the term "$\dot{x}\,dG$" to "$\dot{x}\,dG_x$." See also *Einstein 1908b* (Doc. 49).

[92] See *Planck 1907a*, § 2. The definition of the kinetic potential H is given on p. 549, and its relationship to the energy E on p. 550.

[93] See *Einstein 1908b* (Doc. 49), p. 99, for a clarification of the meaning of constant acceleration.

[94] This paragraph contains Einstein's first formulation of what he later termed the equivalence principle (see *Einstein 1912b*, p. 365).

[95] In a reprint copy (see note 34), Einstein crossed out "nutzbar" and wrote "ersetzbar." See also *Einstein 1908b* (Doc. 49).

[96] See p. 413 of the text above, and note 6. The first equation on p. 457 corresponds to Lorentz's original definition of the local time ("Ortszeit").

[97] In a reprint copy (see note 34), Einstein corrected "gleichen" to "gleicher."

[98] The reference to eq. (29) should be to eq. (12).

[99] See *Einstein 1908b* (Doc. 49), p. 99, for a comment on this equation.

[100] Einstein evidently calculated the gravitational potential at the surface of the sun and divided it by c^2. This is Einstein's first prediction of what came to be known as the gravitational red shift of spectral lines. His next paper on gravitation contains a discussion of this effect that differs from the one given here (see *Einstein 1911g*, p. 905).

[101] In a reprint copy (see note 34), Einstein corrected "$\frac{\partial}{\partial\tau}$" on the right-hand side of this equation to "$\frac{\partial}{\partial\sigma}$." See also *Einstein 1908b* (Doc. 49), p. 99.

[102] In a reprint copy (see note 34), Einstein added "u_ξ" following the closing parenthesis in

this equation. See also *Einstein 1908b* (Doc. 49), p. 99.

[103] In a reprint copy (see note 34), Einstein corrected the second " = " to " − ."

[104] This is Einstein's first discussion of the gravitational deflection of a light ray.

[105] See *Pound and Rebka 1960* for the first terrestrial confirmation of the gravitational red shift of spectral lines.

[106] In a reprint copy (see note 34), Einstein corrected the "u" preceding X to "u_ξ," and the "u" preceding Z to "u_ζ." See also *Einstein 1908b* (Doc. 49), p. 99.

[107] The corrections described in the preceding note apply here as well.

[108] In a reprint copy (see note 34), Einstein corrected " − " to " + " in the equation on this line. See also *Einstein 1908b* (Doc. 49), p. 99. "η^σ" should be "η_σ."

48. "A New Electrostatic Method for the Measurement of Small Quantities of Electricity"

[Einstein 1908a]

DATED Bern, 13 February 1908
RECEIVED 15 February 1908
PUBLISHED 1 April 1908

IN: *Physikalische Zeitschrift* 9 (1908): 216–217.

216   Physikalische Zeitschrift.   9. Jahrgang.   No. 7.

ten für lange Wellen. Sie muß deswegen aufgegeben werden.

Man wird natürlich fragen: welche Hypothese muß an ihre Stelle treten? Die Antwort ist schwer; jedenfalls muß die Strahlung von allgemeineren Elektronenbewegungen, welche sich nicht als Summen einer endlichen Anzahl von Pendelschwingungen darstellen lassen, untersucht worden, und es muß nach Ursachen gesucht worden, welche die Erzeugung von Lichtwellen auf enge, der Feinheit der Spektrallinien entsprechende, Periodenintervalle beschränken, weil der Isochronismus der Schwingungen wegfällt. Auf diese Probleme kann hier nicht weiter eingegangen werden; es möge nur noch bemerkt werden, daß J. J. Thomson schon eine Ursache letzterer Art angegeben hat[1]); allerdings muß man zugeben, daß es dabei fraglich ist, ob die Spektrallinien fein genug ausfallen werden, denn ohne große Rechnungsarbeit ist es voraussichtlich unmöglich, etwas Definitives zu sagen.

Haben wir einmal Elektronenbewegungen gefunden, welche Spektrallinien genügender Feinheit und genügend großer Wellenlänge erzeugen, so können wir hoffen, noch manche andere Schwierigkeiten zu überwinden. Offenbar steht zu erwarten, daß ein einziges Elektron eine Anzahl von Linien erzeugen kann, möglicherweise eine ganze Serie, sei es mittels verschiedener möglicher Bewegungen, sei es während verschiedener Phasen derselben Bewegung; denn wir haben genügend Grund zu glauben, daß die Anzahl Elektronen im Atom ziemlich klein ist, und doch müssen sie die große Zahl beobachtbarer Linien der Spektra des Atomes erzeugen. Vielleicht bestimmt dieselbe Ursache, welche die Breite der Linien begrenzt, auch noch den Zusammenhang zwischen den verschiedenen Linien welche ein Elektron erzeugen kann, und zwar in der Form einer den Serienformeln entsprechenden Gleichung.

---

1) J. J. Thomson, Corpuscular Theory of Matter, S. 158.

Bonn, 17. Febr. 1908.

(Eingegangen 20. Februar 1908.)

Eine neue elektrostatische Methode zur Messung kleiner Elektrizitätsmengen.

Von A. Einstein.

Mit empfindlichen elektrostatischen Quadrantenelektrometern vermag man bekanntlich Spannungen bis gegen 10^{-6} Volt herab zu messen, falls man der Nadel ein genügend [1] großes Hilfspotential gibt. Eine Erhöhung

dieses Hilfspotentials hat nun aber keine Erhöhung, sondern eine Verkleinerung der Empfindlichkeit des Apparates zur Folge, wenn es sich um die Messung elektrischer Mengen handelt. Je höher das Nadelpotential ist, desto kleiner ist vielmehr der Ausschlag, den eine bestimmte Elektrizitätsmenge liefert. Falls das Potential der Nadel einen absoluten Wert hat, der groß ist gegenüber der Spannungsdifferenz zwischen den Quadranten, hängt der Ausschlag nur vom Produkt aus Spannung und zugeführter elektrischer Menge, also von der zugeführten elektrischen Energie ab, und es muß die [2] zur Erzeugung des Ausschlags erforderliche Energie derjenigen des messend zu verfolgenden Systems entnommen werden. Die Grenze der praktisch erzielbaren Empfindlichkeit des Quadrantenelektrometers und analoger Apparate bezüglich der Messung von elektrischen Mengen bzw. Energiemengen ist durch diesen Umstand bedingt.

Es ist nun aber möglich, Meßapparate zu konstruieren, bei welchen die zur Erzeugung des Ausschlags erforderliche Energie nicht dem messend zu verfolgenden System, sondern einer Hilfsenergiequelle entnommen wird, so daß es möglich sein wird, die erwähnte praktische Empfindlichkeitsgrenze zu überschreiten. Im folgenden beschreibe ich das Schema einer Influenzmaschine, mit welcher dies Ziel nach [3] meiner Meinung erreicht werden kann.

A_1 und A'_1 (Fig. 1) seien zwei fest ange-

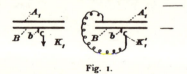

Fig. 1.

ordnete Leiter, an denen sich die starr miteinander verbundenen, etwa auf einem Rädchen befestigten Metallblättchen B vorbeibewegen. Letztere besitzen Kontaktstifte b, die im Bereich von fest angeordneten Kontaktfedern K_1 bzw. K'_1 stehen. K_1 sei geerdet, K'_1 mit A'_1 leitend verbunden.

A^1 werde nun konstant auf einem positiven Potential P_1 gehalten. Wenn das gerade passierende Blättchen K berührt, induziert die auf A_1 befindliche elektrische Ladung auf b eine entgegengesetzte Ladung $-e$. Gelangt dieses Blättchen A'_1 gegenüber, so daß es K' berührt, so gibt es negative Elektrizität an A'_1 ab. Jedes passierende Blättchen wird in dieser Weise die auf A'_1 befindliche Elektrizitätsmenge ändern, so lange, bis ein stationärer Zustand erreicht wird. Nennen wir P'_1 den absoluten Betrag des negativen Potentials, welchen A'_1 im stationären Zustand besitzt, so muß

Physikalische Zeitschrift. 9. Jahrgang. No. 7. 217

$$\frac{P'_1}{P_1} = a_1$$

sein, wobei a_1 eine von P_1 unabhängige Konstante, das Transformationsverhältnis ist. Wenn A_1 und A'_1 die Form von Blättchen haben, wird a_1 ein echter Bruch sein. Wenn wir A_1 und A'_1 aber die Form von Bügeln geben, welchen in den Augenblicken der Kontaktbildung die Blättchen B auf beiden Seiten umgeben, so können wir leicht erreichen, daß $a_1 > 1$ z. B. $a_1 = 10$. Dies sei im folgenden vorausgesetzt.

Wir denken uns nun mehrere solcher Elemente hintereinander geschaltet nach dem nachstehend abgebildeten Schema. Der sekun-

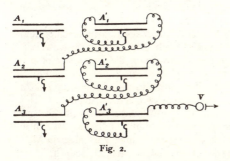

Fig. 2.

däre Leiter A'_1 des ersten Elementes sei mit dem primären A_2 des zweiten Elementes, der sekundäre Leiter A'_2 des zweiten Elementes mit dem primären A_3 des dritten Elementes verbunden usw. Der sekundäre Leiter des letzten Elementes sei an ein Elektrometer V angeschlossen.

Denkt man sich nun dem primären Leiter A_1 eine bestimmte Spannung P_1 mitgeteilt, so wird sich nach einiger Zeit ein stationärer Zustand der ganzen Einrichtung herstellen. Für diesen gelten die folgenden Gleichungen:

$$P_2 = P'_1 = P_1 \cdot a_1$$
$$P_3 = P'_2 = P_2 \cdot a_2 = P_1 \cdot a_1 \cdot a_2$$
$$P'_3 = P_3 \cdot a_3 = P_1 \cdot a_1 \cdot a_2 \cdot a_3.$$

Sind n-Elemente vorhanden, welche alle das gleiche Übersetzungsverhältnis a haben, so ist

$$P'_n = P_1 \cdot a^n.$$

Man sieht, daß die Energie, welche dem beweglichen System des Elektrometers zuzuführen ist, der mechanischen Energie entnommen wird, die den Blättchen B mitgeteilt wird, nicht aber dem an das Blättchen A angeschlossenen, messend zu verfolgenden System. Die Empfindlichkeitsgrenze des Verfahrens ist lediglich durch die äußeren Fehlerquellen begrenzt, da man durch Vergrößern von n a^n be- [4] liebig groß machen kann.

Man kann natürlich die ganze Einrichtung

zweipolig ausführen. Jedes Element erhält dann zwei primäre und zwei sekundäre Leiter.

Da die Erhöhung der Empfindlichkeit der elektrostatischen Meßmethoden von Bedeutung ist für die Erforschung der Radioaktivität, [5] hoffe ich, daß sich ein Physiker für diese Sache interessieren wird. Meine weiteren Über- [6] legungen über den Gegenstand würde ich demselben gerne mitteilen. Auf den vorliegenden Plan wurde ich geführt durch Nachdenken darüber, wie die von der Molekulartheorie der Wärme geforderten, der Brownschen Bewegung analogen spontanen Ladungen von Leitern[1]) konstatiert und gemessen werden könnten. Auch dieses Problem hoffe ich mit dem geschilderten Plane seiner Lösung um einen Schritt näher gebracht zu haben.

1) A. Einstein, Ann. d. Phys. (4) **22**, 569, 1907. [7]

Bern, 13. Februar 1908.

(Eingegangen 15. Februar 1908.)

Wassertropfkollektor mit kontinuierlicher Wasserzufuhr.

Von Wilhelm Schmidt (Wien).

Einer der größten Mängel, die der Wassertropfkollektor in der Form, in welcher er heute gebraucht wird, aufweist, ist der, daß er immer nur für eine begrenzte Zeit in Betrieb bleiben kann, nach deren Ablauf eine Neufüllung notwendig wird, die mit einer Unterbrechung der Beobachtungen verbunden ist. Um noch bequeme Beobachtungszeiten zu gewinnen, hat man sich also zu entscheiden zwischen geringerer Aufladegeschwindigkeit oder Vergrößerung der Vorratsgefäße, welch letztere aber auch wieder nachteilig wirkt durch die mit ihr verbundene Vergrößerung der Kapazität.

Es gibt jedoch ein einfaches Mittel, diesen Unannehmlichkeiten auszuweichen. Man braucht bloß auch die Wasserzufuhr in einzelnen Tropfen vor sich gehen zu lassen und dabei dafür zu sorgen, daß jeder Tropfen ohne Ladung ist. Zu diesem Zwecke ist die eine Tropfstelle, die direkt mit der Wasserleitung in Verbindung stehen kann, von einem Stück eines weiten Metallrohres umgeben, welches gleich wie das abtropfende Wasser geerdet ist. Weiter unten ist der Auffangtrichter aufgestellt, von dem aus die Rohrleitung zur eigentlichen Kollektortropfstelle führt. Auch der Auffangtrichter wird zweckmäßigerweise aus Metall gefertigt, damit etwaige Änderungen des Wasserstandes in ihm nicht zu großen Einfluß auf die Kapazität des Systems nehmen. Um ja allen den Tropfen der ersten (der geerdeten) Tropfstelle noch anhaftenden Eigenladungen ihre Wirksam-

Published in *Physikalische Zeitschrift* 9 (1908): 216–217. Dated Bern, 13 February 1908, received 15 February 1908, published 1 April 1908.

[1] An explanation of the quadrant electrometer is found in Einstein's ETH notes, H. F. Weber's Lectures on Physics, ca. December 1897–ca. June 1898 (Vol. 1, Doc. 37), pp. 156–158. The quadrant electrometer was first proposed by William Thomson (see, e.g., *Thomson, W. 1867*); later versions of this instrument are discussed in *Cermak 1918*, pp. 110–123. At the time of this paper, a sensitivity of 10^{-6} volt had only been achieved with the instrument proposed by Kleiner (see *Kleiner 1906* and *Cermak 1918*, pp. 141–142). For the order of magnitude of the effects Einstein hoped to measure, see *Einstein 1907b* (Doc. 39), pp. 571–572.

[2] Einstein's conclusion follows from the following expression for the deflection φ of the needle in Thomson's quadrant electrometer: $\varphi = K (A - B) \left(C - 1/2 (A + B)\right)$, if C, the auxiliary potential of the needle, is assumed to be high; A, B are the potentials of the two quadrants, and K is a constant. For a discussion of this formula, see, e.g., *Maxwell 1891*, part I, chap. 13, § 219; it is also mentioned in Einstein's ETH notes, H. F. Weber's Lectures on Physics, ca. December 1897–ca. June 1898 (Vol. 1, Doc. 37), p. 157.

[3] Inductive machines were further developments of potential multipliers, which in fact were intended to measure small quantities of electricity, in a way similar to the device proposed by Einstein (see, e.g., *Schmidt 1918*, pp. 33–40). However, at the time of this paper, inductive machines were used mainly for the generation of high potentials, and not as measuring instruments. For a description that includes a historical account of their development, see, e.g., *Schmidt 1918*, pp. 40–67.

[4] The method proposed by Einstein was later severely criticized for not taking into account the phenomenon of self-induction, caused by the different metal parts of the instrument that have to be brought into contact with each other. See, e.g., *Schmidt 1918*, p. 35; see also Vol. 5, the editorial note, "Einstein's 'Maschinchen' for the Measurement of Small Quantities of Electricity."

[5] For a discussion of the use of electrometers in contemporary investigations of radioactivity, see, e.g., *Rutherford 1906*, pp. 23–36 (a German edition appeared in 1907, *Rutherford 1907*). The significance of the instrument proposed by Einstein for research on radioactivity is also emphasized in *Habicht and Habicht 1910*, p. 535.

[6] Einstein's paper aroused immediate interest. On 30 March 1908, Joseph de Kowalski, Professor of Physics at the University of Fribourg, wrote to Einstein: "Today, in the Physikalische Zeitschrift, I read of your proposal for the measurement of electrical quantities. This interested me greatly and I would like to have such an instrument built" ("Heute lese ich in der Physik. Zeitschft. Ihren Vorschlag zur Messung der Elektricitätsmengen. Derselbe hat mich sehr interessirt und ich möchte gern ein entsprechendes Instrument ausführen lassen."). Before the publication of Einstein's paper, his friends, Conrad and Paul Habicht, had started working on the measuring device proposed by Einstein; see the editorial note, "Einstein on Brownian Motion," pp. 221–222, and Vol. 5, the editorial note, "Einstein's 'Maschinchen' for the Measurement of Small Quantities of Electricity."

[7] *Einstein 1907b* (Doc. 39).

49. ''Corrections to the Paper: 'On the Relativity Principle and the Conclusions Drawn from It' ''

[*Einstein 1908b*]

Submitted with a covering letter of 29 February 1908 to Johannes Stark, the editor of the *Jahrbuch*
RECEIVED 3 March 1908
PUBLISHED 1908

IN: *Jahrbuch der Radioaktivität und Elektronik* 5 (1908): 98–99.

98   Berichtigungen.

Die in den verschiedenen Versuchen benutzten Geschwindigkeiten der Ladungen übertreffen nicht $1,5 \cdot 10^4$ cm/sec.; das Verhältnis $\beta = \dfrac{v}{c}$ war also höchstens $0,5 \cdot 10^{-6}$. Die besten Versuche waren mit einem möglichen Fehler von etwa 3 Proz. behaftet. Wenn also die Versuche die Abhängigkeit des Magnetfeldes der elektrischen Konvektion von der ersten Potenz des β festzustellen erlauben, so sind sie gar nicht imstande, den etwaigen Einfluß des β^2 zu entdecken.

Dasselbe gilt für den Versuch von Fizeau über den Einfluß des strömenden Wassers auf die Lichtgeschwindigkeit.

Wenn also alle diese Versuche gegen die Maxwell-Hertzschen elektrodynamischen Gleichungen bewegter Körper und zugunsten der Lorentzschen Elektronentheorie sprechen, so können sie dennoch zwischen den neueren Theorien von H. Lorentz und E. Cohn nicht entscheiden.

Moskau, Ingenieur-Hochschule. 9. Februar 1908.

(Eingegangen 12. Februar 1908.)

Berichtigungen

zu der Arbeit: „Über das Relativitätsprinzip und die aus demselben gezogenen Folgerungen".[1)]

Von A. Einstein.

Bei Durchsicht der Korrekturbogen der genannten Arbeit ist mir leider eine Anzahl Fehler entgangen, die ich berichtigen muß, weil sie das Lesen der Arbeit erschweren.

[2]

Formel 15b (S. 435) sollte lauten:

$$\frac{d}{dt}\left[\int \frac{1}{4\pi c}(YN - ZM)d\omega\right] + \Sigma \frac{\mu\dot{x}}{\sqrt{1 - \dfrac{q^2}{c^2}}} = 0.$$

Die zweite Formel auf S. 451 hat fälschlich den Faktor $\dfrac{4}{3}$; es sollte heißen:

$$G = \frac{q}{\sqrt{1 - \dfrac{q^2}{c^2}}} \frac{E_0}{c^2}.$$

Formel 28 auf S. 453 lautet richtig:

$$dE = F_x dx + F_y dy + F_z dz - p\,dV + T\,d\eta.$$

Einige Zeilen weiter unten ist der Index bei G_x zu ergänzen. In der vorletzten Zeile der S. 455 sollte es heißen „ersetzbar" statt „nutzbar".

[1]   1) Dieses Jahrbuch 4, 411, 1907.

Berichtigungen.   99

Auf S. 461 sollte es heißen:

$$\frac{\delta}{\delta \tau} = \left(1 + \frac{\gamma \xi}{c^2}\right)\frac{\delta}{\delta \sigma}$$

und

$$w_\xi = \left(1 + \frac{\gamma \xi}{c^2}\right)u_\xi.$$

Auf S. 462 sind ferner bei den Größen u_ξ und w_ξ die Indizes zu er-gänzen. Außerdem ist etwa in der Mitte dieser Seite ein Zeichenfehler zu berichtigen; es sollte heißen:

$$\eta_\sigma = \eta_\tau \left(1 - \frac{\gamma \xi}{c^2}\right).$$

Eine briefliche Mitteilung von Herrn Planck veranlaßt mich dazu, zur Vermeidung eines naheliegenden Mißverständnisses eine ergänzende Bemerkung beizufügen.

Im Abschnitt „Relativitätsprinzip und Gravitation" wird ein ruhendes, in einem zeitlich konstanten, homogenen Schwerefeld gelegenes Bezugs-system als physikalisch gleichwertig behandelt mit einem gleichförmig beschleunigten, gravitationsfreien Bezugssystem. Der Begriff „gleichförmig beschleunigt" bedarf noch einer Erläuterung.

Wenn es sich — wie in unserem Falle — um eine gradlinige Bewegung (des Systems Σ) handelt, so ist die Beschleunigung durch den Ausdruck $\frac{dv}{dt}$ gegeben, wobei v die Geschwindigkeit bedeutet. Nach der bisher ge-bräuchlichen Kinematik ist $\frac{dv}{dt}$ eine vom Bewegungszustande des (beschleu-nigungsfreien) Bezugssystems unabhängige Größe, so daß man, wenn die Bewegung in einem bestimmten Zeitteilchen gegeben ist, ohne weiteres von der (momentanen) Beschleunigung reden kann. Gemäß der von uns angewendeten Kinematik hängt $\frac{dv}{dt}$ vom Bewegungszustande des (beschleu-nigungsfreien) Bezugssystems ab. Unter allen Beschleunigungswerten, die man so für eine bestimmte Bewegungsepoche erhalten kann, ist aber der-jenige ausgezeichnet, welcher einem Bezugssystem entspricht, demgegen-über der betrachtete Körper die Geschwindigkeit $v = 0$ besitzt. Dieser Beschleunigungswert ist es, der bei unserem „gleichförmig beschleunigten" System konstant bleiben soll. Die auf S. 457 gebrauchte Beziehung $v = \gamma t$ gilt also nur in erster Annäherung; dies genügt aber, weil in der Be-trachtung nur bezüglich t bezw. τ lineare Glieder zu berücksichtigen sind.

[3]

(Eingegangen 3. März 1908.)

Published in *Jahrbuch der Radioaktivität und Elektronik* 5 (1908): 98–99. Submitted 29 February 1908, received 3 March 1908.

[1] *Einstein 1907j* (Doc. 47).

[2] Many of these corrections, in Einstein's hand, are in a reprint copy of *Einstein 1907j*

(Doc. 47), in the Holton collection of Einstein's reprints (see the Acknowledgments, p. xxxiv); they are indicated in the notes to that document.

[3] The quantity here defined is now often re-ferred to as the proper acceleration.

50. "Elementary Theory of Brownian Motion"

[*Einstein 1908c*]

RECEIVED 1 April 1908
PUBLISHED 24 April 1908

IN: *Zeitschrift für Elektrochemie und angewandte physikalische Chemie* 14 (1908): 235–239.

Republished in *Einstein 1922*, pp. 43–53.

1908.] ZEITSCHRIFT FÜR ELEKTROCHEMIE. 235

3. Geringe Mengen Sauerstoff in einem anderen Gase steigern die kathodische Zerstäubung nicht abnorm, wie die eines glühenden Drahtes.

4. Bei Anwendung von Induktionsstrom ist die Zerstäubung nicht gleichmäßig auf der ganzen Oberfläche einer kreisrunden Kathode.

5. Bei regelmäßiger Entladung wurde nur eine Art größerer Druckänderung, nämlich Abnahme, beobachtet. Sie ist bei den edlen Metallen durch Bildung von Verbindungen zwischen dem Gas (Sauerstoff und Stickstoff) und dem zerstäubten Metall verursacht. Bei unedlen Metallen erfolgt Bindung von Gas teilweise durch die Kathode selbst.

Eintritt bestimmter „Nebenreaktionen" kann neben anderen Störungen auch eine Drucksteigerung veranlassen.

6. Die Reihenfolge der Metalle nach ihrer Zerstäubbarkeit ist in allen Gasen die gleiche. Die Gewichtsverluste der Kathoden stehen unter gleichen Bedingungen im Verhältnis der Aequivalentgewichte.

7. Die Reihenfolge der Gase nach ihrer Fähigkeit, die Kathode anzugreifen, ist bei allen Metallen die gleiche; sie ist die der Atomgewichte der Gase.

(Eingegangen: 19. März.)

ELEMENTARE THEORIE DER BROWNSCHEN [1]) BEWEGUNG.

Von Dr. A. Einstein.

[2] Herr Professor R. Lorenz machte mich gesprächsweise darauf aufmerksam, daß eine elementare Theorie der Brownschen Bewegung manchem Chemiker willkommen wäre. Seiner Aufforderung folgend, gebe ich in nachfolgendem eine einfache Theorie dieses Phänomens. Der mitzuteilende Gedankengang ist kurz folgender. Zunächst untersuchen wir, wie der Diffusionsvorgang in einer nicht dissoziierten verdünnten Lösung von der Verteilung des osmotischen Druckes in der Lösung und von der Beweglichkeit des gelösten Stoffes gegenüber dem Lösungsmittel abhängt. Wir erhalten so für den Fall, daß ein Molekül des gelösten Stoffes groß ist gegenüber einem Molekül des Lösungsmittels, einen Ausdruck für den Diffusionskoeffizienten, in welchem keine von der Natur der Lösung abhängige Größen auftreten, außer der Zähigkeit des Lösungsmittels und dem Durchmesser der gelösten Moleküle.

Hierauf führen wir den Diffusionsvorgang auf die ungeordneten Bewegungen der gelösten Moleküle zurück und finden, wie die mittlere [3] Größe dieser ungeordneten Bewegungen der gelösten Moleküle aus dem Diffusionskoeffizienten, also nach dem vorher erwähnten Ergebnis aus der Zähigkeit des Lösungsmittels und der Größe der gelösten Moleküle, berechnet werden kann. Das so ermittelte Resultat gilt dann nicht nur für eigentliche gelöste Moleküle, sondern auch für beliebige, in der Flüssigkeit suspendierte kleine Körperchen.

§ 1. Diffusion und osmotischer Druck.

Das zylindrische Gefäß Z (Fig. 93) sei gefüllt mit einer verdünnten Lösung. Der Innenraum von Z werde durch den, eine semipermeable Wand bildenden, beweglichen Kolben K in zwei Teile A und B geteilt. Ist die Konzentration der Lösung in A größer als in B, so muss man eine äussere, nach links gerichtete Kraft auf den Kolben ausüben, um ihn im Gleichgewicht zu erhalten, und zwar ist diese Kraft gleich der Differenz der beiden osmotischen Drucke, welche die gelöste Substanz von links bezw. von rechts her auf den Kolben

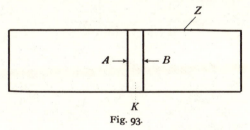

Fig. 93.

ausübt. Läßt man jene äußere Kraft nicht auf den Kolben wirken, so verschiebt er sich unter dem Einflusse des von der in A befindlichen Lösung ausgeübten stärkeren osmotischen Druckes so lange nach rechts, bis die Konzentration in A und B nicht mehr verschieden ist. Aus dieser Betrachtung geht hervor, daß es die osmotischen Druckkräfte sind, welche bei der Diffusion den Ausgleich der Konzentrationen bedingen; denn wir können eben eine Diffusion, d. h. einen Ausgleich der Konzentrationen, dadurch verhindern, daß wir die osmotischen Differenzen, welche den Konzentrationsverschiedenheiten entsprechen, durch äußere, auf semipermeable Wände wirkende Kräfte ausgleichen. Daß der osmotische Druck als bewegende Kraft bei Diffusionsvorgängen aufgefaßt werden kann, ist längst bekannt. Nernst hat bekanntlich hierauf seine Untersuchung über den Zusammenhang zwischen Ionenbeweglich-

---

[1] Man versteht unter Brownscher Bewegung jene ungeordnete Bewegung, welche mikroskopisch kleine, in Flüssigkeit suspendierte Teilchen ausführen. Vergl.
[1] z. B. The Svedberg, Z. f. Elektroch. **12**, 47 u. 51 (1906).

[4] keit, Diffusionskoeffizient und EMK bei Konzentrationselementen gegründet.

Im Innern des Zylinders Z (Fig. 94), dessen Querschnitt = 1 sei, finde ein Diffusionsvorgang längs der Zylinderachse statt. Wir fragen zunächst nach den osmotischen Kräften, welche die Diffusionsbewegung der zwischen den unendlich nahen Ebenen E und E' befindlichen gelösten Substanz bewirken. Von links wirkt auf die Grenzfläche E der Lamelle die osmotische Druckkraft p, von rechts auf die Grenzfläche E' die Kraft p'; die Resultierende der Druckkräfte ist also

$$p - p'.$$

Wir wollen nun die Entfernung der Fläche E vom linken Gefäßende mit x, die Entfernung der Fläche E' von jenem Gefäßende mit $x + dx$ bezeichnen; es ist dann dx zugleich das Volumen der betrachteten Flüssigkeitslamelle. Da $p - p'$ die osmotische Kraft ist, welche auf das Volumen dx gelöster Substanz wirkt, so ist

$$K = \frac{p - p'}{dx} = -\frac{p' - p}{dx} = -\frac{dp}{dx}$$

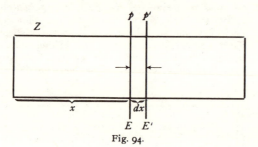

Fig. 94.

die osmotische Kraft, welche auf die in der Volumeneinheit befindliche gelöste Substanz
[5] wirkt. Da ferner der osmotische Druck durch die Gleichung

$$p = RTv$$

gegeben ist, wobei R die Konstante der Gasgleichung $(8{,}31 \cdot 10^7)$, T die absolute Temperatur und v die Anzahl gelöster Grammmoleküle pro Volumeneinheit bedeutet, erhalten wir schließlich für die auf die gelöste Substanz pro Volumeneinheit wirkende osmotische Kraft K den Ausdruck

$$K = -RT\frac{dv}{dx} \quad \ldots \quad (1)$$

Um nun die Diffusionsbewegungen berechnen zu können, welche diese bewegenden Kräfte zu erzeugen vermögen, muß man noch wissen, einen wie großen Widerstand das Lösungsmittel einer Bewegung der gelösten Substanz entgegensetzt. Wirkt auf ein Molekül eine bewegende Kraft k, so wird diese dem Molekül eine proportionale Geschwindigkeit v verleihen, gemäß der Gleichung

$$v = \frac{k}{\Re} \quad \ldots \quad (2)$$

wobei $\Re$ eine Konstante bedeutet, welche wir den Reibungswiderstand des Moleküls nennen wollen. Dieser Reibungswiderstand ist im allgemeinen nicht theoretisch ermittelbar. Wenn aber das gelöste Molekül angenähert als eine Kugel aufgefaßt werden darf, welche groß ist gegenüber einem Molekül des Lösungsmittels, so werden wir den Reibungswiderstand des gelösten Moleküls nach den Methoden der gewöhnlichen Hydrodynamik ermitteln dürfen, welche der molekularen Konstitution der Flüssigkeit nicht Rechnung trägt. Innerhalb des Gültigkeitsbereiches der gewöhnlichen Hydrodynamik gilt nun für eine in einer Flüssigkeit bewegte Kugel die Gleichung (2), wobei gesetzt ist

$$\Re = 6\pi\eta\varrho \quad \ldots \quad (3)$$

Dabei bedeutet η den Viskositätskoeffizienten (Zähigkeit) der Flüssigkeit, ϱ den Radius der Kugel. Falls man annehmen darf, daß die Moleküle einer gelösten Substanz annähernd kugelförmig und groß gegenüber den Molekülen des Lösungsmittels seien, darf Gleichung (3) auf die einzelnen gelösten Moleküle angewendet werden.

Wir können nun die durch einen Querschnitt des Zylinders pro Zeiteinheit diffundierende Menge gelöster Substanz berechnen. In der Volumeneinheit sind v Grammmoleküle, also vN wirkliche Moleküle vorhanden, wobei N die Anzahl der wirklichen Moleküle in einem Grammmolekül bedeutet. Verteilt sich eine Kraft K auf diese vN in der Volumeneinheit enthaltenen Moleküle, so wird sie diesen eine vNmal kleinere Geschwindigkeit erteilen, als sie einem einzigen Molekül zu erteilen imstande wäre, wenn sie nur auf dieses wirkte. Man erhält daher mit Rücksicht auf Gleichung (2) für die Geschwindigkeit v, welche die Kraft K den vN Molekülen zu erteilen vermag, den Ausdruck

$$v = \frac{1}{vN} \cdot \frac{K}{\Re}.$$

Im vorliegenden Falle ist K gleich der auf die vN Moleküle der Volumeneinheit wirkenden, oben berechneten osmotischen Kraft, so daß wir hieraus unter Benutzung von Gleichung (1) erhalten:

$$vv = -\frac{RT}{N} \cdot \frac{1}{\Re} \cdot \frac{dv}{dx} \quad \ldots \quad (4)$$

Auf der linken Seite steht das Produkt aus der Konzentration v der gelösten Substanz und der Geschwindigkeit, mit welcher die gelöste Substanz durch den Prozeß der Diffusion fortbewegt wird. Dies Produkt stellt also die pro Sekunde durch Diffusion durch die Querschnittseinheit hindurchtransportierte Menge gelöster Substanz (in Grammmolekülen) dar. Der Faktor von $\frac{dv}{dx}$ auf der rechten Seite dieser Gleichung

1908.]   ZEITSCHRIFT FÜR ELEKTROCHEMIE.   237

ist daher[1]) nichts anderes als der Diffusions-koeffizient D der betrachteten Lösung. Man hat also allgemein

$$D = \frac{RT}{N} \cdot \frac{1}{\Re} \quad \cdots \quad (5)$$

und, falls die diffundierenden Moleküle als kugel-förmig und als groß gegenüber den Molekülen des Lösungsmittels aufgefaßt werden dürfen, nach Gleichung (3):

$$D = \frac{RT}{N} \cdot \frac{1}{6\pi\eta\varrho} \quad \cdots \quad (5\,\text{a})$$

In dem letzteren Falle hängt also der Diffu-sionskoeffizient von keiner andern für die be-trachteten Substanzen charakteristischen Kon-stanten ab, als von der Viskosität η des Lösungs-mittels und vom Radius ϱ des Moleküls[2]).

§ 2. Diffusion und ungeordnete Bewegung der Moleküle.

Die Molekulartheorie der Wärme eröffnet noch einen zweiten Gesichtspunkt, von welchem aus der Vorgang der Diffusion betrachtet werden kann. Der unregelmäßige Bewegungsprozeß, als welchen wir den Wärmeinhalt einer Substanz aufzufassen haben, wird bewirken, daß die ein-zelnen Moleküle einer Flüssigkeit in denkbar unregelmäßigster Weise ihren Ort ändern. Dieses gewissermaßen planlose Umherirren der Mole-küle gelöster Substanz in einer Lösung wird zur Folge haben, daß eine anfängliche ungleich-mäßige Konzentrationsverteilung der gelösten Substanz allmählich einer gleichmäßigen Platz machen wird.

Wir betrachten nun diesen Vorgang etwas eingehender, indem wir uns wieder auf den in § 1 betrachteten Fall beschränken, daß lediglich die Diffusion in einer Richtung, nämlich in Richtung der Achse (x-Achse) des Zylinders Z ins Auge zu fassen ist. Wir wollen uns vor-stellen, daß wir zu einer bestimmten Zeit t die x-Koordinaten sämtlicher gelöster Moleküle kennen und ebenso zur Zeit $t + \tau$, wobei τ ein so kurzes Zeitintervall bedeutet, daß sich die Kon-zentrationsverhältnisse unserer Lösung während desselben nur sehr wenig ändern. Während dieser Zeit τ wird die x-Koordinate des ersten gelösten Moleküls sich durch die unregelmäßige

---

2) Diese Gleichung erlaubt, aus dem Diffusions-koeffizienten den Radius (großer) Moleküle angenähert zu ermitteln, falls der Diffusionskoeffizient bekannt ist; denn es ist

$$\varrho = \frac{RT}{6\pi N \eta} \cdot \frac{1}{D},$$

[6] wobei $R = 8{,}31 \cdot 10^7$, $N = 6 \cdot 10^{23}$ zu setzen ist. Dem Wert von N haftet allerdings noch eine Unsicherheit von etwa 50 % an. Für die Ermittelung der ungefähren Größe der Moleküle in kolloidalen Lösungen dürfte
[7] diese Beziehung von Bedeutung sein.

Wärmebewegung um eine gewisse Größe Δ_1 ändern, die des zweiten um Δ_2 usw. Diese Verschiebungen Δ_1, Δ_2 usw. werden zum Teil negativ (nach links gerichtet) zum Teil positiv (nach rechts gerichtet) sein. Ferner wird die Größe dieser Verschiebungen bei den einzelnen Molekülen eine verschiedene sein. Aber da wir, wie oben, eine verdünnte Lösung voraussetzen, ist diese Verschiebung nur durch das umgebende Lösungsmittel nicht aber in merklichem Grade durch die übrigen gelösten Moleküle bedingt; deshalb werden diese Verschiebungen Δ in ver-schieden konzentrierten Teilen der Lösung im Mittel gleich groß ebenso oft positiv wie negativ sein.

Wir wollen nun sehen, wie groß die durch die Querschnittseinheit unserer Lösung in der Zeit τ hindurchdiffundierende Substanzmenge anfällt, wenn die Größe der Verschiebungen Δ im Sinne der Zylinderachse bekannt ist, welche

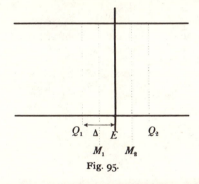

Fig. 95.

die gelösten Moleküle im Durchschnitt erfahren. Um diese Betrachtung zu vereinfachen, rechnen wir nun so, wie wenn alle Moleküle eine gleich große Verschiebung Δ erfahren würden, und zwar die eine Hälfte der Moleküle die Ver-schiebung $+\Delta$ (d. h. nach rechts), die andere Hälfte die Verschiebung $-\Delta$ (d. h. nach links). Wir ersetzen also die einzelnen Verschiebungen Δ_1, Δ_2 usw. durch deren Mittelwert Δ.   [8]

Durch die Ebene E unseres Zylinders (Fig. 95) werden nach unserer vereinfachenden Annahme während der Zeit τ nur solche gelöste Moleküle von links nach rechts hindurchtreten können, welche sich vor Ablauf von τ links von E und in einem Abstand von E befinden, welcher kleiner ist als Δ. Diese Moleküle befinden sich alle zwischen den Ebenen Q_1 und E (Fig. 95). Weil aber von diesen Molekülen nur die Hälfte die Verschiebung $+\Delta$ erfahren, so passieren von ihnen auch nur die Hälfte die Ebene E. Die Hälfte der zwischen Q_1 und E befindlichen gelösten Substanz ist aber, in Grammmolekülen ausgedrückt, gleich

$$\frac{1}{2} \, r_1 \, \Delta,$$

34

wobei v_1 die mittlere Konzentration im Volumen $Q_1 E$, d. h. die Konzentration in der Mittelebene M_1 ist. Denn Δ ist, da der Querschnitt $= 1$ ist, das zwischen Q_1 und E befindliche Volumen, welches, multipliziert mit der mittleren Konzentration, die in diesem Volumen befindliche gelöste Substanz in Grammmolekülen ergibt.

Durch eine analoge Betrachtung ergibt sich, daß die von rechts nach links in der Zeit τ durch E hindurchtretende Menge gelöster Substanz gleich

$$\frac{1}{2} v_2 \Delta$$

ist, wobei v_2 die Konzentration in der Mittelebene M_2 bedeutet. Die während τ durch E von links nach rechts diffundierende Substanzmenge ist nun offenbar gleich der Differenz dieser beiden Werte, also gleich

$$\frac{1}{2} \Delta (v_1 - v_2) \quad \ldots \quad (6)$$

v_1 und v_2 sind die Konzentrationen in zwei Querschnitten, welche den sehr kleinen Abstand Δ haben. Bezeichnet man wieder den Abstand eines Querschnittes vom linken Zylinderende mit x, so ist nach der Definition des Differentialquotienten

$$\frac{v_2 - v_1}{\Delta} = \frac{dv}{dx},$$

folglich

$$v_1 - v_2 = - \Delta \frac{dv}{dx},$$

so daß die während τ durch E diffundierte Substanzmenge auch gleich ist:

$$- \frac{1}{2} \Delta^2 \frac{dv}{dx} \quad \ldots \quad (6\,a)$$

Die in der Zeiteinheit durch E diffundierende, in Grammmolekülen ausgedrückte Substanzmenge ist also gleich

$$- \frac{1}{2} \frac{\Delta^2}{\tau} \frac{dv}{dx}.$$

Wir haben damit einen zweiten Wert für den Diffusionskoeffizienten D gewonnen. Es ist

$$D = \frac{1}{2} \frac{\Delta^2}{\tau} \quad \ldots \quad (7)$$

wobei Δ die im Mittel[1]) von einem gelösten Molekül während der Zeit τ, im Sinne der x-Achse zurückgelegte Wegstrecke bedeutet.

Löst man (7) nach Δ auf, so erhält man:

[10]
$$\Delta = \sqrt{2D}\sqrt{\tau} \quad \ldots \quad (7\,a)$$

---

[9] 1) Genauer genommen, ist Δ gleich der Wurzel aus dem Mittel der Quadrate der Einzelverschiebungen $\Delta_1{}^2$, $\Delta_2{}^2$ usw. Wir sollten deshalb statt Δ genauer $\sqrt{\Delta^2}$ schreiben.

§ 3. Bewegung der einzelnen Moleküle. Brownsche Bewegung.

Setzen wir die in den Gleichungen (5) und (7) für den Diffusionskoeffizienten gegebenen Werte einander gleich, so erhalten wir durch Auflösen nach Δ:

$$\Delta = \sqrt{\frac{2RT}{N\mathfrak{R}}} \sqrt{\tau} \quad \ldots \quad (8)$$

Aus dieser Formel sehen wir, daß der von einem Molekül im Mittel zurückgelegte Weg nicht proportional der Zeit ist[1]), sondern proportional der Quadratwurzel aus der Zeit. Es liegt dies daran, daß sich die in zwei aufeinanderfolgenden Zeiteinheiten zurückgelegten Wege nicht stets addieren, sondern ebenso häufig subtrahieren werden. Man kann die infolge der unregelmäßigen Molekularbewegung im Mittel eintretende Verschiebung des Moleküls vermöge Gleichung (7 a) aus dem Diffusionskoeffizienten, vermöge Gleichung (8) aus der Widerstandskraft $\mathfrak{R}$ berechnen, welche sich einer erzwungenen, mit der Geschwindigkeit $v = 1$ erfolgenden Bewegung entgegenstellt.

Für den Fall, daß das gelöste Molekül groß gegen das Molekül des Lösungsmittels und kugelförmig ist, läßt sich in Gleichung (8) für $\mathfrak{R}$ der in Gleichung (3) angegebene Wert einsetzen, so daß man erhält:

$$\Delta = \sqrt{\frac{RT}{N} \cdot \frac{1}{3\pi\eta\varrho}} \cdot \sqrt{\tau}. \quad \ldots \quad (8\,a)$$

Diese Gleichung erlaubt das Verschiebungsmittel[2]) Δ zu berechnen, aus der Temperatur T, der Zähigkeit des Lösungsmittels η und dem Molekülradius ϱ.

Nach der molekularkinetischen Auffassung existiert nun kein prinzipieller Unterschied zwischen einem gelösten Molekül und einem suspendierten Körperchen. Wir werden daher die Gleichung (8 a) auch dann für gültig zu halten haben, wenn es sich um irgendwelche suspendierten kugelförmigen Teilchen handelt.

Wir berechnen den Weg Δ, welchen ein Teilchen von 1 Mikron Durchmesser in Wasser in 1 Sekunde in einer bestimmten Richtung bei Zimmertemperatur durchschnittlich zurücklegt. Es ist zu setzen:

$$\begin{aligned} R &= 8{,}31 \cdot 10^7, & \eta &= 0{,}0135, \\ T &= 290, & \varrho &= 0{,}5 \cdot 10^{-4}, \\ N &= 6 \cdot 10^{23}. & \tau &= 1. \end{aligned}$$ [12]

Man erhält:

$$\Delta = 0{,}8 \cdot 10^{-4} \text{ cm} = 0{,}8 \text{ Mikron}.$$

Diese Zahl ist wegen der geringen Genauigkeit, mit welcher N bekannt ist, noch mit einer Unsicherheit von etwa $\pm 25\,\%$ behaftet.   [13]

---

1) Vergl. A. Einstein, Z. f. Elektroch. 6 (1907).   [11]
2) Genauer genommen die Wurzel aus dem Mittelwert von Δ^2.

1908.]   ZEITSCHRIFT FÜR ELEKTROCHEMIE.   239

Es ist von Interesse, die eben berechnete mittlere Eigenbewegung mikroskopischer Teilchen mit derjenigen gelöster Moleküle bezw. Ionen zu vergleichen. Für eine nicht dissoziierte gelöste Substanz, deren Diffusionskoeffizient bekannt ist, können wir Δ auf Gleichung (7 a) berechnen. Für Zucker ist bei Zimmertemperatur

[14] $$D = \frac{0{,}33}{24 \cdot 60 \cdot 60}.$$ Hieraus berechnet man aus Gleichung (7 a) für $\tau = 1$:

$$\Delta = 27{,}6 \text{ Mikron}$$

Aus der Zahl N und dem Molekularvolumen des festen Zuckers kann man schließen, daß der Durchmesser eines Zuckermoleküls von der Größenordnung eines tausendstel Mikron, also etwa tausendmal kleiner ist, als der Durchmesser

[15] des vorher betrachteten suspendierten Teilchens. Nach der Gleichung (8 a) ist daher zu erwarten, daß Δ bei Zucker etwa $\sqrt{1000}$ mal größer sei, als bei dem Teilchen von 1 Mikron Durchmesser. Dies ist nun, wie man sieht, wirklich angenähert richtig.

[16] Für Ionen können wir aus ihrer Wanderungsgeschwindigkeit l aus der Gleichung (8) bestimmen. l ist gleich der Elektrizitätsmenge in Coulomb, welche bei der Konzentration $v = 1$ des betreffenden Ions und dem Spannungsgefälle 1 Volt pro Zentimeter durch 1 qcm in 1 Sekunde hindurch ginge. Bei diesem gedachten Vorgang ist die Geschwindigkeit v der Ionenbewegung (in Zentimeter/Sekunde) offenbar durch die Gleichung bestimmt:

[17] $$l = v \cdot 96\,000.$$

Da ferner 1 Volt 10^8 elektromagnetische Einheiten enthält, und die Ladung eines (einwertigen) Ions gleich $\dfrac{9600}{N}$ elektromagnetischen Einheiten ist, so ist die bei dem gedachten Vorgang auf ein Ion wirkende Kraft k:

[18] $$k = \frac{10^8 \cdot 9600}{N}.$$

Setzt man diesen Wert von k und den aus der vorhin gefundenen Gleichung sich ergebenden Wert von v:

$$v = \frac{l}{96\,000}$$

in die Gleichung (2) ein, so erhält man:

$$\Re = \frac{k}{v} = \frac{10^8 \cdot 9600 \cdot 96\,000}{l \cdot N}.$$

Diese Formel gilt bei der üblichen Definition von l auch für mehrwertige Ionen. Durch Einsetzen dieses Wertes für $\Re$ in die Gleichung (8) erhält man:

$$\Delta = 4{,}25 \cdot 10^{-5} \sqrt{l\,T\,\tau}.$$

Die Formel ergibt für Zimmertemperatur und $\tau = 1$:

Bezeichnung des Ions	l	Δ in Mikron	
H	300	125	[19]
K	65	58	
Düsoamylammoniumion $C_{10} H_{24} N$	24	35	

(Eingegangen: 1. April.)

Published in *Zeitschrift für Elektrochemie und angewandte physikalische Chemie* 14 (1908): 235–239. Received 1 April 1908, published 24 April 1908.

[1] *Svedberg 1906a* and *1906b*. Svedberg was the first to attempt to relate Einstein's theoretical studies of Brownian motion to experimental data. Svedberg's analysis of his experiments had been criticized by Einstein (*Einstein 1907c* [Doc. 40]). For a discussion of Einstein's reaction to Svedberg's work, see the editorial note, "Einstein on Brownian Motion," § VII, pp. 219–222.

[2] Richard Lorenz was Professor of Electrochemistry and Physical Chemistry at the ETH. After reading *Einstein 1905k* (Doc. 16), he wrote to Einstein requesting reprints of Einstein's papers on the foundations of statistical

physics (Richard Lorenz to Einstein, 15 November 1907).

[3] The "mittlere Größe dieser ungeordneten Bewegungen" refers to the mean square displacement. Its relationship to diffusion was first analyzed in *Einstein 1905k* (Doc. 16), § 4.

[4] See *Nernst 1888*, *1889*, and *1898*, pp. 357–364, pp. 659–661. Einstein had earlier demonstrated his familiarity with Nernst's theory of electrolytes; see *Einstein 1902a* (Doc. 2), pp. 802–803.

[5] The remainder of this paragraph essentially follows the line of reasoning in *Einstein 1905j* (Doc. 15), § 4.

[6] In *Einstein 1906c* (Doc. 33), on the basis of an erroneous calculation, Einstein calculated N to be 4.15×10^{23} per mole. The values obtained on the basis of the kinetic theory of gases

and of Planck's theory of black-body radiation are close to the value given by Einstein here (see, for e.g., *Meyer, O. E. 1899*, p. 337, and *Planck 1901b*). This is the first time in his papers on Brownian motion that Einstein noted this discrepancy.

[7] For a determination of molecular dimensions from data on diffusion that is based on an argument similar to Einstein's, see *Sutherland 1905*.

[8] This assumption is the main simplification in this paper compared to Einstein's previous work on Brownian motion, in which the displacements are determined by a probability distribution (see *Einstein 1905k* [Doc. 16], p. 556, and *Einstein 1906b* [Doc. 32], p. 377).

[9] The correct expression is "$\sqrt{\overline{\Delta^2}}$."

[10] This equation was first derived in *Einstein 1905k* (Doc. 16), p. 559.

[11] *Einstein 1907c* (Doc. 40).

[12] The value for the viscosity of water is actually given for the temperature 9.5°C (as in *Einstein 1905j* [Doc. 15], p. 21). The particle size is that assumed in *Einstein 1905k* (Doc. 16), p. 559.

[13] See note 6.

[14] The same value for the diffusion constant was used in *Einstein 1906c* (Doc. 33), pp. 305–306.

[15] For the size of a sugar molecule, see also *Einstein 1905j* (Doc. 15).

[16] This sentence should be: "Für Ionen können wir Δ aus ihrer Wanderungsgeschwindigkeit l. . . ." For a contemporary discussion of the concept of migration velocity of ions, see *Nernst 1898*, pp. 352–353; for a comprehensive account, see *Bredig 1894*.

[17] This equation follows from Faraday's law; for a contemporary survey of its history and of its applications, see *Ostwald 1893*, pp. 579–592.

[18] The units of k are dynes.

[19] In *Landolt and Börnstein 1905*, p. 763, the migration velocity for hydrogen at 18°C is given as 318, and that for potassium at the same temperature as 64.7. The migration velocity for the "Diisoamylammoniumion" at 25°C is listed in *Bredig 1894*, p. 228, as 24.2.

EINSTEIN AND LAUB ON THE
ELECTRODYNAMICS OF MOVING MEDIA

At the beginning of this century, the electrodynamics of moving bodies, as usually understood, included not only the microscopic electron theory, but also the macroscopic theory of electromagnetic and optical phenomena in polarizable and magnetizable material media in motion. To distinguish between the two topics, the latter is referred to here as the electrodynamics of moving media. In 1905 Einstein had only applied relativistic kinematics to Lorentz's electron theory. In 1908 Minkowski offered the first solution to the problem of formulating a relativistic electrodynamics of moving media. Since then, the nature of the proper solution to this problem has been a subject of considerable controversy, with a number of questions still in dispute. Einstein, in collaboration with Jakob Laub, discussed the topic in two papers, *Einstein and Laub 1908a* (Doc. 51), *1908b* (Doc. 52). In the first, they rederived Minkowski's relativistic field equations and suggested an experimental test of them. In the second, they disputed his expression for the force exerted by the magnetic field on a volume element of a magnetic medium. They also published two corrections to the first paper, *Einstein and Laub 1908c* (Doc. 53), *1909* (Doc. 54). In *Einstein 1909a* (Doc. 55), Einstein commented negatively on an attempt to show that Lorentz's electrodynamics of moving media, which appears to be a nonrelativistic theory, is actually in accord with relativity.

Starting with Hertz, one approach to the electrodynamics of a (resting or moving) medium was simply to postulate the macroscopic field equations, which may be considered to be fundamental, or to be ultimately derivable from some underlying microscopic model.[1] This approach employs a pair of fields (**E** and **D**) to describe the electric state of the medium, another pair (**B** and **H**) to characterize its magnetic state, as well as charge and current density functions (the latter may include conductive and convective terms). In addition to Maxwell's field equations, which take the same form in all media, constitutive equations must be specified to characterize the relations between the field quantities in a particular medium. An expression for the force exerted by the electric and magnetic fields on a volume element of the medium, then known as the ponderomotive force, must also be specified.

Starting with Lorentz, another approach to the macroscopic equations was to derive them from microscopic models of the structure of material media, both dielectrics and conductors. Lorentz's electron theory, the most influential such program, postulates only one electric and one magnetic field vector in vacuum (ether), and charged particles, the motions of which constitute convection currents.[2] Lorentz also specified an expression for the ponderomotive force on a moving charged particle, which is now called the Lorentz

---

[1] For a review of Hertz's theory, as well as other theories of electrodynamics before 1905, see *Hirosige 1966*, which, however, does not stress the distinction between the macroscopic and microscopic approaches.

[2] See *Lorentz 1904c, 1909b*, and *1915* for contemporary reviews of his work on the electron theory. For further discussion of Lorentz's theory, see the editorial note, "Einstein on the Theory of Relativity," § II, pp. 256–257.

force. Assuming the presence of elastically bound charges in dielectrics, and of free charges in conductors, the electron theory aims to derive the macroscopic field equations for various material media, as well as their constitutive relations, by suitably averaging fields, charges, and currents over macroscopic regions of each medium. Lorentz was able to use such microscopic models of matter to explain a number of electrical, magnetic, and optical phenomena in material media.[3]

Hertz's electrodynamics of moving media obeys the Galileian relativity principle, and hence is able to account for the failure of attempts to detect the motion of the earth by means of optical experiments. But it failed to explain other optical phenomena, and was soon rejected.[4] Emil Cohn postulated another set of phenomenological equations that remedies the defects of Hertz's theory and still is able to account for the failure of attempts to detect the motion of the earth.[5]

By 1899 Einstein was a firm adherent of atomism for both matter and electricity, and expressed himself in favor of the basic concepts of the electron theory a number of times thereafter.[6] In his first paper on the theory of relativity he showed that the basic equations of Lorentz's electron theory are consistent with relativistic kinematics if the transformation properties of the quantities entering the equations are properly defined. But he did not address the problem of formulating a relativistic electrodynamics of moving media.

In 1908, Hermann Minkowski turned to this problem. He had recently discovered a four-dimensional geometrical reformulation of the theory of relativity, in which Lorentz transformations are interpreted as rotations in a space with three real and one imaginary coordinate.[7] He utilized the four-dimensional formalism to facilitate investigation of the invariance of various equations under the Lorentz group. In particular, he found the electrodynamical equations for a moving medium that follow, with the use of the relativity principle, from the form of the equations for a medium at rest.[8] Minkowski showed that Lorentz's macroscopic equations for moving media do not agree with the relativistic equations.[9] He was working on a derivation of the relativistic equations from the electron theory at the time of his death in 1909.[10]

Einstein and Laub wrote their two papers in 1908 as a response to Minkowski's work.

---

[3] See *Lorentz 1904c*, § IV, "Elektromagnetische Vorgänge in ponderablen Körpern."

[4] For a discussion of Hertz's theory and its rejection, see the editorial note, "Einstein on the Theory of Relativity," § II, p. 255.

[5] See *Hirosige 1966* for a discussion of Hertz's and Cohn's theories.

[6] See Vol. 1, the editorial notes, "Einstein on Thermal, Electrical, and Radiation Phenomena," pp. 235–237, and "Einstein on the Electrodynamics of Moving Bodies," pp. 223–225.

[7] His first lecture on the subject (*Minkowski 1907b*) was not published until 1915. His first publication on the four-dimensional approach

was *Minkowski 1908*. In a letter of 9 October 1907, Minkowski informed Einstein that *Einstein 1905r* (Doc. 23) was to be discussed in a Göttingen seminar (see *Pyenson 1985*, p. 83), and requested an offprint. For discussions of Minkowski's work on special relativity, see *Galison 1979* and *Pyenson 1977*.

[8] See *Minkowski 1909*.

[9] See *Minkowski 1908*, § 9. In § 10 Minkowski showed that, up to terms of first order in the velocity, Cohn's equations agree with his own.

[10] His work was completed by Max Born and published as *Minkowski 1910*.

At the time of their collaboration, Laub was *Assistent* to Wilhelm Wien, Professor of Physics at the University of Würzburg.[11] Laub's interest in the theory of relativity had already led to two articles applying the theory to optical problems.[12] He started to correspond with Einstein,[13] and on 2 February 1908 suggested that he come to Bern to work with him. On 1 March Laub wrote that he planned to arrive at the beginning of April.[14] The datelines of their joint papers[15] suggest that both were written during a little over a month of collaboration, after which Laub returned to Würzburg.[16] Laub showed the manuscripts to Wien who, in addition to being Laub's superior, was a co-editor of the *Annalen*; he added several references at Wien's suggestion.[17]

At the time they wrote their papers, Einstein and Laub evidently were not at ease with the four-dimensional approach. They justified publication of a more elementary derivation of Minkowski's field equations in their first paper on the grounds of the "rather great [mathematical] demands" ("ziemlich große Anforderungen") placed on readers by Minkowski's work.[18] The second and final section of the paper applies Minkowski's equations to the Wilson effect, the occurrence of equal and opposite charges on the two surfaces of an uncharged dielectric cylindrical shell when it is rotated in a magnetic field.[19] They showed, for a dielectric with significant magnetic permeability, that Minkowski's equations predict a different charge density than do Lorentz's. The Minkowski prediction was confirmed several years afterward.[20]

The discovery of several incorrect factors of $1/c$ in their equations led Einstein and Laub to publish a correction to the paper.[21] Einstein wrote consolingly to Laub that "well patched is always still better than full of holes" ("gut geflickt ist immer noch besser als verlöchert").[22]

A second correction was soon needed. Max Laue criticized Einstein and Laub's discus-

---

[11] Laub studied at Göttingen from 1902 to 1905, and then went to Würzburg to work with Wien. He submitted a doctoral dissertation on secondary cathode ray emission in 1906, but had already developed a particular interest in the theory of relativity. See *Pyenson 1976* for a discussion of Laub's career that emphasizes his work with Einstein.

[12] See *Laub 1907, 1908*. For Einstein's comment on the first paper, see *Einstein 1907j* (Doc. 47), p. 414. For Laue's comments, see Max Laue to Einstein, 4 September 1907.

[13] The first known letter in their correspondence is Jakob Laub to Einstein, 27 January 1908.

[14] In a letter to Mileva Einstein-Marić of 17 April 1908, Einstein stated that he was working with Laub.

[15] *Einstein and Laub 1908a* (Doc. 51) is dated 29 April, and *Einstein and Laub 1908b*

(Doc. 52) is dated 7 May.

[16] Jakob Laub to Einstein, 18 May 1908, is written from Würzburg.

[17] See Jakob Laub to Einstein 18 May, 19 May, and 30 May 1908 for reports of comments by Wien on the two papers.

[18] *Einstein and Laub 1908a* (Doc. 51), p. 532.

[19] See *Wilson 1904*.

[20] See *Wilson and Wilson 1913*. Soon after the Einstein-Laub paper was written, Wien suggested the possibility of testing the prediction with an iron solution (see Jakob Laub to Einstein, 19 May 1908), but nothing seems to have come of this suggestion.

[21] See *Einstein and Laub 1908c* (Doc. 53).

[22] Einstein to Jakob Laub, mid-October 1908 (excerpt from the auction catalogue of Gerd Rosen, Berlin, April 1961, p. 3, item no. 2365a).

sion of the conditions for the field vectors at the boundary between two media, and derived a different set of boundary conditions.[23] In a supplementary note to their paper,[24] Einstein and Laub acknowledged Laue's criticism and gave their own derivation of his boundary conditions.

Their first paper had rederived and applied Minkowski's version of the field equations. In their second paper,[25] Einstein and Laub criticized his expression for the ponderomotive force density on a conduction current element in a magnetizable medium. Minkowski gave this force density as [s,B], the vector product of the current vector density s and the magnetic induction B. Using arguments based on the electron theory, they rejected this expression and proposed another: [s,H], where H is the magnetic force, which they supported by a simple example.[26]

Minkowski had shown that, from the four-dimensional point of view, the energy density, momentum density, and Maxwell stress tensor are unified in a four-dimensional tensor. The tensor he proposed is now often referred to as Minkowski's form of the stress-momentum-energy tensor of the electromagnetic field in a material medium.[27] By taking the four-dimensional divergence of this tensor, he derived an expression for the total ponderomotive force density exerted on a magnetizable medium by the electromagnetic field. Starting from the standpoint of the electron theory, Einstein and Laub derived a different expression for the total ponderomotive force density that includes a term [s,H], which they interpreted as the force density per unit volume on a conduction current element.[28] They also gave expressions for the components of the stress-momentum-energy tensor of the field in a medium, differing from those of Minkowski, from which their expression for the total ponderomotive force density can be derived.[29]

Soon after he returned to Würzburg, Laub wrote to Einstein setting forth Wien's objections to their expression for the ponderomotive force density.[30] These objections were presumably discussed in letters now missing, for two months later Einstein wrote Laub: "I am also quite firmly convinced that our expressions for the ponderomotive forces are the correct ones" ("Ich bin auch ganz fest überzeugt, dass unsere Ausdrücke für die ponderomotorischen Kräfte die richtigen sind").[31] In 1910, Einstein gave a talk in which he

[23] For their original boundary conditions, see *Einstein and Laub 1908a* (Doc. 51), p. 535. Laue's letter with his comments is mentioned in *Einstein and Laub 1909* (Doc. 54), p. 445. Laue published a discussion of these boundary conditions in *Laue 1911b*, pp. 127–129.

[24] See *Einstein and Laub 1909* (Doc. 54).

[25] See *Einstein and Laub 1908b* (Doc. 52).

[26] See *Einstein and Laub 1908b* (Doc. 52), pp. 545–546.

[27] See *Minkowski 1908*, pp. 92–93, 97. Minkowski, rather than using tensor notation, actually wrote out the matrix corresponding to this tensor. Abbreviated and variant forms of the name "stress-momentum-energy tensor" are common.

[28] See *Einstein and Laub 1908b* (Doc. 52), § 2.

[29] See *Einstein and Laub 1908b* (Doc. 52), § 3. Since they did not use four-dimensional notation in this paper, Einstein and Laub did not combine these expressions into a four-dimensional tensor, nor take its four-dimensional divergence.

[30] See Jakob Laub to Einstein, 30 May 1908.

[31] Einstein to Jakob Laub, 30 July 1908. The letter also indicates that they were contemplating additional calculations involving the ponderomotive forces, but Einstein was not enthusiastic about the prospect.

again defended the Einstein-Laub expression with the help of a new example.[32] However, the expression was soon disputed in print.[33] Controversy continued over this question, as well as over the closely related question of the correct expression for the stress-momentum-energy tensor of the electromagnetic field in a material medium.[34]

In recent years, it has been suggested that these controversies arose from the failure to recognize that only the total force acting on a macroscopic body is uniquely defined. Its division into two parts—an electromagnetic (ponderomotive) force exerted by the field and a mechanical force exerted by ponderable matter—is arbitrary to a large extent. Furthermore, the division of the electromagnetic force density into components acting respectively on charge and conduction current densities and on polarization and magnetization densities is also not unique.[35] It is possible to rewrite Einstein's and Laub's expression for the total electromagnetic force density in such a way that it is consistent with Minkowski's expression for the force density on a conduction current element.[36]

As noted above, Lorentz's equations for moving media differ from Minkowski's relativistic equations. This was considered puzzling, since the underlying equations of the electron theory, from which Lorentz derived the macroscopic equations, are relativistically invariant. Dmitry Mirimanoff, a *Privatdozent* in Mathematics at the University of Geneva, attempted to show that Lorentz's equations can be brought into accord with the principle of relativity by suitably redefining the transformation properties of some of the field variables under Lorentz transformations.[37] *Einstein 1909a* (Doc. 55) shows that this is impossible.[38] Einstein apparently had some difficulty getting this paper accepted by the *Annalen*. In a letter of 19 January 1909, Wien indicated that he was once again returning the paper to Einstein, and asked for a somewhat more detailed exposition. Presumably, Einstein's revision of the manuscript satisfied Wien, since it was accepted three days later.

In the last paragraph of his paper, Einstein drew attention to an electron-theoretical derivation of Minkowski's equations by Frank.[39] This paper established, for nonmagnetic media, that it was Lorentz's failure to take the Lorentz contraction and time dilation fully into account which led to his derivation of nonrelativistic equations.

---

[32] See *Einstein 1910b*.

[33] See, e.g., *Gans 1911* and *Grammel 1913*.

[34] See *Pauli 1921*, pp. 662–668, for a review of these discussions up to 1921 (see p. 216 of *Pauli 1958* for a note added to the English edition). See *De Groot and Suttorp 1972*, chap. V, § 7, for a review that includes subsequent contributions.

[35] See, e.g., *De Groot and Suttorp 1972*, chap. V, § 7, and *Pavlov 1978* for discussions of the nonuniqueness of these divisions.

[36] See, e.g., *Pavlov 1978*, p. 172.

[37] See *Mirimanoff 1909*.

[38] A letter of 12 February 1909 from Mirimanoff to Einstein indicates that they had previously corresponded about Mirimanoff's paper.

[39] See *Frank 1908*. For a review of electron-theoretical derivations of Minkowski's equations up to 1920, see *Pauli 1921*, pp. 659–662.

51. ''On the Fundamental Electromagnetic Equations for Moving Bodies''

[*Einstein and Laub 1908a*]

DATED Bern, 29 April 1908
RECEIVED 2 May 1908
PUBLISHED 7 July 1908

IN: *Annalen der Physik* 26 (1908): 532–540.

532

5. *Über die elektromagnetischen Grundgleichungen für bewegte Körper; von A. Einstein und J. Laub.*

In einer kürzlich veröffentlichten Abhandlung[1]) hat Hr. Minkowski die Grundgleichungen für die elektromagnetischen Vorgänge in bewegten Körpern angegeben. In Anbetracht des Umstandes, daß diese Arbeit in mathematischer Beziehung an den Leser ziemlich große Anforderungen stellt, halten wir es nicht für überflüssig, jene wichtigen Gleichungen im folgenden auf elementarem Wege, der übrigens mit dem Minkowskischen im wesentlichen übereinstimmt, abzuleiten.

§ 1. Ableitung der Grundgleichungen für bewegte Körper.

Der einzuschlagende Weg ist folgender: Wir führen zwei Koordinatensysteme K und K' ein, welche beide beschleunigungsfrei, jedoch relativ zueinander bewegt sind. Ist im Raume Materie vorhanden, die relativ zu K' ruht, gelten in bezug auf K' die Gesetze der Elektrodynamik ruhender Körper, welche durch die Maxwell-Hertzschen Gleichungen dargestellt sind. Transformieren wir diese Gleichungen auf das System K, so erhalten wir unmittelbar die elektrodynamischen Gleichungen bewegter Körper für den Fall, daß die Geschwindigkeit der Materie räumlich und zeitlich konstant ist. Die so erhaltenen Gleichungen gelten offenbar mindestens in erster Annäherung auch dann, wenn die Geschwindigkeitsverteilung der Materie eine beliebige ist. Diese Annahme rechtfertigt sich zum Teil auch dadurch, daß das auf diese Weise erhaltene Resultat streng gilt in dem Falle, daß eine Anzahl von mit verschiedenen Geschwindigkeiten gleichförmig bewegten Körpern vorhanden ist, welche voneinander durch Vakuumzwischenräume getrennt sind.

[2]

---

[1]   1) H. Minkowski, Göttinger Nachr. 1908.

Elektromagnetische Grundgleichungen für bewegte Körper.   **533**

Wir wollen mit Bezug auf das System K' den Vektor der elektrischen Kraft $\mathfrak{E}'$, der magnetischen Kraft $\mathfrak{H}'$, der dielektrischen Verschiebung $\mathfrak{D}'$, der magnetischen Induktion $\mathfrak{B}'$, den des elektrischen Stromes $\mathfrak{s}'$ nennen; ferner bezeichne ϱ' die elektrische Dichte. Es mögen für das Bezugssystem K' die **Maxwell-Hertz**schen Gleichungen gelten:

$$(1) \qquad \operatorname{curl}' \mathfrak{H}' = \frac{1}{c}\left(\frac{\partial \mathfrak{D}'}{\partial t'} + \mathfrak{s}'\right),$$

$$(2) \qquad \operatorname{curl}' \mathfrak{E}' = -\frac{1}{c}\frac{\partial \mathfrak{B}'}{\partial t'},$$

$$(3) \qquad \operatorname{div}' \mathfrak{D}' = \varrho',$$

$$(4) \qquad \operatorname{div}' \mathfrak{B}' = 0.$$

Wir betrachten ein zweites rechtwinkliges Bezugssystem K, dessen Achsen dauernd parallel sind denen von K'. Der Anfangspunkt von K' soll sich mit der konstanten Geschwindigkeit v in der positiven Richtung der x-Achse von K bewegen. Dann gelten bekanntlich bei passend gewähltem Anfangspunkt der Zeit nach der Relativitätstheorie für jedes Punktereignis folgende Transformationsgleichungen [1]):

$$(5) \qquad \begin{cases} x' = \beta\,(x - v\,t), \\ y' = y, \\ z' = z, \\ t' = \beta\left(t - \frac{v}{c^2}\,x\right), \end{cases} \qquad \left(\beta = \frac{1}{\sqrt{1 - \frac{v^2}{c^2}}}\right),$$

wobei x, y, z, t die Raum- und Zeitkoordinaten im System K bedeuten. Führt man die Transformationen aus, so erhält man die Gleichungen:

$$(1\,a) \qquad \operatorname{curl} \mathfrak{H} = \frac{1}{c}\left(\frac{\partial \mathfrak{D}}{\partial t} + \mathfrak{s}\right),$$

$$(2\,a) \qquad \operatorname{curl} \mathfrak{E} = -\frac{1}{c}\frac{\partial \mathfrak{B}}{\partial t},$$

$$(3\,a) \qquad \operatorname{div} \mathfrak{D} = \varrho,$$

$$(4\,a) \qquad \operatorname{div} \mathfrak{B} = 0,$$

[3]   1) A. **Einstein**, Ann. d. Phys. **17.** p. 902. 1905.

534 *A. Einstein u. J. Laub.*

wobei gesetzt ist:

$$(6)\begin{cases} \mathfrak{E}_x = \mathfrak{E}_x', \\[4pt] \mathfrak{E}_y = \beta\left(\mathfrak{E}_y' + \frac{v}{c}\,\mathfrak{B}_z'\right), \\[4pt] \mathfrak{E}_z = \beta\left(\mathfrak{E}_z' - \frac{v}{c}\,\mathfrak{B}_y'\right), \\[4pt] \mathfrak{D}_x = \mathfrak{D}_x', \\[4pt] \mathfrak{D}_y = \beta\left(\mathfrak{D}_y' + \frac{v}{c}\,\mathfrak{H}_z'\right), \\[4pt] \mathfrak{D}_z = \beta\left(\mathfrak{D}_z' - \frac{v}{c}\,\mathfrak{H}_y'\right); \end{cases}$$

$$(7)\begin{cases} \mathfrak{H}_x = \mathfrak{H}_x', \\[4pt] \mathfrak{H}_y = \beta\left(\mathfrak{H}_y' - \frac{v}{c}\,\mathfrak{D}_z'\right), \\[4pt] \mathfrak{H}_z = \beta\left(\mathfrak{H}_z' + \frac{v}{c}\,\mathfrak{D}_y'\right), \\[4pt] \mathfrak{B}_x = \mathfrak{B}_x', \\[4pt] \mathfrak{B}_y = \beta\left(\mathfrak{B}_y' - \frac{v}{c}\,\mathfrak{E}_z'\right), \\[4pt] \mathfrak{B}_z = \beta\left(\mathfrak{B}_z' + \frac{v}{c}\,\mathfrak{E}_y'\right) \end{cases}$$

und

$$(8)\qquad \varrho = \beta\left(\varrho' + \frac{v}{c}\,\mathfrak{s}_x'\right), \qquad [4]$$

$$(9)\begin{cases} \mathfrak{s}_x = \beta\left(\mathfrak{s}_x' + \frac{v}{c}\,\varrho'\right), \\[4pt] \mathfrak{s}_y = \mathfrak{s}_y', \\[4pt] \mathfrak{s}_z = \mathfrak{s}_z'. \end{cases} \qquad [5]$$

Will man die Ausdrücke für die gestrichenen Größen als Funktion der ungestrichenen haben, so vertauscht man die gestrichenen und ungestrichenen Größen und ersetzt v durch $-v$.

Die Gleichungen (1a) bis (4a), welche die elektromagnetischen Vorgänge relativ zum System K beschreiben, haben dieselbe Gestalt, wie die Gleichungen (1) bis (4). *Wir wollen daher die Größen*

$$\mathfrak{E}, \mathfrak{D}, \mathfrak{H}, \mathfrak{B}, \varrho, \mathfrak{s}$$

analog benennen, *wie die entsprechenden Größen relativ zum System K'. Es sind also $\mathfrak{E}, \mathfrak{D}, \mathfrak{H}, \mathfrak{B}, \varrho, \mathfrak{s}$ die elektrische Kraft, die dielektrische Verschiebung, die magnetische Kraft, die magne-*

Elektromagnetische Grundgleichungen für bewegte Körper.   535

tische *Induktion*, *die elektrische Dichte*, *der elektrische Strom* in bezug auf K.

Die Transformationsgleichungen (6) und (7) reduzieren sich für das Vakuum auf die früher gefundenen[1]) Gleichungen für elektrische und magnetische Kräfte.

Es ist klar, daß man durch wiederholte Anwendung solcher Transformationen, wie die soeben durchgeführte, stets auf Gleichungen von derselben Gestalt wie die ursprünglichen (1) bis (4) kommen muß, und daß für solche Transformationen die Gleichungen (6) bis (9) maßgebend sind. Denn es wurde bei der ausgeführten Transformation in formaler Beziehung nicht davon Gebrauch gemacht, daß die Materie relativ zu dem ursprünglichen System K' ruhte.

Die Gültigkeit der transformierten Gleichungen (1a) bis (4a) nehmen wir an auch für den Fall, daß die Geschwindigkeit der Materie räumlich und zeitlich variabel ist, was in erster [6]   Annäherung richtig sein wird.

Es ist bemerkenswert, daß die Grenzbedingungen für die Vektoren $\mathfrak{E}$, $\mathfrak{D}$, $\mathfrak{H}$, $\mathfrak{B}$ an der Grenze zweier Medien dieselben sind, wie für ruhende Körper. Es folgt dies direkt aus den [7]   Gleichungen (1a) bis (4a).

Die Gleichungen (1a) bis (4a) gelten genau wie die Gleichungen (1) bis (4) ganz allgemein für inhomogene und anisotrope Körper. Dieselben bestimmen die elektromagnetischen Vorgänge noch nicht vollständig. Es müssen vielmehr noch Beziehungen gegeben sein, welche die Vektoren $\mathfrak{D}$, $\mathfrak{B}$ und $\mathfrak{s}$ als Funktion von $\mathfrak{E}$ und $\mathfrak{H}$ ausdrücken. Solche Gleichungen wollen wir nun für den Fall angeben, daß die *Materie isotrop* ist. Betrachten wir zunächst wieder den Fall, daß alle Materie relativ zu K' ruht, so gelten in bezug auf K' die Gleichungen:

$$(10) \qquad\qquad \mathfrak{D}' = \varepsilon\,\mathfrak{E}',$$

$$(11) \qquad\qquad \mathfrak{B}' = \mu\,\mathfrak{H}',$$

$$(12) \qquad\qquad \mathfrak{s}' = \sigma\,\mathfrak{E}',$$

wobei $\varepsilon =$ Dielektrizitätskonstante, $\mu =$ Permeabilität, $\sigma =$ elektrische Leitfähigkeit als bekannte Funktionen von x', y', z', t' anzusehen sind. Durch die Transformation von (10) bis (12)

---

1) A. Einstein, l. c. p. 909.

536   *A. Einstein u. J. Laub.*

auf K mittels der Umkehrung unserer Transformationsgleichungen (6) bis (9) erhält man die für das System K
geltenden Beziehungen:

(10a)
$$\begin{cases} \mathfrak{D}_x = \varepsilon\,\mathfrak{E}_x, \\[1mm] \mathfrak{D}_y - \dfrac{v}{c}\,\mathfrak{H}_z = \varepsilon\left(\mathfrak{E}_y - \dfrac{v}{c}\,\mathfrak{B}_z\right), \\[1mm] \mathfrak{D}_z + \dfrac{v}{c}\,\mathfrak{H}_y = \varepsilon\left(\mathfrak{E}_z + \dfrac{v}{c}\,\mathfrak{B}_y\right), \end{cases}$$

(11a)
$$\begin{cases} \mathfrak{B}_x = \mu\,\mathfrak{H}_x, \\[1mm] \mathfrak{B}_y + \dfrac{v}{c}\,\mathfrak{E}_z = \mu\left(\mathfrak{H}_y + \dfrac{v}{c}\,\mathfrak{D}_z\right), \\[1mm] \mathfrak{B}_z - \dfrac{v}{c}\,\mathfrak{E}_y = \mu\left(\mathfrak{H}_z - \dfrac{v}{c}\,\mathfrak{D}_y\right), \end{cases}$$

(12a)
$$\begin{cases} \beta\left(\mathfrak{s}_x - \dfrac{v}{c}\,\varrho\right) = \sigma\,\mathfrak{E}_x, \\[1mm] \mathfrak{s}_y = \sigma\,\beta\left(\mathfrak{E}_y - \dfrac{v}{c}\,\mathfrak{B}_z\right), \\[1mm] \mathfrak{s}_z = \sigma\,\beta\left(\mathfrak{E}_z + \dfrac{v}{c}\,\mathfrak{B}_y\right), \end{cases}$$
[8]

Ist die Geschwindigkeit der Materie nicht der X-Achse
parallel, sondern ist diese Geschwindigkeit durch den Vektor $\mathfrak{v}$
bestimmt, so erhält man die mit den Gleichungen (10a) bis (12a)
gleichartigen vektoriellen Beziehungen:

(13)
$$\begin{cases} \mathfrak{D} + \dfrac{1}{c}\,[\mathfrak{v}\,\mathfrak{H}] = \varepsilon\left\{\mathfrak{E} + \dfrac{1}{c}\,[\mathfrak{v}\,\mathfrak{B}]\right\}, \\[1mm] \mathfrak{B} - \dfrac{1}{c}\,[\mathfrak{v}\,\mathfrak{E}] = \mu\left\{\mathfrak{H} - \dfrac{1}{c}\,[\mathfrak{v}\,\mathfrak{D}]\right\}, \\[1mm] \beta\left(\mathfrak{s}_{\mathfrak{v}} - \dfrac{|\mathfrak{v}|}{c}\,\varrho\right) = \sigma\left\{\mathfrak{E} + \dfrac{1}{c}\,[\mathfrak{v}\,\mathfrak{B}]\right\}_{\mathfrak{v}}, \\[1mm] \mathfrak{s}_{\bar{\mathfrak{v}}} = \sigma\,\beta\left\{\mathfrak{E} + \dfrac{1}{c}\,[\mathfrak{v}\,\mathfrak{B}]\right\}_{\bar{\mathfrak{v}}}, \end{cases}$$
[9]

wobei der Index $\mathfrak{v}$ bedeutet, daß die Komponente nach der
Richtung von $\mathfrak{v}$, der Index $\bar{\mathfrak{v}}$, daß die Komponenten nach den
auf $\mathfrak{v}$ senkrechten Richtungen $\bar{\mathfrak{v}}$ zu nehmen ist.

§ 2. Über das elektromagnetische Verhalten bewegter Dielektrika. Versuch von Wilson.

Im folgenden Abschnitt wollen wir noch an einem einfachen Spezialfall zeigen, wie sich bewegte Dielektrika nach

Elektromagnetische Grundgleichungen für bewegte Körper.   537

der Relativitätstheorie verhalten, und worin sich die Resultate
von den durch die Lorentzsche Theorie gelieferten, unterscheiden.

Es sei S ein im Querschnitt angedeuteter, prismatischer
Streifen (vgl. Figur) aus einem homogenen, isotropen Nicht-
leiter, der sich senkrecht zur Papierebene in beiderlei Sinn
ins Unendliche erstreckt und sich vom Beschauer nach der
Papierebene zu mit der konstanten Geschwindigkeit v zwischen den beiden Kondensatorplatten A_1 und A_2 hindurch-
bewegt. Die Ausdehnung des Streifens S
senkrecht zu den Platten A sei unendlich klein relativ zu dessen Ausdehnung
parallel den Platten und zu beiden Aus-
dehnungen der Platten A; der Zwischenraum zwischen S und den Platten A (im
folgenden kurz Zwischenraum genannt)
sei außerdem gegenüber der Dicke von S zu vernachlässigen.

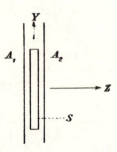

Das betrachtete Körpersystem beziehen wir auf ein relativ zu
den Platten A ruhendes Koordinatensystem, dessen positive
X-Richtung in die Bewegungsrichtung falle, und dessen Y- und
Z-Achsen parallel bzw. senkrecht zu den Platten A sind. Wir
wollen das elektromagnetische Verhalten des zwischen den
Platten A sich befindenden Streifenstückes untersuchen, falls
der elektromagnetische Zustand stationär ist.

Wir denken uns eine geschlossene Fläche, welche gerade
den wirksamen Teil der Kondensatorplatten nebst dem des
dazwischen liegenden Streifenstückes einschließt. Da sich innerhalb dieser Fläche weder bewegte wahre Ladungen, noch
elektrische Leitungsströme befinden, gelten die Gleichungen
(vgl. Gleichungen (1a) bis (4a)):

$$\operatorname{curl} \mathfrak{H} = 0,$$

$$\operatorname{curl} \mathfrak{E} = 0.$$

Innerhalb dieses Raumes sind also sowohl die elektrische, wie
auch die magnetische Kraft von einem Potential ableitbar.
Wir können daher sofort die Verteilung der Vektoren $\mathfrak{E}$ und $\mathfrak{H}$,
falls die Verteilung der freien elektrischen bzw. magnetischen
Dichte bekannt ist. Wir beschränken uns auf die Betrachtung

des Falles, daß die magnetische Kraft $\mathfrak{H}$ parallel der Y-Achse ist, die elektrische $\mathfrak{E}$ parallel der Z-Achse. Dazu, sowie zu der Voraussetzung, daß die in Betracht kommenden Felder innerhalb des Streifens, sowie innerhalb des Zwischenraumes homogen sind, berechtigen uns die oben erwähnten Größenordnungsbedingungen für die Abmessungen des betrachteten Systems. Ebenso schließen wir unmittelbar, daß die an den Enden des Streifenquerschnittes sich befindenden magnetischen Massen nur einen verschwindend kleinen Beitrag zum magnetischen Feld liefern.[1]) Die Gleichungen (13) geben dann für das Innere des Streifens folgende Beziehungen:

$$\mathfrak{D}_z + \frac{v}{c}\,\mathfrak{H}_y = \varepsilon\left(\mathfrak{E}_z + \frac{v}{c}\,\mathfrak{B}_y\right),$$

$$\mathfrak{B}_y + \frac{v}{c}\,\mathfrak{E}_z = \mu\left(\mathfrak{H}_y + \frac{v}{c}\,\mathfrak{D}_z\right).$$

Diese Gleichungen lassen sich auch in folgender Form schreiben:

$$(1) \quad \begin{cases} \left(1 - \varepsilon\,\mu\,\dfrac{v^2}{c^2}\right)\mathfrak{B}_y = \dfrac{v}{c}\,(\varepsilon\,\mu - 1)\,\mathfrak{E}_z + \mu\left(1 - \dfrac{v^2}{c^2}\right)\mathfrak{H}_y, \\[3mm] \left(1 - \varepsilon\,\mu\,\dfrac{v^2}{c^2}\right)\mathfrak{D}_z = \varepsilon\left(1 - \dfrac{v^2}{c^2}\right)\mathfrak{E}_z + \dfrac{v}{c}\,(\varepsilon\,\mu - 1)\,\mathfrak{H}_y. \end{cases}$$

Zur Deutung von (1) bemerken wir folgendes: An der Oberfläche des Streifens erfährt die dielektrische Verschiebung $\mathfrak{D}_z$ keinen Sprung, also ist $\mathfrak{D}_z$ die Ladung der Kondensatorplatten (genauer der Platte A_1) pro Flächeneinheit. Ferner ist $\mathfrak{E}_z \times \delta$ gleich der Potentialdifferenz zwischen den Kondensatorplatten A_1 und A_2, falls δ den Abstand der Platten bezeichnet, denn denkt man sich den Streifen durch einen parallel der XZ-Ebene verlaufenden unendlich engen Spalt getrennt, so ist $\mathfrak{E}$, nach den für diesen Vektor geltenden Grenzbedingungen, gleich der elektrischen Kraft in dem Spalt.

Wir betrachten nun zunächst den Fall, daß ein von außen erregtes Magnetfeld nicht vorhanden ist, d. h. nach dem obigen, daß in dem betrachteten Raume die magnetische Feldstärke $\mathfrak{H}_y$

---

1) Es erhellt dies auch daraus, daß wir ohne wesentliche Änderung der Verhältnisse den Kondensatorplatten und dem Streifen Kreiszylinderform geben könnten, in welchem Falle freie magnetische Massen aus Symmetriegründen überhaupt nicht auftreten könnten.

[10]

Elektromagnetische Grundgleichungen für bewegte Körper. 539

überhaupt verschwindet. Dann haben die Gleichungen (1) folgende Gestalt:

$$\left(1 - \varepsilon\mu\frac{v^2}{c^2}\right)\mathfrak{B}_y = \frac{v}{c}(\varepsilon\mu - 1)\mathfrak{E}_z,$$

$$\left(1 - \varepsilon\mu\frac{v^2}{c^2}\right)\mathfrak{D}_z = \varepsilon\left(1 - \frac{v^2}{c^2}\right)\mathfrak{E}_z.$$

Da $v < c$ sein muß, so sind, falls $\varepsilon\mu - 1 > 0$ ist, die Koeffizienten von $\mathfrak{E}_z$ in den beiden letzten Gleichungen positiv. Die Koeffizienten von $\mathfrak{B}_y$ und $\mathfrak{D}_z$ sind dagegen größer, gleich bzw. kleiner als Null, je nachdem die Streifengeschwindigkeit kleiner, gleich oder größer als $c/\sqrt{\varepsilon\mu}$, d. h. als die Geschwindigkeit elektromagnetischer Wellen in dem Streifenmedium, ist. Hat also $\mathfrak{E}_z$ einen bestimmten Wert, d. h. legt man an die Kondensatorplatten eine bestimmte Spannung an und variiert man die Streifengeschwindigkeit von kleineren zu größeren Werten, so wächst zunächst sowohl die dem Vektor $\mathfrak{D}$ proportionale Ladung der Kondensatorplatten, wie die magnetische Induktion $\mathfrak{B}$ im Streifen. Erreicht v den Wert $c/\sqrt{\varepsilon\mu}$, so wird sowohl die Ladung des Kondensators, wie auch die magnetische Induktion unendlich groß. Es würde also in diesem Falle eine Zerstörung des Streifens durch beliebig kleine angelegte Potentialdifferenzen stattfinden. Für alle $v > c/\sqrt{\varepsilon\mu}$ resultiert ein negativer Wert für $\mathfrak{D}$ und $\mathfrak{B}$. In dem letzten Falle würde also eine an die Kondensatorplatten gelegte Spannung eine Ladung des Kondensators in dem der Spannungsdifferenz entgegengesetzten Sinne bewirken.

Wir betrachten jetzt noch den Fall, daß ein von außen erregtes magnetisches Feld $\mathfrak{H}_y$ vorhanden ist. Dann hat man die Gleichung:

$$\left(1 - \varepsilon\mu\frac{v^2}{c^2}\right)\mathfrak{D}_z = \varepsilon\left(1 - \frac{v^2}{c^2}\right)\mathfrak{E}_z + \frac{v}{c}(\varepsilon\mu - 1)\mathfrak{H}_y,$$

welche bei gegebenem $\mathfrak{H}_y$ eine Beziehung zwischen $\mathfrak{E}_z$ und $\mathfrak{D}_z$ gibt. Beschränkt man sich nur auf Größen erster Ordnung in v/c, so hat man:

$$(2) \qquad \mathfrak{D}_z = \varepsilon\mathfrak{E}_z + \frac{v}{c}(\varepsilon\mu - 1)\mathfrak{H}_y,$$

während die Lorentzsche Theorie auf den Ausdruck:

[11] $(3) \qquad \mathfrak{D}_z = \varepsilon\mathfrak{E}_z + \frac{v}{c}(\varepsilon - 1)\mu\mathfrak{H}_y$

führt.

35*

540 *A. Einstein u. J. Laub. Elektromagn. Grundgleichungen usw.*

Die letzte Gleichung wurde bekanntlich von H. A. Wilson (Wilsoneffekt) experimentell geprüft. Man sieht, daß sich (2) [12] und (3) in Gliedern erster Ordnung unterscheiden. Hätte man einen dielektrischen Körper von beträchtlicher Permeabilität, so könnte man eine experimentelle Entscheidung zwischen den Gleichungen (2) und (3) treffen. [13]

Verbindet man die Platten A_1 und A_2 durch einen Leiter, so tritt auf den Kondensatorplatten eine Ladung von der Größe $\mathfrak{D}_z$ pro Flächeneinheit auf; man erhält sie aus der Gleichung (2), indem man berücksichtigt, daß bei verbundenen Kondensatorplatten $\mathfrak{E}_z = 0$ ist. Es ergibt sich:

$$\mathfrak{D}_z = \frac{v}{c}(\varepsilon\mu - 1)\mathfrak{H}_y.$$

Verbindet man die Kondensatorplatten A_1 und A_2 mit einem Elektrometer von unendlich kleiner Kapazität, so ist $\mathfrak{D}_z = 0$, und man bekommt für die Spannung $(\mathfrak{E}_z.\delta)$ die Gleichung:

$$0 = \varepsilon\,\mathfrak{E}_z + \frac{v}{c}(\varepsilon\mu - 1)\mathfrak{H}_y.$$

Bern, 29. April 1908.

(Eingegangen 2. Mai 1908.)

Published in *Annalen der Physik* 26 (1908): 532–540. Dated Bern, 29 April 1908, received 2 May 1908, published 7 July 1908.

[1] *Minkowski 1908*.

[2] *Minkowski 1908* arrives at the macroscopic equations for moving media by simply postulating that, in a co-moving frame of reference at any point, the equations must reduce to those for bodies at rest. Einstein and Laub considered this assumption in need of justification by a derivation of the equations from the electron theory (see *Einstein 1909a* [Doc. 55], p. 888).

[3] *Einstein 1905r* (Doc. 23).

[4] *c* in the denominator should be squared. See *Einstein and Laub 1908c* (Doc. 53).

[5] *c* in the denominator should be omitted. See *Einstein and Laub 1908c* (Doc. 53).

[6] See note 2.

[7] This assertion is corrected in *Einstein and Laub 1909* (Doc. 54).

[8] *c* in the denominator should be omitted. See *Einstein and Laub 1908c* (Doc. 53).

[9] On the left-hand side of this equation, *c* in the denominator should be omitted. See *Einstein and Laub 1908c* (Doc. 53).

[10] Such a cylindrical arrangement was used in Wilson's experiment (see *Wilson 1904*), discussed below.

[11] See *Lorentz 1904c*. Lorentz did not explicitly consider the case of a material with permeability different from 1; but the equation given here can be obtained by combining several of Lorentz's equations.

[12] See *Wilson 1904*.

[13] *Wilson and Wilson 1913* reports the results of such an experiment, which confirmed Einstein's and Laub's prediction.

52. "On the Ponderomotive Forces Exerted on Bodies at Rest in the Electromagnetic Field"

[*Einstein and Laub 1908b*]

DATED Bern, 7 May 1908
RECEIVED 13 May 1908
PUBLISHED 7 July 1908

IN: *Annalen der Physik* 26 (1908): 541–550.

541

6. *Über die*
im elektromagnetischen Felde auf ruhende
Körper ausgeübten ponderomotorischen Kräfte;
von A. Einstein und J. Laub.

———

In einer kürzlich erschienenen Abhandlung[1]) hat Hr. Minkowski einen Ausdruck für die auf beliebig bewegte Körper
wirkenden ponderomotorischen Kräfte elektromagnetischen Ursprunges angegeben. Spezialisiert man die Minkowskischen [2]
Ausdrücke auf ruhende, isotrope und homogene Körper, so
erhält man für die X-Komponente der auf die Volumeneinheit
wirkenden Kraft:

(1) $$K_x = \rho\,\mathfrak{E}_x + \mathfrak{z}_y\mathfrak{B}_z - \mathfrak{z}_z\mathfrak{B}_y,$$ [3]

wobei ρ die elektrische Dichte, $\mathfrak{z}$ den elektrischen Leitungsstrom,
$\mathfrak{E}$ die elektrische Feldstärke, $\mathfrak{B}$ die magnetische Induktion bedeuten. Dieser Ausdruck scheint uns aus folgenden Gründen mit
dem elektronentheoretischen Bild nicht in Einklang zu stehen:
Während nämlich ein von einem elektrischen Strom (Leitungsstrom) durchflossener Körper im Magnetfeld eine Kraft erleidet, wäre dies nach Gleichung (1) nicht der Fall, wenn der
im Magnetfeld befindliche Körper statt von einem Leitungs-
strom von einem Polarisationsstrom $(\partial\mathfrak{D}/\partial t)$ durchsetzt wird.
Nach Minkowski besteht also hier ein prinzipieller Unter-
schied zwischen einem Verschiebungsstrom und einem Leitungsstrom derart, daß ein Leiter nicht betrachtet werden kann
als ein Dielektrikum von unendlich großer Dielektrizitätskonstante.

Angesichts dieser Sachlage schien es uns von Interesse
zu sein, die ponderomotorischen Kräfte für beliebige magnetisierbare Körper auf elektronentheoretischem Wege abzuleiten.
Wir geben im folgenden eine solche Ableitung, wobei wir uns
aber auf ruhende Körper beschränken.

———

1) H. Minkowski, Gött. Nachr. 1908. p. 45. [1]

542 *A. Einstein u. J. Laub.*

§ 1. Kräfte, welche nicht von Geschwindigkeiten der Elementarteilchen abhängen.

[4] Wir wollen uns bei der Ableitung konsequent auf den Standpunkt der Elektronentheorie stellen [1]); wir setzen also:

$$(2) \qquad \mathfrak{D} = \mathfrak{E} + \mathfrak{P},$$

$$(3) \qquad \mathfrak{B} = \mathfrak{H} + \mathfrak{Q},$$

wobei $\mathfrak{P}$ den elektrischen, $\mathfrak{Q}$ den magnetischen Polarisationsvektor bedeutet. Die elektrische bzw. die magnetische Polarisation denken wir uns bestehend in räumlichen Verschiebungen von an Gleichgewichtslagen gebundenen, elektrischen bzw. magnetischen Massenteilchen von Dipolen. Außerdem nehmen wir noch das Vorhandensein von nicht an Dipole gebundenen, beweglichen elektrischen Teilchen (Leitungselektronen) an. In dem Raume zwischen den genannten Teilchen mögen die Maxwellschen Gleichungen für den leeren Raum gelten, und es seien, wie bei Lorentz, *die Wechselwirkungen zwischen Materie und elektromagnetischem Felde ausschließlich* [6] *durch diese Teilchen bedingt.* Dementsprechend nehmen wir an, daß die vom elektromagnetischen Felde auf das Volumenelement der Materie ausgeübten Kräfte gleich sind der Resultierenden der ponderomotorischen Kräfte, welche von diesem Felde auf alle in dem betreffenden Volumenelement befindlichen elektrischen und magnetischen Elementarteilchen ausgeübt werden. Unter Volumenelement der Materie verstehen wir stets einen so großen Raum, daß er eine sehr große Zahl von elektrischen und magnetischen Teilchen enthält. Die Grenzen eines betrachteten Volumenelementes muß man sich ferner stets so genommen denken, daß die Grenzfläche keine elektrische bzw. magnetische Dipole schneidet.

Wir berechnen zunächst diejenige auf einen elektrischen Dipol wirkende Kraft, welche daher herrührt, daß die Feldstärke $\mathfrak{E}$ an den Orten, an welchen sich die Elementarmassen des Dipols befinden, nicht genau dieselbe ist. Bezeichnet man

---

[5]   1) Der einfacheren Darstellung halber halten wir aber an der dualen Behandlung der elektrischen und magnetischen Erscheinungen fest.

Ponderomotorische Kräfte.   **543**

mit $\mathfrak{p}$ den Vektor des Dipolmomentes, so erhält man für die X-Komponente der gesuchten Kraft den Ausdruck:

$$\mathfrak{f}_x = \mathfrak{p}_x \frac{\partial \mathfrak{E}_x}{\partial x} + \mathfrak{p}_y \frac{\partial \mathfrak{E}_x}{\partial y} + \mathfrak{p}_z \frac{\partial \mathfrak{E}_x}{\partial z}.$$

Denkt man sich den letzten Ausdruck für alle Dipole in der Volumeneinheit gebildet und summiert, so erhält man unter Berücksichtigung der Beziehung:

$$\sum \mathfrak{p} = \mathfrak{P}$$

die Gleichung:

$$(4) \qquad \mathfrak{F}_{1x} = \left\{ \mathfrak{P}_x \frac{\partial \mathfrak{E}_x}{\partial x} + \mathfrak{P}_y \frac{\partial \mathfrak{E}_x}{\partial y} + \mathfrak{P}_z \frac{\partial \mathfrak{E}_x}{\partial z} \right\}.$$

Wenn die algebraische Summe der positiven und negativen Leitungselektronen nicht verschwindet, dann kommt zum Ausdruck (4) noch ein Term hinzu, den wir nun berechnen wollen. Die X-Komponente der auf ein Leitungselektron von der elektrischen Masse e wirkenden ponderomotorischen Kraft ist $e\,\mathfrak{E}_x$. Summiert man über alle Leitungselektronen der Volumeneinheit, so erhält man:

$$(5) \qquad \mathfrak{F}_{2x} = \mathfrak{E}_x \sum e.$$

Denkt man sich die betrachtete in der Volumeneinheit befindliche Materie von einer Fläche umschlossen, welche keine Dipole schneidet, so erhält man nach dem Gaussschen Satz und nach der Definition des Verschiebungsvektors $\mathfrak{D}$:

$$\sum e = \operatorname{div} \mathfrak{D},$$

so daß

$$(5\,\mathrm{a}) \qquad \mathfrak{F}_{2x} = \mathfrak{E}_x \operatorname{div} \mathfrak{D}$$

wird. Die X-Komponente der von der elektrischen Feldstärke auf die Volumeneinheit der Materie ausgeübten Kraft ist daher gleich:

$$(6) \quad \mathfrak{F}_{e_x} = \mathfrak{F}_{1x} + \mathfrak{F}_{2x} = \mathfrak{P}_x \frac{\partial \mathfrak{E}_x}{\partial x} + \mathfrak{P}_y \frac{\partial \mathfrak{E}_x}{\partial y} + \mathfrak{P}_z \frac{\partial \mathfrak{E}_x}{\partial z} + \mathfrak{E}_x \operatorname{div} \mathfrak{D}.$$

Analog erhalten wir unter Berücksichtigung der Beziehung

$$\operatorname{div} \mathfrak{B} = 0$$

für die X-Komponente der von der magnetischen Feldstärke gelieferten Kraft:

$$(7) \qquad \mathfrak{F}_{mx} = \left\{ \mathfrak{D}_x \frac{\partial \mathfrak{H}_x}{\partial x} + \mathfrak{D}_y \frac{\partial \mathfrak{H}_x}{\partial y} + \mathfrak{D}_z \frac{\partial \mathfrak{H}_x}{\partial z} \right\}.$$

Es ist zu bemerken, daß für die Herleitung der Ausdrücke (6) und (7) keinerlei Voraussetzung gemacht werden muß über die Beziehungen, welche die Feldstärken $\mathfrak{E}$ und $\mathfrak{H}$ mit den Polarisationsvektoren $\mathfrak{P}$ und $\mathfrak{O}$ verbinden.

Hat man es mit anisotropen Körpern zu tun, so liefern die elektrische bzw. die magnetische Feldstärke nicht nur eine Kraft, sondern auch Kräftepaare, welche sich auf die Materie übertragen. Das gesuchte Drehmoment ergibt sich leicht für die einzelnen Dipole und Summation über alle elektrischen und magnetischen Dipole in der Volumeneinheit. Man erhält:

$$(8) \qquad \mathfrak{L} = \{[\mathfrak{P}\,\mathfrak{E}] + [\mathfrak{O}\,\mathfrak{H}]\}.$$

Die Formel (6) liefert diejenigen ponderomotorischen Kräfte, welche bei elektrostatischen Problemen eine Rolle spielen. Wir wollen diese Gleichung für den Fall, daß es sich um isotrope Körper handelt, so umformen, daß sie einen Vergleich gestattet mit demjenigen Ausdrucke für die ponderomotorischen Kräfte, wie er in der Elektrostatik angegeben wird. Setzen wir

$$\mathfrak{P} = (\varepsilon - 1)\,\mathfrak{E},$$

so geht die Gleichung (6) über in:

$$\mathfrak{F}_{e_x} = \mathfrak{E}_x \operatorname{div} \mathfrak{D} - \frac{1}{2}\,\mathfrak{E}^2\,\frac{\partial \varepsilon}{\partial x} + \frac{1}{2}\,\frac{\partial}{\partial x}(\varepsilon - 1)\,\mathfrak{E}^2.$$

Die ersten beiden Glieder dieses Ausdruckes sind identisch mit den aus der Elektrostatik bekannten. Das dritte Glied ist, wie man sieht, von einem Potential ableitbar. Handelt es sich um Kräfte, die auf einen im Vakuum befindlichen Körper wirken, so liefert das Glied bei Integration über den Körper keinen Beitrag. Handelt es sich aber um die ponderomotorische Wirkung auf Flüssigkeiten, so wird der dem dritten Glied entsprechende Anteil der Kraft bei Gleichgewicht durch eine Druckverteilung in der Flüssigkeit kompensiert.

§ 2. Kräfte, welche von den Geschwindigkeiten der Elementarteilchen abhängen.

Wir gehen jetzt über zu demjenigen Anteile der ponderomotorischen Kraft, welcher durch die Bewegungsgeschwindigkeiten der Elementarladungen geliefert wird.

Ponderomotorische Kräfte.   **545**

Wir gehen aus vom Biot-Savartschen Gesetz. Auf ein stromdurchflossenes Volumenelement, welches sich in einem magnetischen Felde befindet, wirkt erfahrungsgemäß pro Volumeneinheit die Kraft:

$$\frac{1}{c}\,[\mathfrak{s}\,\mathfrak{H}],$$

falls die betrachtete, stromdurchflossene Materie nicht magnetisch polarisierbar ist. Für das Innere von magnetisch polarisierbaren Körpern wurde, soviel uns bekannt ist, bis jetzt jene Kraft gleich [1])

$$\frac{1}{c}\,[\mathfrak{s}\,\mathfrak{B}]$$

gesetzt, wobei $\mathfrak{B}$ die magnetische Induktion bedeutet. Wir wollen nun zeigen, daß *auch* im Falle, daß das stromdurchflossene Material *magnetisch polarisierbar ist*, die auf das stromdurchflossene Volumenelement wirkende Kraft erhalten wird, wenn man zu der durch die Gleichung (7) ausgedrückten Kraft noch die Volumenkraft:

(9) $$\mathfrak{F}_s = \frac{1}{c}\,[\mathfrak{s}\,\mathfrak{H}] \qquad\text{[8]}$$

hinzufügt. Wir wollen dies zuerst an einem einfachen Beispiel anschaulich machen.

Der unendlich dünne im Querschnitt gezeichnete Streifen S erstrecke sich senkrecht zur Papierebene nach beiden Seiten ins Unendliche. Er bestehe aus magnetisch polarisierbarem Material und befinde sich in einem homogenen Magnetfelde $\mathfrak{H}_a$, dessen Richtung durch die Pfeile (vgl. Figur) angedeutet ist. Wir fragen nach der auf den Materialstreifen wirkenden Kraft, falls derselbe von einem Strome i durchflossen ist.

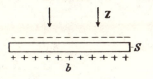

Die Erfahrung lehrt, daß diese Kraft von der magnetischen Permeabilität des Leitermateriales unabhängig ist, und man schloß daraus, daß es nicht die Feldstärke $\mathfrak{H}$, sondern die magnetische Induktion $\mathfrak{B}_i$ sein müsse, welche für die pondero-

---

1) Vgl. z. B. auch M. Abraham, Theorie der Elektrizität 2. p. 319. [7]
1905.

546   *A. Einstein u. J. Laub.*

motorische Kraft maßgebend ist, denn im Innern des Streifens ist die magnetische Induktion $\mathfrak{B}_i$ gleich der außerhalb des Streifens wirkenden Kraft $\mathfrak{H}_a$, unabhängig von dem Werte der Permeabilität des Streifens, während die im Innern des Streifens herrschende Kraft $\mathfrak{H}_i$ bei gegebenem äußeren Felde von μ abhängt. Dieser Schluß ist aber nicht stichhaltig, weil die ins Auge gefaßte ponderomotorische Kraft nicht die einzige ist, welche auf unseren Materialstreifen wirkt. Das äußere Feld $\mathfrak{H}_a$ induziert nämlich auf der Oberseite und Unterseite des Materialstreifens magnetische Belegungen von der Dichte[1]: $\mathfrak{H}_a(1 - 1/\mu)$, und zwar auf der Oberseite eine negative, auf der Unterseite eine positive Belegung. Auf jede dieser Belegungen wirkt eine von dem im Streifen fließenden Strom erzeugte Kraft von der Stärke $i/2\,b$ pro Längeeinheit des Streifens[2], welche magnetische Kraft an der Oberseite und Unterseite verschieden gerichtet ist. Die so resultierenden ponderomotorischen Kräfte addieren sich, so daß wir die ponderomotorische Kraft erhalten: $(1 - 1/\mu)\mathfrak{H}_a\,i$. Diese Kraft scheint bis jetzt nicht berücksichtigt worden zu sein.

Die auf die Längeeinheit unseres Streifens im ganzen ausgeübte Kraft ist nun gleich der Summe der soeben berechneten und der auf die Volumenelemente des Streifens infolge des Stromdurchganges im Magnetfeld wirkenden Kraft R. Da die gesamte auf die Längeeinheit wirkende ponderomotorische Kraft erfahrungsgemäß gleich $i\,\mathfrak{H}_a$ ist, so besteht die Gleichung:

$$\left(1 - \frac{1}{\mu}\right) i\,\mathfrak{H}_a + R = i\,\mathfrak{H}_a$$

oder

$$R = \frac{i\,\mathfrak{H}_a}{\mu} = i\,\mathfrak{H}_i.$$

Man sieht also, daß für die Berechnung der ponderomotorischen Kraft R, welche auf stromdurchflossene Volumenelemente

---

1) Die Dichte ist nämlich gleich:
$$\mathfrak{L}_i = \mathfrak{B}_i - \mathfrak{H}_i = \mathfrak{H}_a\left(1 - \frac{1}{\mu}\right).$$

2) Statt dieser auf die Belegungen wirkenden Kräfte hätten wir streng genommen nach den Resultaten des vorigen Paragraphen allerdings Volumenkräfte einführen müssen, was jedoch ohne Belang ist.

Ponderomotorische Kräfte. 547

wirkt, nicht die Induktion $\mathfrak{B}_i$, sondern die Feldstärke $\mathfrak{H}_i$ maßgebend ist.

Um jeden Zweifel zu beseitigen, wollen wir noch ein Beispiel behandeln, aus welchem man ersieht, daß das Prinzip der Gleichheit von Wirkung und Gegenwirkung den von uns gewählten Ansatz fordert.

Wir denken uns einen zylindrischen, von leerem Raum umgebenen und vom Strom $\mathfrak{z}$ durchflossenen Leiter, welcher sich längs der X-Achse eines Koordinatensystems beiderseits ins Unendliche erstreckt. Die Materialkonstanten des Leiters, sowie die im folgenden auftretenden Feldvektoren seien von x unabhängig, aber Funktionen von y und z. Der Leiter sei ein magnetisch harter Körper und besitze eine Magnetisierung quer zur X-Achse. Wir nehmen an, daß ein äußeres Feld auf den Leiter nicht wirkt, daß also die magnetische Kraft $\mathfrak{H}$ in großen Entfernungen vom Leiter verschwindet.

Es ist klar, daß auf den Leiter als Ganzes keine ponderomotorische Kraft wirkt, denn es würde zu dieser Wirkung keine Gegenwirkung angebbar sein. Wir wollen nun zeigen, daß bei Wahl unseres Ansatzes jene Kraft in der Tat verschwindet. Die gesamte auf die Längeeinheit unseres Leiters in der Richtung der Z-Achse wirkende Kraft läßt sich darstellen gemäß den Gleichungen (7) und (9) in der Form:

$$(10) \qquad R = \int \left(\mathfrak{D}_y \frac{\partial \mathfrak{H}_z}{\partial y} + \mathfrak{D}_z \frac{\partial \mathfrak{H}_z}{\partial z} \right) df + \int \frac{1}{c} \mathfrak{z}_x \mathfrak{H}_y \, df,$$

wobei df ein Flächenelement der YZ-Ebene bedeutet. Wir nehmen an, daß sämtliche in Betracht kommende Größen an der Oberfläche des Leiters stetig sind. Wir behandeln zuerst das erste Integral der Gleichung (10). Es ist:

$$\mathfrak{D}_y \frac{\partial \mathfrak{H}_z}{\partial y} + \mathfrak{D}_z \frac{\partial \mathfrak{H}_z}{\partial z} = \frac{\partial \mathfrak{D}_y \mathfrak{H}_z}{\partial y} + \frac{\partial \mathfrak{D}_z \mathfrak{H}_z}{\partial z} - \mathfrak{H}_z \left(\frac{\partial \mathfrak{D}_y}{\partial y} + \frac{\partial \mathfrak{D}_z}{\partial z} \right).$$

Setzt man die rechte Seite dieser Gleichung in unser Integral ein, so verschwinden bei Integration über die YZ-Ebene die beiden ersten Glieder, da die Kräfte im Unendlichen verschwinden. Das dritte Glied kann unter Berücksichtigung:

$$\mathrm{div}\,\mathfrak{B} = 0$$

548   *A. Einstein u. J. Laub.*

umgeformt werden, so daß unser Integral die Form annimmt:

$$\int \mathfrak{H}_z \left(\frac{\partial \mathfrak{H}_y}{\partial y} + \frac{\partial \mathfrak{H}_z}{\partial z} \right) df.$$

Nun ist:

$$\mathfrak{H}_z \left(\frac{\partial \mathfrak{H}_y}{\partial y} + \frac{\partial \mathfrak{H}_z}{\partial z} \right) = \frac{\partial \mathfrak{H}_y \mathfrak{H}_z}{\partial y} + \frac{1}{2} \frac{\partial \mathfrak{H}_z{}^2}{\partial z} - \mathfrak{H}_y \frac{\partial \mathfrak{H}_z}{\partial y}.$$

Bei der Integration verschwinden aber die beiden Glieder $\frac{\partial \mathfrak{H}_y \mathfrak{H}_z}{\partial y} + \frac{1}{2} \frac{\partial \mathfrak{H}_z{}^2}{\partial z}$. Das Glied $- \mathfrak{H}_y \frac{\partial \mathfrak{H}_z}{\partial y}$ läßt sich umformen mittels der **Maxwell**schen Gleichungen in:

$$- \frac{1}{c} \mathfrak{H}_y \left\{ \mathfrak{z}_x + \frac{\partial \mathfrak{H}_y}{\partial z} \right\},$$

so daß wir endlich die Gleichung (10) schreiben können:

$$R = - \frac{1}{c} \int \mathfrak{H}_y \left\{ \mathfrak{z}_x + \frac{\partial \mathfrak{H}_y}{\partial z} \right\} df + \frac{1}{c} \int \mathfrak{z}_x \mathfrak{H}_y \, df$$

$$= - \frac{1}{c} \int \mathfrak{H}_y \frac{\partial \mathfrak{H}_y}{\partial z} \, df = - \frac{1}{2c} \int \frac{\partial \mathfrak{H}_y{}^2}{\partial z} \, df.$$

Das letzte Integral wird Null, weil im Unendlichen die Kräfte verschwinden. —

Nachdem wir so die Kraft festgestellt haben, welche auf von einem Leitungsstrom durchflossene Materie wirkt, erhalten wir die Kraft, die auf einen von einem Polarisationsstrom durchsetzten Körper wirkt, indem wir beachten, daß Polarisationsstrom und Leitungsstrom in bezug auf elektrodynamische Wirkung vom Standpunkt der Elektronentheorie durchaus äquivalent sein müssen.

Durch Berücksichtigung der Dualität von magnetischen und elektrischen Erscheinungen erhält man auch noch die Kraft, welche auf einen von einem magnetischen Polarisationsstrom durchsetzten Körper im elektrischen Felde ausgeübt wird. Als Gesamtausdruck für diejenigen Kräfte, welche von der Geschwindigkeit der Elementarteilchen abhängen, erhalten wir auf diese Weise die Gleichungen:

$$(11) \qquad \mathfrak{F}_a = \frac{1}{c} [\mathfrak{s} \mathfrak{H}] + \frac{1}{c} \left[\frac{\partial \mathfrak{P}}{\partial t} \mathfrak{H} \right] + \frac{1}{c} \left[\mathfrak{E} \frac{\partial \mathfrak{Q}}{\partial t} \right].$$

Ponderomotorische Kräfte. **549**

§ 3. Gleichheit von actio und reactio.

Addiert man die Gleichungen (6), (7) und (11), so erhält man den Gesamtausdruck für die X-Komponente der pro Volumeneinheit auf die Materie wirkenden ponderomotorischen Kraft in der Form:

$$\mathfrak{F}_x = \mathfrak{E}_x \operatorname{div} \mathfrak{D} + \mathfrak{P}_x \frac{\partial \mathfrak{E}_x}{\partial x} + \mathfrak{P}_y \frac{\partial \mathfrak{E}_x}{\partial y} + \mathfrak{P}_z \frac{\partial \mathfrak{E}_x}{\partial z}$$

$$+ \mathfrak{D}_x \frac{\partial \mathfrak{H}_x}{\partial x} + \mathfrak{D}_y \frac{\partial \mathfrak{H}_x}{\partial y} + \mathfrak{D}_z \frac{\partial \mathfrak{H}_x}{\partial z}$$

$$+ \frac{1}{c}[\mathfrak{s} \mathfrak{H}]_x + \frac{1}{c}\left[\frac{\partial \mathfrak{P}}{\partial t}\mathfrak{H}\right]_x + \frac{1}{c}\left[\mathfrak{E}\frac{\partial \mathfrak{D}}{\partial t}\right]_x.$$

Die Gleichung kann man auch schreiben:

$$\mathfrak{F}_x = \mathfrak{E}_x \operatorname{div} \mathfrak{E} + \frac{1}{c}[\mathfrak{s} \mathfrak{H}]_x + \frac{1}{c}\left[\frac{\partial \mathfrak{D}}{\partial t}\mathfrak{H}\right]_x + \mathfrak{H}_x \operatorname{div}\mathfrak{H} + \frac{1}{c}\left[\mathfrak{E}\frac{\partial \mathfrak{P}}{\partial t}\right]_x$$

$$+ \frac{\partial(\mathfrak{P}_x\mathfrak{E}_x)}{\partial x} + \frac{\partial(\mathfrak{P}_y\mathfrak{E}_x)}{\partial y} + \frac{\partial(\mathfrak{P}_z\mathfrak{E}_x)}{\partial z}$$

$$+ \frac{\partial(\mathfrak{D}_x\mathfrak{H}_x)}{\partial x} + \frac{\partial(\mathfrak{D}_y\mathfrak{H}_x)}{\partial y} + \frac{\partial(\mathfrak{D}_z\mathfrak{H}_x)}{\partial z} - \frac{1}{c}\frac{\partial}{\partial t}[\mathfrak{E}\mathfrak{H}]_x.$$

Ersetzt man

$$\frac{1}{c}\left(\sigma + \frac{\partial \mathfrak{D}}{\partial t}\right) \quad \text{und} \quad \frac{1}{c}\frac{\partial \mathfrak{P}}{\partial t}$$

mittels der **Maxwell**schen Gleichungen durch curl $\mathfrak{H}$ bzw. durch curl $\mathfrak{E}$, so erhält man durch eine einfache Umformung: [9]

$$(12) \qquad \mathfrak{F}_x = \frac{\partial X_x}{\partial x} + \frac{\partial X_y}{\partial y} + \frac{\partial X_z}{\partial z} - \frac{1}{c^2}\frac{\partial \mathfrak{S}_x}{\partial t},$$

wobei gesetzt ist[1]: [10]

$$(13) \qquad \begin{cases} X_x = -\frac{1}{2}(\mathfrak{E}^2 + \mathfrak{H}^2) + \mathfrak{E}_x\mathfrak{D}_x + \mathfrak{H}_x\mathfrak{B}_x, \\ X_y = \quad \mathfrak{E}_x\mathfrak{D}_y + \mathfrak{H}_x\mathfrak{B}_y, \\ X_z = \quad \mathfrak{E}_x\mathfrak{D}_z + \mathfrak{H}_x\mathfrak{B}_z, \\ \mathfrak{S}_x = \quad c\,[\mathfrak{E}\mathfrak{H}]_x. \end{cases}$$

---

1) Hr. Geheimrat **Wien** hatte die Güte, uns darauf aufmerksam zu machen, daß bereits H. A. **Lorentz** die ponderomotorischen Kräfte für nicht magnetisierbare Körper in dieser Form angegeben hat. Enzykl. d. mathem. W. 5. p. 247. [11]

550 *A. Einstein u. J. Laub. Ponderomotorische Kräfte.*

Entsprechende Gleichungen gelten für die beiden anderen Komponenten der ponderomotorischen Kraft.

Integriert man (12) über den unendlichen Raum, so erhält man, falls im Unendlichen die Feldvektoren verschwinden, die Gleichung:

[12] (14)
$$\int \mathfrak{F}_x \, d\tau = -\frac{1}{c^2} \int d\tau \, \frac{d\mathfrak{S}_x}{dt}.$$

Sie sagt aus, daß unsere ponderomotorischen Kräfte bei Einführung der elektromagnetischen Bewegungsgröße dem Satz von der Gleichheit von actio und reactio genügen.

Bern, 7. Mai 1908.

(Eingegangen 13. Mai 1908.)

Published in *Annalen der Physik* 26 (1908): 541–550. Dated Bern, 7 May 1908, received 13 May 1908, published 7 July 1908.

[1] *Minkowski 1908.*

[2] For Minkowski's expression for the ponderomotive force, see *Minkowski 1908*, p. 97.

[3] In the units used in *Einstein and Laub 1908a* (Doc. 51) and later in this paper (see p. 545), the last two terms in this equation should be divided by *c*.

[4] For contemporary expositions of the electron theory, see *Lorentz 1904c, Bucherer 1904*.

[5] For discussions of this duality, also called the Heaviside-Hertz analogy, see *Lorentz 1904b*, p. 99; *Föppl 1894*, pp. 121–122; and *Abraham/Föppl 1904*, p. 211.

[6] See, e.g., *Lorentz 1904c*, pp. 151–155.

[7] *Abraham 1905*. This reference was added at Wien's suggestion (see Jakob Laub to Einstein, 18 May 1908).

[8] For a discussion of criticisms of Einstein's and Laub's expression for the ponderomotive force, see the editorial note, "Einstein and Laub on the Electrodynamics of Moving Media," pp. 506–507.

[9] "Curl $\mathfrak{E}$" should be preceded by a minus sign.

[10] As a consequence of the difference between the Einstein-Laub and the Minkowski expressions for the ponderomotive force density, Einstein's and Laub's expressions for the components of the electromagnetic stress tensor, to which the force density is related by eq. (12), also differ from Minkowski's expressions for the spatial components of the four-dimensional stress-energy-momentum tensor (see *Minkowski 1908*, pp. 92–93). For a discussion of controversies about the electromagnetic stress-energy-momentum tensor, see the editorial note, "Einstein and Laub on the Electrodynamics of Moving Media," p. 507.

[11] *Lorentz 1904c*. Lorentz's expressions for X_x, X_y, X_z differ somewhat from Einstein's and Laub's. For an account of Wien's comments to Laub, indicating Wien did not accept the Einstein-Laub definition of the ponderomotive force density, see Jakob Laub to Einstein, 18 May 1908.

[12] The total time derivative should precede the integral sign.

53. "Correction to the Paper: 'On the Fundamental Electromagnetic Equations for Moving Bodies'"

[*Einstein and Laub 1908c*]

RECEIVED 24 August 1908
PUBLISHED 25 September 1908

IN: *Annalen der Physik* 27 (1908): 232.

232

14. *Berichtigung zur Abhandlung:*
„Über die elektromagnetischen Grundgleichungen
für bewegte Körper";
von A. Einstein und J. Laub.

[1]   In der genannten Abhandlung dieser Zeitschrift **26**. p. 532. 1908 sind zwei Fehler unterlaufen:

p. 534 Formel (8) muß heißen:

$$\varrho = \beta \left(\varrho' + \frac{v}{c^2} \mathfrak{F}_x' \right)$$

statt:

$$\varrho = \beta \left(\varrho' + \frac{v}{c} \mathfrak{F}_x' \right),$$

ferner die erste der Formeln (9):

$$\mathfrak{F}_x = \beta \left(\mathfrak{F}_x' + v \varrho' \right)$$

statt:

$$\mathfrak{F}_x = \beta \left(\mathfrak{F}_x' + \frac{v}{c} \varrho' \right).$$

Die erste der Formeln (12a) sowie die dritte der Formeln (13) müssen ebenfalls heißen:

$$\beta \left(\mathfrak{F}_x - v \varrho \right) = \sigma \mathfrak{E}_x$$

und

$$\beta \left(\mathfrak{F}_v - |\mathfrak{v}| \varrho \right) = \sigma \left\{ \mathfrak{E} + \frac{1}{c} \left[\mathfrak{v} \mathfrak{B} \right] \right\}_v.$$

(Eingegangen 24. August 1908.)

Published in the *Annalen der Physik* 27 (1908): 232. Received 24 August 1908, published 25   September 1908.
[1] *Einstein and Laub 1908a* (Doc. 51).

54. ''Remarks on Our Paper: 'On the Fundamental Electromagnetic Equations for Moving Bodies' '' and ''Supplement''

[*Einstein and Laub 1909*]

DATED Bern and Würzburg, November 1908
RECEIVED 6 December 1908
SUPPLEMENT RECEIVED 19 January 1909
PUBLISHED 4 February 1909

IN: *Annalen der Physik* 28 (1909): 445–447.

445

11. *Bemerkungen zu unserer Arbeit:*
„*Über die elektromagnetischen Grundgleichungen*
für bewegte Körper";
von A. Einstein und J. Laub.

Hr. Laue war so freundlich, uns auf eine in unserer im Titel genannten Arbeit enthaltene Unrichtigkeit hinzuweisen.[1]

[2] Wir sagen dort (Ann. d. Phys. **26**. p. 535. 1908):

„Es ist bemerkenswert, daß die Grenzbedingungen für die Vektoren $\mathfrak{E}$, $\mathfrak{D}$, $\mathfrak{H}$, $\mathfrak{B}$ an der Grenze zweier Medien dieselben sind, wie für ruhende Körper. Es folgt dies direkt aus den Gleichungen (1 a) bis (4 a)."

Abgesehen davon, daß für die Herleitung der Grenzbedingungen die Gleichungen (3 a) und (4 a) nicht in Betracht kommen, ist diese Behauptung nur dann richtig, wenn die Bewegungskomponente normal zur Grenzfläche verschwindet, was bei der im § 2 der genannten Arbeit behandelten Aufgabe tatsächlich zutrifft. Die allgemein gültigen Grenzbedingungen findet man wohl am leichtesten auf folgendem Wege, der dem

[3] von Heinrich Hertz eingeschlagenen entspricht.

Ist die Grenzfläche, oder besser gesagt, die unendlich dünne Grenzübergangsschale, beliebig bewegt, so werden sich in einem momentan in ihr gelegenen ruhenden Punkt die das elektromagnetische Feld bestimmenden Größen im allgemeinen unstetig bzw. unendlich rasch mit der Zeit ändern; diese Änderungen werden aber stetig sein für einen Punkt, der sich *mit der Materie bewegt*. Es wird also die Anwendung des Operators

$$\frac{\partial}{\partial t} + (\mathfrak{v} \nabla)$$

an einem Skalar oder einem Vektor auch in der Grenzfläche

---

[1]   1) Hr. Laue hat uns in seinem Briefe bereits die Grenzbedingungen richtig angegeben und uns eine andere Ableitung derselben mitgeteilt.

446   *A. Einstein u. J. Laub.*

nicht zu unendlich großen Werten führen. Schreiben wir nun die Gleichung (1 a)[1] in der Form:

$$\frac{1}{c}\left\{\frac{\partial \mathfrak{D}}{\partial t} + (\mathfrak{v} \, \nabla) \, \mathfrak{D}\right\} + \mathfrak{s} = \mathrm{curl}\, \mathfrak{H} + \frac{1}{c}\,(\mathfrak{v} \, \nabla) \, \mathfrak{D}$$

und nehmen wir an, daß die Stromdichte $\mathfrak{s}$ auch in der Grenzschicht endlich sei, so ist die linke Seite dieser Gleichung in der Grenzschicht endlich. Dasselbe gilt also auch für die rechte Seite der Gleichung.

Zur leichten Interpretation dieses Resultates denken wir uns das Koordinatensystem so gelegt, daß ein bestimmtes, unendlich kleines Stück der Grenzfläche, das wir nun betrachten wollen, der YZ-Ebene parallel sei. Dann ist klar, daß die Ableitungen aller Größen nach y und z in dem betrachteten Stück der Grenzfläche endlich bleiben. Es muß also auch der Inbegriff derjenigen Glieder der rechten Seite obiger Gleichung, die Differentiationen nach x enthalten, etwas Endliches liefern. Durch einfaches Entwickeln der rechten Seite und Weglassen der nach y und z differenzierten Glieder gelangt man zu dem Resultate, daß in der Grenzschicht die Ausdrücke:

$$\frac{\mathfrak{v}_x}{c}\,\frac{\partial \mathfrak{T}_x}{\partial x}\,,$$

$$\frac{\partial \mathfrak{H}_z}{\partial x} - \frac{\mathfrak{v}_x}{c}\,\frac{\partial \mathfrak{T}_y}{\partial x}\,,$$

$$\frac{\partial \mathfrak{H}_y}{\partial x} + \frac{\mathfrak{v}_x}{c}\,\frac{\partial \mathfrak{T}_x}{\partial x}$$

endlich bleiben. Setzen wir noch voraus, daß die Geschwindigkeitskomponenten an der Grenzfläche keinen Sprung erleiden, so folgt daraus, daß die Ausdrücke:

$$\mathfrak{D}_x\,,$$

$$\mathfrak{H}_y + \frac{\mathfrak{v}_x}{c}\,\mathfrak{D}_z\,,$$

$$\mathfrak{H}_z - \frac{\mathfrak{v}_x}{c}\,\mathfrak{D}_y$$

auf beiden Seiten der Grenzfläche (YZ-Ebene) denselben Wert

---

1) l. c.

Bemerkungen zu unserer Arbeit.   **447**

haben. Da $\mathfrak{D}_x$ und die Komponenten von $\mathfrak{v}$ stetig sind, können wir die beiden letzten Ausdrücke auch ersetzen durch:

$$\mathfrak{H}_y - \frac{1}{c}\,(\mathfrak{v}_z\,\mathfrak{D}_x - \mathfrak{v}_x\,\mathfrak{D}_z),$$

$$\mathfrak{H}_z - \frac{1}{c}\,(\mathfrak{v}_x\,\mathfrak{D}_y - \mathfrak{v}_y\,\mathfrak{D}_x).$$

Von der speziellen Wahl der Lage der Koordinatenachsen relativ zum betrachteten Element der Grenzfläche machen wir uns frei, indem wir das Resultat in den Bezeichnungen der Vektoranalysis schreiben. Bezeichnen wir durch die Indizes n bzw. $\bar{n}$ die Komponente des betreffenden Vektors im Sinne bzw. senkrecht zur Normale der Unstetigkeitsfläche, so folgt, daß

$$\mathfrak{D}_n,$$

$$\left\{\mathfrak{H} - \frac{1}{c}\,[\mathfrak{v}\,\mathfrak{D}]\right\}_{\bar{n}}$$

an der Grenzfläche stetig sein müssen.

In gleicher Weise schließt man aus der Gleichung (2a) [1]) die Stetigkeit der Komponenten:

$$\mathfrak{B}_n,$$

$$\left\{\mathfrak{E} + \frac{1}{c}\,[\mathfrak{v}\,\mathfrak{B}]\right\}_{\bar{n}}.$$

Bern und Würzburg, November 1908.

----

1) l. c.

(Eingegangen 6. Dezember 1908.)

Nachtrag. Wenn an der betrachteten Grenzfläche eine Schicht wahrer Elektrizität ($\int \varrho\,d\tau$) von der Flächendichte η sich befindet, so wird $\mathfrak{s}$ unendlich. Es ist dann

[4]
$$\operatorname{curl}\mathfrak{H} + \frac{1}{c}\,(\mathfrak{v}\,\triangle)\,\mathfrak{D} - \mathfrak{s}$$

in der Grenzschicht endlich, wobei $\mathfrak{s}$ durch $(\mathfrak{v}/c)\,\varrho$ ersetzt werden kann. Für diesen Fall findet man ebenfalls die obigen Grenzbedigungen, mit dem Unterschiede, daß die erste derselben durch

$$\mathfrak{D}_{n2} - \mathfrak{D}_{n1} = \eta$$

zu ersetzen ist.

(Eingegangen 19. Januar 1909.)

Published in *Annalen der Physik* 28 (1909): 445–447. Dated Bern and Würzburg, November 1908, received 6 December 1908. "Supplement" received 19 January 1909. Both published 4 February 1909.

[1] Laue discussed these boundary conditions in *Laue 1911b*, pp. 127–129.

[2] *Einstein and Laub 1908a* (Doc. 51).

[3] See *Hertz, H. 1890b*, as reprinted in *Hertz, H. 1892*, pp. 271–275.

[4] The "Δ" should be "∇."

55. ''Comment on the Paper of D. Mirimanoff: 'On the Fundamental Equations . . .' ''

[*Einstein 1909a*]

DATED Bern, January 1909
RECEIVED 22 January 1909
PUBLISHED 16 March 1909

IN: *Annalen der Physik* 28 (1909): 885–888.

885

5. *Bemerkung zu der Arbeit von D. Mirimanoff* *„Über die Grundgleichungen . . .“;* *von A. Einstein.*

1. Das in dieser Arbeit[1]) angegebene System von Differentialgleichungen und Transformationsgleichungen unterscheidet sich von dem Minkowskis in keiner Weise bzw. *nur* dadurch, daß derjenige Vektor, welcher gewöhnlich mit $\mathfrak{H}$ bezeichnet wird (magnetische Kraft), vom Verfasser mit [2]

$$\mathfrak{Q} = \mathfrak{\check{H}} - \frac{1}{c} \, [\mathfrak{P} \, \mathfrak{w}]$$

bezeichnet wurde.

Differentialgleichung (I) ist nämlich bei Einführung von $\mathfrak{Q}$, [3] wie der Verfasser selbst zeigt, identisch mit der betreffenden Gleichung Minkowskis, während die übrigen drei Differentialgleichungen $\mathfrak{H}$ nicht enthalten und bereits die Form der entsprechenden Gleichungen Minkowskis haben. Der Verfasser sagt auch selbst, daß sich seine Vektoren $\mathfrak{E}, \mathfrak{D}, \mathfrak{Q}, \mathfrak{B}$ transformieren, wie die gewöhnlich mit $\mathfrak{E}, \mathfrak{D}, \mathfrak{H}, \mathfrak{B}$ bezeichneten Vektoren.

2. Auch die Beziehungen zwischen den Vektoren, welche Materialkonstanten (ε, μ und σ) enthalten, unterscheiden sich nicht von den entsprechenden Minkowskis. Der Verfasser geht nämlich davon aus, daß für ein relativ zu dem betrachteten Systempunkt momentan ruhendes Koordinatensystem die Gleichungen

$$\mathfrak{D} = \varepsilon \, \mathfrak{E}, \quad \mathfrak{H} = \frac{1}{\mu} \, \mathfrak{B}, \quad \mathfrak{J} = \sigma \, \mathfrak{E}$$

gelten sollen; bedenkt man nun, daß der Vektor $\mathfrak{H}$ (des Verfassers) für $\mathfrak{w} = 0$ mit dem Vektor $\mathfrak{Q}$ identisch ist, und daß $\mathfrak{Q}$ in den Differentialgleichungen des Verfassers und in dessen Transformationsgleichungen genau dieselbe Rolle spielt, wie $\mathfrak{m}$ in Minkowskis Gleichungen (gewöhnlich mit $\mathfrak{H}$ bezeichnet).

---

1) D. Mirimanoff, Ann. d. Phys. **28**. p. 192. 1909. [1]

886 *A. Einstein.*

so ersieht man, daß auch diese Gleichungen mit den entsprechenden Minkowskis übereinstimmen, bis auf den Umstand, daß die Bezeichnung $\mathfrak{H}$ durch die Bezeichnung $\mathfrak{Q}$ ersetzt ist.

3. Es ist also gezeigt, daß die Größe $\mathfrak{Q}$ Mirimanoffs in dessen sämtlichen Gleichungen dieselbe Rolle spielt wie diejenige Größe, welche man gewöhnlich mit $\mathfrak{H}$ bezeichnet und „magnetische Kraft" oder „magnetische Feldstärke" nennt. Trotzdem hätten die Gleichungen Mirimanoffs einen anderen Inhalt als die Gleichungen Minkowskis, wenn die Größe $\mathfrak{Q}$ Mirimanoffs definitionsgemäß eine andere physikalische Bedeutung hätte als die gewöhnlich mit $\mathfrak{H}$ bezeichnete Größe.

Um hierüber ein Urteil zu gewinnen, fragen wir uns zunächst, was in den Minkowskischen Gleichungen

$$(A) \begin{cases} \operatorname{curl} \mathfrak{H} = \dfrac{1}{c}\dfrac{\partial \mathfrak{D}}{\partial t} + \mathfrak{i}, \\[2mm] \operatorname{curl} \mathfrak{E} = -\dfrac{1}{c}\dfrac{\partial \mathfrak{B}}{\partial t}, \\[2mm] \operatorname{div} \mathfrak{D} = \varrho, \\[2mm] \operatorname{div} \mathfrak{B} = 0 \end{cases}$$

die Vektoren $\mathfrak{E}$, $\mathfrak{D}$, $\mathfrak{H}$, $\mathfrak{B}$ für eine Bedeutung haben. Man muß zugeben, daß diese Vektoren für den Fall, daß die Geschwindigkeit $\mathfrak{w}$ der Materie von Null abweicht, bisher nicht eigens definiert worden sind; Definitionen, auf welchen (ideale) Messungen dieser Größen basiert werden könnten, besitzen wir nur für den Fall, daß $\mathfrak{w}$ verschwindet, und zwar denke ich an jene Definitionen, welche aus der Elektrodynamik ruhender Körper wohlbekannt sind. Wenn daher unter Benutzung der Minkowskischen Gleichungen gefunden ist, daß in einem bestimmten, mit der Geschwindigkeit $\mathfrak{w}$ bewegten Volumelement des Körpers die Feldvektoren zu einer gewissen Zeit die bestimmten (Vektor-)Werte $\mathfrak{E}$, $\mathfrak{D}$, $\mathfrak{H}$, $\mathfrak{B}$ haben, so müssen wir diese Feldvektoren erst auf ein mit Bezug auf das betreffende Volumelement ruhendes Bezugssystem transformieren. Die so erhaltenen Vektoren $\mathfrak{E}'$, $\mathfrak{D}'$, $\mathfrak{H}'$, $\mathfrak{B}'$ haben erst eine bestimmte physikalische Bedeutung, die aus der Elektrodynamik ruhender Körper bekannt ist.

Bemerkung zu der Arbeit von D. Mirimanoff. 887

Die Minkowskischen Differentialgleichungen sagen also für Punkte, in denen $\mathfrak{w} \neq 0$ ist, für sich allein noch gar nichts aus, wohl aber die Minkowskischen Differentialgleichungen zusammen mit den Minkowskischen Transformationsgleichungen und mit der Bestimmung, daß für den Fall $\mathfrak{w} = 0$ die Definitionen der Elektrodynamik ruhender Körper für die Feldvektoren gelten sollen.

Wir haben nun zu fragen: Ist der Vektor $\mathfrak{L}$ Mirimanoffs in anderer Weise definiert als der von uns soeben mit $\mathfrak{H}$ bezeichnete Vektor? Dies ist nicht der Fall, und zwar aus folgenden Gründen:

1. Für die Feldvektoren $\mathfrak{E}$, $\mathfrak{D}$, $\mathfrak{L}$, $\mathfrak{B}$ Mirimanoffs gelten dieselben Differentialgleichungen und Transformationsgleichungen wie für die Vektoren $\mathfrak{E}$, $\mathfrak{D}$, $\mathfrak{H}$, $\mathfrak{B}$ der Minkowskischen Gleichungen (A).

2. Sowohl Mirimanoffs Vektor $\mathfrak{L}$ als auch der Vektor $\mathfrak{H}$ von (A) sind nur für den Fall $\mathfrak{w} = 0$ definiert. In diesem Falle ist aber wegen Mirimanoffs Gleichung

$$\mathfrak{L} = \mathfrak{H} - \frac{1}{e}\,[\mathfrak{B}\,\mathfrak{w}]$$

$\mathfrak{L} = \mathfrak{H} =$ Feldstärke zu setzen; für den Vektor $\mathfrak{H}$ der Gleichungen (A) gilt genau in gleicher Weise, daß er im Falle $\mathfrak{w} = 0$ mit der Feldstärke im Sinne der Elektrodynamik ruhender Körper gleichbedeutend ist.

Aus diesen beiden Argumenten folgt, daß der Vektor $\mathfrak{L}$ Mirimanoffs und der Vektor $\mathfrak{H}$ von (A) durchaus gleichwertig sind.

4. Um seine Resultate bezüglich der Wilsonschen Anordnung mit den von Hrn. Laub und mir erhaltenen zu vergleichen, hätte der Verfasser die Betrachtung so weit durchführen müssen, daß er zu Beziehungen zwischen definierten. d. h. wenigstens prinzipiell der Erfahrung zugänglichen Größen gelangt wäre. Er hätte zu diesem Zwecke nur die seinem Gleichungssystem entsprechenden Grenzbedingungen anzuwenden gehabt. Nach dem Vorigen hätte er so zu genau denselben Folgerungen gelangen müssen wie wir, da seine Theorie mit der von Minkowski identisch ist. [4]

Schließlich möchte ich noch hinweisen auf die Bedeutung

888   *A. Einstein. Bemerkung zu der Arbeit von D. Mirimanoff.*

der neulich erschienenen Arbeit von Ph. Frank[1]), welche die Übereinstimmung zwischen der Lorentzschen elektronen-
[6]   theoretischen und der Minkowskischen Behandlung der Elektrodynamik bewegter Körper durch Berücksichtigung der Lorentzkontraktion wiederherstellt. Der Vorzug der elektronen-theoretischen Behandlungsweise liegt einerseits darin, daß sie eine anschauliche Deutung der Feldvektoren liefert, anderer-seits darin, daß sie auskommt ohne die willkürliche Voraus-setzung, daß die Differentialquotienten der Geschwindigkeit der
[7]   Materie in den Differentialgleichungen nicht auftreten.

[8]   **Bern, Januar 1909.**

[5]   1) **Ph. Frank**, Ann. d. Phys. **27.** p. 1059. 1908.

[9]   **(Eingegangen 22. Januar 1909.)**

Published in *Annalen der Physik* 28 (1909): 885–888. Dated Bern, January 1909, received 22 January 1909, published 16 March 1909.

[1] *Mirimanoff 1909.*

[2] Sec *Minkowski 1908.*

[3] Eq. (I) in Mirimanoff's paper (see note 1) is:

$$\text{curl } \mathfrak{H} = \frac{1}{c}\left(\frac{\partial \mathfrak{D}}{\partial t} + J + \rho\, \mathfrak{w} + \text{curl }[\mathfrak{P}\,\mathfrak{w}\,]\right).$$

Minkowski's equations are given on p. 886 of this paper.

[4] See *Einstein and Laub 1908a*, pp. 539–540.

[5] *Frank 1908.*

[6] See *Lorentz 1904c.*

[7] Minkowski simply postulated this assumption (see *Minkowski 1908*, pp. 368–369). See *Einstein and Laub 1908a* (Doc. 51), pp. 532, 535, for earlier comments on this assumption.

[8] Einstein evidently sent a copy of this paper to Mirimanoff before its publication (see Dmitry Mirimanoff to Einstein, 12 February 1909).

[9] Einstein evidently had submitted this paper twice before to Wien, editor of the *Annalen*, who returned it to Einstein for revisions (see Wilhelm Wien to Einstein, 19 January 1909).

56. ''On the Present Status of the Radiation Problem''

[Einstein 1909b]

DATED Bern, January 1909
RECEIVED 23 January 1909
PUBLISHED 15 March 1909

IN: *Physikalische Zeitschrift* 10 (1909): 185–193.

No. 6.   15. März 1909.   10. Jahrgang.
                   Redaktionsschluß für No. 7 am 22. März 1909.

ORIGINALMITTEILUNGEN.

Zum gegenwärtigen Stand des Strahlungsproblems.

Von A. Einstein.

In der letzten Zeit sind in dieser Zeitschrift von den Herren H. A. Lorentz[1]), Jeans[2]) und Ritz[3]) Meinungsäußerungen erschienen, die geeignet sind, den heutigen Stand dieses ungemein wichtigen Problems erkennen zu lassen. In der Meinung, daß es von Vorteil sei, wenn alle, die über diese Sache ernsthaft nachgedacht haben, ihre Ansichten mitteilen, auch wenn sie zu endgültigem Resultat nicht haben vordringen können, teile ich das Folgende mit.

1. Die einfachste Form, in der wir die bisher erkannten Gesetzmäßigkeiten der Elektrodynamik ausdrücken können, ist durch die Maxwell-Lorentzschen partiellen Differentialgleichungen gegeben. Diejenigen Formen, in denen retardierte Funktionen vorkommen, sehe ich, im Gegensatz zu Herrn Ritz[3]), nur als mathematische Hilfsformen an. Ich sehe mich dazu in erster Linie dadurch gezwungen, daß jene Formen das Energieprinzip nicht in sich schließen, indem ich glaube, daß wir an der strengen Gültigkeit des Energieprinzips so lange festhalten sollen, bis wir gewichtige Gründe gefunden haben, auf diesen Leitstern zu verzichten. Es ist ja gewiß richtig, daß die Maxwellschen Gleichungen für den leeren Raum, für sich allein genommen, gar nichts aussagen, daß sie nur eine Zwischenkonstruktion darstellen; genau das gleiche läßt sich ja bekanntlich auch von den Newtonschen Bewegungsgleichungen sagen, sowie von jeder

Theorie, die noch der Ergänzung durch andere Theorien bedarf, um ein Bild für einen Komplex von Erscheinungen liefern zu können. Was die Maxwell-Lorentzschen Differentialgleichungen gegenüber Formen, welche retardierte Funktionen enthalten, auszeichnet, das ist der Umstand, daß sie für jeden Augenblick, und zwar relativ zu jedem unbeschleunigten Koordinatensystem, einen Ausdruck für die Energie und für die Bewegungsgröße des betrachteten Systems liefern. Bei einer Theorie, die mit retardierten Kräften operiert, kann man den Momentanzustand eines Systems überhaupt nicht beschreiben, ohne für diese Beschreibung frühere Zustände des Systems zu benützen. Hat z. B. eine Lichtquelle A einen Lichtkomplex gegen den Schirm B hin abgesandt, dieser den Schirm B aber noch nicht erreicht, so ist nach den mit retardierten Kräften operierenden Theorien der Lichtkomplex durch nichts repräsentiert als durch die Vorgänge, welche bei der vorhergegangenen Emission im aussendenden Körper stattgefunden haben. Energie und die Bewegungsgröße müssen dann — wenn man auf diese Größen nicht überhaupt verzichten will — als Zeitintegrale dargestellt werden.

Herr Ritz behauptet nun zwar, daß wir durch die Erfahrung dazu gezwungen seien, die Differentialgleichungen zu verlassen und die retardierten Potentiale einzuführen. Indessen scheint mir seine Begründung nicht stichhaltig zu sein.

Setzt man mit Ritz:

$$f_1 = \frac{1}{4\pi} \int \frac{\varphi\left(x', y', z', t - \dfrac{r}{c}\right)}{r}\, dx'\, dy'\, dz'$$

und

---

[1]   1) H. A. Lorentz, diese Zeitschr. **9**, 562—563, 1908.
[2]   2) J. H. Jeans, diese Zeitschr. **9**, 853—855, 1908.
[3]   3) W. Ritz, diese Zeitschr. **9**, 903—907, 1908.

$$f_2 = \frac{1}{4\pi} \int \frac{\varphi\left(x', y', z', t + \frac{r}{c}\right)}{r} dx', dy', dz,$$

so ist sowohl f_1 wie f_2 eine Lösung der Gleichung

$$\frac{1}{c^2} \frac{\partial^2 f}{\partial t^2} - \Delta f = \varphi(x\, y\, z\, t),$$

es ist also auch

$$f_3 = a_1 f_1 + a_2 f_2$$

eine Lösung, wenn $a_1 + a_2 = 1$. Es ist aber nicht richtig, daß die Lösung f_3 eine allgemeinere Lösung ist als f_1, und daß man die Theorie spezialisiert, indem man $a_1 = 1$, $a_2 = 0$ setzt. Setzt man

$$f(x, y, z, t) = f_1,$$

so kommt dies darauf hinaus, daß man die elektromagnetische Wirkung im Punkte x, y, z berechnet aus denjenigen Bewegungen und Konfigurationen der elektrischen Mengen, welche vor dem Augenblick t stattgefunden haben. Setzt man

$$f(x, y, z, t) = f_2,$$

so benützt man zur Bestimmung jener elektromagnetischen Wirkung diejenigen Bewegungen und Konfigurationen, welche nach dem Augenblick t stattfinden.

Im ersteren Fall berechnet man das elektromagnetische Feld aus der Gesamtheit der es erzeugenden, im zweiten Fall aus der Gesamtheit der es absorbierenden Vorgänge. Wenn der ganze Vorgang in einem allseitig begrenzten (endlichen) Raume vor sich geht, kann man ihn ebensowohl in der Form

$$f = f_1$$

wie in der Form

$$f = f_2$$

darstellen. Wenn nun ein vom Endlichen ins Unendliche emittiertes Feld betrachtet wird, kann man naturgemäß nur die Form

$$f = f_1$$

anwenden, weil eben die Gesamtheit der absorbierenden Vorgänge nicht in Betracht gezogen wird. Aber es handelt sich hier um ein irreführendes Paradoxon des Unendlichen. Es lassen sich stets beide Darstellungsweisen anwenden, wie entfernt man sich auch die absorbierenden Körper denken mag. Man kann also nicht schließen, daß die Lösung $f = f_1$ spezieller sei, als die Lösung $a_1 f_1 + a_2 f_2$, wobei $a_1 + a_2 = 1$.

[6] Daß ein Körper nicht „Energie aus dem Unendlichen empfängt, ohne daß irgend ein anderer Körper ein entsprechendes Quantum Energie verliert", kann nach meiner Meinung ebenfalls nicht als Argument angeführt werden. Zunächst können wir, wenn wir bei der Erfahrung bleiben wollen, nicht vom Unendlichen reden, sondern nur von Räumen, die außerhalb

des betrachteten Raumes liegen. Ferner aber kann aus der Nichtbeobachtbarkeit eines derartigen Vorgangs eine Nichtumkehrbarkeit der elektromagnetischen Elementarvorgänge ebensowenig geschlossen werden, als eine Nichtumkehrbarkeit der elementaren Bewegungsvorgänge der Atome aus dem zweiten Hauptsatz der Thermodynamik gefolgert werden darf. [7]

2. Man kann der Jeansschen Auffassung entgegenhalten, daß es vielleicht unzulässig sei, die allgemeinen Ergebnisse der statistischen Mechanik auf mit Strahlung gefüllte Hohlräume [8] anzuwenden. Indessen kann man auch auf folgendem Wege zu dem von Jeans gefolgerten Gesetz gelangen[1].

Ein Ion, welches um eine Gleichgewichtslage in Richtung der X-Achse zu oszillieren vermag, emittiert und absorbiert nach der Maxwellschen Theorie nur dann im Mittel gleiche Mengen Strahlung pro Zeiteinheit, wenn zwischen der mittleren Schwingungsenergie E_ν und der Energiedichte der Strahlung ϱ_ν bei der Eigenfrequenz ν des Oszillators die Beziehung

$$E_\nu = \frac{c^3}{8\pi \nu^2} \varrho_\nu, \qquad \text{(I)} \quad [10]$$

besteht, wobei c die Lichtgeschwindigkeit bedeutet. Wenn das oszillierende Ion auch mit Gasmolekülen (oder überhaupt mit einem mittels der Molekulartheorie darstellbaren System) in Wechselwirkung zu treten vermag, so muß nach der statistischen Theorie der Wärme notwendig

$$\overline{E_\nu} = \frac{R T}{N} \qquad \text{(II)}$$

sein ($R =$ Konstante der Gasgleichung, $N =$ Zahl der Atome in einem Grammatome, $T =$ absolute Temperatur), wenn im Mittel keine Energie vom Gas durch den Oszillator auf den Strahlungsraum übertragen wird[2].

Aus diesen beiden Gleichungen folgt

$$\varrho_\nu = \frac{R}{N} \frac{8\pi}{c^3} \nu^2 T, \qquad \text{(III)}$$

also genau das auch von den Herren Jeans und H. A. Lorentz gefundene Gesetz[3]). [12]

3. Daran, daß unsere heutigen theoretischen Ansichten zu dem von Herrn Jeans vertretenen Gesetz mit Notwendigkeit führen, ist nach meiner Meinung nicht zu zweifeln. Aber als

---

1) Vgl. A. Einstein, Ann. d. Phys. (4) 17, 133—136, 1905. [9]

2) M. Planck, Ann. d. Phys. 1, 99, 1900. M. Planck, Vorlesungen über die Theorie der Wärmestrahlung. III. [11] Kapitel.

3) Es sei ausdrücklich bemerkt, daß diese Gleichung eine unabweisbare Konsequenz der statistischen Theorie der Wärme ist. Der im soeben zitierten Planckschen Buche auf S. 178 enthaltene Versuch, die Allgemeingültigkeit der Gleichung II in Frage zu stellen, beruht — wie mir scheint — [13] nur auf einer Lücke in Boltzmanns Betrachtungen, welche unterdessen durch die Gibbsschen Untersuchungen ausgefüllt wurde. [14]

Physikalische Zeitschrift. 10. Jahrgang. No. 6. 187

nicht viel weniger sicher erwiesen können wir es ansehen, daß die Formel (III) nicht mit den Tatsachen vereinbar ist. Warum senden denn die festen Körper nur von einer gewissen, ziemlich scharf ausgesprochenen Temperatur an sichtbares Licht aus? Warum wimmelt es nicht überall von ultravioletten Strahlen, wenn doch beständig solche bei gewöhnlicher Temperatur erzeugt werden? Wie ist es möglich, höchst empfindliche photographische Platten lange Zeit in Kassetten aufzubewahren, wenn diese beständig kurzwellige Strahlen erzeugen? Bezüglich weiterer Argumente verweise ich auf § 166 des mehrfach zitierten Planckschen Werkes. Wir werden also wohl sagen müssen, daß uns die Erfahrung dazu zwingt, entweder die von der elektromagnetischen Theorie geforderte Gleichung (I) oder die von der statistischen Mechanik geforderte Gleichung (II) oder endlich beide Gleichungen zu verwerfen.

4. Wir müssen uns fragen, in welcher Beziehung steht die Plancksche Strahlungstheorie zu der unter 2. angedeuteten, auf unseren gegenwärtig anerkannten theoretischen Grundlagen ruhenden Theorie? Die Antwort auf diese Frage wird nach meiner Meinung dadurch erschwert, daß der Planckschen Darstellung seiner eigenen Theorie eine gewisse logische Unvollkommenheit anhaftet. Ich will im folgenden dies kurz auseinander zu setzen versuchen.

a) Wenn man sich auf den Standpunkt stellt, daß die Nichtumkehrbarkeit der Naturvorgänge nur eine scheinbare ist, und daß der nichtumkehrbare Vorgang in einem Übergang zu einem wahrscheinlicheren Zustand bestehe, so muß man zunächst eine Definition der Wahrscheinlichkeit W eines Zustandes geben. Die einzige solche Definition, die nach meiner Meinung in Betracht kommen kann, wäre die folgende:

[17] Es seien A_1, A_2 ····A_l alle Zustände, welche ein nach außen abgeschlossenes System bei bestimmtem Energieinhalt anzunehmen vermag, bzw. genauer gesagt, alle Zustände, welche wir an einem solchen System mit gewissen Hilfsmitteln zu unterscheiden vermögen. Nach der klassischen Theorie nimmt das System nach einer bestimmten Zeit einen bestimmten dieser Zustände (z. B. A_l) an, und verharrt darauf in diesem Zustand (thermodynamisches Gleichgewicht). Nach der statistischen Theorie nimmt aber das System in unregelmäßiger Folge alle Zustände A_1 ....A_l immer wieder an[1]). Beobachtet man das System eine sehr lange Zeit Θ hindurch, so wird es einen gewissen Teil τ_ν dieser Zeit geben, so daß das System während τ_ν und

zwar nur während τ_ν den Zustand A_ν inne hat.

Es wird $\dfrac{\tau_\nu}{\Theta}$ einen bestimmten Grenzwert besitzen, den wir die Wahrscheinlichkeit W des betreffenden Zustandes A_ν nennen.

Ausgehend von dieser Definition kann man zeigen, daß für die Entropie S die Gleichung bestehen muß

$$S = \frac{R}{N} \lg W + \text{konst},$$

wobei die Konstante für alle Zustände gleicher Energie dieselbe ist.

b) Weder Herr Boltzmann noch Herr Planck haben eine Definition von W gegeben. Sie setzen rein formal $W =$ Anzahl der Komplexionen des betrachteten Zustandes. Verlangt man nun, daß diese Komplexionen gleich wahrscheinlich sein sollen, wobei man die Wahrscheinlichkeit der Komplexion analog definiert, wie wir unter a) die Wahrscheinlichkeit des Zustandes definiert haben, so gelangt man genau zu der unter a) gegebenen Definition der Zustandswahrscheinlichkeit; man hat nur das logisch unnötige Element Komplexion in der Definition mit verwendet.

Obwohl nun die angegebene Beziehung zwischen S und W nur gilt, wenn die Komplexionswahrscheinlichkeit in der angegebenen oder in gleichbedeutender Weise definiert wird, hat weder Herr Boltzmann noch Herr Planck die Komplexionswahrscheinlichkeit definiert. Aber Herr Boltzmann hatte doch klar erkannt, daß das von ihm gewählte molekulartheoretische Bild ihm die von ihm getroffene Wahl der Komplexionen in ganz bestimmter Weise vorschrieb; er hat dies in seiner in den Wiener Sitzungsberichten des Jahres 1877 erschienenen Arbeit „Über die Beziehung ....." auf S. 404 und 405 dargelegt[1]). Auch bei der Resonatorentheorie der Strahlung wäre Herr Planck in der Wahl der Komplexionen nicht frei gewesen. Er hätte das Gleichungspaar

$$S = \frac{R}{N} \lg W$$

und

$$W = \text{Zahl der Komplexionen}$$

nur ansetzen dürfen, wenn er die Bedingung hinzugefügt hätte, daß die Komplexionen so gewählt werden müssen, daß sie in dem von ihm gewählten theoretischen Bilde auf Grund statistischer Betrachtungen als gleich wahrscheinlich befunden werden. Er wäre auf diesem Wege zu der von Jeans verteidigten Formel gelangt. So sehr sich jeder Physiker darüber freuen muß, daß sich Herr Planck in so glücklicher Weise über diese Forderung hinweg-

[15]
[16]

[18]

[20]

[21]

[22]

[23]

[25]

---

[19] 1) Daß diese letztere Auffassung die allein haltbare ist, geht unmittelbar aus den Eigenschaften der Brownschen Bewegung hervor.

1) Vgl. auch L. Boltzmann, Vorlesungen über Gastheorie, 1. Bd., S. 40, Zeile 9—23. [24]

188 Physikalische Zeitschrift. 10. Jahrgang. No. 6.

setze, so wenig wäre es angebracht, zu vergessen, daß die Plancksche Strahlungsformel mit der theoretischen Grundlage, von welcher Herr Planck ausgegangen ist, unvereinbar ist.

5. Es ist einfach zu sehen, in welcher Weise die Grundlagen der Planckschen Theorie abgeändert werden können, damit die Plancksche Strahlungsformel wirklich als Konsequenz der theoretischen Grundlagen resultiert. Ich gebe hier nicht die betreffenden Ableitungen, sondern verweise nur auf meine diesbezüglichen Abhandlungen [1]). Das Resultat ist folgendes: Man gelangt zur Planckschen Strahlungsformel, wenn man

1. an der von Planck aus der Maxwellschen Theorie hergeleiteten Gleichung (I) zwischen Resonatorenergie und Strahlungsdichte festhält [2]),
2. die statistische Theorie der Wärme durch folgende Annahme modifiziert: Ein Gebilde, welches mit der Frequenz v Schwingungen auszuführen vermag, und welches dadurch, daß es eine elektrische Ladung besitzt, Strahlungsenergie in Energie der Materie und umgekehrt zu verwandeln vermag, vermag nicht Schwingungszustände jeder beliebigen Energie anzunehmen, sondern nur solche Schwingungszustände, deren Energie ein Vielfaches von $h \cdot v$ ist. h ist dabei die von Planck so benannte, in seiner Strahlungsgleichung auftretende Konstante.

[28]

6. Da die soeben mitgeteilte Modifikation der Grundlagen der Planckschen Theorie zu sehr tiefgreifenden Änderungen unserer physikalischen Theorien mit Notwendigkeit hinführt, ist es sehr wichtig, möglichst einfache, voneinander unabhängige Interpretationen der Planckschen Strahlungsformel sowie überhaupt des Strahlungsgesetzes, soweit dasselbe als bekannt vorausgesetzt werden darf, aufzusuchen. Zwei diesbezügliche Betrachtungen, die sich durch ihre Einfachheit auszeichnen, seien im folgenden kurz mitgeteilt.

Die Gleichung $S = \frac{R}{N} \lg W$ wurde bisher hauptsächlich derart angewendet, daß man auf Grund einer mehr oder weniger vollständigen Theorie die Größe W und aus dieser die Entropie berechnete. Man kann diese Gleichung aber auch umgekehrt dazu benutzen, um aus den mit Hilfe der Erfahrung ermittelten Entropiewerten S_r die statistische Wahrscheinlichkeit der einzelnen Zustände A_r eines nach außen

[26]   [1]) A. Einstein, Ann. d. Phys. (4) **20**, 1906 und Ann. d. Phys (4) **22**, 1907, § 1.
[27]   [2]) Es kommt dies darauf hinaus, daß man annimmt, daß die elektromagnetische Theorie der Strahlung wenigstens richtige zeitliche Mittelwerte liefert. Daran läßt sich aber angesichts der Brauchbarkeit der Theorie in der Optik kaum zweifeln.

abgeschlossenen Systems zu ermitteln. Eine Theorie, welche andere als die so ermittelten Werte für die Zustandswahrscheinlichkeit liefert, ist offenbar zu verwerfen.

Eine Betrachtung der angedeuteten Art zur Ermittlung gewisser statistischer Eigenschaften von in einen Hohlraum eingeschlossener Wärmestrahlung habe ich bereits in einer früheren Arbeit [1]) durchgeführt, in der ich die Theorie der Lichtquanten zuerst darlegte. Da ich aber damals von der nur in der Grenze (für kleine Werte von $\frac{v}{T}$) gültigen Wienschen Strahlungs- [30] formel ausging, will ich hier eine ähnliche Betrachtung angeben, welche eine einfache Deutung des Inhalts der Planckschen Strahlungsformel liefert.

Es seien V und v zwei miteinander kommunizierende Räume, die durch diffus vollkommen reflektierende Wände begrenzt seien. In diese Räume sei Wärmestrahlung vom Frequenzbereich dv eingeschlossen. H sei die momentan in V, η die momentan in v befindliche Strahlungsenergie. Nach einiger Zeit gilt dann mit gewisser Annäherung dauernd die Proportion $H_0 : \eta_0 = V : v$. In einem beliebig herausgegriffenen Zeitpunkt wird η von η_0 abweichen nach [31] einem statistischen Gesetz, das sich aus der Beziehung zwischen S und W unmittelbar ergibt, indem man zu den Differentialen übergeht

$$dW = \text{konst } e^{\frac{N}{R} \cdot S} d\eta.$$

Bezeichnet man mit Σ bzw. σ die Entropie der in den beiden Räumen befindlichen Strahlung und setzt man $\eta = \eta_0 + \varepsilon$, so hat man
$$d\eta = d\varepsilon$$
und
$$S = \Sigma + \sigma = \Sigma_0 + \sigma_0 + \left\{\frac{d(\Sigma+\sigma)}{d\varepsilon}\right\}_0 \varepsilon + \frac{1}{2}\left\{\frac{d^2(\Sigma+\sigma)}{d\varepsilon^2}\right\}_0 \varepsilon^2 \cdots$$

Die letztere Gleichung geht wegen
$$\left\{\frac{d(\Sigma+\sigma)}{d\varepsilon}\right\}_0 = 0,$$
wenn man annimmt, daß V sehr groß ist gegen v, über in
$$S = \text{konst} + \frac{1}{2}\left\{\frac{d^2\sigma}{d\varepsilon^2}\right\}_0 \varepsilon^2 + \cdots \quad [32]$$

Begnügt man sich mit dem ersten nicht verschwindenden Glied der Entwicklung, was einen um so kleineren Fehler bedingt, je größer v, gegenüber dem Kubus der Strahlungswellenlänge ist, so erhält man
$$dW = \text{konst} \cdot e^{-\frac{1}{2}\frac{N}{R}\left(\frac{d^2\sigma}{d\varepsilon^2}\right)_0 \varepsilon^2} \cdot d\varepsilon. \quad [33]$$

[1]) Ann. d. Phys. (4) **17**, 132—148, 1905.   [29]

Physikalische Zeitschrift. 10. Jahrgang. No. 6.   189

Für den Mittelwert ε^2 des Quadrates der Energieschwankung der in v befindlichen Strahlung erhält man daraus

[34]
$$\varepsilon^2 = \cfrac{1}{\cfrac{N}{R}\left(\cfrac{d^2\sigma}{d\varepsilon^2}\right)_0}.$$

Ist die Strahlungsformel bekannt, so kann man σ aus derselben berechnen[1]). Betrachtet man als Ausdruck der Erfahrung die Plancksche Strahlungsformel, so erhält man nach einfacher Rechnung

[36]
$$\varepsilon^2 = \frac{R}{Nk}\left\{v\,h\,\eta_0 + \frac{c^3}{8\pi\,v^2\,dv}\cdot\frac{\eta_0^2}{v}\right\}.$$

Wir haben so einen leicht zu interpretierenden Ausdruck für die mittlere Größe der Schwankungen der in v befindlichen Strahlungsenergie erlangt. Wir wollen nun zeigen, daß die jetzige Theorie der Strahlung mit diesem Resultat unvereinbar ist.

Nach der jetzigen Theorie rühren die Schwankungen lediglich daher, daß die unendlich vielen, den Raum durchsetzenden Strahlen, welche die Strahlung von v konstituieren, miteinander interferieren und so einen Wert der Momentanenergie liefern, der bald größer, bald kleiner ist, als die Summe der Energie, welche die einzelnen Strahlen liefern würden, wenn sie gar nicht miteinander interferierten. Man könnte so die Größe ε^2 durch eine mathematisch etwas komplizierte Betrachtung exakt ermitteln. Wir begnügen uns hier mit einer einfachen Dimensionalbetrachtung. Es müssen folgende Bedingungen erfüllt sein:

1. Die Größe der mittleren Schwankung hängt nur von λ (Wellenlänge), $d\lambda$, σ und v ab, wobei σ die auf Wellenlängen bezogene Strahlungsdichte bedeutet ($\sigma\,d\lambda = \varrho\,dv$).

[37]

2. Da sich die Strahlenenergien benachbarter Wellenlängenbereiche und Volumina[2]) einfach addieren, und die betreffenden Schwankungen voneinander unabhängig sind, muß ε^2 bei bestimmtem λ und ϱ den Größen $d\lambda$ und v proportional sein.

3. $\overline{\varepsilon^2}$ hat die Dimension des Quadrates einer Energie.

Dadurch ist der Ausdruck für ε^2 bis auf einen Zahlenfaktor (von der Größenordnung 1) vollkommen bestimmt. Man gelangt auf diese Weise zum Ausdruck $\sigma^2\lambda^4 v\,d\lambda$, der bei Einführung der oben benützten Variabeln in den zweiten Term der vorhin für $\overline{\varepsilon^2}$ entwickelten Formel übergeht. Diesen zweiten Term aber hätten wir allein für ε^2 erhalten, wenn wir von der Jeansschen Formel ausgegangen wären.

[38]

[35]
1) Vgl. z. B. das mehrfach zitierte Plancksche Werk Gleichung (230).
2) Natürlich nur, wenn diese genügend groß sind.

Man hätte dann noch $\dfrac{R}{Nk}$ gleich einer Konstanten von der Größenordnung 1 zu setzen, was der Planckschen Bestimmung des Elementarquantums entspricht[1]). Das erste Glied des obigen Ausdrucks für ε^2, das bei der sichtbaren Strahlung, die uns allenthalben umgibt, einen weitaus größeren Beitrag liefert als das zweite, ist also mit der jetzigen Theorie nicht vereinbar.

[39]

Setzt man mit Planck $\dfrac{R}{Nk} = 1$, so würde das erste Glied, wenn es allein vorhanden wäre, eine solche Schwankung der Strahlungsenergie liefern, wie wenn die Strahlung aus voneinander unabhängig beweglichen, punktförmigen Quanten von der Energie $h\,v$ bestünde. Es läßt sich dies durch eine einfache Rechnung zeigen. Es sei ausdrücklich daran erinnert, daß das erste Glied einen um so größeren Beitrag zur mittleren prozentischen Energieschwankung

[40]

$$\left(\frac{\sqrt{\varepsilon^2}}{\eta_0^2}\right)$$

liefert, je kleiner die Energie η_0 ist, und daß die Größe dieser vom ersten Glied gelieferten prozentischen Schwankung davon unabhängig ist, über einen wie großen Raum v die Strahlung verteilt ist; ich erwähne dies, um zu zeigen, wie grundverschieden die tatsächlichen statistischen Eigenschaften der Strahlung sind von denjenigen, welche wir nach unserer jetzigen Theorie, die sich auf lineare, homogene Differentialgleichungen stützt, erwarten sollten.

[41]

7. Im vorigen haben wir die Schwankungen der Energieverteilung berechnet, um Aufschlüsse über die Natur der Wärmestrahlung zu erhalten. Im folgenden soll kurz gezeigt werden, wie man durch Berechnen der Schwankungen des Strahlungsdruckes also von Schwankungen der Bewegungsgröße zu ganz entsprechenden Resultaten gelangen kann.

[42]

Es befinde sich in einem allseitig von Materie von der absoluten Temperatur T umgebenen Hohlraum ein in Richtung senkrecht zu seiner Normalen frei beweglicher Spiegel[2]. Denken wir uns diesen von Anfang an mit einer gewissen Geschwindigkeit bewegt, so wird infolge dieser Bewegung an seiner Vorderseite mehr Strahlung reflektiert, als an seiner Rückseite; es ist daher der auf seiner Vorderseite wirkende Strahlungsdruck größer als der auf die Rückseite wirkende. Es wird also auf den Spiegel infolge seiner Bewegung relativ zur

1) Bei Durchführung der oben angedeuteten Interferenzbetrachtung würde man wohl $\dfrac{R}{Nk} = 1$ erhalten.

2) Die Bewegungen des Spiegels, von denen hier die Rede ist, sind den sogenannten Brownschen Bewegungen suspendierter Teilchen durchaus analog.

Hohlraumstrahlung eine der Reibung vergleichbare Kraft wirken, welche nach und nach die Bewegungsgröße des Spiegels aufzehren müßte, wenn nicht andererseits eine bewegende Ursache bestünde, welche die durch jene Reibungskraft verlorene Bewegungsgröße im Mittel gerade ersetzte. Den im vorigen studierten unregelmäßigen Schwankungen der Energie eines Strahlungsraumes entsprechen nämlich auch unregelmäßige Schwankungen der Bewegungsgröße bzw. unregelmäßige Schwankungen der von der Strahlung auf den Spiegel ausgeübten Druckkräfte, die den Spiegel in Bewegung versetzen müßten, auch wenn er anfänglich ruhte. Die mittlere Bewegungsgeschwindigkeit des Spiegels ist nun aus der Entropie-Wahrscheinlichkeitsbeziehung, das Gesetz der obengenannten Reibungskräfte aus dem als bekannt angenommenen Strahlungsgesetz zu ermitteln. Aus diesen beiden Resultaten berechnet man dann die Wirkung der Druckschwankungen und ist in der Lage, aus diesen wieder Schlüsse in betreff der Konstitution der Strahlung oder — genauer gesprochen — in betreff der Elementarvorgänge der Reflexion der Strahlung am Spiegel zu ziehen.

Es sei mit v die Geschwindigkeit des Spiegels zur Zeit t bezeichnet. Infolge der obenerwähnten Reibungskraft nimmt im darauf folgenden Zeitteilchen τ diese Geschwindigkeit um $\dfrac{P v \tau}{m}$ ab, wenn mit m die Masse des Spiegels, mit P die verzögernde Kraft bezeichnet wird, welche der Einheit der Geschwindigkeit des Spiegels entspricht. Wir bezeichnen ferner mit $\varDelta$ diejenige Geschwindigkeitsänderung des Spiegels während τ, welche den unregelmäßigen Schwankungen des Strahlungsdruckes entsprechen. Die Geschwindigkeit des Spiegels zur Zeit $t + \tau$ ist

$$v - \frac{P\tau}{m} v + \varDelta.$$

Als Bedingung dafür, daß v im Mittel, während τ ungeändert bleibt, erhalten wir

[43]
$$\overline{\left(v - \frac{P\tau}{m} v + \varDelta \right)} = \overline{v^2}$$

oder, indem man relativ unendlich kleines wegläßt und berücksichtigt, daß der Mittelwert von $v\,\varDelta$ offenbar verschwindet:

[44]
$$\varDelta^2 = \frac{2 P \tau}{m} \overline{v^2}.$$

In dieser Gleichung läßt sich zunächst $\overline{v^2}$ mittels der aus der Entropie-Wahrscheinlichkeitsgleichung ableitbaren Gleichung

[45]
$$\frac{m \overline{v^2}}{2} = \frac{1}{2} \frac{R T}{N}$$

ersetzen. Bevor wir ferner den Wert der Reibungskonstante P angeben, spezialisieren wir das behandelte Problem durch die Annahme, daß der Spiegel Strahlung von dem bestimmten Frequenzbereich (zwischen v und $v + dv$) vollkommen reflektiere, für Strahlung anderer Frequenz aber vollkommen durchlässig sei. Durch eine Rechnung, welche ich hier der Kürze halber nicht angebe, erhält man durch eine [46] rein elektrodynamische Untersuchung, die für jede beliebige Strahlungsverteilung gültige Gleichung

[47]
$$P = \frac{3}{2 c} \left[\varrho - \frac{1}{3} v \frac{d\varrho}{dv} \right] dv f,$$

falls man mit ϱ wieder die Strahlungsdichte bei der Frequenz v, mit f die Fläche des Spiegels bezeichnet. Durch Einsetzen der für $\overline{v^2}$ und P ermittelten Werte erhält man

$$\frac{\overline{\varDelta^2}}{\tau} = \frac{R T}{N} \cdot \frac{3}{c} \left[\varrho - \frac{1}{3} v \frac{d\varrho}{dv} \right] dv f.$$

Indem wir unter Benutzung der Planckschen Strahlungsformel diesen Ausdruck umformen, erhalten wir

[48]
$$\frac{\overline{\varDelta^2}}{\tau} = \frac{1}{c} \left[h \varrho v + \frac{c^3}{8 \pi} \frac{\varrho^2}{v^2} \right] dv f.$$

Die nahe Verwandtschaft dieser Beziehung mit der im vorigen Abschnitt für die Energieschwankung $\overline{(\varepsilon^2)}$ abgeleiteten ist unmittelbar zu sehen [1]), und man kann an sie genau entsprechende Betrachtungen anknüpfen wie an jene. Wieder müßte sich nach der jetzigen Theorie der Ausdruck auf das zweite Glied reduzieren (Schwankung durch Interferenz). Wäre das erste Glied allein vorhanden, so ließen sich die Schwankungen des Strahlungsdruckes vollständig erklären durch die Annahme, daß die Strahlung aus voneinander unabhängig beweglichen, wenig ausgedehnten Komplexen von der Energie $h v$ bestehe. Auch hier besagt die Formel, daß [49] nach der Planckschen Formel die Wirkungen der beiden genannten Schwankungsursachen sich verhalten wie Schwankungen (Fehler), welche voneinander unabhängigen Ursachen entspringen (additive Verknüpfung der Terme, aus denen sich das Schwankungsquadrat zusammensetzt).

8. Aus den letzten beiden Betrachtungen geht nach meiner Meinung unwiderlegbar hervor, daß die Konstitution der Strahlung eine andere sein muß, als wir gegenwärtig meinen. Unsere gegenwärtige Theorie liefert zwar, wie die treffliche Übereinstimmung von Theorie und Experiment in der Optik beweist, die

---

1) Man kann jene in der Form schreiben (wobei $\dfrac{R}{Nk} = 1$ gesetzt ist:

$$\varepsilon^2 = \left\{ h \varrho v + \frac{c^3 \varrho^2}{8 \pi v^2} \right\} v \, dv.$$

Physikalische Zeitschrift. 10. Jahrgang. No. 6. 191

allein direkt wahrnehmbaren zeitlichen Mittelwerte in richtiger Weise, führt aber mit Notwendigkeit zu mit der Erfahrung unvereinbaren Gesetzen über die thermischen Eigenschaften der Strahlung, sobald man nur an der Entropie-Wahrscheinlichkeit-Beziehung festhält. Die Abweichung der Erscheinungen von der Theorie tritt desto stärker hervor, je größer v und je kleiner ϱ ist. Es sind bei kleinem ϱ die zeitlichen Schwankungen der Strahlungsenergie eines bestimmten Raumes bzw. der Druckkraft der Strahlung auf eine bestimmte Fläche viel größer als unsere jetzige Theorie erwarten läßt.

Wir haben gesehen, daß das Plancksche Strahlungsgesetz sich begreifen läßt unter Heranziehung der Annahme, daß Oszillationsenergie von der Frequenz v nur auftreten kann in Quanten von der Größe $h\,v$. Es genügt nach dem Vorigen nicht die Annahme, daß Strahlung nur in Quanten von dieser Größe emittiert und absorbiert werden könne, daß es sich also lediglich um eine Eigenschaft der emittierenden bzw. absorbierenden Materie [50] handle; die Betrachtungen 6 und 7 zeigen, daß auch die Schwankungen in der räumlichen Verteilung der Strahlung und diejenigen des Strahlungsdruckes derart erfolgen, wie wenn die Strahlung aus Quanten von der angegebenen Größe bestünden. Es kann nun zwar nicht behauptet werden, daß die Quantentheorie aus dem Planckschen Strahlungsgesetze als Konsequenz folge, und daß andere Interpretationen ausgeschlossen seien. Man kann aber wohl behaupten, daß die Quantentheorie die einfachste Interpretation der Planckschen Formel liefert.

Es ist hervorzuheben, daß die angegebenen Überlegungen im wesentlichen keineswegs ihren Wert verlieren würden, falls sich die Plancksche Formel noch als ungültig erweisen sollte; gerade der von der Erfahrung genügend bestätigte Teil der Planckschen Formel (das für große $\dfrac{v}{T}$ in der Grenze gültige Wiensche Strahlungsgesetz) ist es, welcher zur Lichtquan[51] tentheorie führt.

9. Die experimentelle Erforschung der Konsequenzen der Lichtquantentheorie ist nach meiner Meinung eine der wichtigsten Aufgaben, welche die Experimentalphysik der Gegenwart zu lösen hat. Die bis jetzt gezogenen Konsequenzen kann man in drei Gruppen ordnen.

a) Es ergeben sich Anhaltspunkte für die Energie derjenigen Elementarvorgänge, welche mit Absorption bzw. Emission von Strahlung bestimmter Frequenz verknüpft sind (Stokes[52] sche Regel; Geschwindigkeit der durch Licht oder Röntgenstrahlen erzeugten Kathodenstrahlen; Kathodolumineszenz usw.). Hierher

gehört auch die interessante Anwendung, die Herr Stark von der Lichtquantentheorie gemacht hat, um die eigentümliche Energieverteilung im Spektrum einer von Kanalstrahlen emittierten Spektrallinie zu erklären [1]).

Die Schlußweise ist hier immer folgende: Erzeugt ein Elementarvorgang einen andern, so ist die Energie des letzteren nicht größer als die des ersteren. Die Energie eines der beiden Elementarvorgänge ist aber bekannt (von der Größe $h\,v$), wenn letzterer in der Absorption oder Emission von Strahlung bestimmter Frequenz besteht.

Besonders interessant wäre das Studium der Ausnahmen vom Stokesschen Gesetz. Zur [54] Erklärung dieser Ausnahmen muß angenommen werden, daß ein Lichtquant erst dann emittiert wird, wenn das betreffende Emissionszentrum zwei Lichtquanten absorbiert hat. Die Häufigkeit eines derartigen Ereignisses, also auch die Intensität des emittierten Lichtes von kleinerer Wellenlänge als das erzeugende, wird in diesem Falle bei schwacher Bestrahlung (nach dem Massenwirkungsgesetz) dem Quadrat der erregenden Lichtstärke proportional sein müssen, während bei Gültigkeit der Stokesschen Regel bei schwacher Bestrahlung Proportionalität mit der ersten Potenz der erregenden Lichtintensität zu erwarten ist. [55]

b) Wird bei Absorption [2]) jedes Lichtquants ein Elementarvorgang gewisser Art bewirkt, so ist $\dfrac{E}{h\,v}$ die Anzahl dieser Elementarvorgänge, falls die Energiemenge E von Strahlung der Frequenz v absorbiert wird.

Wird also z. B. die Menge E einer Strahlung von der Frequenz v unter Ionisierung eines Gases von diesem absorbiert, so ist zu erwarten, daß $\dfrac{E}{N\,h\,v}$ Grammoleküle des Gases dabei ionisiert werden. Diese Beziehung setzt nur scheinbar die Kenntnis von N voraus; schreibt man nämlich die Plancksche Strahlungsformel in der Form

$$\varrho = \alpha\,v^3\,\dfrac{1}{e^{\frac{\beta v}{T}} - 1},$$

so ist $\dfrac{E}{R\,\beta\,v}$ die Anzahl der ionisierten Grammmoleküle.

Diese Beziehung, welche ich bereits in meiner ersten Arbeit [3]) über diesen Gegenstand angab, ist leider bisher unbeachtet geblieben. [58]

c) Das unter 5 Mitgeteilte führt zu einer

[53] 1) J. Stark, diese Zeitschr. 9, 767, 1908.
[56] 2) Die analoge Betrachtung gilt natürlich auch umgekehrt für die Lichterzeugung durch Elementarvorgänge (z. B. durch Ionenstöße).
[57] 3) Ann. d. Phys. (4) 17, 132—148, 1905. § 9.

Modifikation der kinetischen Theorie der spezifischen Wärme [1]) und zu gewissen Beziehungen zwischen optischem und thermischem Verhalten der Körper.

10. Es erscheint schwierig, ein theoretisches System aufzustellen, welches die Lichtquanten in vollständiger Weise deutet, wie unsere heutige Molekularmechanik in Verbindung mit der Maxwell-Lorentzschen Theorie die von Herrn Jeans vertretene Strahlungsformel zu deuten vermag. Daß es sich nur um eine Modifikation unserer heutigen Theorien, nicht um ein vollständiges Verlassen derselben handeln wird, scheint schon daraus hervorzugehen, daß das Jeanssche Gesetz in der Grenze (für kleine $\frac{v}{T}$) gültig zu sein scheint. Einen Hinweis darauf, wie jene Modifikation durchzuführen sein dürfte, liefert eine von Herrn Jeans vor einigen Jahren durchgeführte, nach meiner Meinung höchst wichtige Dimensionalbetrachtung, die ich im folgenden — in einigen Punkten modifiziert — kurz wiedergebe.

Wir denken uns, daß in einem abgeschlossenen Raume ein ideales Gas, Strahlung sowie Ionen vorhanden seien, welch letztere vermöge ihrer Ladung einen Energieaustausch zwischen Gas und Strahlung zu vermitteln vermögen. Es ist zu erwarten, daß in einer an die Betrachtung dieses Systems geknüpften Strahlungstheorie folgende Größen eine Rolle spielen, also in dem zu ermittelnden Ausdruck für die Strahlungsdichte ϱ auftreten werden:

a) die mittlere Energie η eines molekularen Gebildes (bis auf einen unbenannten Zahlenfaktor gleich $\frac{RT}{N}$),

b) die Lichtgeschwindigkeit c,

c) das Elementarquantum ε der Elektrizität,

d) die Frequenz v.

Aus der Dimension von ϱ kann man nun unter ausschließlicher Berücksichtigung der Dimensionen der vier ebengenannten Größen in einfacher Weise ermitteln, welche Gestalt der Ausdruck für ϱ haben muß. Man erhält, indem man für η den Wert $\frac{RT}{N}$ setzt:

$$\varrho = \frac{\varepsilon^2}{c^4} v^3 \psi(\alpha),$$

wobei

[62]
$$\alpha = \frac{R\,\varepsilon^2}{Nc}\frac{v}{T},$$

wobei ψ eine unbestimmt bleibende Funktion bezeichnet. Diese Gleichung enthält das Wiensche Verschiebungsgesetz, dessen Gültigkeit kaum mehr bezweifelt werden kann. Man

[59]   [1]) A. Einstein, Ann. d. Phys. (4) **22**, 1907, S. 180—190 und S. 800.

hat dies als eine Bestätigung dafür aufzufassen, daß außer den oben eingeführten vier Größen in dem Strahlungsgesetz keine weiteren Größen eine Rolle spielen, die eine Dimension haben.

Daraus schließen wir, daß die in der Gleichung für ϱ auftretenden Koeffizienten $\frac{\varepsilon^2}{c^4}$ und $\frac{R\,\varepsilon^2}{Nc}$ bis auf bei theoretischen Entwicklungen auftretende dimensionslose Zahlenfaktoren, welche sich natürlich aus einer Dimensionalbetrachtung nicht ergeben können, numerisch gleich sein sollen den in der Planckschen (oder Wienschen) Strahlungsformel auftretenden Koeffizienten. Da jene sich nicht ergebenden dimensionslosen Zahlenfaktoren die Größenordnung kaum wesentlich ändern dürften, so kann man der Größenordnung nach setzen [1]):

[65]
$$\frac{h}{c^3} = \frac{\varepsilon^2}{c^4} \quad \text{und} \quad \frac{h}{k} = \frac{R}{N}\frac{\varepsilon^2}{c},$$

also

$$h = \frac{\varepsilon^2}{c} \quad \text{und} \quad k = \frac{N}{R}.$$

Die zweite dieser Gleichungen ist die, mittels welcher Herr Planck die Elementarquanta der Materie oder Elektrizität bestimmt hat. Zum Ausdruck für h ist zu bemerken, daß

$$h = 6 \cdot 10^{-27}$$

und

$$\frac{\varepsilon^2}{c} = 7 \cdot 10^{-30}.$$

Es fehlt ja hier um 3 Dezimalen. Aber dies dürfte wohl darauf zurückzuführen sein, daß die dimensionslosen Faktoren unbekannt sind.

Das Wichtigste dieser Ableitung liegt darin, daß durch sie die Lichtquantenkonstante h auf das Elementarquantum ε der Elektrizität zurückgeführt wird. Es ist nun daran zu erinnern, daß das Elementarquantum ε ein Fremdling ist in der Maxwell-Lorentzschen Elektrodynamik [2]). Man muß fremde Kräfte heranziehen, um in der Theorie das Elektron zu konstruieren; man pflegt ein starres Gerüst einzuführen, das verhindern soll, daß die elektrischen Massen des Elektrons unter dem Einfluß ihrer elektrischen Wechselwirkung auseinanderfahren. Es scheint mir nun aus der Beziehung $h = \frac{\varepsilon^2}{c}$ hervorzugehen, daß die gleiche Modifikation der Theorie, welche das Elementarquantum ε als Konsequenz enthält, auch die Quantenstruktur der Strahlung als Konsequenz

[1]) Die Plancksche Formel lautet:
$$\varrho = \frac{8\pi h v^3}{c^3} \frac{1}{e^{\frac{hv}{kT}} - 1}$$

[2]) Vgl. Levi-Civita. Comptes Rendus 1907: „Sur le mouvement etc.".

[60]
[61]

[64]

[63]

[66]

[67]

[69]

[68]

enthalten wird. Es wird die Fundamental-
gleichung der Optik

$$D(\varphi) = \frac{1}{c^2}\frac{\partial^2 \varphi}{\partial^2 t} - \left(\frac{\partial^2 \varphi}{\partial x^2} + \frac{\partial^2 \varphi}{\partial y^2} + \frac{\partial^2 \varphi}{\partial z^2}\right) = 0$$

zu ersetzen sein durch eine Gleichung, in der
auch die universelle Konstante ε (wahrscheinlich das Quadrat derselben) in einem Koeffizienten auftritt. Die gesuchte Gleichung (bzw.
das gesuchte Gleichungssystem) muß in den
Dimensionen homogen sein. Es muß bei Anwendung der Lorentz-Transformation in sich
selbst übergehen. Sie kann nicht linear und
homogen sein. Sie muß — wenigstens falls
[70] das Jeanssche Gesetz wirklich in der Grenze
für kleine $\frac{\nu}{T}$ gültig ist — für große Ampli-
tuden in der Grenze auf die Form $D(\varphi) = 0$
führen.

Es ist mir noch nicht gelungen, ein diesen
Bedingungen entsprechendes Gleichungssystem
zu finden, von dem ich hätte einsehen können,
daß es zur Konstruktion des elektrischen Elementarquantums und der Lichtquanten geeignet
[71] sei. Die Mannigfaltigkeit der Möglichkeiten
scheint aber nicht so groß zu sein, daß man
vor der Aufgabe zurückschrecken müßte.

Nachtrag.

Aus dem unter 4. in der vorstehenden Abhandlung Gesagten könnte der Leser leicht
einen unzutreffenden Eindruck gewinnen über
den Standpunkt, welchen Herr Planck seiner
eigenen Theorie der Temperaturstrahlung gegenüber einnimmt. Deshalb halte ich es für angezeigt, das Folgende zu bemerken.

Herr Planck hat in seinem Buche an mehreren Stellen hervorgehoben, daß seine Theorie
noch nicht als etwas Fertiges, Abgeschlossenes
aufzufassen sei. Er sagt z. B. am Schluß der
Vorrede wörtlich: „Es liegt mir aber daran,
auch an dieser Stelle noch besonders hervorzuheben, was sich im letzten Paragraphen des
Buches näher ausgeführt findet, daß die hier
entwickelte Theorie keineswegs den Anspruch
erhebt, als vollkommen abgeschlossen zu gelten,
wenn sie auch, wie ich glaube, einen gangbaren
Weg eröffnet, um die Vorgänge der Energie-
strahlung von dem nämlichen Gesichtspunkt
aus zu überblicken wie die der Molekular-
[72] bewegung.“

Die betreffenden Auseinandersetzungen in
meiner Abhandlung sind nicht als ein Einwand
(im eigentlichen Sinne des Wortes) gegen die
Plancksche Theorie aufzufassen, sondern lediglich als ein Versuch, das Entropie-Wahrschein-
lichkeitsprinzip etwa schärfer zu fassen und anzuwenden, als man es bisher getan hat. Eine
schärfere Fassung dieses Prinzips war not-

wendig, weil ohne eine solche die folgenden
Entwicklungen in der Abhandlung, in welchen
auf die molekulare Struktur der Strahlung geschlossen wird, nicht genügend begründet gewesen wären. Damit meine Fassung des
Prinzips nicht als etwas ad hoc Gewähltes,
Willkürliches erscheine, mußte ich zeigen, warum
mich die bisherige Formulierung des Prinzips
noch nicht vollkommen befriedigte.

Bern, Januar 1909.

(Eingegangen 23. Januar 1909.)

Über das Plancksche Strahlungsgesetz.

(Vorläufige Mitteilung.)

Von J. Weiß.

§ 1. Das Plancksche Strahlungsgesetz bezieht sich auf einen von absolut spiegelnden
Wänden eingeschlossenen, strahlungserfüllten
Hohlraum; er soll die Form eines Würfels
haben, dessen Kanten parallel den Koordinatenachsen seien und die Länge l besitzen. Dann
ist die Grenzbedingung für eine ebene elektrische Welle, die ihn durchsetzt, die, daß die
Tangentialkomponente der elektrischen Feldstärke an den Wänden gleich Null ist. Betrachten wir z. B. die zu einer Kante senkrechten Flächen, so wird die Bedingung überall
auf ihnen erfüllt sein, wenn sie in den Ecken
gilt, welche die Kante begrenzen. α, β, γ seien
die Richtungskosinusse der Wellennormalen.
Das Stück der Kante, das zwischen entsprechenden Punkten der Welle liegt, hat die
Länge $\frac{\alpha}{\lambda}$; dabei haben wir angenommen, daß
die betrachtete Kante in der X Achse liege.
Es müssen α solche Stücke auf $2l$ liegen, wo α
eine ganze positive Zahl ist, wenn in den Ecken,
die l begrenzen, Knoten der Welle sein sollen.

§ 2. Wir wollen die Anzahl der in dem
Würfelraum l^3 enthaltenen Parameter berechnen,
auf welche sich die elektromagnetische Energie
des Volumens verteilt.

Eine elektromagnetische Schwingung in
einem Punkte kann als Superposition von ebenen
Wellen aller möglichen Richtungen betrachtet
werden. Die Polarisation derselben kommt
bei der Berechnung der Energie nicht in Betracht. In einem bestimmten Moment ist an
einem bestimmten Punkte die Energie durch
die jeweilige elektrische und magnetische Elongation gegeben. Die Energie einer einzigen
Welle wird also durch 2 Parameter dargestellt.
und es kommt uns darauf an, die Zahl aller
Wellen im Volumen l^3 zu finden, welche die
Frequenz ν bis $\nu + d\nu$ besitzen. Doppelt so
groß ist die Zahl der Parameter, auf welche
Energie entfällt.

Published in *Physikalische Zeitschrift* 10 (1909): 185–193. Dated Bern, January 1909, received 23 January 1909, published 15 March 1909.

[1] *Lorentz 1908b*.

[2] *Jeans 1908*. See also *Jeans 1905c*.

[3] *Ritz 1908b*.

[4] For a discussion of the works cited, see the editorial note, "Einstein's Early Work on the Quantum Hypothesis," pp. 144–146, and *Kuhn 1978*, pp. 189–205. All of the works cited by Einstein refer to *Lorentz 1908a*, to which Einstein refers in the slightly revised version, *Lorentz 1909a* (see Einstein to Hendrik Lorentz, 13 April 1909).

[5] See *Ritz 1908b*, p. 904.

[6] *Ritz 1908b*, p. 904. Beginning with "ohne," Ritz indicated emphasis by using spaced type ("Sperrdruck") in the remainder of the quotation.

[7] Ritz offered a rebuttal to Einstein's argument (see *Ritz 1909*). These and other points of difference between Einstein and Ritz are noted in *Ritz and Einstein 1909* (Doc. 57).

[8] Einstein and Hopf disposed of this doubt in *Einstein and Hopf 1910a* and *1910b*.

[9] *Einstein 1905i* (Doc. 14).

[10] "E_ν" should be "$\overline{E}_\nu$". This equation appears in *Planck 1900e*, p. 241.

[11] *Planck 1900a* and *1906c*.

[12] *Lorentz 1908a*; Jeans restated his law in *Jeans 1908*, p. 853.

[13] *Planck 1906c*.

[14] For a discussion of the nature of the "Lücke" ("gap"), see the editorial note, "Einstein on the Foundations of Statistical Physics," pp. 48–50. The references to Gibbs are *Gibbs 1902* and *1905*.

[15] The disagreement between the "Jeans-Lorentz" law and "all radiation observations," as well as the contradiction with everyday experience, were especially emphasized in *Lummer and Pringsheim 1908*. Jeans responded in *Jeans 1908*, p. 853, that the formula does agree with observations of long wavelength thermal radiation.

[16] *Planck 1906c*, pp. 177–179.

[17] *Klein 1974b*, p. 190, suggests that the following discussion is related to Einstein's promise to show that statistical probability is sufficient for the discussion of thermal processes (see *Einstein 1905i* [Doc. 14], p. 140). For an earlier version of the following definition, see *Einstein 1903* (Doc. 4), pp. 171–172.

[18] For further discussion of this point, see *Einstein 1910c*, pp. 1276–1277.

[19] See *Einstein 1906b* (Doc. 32), pp. 372–373, where this assumption leads to an expression that accounts for Brownian motion. See also the editorial note, "Einstein on Brownian Motion," pp. 213–214.

[20] The bar above τ_ν appears to be a printer's error.

[21] For an earlier derivation of this formula, see *Einstein 1903* (Doc. 4); *Einstein 1904* (Doc. 5), pp. 354–355; and *Einstein 1905i* (Doc. 14), pp. 140–141.

[22] See, e.g., *Boltzmann 1877*; *Planck 1900d*, pp. 242–243; *Planck 1906c*, pp. 137–140.

[23] *Boltzmann 1877*.

[24] *Boltzmann 1896*.

[25] See *Einstein 1906d* (Doc. 34), pp. 201–202. *Planck 1906c*, pp. 178–179, discusses this difficulty.

[26] *Einstein 1906d* (Doc. 34); *Einstein 1907a* (Doc. 38).

[27] See *Einstein 1905i* (Doc. 14), pp. 132–133, and *Einstein 1907h* (Doc. 45), pp. 372–373.

[28] This is Einstein's first use of h to denote what is now called "Planck's constant."

[29] *Einstein 1905i* (Doc. 14), especially pp. 139–144.

[30] Wien's formula is valid for *large* values of ν/T.

[31] H_0 and η_0 are the equilibrium values of H and η, respectively.

[32] For an earlier version of this argument, see *Einstein 1907b* (Doc. 39), pp. 570–571.

[33] In this equation, the absolute value of $\left(\dfrac{d^2\sigma}{d\epsilon^2}\right)_0$ should be taken, since this quantity is negative because the entropy is a maximum.

[34] In this equation, the absolute value of $\left(\dfrac{d^2\sigma}{d\epsilon^2}\right)_0$ should be taken, and in this and the following equation ϵ^2 should be $\overline{\epsilon^2}$. This expression for the mean square energy fluctuation is equivalent to the one given in *Einstein 1904* (Doc. 5), p. 360.

[35] Eq. (230) on p. 156 of *Planck 1906c* is an expression for the entropy density $\mathfrak{s}$ per unit volume per unit frequency. The equivalent expression for the entropy σ, defined by Einstein above, may be written

$$\sigma = \frac{8\pi k\nu^2}{c^3}\left\{\left(1 + \frac{c^3\rho}{8\pi h\nu^3}\right)\log\left(1 + \frac{c^3\rho}{8\pi h\nu^3}\right) - \frac{c^3\rho}{8\pi h\nu^3}\log\frac{c^3\rho}{8\pi h\nu^3}\right\},$$

where ρ is the energy density per unit frequency.

[36] In this equation, $\eta_0 = \rho v dv$, as follows from the definition of η on p. 188. See also fn. 1 on p. 190.

[37] The equation in parentheses should read "$\sigma d\lambda = -\rho dv$".

[38] Except for the factor $1/8\pi$. The second term was derived from electromagnetic theory in *Lorentz 1916*.

[39] See *Planck 1901b* and *Planck 1906c*, pp. 162–163.

[40] The first term can be obtained directly from Einstein's expression for $\overline{\epsilon^2}$ above by using the expression for entropy corresponding to Wien's law given in *Einstein 1905i* (Doc. 14), p. 139.

[41] Einstein wrote Hendrik Lorentz, 30 March 1909: "In particular the argument given in section 7 of the paper convinces me" ("Insbesondere wirkt die unter 7. in der Arbeit mitgeteilte Überlegung auf mich überzeugend"). The argument given is closely related to his prior work on Brownian motion, especially *Einstein 1906b* (Doc. 32).

[42] On 17 January 1952 Einstein wrote to Max von Laue: "In 1905 I already knew for certain that it [Maxwell's theory] leads to false fluctuations of the radiation pressure and thus to an incorrect Brownian motion of a mirror in a Planck radiation cavity" ("1905 wusste ich schon sicher, dass sie zu falschen Schwankungen des Strahlungsdruckes führt und damit zu einer unrichtigen Brown'schen Bewegung eines Spiegels in einem Planck'schen Strahlungshohlraum").

[43] The parenthetical expression on the left-hand side of this equation should be squared.

[44] "Δ^2" should be "$\overline{\Delta^2}$".

[45] The motion of the mirror is assumed to be one-dimensional.

[46] A calculation of P is given in *Einstein and Hopf 1910b*. After citing *Einstein 1909b* (Doc. 56), the authors state: "The essentially new element in the present paper lies in the fact that the momentum fluctuations were, for the first time, calculated exactly" ("Das wesentlich Neue der vorliegenden Arbeit besteht darin, daß die Impulsschwankungen zum erstenmal exakt ausgerechnet wurden") (*Einstein and Hopf 1910b*, p. 1115).

[47] A factor of $1/m^2$ is missing from the right-hand side of this equation.

[48] The left-hand side of this equation should be "$m^2\Delta^2/\tau$".

[49] Planck criticized this interpretation of the first term in *Planck 1910*. For Einstein's response, see "Antwort auf Planks Manuskript," ca. January 1910. In addition, *Einstein and Hopf 1910b* shows that the momentum fluctuations derived from Maxwell's theory yield a differential equation for the energy density, the only solution of which is the Rayleigh-Jeans law.

[50] Laue and Planck held this view (see Max Laue to Einstein, 2 June 1906; Max Planck to Einstein, 6 July 1907; and Planck's comment in *Einstein et al. 1909c* [Doc. 61], pp. 825–826).

[51] See *Einstein 1905i* (Doc. 14).

[52] For treatments of these phenomena, see *Einstein 1905i* (Doc. 14), pp. 144–147; *Ladenburg 1907*; *Stark 1908*; and *Lenard and Saeland 1909*.

[53] *Stark 1908*.

[54] For the predicted exceptions, see *Einstein 1905i* (Doc. 14), p. 145.

[55] Direct proportionality between the intensities of emitted and absorbed fluorescent light is reported in *Knoblauch 1895*. It is confirmed in *Kowalski 1910*.

[56] *Einstein 1909c* (Doc. 60), p. 492, considers a similar process, in which decelerated cathode rays generate X-rays.

[57] *Einstein 1905i* (Doc. 14). See p. 148.

[58] Stark reported tests of this relationship in *Stark 1909b*, which appeared shortly after this paper.

[59] *Einstein 1907a* (Doc. 38); *Einstein 1907d* (Doc. 42).

[60] This is probably a reference to *Jeans 1905d*. Jeans does not, however, use the constant h, nor does he consider the ratio e^2/c.

[61] Einstein wrote nearly the same words to Hendrik Lorentz, 30 March 1909. He then continued: "I hope that you can find the correct way, if you find the reasons given in the paper for the untenability of the current foundations at all valid" ("Ich hege nun die Hoffnung, dass Sie den richtigen Weg finden könnten, wenn Sie überhaupt die in der Arbeit angegebenen Gründe für die Haltlosigkeit der heutigen Grundlagen stichhaltig finden").

[62] "R/N" should be "N/R".

[63] See *Wien 1893*. Einstein wrote Hendrik Lorentz, 23 May 1909: "It seems to me indeed quite significant that one arrives by means of a dimensional consideration at the Wien displacement law and at the Planck determination of the elementary quantum, which is certainly correct in order of magnitude" ("Aber die Thatsache, dass man durch die Dimensionalbetrachtung

zum Wien'schen Verschiebungsgesetz und zur in der Grössenordnung sicher richtigen Planck'schen Bestimmung des Elementarquantums gelangt, erscheint mir doch recht bedeutsam'').

[64] See note 62.

[65] See note 62.

[66] See note 39.

[67] Because of the size of the dimensionless factor, Lorentz rejected Einstein's argument: "I cannot declare myself in agreement with your opinion that h is probably related to ϵ (charge of the electron); in any case I doubt it very much, since the three missing decimals are no trifle" ("Mit Ihrer Meinung, daß h wohl mit ϵ (Ladung des Elektrons) zusammenhängt, kann ich mich nicht einverstanden erklären; jedenfalls zweifle ich sehr. Denn die drei fehlenden Dezimalen sind keine Kleinigkeit") (Hendrik Lorentz to Einstein, 6 May 1909). Today e^2/hc is known as the fine structure constant.

[68] In *Levi-Civita 1907*, Levi-Civita found a solution of the field equations corresponding to the motion of a stable, isolated charge moving at the speed of light.

[69] The notion of a rigid electron was introduced in *Abraham 1902b, 1903*. Poincaré introduced a deformable electron possessing internal, nonelectromagnetic stresses (see *Poincaré 1906*).

[70] In a letter to Hendrik Lorentz of 23 May 1909, Einstein considered the possibility that the differential equations could be both linear and homogeneous if one regarded electrons and light quanta as point singularities. But linear, homogeneous equations would not be possible "if one wants to avoid the introduction of singular points, which would probably be the most satisfactory" ("wenn man ohne die Einführung singulärer Punkte auskommen möchte, was wohl am befriedigendsten wäre").

[71] For a discussion of Einstein's early attempt to construct a unified field theory of matter and light, see the editorial note, "Einstein's Early Work on the Quantum Hypothesis," pp. 147–148, and *McCormmach 1970a*.

[72] *Planck 1906c*, p. vi (the last word is "Molekularbewegungen"); see also pp. 220–221.

57. ''On the Present Status of the Radiation Problem''

[Ritz and Einstein 1909]

DATED Zurich, April 1909
RECEIVED 13 April 1909
PUBLISHED 1 May 1909

IN: *Physikalische Zeitschrift* 10 (1909): 323–324.

Physikalische Zeitschrift.   10. Jahrgang.   No. 9.   323

Zum gegenwärtigen Stand des Strahlungsproblems.

Von W. Ritz und A. Einstein.   [1]

Zur Aufklärung der Meinungsverschiedenheiten, welche in unseren beiderseitigen Publikationen[1]) zutage getreten sind, bemerken wir folgendes.

In den speziellen Fällen, in denen ein elektro-

---

1) W. Ritz, diese Zeitschr. **9**, 903—907, 1908 und A. Einstein, diese Zeitschr. **10**, 185—193, 1909.   [2]

324   Physikalische Zeitschrift.   10. Jahrgang.   No. 9.

magnetischer Vorgang auf einen endlichen Raum beschränkt bleibt, ist die Darstellung des Vorganges sowohl in der Form

$$f = f_1 = \frac{1}{4\pi} \int \frac{\varphi\left(x', y', z', t - \frac{r}{c}\right)}{r} \, dx' \, dy' \, dz'$$

als auch in der Form

$$f = f_2 = \frac{1}{4\pi} \int \frac{\varphi\left(x', y', z', t + \frac{r}{c}\right)}{r} \, dx' \, dy' \, dz'$$

und in anderen Formen möglich.

Während Einstein glaubt, daß man sich auf diesen Fall beschränken könne, ohne die Allgemeinheit der Betrachtung wesentlich zu beschränken, betrachtet Ritz diese Beschränkung als eine prinzipiell nicht erlaubte. Stellt man sich auf diesen Standpunkt, so nötigt die Erfahrung dazu, die Darstellung mit Hilfe der retardierten Potentiale als die einzig mögliche zu betrachten, falls man der Ansicht zuneigt, daß die Tatsache der Nichtumkehrbarkeit der Strahlungsvorgänge bereits in den Grundgesetzen ihren Ausdruck zu finden habe. Ritz betrachtet die Einschränkung auf die Form der retardierten Potentiale als eine der Wurzeln des zweiten Hauptsatzes, während Einstein glaubt, daß die Nichtumkehrbarkeit ausschließlich auf Wahrscheinlichkeitsgründen beruhe.   [3]

Zürich, April 1909.

(Eingegangen 13. April 1909.)

Published in *Physikalische Zeitschrift* 10 (1909): 323–324. Dated Zurich, April 1909, received 13 April 1909, published 1 May 1909.

[1] Walter Ritz was a Privatdozent in Göttingen. For a discussion of Ritz's work, emphasizing his relationship with Einstein, see *Pyenson 1976*, pp. 100–104.

[2] *Ritz 1908b* and *Einstein 1909b* (Doc. 56). See also *Ritz 1909* and the editorial note, "Einstein's Early Work on the Quantum Hypothesis," pp. 145–146.

[3] See *Ritz 1908b*, p. 904, and the editorial note, "Einstein on the Foundations of Statistical Physics," pp. 48–53.

58. "Discussion" following lecture version of Henry Siedentopf, "On Ultramicroscopic Images"

[*Einstein et al. 1909a*]

Excerpt, printed version of the discussion held 20 September 1909 following presentation of Siedentopf's paper at the 81st meeting of the Gesellschaft Deutscher Naturforscher und Ärzte in Salzburg.

PUBLISHED 10 November 1909

IN: *Physikalische Zeitschrift* 10 (1909): 779–780.

Physikalische Zeitschrift. 10. Jahrgang. No. 22. 779

densoren, dem Paraboloidkondensor und noch besser dem neuen aplanatischen Dunkelfeldkondensor von Zeiß lassen sich eine Reihe von Eigentümlichkeiten ultramikroskopischer Abbildung bequem studieren, wie die Veränderung der Beugungsscheiben durch Diaphragmierung der Öffnung des Mikroskopobjektivs, und die praktisch wichtigen Erscheinungen nichtsphärischer und asymmetrischer Wellen.

Von besonderem physikalischen Interesse sind die Anzeichen, die auf Doppelbrechung in den Beugungsscheiben hindeuten, so daß wir bei den Ultramikronen isotrope und anisotrope unterscheiden müssen.

Relativ einfach liegen die Verhältnisse noch bei Goldteilchen, welche das Goldrubinglas färben. Hier kommt hinsichtlich des Polarisationszustandes bis zu Größen, die 100 $\mu\mu$ nicht sehr übersteigen, im wesentlichen nur die Rayleighsche Welle zur Geltung. Entsprechend zeigt sich bei Beleuchtung mit linear polarisiertem Licht in der hinteren Brennebene des Mikroskopobjektives jedesmal in demjenigen Punkte Dunkelheit, welcher einer zu der Schwingungsrichtung im Polarisator parallelen Richtung im Fokus des Objektives entspricht.

Jedes Goldteilchen verhält sich also wie eine linear polarisierte Lichtquelle, deren Schwingungen zur Schwingungsebene des Polarisators parallel liegen. In der Richtung dieser Schwingungen kann kein Licht emittiert werden — daher der dunkle Fleck — weil das ja sonst auf longitudinale Schwingungen führen würde.

Viel verwickelter sind die Erscheinungen bei Silberteilchen, die sich aus kolloidaler Lösung durch Adsorption am Glase absetzen. Hier besteht keine Richtung verschwindender Intensität. Die Teilchen verhalten sich bei ringförmiger Seitenbeleuchtung wie kleine Lichtquellen, in denen nach zwei zueinander senkrechten Richtungen das Licht schwingen kann. Die Mannigfaltigkeit der Erscheinungen ist sehr groß, weil die Teilchen ungeordnet liegen, in allen Farben, wenn auch vorwiegend violett auftreten, und dazu noch pleochroitisch sind.

Die lichtstarken Dunkelfeldkondensoren eignen sich schließlich auch gut zur Momentaufnahme schnell ablaufender mikroskopischer Vorgänge. Ich zeige als Beispiele Bilder lebender Bakterien, wie die Spirochaete pallida und lebender Spermatozoen des Menschen. Für den Physiker sind besonders interessant Aufnahmen der Brownschen Molekularbewegung. Eine bemerkenswerte Bestätigung der kinetischen Theorie bieten Momentaufnahmen einer kolloidalen Silberlösung nach Carey Lea auf fallender Platte. Die Teilchen beschreiben eine in der Fallrichtung der Platte auseinander gezogene Kurve auf der Platte; die ganz unregelmäßigen Schwingungen der Teilchen zeigen,

daß auch in den kleinen Zeitelementen von $^1/_{100}$ Sekunde und darunter der von der kinetischen Theorie geforderte Zufall die Schwingungen regiert (vergl. Figur auf Tafel XIII).

Wie das ausgestellte Demonstrationspräparat und der zugehörige Apparat zeigt, ist die Zusammensetzung des neuen lichtstarken Ultramikroskops von Zeiß so einfach, daß diese wirkungsvolle Demonstration der Brownschen Bewegung und kolloidaler Lösungen allgemeinere Aufnahme in den physikalischen und chemischen Experimentalvorlesungen finden sollte. [1]

Diskussion.

v. Ignatowsky: Ich möchte bemerken, daß mein Spiegelkondensor schon seit etwa Januar 1909 von der Firma E. Leitz in Wetzlar genau in der Weise ausgeführt wird, wie Dr. Siedentopf es hier in der Figur 13 zeigte. Die Trennungsfläche zwischen den beiden Teilen des Glaskörpers besteht nicht wie früher aus einer Ebene, sondern aus einer sphärischen Fläche, weshalb die äußere spiegelnde Fläche tatsächlich eine einzige Kugelfläche bildet. Der Strahlengang ist dadurch sehr erheblich verbessert worden und entspricht dem theoretisch berechneten [1]).

Vortragender: Ich freue mich jedenfalls, daß die Konstruktion richtig ist und das ist doch der Fall, wenn dieselbe Methode von zwei voneinander unabhängigen und sich in gewissem Sinne gegenseitig kontrollierenden Stellen gefunden ist. Wesentlich neu ist meine Erklärung der aplanatischen Natur dieser Kondensoren aus einer bisher unbekannten Eigenschaft der Kardioide. Übrigens sind diese Kondensoren nach meinen Angaben schon 1908 von Zeiß ausgeführt (Anm. bei der Korrektur), auch hat damals schon Verf. auf den Konstruktionsfehler in dem Leitzschen Spiegelkondensor öffentlich hingewiesen (vgl. Ztschr. wiss. Mikroskopie 25, Seite 273 ff., Fig. 13 u. 11, 1908).

Rubens: Ich möchte Herrn Dr. Siedentopf fragen, ob er diese schöne Methode vielleicht zur quantitativen Messung der Brownschen Molekularbewegung benutzt hat. Nach dem Vorgang des Herrn Perrin lassen sich solche Versuche zur Ermittlung der Loschmidtschen Zahl verwenden. [2]

Vortragender: Ich möchte das den Physikern im Laboratorium überlassen. [3]

Rubens: Mir fällt auf, daß die Methode den großen Vorteil bietet, daß man die in Betracht kommenden Größen in Ruhe auf dem Photogramm ausmessen kann. [4]

Vortragender: Ich bitte die Herren, das [5]

1) In dem nächsten Heft der Zeitschrift f. wiss. Mikroskopie erscheint die genaue Beschreibung dieses Spiegelkondensors und auch die Abbildung des Strahlenganges.

780 **Physikalische Zeitschrift. 10. Jahrgang. No. 22.**

selbst in die Hand zu nehmen; ich bin beruf-lich so in Anspruch genommen, daß ich keine Zeit dazu habe.

[6] **Einstein:** Die Hauptschwierigkeit liegt in der Temperatur; die ist nicht konstant zu halten. Und das ist auch bei den französischen Messungen der Fall.

[7] **Seddig:** Diesen letzteren Übelstand, den der Vorredner erwähnte und der darin besteht, daß während der Beobachtung Temperatur-änderungen des Präparats durch Strahlungs-absorption auftreten und das Resultat in nicht kontrollierbarer Weise beeinflussen, habe ich bei ähnlichen Untersuchungen auch empfunden, und deshalb bei meinen Versuchen, die Braun-sche Molekularbewegung messend zu verfolgen, keine andauernde Beleuchtung benutzt, sondern immer nur zwei ganz kurze Lichtblitze durch das Präparat zur photographischen Markierung der jeweiligen Lage der Teilchen hindurchgehen

[8] lassen. Übrigens ist eine etwas ähnliche Methode wie die eben vorgeführte, vor 2 oder 3 Jahren von **The Svedberg** gemacht worden; er ließ die kolloidale Lösung durch die Beobachtungs-küvette langsam strömen und er bekam dann ähnliche Elongationen, von der geradlinigen Be-wegung, die er freilich nicht photographierte,

[9] sondern mit einem Okularmikrometer beobach-tete.

Vortragender: Die Methode von **The Svedberg** hat den Fehler, daß man auf stö-rende Strömungen keinen Einfluß hat, die noch dazu mit vergrößert werden. Außerdem nahm er fälschlich an, daß die Bewegung sinus-

[10] förmig sei.

F. Paul Liesegang (Düsseldorf), **Einige neue Versuchsanordnungen mit dem Projektions-apparat. (Mit Demonstrationen.)**

Die im folgenden beschriebenen Demon-strationen stellen Abänderungen oder Ergän-zungen bekannter Versuchsanordnungen dar; sie lassen sich z. T. mit den meist vorhandenen Hilfsmitteln ausführen.

1. **Schattenversuche.** Zur Darstellung der Schattenversuche wird dicht vor eine Matt-glasscheibe, die vor die Projektionslaterne ge-setzt ist und als lichtspendender Körper dient, eine große Irisblende gebracht, mittels der man die Größe der zur Wirkung kommenden leuch-tenden Fläche verändern kann. Zunächst stellt man die Irisblende ganz klein, so daß man annähernd einen leuchtenden Punkt erhält, der von dem schattenwerfenden Körper (es wurde dazu bei der Demonstration eine weiße Kugel von 5 cm Durchmesser benutzt) auf dem weißen Schirm einen scharfen Schlagschatten gibt (Fig. 1). Bei größerer Öffnung der Irisblende macht sich um den kleiner gewordenen Kern-

schatten der Halbschatten bemerkbar. Ordnet man nun den Schirm in der Richtung der op-tischen Achse an, so daß der Weg des Schat-tens darauf sichtbar wird, und gibt der Iris-blende den Durchmesser des schattenwerfenden Körpers, so kann man zeigen, wie der Kern-schatten eine zylindrische Form annimmt. Ein zweiter quergestellter Schirm fängt den Schatten auf und gestattet, ihn nochmals zu beobachten (Fig. 2). Mit völlig geöffneter Irisblende de-monstriert man die Wirkung eines leuchtenden Körpers, der größer ist als der schattenwer-fende. Durch einen vorgesetzten, horizontalen Spalt läßt sich eine einseitige Bildung des Halbschattens veranschaulichen.

2. **Mischung prismatischer Farben.** Zur Mischung der Farben zweier Spektren dient eine Platte mit zwei übereinander angeordneten Hochspalten, die sich einzeln in horizontaler Richtung verschieben lassen. Von den zwei Spektren, die Objektiv und Prisma entwerfen, wird das untere durch ein Ablenkungsprisma mit dem oberen ganz oder teilweise zur Deckung gebracht und nun wird durch Verschieben des unteren Spaltes das eine Spektrum über das andere gezogen (Fig. 3). (Um den durch die Ablenkung bewirkten Lichtverlust auszugleichen, nimmt man den oberen Spalt etwas breiter.)

Unter Benutzung eines einfachen Spaltes kann man diesen Versuch mit Hilfe eines achromatisierten Kalkspatprismas ausführen, welches das Spaltbild bezw. Spektrum verdop-pelt: eine Anordnung, die für subjektive Beob-achtung bekannt ist. Das Kalkspatprisma wird zwischen Spalt und Objektiv verschoben; da-bei gibt es eine um so stärkere Ablenkung, je näher es an das Objektiv kommt (Fig. 4). Durch Drehen des Prismas kann man die bei-den Spektren beliebig stark gegeneinander ver-setzen.

3. **Additive und subtraktive Farben-mischung.** Zur Mischung der Farben zweier oder dreier Farbfilter dient ein kleiner Apparat, der eine Platte mit sechs runden Öffnungen besitzt, die in zwei Reihen übereinander ange-ordnet sind. Eine Linse entwirft davon sechs Bildfelder. Vor der Linse befindet sich ein nach beiden Seiten prismatisch geschliffener Glaskörper, der die beiden äußeren Felder der unteren Reihe nach innen ablenkt und mit dem mittleren zur Deckung bringt. Die oberen werden nicht abgelenkt. Spindelschrauben rechts und links an der Platte gestatten, die äußeren Lochpaare seitlich etwas zu verschieben und so die Einstellung wenn nötig zu korrigieren. Für die Filterscheiben sind vor und hinter der Platte Nuten vorgesehen.

Setzt man nun in die vertikalen Nuten drei Farbfilter ein, so bekommt man auf dem Schirme unten die additive Mischung, während oben

Published in *Physikalische Zeitschrift* 10 (1909): 779–780. Discussion following *Siedentopf 1909*, held on 20 September 1909, published 10 November 1909.

[1] In his talk, *Siedentopf 1909*, which he accompanied by slides and demonstrations, Siedentopf discussed various methods for dark-field illumination in ultramicroscopic observations. He emphasized the advantages offered by the new powerful dark-field condensers, produced by Zeiss, for various applications, including the analysis of Brownian motion.

[2] Heinrich Rubens was Professor of Experimental Physics at the University of Berlin. His research interests included questions of optics and black-body radiation.

[3] For an account of Perrin's experiments, see the editorial note, "Einstein on Brownian Motion," § VII, pp. 219–222.

[4] Siedentopf was director of the Zeiss company's microscopy division. It appears from his earlier publications (see, e.g., *Siedentopf and Zsigmondy 1903*) as well as from a remark in *Einstein 1906b* (Doc. 32), p. 371, that he had earlier been interested in the observation as well as in the interpretation of Brownian motion.

[5] In his lecture, Siedentopf had earlier pointed out that "finally, dark-field condensers of high intensity are also appropriate for high-speed photography of rapid microscopic processes. . . . Photographs of Brownian molecular motion are particularly interesting for the physicist" ("[d]ie lichtstarken Dunkelfeldkondensatoren eignen sich schließlich auch gut zur Momentaufnahme schnell ablaufender mikroskopischer Vorgänge. . . . Für den Physiker sind besonders interessant Aufnahmen der Brownschen Molekularbewegung" (see *Siedentopf 1909*, p. 779). Photographs of Brownian motion had been taken previously by Henri (*Henri 1908*) and by Seddig (*Seddig 1907*).

[6] Einstein was probably referring to the measurements of Henri and possibly also to some of the experiments performed by Perrin and his group. The cinematographic studies of Henri (*Henri 1908*) had earlier been criticized because the large quantity of light used for obtaining photographs prevented satisfactory control of the temperature (*Cotton 1908*). A more careful control of the temperature was achieved in the experiments on the displacement formula that Perrin and his coworker Dabrowski published in 1909 (*Perrin and Dabrowski 1909*; see also *Perrin 1909b*, p. 79).

[7] Max Seddig was at that time at the University of Marburg; he later became Professor of Physics at the University of Frankfurt. His research interests included scientific photography.

[8] In his work on the temperature dependence of Brownian motion, Seddig had—in spite of using short exposures for his photographs—encountered serious difficulties in measuring the correct temperature of the liquids he studied; this was one of the reasons for the lack of agreement between his observations and Einstein's predictions (see *Seddig 1908*).

[9] See *Svedberg 1906a*.

[10] Svedberg described the trajectories of the suspended particles as "sinusoidlike" ("sinusoidähnlich") (*Svedberg 1906a*, p. 854). Svedberg's work on Brownian motion had been discussed by Einstein (see *Einstein 1907c* [Doc. 40]).

59. "Discussion" following lecture version of Arthur Szarvassi, "The Theory of Electromagnetic Phenomena in Moving Bodies and the Energy Principle"

[*Einstein et al. 1909b*]

Excerpt, printed version of the discussion held 21 September 1909 following presentation of Szarvassi's paper at the 81st meeting of the Gesellschaft Deutscher Naturforscher und Ärzte in Salzburg.

PUBLISHED 10 November 1909

IN: *Physikalische Zeitschrift* 10 (1909): 813.

Physikalische Zeitschrift. 10. Jahrgang. No. 22.   813

im allgemeinen nicht Null war, sondern gleich dem Ausdruck

$$\frac{v_0 w'}{c^2} \int_{z_0}^{z_1} (G^2 - F^2)\, dz.$$

Hier bedeuten $w' = \dfrac{dw}{dt}$ die Beschleunigung der Platte, c die Lichtgeschwindigkeit im Vakuum, v_0 jene in der dielektrischen Platte, falls dieselbe ruhte, F und G die Wellenfunktionen des im Dielektrikum gegen den Spiegel wandernden, bezw. reflektierten Lichtes, endlich z_0 und z_1 die Koordinaten der inneren und äußeren Grenzfläche der Platte, bezogen auf ein ruhendes Koordinatensystem, dessen positive z-Achse vom Boden des Zylinders gegen den Spiegel gerichtet ist. Dieser Ausdruck ist sicher nicht Null; denn das Licht kehrt im allgemeinen nach der Reflexion am bewegten Spiegel mit geänderter Amplitude zurück. Ist z. B. w positiv, d. h. bewegt sich der Spiegel im Sinne der Lichtfortpflanzung, so kehrt das Licht geschwächt zurück, und es ist $G^2 < F^2$; ist also auch w' positiv, so ist der genannte Ausdruck wesentlich negativ; es findet demnach ein unkompensierter Energiegewinn statt.

Man könnte, anknüpfend an neuere Untersuchungen der Herren Planck, Hasenöhrl usw., meinen, daß der hier gewählte spezielle Mechanismus die Verletzung des Energieprinzips durch die Lorentzsche Theorie deshalb nicht erweise, weil für einen bewegten Körper, in dessen Innerem eine elektromagnetische Strahlung herrscht, nicht die gewöhnliche Energiebeziehung gelte. Es ist aber klar, daß dieser Einwand nicht stichhaltig wäre: denn die betreffenden Untersuchungen gründen sich auf die Theorie der Hohlraumstrahlung, diese aber ist fundiert auf einer bestimmten Theorie des Elektromagnetismus, nämlich gerade auf jener, deren Unzulänglichkeit in energetischer Hinsicht hier nachgewiesen worden ist.

(Eingegangen 1. Oktober 1909.)

Diskussion.

Mie: Es hat schon früher einmal Abraham einen ähnlichen Einwand gegen die Lorentzsche Relativitätstheorie erhoben, indem er sagte, nach der Lorentzschen Relativitätstheorie wären die Naturerscheinungen nicht rein elektromagnetisch zu erklären, weil ein Teil der Energie, die ein Elektron bei beschleunigter Bewegung aufnimmt, nicht dem elektromagnetischen Feld zugute kommt, sondern, wenn man so sagen will, als Deformationsarbeit zur Deformation des Elektrons verbraucht wird. Ich möchte den Vortragenden fragen, ob der Widerspruch, den er zwischen der Relativitäts-theorie und dem Energieprinzip gefunden hat, nicht vielleicht in ähnlicher Weise gelöst werden kann, daß man nämlich nicht so rechnen darf, als ob die ganze Energiemenge, die ein Elektron aufnimmt, rein elektromagnetisch sei, sondern noch eine neue Energiemenge in dem bewegten Elektron selber hinzufügen muß.

Einstein: Ich meine, daß ein Körper, [1] welcher Kräften unterworfen ist, daß dieser, wenn er von einem relativ bewegten Koordinatensystem betrachtet wird, deshalb, weil er Kräften unterworfen ist, eine Energie repräsentiert. Macht man diese Annahme nicht, so [2] tritt eine Verletzung des Energieprinzips ein. Glauben Sie nicht, daß das von Ihnen behandelte Beispiel vielleicht darauf beruhen könnte? Ist Ihnen klar, was ich meine?

Vortragender: Nicht vollständig.

Einstein: Man kann zeigen, daß ein bewegter Körper, welcher Kräften unterworfen ist, deren Resultierende nicht verschwindet, dadurch in gewissen Fällen nicht beschleunigt wird. Man muß in der Relativitätstheorie deshalb annehmen, daß der bewegte, Kräften unterworfene (starre) Körper einen gewissen Energieinhalt besitzt; sonst kommt man zu einer Verletzung des Energieprinzips.

Vortragender: Das würde bedeuten, daß außer der sogenannten Bewegungsenergie des Systems im landläufigen Sinn und außer der sogenannten gewöhnlichen potentiellen elektrischen Energie des Systems noch ein Energieteil hinzutritt. Dieser Teil müßte, weil ich über die Größe ψ nichts vorausgesetzt habe, [3] in dieser Funktion drin stecken. Ich habe ja gar nichts über die Form der Funktion ψ gesagt. Die Energiegleichung spricht das Energieprinzip ganz allgemein aus. Diese weitere Energiegröße kann sehr wohl in ψ enthalten sein.

Einstein: Darauf kann ich nichts sagen, weil ich nicht genug in den Geist dieser Überlegung eingedrungen bin.

Vortragender: Was die andere Bemerkung betrifft, so wäre es ja sehr erwünscht, wenn schon Abraham auf diese Inkongruenzen der gegenwärtig herrschenden elektromagnetischen Theorien hingewiesen hätte. Aber ich glaube, Herr Mie bezieht sich auf die Vorgänge beim Elektron, wo Abraham meint, daß ein nicht starres Elektron fremde Energie benötige. Das hat aber hier keine Anwendung; denn ich habe nichts vorausgesetzt, als die Feldgleichungen, und es ist gleichgültig, aus welcher theoretischen Vorstellung sie fließen.

Published in *Physikalische Zeitschrift* 10 (1909): 813. Discussion following *Szarvassi 1909*, held on 21 September 1909, published 10 November 1909.

[1] Arthur Szarvassi's paper, *Szarvassi 1909*, claims to show that Lorentz's equations for the electromagnetic field lead to a violation of the principle of conservation of energy in the for-mulation of that principle that Szarvassi gives.

[2] Einstein here briefly summarized his argument in *Einstein 1907h* (Doc. 45), § 1, pp. 373–377.

[3] Szarvassi defined ψ as the density of potential energy of the system to which the principle of conservation of energy is applied (see *Szarvassi 1909*, p. 811).

60. "On the Development of Our Views Concerning the Nature and Constitution of Radiation"

[*Einstein 1909c*]

Printed version of the paper presented 21 September 1909 at the 81st meeting of the Gesellschaft Deutscher Naturforscher und Ärzte in Salzburg, probably under the title: "On the More Recent Changes Which Our Views Concerning the Nature of Light Have Experienced" ("Über die neueren Umwandlungen, welche unsere Anschauungen über die Natur des Lichtes erfahren haben") (*Verhandlungen 1910*, Part 2, p. 41).

PUBLISHED 30 October 1909

IN: *Deutsche Physikalische Gesellschaft, Verhandlungen* 7 (1909): 482–500.

The same version was received 14 October 1909 and published 10 November 1909

IN: *Physikalische Zeitschrift* 10 (1909): 817–825.

The occurrence of page breaks in the version published in the *Physikalische Zeitschrift*—a version which differs only in orthography from this document—is noted on this document in the margins: the page indication is followed by the first word or part of a word, at the beginning of the page cited.

482

Über die Entwickelung unserer Anschauungen über das Wesen und die Konstitution der Strahlung; von A. Einstein.

(Vorgetragen in der Sitzung der physikalischen Abteilung der 81. Versammlung
[1]   Deutscher Naturforscher und Ärzte zu Salzburg am 21. September 1909.)
(Vgl. oben S. 417.)

Als man erkannt hatte, daß das Licht die Erscheinungen der Interferenz und Beugung zeige, da erschien es kaum mehr bezweifelbar, daß das Licht als eine Wellenbewegung aufzufassen sei. Da das Licht sich auch durch das Vakuum fortzupflanzen vermag, so mußte man sich vorstellen, daß auch in diesem eine Art besonderer Materie vorhanden sei, welche die Fortpflanzung der Lichtwellen vermittelt. Für die Auffassung der Gesetze der Ausbreitung des Lichtes in ponderabeln Körpern war es nötig, anzunehmen, daß jene Materie, welche man Lichtäther nannte, auch in diesen vorhanden sei, und daß es auch im Innern der ponderabeln Körper im wesentlichen der Lichtäther sei, welcher die Ausbreitung des Lichtes vermittelt. Die Existenz jenes Lichtäthers schien unbezweifelbar. In dem 1902 erschienenen ersten Bande des vortrefflichen Lehrbuches der Physik von CHWOLSON findet sich in der Einleitung über den Äther der Satz: „Die Wahrscheinlichkeit der Hypothese von der Existenz dieses einen [2] Agens grenzt außerordentlich nahe an Gewißheit."

Heute aber müssen wir wohl die Ätherhypothese als einen überwundenen Standpunkt ansehen. Es ist sogar unleugbar, daß es eine ausgedehnte Gruppe von die Strahlung betreffenden Tatsachen gibt, welche zeigen, daß dem Lichte gewisse fundamentale Eigenschaften zukommen, die sich weit eher vom Standpunkte der NEWTONschen Emissionstheorie des Lichtes als vom Standpunkte der Undulationstheorie begreifen lassen. Deshalb ist es [3] meine Meinung, daß die nächste Phase der Entwickelung der theoretischen Physik uns eine Theorie des Lichtes bringen wird, welche sich als eine Art Verschmelzung von Undulations- und Emissionstheorie des Lichtes auffassen läßt. Diese Meinung zu begründen, und zu zeigen, daß eine tiefgehende Änderung unserer Anschauungen vom Wesen und von der Konstitution des Lichtes unerläßlich ist, das ist der Zweck der folgenden Ausführungen.

Der größte Fortschritt, welchen die theoretische Optik seit der Einführung der Undulationstheorie gemacht hat, besteht wohl in MAXWELLs genialer Entdeckung von der Möglichkeit, das Licht als einen elektromagnetischen Vorgang aufzufassen. Diese Theorie führt statt der mechanischen Größen, nämlich Deformation und Geschwindigkeit der Teile des Äthers, die elektromagnetischen Zustände des Äthers und der Materie in die Betrachtung ein und reduziert dadurch die optischen Probleme auf elektromagnetische. Je mehr sich die elektromagnetische Theorie entwickelte, desto mehr trat die Frage, ob sich die elektromagnetischen Vorgänge auf mechanische zurückführen lassen, in den Hintergrund; man gewöhnte sich daran, die Begriffe elektrische und magnetische Feldstärke, elektrische Raumdichte usw. als elementare Begriffe zu behandeln, die einer mechanischen Interpretation nicht bedürfen. [4]

Durch die Einführung der elektromagnetischen Theorie wurden die Grundlagen der theoretischen Optik vereinfacht, die Anzahl der willkürlichen Hypothesen vermindert. Die alte Frage nach der Schwingungsrichtung des polarisierten Lichtes wurde gegenstandslos. [5] Die Schwierigkeiten, betreffend die Grenzbedingungen an der Grenze zweier Media ergaben sich aus dem Fundament der Theorie. Es bedurfte keiner willkürlichen Hypothese mehr, um longitudinale Lichtwellen anzuschließen. Der erst in neuerer Zeit experimentell konstatierte Lichtdruck, welcher in der Theorie [6] der Strahlung eine so wichtige Rolle spielt, ergab sich als Konsequenz der Theorie. Ich will gar keine erschöpfende Aufzählung der wohlbekannten Errungenschaften hier versuchen, sondern einen Hauptpunkt ins Auge fassen, in bezug auf welchen die elektromagnetische Theorie mit der kinetischen Theorie übereinstimmt oder, besser gesagt, übereinzustimmen scheint.

Nach beiden Theorien erscheinen nämlich die Lichtwellen im wesentlichen als ein Inbegriff von Zuständen eines auch bei Ab-

P.818[/Hypothese]

wesenheit von Strahlung allenthalben vorhandenen hypothetischen Mediums, des Äthers. Es war daher anzunehmen, daß Bewegungen dieses Mediums auf die optischen und elektromagnetischen Erscheinungen von Einfluß sein müssen. Das Suchen nach den Gesetzen, welche dieser Einfluß unterliege, veranlaßte eine Wandlung in den die Natur der Strahlung betreffenden Grundanschauungen, deren Verlauf wir kurz betrachten wollen.

Die Grundfrage, die sich da aufdrängte, war folgende: Macht der Lichtäther die Bewegungen der Materie mit, oder ist er im Innern bewegter Materie anders bewegt als diese, oder endlich nimmt er vielleicht an den Bewegungen der Materie überhaupt gar nicht Anteil, sondern bleibt stets in Ruhe. Um diese Frage zu entscheiden, stellte FIZEAU einen wichtigen Interferenzversuch [7] an, der auf folgender Überlegung beruht. Es breite sich das Licht in einem Körper mit der Geschwindigkeit V aus, falls dieser ruht. Falls dieser Körper, wenn er bewegt ist, seinen Äther vollkommen mitnimmt, so wird sich in diesem Falle das Licht relativ zum Körper ebenso ausbreiten, wie wenn der Körper ruhte. Die Ausbreitungsgeschwindigkeit relativ zum Körper wird also auch in diesem Falle V sein. Absolut genommen, d. h. relativ zu einem nicht mit dem Körper bewegten Beobachter, wird aber die Fortpflanzungsgeschwindigkeit eines Lichtstrahles gleich sein der geometrischen Summe aus V und der Bewegungsgeschwindigkeit v des Körpers. Falls Fortpflanzungs- und Bewegungsgeschwindigkeit gleichgerichtet und gleichsinnig sind, ist V_{abs} einfach gleich der Summe der beiden Geschwindigkeiten, d. h.

$$V_{abs} = V + v.$$

Um zu prüfen, ob diese Konsequenz aus der Hypothese des [8] vollkommen mitbewegten Lichtäthers zutreffe, ließ FIZEAU zwei kohärente monochromatische Lichtbündel je eine mit Wasser gefüllte Röhre axial passieren und nachher zur Interferenz gelangen. Ließ er nun gleichzeitig das Wasser in den Röhren sich axial durch diese hindurchbewegen, und zwar durch die eine im Sinne des Lichtes, durch die andere im entgegengesetzten Sinne, so ergab sich eine Verschiebung der Interferenzfransen, aus denen er einen Rückschluß ziehen konnte auf den Einfluß der Körpergeschwindigkeit auf die Absolutgeschwindigkeit.

Es ergab sich bekanntlich, daß ein Einfluß der Körpergeschwindigkeit in dem zu erwartenden Sinne vorhanden ist, daß
er aber stets kleiner ist, als der Hypothese von der vollständigen
Mitführung entspricht. Es ist

$$V_{abs} = V + \alpha v,$$

wobei α stets kleiner als 1 ist. Unter Vernachlässigung der Dispersion ist

$$\alpha = 1 - \frac{1}{n^2}.$$ [9]

Aus diesem Experiment folgte, daß eine vollständige Mitführung des Äthers durch die Materie nicht stattfinde, daß also
eine Relativbewegung des Äthers gegen die Materie im allgemeinen
vorhanden sei. Nun ist aber die Erde ein Körper, der in bezug
auf das Sonnensystem im Laufe des Jahres Geschwindigkeiten
verschiedener Richtung hat, und es war anzunehmen, daß der
Äther in unseren Laboratorien ebensowenig diese Bewegung der
Erde vollkommen mitmache, wie er beim FIZEAUschen Versuch
die Bewegung des Wassers vollkommen mitzumachen schien. Es war
also zu folgern, daß eine mit der Tages- und Jahreszeit wechselnde
Relativbewegung des Äthers gegen unsere Apparate existiere, und
man mußte erwarten, daß diese Relativgeschwindigkeit bei optischen Versuchen eine scheinbare Anisotropie des Raumes herbei-
führe, d. h. daß die optischen Erscheinungen von der Orientierung
der Apparate abhängig seien. Die verschiedensten Experimente   [10]
zur Konstatierung einer solchen Anisotropie wurden ausgeführt,
ohne daß man die erwartete Abhängigkeit der Erscheinungen von
der Orientierung der Apparate hätte konstatieren können.

Dieser Widerspruch wurde zum größten Teil beseitigt durch
die bahnbrechende Arbeit von H. A. LORENTZ vom Jahre 1895.   [11]
LORENTZ zeigte, daß man unter Zugrundelegung eines ruhenden,
an den Bewegungen der Materie nicht teilnehmenden Äthers ohne
Aufstellung sonstiger Hypothesen zu einer Theorie gelangt, welche
fast allen Erscheinungen gerecht wird. Insbesondere erklärten
sich die Ergebnisse des oben angedeuteten Versuches von FIZEAU
sowie das negative Ergebnis der erwähnten Versuche, die Bewegung der Erde gegen den Äther zu konstatieren. Nur mit einem
einzigen Experiment schien die LORENTZsche Theorie nicht ver-

einbar zu sein, nämlich mit dem Interferenzversuch von MICHELSON

[12] und MORLEY.

P.819[/Theorie,]   LORENTZ hatte gezeigt, daß nach seiner Theorie, abgesehen von Gliedern, welche den Quotienten $\dfrac{\text{Körpergeschwindigkeit}}{\text{Lichtgeschwindigkeit}}$ in der zweiten oder einer höheren Potenz als Faktor enthielten, ein Einfluß einer gemeinsamen Translationsbewegung der Apparate auf den Strahlengang bei optischen Versuchen nicht vorhanden sei. Es war aber damals schon der Interferenzversuch von MICHELSON und MORLEY bekannt, welcher dartat, daß in einem speziellen Falle auch Glieder zweiter Ordnung in bezug auf den Quotienten $\dfrac{\text{Körpergeschwindigkeit}}{\text{Lichtgeschwindigkeit}}$ sich nicht bemerkbar machten, trotzdem dies vom Standpunkte der Theorie des ruhenden Lichtäthers aus zu erwarten war. Damit dieser Versuch von der Theorie mit

[13] umfaßt werde, wurde von LORENTZ und FR. GERALD bekanntlich die Annahme eingeführt, daß alle Körper, also auch diejenigen, welche die Bestandteile der Versuchsanordnung von MICHELSON und MORLEY miteinander verbanden, in bestimmter Weise ihre Gestalt ändern, falls sie relativ zum Äther bewegt werden.

Diese Sachlage war nun eine höchst unbefriedigende. Die einzige Theorie, welche brauchbar und in ihren Grundlagen durchsichtig war, war die LORENTZsche Theorie. Diese ruhte auf der Voraussetzung eines absolut unbeweglichen Äthers. Die Erde mußte relativ zu diesem Äther als bewegt angesehen werden. Alle Versuche aber, jene Relativbewegung nachzuweisen, verliefen resultatlos, so daß man zur Aufstellung einer ganz eigentümlichen Hypothese gezwungen wurde, um begreifen zu können, daß jene Relativbewegung sich nicht bemerkbar mache.

Der MICHELSONsche Versuch legte die Voraussetzung nahe, daß alle Erscheinungen relativ zu einem mit der Erde bewegten Koordinatensystem, allgemeiner überhaupt relativ zu jedem beschleunigungsfrei bewegten System, nach genau den gleichen Gesetzen verlaufen. Diese Voraussetzung wollen wir im folgenden

[14] kurz „Relativitätsprinzip" nennen. Bevor wir die Frage berühren, ob es möglich sei, an dem Relativitätsprinzip festzuhalten, wollen wir kurz überlegen, was bei Festhaltung dieses Prinzips aus der Ätherhypothese wird.

Unter Zugrundelegung der Ätherhypothese führte das Experiment dazu, den Äther als unbeweglich anzunehmen. Das Relativitätsprinzip besagt dann, daß alle Naturgesetze in bezug auf ein relativ zum Äther gleichförmig bewegtes Koordinatensystem K' gleich seien den entsprechenden Gesetzen in bezug auf ein relativ zum Äther ruhendes Koordinatensystem K. Ist dem aber so, dann haben wir ebensoviel Grund, uns den Äther als relativ zu K' ruhend vorzustellen wie als relativ zu K ruhend. Es ist dann überhaupt ganz unnatürlich, eines der beiden Koordinatensysteme K, K' dadurch auszuzeichnen, daß man einen relativ zu ihm ruhenden Äther einführt. Daraus folgt, daß man zu einer befriedigenden Theorie nur dann gelangen kann, wenn man auf die Ätherhypothese verzichtet. Die das Licht konstituierenden elektromagnetischen Felder erscheinen dann nicht mehr als Zustände eines hypothetischen Mediums, sondern als selbständige Gebilde, welche von den Lichtquellen ausgesandt werden, gerade wie nach der NEWTONschen Emissionstheorie des Lichtes. Ebenso [15] wie gemäß letzterer Theorie erscheint ein nicht von Strahlung durchsetzter, von ponderabler Materie freier Raum wirklich als leer.

Bei oberflächlicher Betrachtung erscheint es unmöglich, das Wesentliche der LORENTZschen Theorie mit dem Relativitätsprinzip in Einklang zu bringen. Pflanzt sich nämlich ein Lichtstrahl im Vakuum fort, so geschieht dies nach der LORENTZschen Theorie in bezug auf ein im Äther ruhendes Koordinatensystem K stets mit der bestimmten Geschwindigkeit c, unabhängig vom Bewegungszustande des emittierenden Körpers. Wir wollen diesen Satz das Prinzip von der Konstanz der Lichtgeschwindigkeit nennen. Nach dem Additionstheorem der Geschwindigkeiten wird sich derselbe Lichtstrahl in bezug auf ein relativ zum Äther in gleichförmiger Translationsbewegung befindliches Koordinatensystem K' nicht ebenfalls mit der Geschwindigkeit c fortpflanzen. Die Gesetze der Lichtfortpflanzung scheinen also in bezug auf beide Koordinatensysteme verschieden zu sein, und es scheint daraus zu folgen, daß das Relativitätsprinzip mit den Gesetzen der Lichtausbreitung unvereinbar ist.

Das Additionstheorem der Geschwindigkeiten beruht indessen auf den willkürlichen Voraussetzungen, daß Zeitangaben sowie Angaben über die Gestalt von bewegten Körpern eine vom Bewegungszustande des benutzten Koordinatensystems unabhängige Bedeutung haben. Man überzeugt sich aber, daß man zu einer Definition der Zeit und der Gestalt bewegter Körper der Einführung von Uhren bedarf, welche relativ zu dem benutzten Koordinatensystem ruhen. Man muß deshalb jene Begriffe für jedes Koordinatensystem besonders festlegen, und es ist nicht selbstverständlich, daß diese Definitionen für zwei relativ zueinander bewegte Koordinatensysteme K und K' zu gleichen Zeitwerten t und t' für die einzelnen Ereignisse führen; ebensowenig läßt sich a priori sagen, daß jegliche Aussage über die Gestalt von Körpern, welche in bezug auf das Koordinatensystem K gilt, auch in bezug auf das relativ zu K bewegte Koordinatensystem K' gelte.

P.820[/lichen]   Daraus geht hervor, daß die bisher gebräuchlichen Transformationsgleichungen für den Übergang von einem Koordinatensystem zu einem relativ zu ihm gleichförmig bewegten Koordinatensystem auf willkürlichen Annahmen beruhen. Läßt man diese fallen, so zeigt sich, daß man das Fundament der LORENTZschen Theorie bzw. allgemeiner das Prinzip der Konstanz der Lichtgeschwindigkeit mit dem Relativitätsprinzip in Einklang bringen kann. Man gelangt so zu neuen, durch die beiden Prinzipe eindeutig bestimmten Gleichungen der Koordinatentransformation, welche bei passender Wahl der Anfangspunkte von Koordinaten und Zeiten dadurch charakterisiert sind, daß durch sie die Gleichung

$$x^2 + y^2 + z^2 - c^2 t^2 = x'^2 + y'^2 + z'^2 - c^2 t'^2$$

zu einer Identität wird. Hierbei bedeutet c die Lichtgeschwindigkeit im Vakuum. x, y, z, t sind Raum-Zeit-Koordinaten in bezug auf K, x', y', z', t' in bezug auf K'.

Dieser Weg führt zu der sogenannten Relativitätstheorie, von deren Konsequenzen ich hier nur eine einzige anführen möchte, weil sie eine gewisse Modifikation der Grundanschauungen auf dem Gebiete der Physik mit sich bringt. Es zeigt sich nämlich, daß die träge Masse eines Körpers um $L c^2$ abnimmt, wenn derselbe die Strahlungsenergie L emittiert. Man kann dazu auf folgendem Wege gelangen.

[16]

Wir betrachten einen unbewegten, frei schwebenden Körper, welcher nach zwei entgegengesetzten Richtungen die gleiche

Energiemenge in Form von Strahlung aussendet. Dabei bleibt der Körper in Ruhe. Bezeichnen wir mit E_0 die Energie des Körpers vor der Emission, mit E_1 dessen Energie nach der Emission, mit L die Menge der emittierten Strahlung, so hat man nach dem Energieprinzip

$$E_0 = E_1 + L.$$

Wir betrachten nun den Körper sowie die von demselben emittierte Strahlung von einem Koordinatensystem aus, relativ zu welchem sich der Körper mit der Geschwindigkeit v bewegt. Es liefert dann die Relativitätstheorie die Mittel, um die Energie der ausgesandten Strahlung bezüglich des neuen Koordinatensystems zu berechnen. Man erhält hierfür den Wert

$$L' = L \cdot \frac{1}{\sqrt{1 - \dfrac{v^2}{c^2}}}.$$

Da in bezug auf das neue Koordinatensystem ebenfalls das Prinzip von der Erhaltung der Energie gelten muß, erhält man in analoger Bezeichnungsweise

$$E_0' = E_1' + L \frac{1}{\sqrt{1 - \dfrac{v^2}{c^2}}}.$$

Durch Subtraktion erhält man unter Weglassung der Glieder, welche in v/c von vierter und höherer Ordnung sind:

$$(E_0' - E_0) = (E_1' - E_1) + \frac{1}{2} \frac{L}{c^2} v^2.$$

Nun ist aber $E_0' - E_0$ nichts anderes als die kinetische Energie des Körpers vor der Lichtaussendung, $E_1' - E_1$ nichts anderes als dessen kinetische Energie nach der Lichtaussendung. Nennt man M_0 die Masse des Körpers vor der Aussendung, M_1 dessen Masse nach der Lichtaussendung, so kann man unter Vernachlässigung der Glieder höheren als zweiten Grades setzen:

$$\frac{1}{2} M_0 v^2 = \frac{1}{2} M_1 v^2 + \frac{1}{2} \frac{L}{c^2} v^2$$

oder

$$M_0 = M_1 + \frac{L}{c^2}.$$

490   Verhandlungen der Deutschen Physikalischen Gesellschaft.   [Nr. 20.

Es vermindert sich also die träge Masse eines Körpers bei Lichtaussendung. Die abgegebene Energie figuriert als Teil der Masse des Körpers. Man kann hieraus weiter schließen, daß jegliche Energieaufnahme bzw. -abgabe eine Zu- bzw. Abnahme der Masse des betreffenden Körpers mit sich bringt. Es scheinen Energie und Masse ebenso als äquivalente Größen wie Wärme und mechanische Energie.

Die Relativitätstheorie hat also unsere Anschauungen über die Natur des Lichtes insofern geändert, als sie das Licht nicht als Folge von Zuständen eines hypothetischen Mediums auffaßt, sondern als etwas wie die Materie selbständig Bestehendes. Es hat ferner nach dieser Theorie mit einer Korpuskulartheorie des Lichtes das Merkmal gemeinsam, träge Masse vom emittierenden zum absorbierenden Körper zu übertragen. An unserer Auffassung von der Struktur der Strahlung, insbesondere von der Verteilung der Energie in dem durchstrahlten Raume änderte die Relativitätstheorie nichts. Es ist jedoch meine Meinung, daß wir in bezug auf diese Seite der Frage am Anfange einer noch nicht übersehbaren, jedoch zweifellos höchst bedeutsamen Entwickelung stehen. Was ich im folgenden vorbringen werde, ist großenteils bloße persönliche Meinung bzw. Ergebnis von Überlegungen, welche eine genügende Nachprüfung durch andere noch nicht erfahren haben. Wenn ich dieselben trotzdem hier vorbringe, so ist dies nicht auf übermäßiges Vertrauen in die eigenen Ansichten zurückzuführen, sondern auf die Hoffnung, den einen oder anderen von Ihnen dazu veranlassen zu können, sich mit den in Betracht kommenden Fragen abzugeben.

P.821[/klären]
[17]

Auch ohne tiefer in irgend welche theoretische Betrachtung einzugehen, bemerkt man, daß unsere Lichttheorie gewisse fundamentale Eigenschaften der Lichtphänomene nicht zu erklären vermag. Warum hängt es nur von der Farbe, nicht aber von der Intensität des Lichtes ab, ob eine bestimmte photochemische Reaktion eintritt oder nicht? Warum sind die kurzwelligen Strahlen im allgemeinen chemisch wirksamer als die langwelligen? Warum ist die Geschwindigkeit der lichtelektrisch erzeugten Kathodenstrahlen von der Intensität des Lichtes unabhängig? Weshalb bedarf es hoher Temperaturen, also hoher Molekularenergien,

damit die durch die Körper emittierte Strahlung kurzwellige Bestandteile enthalte?

Auf alle diese Fragen gibt die Undulationstheorie in ihrer
heutigen Fassung keine Antwort. Insbesondere ist es durchaus
nicht begreiflich, warum die lichtelektrisch oder durch Röntgenstrahlen erzeugten Kathodenstrahlen eine so bedeutende, von der
Strahlintensität unabhängige Geschwindigkeit erlangen. Das Auf-   [18]
treten so großer Energiemengen an einem molekularen Gebilde
unter dem Einfluß einer Quelle, in welcher die Energie so wenig
dicht verteilt ist, wie wir dies bei der Licht- und Röntgenstrahlung
nach der Undulationstheorie voraussetzen müssen, veranlaßte
tüchtige Physiker dazu, ihre Zuflucht zu einer recht fernliegenden
Hypothese zu nehmen. Sie nahmen an, daß das Licht bei dem
Vorgang lediglich eine auslösende Rolle spiele, die zum Vorschein
kommenden Molekularenergien aber radioaktiver Natur seien.   [19]
Weil diese Hypothese bereits wieder so ziemlich verlassen ist,
will ich gegen sie keine Gründe vorbringen.

Die Grundeigenschaft der Undulationstheorie, welche diese
Schwierigkeiten mit sich bringt, scheint mir im folgenden zu
liegen. Während in der kinetischen Molekulartheorie zu jedem
Vorgang, bei welchem nur wenige Elementarteilchen beteiligt sind,
z. B. zu jedem molekularen Zusammenstoß, der inverse Vorgang
existiert, ist dies nach der Undulationstheorie bei den elementaren Strahlungsvorgängen nicht der Fall. Ein oszillierendes Ion
erzeugt nach der uns geläufigen Theorie eine nach außen sich
fortpflanzende Kugelwelle. Der umgekehrte Prozeß existiert als
Elementarprozeß nicht. Die nach innen fortschreitende Kugelwelle ist nämlich zwar mathematisch möglich; aber es bedarf zu
deren angenäherter Realisierung einer ungeheuren Menge von
emittierenden Elementargebilden. Dem Elementarprozeß der Lichtemission als solchem kommt also der Charakter der Umkehrbarkeit nicht zu. Hierin trifft, glaube ich, unsere Undulationstheorie
nicht das Richtige. Es scheint, daß in bezug auf diesen Punkt
die Emissionstheorie des Lichtes von NEWTON mehr Wahres enthält als die Undulationstheorie, da nach ersterer die Energie,
welche einem Lichtteilchen bei der Aussendung verliehen wird,
nicht über den unendlichen Raum zerstreut wird, sondern für
einen Elementarprozeß der Absorption disponibel bleibt. Man

denke an die Gesetze der Erzeugung der sekundären Kathodenstrahlung durch Röntgenstrahlen.

[20]

Fallen primäre Kathodenstrahlen auf eine Metallplatte P_1,
so erzeugen sie Röntgenstrahlen. Fallen diese auf eine zweite
Metallplatte P_2, so werden wieder Kathodenstrahlen erzeugt, deren
Geschwindigkeit von derselben Größenordnung ist wie die Geschwindigkeit der primären Kathodenstrahlen. Die Geschwindigkeit der sekundären Kathodenstrahlen hängt, soviel wir heute
wissen, weder vom Abstand der Platten P_1 und P_2 noch von der
Intensität der primären Kathodenstrahlen, sondern ausschließlich
von der Geschwindigkeit der primären Kathodenstrahlen ab.
Nehmen wir einmal an, dies sei streng richtig. Was wird geschehen, wenn wir die Intensität der primären Kathodenstrahlen
oder die Größe der Platte P_1, auf die sie fallen, derart abnehmen
lassen, daß man das Auftreffen eines Elektrons der primären
Kathodenstrahlen als einen isolierten Prozeß auffassen kann?
Wenn das Vorhergehende wirklich richtig ist, so werden wir
wegen der Unabhängigkeit der Geschwindigkeit der Sekundärstrahlen von der Intensität der primären Kathodenstrahlen anzunehmen haben, daß an P_2 (infolge des Auftreffens jenes Elektrons
auf P_1) entweder gar nichts erzeugt wird, oder aber an P_2 eine
sekundäre Emission eines Elektrons erfolgt mit einer Geschwindigkeit, die von derselben Größenordnung ist, wie diejenige des
auf P_1 auffallenden Elektrons gewesen ist. Mit anderen Worten,
der elementare Strahlungsprozeß scheint derart zu verlaufen, daß
er nicht, wie die Undulationstheorie verlangt, die Energie des
primären Elektrons durch eine nach allen Seiten sich fortpflanzende Kugelwelle verteilt und zerstreut, sondern es scheint wenigstens ein großer Teil dieser Energie an irgend einer Stelle von
P_2 oder anderswo disponibel zu sein. Der Elementarvorgang

[21]

der Strahlungsemission scheint **gerichtet** zu sein. Es
macht ferner den Eindruck, daß der Prozeß der Erzeugung des
Röntgenstrahles in P_1 und die Erzeugung des sekundären Kathoden-
strahles in P_2 im wesentlichen inverse Prozesse seien.

Die Konstitution der Strahlung scheint also eine andere zu
sein, als unsere Undulationstheorie folgern läßt. Wichtige Anhaltspunkte hierüber hat die Theorie der Temperaturstrahlung geliefert, und zwar zuerst und in erster Linie diejenige Theorie,

durch welche Herr PLANCK seine Strahlungsformel begründet hat.
Da ich diese Theorie wohl nicht als allgemein bekannt voraussetzen darf, will ich das Notwendigste über dieselbe kurz angeben.

[22]

Im Innern eines Hohlraumes von der Temperatur T befindet
sich Strahlung von bestimmter, von der Natur des Körpers unab-
hängiger Zusammensetzung. Pro Volumeneinheit ist in dem Hohlraum die Strahlungsmenge $\varrho\,d\nu$ vorhanden, deren Frequenz zwischen
ν und $\nu + d\nu$ liegt. Das Problem besteht darin, daß ϱ in Funktion von ν und T gesucht wird. Befindet sich in dem Hohlraum
ein elektrischer Resonator von der Eigenfrequenz ν_0 und geringer
Dämpfung, so erlaubt die elektromagnetische Theorie der Strahlung,
das zeitliche Mittel der Energie $(\overline{E})$ des Resonators in Funktion
von $\varrho\,(\nu_0)$ zu berechnen. Das Problem ist dadurch auf dasjenige
reduziert, $\overline{E}$ in Funktion der Temperatur zu ermitteln. Das letztere
Problem läßt sich aber wieder auf das folgende reduzieren. Es
seien in dem Hohlraum sehr viele (N) Resonatoren von der Frequenz ν_0 vorhanden. Wie hängt die Entropie dieses Resonatorensystems von dessen Energie ab?

P.822[/von der]

Um diese Frage zu lösen, wendet Herr PLANCK die allgemeine
Beziehung zwischen Entropie und Zustandswahrscheinlichkeit an,
wie sie von BOLTZMANN aus seinen gastheoretischen Untersuchungen
gefolgert worden ist. Es ist allgemein

$$\text{Entropie} = k.\log W,$$

wobei k eine universelle Konstante und W die Wahrscheinlichkeit des ins Auge gefaßten Zustandes bedeutet. Diese Wahrscheinlichkeit wird gemessen durch die „Anzahl der Komplexionen", eine Zahl, die angibt, auf wieviele verschiedene Weisen
der ins Auge gefaßte Zustand sich realisieren läßt. Im Falle der
obigen Fragestellung ist der Zustand des Resonatorensystems durch
die Gesamtenergie desselben definiert, so daß die zu lösende
Frage lautet: Auf wieviele verschiedene Arten kann die gegebene Gesamtenergie unter die N Resonatoren verteilt werden?
Um dies zu finden, teilt Herr PLANCK die gesamte Energie in
gleiche Teilchen von bestimmter Größe ε. Eine Komplexion wird
dadurch bestimmt, daß angegeben wird, wieviele Teilchen ε auf
jeden Resonator entfallen. Die Anzahl solcher Komplexionen,

welche die gegebene Gesamtenergie ergeben, wird bestimmt und gleich W gesetzt.

Herr PLANCK folgert dann weiter aus dem auf thermodynamischer Grundlage ableitbaren WIENschen Verschiebungsgesetz, daß $\varepsilon = h\nu$ gesetzt werden müsse, wobei h eine von ν unabhängige Zahl bedeutet. Er findet so seine mit aller bisherigen Erfahrung übereinstimmende Strahlungsformel

$$\varrho = \frac{8\pi h\nu^3}{c^3} \cdot \frac{1}{e^{\frac{h\nu}{kT}} - 1}.$$

[23] Es könnte scheinen, daß gemäß dieser Ableitung die PLANCKsche Strahlungsformel als eine Konsequenz der heutigen elektromagnetischen Theorie der Strahlung anzusehen sei. Dies ist jedoch insbesondere aus folgendem Grunde nicht der Fall. Man könnte die Anzahl der Komplexionen, von welcher soeben die Rede war, nur dann als einen Ausdruck für die Mannigfaltigkeit der Verteilungsmöglichkeiten für die Gesamtenergie unter die N Resonatoren ansehen, wenn jede denkbare Verteilung der Energie wenigstens mit gewisser Annäherung unter den zur Berechnung von W benutzten Komplexionen vorkäme. Hierfür ist notwendig, daß für alle ν, denen eine merkbare Energiedichte ϱ entspricht, das Energiequantum ε klein sei gegenüber der mittleren Resonatorenergie $\overline{E}$. Nun findet man aber durch einfache Rechnung, daß $\varepsilon/\overline{E}$ für die Wellenlänge $0{,}5\,\mu$ und eine absolute Temperatur $T = 1700$ nicht nur nicht klein gegen 1, sondern sogar sehr [24] groß gegen 1 ist. Es hat den Wert etwa $6{,}5 \cdot 10^7$. Es wird also bei dem gegebenen Zahlenbeispiel bei der Zählung der Komplexionen so verfahren, wie wenn die Energie des Resonators nur den Wert Null, das $6{,}5 \cdot 10^7$fache seines mittleren Energiewertes oder ein Vielfaches davon anzunehmen vermöchte. Es ist klar, daß bei dieser Art des Vorgehens nur ein ganz verschwindend kleiner Teil derjenigen Verteilungen der Energie, welche wir nach den Grundlagen der Theorie als möglich ansehen müssen, zur Berechnung der Entropie herangezogen wird. Die Anzahl dieser Komplexionen ist also nach den Grundlagen der Theorie kein Ausdruck für die Wahrscheinlichkeit des Zustandes im BOLTZMANNschen Sinne. Die PLANCKsche Theorie annehmen heißt nach meiner

Meinung geradezu die Grundlagen unserer Strahlungstheorie verwerfen.

Daß unsere jetzigen Grundlagen der Strahlungstheorie verlassen werden müssen, habe ich schon vorher zu zeigen versucht.
Jedenfalls kann man nicht daran denken, die PLANCKsche Theorie
zu refüsieren, weil sie zu jenen Grundlagen nicht paßt. Diese
Theorie hat zu einer Bestimmung der Elementarquanta geführt,
welche durch die neuesten Messungen dieser Größen auf Grund
der Zählung der α-Teilchen glänzend bestätigt worden ist. Für
das Elementarquantum der Elektrizität erhielten RUTHERFORD und
GEIGER im Mittel den Wert $4,65 \cdot 10^{-10}$, REGENER $4,79 \cdot 10^{-10}$,   [25]
während Herr PLANCK mit Hilfe seiner Strahlungstheorie aus den
Konstanten der Strahlungsformel den dazwischen liegenden Wert
$4,69 \cdot 10^{-10}$ ermittelte.   [26]

Die PLANCKsche Theorie führt auf folgende Vermutung. Wenn
es wirklich wahr ist, daß ein Strahlungsresonator nur solche
Energiewerte aufzunehmen vermag, welche Vielfache von $h\nu$ sind,
so liegt die Annahme nahe, daß Emission und Absorption von
Strahlung überhaupt nur in Quanten von dieser Energiegröße   P.823[/nur]
stattfinde. Auf Grund dieser Hypothese, der Lichtquantenhypo-   [27]
these, kann man die oben aufgeworfenen, die Absorption und
Emission von Strahlung betreffenden Fragen beantworten. Soweit
unsere Kenntnisse reichen, werden auch die Folgerungen quantitativen Inhaltes dieser Lichtquantenhypothese bestätigt. Es erhebt sich nun folgende Frage. Wäre es nicht denkbar, daß zwar
die von PLANCK gegebene Strahlungsformel richtig wäre, daß aber
eine Ableitung derselben gegeben werden könnte, die nicht auf
einer so ungeheuerlich erscheinenden Annahme beruht wie die
PLANCKsche Theorie? Wäre es nicht möglich, die Lichtquantenhypothese durch eine andere Annahme zu ersetzen, mit welcher
man ebenso den bekannten Erscheinungen gerecht werden könnte?
Wenn es nötig ist, die Elemente der Theorie zu modifizieren,
könnte man nicht wenigstens die Gleichungen der Ausbreitung
der Strahlung beibehalten und nur die Elementarvorgänge der
Emission und Absorption anders auffassen als bisher?

Um uns hierüber klar zu werden, wollen wir versuchen, in
umgekehrter Richtung vorzugehen wie Herr PLANCK in seiner
Strahlungstheorie. Wir sehen die PLANCKsche Strahlungsformel

*

als richtig an und fragen uns, ob aus ihr etwas gefolgert werden kann bezüglich der Konstitution der Strahlung. Von zwei Betrachtungen, die ich in diesem Sinne ausgeführt habe, will ich Ihnen nur eine hier skizzieren, die mir wegen ihrer Anschaulichkeit besonders überzeugend erscheint.

[28]

In einem Hohlraume befinde sich ein ideales Gas sowie eine Platte aus fester Substanz, welche lediglich senkrecht zu ihrer Ebene frei beweglich sei. Infolge der Unregelmäßigkeit der Zusammenstöße zwischen Gasmolekülen und Platte wird letztere in Bewegung geraten, und zwar derart, daß ihre mittlere kinetische Energie gleich ist dem dritten Teil der mittleren kinetischen Energie eines einatomigen Gasmoleküls. Es ist dies eine Folgerung aus der statistischen Mechanik. Wir nehmen nun an, daß außer dem Gas, welches wir uns als aus wenigen Molekülen bestehend denken können, in dem Hohlraume Strahlung vorhanden ist, und zwar sei diese Strahlung sogenannte Temperaturstrahlung von der nämlichen Temperatur wie das Gas. Dies wird der Fall sein, wenn die Wände des Hohlraumes die bestimmte Temperatur T besitzen, für Strahlung nicht durchlässig und gegen den Hohlraum hin nicht überall vollkommen reflektierend sind. Wir nehmen ferner vorläufig an, daß unsere Platte auf beiden Seiten vollkommen reflektierend sei. Bei dieser Sachlage wird nicht nur das Gas, sondern auch die Strahlung auf die Platte einwirken. Die Strahlung wird nämlich auf beide Seiten der Platte einen Druck ausüben. Die auf die beiden Seiten wirkenden Druckkräfte sind einander gleich, wenn die Platte ruht. Ist sie aber bewegt, so wird an der bei der Bewegung vorangehenden Fläche (Vorderfläche) mehr Strahlung reflektiert als an der Rückfläche. Die auf die Vorderfläche nach rückwärts wirkende Druckkraft ist also größer als die auf die Rückfläche wirkende Druckkraft. Es bleibt also als Resultierende der beiden eine Kraft übrig, welche der Bewegung der Platte entgegenwirkt und mit der Geschwindigkeit der Platte wächst. Wir wollen diese Resultierende kurz „Strahlungsreibung" nennen.

Nehmen wir nun für einen Augenblick an, wir hätten damit die ganze mechanische Einwirkung der Strahlung auf die Platte berücksichtigt, so gelangen wir zu folgender Auffassung. Durch Zusammenstöße mit Gasmolekülen werden der Platte in unregelmäßigen Intervallen Impulse unregelmäßiger Richtung erteilt. Die Geschwindigkeit der Platte zwischen zwei solchen Stößen nimmt infolge der Strahlungsreibung stets ab, wobei kinetische Energie der Platte in Strahlungsenergie verwandelt wird. Die Konsequenz wäre die, daß unausgesetzt Energie der Gasmoleküle durch die Platte in Energie der Strahlung verwandelt wird, so lange, bis alle vorhandene Energie in Energie der Strahlung übergegangen ist. Es gäbe also kein Temperaturgleichgewicht zwischen Gas und Strahlung.

Diese Betrachtung ist deshalb fehlerhaft, weil man die von der Strahlung auf die Platte ausgeübten Druckkräfte ebensowenig als zeitlich konstant und als frei von unregelmäßigen Schwankungen ansehen darf wie die vom Gase auf die Platte ausgeübten Druckkräfte. Jene Schwankungen der Druckkräfte der Strahlung müssen nun, damit thermisches Gleichgewicht möglich sei, so beschaffen sein, daß sie im Mittel die Geschwindigkeitsverluste der Platte durch Strahlungsreibung kompensieren, wobei die mittlere kinetische Energie der Platte gleich ist dem dritten Teil der mittleren kinetischen Energie eines einatomigen Gasmoleküls. Wenn das Strahlungsgesetz bekannt ist, so kann man die Strahlungsreibung berechnen, und hieraus die mittlere Größe der Impulse, welche die Platte infolge der Schwankungen des Strahlungsdruckes erhalten muß, damit statistisches Gleichgewicht bestehen könne.

Noch interessanter wird die Betrachtung dadurch, daß man die Platte so wählt, daß sie nur Strahlung von dem Frequenzbereich $d\nu$ vollkommen reflektiert, Strahlung anderer Frequenz aber ohne Absorption durchläßt; man erhält dann die Schwankungen des Strahlungsdruckes der Strahlung vom Frequenzbereich $d\nu$. Für diesen Fall will ich nun das Resultat der Rechnung angeben. Bezeichnet man mit Δ die Bewegungsgröße, welche in der Zeit τ infolge der unregelmäßigen Schwankungen des Strahlungsdruckes auf die Platte übertragen wird, so erhält man für den Mittelwert des Quadrates von Δ den Ausdruck:

P.824[/Für]

$$\overline{\Delta^2} = \frac{1}{c}\left[h\varrho\nu + \frac{c^3}{8\pi}\frac{\varrho^2}{\nu^2}\right]d\nu\,f\tau.$$

Zunächst fällt die Einfachheit dieses Ausdruckes auf; es dürfte keine mit der Erfahrung innerhalb der Grenzen der Beob-

achtungsfehler übereinstimmende Strahlungsformel geben, welche
einen so einfachen Ausdruck für die statistischen Eigenschaften
des Strahlungsdruckes liefert wie die PLANCKsche.

[29]

Zur Interpretation ist zunächst zu bemerken, daß der Ausdruck für das mittlere Schwankungsquadrat eine Summe von zwei
Termen ist. Es ist also so, wie wenn zwei voneinander unabhängige verschiedene Ursachen vorhanden wären, welche ein
Schwanken des Strahlungsdruckes verursachen. Daraus, daß
$\overline{\varDelta^2}$ proportial f ist, schließt man, daß die Druckschwankungen
für nebeneinander liegende Teile der Platte, deren Linearabmes-
sungen groß sind gegen die Wellenlänge der Reflexionsfrequenz,
voneinander unabhängige Ereignisse sind.

[30]

Die Undulationstheorie liefert nun nur für das zweite Glied
des für $\overline{\varDelta^2}$ gefundenen Ausdruckes eine Erklärung. Nach der
Undulationstheorie müssen nämlich Strahlenbündel von wenig verschiedener Richtung, wenig verschiedener Frequenz und wenig
verschiedenem Polarisationszustand miteinander interferieren, und
es muß der Gesamtheit dieser in ungeordnetster Weise eintretenden Interferenzen ein Schwanken des Strahlungsdruckes entsprechen.
Daß diese Schwankung durch einen Ausdruck von der Gestalt
des zweiten Gliedes unserer Formel sein muß, läßt sich durch
eine einfache Dimensionalbetrachtung einsehen. Man sieht, daß
die Undulationsstruktur der Strahlung in der Tat zu den aus ihr
zu erwartenden Schwankungen des Strahlungsdruckes Veranlassung gibt.

Wie ist aber das erste Glied der Formel zu erklären? Dieses
ist keineswegs zu vernachlässigen, sondern ist im Gültigkeitsbereich des sogenannten WIENschen Strahlungsgesetzes sozusagen
allein maßgebend. So ist für $\lambda = 0{,}5\,\mu$ und $T = 1700$ dieses
Glied etwa $6{,}5 \cdot 10^7$ mal größer als das zweite. Bestände die Strahlung aus sehr wenig ausgedehnten Komplexen von der Energie
$h\nu$, welche sich unabhängig voneinander durch den Raum bewegen
und unabhängig voneinander reflektiert werden — eine Vorstellung, welche die roheste Veranschaulichung der Lichtquantenhypothese darstellt —, so würden infolge Schwankungen des
Strahlungsdruckes derartige Impulse auf unsere Platte wirken,
wie sie durch das erste Glied unserer Formel allein dargestellt
werden.

Nach meiner Meinung muß also aus obiger Formel, welche ihrerseits eine Konsequenz aus der PLANCK schen Strahlungsformel ist, folgendes geschlossen werden. Außer den räumlichen Ungleichmäßigkeiten in der Verteilung der Bewegungsgröße der Strahlung, die aus der Undulationstheorie hervorgehen, sind noch andere Ungleichmäßigkeiten in der räumlichen Verteilung der Bewegungsgröße vorhanden, welche bei geringer Energiedichte der Strahlung die erstgenannten Ungleichmäßigkeiten an Einfluß weit überragen. Ich füge hinzu, daß eine andere Betrachtung bezüglich der räumlichen Verteilung der Energie ganz entsprechende Resultate liefert, wie die im vorstehenden angedeutete bezüglich der räumlichen Verteilung der Bewegungsgröße.

Soviel mir bekannt ist, ist die Aufstellung einer mathematischen Theorie der Strahlung, welche der Undulationsstruktur und der aus dem ersten Glied der obigen Formel zu folgernden Struktur (Quantenstruktur) zusammen gerecht wird, noch nicht gelungen. Die Schwierigkeit liegt hauptsächlich darin, daß die Schwankungseigenschaften der Strahlung, wie sie durch obige Formel ausgedrückt werden, wenig formale Anhaltspunkte für die Aufstellung einer Theorie bieten. Man denke sich, es wären die Beugungs- und Inferenzerscheinungen noch unbekannt, aber man wüßte, daß die mittlere Größe der unregelmäßigen Schwankungen des Strahlungsdruckes durch das zweite Glied der obigen Gleichung bestimmt sei, wobei v ein die Farbe bestimmender Parameter unbekannter Bedeutung ist. Wer hätte genug Phantasie, um auf dieser Grundlage die Indulationstheorie des Lichtes aufzubauen?

Immerhin erscheint mir vor der Hand die Auffassung die natürlichste, daß das Auftreten der elektromagnetischen Felder des Lichtes ebenso an singuläre Punkte gebunden sei wie das Auftreten elektrostatischer Felder nach der Elektronentheorie. Es [31] ist nicht ausgeschlossen, daß in einer solchen Theorie die ganze Energie des elektromagnetischen Feldes als in diesen Singularitäten lokalisiert angesehen werden könnte, ganz wie bei der alten Fernwirkungstheorie. Ich denke mir etwa jeden solchen singu- [32] lären Punkt von einem Kraftfeld umgeben, das im wesentlichen den Charakter einer ebenen Welle besitzt, und dessen Amplitude mit der Entfernung vom singulären Punkte abnimmt. Sind solcher

Singularitäten viele in Abständen vorhanden, die klein sind gegenüber den Abmessungen des Kraftfeldes eines singulären Punktes,
so werden die Kraftfelder sich übereinanderlagern und in ihrer
Gesamtheit ein undulatorisches Kraftfeld ergeben, das sich von
einem undulatorischen Felde im Sinne der gegenwärtigen elektromagnetischen Lichttheorie vielleicht nur wenig unterscheidet. Daß
einem derartigen Bilde, solange dasselbe nicht zu einer exakten
Theorie führt, kein Wert beizumessen ist, braucht wohl nicht besonders hervorgehoben zu werden. Ich wollte durch dasselbe nur
kurz veranschaulichen, daß die beiden Struktureigenschaften
(Undulationsstruktur und Quantenstruktur), welche gemäß der
PLANCKschen Formel beide der Strahlung zukommen sollen, nicht
als miteinander unvereinbar anzusehen sind.

P.825[/ein-
ander-
lagern]

Published in *Deutsche Physikalische Gesellschaft, Verhandlungen* 7 (1909): 482–500. Lecture, held on 21 September 1909, published 30
October 1909.

[1] The session was chaired by Woldemar
Voigt (see *Versammlung 1909a*, p. 417, and
1909b, p. 778). A list of some of those in atten-
dance is given in *Mehra and Rechenberg 1982*,
p. 121.

[2] *Khvolson 1902*, p. 9.

[3] See, e.g., *Einstein 1905i* (Doc. 14).

[4] See, e.g., *Lorentz 1895*.

[5] See, e.g., *Wiener 1890* and *Drude 1891*.

[6] The expression for the pressure exerted by
light was first derived in *Maxwell 1873*, § 792.
It was experimentally confirmed in *Lebedev
1901* and *Nichols and Hull 1903*.

[7] See *Fizeau 1851*.

[8] Fizeau actually considered three possibilities: the ether is totally dragged by the moving
body; it is not dragged at all; or, in accord with
Fresnel's hypothesis, it is partially dragged. His
results are in accord with Fresnel's hypothesis.

[9] In this equation, *n* is the index of refraction
of the body through which light is propagating.

[10] For a discussion of this problem, see *Hirosige 1976* and *Swenson 1972*.

[11] *Lorentz 1895*.

[12] *Michelson and Morley 1887*.

[13] See *Lorentz 1892b*; *FitzGerald 1889*.

[14] See *Einstein 1905r* (Doc. 23) and the editorial note, "Einstein on the Theory of Relativity," pp. 253–274. For details of the following summary of the nature of light in the
relativity theory, see *Einstein 1905r* (Doc. 23)
and *Einstein 1905s* (Doc. 24).

[15] For a discussion of the connection between
Einstein's work on relativity and on light
quanta, see the Introduction, pp. xvii–xviii.

[16] See *Einstein 1905s* (Doc. 24).

[17] See *Einstein 1905i* (Doc. 14), *Einstein
1907a* (Doc. 38), and *Einstein 1909b* (Doc. 56),
p. 191.

[18] See *Lenard 1902*, pp. 166–168.

[19] This hypothesis, now known as the "triggering hypothesis," appeared in *Lenard 1902*,
pp. 150 and 170. It is also used in *Thomson,
J. J. 1905*, pp. 588–589; *Wien 1905*; *Ladenburg
1907*; and in a postscript to Hendrik Lorentz to
Einstein, 6 May 1909. In his reply to the last,
Einstein to Hendrik Lorentz, 23 May 1909, Einstein refuted Lorentz's version of the hypothesis, whereby the electrons are ejected with velocities equal to their thermal motions in the

[20] metal. For a discussion of contemporary criticisms of the triggering hypothesis, see *Wheaton 1978a*, pp. 146–150.

[20] See *Stark 1909a*, p. 583; and Stark's comment on this paper, in *Einstein et al. 1909c* (Doc. 61), p. 826.

[21] Einstein's conclusion is similar to Stark's assumption that light quanta are endowed with both energy and momentum (see *Stark 1909c*). Stark's paper was received on 16 November 1909, after Einstein's lecture. Einstein discussed the directed emission of light quanta in Einstein to Arnold Sommerfeld, 19 January 1910. He later elaborated on the idea in *Einstein 1916b*.

[22] For details of Planck's theory of thermal radiation and the derivation of his formula, see *Planck 1900a, 1900e*, and *1906c*.

[23] For Einstein's critique of Planck's theory, see *Einstein 1905i* (Doc. 14), *Einstein 1906d* (Doc. 34), *Einstein 1907a* (Doc. 38), and *Einstein 1909b* (Doc. 56).

[24] This figure can be calculated from the formula for $\overline{E}_\nu/\epsilon$ in *Einstein 1906d* (Doc. 34), p. 203.

[25] See *Rutherford and Geiger 1908* (German translation, *Rutherford and Geiger 1909*), which gives the charge of an α-particle ($2e$) as 9.3×10^{-10} electrostatic units (pp. 168, 173). For Regener's result, see *Regener 1909*, p. 965.

[26] *Planck 1901b*, p. 566; *Planck 1906c*, p. 163.

[27] See *Einstein 1905i* (Doc. 14).

[28] For further details of the following argument, see *Einstein 1909b* (Doc. 56), pp. 189–190.

[29] In *Planck 1910*, Planck criticized Einstein's interpretation of the expression for Δ^2. In particular, he did not agree with Einstein's interpretation of the first term. For Einstein's response, see "Antwort auf Planks Manuskript," ca. January 1910.

[30] For an earlier discussion of the following implications of the momentum fluctuation formula, see *Einstein 1909b* (Doc. 56), pp. 190–191.

[31] See also Einstein to Hendrik Lorentz, 23 May 1909. He may have had this idea when he wrote Arnold Sommerfeld on 19 January 1910 that "the localization which we give the electromagnetic energy in Maxwell's theory is entirely arbitrary; but so far this conception has helped me little in clarifying the question" ("die Lokalisation, welche wir der elektromagnetischen Energie in der Maxwell'schen Theorie geben, [ist] eine ganz willkürliche; bis jetzt hat mir diese Erkenntnis aber zur Aufklärung der Frage wenig genützt").

[32] See also Einstein's comment in *Einstein et al. 1909c* (Doc. 61), p. 826.

61. "Discussion" following lecture version of "On the Development of Our Views Concerning the Nature and Constitution of Radiation"

[*Einstein et al. 1909c*]

Printed version of the discussion held 21 September 1909 following presentation of Einstein's paper (Document 60) at the 81st meeting of the Gesellschaft Deutscher Naturforscher und Ärzte in Salzburg.

PUBLISHED 10 November 1909

IN: *Physikalische Zeitschrift* 10 (1909): 825–826.

Physikalische Zeitschrift. 10. Jahrgang. No. 22. 825

einanderlagern und in ihrer Gesamtheit ein undulatorisches Kraftfeld ergeben, das sich von einem undulatorischen Felde im Sinne der gegenwärtigen elektromagnetischen Lichttheorie vielleicht nur wenig unterscheidet. Daß einem derartigen Bilde, solange dasselbe nicht zu einer exakten Theorie führt, kein Wert beizumessen ist, braucht wohl nicht besonders hervorgehoben zu werden. Ich wollte durch dasselbe nur kurz veranschaulichen, daß die beiden Struktureigenschaften (Undulationsstruktur und Quantenstruktur), welche gemäß der Planckschen Formel beide der Strahlung zukommen sollen, nicht als miteinander unvereinbar anzusehen sind.

(Eingegangen 14. Oktober 1909.)

Diskussion.

Planck: Wenn ich mir erlaube, einige Worte zu dem Vortrage zu bemerken, so kann ich mich zunächst nur dem Danke der ganzen Versammlung anschließen, welche mit größtem Interesse das, was Herr Einstein vorgebracht hat, angehört hat und auch da, wo vielleicht ein Widerspruch auftauchte, zu weiterem Nachdenken angeregt wurde. Ich werde mich naturgemäß auf das beschränken, worin ich anderer Meinung bin als der Vortragende. Das meiste, was der Vortragende ausgeführt hat, wird ja nicht auf Widerspruch stoßen. Auch ich betone die Notwendigkeit der Einführung von gewissen Quanten. Wir kommen mit der ganzen Strahlungstheorie nicht weiter, ohne daß wir die Energie in gewissem Sinne in Quanten teilen, die als Wirkungsatome zu denken sind. Es fragt sich nun, wo man diese Quanten suchen soll. Nach den letzten Ausführungen von Herrn Einstein wäre es notwendig, die freie Strahlung im Vakuum, also die Lichtwellen selber, als atomistisch konstituiert anzunehmen, mithin die Maxwellschen Gleichungen aufzugeben. Das scheint mir ein Schritt, der in meiner Auffassung noch nicht als notwendig geboten ist. Ich will nicht ins einzelne eingehen, sondern nur folgendes bemerken. In der letzten Betrachtung des Herrn Einstein wird von der Bewegung der Materie auf die Schwankungen der freien Strahlung im reinen Vakuum geschlossen. Dieser Schluß scheint mir nur dann ganz einwurfsfrei, wenn man die Wechselwirkungen zwischen der Strahlung im Vakuum und der Bewegung der Materie vollständig kennt; wenn das nicht der Fall ist, fehlt die Brücke, die notwendig ist, um von der Bewegung des Spiegels auf die Intensität der auffallenden Strahlung überzugehen. Nun scheint mir diese Wechselwirkung zwischen freier elektrischer Energie im Vakuum und der Bewegung der Atome der Materie doch sehr wenig bekannt zu sein. Sie beruht im wesentlichen auf Emission und Absorption des Lichtes. Auch der Strahlungsdruck besteht im wesentlichen darin, wenigstens nach der allgemein als gültig angenommenen Dispersionstheorie, welche auch die Reflexion auf Absorption und Emission zurückführt. Nun ist gerade die Emission und die Absorption der dunkle Punkt, über den wir sehr wenig wissen. Über die Absorption wissen wir vielleicht noch eher etwas, aber wie steht es mit der Emission? Man stellt sie sich vor als hervorgebracht durch Beschleunigung von Elektronen. Aber dieser Punkt ist der schwächste der ganzen Elektronentheorie. Man stellt sich vor, das Elektron besitzt ein bestimmtes Volumen und eine bestimmte endliche Ladungsdichte sei es räumliche oder flächenhafte Ladung ohne das kommt man nicht aus; das widerspricht aber wieder im gewissen Sinne der atomistischen Auffassung der Elektrizität. Das sind nicht Unmöglichkeiten, aber Schwierigkeiten, und ich wundere mich fast, daß sich nicht mehr Widerspruch dagegen erhoben hat.

An diesem Punkt kann, glaube ich, mit Nutzen die Quantentheorie einsetzen. Wir können nur für große Zeiten die Gesetze aussprechen. Aber für kleine Zeiten und für große Beschleunigungen steht man einstweilen noch vor einer Lücke, deren Ausfüllung neue Hypothesen erfordert. Vielleicht darf man annehmen, daß ein schwingender Resonator nicht eine stetig veränderliche Energie besitzt sondern daß seine Energie ein einfaches Vielfaches eines Elementarquantums ist. Ich glaube, wenn man diesen Satz benutzt, kann man zu einer befriedigenden Strahlungstheorie kommen. Nun ist die Frage immer: wie stellt man sich so etwas vor? Das heißt, man verlangt ein mechanisches oder elektrodynamisches Modell eines solchen Resonators. Aber in der Mechanik und in der jetzigen Elektrodynamik haben wir keine diskreten Wirkungselemente und daher können wir auch ein mechanisches oder elektrodynamisches Modell nicht herstellen. Mechanisch erscheint das also unmöglich und man wird sich daran gewöhnen müssen. Auch unsere Versuche, den Lichtäther mechanisch darzustellen, sind ja vollständig gescheitert. Auch den elektrischen Strom hat man sich mechanisch vorstellen wollen und hat an den Vergleich mit einer Wasserströmung gedacht, aber auch das hat man aufgeben müssen, und wie man sich hieran gewöhnt hat, wird man sich auch an einen solchen Resonator gewöhnen müssen. Selbstverständlich müßte diese Theorie noch viel weiter im einzelnen ausgearbeitet werden als bisher geschehen; vielleicht ist da ein anderer glücklicher als ich. Jedenfalls meine ich, man müßte zunächst versuchen, die ganze Schwierigkeit der Quantentheorie zu verlegen in das Gebiet der Wechselwirkung

[1]

826   Physikalische Zeitschrift. 10. Jahrgang. No. 22.

zwischen der Materie und der strahlenden Energie; die Vorgänge im reinen Vakuum könnte man dann vorläufig noch mit den Maxwellschen Gleichungen erklären.

H. Ziegler: Wenn man sich die Uratome der Materie als unsichtbare Kügelchen vorstellt, welche unveränderliche Lichtgeschwindigkeit besitzen, so lassen sich alle Wechselwirkungen von körperlichen Zuständen und elektromagnetischen Erscheinungen darstellen und damit wäre auch die von Herrn Planck noch vermißte Brücke zwischen Materiellem und Nichtmateriellem geschlagen.

Stark: Herr Planck hat darauf hingewiesen, daß wir vorläufig keinen Anlaß haben, zu der Einsteinschen Konsequenz überzugehen, die Strahlung im Raume, wo sie losgelöst von Materie auftritt, als konzentriert anzusehen. Ursprünglich war ich auch der Ansicht, daß man sich vorderhand darauf beschränken könnte, das Elementargesetz zurückzuführen auf eine bestimmte Wirkungsweise der Resonatoren. Aber ich glaube doch, daß es eine Erscheinung gibt, die dazu führt, die elektromagnetische Strahlung losgelöst von Materie, im Raum als konzentriert sich vorstellen zu müssen. Das ist [2] nämlich die Erscheinung, daß die elektromagnetische Strahlung, die von einer Röntgenröhre in den umgebenden Raum weggeht, selbst in großen Distanzen, bis zu 10 m, noch konzentriert zur Wirkung kommen kann an einem einzelnen Elektron. Ich glaube, daß diese Erscheinung doch ein Anlaß ist, die Frage ins Auge zu fassen, ob die elektromagnetische Strahlungsenergie nicht als konzentriert aufzufassen ist, auch da, wo sie losgelöst von Materie auftritt.

Rubens: Aus der von Herrn Einstein vertretenen Anschauung würde sich eine praktische Folgerung ergeben, die sich experimentell prüfen läßt. Bekanntlich rufen nicht nur die α-Strahlen, sondern auch die β-Strahlen eine szintillierende Leuchtwirkung auf dem Fluoreszenzschirm hervor. Nach der entwickelten Anschauung müßte sich das gleiche auch für die γ-Strahlen und für die Röntgenstrahlung ergeben.

Planck: Mit den Röntgenstrahlen ist es eine eigene Sache; ich möchte da nicht zuviel behaupten. — Stark hat etwas für die Quantentheorie angeführt, ich will etwas dagegen anführen; das sind die Interferenzen bei den kolossalen Gangunterschieden von Hunderttausenden von Wellenlängen. Wenn ein Quantum mit sich interferiert, müßte es eine Ausdehnung von Hunderttausenden von Wellenlängen haben. Das ist auch eine gewisse Schwierigkeit.

Stark: Die Interferenzerscheinungen können der Quantenhypothese leicht entgegengestellt werden. Wenn man sie aber mit größerem Wohlwollen für die Quantenhypothese behandeln wird, so wird man auch eine Erklärung dafür gewinnen, das möchte ich als Hoffnung aussprechen. Was die experimentelle Seite betrifft, so muß doch betont werden, daß die Experimente, auf die Herr Planck angespielt hat, mit sehr dichter Strahlung angestellt sind, so daß sehr viele Quanten der gleichen Frequenz in dem Lichtbündel konzentriert waren; das muß bei der Behandlung jener Interferenzerscheinungen wohl berücksichtigt werden. Mit sehr dünner Strahlung würden die Interferenzerscheinungen wohl anders sein.   [3]

Einstein: Die Interferenzerscheinungen würden wohl nicht so schwierig einzureihen sein als man sich vorstellt, und zwar aus folgendem [4] Grunde: man darf nicht annehmen, daß die Strahlungen bestehen aus Quanten, die nicht in Wechselwirkung stehen; das würde unmöglich sein für die Erklärung der Interferenzerscheinungen. Ich denke mir ein Quantum als eine Singularität umgeben von einem großen Vektorfeld. Durch eine große Zahl von Quanten läßt sich ein Vektorenfeld zusammensetzen, das sich wenig von einem solchen unterscheidet, wie wir es bei Strahlungen annehmen. Ich kann mir denken, daß bei Auftreffen von Strahlen an einer Grenzfläche durch Wirkung an der Grenzfläche Separierung der Quanten stattfindet, etwa je nach der Phase des resultierenden Feldes, bei welcher die Quanten die Trennungsfläche erreichen. Die Gleichungen für das resultierende Feld würden sich wenig von denjenigen der bisherigen Theorie unterscheiden. Es ist nicht gesagt, daß wir bezüglich der Interferenzerscheinungen viel zu ändern haben würden an den Auffassungen, wie sie jetzt vorhanden sind. Ich möchte das vergleichen mit dem Vorgange der Molekularisierung der Träger des elektrostatischen Feldes. Das Feld als hervorgebracht von atomisierten elektrischen Teilchen ist nicht sehr wesentlich von der früheren Auffassung unterschieden, und es ist nicht ausgeschlossen, daß in der Strahlungstheorie etwas Ähnliches statthaben wird. Ich sehe eine prinzipielle Schwierigkeit in den Interferenzerscheinungen nicht.

A. Sommerfeld (München), Über die Zusammensetzung der Geschwindigkeiten in der Relativtheorie.

Minkowski hat uns gelehrt, die Lorentz-Einsteinsche Transformation aufzufassen als „Raumzeitdrehung", d. h. als eine Transformation vom Charakter der gewöhnlichen Drehung, aber nicht im Raum xyz, sondern in der vierdimensionalen Mannigfaltigkeit der Größen

Published in *Physikalische Zeitschrift* 10 (1909): 825–826. Discussion following *Einstein 1909c* (Doc. 60), held on 21 September 1909, published 10 November 1909.

[1] For Planck's criticisms see *Planck 1910*; see also Planck to Einstein, 6 July 1907. Einstein gave an answer in his "Antwort auf Planks Manuskript," ca. January 1910.

[2] See *Einstein 1909c* (Doc. 60), p. 821, and *Stark 1909a*, p. 583.

[3] *Taylor 1909*, however, reports the usual interference effects even with radiation of very low intensity.

[4] Lorentz had earlier raised the problem of interference of low-intensity radiation (Hendrik Lorentz to Einstein, 6 May 1909), to which Einstein responded with comments similar to those here (Einstein to Hendrik Lorentz, 23 May 1909).

62. "Discussion" following lecture version of Fritz Hasenöhrl, "On the Transformation of Kinetic Energy into Radiation"

[*Einstein et al. 1909d*]

Printed version of the discussion held 21 September 1909 following presentation of Hasenöhrl's paper at the 81st meeting of the Gesellschaft Deutscher Naturforscher und Ärzte in Salzburg.

PUBLISHED 10 November 1909

IN: *Physikalische Zeitschrift* 10 (1909): 830.

830 **Physikalische Zeitschrift. 10. Jahrgang. No. 22.**

nicht eine abnorme Intensität hat, und daher die Bewegung als quasi stationär anzusehen ist) zu $\frac{8\pi}{3}\frac{e^4}{m^2c^4}w$ ergibt, wo e und m Ladung und Masse des Elektrons sind. (Diese Kraft ist verhältnismäßig nicht allzu gering; hätte die einfallende Strahlung die Intensität der Sonnenstrahlung, so würde das Elektron eine Beschleunigung von ca. 0,05 cm sec^{-2} erfahren.) Der Widerstand, den das bewegte Elektron erfährt, ergibt sich zu:

$$R = \frac{32\pi}{9}\frac{e^4}{m^2c^4}\cdot u\cdot\beta.$$

Um uns eine Vorstellung von der Größenordnung dieses Effektes zu bilden, wollen wir in Beispiel 1 und 2 $a = \frac{1}{2}10^{-7}$ cm setzen, welche Größe ungefähr dem Molekülradius entspricht. Für die Dichte der Strahlung u setzen wir den Betrag der Hohlraumstrahlung bei 0^0 C ein. Bei Beispiel 2 muß noch eine Voraussetzung über die spektrale Verteilung der Energie gemacht werden; wir nehmen die durch das Plancksche Gesetz geforderte Verteilung, ebenfalls für 0^0 C an.

Wir können dann die Abnahme der lebendigen Kraft L berechnen, die der oben angegebene Widerstand bei der Bewegung eines oder vieler solcher Körper (solange sie sich gegenseitig nicht beeinflussen) zur Folge hat.

Es ergibt sich bei den drei betrachteten Beispielen für $-\frac{1}{L}\frac{dL}{dt}$ bezw. der Wert

$$1,4\cdot10^{-5}\text{ sec}^{-1};\quad 1,2\cdot10^{-18}\text{ sec}^{-1};$$
$$1,9\cdot10^{-12}\text{ sec}^{-1}.$$

Fall 1 ist nur wegen seiner Einfachheit mit in Betracht gezogen; die Annahmen, daß $a = \frac{1}{2}10^{-7}$ cm und groß gegen die Wellenlänge der Strahlung bei 0^0 C sei, widersprechen einander. Wir beschäftigen uns daher nur mit den zwei letzten Zahlen; sie zeigen, daß die gesuchte Abnahme der lebendigen Kraft eine außerordentlich geringe ist. Dieselbe würde erst in $3\cdot10^8$ bezw. in 200 Jahren um 1 Proz. sinken. Auch wenn die anfängliche Dichte der Strahlung eine größere ist, wenn sie etwa der Hohlraumstrahlung bei fünffacher absoluter Temperatur, also bei 1092^0 C entspräche, wären die obigen Zahlen 770 Jahre bezw. $^1/_3$ Jahr.

Denken wir uns nun einen nach außen ganz abgeschlossenen Raum, in dem sich Strahlungsenergie und ein „Gas" befinden, dessen Moleküle aus einer reflektierenden Kugel, wie wir sie etwa in Beispiel 2 behandelt haben, bestehen. (Es kann sich hier natürlich höchstens um ein ganz rohes Bild tatsächlicher Vorgänge handeln; die Ausdrücke „Gas" und „Molekül" sind durchaus nicht wörtlich zu verstehen.) Es wird sich nach dem Obigen die kinetische Energie der Moleküle allmählich in Strahlung verwandeln — diese Umwandlung wird aber so langsam vor sich gehen, daß die gewöhnlichen Gasgesetze (das Verteilungsgesetz der Geschwindigkeiten etwa, oder das Verhältnis der spezifischen Wärmen) nicht in merkbarer Weise alteriert würden.

Die Veränderung der Strahlung wird im allgemeinen prozentuell viel rascher vor sich gehen; mit der Vermehrung der Strahlungsenergie ist in den meisten Fällen eine Verkürzung der Wellenlänge verbunden.

Das genauere Studium der Veränderung der Strahlung bezüglich Gesamtintensität und spektraler Verteilung, sowie des Einflusses der Eigenschwingungen der bewegten Körperchen behalte ich einer späteren Untersuchung vor. Nach Herrn Einsteins Hypothese würde der hier statuierte Effekt durch die Unregelmäßigkeiten der Strahlung gerade aufgehoben. (Die ausführliche Begründung der hier angegebenen Resultate wird demnächst an anderer Stelle publiziert werden.)

Wien, im Oktober 1909.

(Eingegangen 13. Oktober 1909.)

Diskussion. [1]

Einstein: Natürlich würden in diesem Falle die unregelmäßigen Schwankungen gerade so sein, daß das Maxwellsche Verteilungsgesetz aufrecht erhalten wird, d. h. daß die Dämpfung kompensiert wird durch die unregelmäßigen Stöße. [2]

Vortragender: Ich weiß nicht, ob ich Sie richtig verstehe. Sie meinen, wenn Sie ein abgeschlossenes Gefäß sich denken und Körperchen sich darin bewegen, daß die überhaupt nicht durch die Strahlung gedämpft würden.

Einstein: Jawohl.

Vortragender: Ich bekomme merkbare Dämpfung der Bewegung erst in praktisch unendlich langer Zeit.

Planck: Die Voraussetzungen, von denen die beiden Herren ausgehen, sind wohl verschieden. Der Herr Vortragende betrachtet eine vollkommen gleichmäßige Strahlungsintensität, während Herr Einstein Schwankungen der Strahlung betrachtet und dadurch auch Schwankungen in den resultierenden Wirkungen, d. h. keine vollständige Dämpfung bekommt. [3]

W. Seitz (Aachen), Über eine neue Röntgenröhre von konzentrierter Wirkung.

Die Intensität der Röntgenstrahlen ist bekanntlich innerhalb gewisser Grenzen umgekehrt proportional dem Quadrat der Entfernung von

Published in *Physikalische Zeitschrift* 10 (1909): 830. Discussion following *Hasenöhrl 1909a*, held on 21 September 1909, published 10 November 1909.

[1] Friedrich Hasenöhrl was Professor of Physics at the University of Vienna. He was well known for his studies of electromagnetic radiation, especially *Hasenöhrl 1904*. In his talk, *Hasenöhrl 1909a*, Hasenöhrl reported the results of calculations of the resistance experienced by various models of molecules when they move through a field of radiation uniformly distributed in all directions. He concluded that, if a gas of such molecules were enclosed in a container filled with radiation, "the kinetic energy of the molecules will . . . be transformed gradually into radiation—this transformation, however, will take place so slowly that the usual gas laws . . . are not altered in a noticeable way. The change of the radiation, in terms of percentage, will in general take place much more rapidly" ("Es wird sich . . . die kinetische Energie der Moleküle allmählich in Strahlung verwandeln—diese Umwandlung wird aber so langsam vor sich gehen, daß die gewöhnlichen Gasgesetze . . . nicht in merkbarer Weise alteriert werden. Die Veränderung der Strahlung wird im allgemeinen prozentuell viel rascher vor sich gehen") (p. 830). *Hasenöhrl 1910* is a fuller account of this work.

[2] Einstein had considered this phenomenon in *Einstein 1909b* (Doc. 56), pp. 189–190, and had just summarized his views in his Salzburg lecture earlier in the same session in which Hasenöhrl spoke (see *Einstein 1909c* [Doc. 60], pp. 496–497).

[3] The printed version of Hasenöhrl's text concludes with the sentence: "According to Einstein's hypothesis, the effect established here would be just canceled by the irregularities of the radiation" ("Nach Herrn Einsteins Hypothese würde der hier statuierte Effekt durch die Unregelmäßigkeiten der Strahlung gerade aufgehoben") (*Hasenöhrl 1909a*, p. 830).

LITERATURE CITED

Abraham 1902a   Abraham, Max. "Dynamik des Elektrons." *Königliche Gesellschaft der Wissenschaften zu Göttingen. Mathematisch-physikalische Klasse. Nachrichten* (1902): 20–41.

Abraham 1902b   ———. "Prinzipien der Dynamik des Elektrons." *Physikalische Zeitschrift* 4 (1902): 57–62.

Abraham 1903   ———. "Prinzipien der Dynamik des Elektrons." *Annalen der Physik* 10 (1903): 105–179.

Abraham 1904a   ———. "Der Lichtdruck auf einen bewegten Spiegel und das Gesetz der schwarzen Strahlung." In *Meyer, S. 1904*, pp. 85–93.

Abraham 1904b   ———. "Zur Theorie der Strahlung und des Strahlungsdruckes." *Annalen der Physik* 14 (1904): 236–287.

Abraham 1905   ———. *Theorie der Elektrizität*. Vol. 2, *Elektromagnetische Theorie der Strahlung*. Leipzig: B. G. Teubner, 1905.

Abraham/Föppl 1904   Abraham, Max. *Theorie der Elektrizität*. Vol. 1, August Föppl, *Einführung in die Maxwellsche Theorie der Elektrizität*. 2nd rev. ed. Max Abraham, ed. Leipzig: B. G. Teubner, 1904.

Ampère 1834   Ampère, André-Marie. *Essai sur la philosophie des sciences ou exposition analytique d'une classification naturelle de toutes les connaissances humaines*. Paris: Bachelier, 1834.

Andenken/Kleiner 1916   "Zum Andenken an Dr. Alfred Kleiner. 1879/1916 Professor für Physik und Direktor des Physikalischen Instituts an der Universität Zürich." *Physikalische Gesellschaft Zürich. Mitteilungen*, no. 18 (1916).

Annuaire 1902   *Annuaire pour l'an 1902*. Publié par le Bureau des Longitudes. Paris: Gauthier-Villars, 1902.

Asquith and Nickles 1983   Asquith, Peter D., and Nickles, Thomas, eds. *PSA 1982: Proceedings of the 1982 Biennial Meeting of the Philosophy of Science Association*. Vol. 2. East Lansing, Michigan: Philosophy of Science Association, 1983.

Auerbach 1893   Auerbach, Felix. "Berührungselektricität." In *Winkelmann 1893*, pp. 106–136.

Avenarius 1888   Avenarius, Richard. *Kritik der reinen Erfahrung*. Vol. 1. Leipzig: O. R. Reisland, 1888.

Avenarius 1890   ———. *Kritik der reinen Erfahrung*. Vol. 2. Leipzig: O. R. Reisland, 1890.

Bancelin 1911a   Bancelin, Jacques. "La viscosité des émulsions." *Académie des sciences* (Paris). *Comptes rendus* 152 (1911): 1382–1384.

Bancelin 1911b   ———. "Ueber die Viskosität von Suspensionen und die Bestimmung der Avogadro'schen Zahl." *Zeitschrift für Chemie und Industrie der Kolloide* 9 (1911): 154–156.

Battelli 1889   Battelli, Angelo. "Sulle proprietà termiche dei vapori. Parte I. Studio del vapore d'etere rispetto alle leggi di Boyle e di Gay-Lussac." *Reale Accademia delle Scienze di Torino. Memorie* 40 (1889): 21–130.

Belluzzo 1904   Belluzzo, Giuseppe. "Principi di termodinamica grafica." *Il Nuovo Cimento* 8 (1904): 196–222, 241–263.

Berlin Verzeichnis 1915b   *Königliche Friedrich-Wilhelms-Universität zu Berlin. Verzeichnis der Vorlesungen Winter-Semester 1915/16*. Berlin: Gustav Schade, 1915.

Berlin Verzeichnis 1917b   *Königliche Friedrich-Wilhelms-Universität zu Berlin. Verzeichnis der Vorlesungen Winter-Semester 1917/18*. Berlin: Arthur Scholem, 1917.

Bernhardt 1971   Bernhardt, Hannelore.

''Über die Entwicklung und Bedeutung der Ergodenhypothese in den Anfängen der statistischen Mechanik.'' *NTM-Schriftenreihe zur Geschichte der Naturwissenschaft, Technik und Medizin* 8 (1971): 13–25.

Bernstein 1853–1857   Bernstein, Aaron. *Aus dem Reiche der Naturwissenschaft. Für jedermann aus dem Volke.* 12 vols. Berlin: Besser, 1853–1857. Reissued as: *Naturwissenschaftliche Volksbücher.* Wohlfeile Gesammt-Ausgabe. 20 vols. Berlin: Franz Duncker, 1867–1869.

Bertrand 1887   Bertrand, Joseph. *Thermodynamique.* Paris: Gauthier-Villars, 1887.

Bikerman 1975   Bikerman, Jacob J. ''Theories of Capillary Attraction.'' *Centaurus* 19 (1975): 182–206.

Bikerman 1978   ———. ''Capillarity before Laplace: Clairaut, Segner, Monge, Young.'' *Archive for History of Exact Sciences* 18 (1978): 103–122.

Birven 1905   Birven, Heinrich. *Grundzüge der mechanischen Wärmetheorie.* Stuttgart: F. Grub, 1905.

Blackmore 1972   Blackmore, John T. *Ernst Mach: His Work, Life, and Influence.* Berkeley: University of California Press, 1972.

Blumenthal 1913   Blumenthal, Otto, ed. *Das Relativitätsprinzip. Eine Sammlung von Abhandlungen.* Leipzig: B. G. Teubner, 1913.

Böhi 1911   Böhi, Paul. ''Eine neue Methode der Bestimmung der Avogadroschen Zahl N.'' *Naturforschende Gesellschaft in Zürich. Vierteljahrsschrift* 56 (1911): 183–212.

Bohlin 1904   Bohlin, Karl. ''Sur le choc, considéré comme fondement des théories cinétiques de la pression des gaz et de la gravitation universelle.'' *Arkiv för Matematik, Astronomi och Fysik* 1 (1904): 529–540.

Boltzmann 1868   Boltzmann, Ludwig. ''Studien über das Gleichgewicht der lebendigen Kraft zwischen bewegten materiellen Punkten.'' *Kaiserliche Akademie der Wissenschaften* (Vienna). *Mathematisch-naturwissenschaftliche Classe. Zweite Abtheilung. Sitzungsberichte* 58 (1868): 517–560. Reprinted in *Boltzmann 1909*, vol. 1, pp. 49–96.

Boltzmann 1871a   ———. ''Einige allgemeine Sätze über Wärmegleichgewicht.'' *Kaiserliche Akademie der Wissenschaften* (Vienna). *Mathematisch-naturwissenschaftliche Classe. Zweite Abtheilung. Sitzungsberichte* 63 (1871): 679–711. Reprinted in *Boltzmann 1909*, vol. 1, pp. 259–287.

Boltzmann 1871b   ———. ''Analytischer Beweis des 2. Hauptsatzes der mechanischen Wärmetheorie aus den Sätzen über das Gleichgewicht der lebendigen Kraft.'' *Kaiserliche Akademie der Wissenschaften* (Vienna). *Mathematisch-naturwissenschaftliche Classe. Zweite Abtheilung. Sitzungsberichte* 63 (1871): 712–732. Reprinted in *Boltzmann 1909*, vol. 1, pp. 288–308.

Boltzmann 1872   ———. ''Weitere Studien über das Wärmegleichgewicht unter Gasmolekülen.'' *Kaiserliche Akademie der Wissenschaften* (Vienna). *Mathematisch-naturwissenschaftliche Classe. Zweite Abtheilung. Sitzungsberichte* 66 (1872): 275–370. Reprinted in *Boltzmann 1909*, vol. 1, pp. 316–402.

Boltzmann 1876   ———. ''Über die Natur der Gasmolecüle.'' *Kaiserliche Akademie der Wissenschaften* (Vienna). *Mathematisch-naturwissenschaftliche Classe. Zweite Abtheilung. Sitzungsberichte* 74 (1876): 553–560. Reprinted in *Boltzmann 1909*, vol. 2, pp. 103–110.

Boltzmann 1877   ———. ''Über die Beziehung zwischen dem zweiten Hauptsatze der mechanischen Wärmetheorie und der Wahrscheinlichkeitsrechnung, respective den Sätzen über das Wärmegleichgewicht.'' *Kaiserliche Akademie der Wissenschaften* (Vienna). *Mathematisch-naturwissenschaftliche Classe. Zweite Abtheilung. Sitzungsberichte* 76 (1877): 373–435. Reprinted in *Boltzmann 1909*, vol. 2, pp. 164–223.

Boltzmann 1878a   ———. ''Weitere Bemerkungen über einige Probleme der mechanischen Wärmetheorie.'' *Kaiserliche Akademie der Wissenschaften* (Vienna). *Mathematisch-naturwissenschaftliche Classe. Zweite Abtheilung. Sitzungsberichte* 78 (1878): 7–46. Reprinted in *Boltzmann 1909*, vol. 2, pp. 250–288.

Boltzmann 1878b   ———. ''Über die Beziehung der Diffusionsphänomene zum zweiten Hauptsatze der mechanischen Wärmetheorie.'' *Kaiserliche Akademie der Wissenschaften* (Vienna). *Mathematisch-naturwissenschaftliche Classe. Zweite Abtheilung. Sitzungsberichte* 78 (1878): 733–763. Reprinted in *Boltzmann 1909*, vol. 2, pp. 289–317.

Boltzmann 1884 ———. ''Ableitung des Stefan'schen Gesetzes, betreffend die Abhängigkeit der Wärmestrahlung von der Temperatur aus der electromagnetischen Lichttheorie.'' *Annalen der Physik und Chemie* 22 (1884): 291–294. Reprinted in *Boltzmann 1909*, vol. 3, pp. 118–121.

Boltzmann 1885 ———. ''Über die Eigenschaften monocyklischer und anderer damit verwandter Systeme.'' *Journal für die reine und angewandte Mathematik* 98 (1885): 68–94. Reprinted in *Boltzmann 1909*, vol. 3, pp. 122–152.

Boltzmann 1887 ———. ''Ueber die mechanischen Analogien des zweiten Hauptsatzes der Thermodynamik.'' *Journal für die reine und angewandte Mathematik* 100 (1887): 201–212. Reprinted in *Boltzmann 1909*, vol. 3, pp. 258–271.

Boltzmann 1896 ———. *Vorlesungen über Gastheorie*. Part 1, *Theorie der Gase mit einatomigen Molekülen, deren Dimensionen gegen die mittlere Weglänge verschwinden.* Leipzig: Johann Ambrosius Barth, 1896.

Boltzmann 1898a ———. *Vorlesungen über Gastheorie*. Part 2, *Theorie Van der Waals'; Gase mit zusammengesetzten Molekülen; Gasdissociation; Schlussbemerkungen.* Leipzig: Johann Ambrosius Barth, 1898.

Boltzmann 1898b ———. ''Vorschlag zur Festlegung gewisser physikalischer Ausdrücke.'' In *Verhandlungen der Gesellschaft Deutscher Naturforscher und Ärzte. 70. Versammlung zu Düsseldorf. 19.–24. September 1898.* Part 2, 1st half, *Naturwissenschaftliche Abtheilungen.* Leipzig: F.C.W. Vogel, 1899, pp. 67–68. Reprinted in *Boltzmann 1909*, vol. 3, p. 642.

Boltzmann 1909 ———. *Wissenschaftliche Abhandlungen.* Fritz Hasenöhrl, ed. 3 vols. Leipzig: Johann Ambrosius Barth, 1909.

Boltzmann and Nabl 1907 Boltzmann, Ludwig, and Nabl, Josef. ''Kinetische Theorie der Materie.'' In *Encyklopädie der mathematischen Wissenschaften, mit Einschluss ihrer Anwendungen.* Vol. 5, *Physik*, part 1, pp. 493–557. Arnold Sommerfeld, ed. Leipzig: B. G. Teubner, 1903–1921. Issued 25 April 1907.

Born 1909 Born, Max. ''Die Theorie des starren Elektrons in der Kinematik des Relativitätsprinzips.'' *Annalen der Physik* 30 (1909): 1–56.

Bosscha 1900 Bosscha, Johannes, ed. *Recueil de travaux offerts par les auteurs à H. A. Lorentz, professeur de physique à l'université de Leiden, à l'occasion du 25me anniversaire de son doctorat le 11 décembre 1900.* The Hague: Martinus Nijhoff, 1900. *Archives Néerlandaises des sciences exactes et naturelles* 5 (1900).

Bousfield 1905a Bousfield, William Robert. ''Ionic Sizes in Relation to the Conductivity of Electrolytes. (Abstract.)'' *Royal Society of London. Proceedings* 74 (1905): 563–564.

Bousfield 1905b ———. ''Ionengrössen in Beziehung zur Leitfähigkeit von Elektrolyten.'' *Zeitschrift für physikalische Chemie, Stöchiometrie und Verwandtschaftslehre* 53 (1905): 257–313.

Bredig 1894 Bredig, Georg. ''Beiträge zur Stöchiometrie der Ionenbeweglichkeit.'' *Zeitschrift für physikalische Chemie, Stöchiometrie und Verwandtschaftslehre* 13 (1894): 191–288.

Brown 1828 Brown, Robert. ''A Brief Account of Microscopical Observations Made in the Months of June, July, and August 1827, on the Particles Contained in the Pollen of Plants; and on the General Existence of Active Molecules in Organic and Inorganic Bodies.'' *Edinburgh New Philosophical Journal* 5 (1828): 358–371. Reprinted in *Philosophical Magazine* 4 (1828): 161–173.

Brush 1968 Brush, Stephen G. ''A History of Random Processes: I. Brownian Movement from Brown to Perrin.'' *Archive for History of Exact Sciences* 5 (1968): 1–36.

Brush 1976 ———. *The Kind of Motion We*

Call Heat: A History of the Kinetic Theory of Gases in the 19th Century. Book 1, *Physics and the Atomists.* Book 2, *Statistical Physics and Irreversible Processes.* Amsterdam: North-Holland, 1976.

Bryan 1904 Bryan, George Hartley. "The Law of Degradation of Energy as the Fundamental Principle of Thermodynamics." In *Meyer, S. 1904*, pp. 123–136.

Bucherer 1903 Bucherer, Alfred Heinrich. "Über den Einfluß der Erdbewegung auf die Intensität des Lichtes." *Annalen der Physik* 11 (1903): 270–283.

Bucherer 1904 ———. *Mathematische Einführung in die Elektronentheorie.* Leipzig: B. G. Teubner, 1904.

Büchner 1855 Büchner, Ludwig. *Kraft und Stoff. Empirisch-naturphilosophische Studien.* Frankfurt am Main: Meidinger Sohn & Cie, 1855.

Buckingham 1905 Buckingham, Edgar. "On Certain Difficulties Which Are Encountered in the Study of Thermodynamics." *Philosophical Magazine and Journal of Science* 9 (1905): 208–214.

Bundesgesez 1854 *Bundesgesez, betreffend die Errichtung einer eidgenössischen polytechnischen Schule. (Vom 7. Hornung 1854.)* [Bern, 1854].

Burbury 1894 Burbury, Samuel Hawksley. "Boltzmann's Minimum Function." *Nature* 51 (1894): 78.

Byrne 1980 Byrne, Patrick H. "Statistical and Causal Concepts in Einstein's Early Thought." *Annals of Science* 37 (1980): 215–228.

Byrne 1981 ———. "The Origins of Einstein's Use of Formal Asymmetries." *Annals of Science* 38 (1981): 191–206.

Cawkell and Garfield 1980 Cawkell, Tony, and Garfield, Eugene. "Assessing Einstein's Impact on Today's Science by Citation Analysis." In *Einstein: The First Hundred Years*, pp. 31–40. Maurice Goldsmith, Alan Mackay, and James Woudhuysen, eds. Oxford: Pergamon, 1980.

Cermak 1918 Cermak, Paul. "Elektrostatische Meßapparate und Messung elektrostatischer Größen." In *Graetz 1918*, pp. 94–156.

Chaudesaigues 1908 Chaudesaigues, ———. "Le mouvement brownien et la formule d'Einstein." *Académie des sciences* (Paris). *Comptes rendus* 147 (1908): 1044–1046.

Clausius 1858 Clausius, Rudolf. "Ueber die mittlere Länge der Wege, welche bei der Molecularbewegung gasförmiger Körper von den einzelnen Molecülen zurückgelegt werden; nebst einigen anderen Bemerkungen über die mechanische Wärmetheorie." *Annalen der Physik und Chemie* 15 (1858): 239–258.

Clausius 1864 ———. "Ueber die Concentration von Wärme- und Lichtstrahlen und die Gränzen ihrer Wirkung." *Annalen der Physik und Chemie* 1 (1864): 1–44.

Clausius 1879–1891 ———. *Die mechanische Wärmetheorie.* 3 vols. 3rd rev. ed. Braunschweig: Friedrich Vieweg und Sohn, 1879–1891.

Clifford 1903 Clifford, William Kingdon. *Von der Natur der Dinge an sich.* Hans Kleinpeter, trans. and ed. Leipzig: Johann Ambrosius Barth, 1903.

Cockcroft and Walton 1932 Cockcroft, John, and Walton, Ernest T. S. "Experiments with High Velocity Positive Ions. II. The Disintegration of Elements by High Velocity Protons." *Royal Society of London. Proceedings A* 137 (1932): 229–242.

Cohn 1900 Cohn, Emil. "Über die Gleichungen der Electrodynamik für bewegte Körper." In *Bosscha 1900*, pp. 516–523.

Cohn 1902 ———. "Ueber die Gleichungen des elektromagnetischen Feldes für bewegte Körper." *Annalen der Physik* 7 (1902): 29–56.

Cohn 1904a ———. "Zur Elektrodynamik bewegter Systeme." *Königlich Preussische Akademie der Wissenschaften* (Berlin). *Sitzungsberichte* (1904): 1294–1303.

Cohn 1904b ———. "Zur Elektrodynamik bewegter Systeme. II." *Königlich Preussische Akademie der Wissenschaften* (Berlin). *Sitzungsberichte* (1904): 1404–1416.

Cotton 1908 Cotton, Aimé. "Recherches récentes sur les mouvements browniens." *Revue du mois* 5 (1908): 737–741.

Cotton and Mouton 1906 Cotton, Aimé, and Mouton, Henri. *Les ultramicroscopes et les objets ultramicroscopiques*. Paris: Masson, 1906.

Cushing 1981 Cushing, James T. "Electromagnetic Mass, Relativity, and the Kaufmann Experiments." *American Journal of Physics* 49 (1981): 1133–1149.

Debus 1968 Debus, Allen G., ed. *World Who's Who in Science: A Biographical Dictionary of Notable Scientists from Antiquity to the Present*. Chicago: Marquis-Who's Who, 1968.

Dedekind 1893 Dedekind, Richard. *Was sind und was sollen die Zahlen?* 2nd ed. Braunschweig: Friedrich Vieweg und Sohn, 1893.

Defay and Prigogine 1951 Defay, Raymond, and Prigogine, Ilya. *Tension superficielle et adsorption*. Liège: Desoer, 1951.

De Groot and Suttorp 1972 De Groot, Sybren Ruurds, and Suttorp, Leendert G. *Foundations of Electrodynamics*. Amsterdam: North-Holland, 1972.

De Haas-Lorentz 1913 De Haas-Lorentz, Geertruida L. *Die Brownsche Bewegung und einige verwandte Erscheinungen*. Braunschweig: Friedrich Vieweg und Sohn, 1913.

Deltete 1983 Deltete, Robert John. "The Energetics Controversy in Late Nineteenth-Century Germany: Helm, Ostwald and Their Critics." 2 vols. Ph.D. dissertation. Yale University, 1983.

Dhar 1914 Dhar, Nilratan. "Verbindung des gelösten Körpers und des Lösungsmittels in der Lösung." *Zeitschrift für Elektrochemie und angewandte physikalische Chemie* 20 (1914): 57–81.

Discussion/Einstein 1911 "Discussion du rapport de M. Einstein." In *Solvay 1911*, pp. 436–450.

Discussion/Planck 1906 "Diskussion" following *Planck 1906b*. *Physikalische Zeitschrift* 7 (1906): 759–761.

Discussion/Planck 1911 "Discussion du rapport de M. Planck." In *Solvay 1911*, pp. 115–132.

Dolezalek 1901 Dolezalek, Fritz. "Ueber ein einfaches und empfindliches Quadrantenelektrometer." *Zeitschrift für Instrumentenkunde* 21 (1901): 345–350.

Drude 1891 Drude, Paul. "Zur Schwingungsrichtung des polarisirten Lichtes." *Annalen der Physik und Chemie* 43 (1891): 177–180.

Drude 1894 ———. *Physik des Aethers auf elektromagnetischer Grundlage*. Stuttgart: Ferdinand Enke, 1894.

Drude 1900a ———. "Zur Elektronentheorie der Metalle. I. Teil." *Annalen der Physik* 1 (1900): 566–613.

Drude 1900b ———. "Zur Elektronentheorie der Metalle. II. Teil. Galvanomagnetische und thermomagnetische Effecte." *Annalen der Physik* 3 (1900): 369–402.

Drude 1900c ———. *Lehrbuch der Optik*. Leipzig: S. Hirzel, 1900.

Drude 1904a ———. "Optische Eigenschaften und Elektronentheorie. I. Teil." *Annalen der Physik* 14 (1904): 677–725.

Drude 1904b ———. "Optische Eigenschaften und Elektronentheorie. II. Teil." *Annalen der Physik* 14 (1904): 936–961.

Drude 1906a ———. *Lehrbuch der Optik*. 2nd enl. ed. Leipzig: S. Hirzel, 1906.

Drude 1906b ———. "Die Natur des Lichtes." In *Winkelmann 1906d*, pp. 1120–1387.

Earman et al. 1982 Earman, John; Glymour, Clark; and Rynasiewicz, Robert. "On Writing the History of Special Relativity." In *Asquith and Nickles 1983*, pp. 403–416.

Eckert and Pricha 1984 Eckert, Michael, and Pricha, Willibald. "Die ersten Briefe Albert Einsteins an Arnold Sommerfeld." *Physikalische Blätter* 40 (1984): 29–34.

Ehrenfest 1906 Ehrenfest, Paul. "Zur Planckschen Strahlungstheorie." *Physikalische Zeitschrift* 7 (1906): 528–532.

Ehrenfest 1907 ———. "Die Translation deformierbarer Elektronen und der Flächensatz." *Annalen der Physik* 23 (1907): 204–205.

Ehrenfest 1911 ———. "Welche Züge der Lichtquantenhypothese spielen in der Theorie der Wärmestrahlung eine wesentliche Rolle?" *Annalen der Physik* 36 (1911): 91–118.

Ehrenfest 1912 ———. "Zur Frage der Entbehrlichkeit des Lichtäthers." *Physikalische Zeitschrift* 13 (1912): 317–319.

Ehrenfest and Ehrenfest 1906   Ehrenfest, Tatiana, and Ehrenfest, Paul. "Bemerkung zur Theorie der Entropiezunahme in der 'Statistischen Mechanik' von W. Gibbs." *Kaiserliche Akademie der Wissenschaften* (Vienna). *Mathematisch-naturwissenschaftliche Klasse. Abteilung IIa. Sitzungsberichte* 115 (1906): 89–98.

Ehrenfest and Ehrenfest 1911   Ehrenfest, Paul, and Ehrenfest, Tatiana. "Begriffliche Grundlagen der statistischen Auffassung in der Mechanik." In *Encyklopädie der mathematischen Wissenschaften, mit Einschluss ihrer Anwendungen.* Vol. 4, *Mechanik*, part 4, pp. 1–90 (separately paginated). Felix Klein and Conrad Müller, eds. Leipzig: B. G. Teubner, 1907–1914. Issued 12 December 1911.

Ehrenhaft 1907   Ehrenhaft, Felix. "Über eine der Brown'schen Molekularbewegung in den Flüssigkeiten gleichartige Molekularbewegung in den Gasen und deren molekularkinetischer Erklärungsversuch." *Kaiserliche Akademie der Wissenschaften* (Vienna). *Mathematisch-naturwissenschaftliche Klasse. Abteilung IIa. Sitzungsberichte* 116 (1907): 1139–1149.

Einstein 1901   Einstein, Albert. "Folgerungen aus den Capillaritätserscheinungen." *Annalen der Physik* 4 (1901): 513–523.

Einstein 1902a ——. "Ueber die thermodynamische Theorie der Potentialdifferenz zwischen Metallen und vollständig dissociirten Lösungen ihrer Salze und über eine elektrische Methode zur Erforschung der Molecularkräfte." *Annalen der Physik* 8 (1902): 798–814.

Einstein 1902b ——. "Kinetische Theorie des Wärmegleichgewichtes und des zweiten Hauptsatzes der Thermodynamik." *Annalen der Physik* 9 (1902): 417–433.

Einstein 1903 ——. "Eine Theorie der Grundlagen der Thermodynamik." *Annalen der Physik* 11 (1903): 170–187.

Einstein 1904 ——. "Zur allgemeinen molekularen Theorie der Wärme." *Annalen der Physik* 14 (1904): 354–362.

Einstein 1905a ——. Review of: Giuseppe Belluzzo, "Principi di termodinamica grafica." *Il Nuovo Cimento* 8 (1904): 196–222, 241–263. [*Belluzzo 1904.*] *Beiblätter zu den Annalen der Physik* 29 (1905): 235–236.

Einstein 1905b ——. Review of: Albert Fliegner, "Über den Clausius'schen Entropiesatz." *Naturforschende Gesellschaft in Zürich. Vierteljahrsschrift* 48 (1903): 1–48. [*Fliegner 1903.*] *Beiblätter zu den Annalen der Physik* 29 (1905): 236–237.

Einstein 1905c ——. Review of: William McFadden Orr, "On Clausius' Theorem for Irreversible Cycles, and on the Increase of Entropy." *Philosophical Magazine and Journal of Science* 8 (1904): 509–527. [*Orr 1904.*] *Beiblätter zu den Annalen der Physik* 29 (1905): 237.

Einstein 1905d ——. Review of: George Hartley Bryan, "The Law of Degradation of Energy as the Fundamental Principle of Thermodynamics." In *Meyer, S. 1904*, pp. 123–136. [*Bryan 1904.*] *Beiblätter zu den Annalen der Physik* 29 (1905): 237.

Einstein 1905e ——. Review of: Nikolay Nikolayevich Schiller, "Einige Bedenken betreffend die Theorie der Entropievermehrung durch Diffusion der Gase bei einander gleichen Anfangsspannungen der letzteren." In *Meyer, S. 1904*, pp. 350–366. [*Schiller 1904.*] *Beiblätter zu den Annalen der Physik* 29 (1905): 237–238.

Einstein 1905f ——. Review of: Jakob Johann Weyrauch, "Ueber die spezifischen Wärmen des überhitzten Wasserdampfes." *Zeitschrift des Vereines deutscher Ingenieure* 48 (1904): 24–28, 50–54. [*Weyrauch 1904.*] *Beiblätter zu den Annalen der Physik* 29 (1905): 240.

Einstein 1905g ——. Review of: Jacobus Henricus Van 't Hoff, "Einfluß der Änderung der spezifischen Wärme auf die Umwandlungsarbeit." In *Meyer, S. 1904*, pp. 233–241. [*Van 't Hoff 1904.*] *Beiblätter zu den Annalen der Physik* 29 (1905): 240–242.

Einstein 1905h ——. Review of: Arturo Giammarco, "Un caso di corrispondenza in termodinamica." *Il Nuovo Cimento* 5 (1903): 377–391. [*Giammarco 1903.*] *Bei-*

blätter zu den Annalen der Physik 29 (1905): 246–247.

Einstein 1905i ———. "Über einen die Erzeugung und Verwandlung des Lichtes betreffenden heuristischen Gesichtspunkt." *Annalen der Physik* 17 (1905): 132–148.

Einstein 1905j ———. *Eine neue Bestimmung der Moleküldimensionen*. Bern: K. J. Wyss, 1905.

Einstein 1905k ———. "Über die von der molekularkinetischen Theorie der Wärme geforderte Bewegung von in ruhenden Flüssigkeiten suspendierten Teilchen." *Annalen der Physik* 17 (1905): 549–560.

Einstein 1905l ———. Review of: Karl Fredrik Slotte, "Über die Schmelzwärme." *Finska Vetenskaps-Societeten. Öfversigt af Förhandlingar* 47, no. 7 (1904): 1–8. [*Slotte 1904a.*] *Beiblätter zu den Annalen der Physik* 29 (1905): 623–624.

Einstein 1905m ———. Review of: Karl Fredrik Slotte, "Folgerungen aus einer thermodynamischen Gleichung." *Finska Vetenskaps-Societeten. Öfversigt af Förhandlingar* 47, no. 8 (1904): 1–3. [*Slotte 1904b.*] *Beiblätter zu den Annalen der Physik* 29 (1905): 629.

Einstein 1905n ———. Review of: Emile Mathias, "La constante a des diamètres rectilignes et les lois des états correspondants [2ᵉ mémoire]." *Journal de physique théorique et appliquée* 4 (1905): 77–91. [*Mathias 1905.*] *Beiblätter zu den Annalen der Physik* 29 (1905): 634–635.

Einstein 1905o ———. Review of: Max Planck, "On Clausius' Theorem for Irreversible Cycles, and on the Increase of Entropy." *Philosophical Magazine and Journal of Science* 9 (1905): 167–168. [*Planck 1905.*] *Beiblätter zu den Annalen der Physik* 29 (1905): 635.

Einstein 1905p ———. Review of: Edgar Buckingham, "On Certain Difficulties Which Are Encountered in the Study of Thermodynamics." *Philosophical Magazine and Journal of Science* 9 (1905): 208–214. [*Buckingham 1905.*] *Beiblätter zu den Annalen der Physik* 29 (1905): 635–636.

Einstein 1905q ———. Review of: Paul Langevin, "Sur une formule fondamentale de la théorie cinétique." *Académie des sciences* (Paris). *Comptes rendus* 140 (1905): 35–38. [*Langevin 1905a.*] *Beiblätter zu den Annalen der Physik* 29 (1905): 640–641.

Einstein 1905r ———. "Zur Elektrodynamik bewegter Körper." *Annalen der Physik* 17 (1905): 891–921.

Einstein 1905s ———. "Ist die Trägheit eines Körpers von seinem Energieinhalt abhängig?" *Annalen der Physik* 18 (1905): 639–641.

Einstein 1905t ———. Review of: Heinrich Birven, *Grundzüge der mechanischen Wärmetheorie*. Stuttgart: F. Grub, 1905. [*Birven 1905.*] *Beiblätter zu den Annalen der Physik* 29 (1905): 950.

Einstein 1905u ———. Review of: Auguste Ponsot, "Chaleur dans le déplacement de l'équilibre d'un système capillaire." *Académie des sciences* (Paris). *Comptes rendus* 140 (1905): 1176–1179. [*Ponsot 1905.*] *Beiblätter zu den Annalen der Physik* 29 (1905): 952.

Einstein 1905v ———. Review of: Karl Bohlin, "Sur le choc, considéré comme fondement des théories cinétiques de la pression des gaz et de la gravitation universelle." *Arkiv för Matematik, Astronomi och Fysik* 1 (1904): 529–540. [*Bohlin 1904.*] *Beiblätter zu den Annalen der Physik* 29 (1905): 952–953.

Einstein 1905w ———. Review of: Georges Meslin, "Sur la constante de la loi de Mariotte et Gay-Lussac." *Journal de physique théorique et appliquée* 4 (1905): 252–256. [*Meslin 1905.*] *Beiblätter zu den Annalen der Physik* 29 (1905): 1114.

Einstein 1905x ———. Review of: Albert Fliegner, "Das Ausströmen heissen Wassers aus Gefässmündungen." *Schweizerische Bauzeitung* 45 (1905): 282–285, 306–308. [*Fliegner 1905a.*] *Beiblätter zu den Annalen der Physik* 29 (1905): 1115.

Einstein 1905y ———. Review of: Jakob Johann Weyrauch, *Grundriss der Wärmetheorie. Mit zahlreichen Beispielen und Anwendungen*. Part 1. Stuttgart: Konrad Wittwer, 1905. [*Weyrauch 1905.*] *Beiblätter zu den*

Annalen der Physik 29 (1905): 1152–1153.

Einstein 1905z ———. Review of: Albert Fliegner, "Über den Wärmewert chemischer Vorgänge." *Naturforschende Gesellschaft in Zürich. Vierteljahrsschrift* 50 (1905): 201–212. [*Fliegner 1905b.*] *Beiblätter zu den Annalen der Physik* 29 (1905): 1158.

Einstein 1906a ———. "Eine neue Bestimmung der Moleküldimensionen." *Annalen der Physik* 19 (1906): 289–305.

Einstein 1906b ———. "Zur Theorie der Brownschen Bewegung." *Annalen der Physik* 19 (1906): 371–381.

Einstein 1906c ———. "Nachtrag" to *Einstein 1906a. Annalen der Physik* 19 (1906): 305–306.

Einstein 1906d ———. "Zur Theorie der Lichterzeugung und Lichtabsorption." *Annalen der Physik* 20 (1906): 199–206.

Einstein 1906e ———. "Das Prinzip von der Erhaltung der Schwerpunktsbewegung und die Trägheit der Energie." *Annalen der Physik* 20 (1906): 627–633.

Einstein 1906f ———. Review of: Max Planck, *Vorlesungen über die Theorie der Wärmestrahlung.* Leipzig: Johann Ambrosius Barth, 1906. [*Planck 1906c.*] *Beiblätter zu den Annalen der Physik* 30 (1906): 764–766.

Einstein 1906g ———. "Über eine Methode zur Bestimmung des Verhältnisses der transversalen und longitudinalen Masse des Elektrons." *Annalen der Physik* 21 (1906): 583–586.

Einstein 1907a ———. "Die Plancksche Theorie der Strahlung und die Theorie der spezifischen Wärme." *Annalen der Physik* 22 (1907): 180–190.

Einstein 1907b ———. "Über die Gültigkeitsgrenze des Satzes vom thermodynamischen Gleichgewicht und über die Möglichkeit einer neuen Bestimmung der Elementarquanta." *Annalen der Physik* 22 (1907): 569–572.

Einstein 1907c ———. "Theoretische Bemerkungen über die Brownsche Bewegung." *Zeitschrift für Elektrochemie und*

angewandte physikalische Chemie 13 (1907): 41–42.

Einstein 1907d ———. "Berichtigung zu meiner Arbeit: 'Die Plancksche Theorie der Strahlung etc.' " *Annalen der Physik* 22 (1907): 800.

Einstein 1907e ———. "Über die Möglichkeit einer neuen Prüfung des Relativitätsprinzips." *Annalen der Physik* 23 (1907): 197–198.

Einstein 1907f ———. "Ueber die Natur der Bewegungen mikroskopisch kleiner, in Flüssigkeiten suspendierter Teilchen." *Naturforschende Gesellschaft Bern. Mitteilungen,* no. 1038 (1907): vii. Report on a lecture held at the meeting of 23 March 1907.

Einstein 1907g ———. "Bemerkungen zu der Notiz von Hrn. Paul Ehrenfest: 'Die Translation deformierbarer Elektronen und der Flächensatz.' " *Annalen der Physik* 23 (1907): 206–208.

Einstein 1907h ———. "Über die vom Relativitätsprinzip geforderte Trägheit der Energie." *Annalen der Physik* 23 (1907): 371–384.

Einstein 1907i ———. Review of: Jakob Johann Weyrauch, *Grundriss der Wärmetheorie. Mit zahlreichen Beispielen und Anwendungen.* Part 2. Stuttgart: Konrad Wittwer, 1907. [*Weyrauch 1907.*] *Beiblätter zu den Annalen der Physik* 31 (1907): 777–778.

Einstein 1907j ———. "Über das Relativitätsprinzip und die aus demselben gezogenen Folgerungen." *Jahrbuch der Radioaktivität und Elektronik* 4 (1907): 411–462. Issued 22 January 1908.

Einstein 1908a ———. "Eine neue elektrostatische Methode zur Messung kleiner Elektrizitätsmengen." *Physikalische Zeitschrift* 9 (1908): 216–217.

Einstein 1908b ———. "Berichtigungen zu der Arbeit: 'Über das Relativitätsprinzip und die aus demselben gezogenen Folgerungen.' " *Jahrbuch der Radioaktivität und Elektronik* 5 (1908): 98–99.

Einstein 1908c ———. "Elementare Theorie der Brownschen Bewegung." *Zeitschrift für Elektrochemie und angewandte physika-*

lische Chemie 14 (1908): 235–239.

Einstein 1909a ———. "Bemerkung zu der Arbeit von D. Mirimanoff 'Über die Grundgleichungen. . . .'" *Annalen der Physik* 28 (1909): 885–888.

Einstein 1909b ———. "Zum gegenwärtigen Stand des Strahlungsproblems." *Physikalische Zeitschrift* 10 (1909): 185–193.

Einstein 1909c ———. "Über die Entwickelung unserer Anschauungen über das Wesen und die Konstitution der Strahlung." *Deutsche Physikalische Gesellschaft. Verhandlungen* 11 (1909): 482–500. Reprinted in *Physikalische Zeitschrift* 10 (1909): 817–825.

Einstein 1910a ———. "Le principe de relativité et ses conséquences dans la physique moderne." *Archives des sciences physiques et naturelles* 29 (1910): 5–28, 125–144.

Einstein 1910b ———. "Sur les forces pondéromotrices qui agissent sur des conducteurs ferromagnétiques disposés dans un champ magnétique et parcourus par un courant." *Archives des sciences physiques et naturelles* 30 (1910): 323–324.

Einstein 1910c ———. "Theorie der Opaleszenz von homogenen Flüssigkeiten und Flüssigkeitsgemischen in der Nähe des kritischen Zustandes." *Annalen der Physik* 33 (1910): 1275–1298.

Einstein 1911a ———. "Bemerkung zu dem Gesetz von Eötvös." *Annalen der Physik* 34 (1911): 165–169.

Einstein 1911b ———. "Eine Beziehung zwischen dem elastischen Verhalten und der spezifischen Wärme bei festen Körpern mit einatomigem Molekül." *Annalen der Physik* 34 (1911): 170–174.

Einstein 1911c ———. "Bemerkungen zu den P. Hertzschen Arbeiten: 'Über die mechanischen Grundlagen der Thermodynamik'." *Annalen der Physik* 34 (1911): 175–176.

Einstein 1911d ———. "Berichtigung zu meiner Arbeit: 'Eine neue Bestimmung der Moleküldimensionen.'" *Annalen der Physik* 34 (1911): 591–592.

Einstein 1911e ———. "Die Relativitäts-Theorie." *Naturforschende Gesellschaft in Zürich. Vierteljahrsschrift* 56 (1911): 1–14.

Einstein 1911f ———. "Elementare Betrachtungen über die thermische Molekularbewegung in festen Körpern." *Annalen der Physik* 35 (1911): 679–694.

Einstein 1911g ———. "Über den Einfluß der Schwerkraft auf die Ausbreitung des Lichtes." *Annalen der Physik* 35 (1911): 898–908.

Einstein 1911h ———. "L'état actuel du problème des chaleurs spécifiques." In *Solvay 1911*, pp. 407–435.

Einstein 1912a ———. "Thermodynamische Begründung des photochemischen Äquivalentgesetzes." *Annalen der Physik* 37 (1912): 832–838.

Einstein 1912b ———. "Lichtgeschwindigkeit und Statik des Gravitationsfeldes." *Annalen der Physik* 38 (1912): 355–369.

Einstein 1913 ———. "Max Planck als Forscher." *Die Naturwissenschaften* 1 (1913): 1077–1079.

Einstein 1914a ———. "Méthode pour la détermination de valeurs statistiques d'observations concernant des grandeurs soumises à des fluctuations irrégulières." *Archives des sciences physiques et naturelles* 37 (1914): 254–256.

Einstein 1914b ———. "Antrittsrede des Hrn. Einstein." *Königlich Preussische Akademie der Wissenschaften* (Berlin). *Sitzungsberichte* (1914): 739–742.

Einstein 1915a ———. "Theoretische Atomistik." In *Die Kultur der Gegenwart. Ihre Entwicklung und ihre Ziele*. Paul Hinneberg, ed. Part 3, sec. 3, vol. 1, *Physik*, pp. 251–263. Emil Warburg, ed. Leipzig: B. G. Teubner, 1915.

Einstein 1915b ———. "Zur allgemeinen Relativitätstheorie." *Königlich Preussische Akademie der Wissenschaften* (Berlin). *Sitzungsberichte* (1915): 778–786.

Einstein 1916a ———. "Strahlungs-Emission und -Absorption nach der Quantentheorie." *Deutsche Physikalische Gesellschaft. Verhandlungen* 18 (1916): 318–323.

Einstein 1916b ———. "Zur Quantentheorie

der Strahlung." *Physikalische Gesellschaft Zürich. Mitteilungen* (1916): 47–62. Reprinted in *Physikalische Zeitschrift* 18 (1917): 121–128.

Einstein 1916c ———. "Ernst Mach." *Physikalische Zeitschrift* 17 (1916): 101–104.

Einstein 1917a ———. *Über die spezielle und die allgemeine Relativitätstheorie. (Gemeinverständlich)*. Braunschweig: Friedrich Vieweg und Sohn, 1917.

Einstein 1917b ———. "Marian von Smoluchowski." *Die Naturwissenschaften* 5 (1917): 737–738.

Einstein 1918 ———. "Bemerkung zu E. Gehrkes Notiz 'Über den Äther.' " *Deutsche Physikalische Gesellschaft. Verhandlungen* 20 (1918): 261.

Einstein 1919 ———. "Time, Space and Gravitation." *Times* (London). 28 November 1919, p. 13.

Einstein 1920 ———. "Bemerkung zu der Abhandlung von W. R. Heß 'Beitrag zur Theorie der Viskosität heterogener Systeme'." *Kolloidzeitschrift* 27 (1920): 137.

Einstein 1921a ———. "A Brief Outline of the Development of the Theory of Relativity." *Nature* 106 (1921): 782–784.

Einstein 1921b ———. *The Meaning of Relativity: Four Lectures Delivered at Princeton University, May, 1921*. London: Methuen, 1922.

Einstein 1922 ———. *Untersuchungen über die Theorie der 'Brownschen Bewegung'*. Reinhold Fürth, ed. Ostwald's Klassiker der exakten Wissenschaften, no. 199. Leipzig: Akademische Verlagsgesellschaft, 1922.

Einstein 1924 ———. "Quantentheorie des einatomigen idealen Gases." *Preussische Akademie der Wissenschaften* (Berlin). *Physikalisch-mathematische Klasse. Sitzungsberichte* (1924): 261–267.

Einstein 1925a ———. "Quantentheorie des einatomigen idealen Gases. Zweite Abhandlung." *Preussische Akademie der Wissenschaften* (Berlin). *Physikalisch-mathematische Klasse. Sitzungsberichte* (1925): 3–14.

Einstein 1925b ———. "Zur Quantentheorie

des idealen Gases." *Preussische Akademie der Wissenschaften* (Berlin). *Physikalisch-mathematische Klasse. Sitzungsberichte* (1925): 18–25.

Einstein 1926 ———. *Investigations on the Theory of the Brownian Movement*. Reinhold Fürth, ed., A. D. Cowper, trans. London: Methuen, 1926.

Einstein 1933 ———. *The Origins of the General Theory of Relativity: Being the First Lecture on the George A. Gibson Foundation in the University of Glasgow, Delivered on June 20th, 1933*. Glasgow University Publications, vol. 30. Glasgow: Jackson, Wylie and Co., 1933.

Einstein 1934 ———. "Einiges über die Entstehung der allgemeinen Relativitätstheorie." In Albert Einstein, *Mein Weltbild*, pp. 248–256. Amsterdam: Querido, 1934. (Original German text of *Einstein 1933*.)

Einstein 1935 ———. "Elementary Derivation of the Equivalence of Mass and Energy." *American Mathematical Society. Bulletin* 41 (1935): 223–230.

Einstein 1955 ———. "Erinnerungen—Souvenirs." *Schweizerische Hochschulzeitung* 28 (*Sonderheft*) (1955): 145–153. Reprinted as "Autobiographische Skizze." In *Seelig 1956*, pp. 9–17.

Einstein 1979 ———. *Autobiographical Notes: A Centennial Edition*. Paul Arthur Schilpp, trans. and ed. La Salle, Illinois: Open Court, 1979. Parallel English and German texts. Corrected version of "Autobiographisches—Autobiographical Notes." In *Albert Einstein: Philosopher-Scientist*, pp. 1–94. Paul Arthur Schilpp, ed. Evanston, Illinois: The Library of Living Philosophers, 1949.

Einstein/Besso 1972 Einstein, Albert, and Besso, Michele. *Correspondance 1903–1955*. Pierre Speziali, trans. and ed. Paris: Hermann, 1972.

Einstein and Hopf 1910a Einstein, Albert, and Hopf, Ludwig. "Über einen Satz der Wahrscheinlichkeitsrechnung und seine Anwendung in der Strahlungstheorie." *Annalen der Physik* 33 (1910): 1096–1104.

Einstein and Hopf 1910b ———. "Statistische Untersuchung der Bewegung eines Resonators in einem Strahlungsfeld.'' *Annalen der Physik* 33 (1910): 1105–1115.

Einstein and Laub 1908a Einstein, Albert, and Laub, Jakob. ''Über die elektromagnetischen Grundgleichungen für bewegte Körper.'' *Annalen der Physik* 26 (1908): 532–540.

Einstein and Laub 1908b ———. ''Über die im elektromagnetischen Felde auf ruhende Körper ausgeübten ponderomotorischen Kräfte.'' *Annalen der Physik* 26 (1908): 541–550.

Einstein and Laub 1908c ———. ''Berichtigung zur Abhandlung: 'Über die elektromagnetischen Grundgleichungen für bewegte Körper.' '' *Annalen der Physik* 27 (1908): 232.

Einstein and Laub 1909 ———. ''Bemerkungen zu unserer Arbeit: 'Über die elektromagnetischen Grundgleichungen für bewegte Körper.' '' *Annalen der Physik* 28 (1909): 445–447.

Einstein et al. 1909a Einstein, Albert, et al. ''Diskussion'' (following Henry Siedentopf, ''Über ultramikroskopische Abbildungen (Vorläufige Mitteilung)'' [*Siedentopf 1909*]). *Physikalische Zeitschrift* 10 (1909): 779–780.

Einstein et al. 1909b Einstein, Albert, et al. ''Diskussion'' (following Arthur Szarvassi, ''Die Theorie der elektromagnetischen Erscheinungen in bewegten Körpern und das Energieprinzip'' [*Szarvassi 1909*]). *Physikalische Zeitschrift* 10 (1909): 813.

Einstein et al. 1909c Einstein, Albert, et al. ''Diskussion'' (following Albert Einstein, ''Über die Entwickelung unserer Anschauungen über das Wesen und die Konstitution der Strahlung'' [*Einstein 1909c*]). *Physikalische Zeitschrift* 10 (1909): 825–826.

Einstein et al. 1909d Einstein, Albert, et al. ''Diskussion'' (following Fritz Hasenöhrl, ''Über die Umwandlung kinetischer Energie in Strahlung'' [*Hasenöhrl 1909a*]). *Physikalische Zeitschrift* 10 (1909): 830.

Elster and Geitel 1891 Elster, Julius, and Geitel, Hans. ''Ueber die Abhängigkeit der durch das Licht bewirkten Electricitätszerstreuung von der Natur der belichteten Oberfläche.'' *Annalen der Physik und Chemie* 43 (1891): 225–240.

Eötvös 1890 Eötvös, Loránd. ''A föld vonzása különböző anyagokra.'' *Akadémiai Értesítő* 1 (1890): 108–110. Reprinted in translation as: Roland Eötvös, ''Über die Anziehung der Erde auf verschiedene Substanzen.'' *Mathematische und naturwissenschaftliche Berichte aus Ungarn* 8 (1890): 65–68.

ETH Programm 1913a *Programm der Eidgenössischen Technischen Hochschule für das Sommersemester 1913*. Zurich: Buchdruckerei Berichthaus, 1913.

Exner 1900 Exner, Felix M. ''Notiz zu Brown's Molecularbewegung.'' *Annalen der Physik* 2 (1900): 843–847.

Faraday 1839 Faraday, Michael. *Experimental Researches in Electricity*. Vol. 1. First Series. London: Taylor and Francis, 1839.

Fick 1855 Fick, Adolf. ''Ueber Diffusion.'' *Annalen der Physik und Chemie* 4 (1855): 59–86.

FitzGerald 1889 FitzGerald, George Francis. ''The Ether and the Earth's Atmosphere.'' *Science* 13 (1889): 390.

Fizeau 1851 Fizeau, Armand. ''Sur les hypothèses relatives à l'éther lumineux, et sur une expérience qui paraît démontrer que le mouvement des corps change la vitesse avec laquelle la lumière se propage dans leur intérieur (Extrait par l'auteur).'' *Académie des sciences* (Paris). *Comptes rendus* 33 (1851): 349–355.

Fliegner 1903 Fliegner, Albert. ''Über den Clausius'schen Entropiesatz.'' *Naturforschende Gesellschaft in Zürich. Vierteljahrsschrift* 48 (1903): 1–48.

Fliegner 1905a ———. ''Das Ausströmen heissen Wassers aus Gefässmündungen.'' *Schweizerische Bauzeitung* 45 (1905): 282–285, 306–308.

Fliegner 1905b ———. ''Über den Wärmewert chemischer Vorgänge.'' *Naturforschende Gesellschaft in Zürich. Vierteljahrsschrift* 50 (1905): 201–212.

Flückiger 1974 Flückiger, Max. *Albert Ein-*

stein in Bern. Das Ringen um ein neues Weltbild. Eine dokumentarische Darstellung über den Aufstieg eines Genies. Bern: Paul Haupt, 1974.

Föppl 1894   Föppl, August. *Einführung in die Maxwell'sche Theorie der Elektricität*. Leipzig: B. G. Teubner, 1894. (2nd ed., *Abraham/Föppl 1904*.)

Fox 1974   Fox, Robert. "The Rise and Fall of Laplacian Physics." *Historical Studies in the Physical Sciences* 4 (1974): 89–136.

Frank 1908   Frank, Philipp. "Relativitätstheorie und Elektronentheorie in ihrer Anwendung zur Ableitung der Grundgleichungen für die elektromagnetischen Vorgänge in bewegten ponderablen Körpern." *Annalen der Physik* 27 (1908): 1059–1065.

Frank 1979   ———. *Einstein. Sein Leben und seine Zeit*. Braunschweig: Friedrich Vieweg und Sohn, 1979.

Freundlich 1909   Freundlich, Herbert. *Kapillarchemie. Eine Darstellung der Chemie der Kolloide und verwandter Gebiete*. Leipzig: Akademische Verlagsgesellschaft, 1909.

Fürth 1922   Fürth, Reinhold. "Vorwort" and "Anmerkungen." In *Einstein 1922*, pp. 3 and 54–72.

Fürth 1980   ———. "Personal Reminiscences." In *Einstein: The First Hundred Years*, pp. 19–21. Maurice Goldsmith, Alan Mackay, and James Woudhuysen, eds. Oxford: Pergamon, 1980.

Galison 1979   Galison, Peter. "Minkowski's Space-Time: From Visual Thinking to the Absolute World." *Historical Studies in the Physical Sciences* 10 (1979): 85–121.

Gans 1905   Gans, Richard. "Zur Elektrodynamik in bewegten Medien." *Annalen der Physik* 16 (1905): 516–534.

Gans 1911   ———. "Über das Biot-Savartsche Gesetz." *Physikalische Zeitschrift* 12 (1911): 806–811.

Garber 1976   Garber, Elizabeth. "Some Reactions to Planck's Law, 1900–1914." *Studies in History and Philosophy of Science* 7 (1976): 89–126.

Giammarco 1903   Giammarco, Arturo. "Un caso di corrispondenza in termodinamica." *Il Nuovo Cimento* 5 (1903): 377–391.

Gibbs 1902   Gibbs, Josiah Willard. *Elementary Principles in Statistical Mechanics Developed with Especial Reference to the Rational Foundation of Thermodynamics*. Yale Bicentennial Publications. New York: Charles Scribner's Sons, 1902.

Gibbs 1905   ———. *Elementare Grundlagen der statistischen Mechanik*. Ernst Zermelo, trans. Leipzig: Johann Ambrosius Barth, 1905.

Gillispie 1970–1980   Gillispie, Charles Coulston, ed. *Dictionary of Scientific Biography*. 16 vols. New York: Charles Scribner's Sons, 1970–1980.

Glick 1987   Glick, Thomas F., ed. *The Comparative Reception of Relativity*. Boston: D. Reidel, 1987.

Glitscher 1917   Glitscher, Karl. "Spektroskopischer Vergleich zwischen den Theorien des starren und des deformierbaren Elektrons." *Annalen der Physik* 52 (1917): 608–630.

Goldberg 1983   Goldberg, Stanley. "Albert Einstein and the Creative Act: The Case of Special Relativity." In *Springs of Scientific Creativity: Essays on Founders of Modern Science*, pp. 232–253. Aris H. Rutherford, Ted Davis, and Roger H. Stuewer, eds. Minneapolis: University of Minnesota Press, 1983.

Goldberg 1984   ———. *Understanding Relativity: Origin and Impact of a Scientific Revolution*. Boston: Birkhäuser, 1984.

Gouy 1888   Gouy, Louis-Georges. "Note sur le mouvement brownien." *Journal de physique théorique et appliquée* 7 (1888): 561–564.

Graetz 1906   Graetz, Leo. "Wärmestrahlung." In *Winkelmann 1906c*, pp. 241–435.

Graetz 1918   Graetz, Leo, ed. *Handbuch der Elektrizität und des Magnetismus*. Vol. 1, *Elektrizitätserregung und Elektrostatik*. Leipzig: Johann Ambrosius Barth, 1918.

Grammel 1913   Grammel, Richard. "Zur relativitätstheoretischen Elektrodynamik bewegter Körper." *Annalen der Physik* 41 (1913): 570–580.

Guggenbühl 1955   Guggenbühl, Gottfried.

Geschichte der Eidgenössischen Technischen Hochschule in Zürich. Zurich: Neue Zürcher Zeitung, 1955.

Guye and Lavanchy 1916 Guye, Charles-Eugène, and Lavanchy, Charles. "Vérification expérimentale de la formule de Lorentz-Einstein par les rayons cathodiques de grande vitesse." *Archives des sciences physiques et naturelles* 42 (1916): 286–299, 353–373, 441–448.

Habicht and Habicht 1910 Habicht, Conrad, and Habicht, Paul. "Elektrostatischer Potentialmultiplikator nach A. Einstein." *Physikalische Zeitschrift* 11 (1910): 532–535.

Hasenöhrl 1904 Hasenöhrl, Fritz. "Zur Theorie der Strahlung in bewegten Körpern." *Annalen der Physik* 15 (1904): 344–370.

Hasenöhrl 1905 ———. "Zur Theorie der Strahlung in bewegten Körpern. Berichtigung." *Annalen der Physik* 16 (1905): 589–592.

Hasenöhrl 1909a ———. "Über die Umwandlung kinetischer Energie in Strahlung." *Physikalische Zeitschrift* 10 (1909): 829–830.

Hasenöhrl 1909b ———. "Bericht über die Trägheit der Energie." *Jahrbuch der Radioaktivität und Elektronik* 6 (1909): 485–502.

Hasenöhrl 1910 ———. "Über den Widerstand, welchen die Bewegung kleiner Körperchen in einem mit Hohlraumstrahlung erfüllten Raume erleidet." *Kaiserliche Akademie der Wissenschaften* (Vienna). *Mathematisch-naturwissenschaftliche Klasse. Sitzungsberichte* 119 (1910): 1327–1349.

Heaviside 1892 Heaviside, Oliver. *Electrical Papers*. Vol. 1. London: Macmillan, 1892.

Helmholtz 1882 Helmholtz, Hermann von. *Wissenschaftliche Abhandlungen*. Vol. 1. Leipzig: Johann Ambrosius Barth, 1882.

Helmholtz 1883 ———. *Wissenschaftliche Abhandlungen*. Vol. 2. Leipzig: Johann Ambrosius Barth, 1883.

Helmholtz 1884 ———. *Vorträge und Reden*. 2 vols. Braunschweig: Friedrich Vieweg und Sohn, 1884.

Helmholtz 1892 ———. "Das Princip der kleinsten Wirkung in der Electrodynamik." *Annalen der Physik und Chemie* 47 (1892): 1–26. Reprinted in *Helmholtz 1895*, pp. 476–504.

Helmholtz 1895 ———. *Wissenschaftliche Abhandlungen*. Vol. 3. Leipzig: Johann Ambrosius Barth, 1895.

Helmholtz 1897 ———. *Vorlesungen über die elektromagnetische Theorie des Lichts*. Arthur König and Carl Runge, eds. Hamburg: Leopold Voss, 1897.

Helmholtz 1903 ———. *Vorlesungen über Theorie der Wärme*. Franz Richarz, ed. Leipzig: Johann Ambrosius Barth, 1903.

Henri 1908 Henri, Victor. "Etude cinématographique des mouvements browniens." *Académie des sciences* (Paris). *Comptes rendus* 146 (1908): 1024–1026.

Herglotz 1910 Herglotz, Gustav. "Über den vom Standpunkt des Relativitätsprinzips aus als 'starr' zu bezeichnenden Körper." *Annalen der Physik* 31 (1910): 393–415.

Hermann 1966 Hermann, Armin. "Albert Einstein und Johannes Stark: Briefwechsel und Verhältnis der beiden Nobelpreisträger." *Sudhoffs Archiv. Vierteljahrsschrift für Geschichte der Medizin und der Naturwissenschaften, der Pharmazie und der Mathematik* 50 (1966): 267–285.

Hermann 1969 ———. *Frühgeschichte der Quantentheorie (1899–1913)*. Mosbach/Baden: Physik Verlag, 1969.

Herneck 1966a Herneck, Friedrich. "Zwei Tondokumente Einsteins zur Relativitätstheorie." *Forschungen und Fortschritte* 40 (1966): 133–135. Reprinted as "Zwei Lautdokumente Einsteins zur Relativitätstheorie." In *Herneck 1976*, pp. 103–108.

Herneck 1966b ———. "Die Beziehungen zwischen Einstein und Mach, dokumentarisch dargestellt." *Wissenschaftliche Zeitschrift der Friedrich-Schiller-Universität Jena. Mathematisch-naturwissenschaftliche Reihe* 15 (1966): 1–14.

Herneck 1976 ———. *Einstein und sein Weltbild. Aufsätze und Vorträge*. Berlin: Buchverlag Der Morgen, 1976.

Hertz, H. 1884 Hertz, Heinrich. "Ueber die Beziehungen zwischen den Maxwell'schen electrodynamischen Grundgleichungen und den Grundgleichungen der gegnerischen Electrodynamik.'' *Annalen der Physik und Chemie* 23 (1884): 84–103.

Hertz, H. 1890a ———. ''Ueber die Grundgleichungen der Electrodynamik für ruhende Körper.'' *Annalen der Physik und Chemie* 40 (1890): 577–624. Reprinted in *Hertz, H. 1892*, pp. 208–255.

Hertz, H. 1890b ———. ''Ueber die Grundgleichungen der Electrodynamik für bewegte Körper.'' *Annalen der Physik und Chemie* 41 (1890): 369–399. Reprinted in *Hertz, H. 1892*, pp. 256–285.

Hertz, H. 1892 ———. *Untersuchungen ueber die Ausbreitung der elektrischen Kraft.* Leipzig: Johann Ambrosius Barth, 1892.

Hertz, H. 1894 ———. *Gesammelte Werke.* Vol. 3, *Die Prinzipien der Mechanik. In neuem Zusammenhange dargestellt.* Philipp Lenard, ed. Leipzig: Johann Ambrosius Barth (Arthur Meiner), 1894.

Hertz, P. 1910a Hertz, Paul. ''Über die mechanischen Grundlagen der Thermodynamik.'' *Annalen der Physik* 33 (1910): 225–274, 537–552.

Hertz, P. 1910b ———. ''Ueber die kanonische Gesamtheit.'' *Koninklijke Akademie van Wetenschappen te Amsterdam. Wis- en Natuurkundige Afdeeling. Verslagen van de Gewone Vergaderingen* 19 (1910): 824–848.

Hertz, P. 1912 ———. ''Ueber einen Boltzmannschen Beweis des zweiten Hauptsatzes.'' *Königliche Gesellschaft der Wissenschaften zu Göttingen. Mathematisch-physikalische Klasse. Nachrichten* (1912): 566–576.

Hertz, P. 1913a ———. ''Über die statistische Mechanik der Raumgesamtheit und die Wahrscheinlichkeit der Komplexion.'' *Königliche Gesellschaft der Wissenschaften zu Göttingen. Mathematisch-physikalische Klasse. Nachrichten* (1913): 177–196.

Hertz, P. 1913b ———. ''Über die statistische Mechanik der Raumgesamtheit und den Begriff der Komplexion.'' *Mathematische Annalen* 74 (1913): 153–203.

Hertz, P. 1916 ———. ''Statistische Mechanik.'' In *Weber and Hertz 1916*, pp. 436–600.

Herzfeld 1913 Herzfeld, Karl Ferdinand. ''Bemerkungen zum Boltzmann'schen Prinzip.'' *Kaiserliche Akademie der Wissenschaften* (Vienna). *Mathematisch-naturwissenschaftliche Klasse. Abteilung IIa. Sitzungsberichte* 122 (1913): 1553–1561.

Herzfeld 1921 ———. ''Physikalische und Elektrochemie.'' In *Encyklopädie der mathematischen Wissenschaften, mit Einschluss ihrer Anwendungen.* Vol. 5, *Physik*, part 1, pp. 947–1112. Arnold Sommerfeld, ed. Leipzig: B. G. Teubner, 1903–1921. Issued 1 November 1921.

Hirosige 1966 Hirosige, Tetu. ''Electrodynamics before the Theory of Relativity, 1890–1905.'' *Japanese Studies in the History of Science* 5 (1966): 1–49.

Hirosige 1976 ———. ''The Ether Problem, the Mechanistic Worldview, and the Origins of the Theory of Relativity.'' *Historical Studies in the Physical Sciences* 7 (1976): 3–82.

Holton 1967 Holton, Gerald. ''Influences on Einstein's Early Work in Relativity Theory.'' *American Scholar* 37 (1967): 59–79.

Holton 1973 ———. *Thematic Origins of Scientific Thought: Kepler to Einstein.* Cambridge, Massachusetts: Harvard University Press, 1973. (Rev. ed., 1988.)

Holton 1980 ———. ''Einstein's Scientific Program: The Formative Years.'' In *Woolf 1980*, pp. 49–65.

Holton 1986 ———. *The Advancement of Science, and Its Burdens.* Cambridge: Cambridge University Press, 1986.

Hughes 1912 Hughes, Arthur Llewelyn. ''On the Emission Velocities of Photo-Electrons.'' *Royal Society of London. Philosophical Transactions A* 212 (1912): 205–226.

Humboldt 1845–1862 Humboldt, Alexander von. *Kosmos. Entwurf einer physischen Weltbeschreibung.* 5 vols. Stuttgart and Tübingen: J. G. Cotta, 1845–1862.

Hume 1739 Hume, David. *A Treatise of Human Nature: Being an Attempt to Introduce* the Experimental Method of Reasoning into Moral Subjects. Book 1, Of the Understanding. London: John Noon, 1739.

Hume 1895 ———. Ein Traktat über die menschliche Natur. Part 1, Über den Verstand. E. Köttgen, trans., Theodor Lipps, ed. Hamburg: Leopold Voss, 1895. (2nd ed., 1904; trans. of Hume 1739.)

Infeld 1940 Infeld, Leopold. On the Theory of Brownian Motion. University of Toronto Studies. Applied Mathematics Series, no. 4. Toronto: The University of Toronto Press, 1940.

Ishiwara 1971 Ishiwara, Jun. Einstein Kyôzyu-Kôen-roku. Tokyo: Kabushika Kaisha, 1971.

Ives and Stilwell 1938 Ives, Herbert E., and Stilwell, G. R. "An Experimental Study of the Rate of a Moving Atomic Clock." Journal of the Optical Society of America 28 (1938): 215–226.

Jammer 1966 Jammer, Max. The Conceptual Development of Quantum Mechanics. New York: McGraw-Hill, 1966.

Jeans 1905a Jeans, James Hopwood. "On the Partition of Energy between Matter and Aether." Philosophical Magazine 10 (1905): 91–98.

Jeans 1905b ———. "The Dynamical Theory of Gases and of Radiation." Nature 72 (1905): 101–102.

Jeans 1905c ———. "A Comparison between Two Theories of Radiation." Nature 72 (1905): 293–294.

Jeans 1905d ———. "On the Laws of Radiation." Royal Society of London. Proceedings A 76 (1905): 545–552.

Jeans 1908 ———. "Zur Strahlungstheorie." Physikalische Zeitschrift 9 (1908): 853–855.

Jeans 1914 ———. Report on Radiation and the Quantum-Theory. London: The Physical Society of London ("The Electrician" Printing and Publishing Co.), 1914.

Jungnickel and McCormmach 1986a Jungnickel, Christa, and McCormmach, Russell. Intellectual Mastery of Nature: Theoretical Physics from Ohm to Einstein. Vol. 1, The Torch of Mathematics 1800–1870. Chicago: University of Chicago Press, 1986.

Jungnickel and McCormmach 1986b ———. Intellectual Mastery of Nature: Theoretical Physics from Ohm to Einstein. Vol. 2, The Now Mighty Theoretical Physics 1870–1925. Chicago: University of Chicago Press, 1986.

Kaiser 1987 Kaiser, Walter. "Early Theories of the Electron Gas." Historical Studies in the Physical and Biological Sciences 17 (1987): 271–272.

Kamerlingh Onnes and Keesom 1912 Kamerlingh Onnes, Heike, and Keesom, Willem Hendrik. "Die Zustandsgleichung." In Encyklopädie der mathematischen Wissenschaften, mit Einschluss ihrer Anwendungen. Vol. 5, Physik, part 1, pp. 615–945. Arnold Sommerfeld, ed. Leipzig: B. G. Teubner, 1903–1921. Issued 12 September 1912.

Kangro 1976 Kangro, Hans. Early History of Planck's Radiation Law. London: Taylor & Francis, 1976.

Kaufmann 1901 Kaufmann, Walter. "Die magnetische und electrische Ablenkbarkeit der Bequerelstrahlen und die scheinbare Masse der Elektronen." Königliche Gesellschaft der Wissenschaften zu Göttingen. Mathematisch-physikalische Klasse. Nachrichten (1901): 143–155.

Kaufmann 1905 ———. "Über die Konstitution des Elektrons." Königlich Preussische Akademie der Wissenschaften (Berlin). Sitzungsberichte (1905): 949–956.

Kaufmann 1906a ———. "Über die Konstitution des Elektrons." Annalen der Physik 19 (1906): 487–553.

Kaufmann 1906b ———. "Nachtrag zu der Abhandlung: 'Über die Konstitution des Elektrons.'" Annalen der Physik 20 (1906): 639–640.

Kayser 1930 Kayser, Rudolf [Anton Reiser, pseud.]. Albert Einstein: A Biographical Portrait. New York: Albert and Charles Boni, 1930.

Kerker 1976 Kerker, Milton. "The Svedberg and Molecular Reality." Isis 67 (1976): 190–216.

Khvolson 1902   Khvolson [Chwolson], Orest Daniylovich. *Lehrbuch der Physik*. Vol. 1, *Einleitung–Mechanik–Einige Messinstrumente und Messmethoden–Die Lehre von den Gasen, Flüssigkeiten und festen Körpern*. H. Pflaum, trans. Braunschweig: Friedrich Vieweg und Sohn, 1902.

Kirchhoff 1860   Kirchhoff, Gustav Robert. ''Ueber das Verhältniss zwischen dem Emissionsvermögen und dem Absorptionsvermögen der Körper für Wärme und Licht.'' *Annalen der Physik und Chemie* 109 (1860): 275–301. Reprinted in *Kirchhoff 1882*, pp. 571–598.

Kirchhoff 1882   ———. *Gesammelte Abhandlungen*. Leipzig: Johann Ambrosius Barth, 1882.

Kirchhoff 1894   ———. *Vorlesungen über mathematische Physik*. Vol. 4, *Theorie der Wärme*. Max Planck, ed. Leipzig: B. G. Teubner, 1894.

Kirchhoff 1897   ———. *Vorlesungen über mathematische Physik*. Vol. 1, *Mechanik*. 4th ed. Wilhelm Wien, ed. Leipzig: B. G. Teubner, 1897.

Kirchner 1962   Kirchner, Joachim. *Das deutsche Zeitschriftenwesen. Seine Geschichte und seine Probleme*. Vol. 2, *Vom Wiener Kongress bis zum Ausgange des 19. Jahrhunderts*. Wiesbaden: Otto Harrasowitz, 1962.

Kleeman 1909   Kleeman, Richard D. ''Some Relations in Capillarity.'' *Philosophical Magazine and Journal of Science* 18 (1909): 491–510.

Klein, F. 1910   Klein, Felix. ''Über die geometrischen Grundlagen der Lorentzgruppe.'' *Deutsche Mathematiker-Vereinigung. Jahresbericht* 19 (1910): 281–300.

Klein 1962   Klein, Martin J. ''Max Planck and the Beginnings of the Quantum Theory.'' *Archive for History of Exact Sciences* 1 (1962): 459–479.

Klein 1963a   ———. ''Planck, Entropy, and Quanta, 1901–1906.'' *The Natural Philosopher* 1 (1963): 83–108.

Klein 1963b   ———. ''Einstein's First Paper on Quanta.'' *The Natural Philosopher* 2 (1963): 59–86.

Klein 1965   ———. ''Einstein, Specific Heats, and the Early Quantum Theory.'' *Science* 148 (1965): 173–180.

Klein 1966   ———. ''Thermodynamics and Quanta in Planck's Work.'' *Physics Today* 19, no. 11 (1966): 23–32.

Klein 1967   ———. ''Thermodynamics in Einstein's Thought.'' *Science* 157 (1967): 509–516.

Klein 1970   ———. *Paul Ehrenfest*. Vol. 1, *The Making of a Theoretical Physicist*. Amsterdam: North-Holland; New York: American Elsevier, 1970.

Klein 1972   ———. ''Mechanical Explanation at the End of the Nineteenth Century.'' *Centaurus* 17 (1972): 58–82.

Klein 1973   ———. ''The Development of Boltzmann's Statistical Ideas.'' In *The Boltzmann Equation: Theory and Applications*, pp. 53–106. E.G.D. Cohen and Walter Thirring, eds. Vienna: Springer-Verlag, 1973.

Klein 1974a   ———. ''The Historical Origins of the Van der Waals Equation.'' *Physica* 73 (1974): 28–47.

Klein 1974b   ———. ''Einstein, Boltzmann's Principle, and the Mechanical World View.'' In *XIVth International Congress of the History of Science. Tokyo & Kyoto Japan 19–27 August, 1974. Texts of Symposia (Proceedings, no. 1)*, pp. 183–194. N.p.: Science Council of Japan, n.d.

Klein 1977   ———. ''The Beginnings of the Quantum Theory.'' In *History of Twentieth Century Physics*. Proceedings of the International School of Physics ''Enrico Fermi,'' Course 57, pp. 1–39. C. Weiner, ed. New York: Academic Press, 1977.

Klein 1979   ———. ''Einstein and the Development of Quantum Physics.'' In *Einstein: A Centenary Volume*, pp. 133–151. Anthony P. French, ed. Cambridge, Massachusetts: Harvard University Press, 1979.

Klein 1980   ———. ''No Firm Foundation: Einstein and the Early Quantum Theory.'' In *Woolf 1980*, pp. 161–185.

Klein 1982a   ———. ''Fluctuations and Statistical Physics in Einstein's Early Work.'' In *Albert Einstein: Historical and Cultural*

Perspectives. The Centennial Symposium in Jerusalem, pp. 39–58. Gerald Holton and Yehuda Elkana, eds. Princeton, New Jersey: Princeton University Press, 1982.

Klein 1982b ———. "Some Turns of Phrase in Einstein's Early Papers." In *Physics and Natural Philosophy: Essays in Honor of Laszlo Tisza on His Seventy-Fifth Birthday*, pp. 364–375. Abner Shimony and Herman Feshbach, eds. Cambridge, Massachusetts: MIT Press, 1982.

Klein 1986 ———. "Ernst Mach's Principles of the Theory of Heat." Introduction to *Mach 1986*, pp. ix–xx.

Klein and Needell 1977 Klein, Martin J., and Needell, Allan. "Some Unnoticed Publications by Einstein." *Isis* 68 (1977): 601–604.

Kleiner 1901 Kleiner, Alfred. "Ueber die Wandlungen in den physikalischen Grundanschauungen." In *Verhandlungen der Schweizerischen Naturforschenden Gesellschaft bei ihrer Versammlung zu Zofingen den 4., 5. und 6. August 1901 (84. Jahresversammlung)*, pp. 3–31. Zofingen: P. Ringier, 1902.

Kleiner 1906 ———. "Über Elektrometer von hoher Empfindlichkeit." *Naturforschende Gesellschaft in Zürich. Vierteljahrsschrift* 51 (1906): 226–228.

Knoblauch 1895 Knoblauch, Oscar. "Ueber die Fluorescenz von Lösungen." *Annalen der Physik und Chemie* 54 (1895): 193–220.

Kollros 1956 Kollros, Louis. "Erinnerungen eines Kommilitonen." In *Seelig 1956*, pp. 17–31.

Kopp 1864 Kopp, Hermann. "Untersuchungen über die specifische Wärme der starren und tropfbar-flüssigen Körper." *Annalen der Chemie und Pharmacie* 3 (suppl. vol. 3) (1864): 1–126.

Kowalski 1910 Kowalski, Joseph de. "Influence de la température sur la fluorescence et la loi de Stokes." *Le Radium* 7 (1910): 56–58.

Krist 1891 Krist, Josef. *Anfangsgründe der Naturlehre für die Unterclassen der Realschulen*. 6th ed. Vienna: Wilhelm Braumüller, 1891.

Kuhn 1978 Kuhn, Thomas S. *Black-body Theory and the Quantum Discontinuity, 1894–1912*. Oxford: Clarendon Press; New York: Oxford University Press, 1978.

Kunitz 1926 Kunitz, Moses. "An Empirical Formula for the Relation between Viscosity of Solution and Volume of Solute." *Journal of General Physiology* 9 (1926): 715–725.

Kurlbaum 1898 Kurlbaum, Ferdinand. "Ueber eine Methode zur Bestimmung der Strahlung in absolutem Maass und die Strahlung des schwarzen Körpers zwischen 0 und 100 Grad." *Annalen der Physik und Chemie* 65 (1898): 746–760.

Ladenburg 1907 Ladenburg, Erich. "Über Anfangsgeschwindigkeit und Menge der photoelektrischen Elektronen in ihrem Zusammenhange mit der Wellenlänge des auslösenden Lichtes. (Vorläufige Mitteilung)." *Physikalische Zeitschrift* 8 (1907): 590–594.

Landolt and Börnstein 1894 Landolt, Hans, and Börnstein, Richard, eds. *Physikalisch-chemische Tabellen*. 2nd ed. Berlin: Julius Springer, 1894.

Landolt and Börnstein 1905 Börnstein, Richard, and Meyerhoffer, Wilhelm, eds. *Landolt-Börnstein physikalisch-chemische Tabellen*. 3rd ed. Berlin: Julius Springer, 1905.

Langevin 1905a Langevin, Paul. "Sur une formule fondamentale de la théorie cinétique." *Académie des sciences* (Paris). *Comptes rendus* 140 (1905): 35–38.

Langevin 1905b ———. "Une formule fondamentale de théorie cinétique." *Annales de chimie et de physique* 5 (1905): 245–288.

Langevin 1905c ———. "La physique des électrons." *Revue générale des sciences pures et appliquées* 16 (1905): 257–276.

Langevin 1908 ———. "Sur la théorie du mouvement brownien." *Académie des sciences* (Paris). *Comptes rendus* 146 (1908): 530–533.

Langevin 1911 ———. "L'évolution de l'espace et du temps." *Scientia* 10 (1911): 31–54.

Laplace 1806 Laplace, Pierre-Simon. *Théorie de l'action capillaire*. Paris: Courcier, 1806. Incorporated in some editions of *Traité de mécanique céleste*. Vol. 4, as *Supplément au dixième livre. Sur l'action capillaire* (separately paginated). Reprinted in *Oeuvres complètes de Laplace*. Vol. 4, pp. 349–417. Paris: Gauthier-Villars, 1880.

Larmor 1894 Larmor, Joseph. "A Dynamical Theory of the Electric and Luminiferous Medium." *Royal Society of London. Philosophical Transactions A* 185 (1894): 719–822.

Larmor 1895 ———. "A Dynamical Theory of the Electric and Luminiferous Medium. Part II: Theory of Electrons." *Royal Society of London. Philosophical Transactions A* 186 (1895): 695–743.

Larmor 1897 ———. "A Dynamical Theory of the Electric and Luminiferous Medium. Part III: Relations with Material Media." *Royal Society of London. Philosophical Transactions A* 190 (1897): 205–300.

Larmor 1900 ———. *Aether and Matter*. Cambridge: Cambridge University Press, 1900.

Laub 1907 Laub, Jakob Johann. "Zur Optik der bewegten Körper." *Annalen der Physik* 23 (1907): 738–744.

Laub 1908 ———. "Zur Optik der bewegten Körper. II." *Annalen der Physik* 25 (1908): 175–184.

Laub 1910 ———. "Über die experimentellen Grundlagen des Relativitätsprinzips." *Jahrbuch der Radioaktivität und Elektronik* 7 (1910): 405–463.

Laue 1906 Laue, Max. Review of: *Einstein 1905i. Fortschritte der Physik* 61 (1906): 349–350.

Laue 1907 ———. "Die Mitführung des Lichtes durch bewegte Körper nach dem Relativitätsprinzip." *Annalen der Physik* 23 (1907): 989–990.

Laue 1911a ———. "Zur Diskussion über den starren Körper in der Relativitätstheorie." *Physikalische Zeitschrift* 12 (1911): 85–87.

Laue 1911b ———. *Das Relativitätsprinzip*. Braunschweig: Friedrich Vieweg und Sohn, 1911.

Laue 1952 Laue, Max von. "Mein physikalischer Werdegang. Eine Selbstdarstellung." In *Schöpfer des neuen Weltbildes*, pp. 178–210. Hans Hartmann, ed. Bonn: Athenäum-Verlag, 1952. Reprinted in: Max von Laue. *Gesammelte Schriften und Vorträge*. Vol. 3, pp. v–xxxiv. Braunschweig: Friedrich Vieweg und Sohn, 1961.

Lebedev 1901 Lebedev, Pëtr Nikolayevich. "Untersuchungen über die Druckkräfte des Lichtes." *Annalen der Physik* 6 (1901): 433–458.

Lenard 1900a Lenard, Philipp. "Ueber Wirkungen des ultravioletten Lichtes auf gasförmige Körper." *Annalen der Physik* 1 (1900): 486–507.

Lenard 1900b ———. "Erzeugung von Kathodenstrahlen durch ultraviolettes Licht." *Annalen der Physik* 2 (1900): 359–375.

Lenard 1900c ———. "Ueber die Elektricitätszerstreuung in ultraviolett durchstrahlter Luft." *Annalen der Physik* 3 (1900): 298–319.

Lenard 1902 ———. "Ueber die lichtelektrische Wirkung." *Annalen der Physik* 8 (1902): 149–198.

Lenard 1903 ———. "Über die Beobachtung langsamer Kathodenstrahlen mit Hilfe der Phosphoreszenz und über Sekundärentstehung von Kathodenstrahlen." *Annalen der Physik* 12 (1903): 449–490.

Lenard and Saeland 1909 Lenard, Philipp, and Saeland, Sem. "Über die lichtelektrische und aktinodielektrische Wirkung bei den Erdalkaliphosphoren." *Annalen der Physik* 28 (1909): 476–502.

Levi-Civita 1907 Levi-Civita, Tullio. "Sur le mouvement de l'électricité sans liaisons ni forces extérieures." *Académie des sciences* (Paris). *Comptes rendus* 145 (1907): 417–420.

Lindemann 1910 Lindemann, Frederick Alexander. "Über die Berechnung molekularer Eigenfrequenzen." *Physikalische Zeitschrift* 11 (1910): 609–612.

Lorentz 1886 Lorentz, Hendrik Antoon. "Over den invloed, dien de beweging der aarde op de lichtverschijnselen uitoefent." *Koninklijke Akademie van Wetenschappen* (Amsterdam). *Afdeeling Natuurkunde. Ver-*

slagen en Mededeelingen 2 (1886): 297–372.

Lorentz 1892a ———. "Over de terugkaatsing van licht door lichamen die zich bewegen." *Koninklijke Akademie van Wetenschappen* (Amsterdam). *Wis- en Natuurkundige Afdeeling. Verslagen der Zittingen* 1 (1892): 28–31.

Lorentz 1892b ———. "De relatieve beweging van de aarde en den aether." *Koninklijke Akademie van Wetenschappen* (Amsterdam). *Wis- en Natuurkundige Afdeeling. Verslagen der Zittingen* 1 (1892): 74–79.

Lorentz 1892c ———. "La théorie électromagnétique de Maxwell et son application aux corps mouvants." *Archives Néerlandaises des sciences exactes et naturelles* 25 (1892): 363–552. Reprint, Leiden: E. J. Brill, 1892.

Lorentz 1895 ———. *Versuch einer Theorie der electrischen und optischen Erscheinungen in bewegten Körpern.* Leiden: E. J. Brill, 1895.

Lorentz 1900 ———. "Über die scheinbare Masse der Ionen." *Physikalische Zeitschrift* 2 (1900): 78–79.

Lorentz 1904a ———. "Electromagnetische verschijnselen in een stelsel dat zich met willekeurige snelheid, kleiner dan die van het licht, beweegt." *Koninklijke Akademie van Wetenschappen te Amsterdam. Wis- en Natuurkundige Afdeeling. Verslagen van de Gewone Vergaderingen* 12 (1904): 986–1009. Reprinted in translation as: "Electromagnetic Phenomena in a System Moving with Any Velocity Smaller Than That of Light." *Koninklijke Akademie van Wetenschappen te Amsterdam. Section of Sciences. Proceedings* 6 (1904): 809–831.

Lorentz 1904b ———. "Maxwells elektromagnetische Theorie." In *Encyklopädie der mathematischen Wissenschaften, mit Einschluss ihrer Anwendungen.* Vol. 5, *Physik,* part 2, pp. 63–144. Arnold Sommerfeld, ed., Leipzig: B. G. Teubner, 1904–1922. Issued 16 June 1904.

Lorentz 1904c ———. "Weiterbildung der Maxwellschen Theorie. Elektronentheorie." In *Encyklopädie der mathematischen Wissenschaften, mit Einschluss ihrer Anwendungen.* Vol. 5, *Physik,* part 2, pp. 145–280. Arnold Sommerfeld, ed. Leipzig: B. G. Teubner, 1904–1922. Issued 16 June 1904.

Lorentz 1908a ———. *Le partage de l'énergie entre la matière pondérable et l'éther.* Rome: R. Accademia dei Lincei, 1908.

Lorentz 1908b ———. "Zur Strahlungstheorie." *Physikalische Zeitschrift* 9 (1908): 562–563.

Lorentz 1909a ———. "Le partage de l'énergie entre la matière pondérable et l'éther." *Revue générale des sciences pures et appliquées* 20 (1909): 14–26.

Lorentz 1909b ———. *The Theory of Electrons and Its Applications to the Phenomena of Light and Radiant Heat.* Leipzig: B. G. Teubner, 1909.

Lorentz 1910 ———. "Alte und neue Fragen der Physik." *Physikalische Zeitschrift* 11 (1910): 1234–1257.

Lorentz 1915 ———. "Die Maxwellsche Theorie und die Elektronentheorie." In *Die Kultur der Gegenwart. Ihre Entwicklung und ihre Ziele.* Paul Hinneberg, ed. Part 3, sec. 3, vol. 1, *Physik,* pp. 311–333. Emil Warburg, ed. Leipzig: B. G. Teubner, 1915.

Lorentz 1916 ———. *Les théories statistiques en thermodynamique. Conférences faits au Collège de France en novembre 1912.* L. Dunoyer, ed. Leipzig: B. G. Teubner, 1916.

Lorentz 1922 ———. *Lessen over theoretische natuurkunde aan de Rijks-Universiteit te Leiden gegeven.* Vol. 6, *Het relativiteitsbeginsel voor eenparige translaties (1910–1912).* Adriaan D. Fokker, ed. Leiden: E. J. Brill, 1922.

Lorenz 1904 Lorenz, Hans. *Lehrbuch der technischen Physik.* Vol. 2, *Technische Wärmelehre.* Munich: R. Oldenbourg, 1904.

Loschmidt 1865 Loschmidt, Josef. "Zur Grösse der Luftmolecüle." *Kaiserliche Akademie der Wissenschaften* (Vienna). *Mathematisch-naturwissenschaftliche Classe. Zweite Abtheilung. Sitzungsberichte* 52 (1865): 395–413.

Lummer and Pringsheim 1899 Lummer, Otto,

and Pringsheim, Ernst. "Die Vertheilung der Energie im Spectrum des schwarzen Körpers und des blanken Platins." *Deutsche Physikalische Gesellschaft. Verhandlungen* 1 (1899): 215–230.

Lummer and Pringsheim 1908 ———. "Über die Jeans-Lorentzsche Strahlungsformel." *Physikalische Zeitschrift* 9 (1908): 449–450.

McCausland 1984 McCausland, Ian. "Einstein and Special Relativity: Who Wrote the Added Footnotes?" *British Journal for the Philosophy of Science* 35 (1984): 60–61.

McCormmach 1967 McCormmach, Russell. "J. J. Thomson and the Structure of Light." *British Journal for the History of Science* 3 (1967): 362–387.

McCormmach 1970a ———. "Einstein, Lorentz, and the Electron Theory." *Historical Studies in the Physical Sciences* 2 (1970): 41–87.

McCormmach 1970b ———. "H. A. Lorentz and the Electromagnetic View of Nature." *Isis* 61 (1970): 459–497.

McCormmach 1976 ———. "Editor's Foreword." *Historical Studies in the Physical Sciences* 7 (1976): xi–xxxv.

Mach 1872 Mach, Ernst. *Die Geschichte und die Wurzel des Satzes von der Erhaltung der Arbeit. Vortrag.* Prague: J. G. Calve, 1872. Reprint, Leipzig: Johann Ambrosius Barth, 1909.

Mach 1886 ———. *Beiträge zur Analyse der Empfindungen.* Jena: Gustav Fischer, 1886.

Mach 1896 ———. *Die Principien der Wärmelehre. Historisch-kritisch entwickelt.* Leipzig: Johann Ambrosius Barth, 1896.

Mach 1897 ———. *Die Mechanik in ihrer Entwickelung. Historisch-kritisch dargestellt.* 3rd ed. Leipzig: F. A. Brockhaus, 1897.

Mach 1900a ———. *Die Analyse der Empfindungen und das Verhältnis des Physischen zum Psychischen.* 2nd enl. ed. of *Mach 1886.* Jena: Gustav Fischer, 1900.

Mach 1900b ———. *Die Principien der Wärmelehre. Historisch-kritisch entwickelt.* 2nd ed. Leipzig: Johann Ambrosius Barth, 1900.

Mach 1901 ———. *Die Mechanik in ihrer Entwickelung. Historisch-kritisch dargestellt.* 4th ed. Leipzig: F. A. Brockhaus, 1901.

Mach 1902 ———. *Die Analyse der Empfindungen und das Verhältnis des Physischen zum Psychischen.* 3rd enl. ed. Jena: Gustav Fischer, 1902.

Mach 1903 ———. *Die Analyse der Empfindungen und das Verhältnis des Physischen zum Psychischen.* 4th enl. ed. Jena: Gustav Fischer, 1903.

Mach 1904 ———. *Die Mechanik in ihrer Entwickelung. Historisch-kritisch dargestellt.* 5th ed. Leipzig: F. A. Brockhaus, 1904.

Mach 1910 ———. "Die Leitgedanken meiner naturwissenschaftlichen Erkenntnislehre und ihre Aufnahme durch die Zeitgenossen." *Scientia* 7 (1910): 225–240.

Mach 1986 ———. *Principles of the Theory of Heat: Historically and Critically Elucidated.* Brian McGuinness, ed. Dordrecht and Boston: D. Reidel, 1986. (Trans. of *Mach 1900b.*)

Marx 1924 Marx, Erich, ed. *Handbuch der Radiologie.* Vol. 6, *Die Theorien der Radiologie.* Leipzig: Akademische Verlagsgesellschaft, 1924.

Mathias 1893 Mathias, Emile. "Sur la densité critique et le théorème des états correspondants." *Journal de physique théorique et appliquée* 2 (1893): 5–22.

Mathias 1899 ———. "La constante a des diamètres rectilignes et les lois des états correspondants." *Journal de physique théorique et appliquée* 8 (1899): 407–413.

Mathias 1905 ———. "La constante a des diamètres rectilignes et les lois des états correspondants [2ᵉ mémoire]." *Journal de physique théorique et appliquée* 4 (1905): 77–91.

Maxwell 1860 Maxwell, James Clerk. "Illustrations of the Dynamical Theory of Gases. Part I. On the Motions and Collisions of Perfectly Elastic Spheres." *Philosophical Magazine and Journal of Science* 19 (1860): 19–32.

Maxwell 1867 ———. "On the Dynamical

Theory of Gases." *Royal Society of London. Philosophical Transactions* 157 (1867): 49–88. Reprinted in *Maxwell 1890*, vol. 2, pp. 26–78.

Maxwell 1871 ———. *Theory of Heat*. London: Longmans, Green, 1871.

Maxwell 1873 ———. *A Treatise on Electricity and Magnetism*. 2 vols. Oxford: Clarendon Press, 1873.

Maxwell 1877 ———. *Theorie der Wärme*. F. Auerbach, trans. Breslau: Maruschke & Berendt, 1877. (Trans. from the 4th English ed., 1875.)

Maxwell 1878 ———. *Theorie der Wärme*. F. Neeson, trans. Braunschweig: Friedrich Vieweg und Sohn, 1878. (Authorized trans. from the 4th English ed., 1875.)

Maxwell 1879 ———. "On Boltzmann's Theorem on the Average Distribution of Energy in a System of Material Points." *Cambridge Philosophical Society. Transactions* 12 (1879): 547–570.

Maxwell 1890 ———. *The Scientific Papers of James Clerk Maxwell*. W. D. Niven, ed. 2 vols. Cambridge: Cambridge University Press, 1890.

Maxwell 1891 ———. *A Treatise on Electricity and Magnetism*. 2 vols. 3rd ed. Oxford: Clarendon Press, 1891. Reprint, New York: Dover, 1954.

Mayer 1893a Mayer, Julius Robert. *Die Mechanik der Wärme in gesammelten Schriften*. 3rd ed. Jakob Johann Weyrauch, ed. Stuttgart: J. G. Cotta, 1893.

Mayer 1893b ———. *Kleinere Schriften und Briefe von Robert Mayer*. Jakob Johann Weyrauch, ed. Stuttgart: J. G. Cotta, 1893.

Mehra and Rechenberg 1982 Mehra, Jagdish, and Rechenberg, Helmut. *The Historical Development of Quantum Theory*. Vol. 1. New York: Springer-Verlag, 1982.

Meslin 1905 Meslin, Georges. "Sur la constante de la loi de Mariotte et Gay-Lussac." *Journal de physique théorique et appliquée* 4 (1905): 252–256.

Meyer, O. E. 1877 Meyer, Oskar Emil. *Die kinetische Theorie der Gase. In elementarer Darstellung mit mathematischen Zusätzen*. Breslau: Maruschke & Berendt, 1877.

Meyer, O. E. 1895 ———. *Die kinetische Theorie der Gase. In elementarer Darstellung mit mathematischen Zusätzen*. 2nd ed. Part 1. Breslau: Maruschke & Berendt, 1895.

Meyer, O. E. 1899 ———. *Die kinetische Theorie der Gase. In elementarer Darstellung mit mathematischen Zusätzen*. 2nd ed. Part 2. Breslau: Maruschke & Berendt, 1899.

Meyer, S. 1904 Meyer, Stefan, ed. *Festschrift. Ludwig Boltzmann gewidmet zum sechzigsten Geburtstage 20. Februar 1904*. Leipzig: Johann Ambrosius Barth, 1904.

Michelson 1881 Michelson, Albert A. "The Relative Motion of the Earth and the Luminiferous Ether." *American Journal of Science* 22 (1881): 120–129.

Michelson and Morley 1887 Michelson, Albert A., and Morley, Edward W. "On the Relative Motion of the Earth and the Luminiferous Ether." *American Journal of Science* 34 (1887): 333–345.

Mill 1872 Mill, John Stuart. *A System of Logic Ratiocinative and Inductive: Being a Connected View of the Principles of Evidence and the Methods of Scientific Investigation*. 2 vols. 8th ed. London: Longmans, Green, Reader, and Dyer, 1872. (1st ed., 1843.)

Mill 1877 ———. *System der deductiven und inductiven Logik. Eine Darlegung der Principien wissenschaftlicher Forschung, insbesondere der Naturforschung*. 4th ed. J. Schiel, trans. Braunschweig: Friedrich Vieweg und Sohn, 1877. (Trans. from 8th English ed., *Mill 1872*.)

Mill 1884–1887 ———. *System der deductiven und inductiven Logik. Eine Darlegung der Grundsätze der Beweislehre und der Methoden wissenschaftlicher Forschung*. 3 vols. 2nd ed. Theodor Gomperz, trans. Leipzig: Fues, 1884–1887.

Miller 1980 Miller, Arthur I. "On Some Other Approaches to Electrodynamics in 1905." In *Woolf 1980*, pp. 66–91.

Miller 1981a ———. "Unipolar Induction: A Case Study of the Interaction between Science and Technology." *Annals of Science* 38 (1981): 155–189.

Miller 1981b ———. *Albert Einstein's Special Theory of Relativity: Emergence (1905) and Early Interpretation (1905–1911)*. Reading, Massachusetts: Addison-Wesley, 1981.

Miller 1986 ———. *Frontiers of Physics: 1900–1911*. Boston: Birkhäuser, 1986.

Millikan 1916a Millikan, Robert A. "Einstein's Photoelectric Equation and Contact Electromotive Force." *Physical Review* 7 (1916): 18–32.

Millikan 1916b ———. "A Direct Photoelectric Determination of Planck's *h*." *Physical Review* 7 (1916): 355–388.

Minkowski 1907a Minkowski, Hermann. "Kapillarität." In *Encyklopädie der mathematischen Wissenschaften, mit Einschluss ihrer Anwendungen*. Vol. 5, *Physik*, part 1, pp. 558–613. Arnold Sommerfeld, ed. Leipzig: B. G. Teubner, 1903–1921. Issued 25 April 1907. Reprinted in *Minkowski 1911*, vol. 2, pp. 298–351.

Minkowski 1907b ———. "Das Relativitätsprinzip." *Annalen der Physik* 47 (1915): 927–938. Lecture, 5 November 1907, Mathematische Gesellschaft, Göttingen.

Minkowski 1908 ———. "Die Grundgleichungen für die elektromagnetischen Vorgänge in bewegten Körpern." *Königliche Gesellschaft der Wissenschaften zu Göttingen. Mathematisch-physikalische Klasse. Nachrichten* (1908): 53–111. Reprinted in *Minkowski 1911*, vol. 2, pp. 352–404.

Minkowski 1909 ———. "Raum und Zeit." *Physikalische Zeitschrift* 10 (1909): 104–111. Reprinted in *Minkowski 1911*, vol. 2, pp. 431–444.

Minkowski 1910 ———. "Eine Ableitung der Grundgleichungen für die elektromagnetischen Vorgänge in bewegten Körpern vom Standpunkte der Elektronentheorie" [prepared for publication by Max Born]. *Mathematische Annalen* 68 (1910): 526–551. Reprinted in *Minkowski 1911*, vol. 2, pp. 405–430.

Minkowski 1911 ———. *Gesammelte Abhandlungen*, David Hilbert, ed. 2 vols. Leipzig: B. G. Teubner, 1911.

Mirimanoff 1909 Mirimanoff, Dmitry. "Über die Grundgleichungen der Elektrodynamik bewegter Körper von Lorentz und das Prinzip der Relativität." *Annalen der Physik* 28 (1909): 192–198.

Mosengeil 1907 Mosengeil, Kurd von. "Theorie der stationären Strahlung in einem gleichförmig bewegten Hohlraum." *Annalen der Physik* 22 (1907): 867–904.

Nägeli 1879 Nägeli, Karl von. "Ueber die Bewegungen kleinster Körperchen." *Königlich Bayerische Akademie der Wissenschaften zu München. Mathematisch-physikalische Classe. Sitzungsberichte* 9 (1879): 389–453.

Needell 1980 Needell, Allan A. "Irreversibility and the Failure of Classical Dynamics: Max Planck's Work on the Quantum Theory 1900–1915." Ph.D. dissertation. Yale University, 1980.

Nernst 1888 Nernst, Walther. "Zur Kinetik der in Lösung befindlichen Körper. I. Theorie der Diffusion." *Zeitschrift für physikalische Chemie, Stöchiometrie und Verwandtschaftslehre* 2 (1888): 613–637.

Nernst 1889 ———. "Die elektromotorische Wirksamkeit der Ionen." *Zeitschrift für physikalische Chemie, Stöchiometrie und Verwandtschaftslehre* 4 (1889): 129–181.

Nernst 1898 ———. *Theoretische Chemie vom Standpunkte der Avogadro'schen Regel und der Thermodynamik*. 2nd ed. Stuttgart: Ferdinand Enke, 1898.

Nernst 1911a ———. "Über neuere Probleme der Wärmetheorie." *Königlich Preussische Akademie der Wissenschaften* (Berlin). *Sitzungsberichte* (1911): 65–90.

Nernst 1911b ———. "Zur Theorie der spezifischen Wärme und über die Anwendung der Lehre von den Energiequanten auf physikalisch-chemische Fragen überhaupt." *Zeitschrift für Elektrochemie und angewandte physikalische Chemie* 17 (1911): 265–275.

Nernst 1911c ———. "Untersuchungen über die spezifische Wärme bei tiefen Temperaturen. III." *Königlich Preussische Akademie der Wissenschaften* (Berlin). *Sitzungs-*

berichte (1911): 306–315.

Neumann, F. E. *1831* Neumann, Franz Ernst. "Untersuchung über die spezifische Wärme der Mineralien." *Annalen der Physik und Chemie* 23 (1831): 1–39.

Nichols and Hull 1903 Nichols, Ernest F., and Hull, Gordon F. "Über Strahlungsdruck." *Annalen der Physik* 12 (1903): 225–263.

Noether 1910 Noether, Fritz. "Zur Kinematik des starren Körpers in der Relativtheorie." *Annalen der Physik* 31 (1910): 919–944.

Nordmeyer 1903 Nordmeyer, Paul. "Über den Einfluß der Erdbewegung auf die Verteilung der Intensität der Licht- und Wärmestrahlung." *Annalen der Physik* 11 (1903): 284–302.

Nye 1972 Nye, Mary Jo. *Molecular Reality: A Perspective on the Scientific Work of Jean Perrin.* London: Macdonald; New York: American Elsevier, 1972.

Oppolzer 1902 Oppolzer, Egon R. von. "Erdbewegung und Aether." *Annalen der Physik* 8 (1902): 898–907.

Ornstein 1910 Ornstein, Leonard Salomon. "Eenige opmerkingen over de mechanische grondslagen der warmteleer." I. *Koninklijke Akademie van Wetenschappen te Amsterdam. Wis- en Natuurkundige Afdeeling. Verslagen van de Gewone Vergaderingen* 19 (1910): 809–823. Reprinted in translation as: "Some Remarks on the Mechanical Foundation of Thermodynamics." I. *Koninklijke Akademie van Wetenschappen te Amsterdam. Section of Sciences. Proceedings* 13 (1910): 804–817. Page numbers are cited from the English translation.

Ornstein 1911 ———. "Eenige opmerkingen over de mechanische grondslagen der warmteleer." II. *Koninklijke Akademie van Wetenschappen te Amsterdam. Wis- en Natuurkundige Afdeeling. Verslagen van de Gewone Vergaderingen* 19 (1911): 947–954. Reprinted in translation as: "Some Remarks on the Mechanical Foundation of Thermodynamics." II. *Koninklijke Akademie van Wetenschappen te Amsterdam. Section of Sciences. Proceedings* 13 (1911):

858–865. Page numbers are cited from the English translation.

Orr 1904 Orr, William McFadden. "On Clausius' Theorem for Irreversible Cycles, and on the Increase of Entropy." *Philosophical Magazine and Journal of Science* 8 (1904): 509–527.

Ostwald 1891 Ostwald, Wilhelm. *Lehrbuch der allgemeinen Chemie.* 2nd rev. ed. Vol. 1, *Stöchiometrie.* Leipzig: Wilhelm Engelmann, 1891.

Ostwald 1893 ———. *Lehrbuch der allgemeinen Chemie.* 2nd rev. ed. Vol. 2, part 1, *Chemische Energie.* Leipzig: Wilhelm Engelmann, 1893.

Ostwald 1907 ———. Review of: *Zsigmondy 1905. Zeitschrift für physikalische Chemie, Stöchiometrie und Verwandtschaftslehre* 57 (1907): 383.

Ostwald 1909 ———. "Vorbericht." In *Grundriss der allgemeinen Chemie.* 4th rev. ed., pp. iii–iv. Dresden: Theodor Steinkopff, 1909.

Pais 1982 Pais, Abraham. *'Subtle is the Lord . . .': The Science and the Life of Albert Einstein.* Oxford: Clarendon Press; New York: Oxford University Press, 1982.

Paschen 1901a Paschen, Friedrich. "Ueber das Strahlungsgesetz des schwarzen Körpers." *Annalen der Physik* 4 (1901): 277–298.

Paschen 1901b ———. "Ueber das Strahlungsgesetz des schwarzen Körpers. Entgegnung auf Ausführungen der Herren O. Lummer und E. Pringsheim." *Annalen der Physik* 6 (1901): 646–658.

Pauli 1921 Pauli, Wolfgang. "Relativitätstheorie." In *Encyklopädie der mathematischen Wissenschaften, mit Einschluss ihrer Anwendungen.* Vol. 5, *Physik*, part 2, pp. 539–775. Arnold Sommerfeld, ed. Leipzig: B. G. Teubner, 1904–1922. Issued 15 November 1921.

Pauli 1958 ———. *Theory of Relativity.* G. Field, trans. London: Pergamon, 1958. (Trans. of *Pauli 1921*.)

Pavlov 1978 Pavlov, V. I. "On Discussions Concerning the Problem of Ponderomotive

Forces." *Soviet Physics—USPEKHI* 21 (1978): 171–173.

Pearson 1900 Pearson, Karl. *The Grammar of Science*. 2nd ed. London: Adam & Charles Black, 1900.

Perrin 1908a Perrin, Jean. "L'agitation moléculaire et le mouvement brownien." *Académie des sciences* (Paris). *Comptes rendus* 146 (1908): 967–970.

Perrin 1908b ———. "La loi de Stokes et le mouvement brownien." *Académie des sciences* (Paris). *Comptes rendus* 147 (1908): 475–476.

Perrin 1908c ———. "L'origine du mouvement brownien." *Académie des sciences* (Paris). *Comptes rendus* 147 (1908): 530–532.

Perrin 1908d ———. "Grandeur des molécules et charge de l'électron." *Académie des sciences* (Paris). *Comptes rendus* 147 (1908): 594–596.

Perrin 1909a ———. "Le mouvement brownien de rotation." *Académie des sciences* (Paris). *Comptes rendus* 149 (1909): 549–551.

Perrin 1909b ———. "Mouvement brownien et réalité moléculaire." *Annales de chimie et de physique* 18 (1909): 5–114.

Perrin 1911 ———. "Les preuves de la réalité moléculaire (Etude spéciale des émulsions)." In *Solvay 1911*, pp. 153–250.

Perrin 1914 ———. *Les atomes*. 4th rev. ed. Paris: Félix Alcan, 1914.

Perrin and Dabrowski 1909 Perrin, Jean and Dabrowski, —. "Mouvement brownien et constantes moléculaires." *Académie des sciences* (Paris). *Comptes rendus* 149 (1909): 477–479.

Planck 1887 Planck, Max. *Das Princip der Erhaltung der Energie*. Leipzig: B. G. Teubner, 1887.

Planck 1891 ———. "Allgemeines zur neueren Entwicklung der Wärmetheorie." *Zeitschrift für physikalische Chemie, Stöchiometrie und Verwandtschaftslehre* 8 (1891): 647–656. Reprinted in *Planck 1958*, vol. 1, pp. 372–381.

Planck 1896 ———. "Gegen die neuere Energetik." *Annalen der Physik und Chemie* 57 (1896): 72–78. Reprinted in *Planck 1958*, vol. 1, pp. 459–465.

Planck 1897 ———. *Vorlesungen über Thermodynamik*. Leipzig: Veit & Comp., 1897.

Planck 1898 ———. "Über irreversible Strahlungsvorgänge. Vierte Mittheilung." *Königlich Preussische Akademie der Wissenschaften zu Berlin. Sitzungsberichte* (1898): 449–476. Reprinted in *Planck 1958*, vol. 1, pp. 532–559.

Planck 1899 ———. "Über irreversible Strahlungsvorgänge. Fünfte Mittheilung (Schluss)." *Königlich Preussische Akademie der Wissenschaften zu Berlin. Sitzungsberichte* (1899): 440–480. Reprinted in *Planck 1958*, vol. 1, pp. 560–600.

Planck 1900a ———. "Ueber irreversible Strahlungsvorgänge." *Annalen der Physik* 1 (1900): 69–122. Reprinted in *Planck 1958*, vol. 1, pp. 614–667.

Planck 1900b ———. "Entropie und Temperatur strahlender Wärme." *Annalen der Physik* 1 (1900): 719–737. Reprinted in *Planck 1958*, vol. 1, pp. 668–686.

Planck 1900c ———. "Ueber eine Verbesserung der Wien'schen Spectralgleichung." *Deutsche Physikalische Gesellschaft. Verhandlungen* 2 (1900): 202–204. Reprinted in *Planck 1958*, vol. 1, pp. 687–689.

Planck 1900d ———. "Kritik zweier Sätze des Hrn. W. Wien." *Annalen der Physik* 3 (1900): 764–766. Reprinted in *Planck 1958*, vol. 1, pp. 695–697.

Planck 1900e ———. "Zur Theorie des Gesetzes der Energieverteilung im Normalspectrum." *Deutsche Physikalische Gesellschaft. Verhandlungen* 2 (1900): 237–245. Reprinted in *Planck 1958*, vol. 1, pp. 698–706.

Planck 1901a ———. "Ueber das Gesetz der Energieverteilung im Normalspectrum." *Annalen der Physik* 4 (1901): 553–563. Reprinted in *Planck 1958*, vol. 1, pp. 717–727.

Planck 1901b ———. "Ueber die Elementarquanta der Materie und der Elektricität." *Annalen der Physik* 4 (1901): 564–566. Reprinted in *Planck 1958*, vol. 1, pp. 728–730.

Planck 1903a ———. Review of: *Gibbs 1902*. *Beiblätter zu den Annalen der Physik* 27 (1903): 748–753.

Planck 1903b ———. *Treatise on Thermodynamics*. Alexander Ogg, trans. London: Longmans, Green and Co., 1903.

Planck 1904 ———. "Über die mechanische Bedeutung der Temperatur und der Entropie." In *Meyer, S. 1904*, pp. 113–122. Reprinted in *Planck 1958*, vol. 2, pp. 79–88.

Planck 1905 ———. "On Clausius' Theorem for Irreversible Cycles, and on the Increase of Entropy." *Philosophical Magazine and Journal of Science* 9 (1905): 167–168. Reprinted in *Planck 1958*, vol. 2, pp. 100–101.

Planck 1906a ———. "Das Prinzip der Relativität und die Grundgleichungen der Mechanik." *Deutsche Physikalische Gesellschaft. Verhandlungen* 8 (1906): 136–141. Reprinted in *Planck 1958*, vol. 2, pp. 115–120.

Planck 1906b ———. "Die Kaufmannschen Messungen der Ablenkbarkeit der β-Strahlen in ihrer Bedeutung für die Dynamik der Elektronen." *Deutsche Physikalische Gesellschaft. Verhandlungen* 8 (1906): 418–432. Reprinted in *Physikalische Zeitschrift* 7 (1906): 753–759; and in *Planck 1958*, vol. 2, pp. 121–135.

Planck 1906c ———. *Vorlesungen über die Theorie der Wärmestrahlung*. Leipzig: Johann Ambrosius Barth, 1906.

Planck 1907a ———. "Zur Dynamik bewegter Systeme." *Königlich Preussische Akademie der Wissenschaften* (Berlin). *Sitzungsberichte* (1907): 542–570. Reprinted in *Annalen der Physik* 26 (1908): 1–34; and in *Planck 1958*, vol. 2, pp. 176–209.

Planck 1907b ———. "Nachtrag zu der Besprechung der Kaufmannschen Ablenkungsmessungen." *Deutsche Physikalische Gesellschaft. Verhandlungen* 9 (1907): 301–305. Reprinted in *Planck 1958*, vol. 2, pp. 210–214.

Planck 1910 ———. "Zur Theorie der Wärmestrahlung." *Annalen der Physik* 31 (1910): 758–768. Reprinted in *Planck 1958*, vol. 2, pp. 237–247.

Planck 1958 ———. *Physikalische Abhandlungen und Vorträge*. 3 vols. Braunschweig: Friedrich Vieweg und Sohn, 1958.

Pockels 1908 Pockels, Friedrich. "Kapillarität." In *Winkelmann 1908*, pp. 1119–1234.

Poincaré 1898 Poincaré, Henri. "La mesure du temps." *Revue de métaphysique et de morale* 6 (1898): 1–13.

Poincaré 1900 ———. "La théorie de Lorentz et le principe de la réaction." In *Bosscha 1900*, pp. 252–278.

Poincaré 1902 ———. *La science et l'hypothèse*. Paris: E. Flammarion, 1902.

Poincaré 1904a ———. *Wissenschaft und Hypothese*. Ferdinand and Lisbeth Lindemann, trans. Annotations by Ferdinand Lindemann. Leipzig: B. G. Teubner, 1904. (Trans. of *Poincaré 1902*.)

Poincaré 1904b ———. "L'état actuel et l'avenir de la physique mathématique." *Bulletin des sciences mathématiques* 28 (1904): 302–324.

Poincaré 1905a ———. *La valeur de la science*. Paris: E. Flammarion, 1905.

Poincaré 1905b ———. "Sur la dynamique de l'électron." *Académie des sciences* (Paris). *Comptes rendus* 140 (1905): 1504–1508.

Poincaré 1906 ———. "Sur la dynamique de l'électron." *Circolo Matematico di Palermo. Rendiconti* 21 (1906): 129–175.

Ponsot 1905 Ponsot, Auguste. "Chaleur dans le déplacement de l'équilibre d'un système capillaire." *Académie des sciences* (Paris). *Comptes rendus* 140 (1905): 1176–1179.

Pound and Rebka 1960 Pound, Robert V., and Rebka, Glen A. "Apparent Weight of Photons." *Physical Review Letters* 4 (1960): 337–341.

Prag Ordnung 1912a *Ordnung der Vorlesungen an der K. K. deutschen Karl Ferdinands-Universität zu Prag im Sommersemester 1912*. Herausgegeben vom K. K. Akadem. Senate. Prague: Statthalterei-Buchdruckerei, [1912].

Precht 1906 Precht, Julius. "Strahlungsenergie von Radium." *Annalen der Physik* 21 (1906): 595–601.

Promotionsordnung 1899   Promotions-
ordnung der II. Sektion der philoso-
phischen Fakultät der Hochschule Zürich
(Vom 10. Juni 1899). [Zurich, 1899].

Pyenson 1976   Pyenson, Lewis. "Einstein's
Early Scientific Collaboration." *Historical
Studies in the Physical Sciences* 7 (1976):
84–123.

Pyenson 1977   ———. "Hermann Min-
kowski and Einstein's Special Theory of
Relativity." *Archive for History of Exact
Sciences* 17 (1977): 71–95.

Pyenson 1985   ———. *The Young Einstein:
The Advent of Relativity.* Bristol: Adam
Hilger, 1985.

Ramsay 1882   Ramsay, William. "On Brownian or Pedetic Motion." *Bristol Naturalists'
Society. Proceedings* 3 (1882): 299–302.

Rayleigh 1900   Lord Rayleigh (John William
Strutt). "Remarks upon the Law of Complete Radiation." *Philosophical Magazine
and Journal of Science* 49 (1900): 539–540.
Reprinted in *Rayleigh 1964*, vol. 4, pp.
483–485.

Rayleigh 1905a   ———. "The Dynamical
Theory of Gases and of Radiation." *Nature*
72 (1905): 54–55. Reprinted in *Rayleigh
1964*, vol. 5, pp. 248–252.

Rayleigh 1905b   ———. "The Constant of
Radiation as Calculated from Molecular
Data." *Nature* 72 (1905): 243–244. Re-
printed in *Rayleigh 1964*, vol. 5, p. 253.

Rayleigh 1964   ———. *Scientific Papers.* 6
vols. New York: Dover, 1964.

Regener 1909   Regener, Erich. "Über Zählung der α-Teilchen durch die Szintillation
und über die Größe des elektrischen Elementarquantums." *Königlich Preussische
Akademie der Wissenschaften* (Berlin). *Sitzungsberichte* (1909): 948–965.

Reinganum 1900   Reinganum, Maximilian.
"Theoretische Bestimmung des Verhältnisses von Wärme- und Elektricitätsleitung
der Metalle aus der Drude'schen Elektronentheorie." *Annalen der Physik* 2 (1900):
398–403.

Richards 1902   Richards, Theodore William.
"The Significance of Changing Atomic
Volume: III. The Relation of Changing

Heat Capacity to Change of Free Energy,
Heat of Reaction, Change of Volume, and
Chemical Affinity." *American Academy of
Arts and Sciences. Proceedings* 38 (1902):
293–317.

Richardson and Compton 1912   Richardson,
Owen W., and Compton, Karl T. "The
Photoelectric Effect." *Philosophical Magazine and Journal of Science* 24 (1912): 575–
594.

Riecke 1890   Riecke, Eduard. "Molekulartheorie der Diffusion und Elektrolyse."
*Zeitschrift für physikalische Chemie, Stö-
chiometrie und Verwandtschaftslehre* 6
(1890): 564–572.

Riemann 1854   Riemann, Bernhard. "Ueber
die Hypothesen, welche der Geometrie zu
Grunde liegen." *Königliche Gesellschaft
der Wissenschaften und der Georg-Augusts-
Universität* (Göttingen). *Mathematische
Classe. Abhandlungen* 13 (1867): 133–152.
Lecture, Göttingen, 10 June 1854.

Ritz 1908a   Ritz, Walter. "Recherches cri-
tiques sur l'électrodynamique générale."
Annales de chimie et de physique 13 (1908):
145–275.

Ritz 1908b   ———. "Über die Grundlagen
der Elektrodynamik und die Theorie der
schwarzen Strahlung." *Physikalische Zeitschrift* 9 (1908): 903–907.

Ritz 1909   ———. "Zum gegenwärtigen
Stand des Strahlungsproblems. (Erwiderung
auf den Aufsatz des Herrn A. Einstein)."
Physikalische Zeitschrift 10 (1909): 224–
225.

Ritz and Einstein 1909   Ritz, Walter, and Einstein, Albert. "Zum gegenwärtigen Stand
des Strahlungsproblems." *Physikalische
Zeitschrift* 10 (1909): 323–324.

Roscoe et al. 1898   Roscoe, Henry E.; Schor-
lemmer, Carl; and Classen, Alexander.
*Roscoe-Schorlemmer's kurzes Lehrbuch der
Chemie.* 11th ed. Braunschweig: Friedrich
Vieweg und Sohn, 1898.

Rowlinson 1973   Rowlinson, John Shipley.
"Legacy of van der Waals." *Nature* 244
(1973): 414–417.

Rowlinson and Widom 1982   Rowlinson, John
Shipley, and Widom, B. *Molecular Theory*

of Capillarity. Oxford: Clarendon Press, 1982.

Rubens and Kurlbaum 1901 Rubens, Heinrich, and Kurlbaum, Ferdinand. "Anwendung der Methode der Reststrahlen zur Prüfung des Strahlungsgesetzes." *Annalen der Physik* 4 (1901): 649–666.

Rüger 1985 Rüger, Alexander. "Die Molekularhypothese in der Theorie der Kapillarerscheinungen (1805–1873)." *Centaurus* 28 (1985): 244–276.

Rutherford 1906 Rutherford, Ernest. *Radioactive Transformations*. New York: Charles Scribner's Sons, 1906.

Rutherford 1907 ———. *Radioaktive Umwandlungen*. Max Levin, trans. Braunschweig: Friedrich Vieweg und Sohn, 1907. (Trans. of *Rutherford 1906*.)

Rutherford and Geiger 1908 Rutherford, Ernest, and Geiger, Hans. "The Charge and Nature of the α-Particle." *Royal Society of London. Proceedings A* 81 (1908): 162–173.

Rutherford and Geiger 1909 ———. "Die Ladung und Natur des α-Teilchens." *Physikalische Zeitschrift* 10 (1909): 42–46. (Trans. of *Rutherford and Geiger 1908*.)

Saint-Venant and Wantzel 1839 Saint-Venant, Adhémar de, and Wantzel, Pierre. "Mémoire et expériences sur l'écoulement de l'air." *Journal de l'Ecole Polytechnique* 16 (1839): 85–122.

Sauter 1901 Sauter, Joseph. "Zur Interpretation der Maxwell'schen Gleichungen des elektromagnetischen Feldes in ruhenden isotropen Medien." *Annalen der Physik* 6 (1901): 331–338.

Sauvage 1892 Sauvage, Eduard. "Ecoulement de l'eau des chaudières." *Annales des Mines. Mémoires* 2 (1892): 192–202.

Schaffner 1982 Schaffner, Kenneth F. "The Historiography of Special Relativity: Comments on the Papers by John Earman, Clark Glymour, and Robert Rynasiewicz and by Arthur Miller." In *Asquith and Nickles 1983*, pp. 417–428.

Schiller 1904 Schiller, Nikolay Nikolayevich. "Einige Bedenken betreffend die Theorie der Entropievermehrung durch Diffusion

der Gase bei einander gleichen Anfangsspannungen der letzteren." In *Meyer, S. 1904*, pp. 350–366.

Schmidt 1918 Schmidt, Heinrich Willy. "Elektrisiermaschinen und Apparate." In *Graetz 1918*, pp. 21–93.

Schottky 1929 Schottky, Walter. *Thermodynamik*. Berlin: Julius Springer, 1929.

Schweidler 1904 Schweidler, Egon von. "Die lichtelektrischen Erscheinungen. (Die Emission negativer Elektronen von belichteten Oberflächen.)" *Jahrbuch der Radioaktivität und Elektronik* 1 (1904): 358–400.

Sciama 1979 Sciama, Dennis W. "Black Holes and Fluctuations of Quantum Particles: An Einstein Synthesis." In *Relativity, Quanta and Cosmology in the Development of the Scientific Thought of Albert Einstein*. Vol. 2, pp. 681–724. F. de Finis, ed. New York: Johnson Reprint, 1979.

Seddig 1907 Seddig, Max. "Abhängigkeit der Brownschen Molekularbewegung von der Temperatur." *Gesellschaft zur Beförderung der gesammten Naturwissenschaften* (Marburg). *Sitzungsberichte* (1907): 182–188.

Seddig 1908 ———. "Über die Messung der Temperaturabhängigkeit der Brownschen Molekularbewegung." *Physikalische Zeitschrift* 9 (1908): 465–468.

Seelig 1956 Seelig, Carl, ed. *Helle Zeit-Dunkle Zeit. In Memoriam Albert Einstein*. Zurich: Europa Verlag, 1956.

Seelig 1960 Seelig, Carl. *Albert Einstein. Leben und Werk eines Genies unserer Zeit*. Zurich: Europa Verlag, 1960.

Sesmat 1937 Sesmat, Augustine. *Systèmes de référence et mouvements (Physique classique)*. Vol. 6, *L'optique des corps en mouvement*. Actualités scientifiques et industrielles, no. 484. Paris: Hermann, 1937.

Siedentopf 1903 Siedentopf, Henry. "On the Rendering Visible of Ultra-Microscopic Particles and of Ultra-Microscopic Bacteria." *Journal of the Royal Microscopical Society. Transactions* (1903): 573–578.

Siedentopf 1909 ———. "Über ultra-mikroskopische Abbildungen (Vorläufige Mitteilung).'' *Physikalische Zeitschrift* 10 (1909): 778–779.

Siedentopf and Zsigmondy 1903   Siedentopf, Henry, and Zsigmondy, Richard A. ''Über Sichtbarmachung und Größenbestimmung ultramikroskopischer Teilchen, mit besonderer Anwendung auf Goldrubingläser.'' *Annalen der Physik* 10 (1903): 1–39.

Simha 1936   Simha, Robert. ''Untersuchungen über die Viskosität von Suspensionen und Lösungen. 7. Über die Viskosität von Kugelsuspensionen. (Suspensionen in Poiseuille'scher Grundströmung.)'' *Kolloidzeitschrift* 76 (1936): 16–19.

Slotte 1900   Slotte, Karl Fredrik. ''Uber die Molecularbewegung fester Körper.'' *Finska Vetenskaps-Societeten. Öfversigt af Förhandlingar* 43 (1900): 49–73.

Slotte 1902   ———. ''Über die thermische Ausdehnung und die specifische Wärme einfacher fester Körper.'' *Finska Vetenskaps-Societeten. Öfversigt af Förhandlingar* 44 (1902): 121–138.

Slotte 1904a   ———. ''Über die Schmelzwärme.'' *Finska Vetenskaps-Societeten. Öfversigt af Förhandlingar* 47, no. 7 (1904): 1–8.

Slotte 1904b   ———. ''Folgerungen aus einer thermodynamischen Gleichung.'' *Finska Vetenskaps-Societeten. Öfversigt af Förhandlingar* 47, no. 8 (1904): 1–3.

Smoluchowski 1904   Smoluchowski, Marian von. ''Über Unregelmäßigkeiten in der Verteilung von Gasmolekülen und deren Einfluß auf Entropie und Zustandsgleichung.'' In *Meyer, S. 1904*, pp. 626–641.

Smoluchowski 1906   ———. ''Zur kinetischen Theorie der Brownschen Molekularbewegung und der Suspensionen.'' *Annalen der Physik* 21 (1906): 756–780.

Smoluchowski 1913   ———. ''Einige Beispiele Brown'scher Molekularbewegung unter Einfluß äußerer Kräfte.'' *Akademie der Wissenschaften in Krakau. Mathematisch-naturwissenschaftliche Klasse. Reihe A. Mathematische Wissenschaften* (1913): 418–434.

Snelders 1976   Snelders, H.A.M. ''Jacobus Henricus van 't Hoff.'' In *Gillispie 1970–1980*, vol. 13, pp. 358–360.

Solovine 1956   Solovine, Maurice, ed. and trans. *Albert Einstein: Lettres à Maurice Solovine*. Paris: Gauthier-Villars, 1956.

Solvay 1911   Langevin, Paul, and de Broglie, Maurice, eds. *La théorie du rayonnement et les quanta. Rapports et discussions de la réunion tenue à Bruxelles, du 30 octobre au 3 novembre 1911. Sous les auspices de M. E. Solvay*. Paris: Gauthier-Villars, 1912.

Spinoza 1677   Spinoza, Baruch. ''Ethica ordine geometrico demonstrata.'' In *Opera postuma*, pp. 1–264. Amsterdam: J. Rieuwertsz, 1677.

Spinoza 1887   ———. *Die Ethik*. J. Stern, trans. Leipzig: Reclam, [1887].

Spinoza 1893   ———. *Benedict von Spinoza's Ethik*. 5th ed. J. H. von Kirchmann, trans. Berlin: Philosophisch-Historischer Verlag, [1893].

Stachel and Torretti 1982   Stachel, John, and Torretti, Roberto. ''Einstein's First Derivation of Mass-Energy Equivalence.'' *American Journal of Physics* 50 (1982): 760–763.

Stark 1902   Stark, Johannes. *Die Elektrizität in Gasen*. Leipzig: Johann Ambrosius Barth, 1902.

Stark 1906   ———. ''Über die Lichtemission der Kanalstrahlen in Wasserstoff.'' *Annalen der Physik* 21 (1906): 401–456.

Stark 1907   ———. ''Elementarquantum der Energie, Modell der negativen und der positiven Elektrizität.'' *Physikalische Zeitschrift* 8 (1907): 881–884.

Stark 1908   ———. ''Neue Beobachtungen an Kanalstrahlen in Beziehung zur Lichtquantenhypothese.'' *Physikalische Zeitschrift* 9 (1908): 767–773.

Stark 1909a   ———. ''Über Röntgenstrahlen und die atomistische Konstitution der Strahlung.'' *Physikalische Zeitschrift* 10 (1909): 579–586.

Stark 1909b   ———. ''Über die Ionisierung von Gasen durch Licht.'' *Physikalische Zeitschrift* 10 (1909): 614–623.

Stark 1909c   ———. ''Zur experimentellen Entscheidung zwischen Ätherwellen- und

Lichtquantenhypothese. I. Röntgenstrahlung.'' *Physikalische Zeitschrift* 10 (1909):
902–913.

Stark and Steubing 1908   Stark, Johannes and
Steubing, Walter. ''Fluoreszenz und lichtelektrische Empfindlichkeit organischer Substanzen.'' *Physikalische Zeitschrift* 9
(1908): 481–495.

Starke 1903   Starke, Hermann. ''Über die
elektrische und magnetische Ablenkung
schneller Kathodenstrahlen.'' *Deutsche
Physikalische Gesellschaft. Verhandlungen*
5 (1903): 241–250.

Stefan 1872   Stefan, Josef. ''Über die dynamische Theorie der Diffusion der Gase.''
Kaiserliche Akademie der Wissenschaften
(Vienna). *Mathematisch-naturwissenschaftliche Classe. Zweite Abtheilung. Sitzungsberichte* 65 (1872): 323–363.

Stefan 1879   ———. ''Über die Beziehung
zwischen der Wärmestrahlung und der
Temperatur.'' *Kaiserliche Akademie der
Wissenschaften* (Vienna). *Mathematisch-
naturwissenschaftliche Classe. Zweite
Abtheilung. Sitzungsberichte* 79 (1879):
391–428.

Stokes 1845   Stokes, George Gabriel. ''On the
Theories of the Internal Friction of Fluids in
Motion, and of the Equilibrium and Motion
of Elastic Solids.'' *Cambridge Philosophical Society. Transactions* 8 (1849): 287–
319.

Strobel 1893   Strobel, Fr. ''Vorwort.'' In
*Beiblätter zu den Annalen der Physik und
Chemie. Namenregister zum 1–15. Bande
(1877–1891)* (1893): vi.

Strobel 1909   ———. ''Vorwort.'' In *Beiblätter zu den Annalen der Physik. Register zu
Band 16 bis 30 (1892–1906)* (1909): iv.

Stuewer 1970   Stuewer, Roger H. ''Non-
Einsteinian Interpretations of the Photoelectric Effect.'' In *Historical and Philosophical Perspectives of Science*, pp. 246–263.
Roger H. Stuewer, ed. Minnesota Studies
in the Philosophy of Science, vol. 5. Herbert Feigl and Grover Maxwell, eds. Minneapolis: University of Minnesota Press,
1970.

Sutherland 1897   Sutherland, William. ''The
Causes of Osmotic Pressure and of the Sim-
plicity of the Laws of Dilute Solutions.''
*Philosophical Magazine and Journal of Sci-
ence* 44 (1897): 493–498.

Sutherland 1902   ———. ''Ionization, Ionic
Velocities, and Atomic Sizes.'' *Philosophical Magazine and Journal of Science* 3
(1902): 161–177.

Sutherland 1905   ———. ''A Dynamical
Theory of Diffusion for Non-Electrolytes
and the Molecular Mass of Albumin.'' *Philosophical Magazine and Journal of Science*
9 (1905): 781–785.

Svedberg 1905   Svedberg, The. ''Ueber die
elektrische Darstellung einiger neuen colloi-
dalen Metalle.'' *Deutsche Chemische Gesellschaft. Berichte* 38 (1905): 3616–3620.

Svedberg 1906a   ———. ''Über die Eigenbewegung der Teilchen in kolloidalen Lösungen.'' *Zeitschrift für Elektrochemie und
angewandte physikalische Chemie* 12
(1906): 853–860.

Svedberg 1906b   ———. ''Über die Eigenbewegung der Teilchen in kolloidalen Lösungen. Zweite Mitteilung.'' *Zeitschrift für
Elektrochemie und angewandte physikalische Chemie* 12 (1906): 909–910.

Svedberg 1910   ———. ''Einige Bemerkungen über die Brownsche Bewegung.''
*Zeitschrift für physikalische Chemie,
Stöchiometrie und Verwandtschaftslehre* 71
(1910): 571–576.

Swenson 1972   Swenson, Loyd S. *The Ethe-
real Aether: A History of the Michelson-
Morley-Miller Aether-Drift Experiments,
1880–1930.* Austin: University of Texas
Press, 1972.

Szarvassi 1909   Szarvassi, Arthur. ''Die Theorie der elektromagnetischen Erscheinungen
in bewegten Körpern und das Energieprinzip.'' *Physikalische Zeitschrift* 10 (1909):
811–813.

Taylor 1909   Taylor, Geoffrey Ingram. ''Interference Fringes with Feeble Light.'' *Cambridge Philosophical Society. Proceedings*
15 (1909): 114–115.

Teske 1969   Teske, Armin. ''Einstein und

Smoluchowski. Zur Geschichte der Brownschen Bewegung und der Opaleszenz.'' *Sudhoffs Archiv. Zeitschrift für Wissenschaftsgeschichte* 53 (1969): 292–305.

Thomson, J. J. 1905   Thomson, Joseph John. ''On the Emission of Negative Corpuscles by the Alkali Metals.'' *Philosophical Magazine and Journal of Science* 10 (1905): 584–590.

Thomson, W. 1858   Thomson, William. ''On the Thermal Effect of Drawing out a Film of Liquid.'' *Royal Society of London. Proceedings* 9 (1858): 255–256.

Thomson, W. 1867   ————. ''Report on Electrometers and Electrostatic Measurements.'' *British Association for the Advancement of Science. Report* (1867): 489–512.

Thomson, W. 1870   ————. ''The Size of Atoms.'' *Nature* 1 (1870): 551–553.

Tisza and Quay 1963   Tisza, Laszlo, and Quay, Paul M. ''The Statistical Thermodynamics of Equilibrium.'' *Annals of Physics* 25 (1963): 48–90.

Torretti 1983   Torretti, Roberto. *Relativity and Geometry*. Oxford: Pergamon, 1983.

Townsend 1920   Townsend, John Sealy. ''Die Ionisation der Gase.'' In *Handbuch der Radiologie*. Vol. 1, pp. 1–398. Erich Marx, ed. Leipzig: Akademische Verlagsgesellschaft, 1920.

Tumlirz 1899   Tumlirz, Ottokar. ''Die Zustandsgleichung des Wasserdampfes.'' *Kaiserliche Akademie der Wissenschaften* (Vienna). *Mathematisch-naturwissenschaftliche Classe. Abtheilung IIa. Sitzungsberichte* 108 (1899): 1058–1069.

Valentiner 1905   Valentiner, Siegfried. Review of: *Planck 1904. Beiblätter zu den Annalen der Physik* 29 (1905): 636–637.

Van der Waals 1873   Van der Waals, Johannes Diderik. *Over de continuiteit van den gasen vloeistoftoestand*. Leiden: A. W. Sijthoff, 1873.

Van 't Hoff 1887   Van 't Hoff, Jacobus Henricus. ''Die Rolle des osmotischen Druckes in der Analogie zwischen Lösungen und Gasen.'' *Zeitschrift für physikalische Chemie, Stöchiometrie und Verwandt-*

schaftslehre 1 (1887): 481–508.

Van 't Hoff 1904   ————. ''Einfluss der Änderung der spezifischen Wärme auf die Umwandlungsarbeit.'' In *Meyer, S. 1904*, pp. 233–241.

Verhandlungen 1904   *Verhandlungen der Schweizerischen Naturforschenden Gesellschaft in Winterthur den 30. und 31. Juli und 1. und 2. August 1904. 87. Jahresversammlung*. Basel: Georg, 1905.

Verhandlungen 1909   ''Sitzung vom 14. Mai 1909.'' *Deutsche Physikalische Gesellschaft. Verhandlungen* 11 (1909): 243–245.

Verhandlungen 1910   *Verhandlungen der Gesellschaft Deutscher Naturforscher und Ärzte. 81. Versammlung zu Salzburg. 19.-25. September 1909*. Parts 1 and 2. Albert Wangerin, ed. Leipzig: F.C.W. Vogel, 1910.

Versammlung 1909a   ''81. Versammlung Deutscher Naturforscher und Ärzte zu Salzburg vom 19. bis 25. September 1909.'' *Deutsche Physikalische Gesellschaft. Verhandlungen* 11 (1909): 415–419.

Versammlung 1909b   ''81. Versammlung Deutscher Naturforscher und Ärzte zu Salzburg, vom 21. bis 25. September 1909.'' *Physikalische Zeitschrift* 10 (1909): 777–778.

Violle 1892   Violle, Jules. *Lehrbuch der Physik*. German edition by E. Gumlich et al. Part 1, *Mechanik*. Vol. 1, *Allgemeine Mechanik und Mechanik der festen Körper*. Berlin: Julius Springer, 1892.

Violle 1893   ————. *Lehrbuch der Physik*. German edition by E. Gumlich et al. Part 1, *Mechanik*. Vol. 2, *Mechanik der flüssigen und gasförmigen Körper*. Berlin: Julius Springer, 1893.

Voigt 1887   Voigt, Woldemar. ''Ueber das Doppler'sche Princip.'' *Königliche Gesellschaft der Wissenschaften und der Georg-Augusts-Universität zu Göttingen. Nachrichten* (1887): 41–51.

Voigt 1896   ————. *Kompendium der theoretischen Physik*. Vol. 2, *Elektricität und Magnetismus. Optik*. Leipzig: Veit & Comp., 1896.

Washburn 1908   Washburn, Edward W. ''Die

neueren Forschungen über die Hydrate in Lösung. I. Teil.'' *Jahrbuch der Radioaktivität und Elektronik* 5 (1908): 493–552.

Washburn 1909 ———. ''Die neueren Forschungen über die Hydrate in Lösung. II. und III. Teil.'' *Jahrbuch der Radioaktivität und Elektronik* 6 (1909): 69–126.

Wassmuth 1915 Wassmuth, Anton. *Grundlagen und Anwendungen der statistischen Mechanik*. Braunschweig: Friedrich Vieweg und Sohn, 1915.

Weber, H. F. 1875 Weber, Heinrich Friedrich. ''Die specifischen Wärmen der Elemente Kohlenstoff, Bor und Silicium.'' *Annalen der Physik und Chemie* 4 (1875): 367–423, 553–582.

Weber, H. F. 1887 ———. ''Die Entwickelung der Lichtemission glühender fester Körper.'' *Königlich Preussische Akademie der Wissenschaften zu Berlin. Sitzungsberichte* (1887): 491–504.

Weber, H. F. 1888 ———. ''Untersuchungen über die Strahlung fester Körper.'' *Königlich Preussische Akademie der Wissenschaften zu Berlin. Sitzungsberichte* (1888): 933–957.

Weber, R. H. 1916 Weber, Rudolf Heinrich. ''Kapillarität.'' In *Weber and Hertz 1916*, pp. 1–122.

Weber and Hertz 1916 Weber, Rudolf Heinrich, and Hertz, Paul, eds. *Kapillarität, Wärme, Wärmeleitung, kinetische Gastheorie und statistische Mechanik*. Part 2, of vol. 1, *Mechanik und Wärme*, of *Repertorium der Physik*. Rudolf Heinrich Weber and Richard Gans, eds. Leipzig and Berlin: B. G. Teubner, 1916.

Weiss 1912 Weiss, Pierre. ''Prof. Dr. Heinrich Friedr. Weber. 1843–1912.'' *Schweizerische Naturforschende Gesellschaft. Verhandlungen* 95 (1912): 44–52.

Weyrauch 1885 Weyrauch, Jakob Johann. *Das Prinzip von der Erhaltung der Energie seit Robert Mayer*. Leipzig: B. G. Teubner, 1885.

Weyrauch 1890 ———. *Robert Mayer. Der Entdecker des Princips von der Erhaltung der Energie*. Stuttgart: Konrad Wittwer, 1890.

Weyrauch 1904 ———. ''Ueber die specifischen Wärmen des überhitzten Wasserdampfes.'' *Zeitschrift des Vereines deutscher Ingenieure* 48 (1904): 24–28, 50–54.

Weyrauch 1905 ———. *Grundriss der Wärmetheorie. Mit zahlreichen Beispielen und Anwendungen*. Part 1. Stuttgart: Konrad Wittwer, 1905.

Weyrauch 1907 ———. *Grundriss der Wärmetheorie. Mit zahlreichen Beispielen und Anwendungen*. Part 2. Stuttgart: Konrad Wittwer, 1907.

Wheaton 1978a Wheaton, Bruce R. ''On the Nature of X and Gamma Rays: Attitudes toward Localization of Energy in the 'New Radiations,' 1896–1922.'' Ph.D dissertation. Princeton University, 1978.

Wheaton 1978b ———. ''Philipp Lenard and the Photoelectric Effect, 1889–1911.'' *Historical Studies in the Physical Sciences* 9 (1978): 299–322.

Wheaton 1983 ———. *The Tiger and the Shark: Empirical Roots of Wave-Particle Dualism*. Cambridge: Cambridge University Press, 1983.

Wiechert 1896 Wiechert, Emil. ''Ueber die Grundlagen der Electrodynamik.'' *Annalen der Physik und Chemie* 59 (1896): 283–323.

Wiedeburg 1901 Wiedeburg, Otto. Review of: *Einstein 1901*. *Zeitschrift für physikalische Chemie, Stöchiometrie und Verwandtschaftslehre* 39 (1901): 378.

Wien 1893 Wien, Wilhelm. ''Eine neue Beziehung der Strahlung schwarzer Körper zum zweiten Hauptsatz der Wärmetheorie.'' *Königlich Preussische Akademie der Wissenschaften zu Berlin. Sitzungsberichte* (1893): 55–62.

Wien 1894 ———. ''Temperatur und Entropie der Strahlung.'' *Annalen der Physik und Chemie* 52 (1894): 132–165.

Wien 1896 ———. ''Ueber die Energievertheilung im Emissionsspectrum eines schwarzen Körpers.'' *Annalen der Physik und Chemie* 58 (1896): 662–669.

Wien 1898 ———. ''Ueber die Fragen, welche die translatorische Bewegung des

Lichtäthers betreffen.'' *Annalen der Physik und Chemie* 65, no. 3 (Beilage) (1898): i–xviii.

Wien 1900 ———. ''Über die Möglichkeit einer elektromagnetischen Begründung der Mechanik.'' In *Bosscha 1900*, pp. 96–107. Reprinted in *Annalen der Physik* 5 (1901): 501–513.

Wien 1904 ———. ''Über die Differentialgleichungen der Elektrodynamik für bewegte Körper.'' *Annalen der Physik* 13 (1904): 641–662.

Wien 1905 ———. ''Über die Energie der Kathodenstrahlen im Verhältnis zur Energie der Röntgen- und Sekundärstrahlen.'' *Annalen der Physik* 18 (1905): 991–1007.

Wien 1909 ———. ''Elektromagnetische Lichttheorie.'' In *Encyklopädie der mathematischen Wissenschaften, mit Einschluss ihrer Anwendungen*. Vol. 5, *Physik*, part 3, pp. 95–198. Arnold Sommerfeld, ed. Leipzig: B. G. Teubner, 1909–1926. Issued 26 January 1909.

Wiener 1890 Wiener, Otto. ''Stehende Lichtwellen und die Schwingungsrichtung polarisirten Lichtes.'' *Annalen der Physik und Chemie* 40 (1890): 203–243.

Wilson 1904 Wilson, Harold Albert. ''On the Electric Effect of Rotating a Dielectric in a Magnetic Field.'' *Royal Society of London. Philosophical Transactions* A 204 (1904): 121–137.

Wilson and Wilson 1913 Wilson, Majorie, and Wilson, Harold Albert. ''On the Electric Effect of Rotating a Magnetic Insulator in a Magnetic Field.'' *Royal Society of London. Proceedings* A 89 (1913): 99–106.

Winkelmann 1893 Winkelmann, Adolph, ed. *Handbuch der Physik*. Vol. 3, part 1, *Elektricität und Magnetismus I*. Breslau: Eduard Trewendt, 1893.

Winkelmann 1906a Winkelmann, Adolph. ''Spezifische Wärme.'' In *Winkelmann 1906c*, pp. 154–240.

Winkelmann 1906b ———. ''Lumineszenz.'' In *Winkelmann 1906d*, pp. 784–813.

Winkelmann 1906c Winkelmann Adolph, ed. *Handbuch der Physik*. 2nd ed. Vol. 3,

Wärme. Leipzig: Johann Ambrosius Barth, 1906.

Winkelmann 1906d ———, ed. *Handbuch der Physik*. 2nd ed. Vol. 6, *Optik*. Leipzig: Johann Ambrosius Barth, 1906.

Winkelmann 1908 ———, ed. *Handbuch der Physik*. 2nd ed. Vol. 1, *Allgemeine Physik*. Leipzig: Johann Ambrosius Barth, 1908.

Winteler-Einstein 1924 Winteler-Einstein, Maja. ''Albert Einstein. Beitrag für sein Lebensbild.'' Typescript. 15 February 1924.

Wolters 1988 Wolters, Gereon. ''Atome und Relativität—Was meinte Mach?'' In *Ernst Mach. Leben-Werk-Wirkung*. Rudolf Haller and Friedrich Stadler, eds. Vienna: Hölder-Pichler-Tempsky, 1988.

Woolf 1980 Woolf, Harry, ed. *Some Strangeness in the Proportion: A Centennial Symposium to Celebrate the Achievements of Albert Einstein*. Reading, Massachusetts: Addison-Wesley, 1980.

Young 1816 Young, Thomas. ''Cohesion.'' In *Supplement to the Fourth, Fifth and Sixth Editions of the Encyclopaedia Britannica*. Vol. 3, pp. 211–222. Edinburgh: Archibald Constable, 1824. Reprinted in: Thomas Young. *Miscellaneous Works*. Vol. 1, pp. 454–484. George Peacock, ed. London: John Murray, 1855.

Zangger 1906 Zangger, Heinrich. ''Ueber Membranen.'' *Naturforschende Gesellschaft in Zürich. Vierteljahrsschrift* 51 (1906): 432–440.

Zangger 1907 ———.''Über Membranen II.'' *Naturforschende Gesellschaft in Zürich. Vierteljahrsschrift* 52 (1907): 500–536.

Zangger 1911 ———. ''Die Bestimmungen der Avogadroschen Zahl N; die untere Teilungsgrenze der Materie (deren Bedeutung für die Biologie und Medizin).'' *Naturforschende Gesellschaft in Zürich. Vierteljahrsschrift* 56 (1911): 168–182.

Zenneck 1903 Zenneck, Johann. ''Gravitation.'' In *Encyklopädie der mathematischen Wissenschaften, mit Einschluss ihrer Anwendungen*. Vol. 5, *Physik*, part 1, pp. 25–

67. Arnold Sommerfeld, ed. Leipzig: B. G. Teubner, 1903–1921. Issued 23 April 1903.

Zermelo 1906 Zermelo, Ernst. Review of: *Gibbs 1902* and *Gibbs 1905*. *Deutsche Mathematiker-Vereinigung. Jahresbericht* 15 (1906): 232–242.

Zeuner 1867a Zeuner, Gustav Anton. "Theorie der überhitzten Wasserdämpfe." *Zeitschrift des Vereins deutscher Ingenieure* 11 (1867): 41–66.

Zeuner 1867b ———. "Ueber das Verhalten der überhitzten und der gemischten Wasserdämpfe." *Civilingenieur* 13 (1867): 343–372.

Zeuner 1905 ———. *Technische Thermodynamik*. 3rd ed. Vol. 1, *Fundamentalsätze der Thermodynamik. Lehre von den Gasen*. Leipzig: Arthur Felix, 1905.

Zeuner 1906 ———. *Technische Thermodynamik*. 3rd ed. Vol. 2, *Die Lehre von den Dämpfen*. Leipzig: Arthur Felix, 1906.

Zsigmondy 1905 Zsigmondy, Richard A. *Zur Erkenntnis der Kolloide. Über irreversible Hydrosole und Ultramikroskopie*. Jena: Gustav Fischer, 1905.

Zürich Verzeichnis 1910a *Verzeichnis der Vorlesungen an der Hochschule Zürich im Sommersemester 1910*. Zurich: Aktien-Buchdruckerei, 1910.

INDEX

Italic page numbers indicate references to front matter, or to editorial notes. Page numbers followed by a lowercase "n" indicate footnotes to Einstein documents. With few exceptions, references are collected under the appropriate English heading. Certain important institutions, organizations, and concepts are also listed under their German designations, with cross references to the corresponding English terms. "Albert Einstein" is abbreviated to "AE" in subentries. Correspondence cited in editorial material is listed alphabetically by correspondent/recipient following the words "correspondence with" (referring to a group of letters exchanged with AE) and/or "letter(s) from/to" (referring to specific letters). Full book titles appear only for AE's readings in the Olympia Academy. A separate index of citations follows the main index.

INDEX OF CITATIONS